NATURAL DISASTERS
NEW EDITION

NATURAL DISASTERS

NEW EDITION

LEE DAVIS

Facts On File
An imprint of Infobase Publishing

Facts On File, Inc.
An imprint of Infobase Publishing
132 West 31st Street
New York NY 10001

Library of Congress Cataloging-in-Publication Data

Davis, Lee (Lee Allyn)
Natural disasters / author, Lee Davis. — New ed.
p. cm.
Includes bibliographical references and index.
ISBN-13: 978-0-8160-7000-8
ISBN-10: 0-8160-7000-8
1. Natural disasters. I. Title.
GB5014.D38 2008
904'.5—dc22   2007050846

Facts On File books are available at special discounts when purchased in bulk quantities for businesses, associations, institutions, or sales promotions. Please call our Special Sales Department in New York at (212) 967-8800 or (800) 322-8755.

You can find Facts On File on the World Wide Web at http://www.factsonfile.com

Text design adapted by Annie O'Donnell
Photo research by Elizabeth H. Oakes

Printed in the United States of America

VB Hermitage 10 9 8 7 6 5 4 3 2 1

This book is printed on acid-free paper and contains 30 percent postconsumer recycled content.

To Christiana, whose sublimely natural life
is at the farthest remove from disaster

CONTENTS

ACKNOWLEDGMENTS
IX

INTRODUCTION
XI

AVALANCHES AND LANDSLIDES
1

EARTHQUAKES
25

FAMINES AND DROUGHTS
105

FLOODS
137

PLAGUES AND EPIDEMICS
203

CYCLONES
231

HURRICANES
245

ICESTORMS AND SNOWSTORMS
319

TORNADOES
335

STORMS
357

TYPHOONS
367

VOLCANIC ERUPTIONS AND NATURAL EXPLOSIONS
381

SELECTED BIBLIOGRAPHY
425

WEB SOURCES
429

INDEX
433

ACKNOWLEDGMENTS

It is the custom, as if it were an awards ceremony, to thank everybody but your dog and your least favorite relative for help in the birthing of a book. That is not going to be the practice on this page. If I were to name all of the people over the passage of three years who, while I was writing about and therefore living through some of the world's worst times, kept me from getting depressed to the point of paralysis, I would compile a cast of—well—hundreds.

So, I won't, but will, instead, express my gratitude to the major players in this drama of disasters:

Particularly to Mary Lou Barber, who forfeited half a summer to help me endlessly and immeasurably in the accumulation of a small mountain range of reference material;

To Diane Johnston, who cut through a continent of red tape at the New York Public Library and made the portion of my life spent there infinitely more productive than it otherwise would have been;

To Jean Kaleda, Edna McCaffery Cichanowicz, Joanne Brooks, Patricia S. Tormey, and the rest of the research staff at the Riverhead Free Library; to Shirley Van Derof, Phyllis Acard, Karen Hewlett, Susan La Vista, Jane Vail, Elva Stanley, Robert Allard, Jan Camarda, and Nancy Foley of the Westhampton Free Library; to Selma Kelson and the research staff of the Patchogue-Medford Library; to the research staff of the library at the Southampton Campus of Long Island University; to the research staff of the print division of the Library of Congress;

To the research staff at the Riverhead Free Library, the Patchogue-Medford library, the library at the Southampton Campus of Long Island University, and the print division of the Library of Congress;

To Elizabeth Hooks of the American Red Cross photo library; to Reynoldo Reyes of the United Nations photo library; to Pedro Soto of the CARE photo library; to Michael Benson, Larry Crabil, and particularly Bill McGruder of the National Transportation Safety Board; to Tim Cronen of the library of the Smithsonian Institution Air and Space museum;

To Tom Deja for his picture research; to Fred Robertson for his patient photography of disintegrating copies of old newspapers; and, to my two faithful research assistants, Colleen Miller and Suzanne Henning;

To my agent, Elizabeth Knappman, for causing this to happen;

To my editor, Frank K. Darmstadt, and Elizabeth Oakes, a fine photo researcher.

And to all of those unnamed friends and spiritual advisers who kept me sane during the creation of this book—my deepest and heartfelt thanks.

Lee Davis
Westhampton, N.Y.

INTRODUCTION

Sometimes it seems as if nature's way is some monumental system of checks and balances, designed to control the world's population. At other times it seems like some cruel cosmic joke. And at still others, it seems like pure happenstance, impure accident, or random violence that proceeds mindlessly out of a universe without a pattern.

In fact, the very randomness and suddenness of some natural disasters are sometimes their most devastating qualities. Out of a freshly polished summer day in Pompeii, death in the form of white hot ashes both struck down and preserved a civilization, in one arrested moment of flight and surprise. In the Alps, families eating breakfast were instantly turned, by sudden avalanches, into recollections of a frozen fraction of time. Japanese fishermen, knitting nets and smoking, literally did not know what hit them when gigantic tsunamis, set in motion by undersea earthquakes hundreds of miles away, curled and broke 40 feet (12.2 m) over their heads, smashing their villages to atoms in an instant.

All of this chaos is of course unsettling to scientists, and so, in an effort to make order out of disorder, they have tried to first identify and then track the patterns in natural disasters. To some extent, they have succeeded. Even the most cursory examination of the world's natural disasters indicates that human fatalities have steadily decreased, largely because humankind, through its scientists, has learned to pay attention to certain warning signs and signals.

Today, volcanologists have deduced that repeated earthquakes are an almost certain prelude to a volcanic eruption.

By comparing the relative heights of specific bodies of water over a fixed period of time, scientists are able to predict, with a fair amount of accuracy, the coming of a flood.

Meteorologists have now identified hurricane and tornado "seasons" and have developed sophisticated ways of tracking and gauging the storms within them.

With increased knowledge about sanitary conditions and the role of germs in the spreading of disease, and the discovery of vaccines and treatments, the unbridled spread of plague and epidemic has come under greater control.

And yet, there are always unpredictable variables.

Flash floods brought on by out of season rainfall can cause unpredictable catastrophes. Tornadoes and hurricanes often choose their own individual paths, confounding both computers and meteorologists. Dormant volcanoes have been known to come to life, inexplicably. These are the vagaries of nature that continue to confound both seers and scientists.

How else to explain city planners locating their creations on faults that will ultimately heave and split apart in some future earthquake? How to explain the inhabitants of the city of Catania, who rebuilt their city on the slopes of the volatile volcano Mount Etna? The statistics books bulge with fatalities that have come about this way, and yet, next year, in some place in the world, it will happen again, as regularly as the arrival of the seasons.

Throughout history, man has attributed natural disasters to a divine cause. For generations, perhaps eons, it was believed that the eruption of volcanoes was the cataclysmic expression of the dissatisfaction of the gods. To prevent such a monstrous eruption, the gods were offered a sacrifice of a goat, a virgin, or a child. When the Black Death stalked Europe in the Middle Ages, it was thought to be a retribution exacted from Earth's sinners. Thus armies of flagellants whipped themselves publicly in atonement for the horror they felt they had brought upon mankind.

As the Age of Reason dawned, scientific evidence was substituted for divine intervention. But, as the world has aged and testimony piles up, it is obvious that natural disasters arise from a complexity of causes, some natural, some sociological. Consider, for instance, poverty-ridden populations in countries like Bangladesh, who, through a multiplicity of economic factors, are forced to live in the direct path of monsoons and typhoons and are put at the mercy of floods caused by the rape of the ecology by uncaring governments—often of other, neighboring countries. Consider corrupt governments that absorb food, medical supplies, life-saving equipment, and the very seed grain that could, for future generations, alleviate the scourge of famine and starvation in Africa, Asia, and the Middle East. Sometimes, mass inefficiency is the culprit. But, more pervasively, these natural disasters are exacerbated by simple, primal greed, and its immediate manifestation, corruption.

Nor should the most dramatic, appalling, and contemporary example of human culpability in the face of a natural disaster be understated. During the first decade of the 21st century, AIDS is still one of the most critical pandemics in the history of the world. Its passage toward eradication continues, however, to be a rocky one. At the end of the 20th century, whole governments ignored the disease. In the 1980s, the decision makers within the U.S. government seemed to be bereft of compassion and intelligence when it came to facing the reality of a nascent epidemic in its midst. AIDS was then perceived as a disease of homosexual men and drug addicts, and the conservative administrations of that time chose to either ignore or condemn, but not treat, the disease. In fact, some of the nation's more conservative representatives, in a disturbing echo of the thinking of the Middle Ages, postulated that AIDS was a form of divine punishment.

Thus, ideology and ignorance conspired at the end of the 20th century—a century in which man walked on the moon—to inhibit intervention in the early stages of a worldwide epidemic, in which millions have died and millions more will die. However, as the 1980s gave way to the 1990s, and as the AIDS epidemic spread among the general population, and as a conservative administration in Washington gave way to a more moderate one, governments in more of the industrialized world began to acknowledge the pandemic proportions of the AIDS scourge and its potential impact. Fortunately, these governments, through the United Nations, are now slowly beginning to undo years of foolish and destructive neglect.

They still, however, have a long way to go. In 2001, the United Nations Declaration of Commitment on HIV/AIDS set the admirable goal of educating 90 percent of young people about preventive measures to take against acquiring AIDS by 2005. This goal has not yet been reached. Less than 50 percent of this populace was reported that year to have, in the United Nations's words, a "comprehensive knowledge" of HIV preventive strategies. The disease has increased apace: Approximately 4.3 million people worldwide contracted HIV in 2006 and an estimated 2.9 million died of AIDS. In 2007, the United Nations reported 15 million AIDS orphans worldwide.

Underfunding by countries that can afford to fund research, education, and treatment is still, in 2007, caused by local and/or national ideologies and priorities. New drugs have been developed but blocked by some governments because of issues with the distribution of condoms voiced by some portions of the populace of these countries. In the United States and other countries, budgetary boundaries and internal needs have frequently taken precedence over international AIDS funding.

As the need continues and increases, the purse strings are loosening, but slowly. Acceleration and concentration have, in the 21st century, become pressing priorities as worldwide awareness of the vastness and steady progression of the disease has increased. Still, there is still much to learn and much to do to stem the rising tide of this, the most virulent and explosive of human pandemics yet.

Before getting on to the second contemporary example of human culpability in the face of a natural—perhaps cataclysmic—disaster, a word is in order regarding the twin phenomena known as El Niño and La Niña.

El Niño in Spanish is "the Christ Child," so named some hundreds of years ago by fishermen along the west coasts of Ecuador and Peru because a particular weather phenomenon occurred every December, around Christmas.

A warm current migrated eastward across the Pacific Ocean at roughly the same time each year, and when it arrived in the particularly bountiful fishing grounds these fishermen frequented, it blocked the rise of nutrient-rich cold water from the bottom of the Pacific and either killed or drove the fish from the fisheries. The fishermen, deprived of a livelihood by the phenomenon, took this time to repair their boats and homes.

El Niño, however, has been found to reach considerably farther than the fishing fields of Ecuador and Peru. Its periodic appearance—usually over a two- to three-year period—disrupts not only the lives of fishermen but also the lives of millions across the globe.

Most particularly, when El Niño joins the so-called Southern Oscillation—the ups and downs in southern Pacific barometers—weather patterns change all over the world. Normally, pressures are high on the eastern side of the Pacific and low on the western side. With a strong El Niño appearance, this is reversed, and easterly surface winds, which extend from the Galápagos Islands to Indonesia, are driven farther eastward.

In 1997 and 1998, a particularly strong El Niño phenomenon occurred, and its effects caused warm winters in the United States; warm waters against the California coast (which enriched the tuna catch there); droughts in Indonesia; Australia, North Korea, and parts of China; increased monsoon rains in Pakistan and India; and decreased rain in Central America.

Two years of El Niño are customarily followed by two years of La Niña (the little child). La Niña is the reverse of El Niño, and its weather effects are the reverse of El Niño also. Thus, weather phenomena in 1999 and 2000 included floods in China and North Korea, droughts in India and Pakistan, and an unusually cold and snowy winter in the United States.

Hundreds of years after their naming, the origins of El Niño and La Niña remain essentially mysteries to modern meteorologists, though some recent volcanic discoveries have presented some clues (see VOLCANIC ERUPTIONS AND NATURAL EXPLOSIONS, Introduction). What is certainly known is that when they occur, trade winds over the equator grow weak and allow warm currents to flow into the central and eastern Pacific, toward Ecuador and Peru. The warming of the ocean then causes the warming of the moist air above it, which in turn forms deep clouds and heavy rain along the equator, which in turn causes the major rain zone over the western Pacific to shift eastward, and the domino effect hits barometers and climates worldwide.

El Niño and La Niña weren't the only culprits causing climatic caprice at the end of the 20th century. In a sotto voce echo of the gigantic volcanic explosions of Java's Tambora in 1815 and Krakatoa in 1883, Mt. Pinatubo's eruption in the Philippines in 1991 set up a global haze of sulfurous aerosols that caught and reflected enough sunlight to cool the Earth about one degree Fahrenheit from shortly after the eruption until early into 1994.

And finally, in contrast to Pinatubo's cooling effect, there is the second most dramatic and appalling contemporary example of human culpability in the onset and sustenance of a major natural disaster: the very present and disturbing phenomenon of global warming. As the evidence of this dramatic deterioration of the planet mounts, the voices who call it a theory that has not yet been proven have grown dimmer. With the 2007 publication of the report of the Intergovernmental Panel on Climate Change (IPCC)—a consortium, established by the United Nations Environmental Programme, of over 130 countries, and prepared by 1,500 scientists from 60 countries—it has become nearly impossible for all but the most conservative local politicians or ideologues to deny the fact, the inevitability, and the implications of accelerating climate change and the role that human beings have played in its causes and its increase.

It is generally agreed that the burning of fossil fuels and the burning off of tropical rain forests has released unacceptable levels of so called "greenhouse gases" into the atmosphere.

The greenhouse effect is caused by the formation of a shield of carbon dioxide and methane—the principal products of the burning of fossil fuel. This shield, in turn, traps radiant heat, much the way glass catches heat inside a greenhouse or in an automobile on a summer day.

It's important for a shield of carbon dioxide to exist. Otherwise, the heat from the sun would bounce back into space and the Earth would cool down to an uninhabitable temperature of minus 65 degrees Fahrenheit ($-54°C$).

For thousands of years, this shield has remained constant, kept that way by the absorption of carbon dioxide by phytoplankton in the seas and trees in the forests. But in the last 200 years, since the Industrial Revolution, forests have been cleared and the Earth's waters have become polluted, and so concentrations of carbon dioxide in the atmosphere have increased dramatically, from 250 parts per million to 370 parts per million. In the same timespan, global temperatures have risen, and the scientists and environmentalists who believe that global warming is with us today see a link between the two. In fact, by 2005, the levels of greenhouse gases—especially carbon dioxide and methane—in the atmosphere were higher than at any time in the past 650,000 years.

Their prognostication: Carbon dioxide levels could reach 500 parts per million by the middle of the 21st century, which could mean a rise in the Earth's temperature of from three to eight degrees Fahrenheit, a temperature rise that has not occurred since the ending of the last ice age, between 10,000 and 18,000 years ago.

The facts seem to bear this out. The 1980s began a trend to set records for the hottest years since these records began to be kept in the 1850s. Then, seven of the hottest years yet occurred in the 1990s, with 1998 establishing a record. And with the coming of the 21st century, new breaching of old records began in earnest: 2005 produced the highest recorded annual average surface temperature worldwide and 2006 became the sixth warmest year since the advent of the documenting of such phenomena. The British Meteorological Office posited that 2006 might have broken all previous records if it had not been for La Niña, which cooled parts of the Pacific Ocean. At any rate, of the 12 hottest years on record, 11 have occurred since 1995, indicating a continuing warming trend. In the measured words of the 2007 report of the IPCC, "Average Northern Hemisphere temperatures during the second half of the 20th century were very likely higher than any other 50-year period in the last 500 years, and likely the highest in at least the past 1,300 years."

As a result of this, the Arctic ice pack has been steadily melting, and sea levels have been rising at the rate of approximately one-tenth of an inch (c. .25 cm) per year.

"You can't say that a hurricane or a flood or a drought is because of global warming," one moderate disaster expert told a meeting of meteorologists in Johannesburg, South Africa, in April 2000. "What you can say is that global warming makes any of these or all of them more likely."

In 2005—the first time in the 52 years that the names given to major storms ran out and Greek letters were used—a team of scientists published a study in the journal *Science* that surveyed global hurricane frequency and intensity from 1960 to 2005. The scientists found that the number of Category 1, 2, and 3 storms fell slightly while the number of Category 4 and 5 storms—which are the most powerful ones—had climbed dramatically. In the 1970s, there had been an average of 10 Category 4 and 5 hurricanes a year worldwide. From 1990 to 2005, the annual number had nearly doubled. Overall, the most severe storms grew from just 20 percent of the global total to 35 percent. Their power, measured by wind speed and duration, jumped 50 percent.

Warming oceans, then, have the ability to increase not only the frequency of storms, but also their power. These steadily warming oceans produced, from 1988 to the present writing, higher levels of vapor than those of cool oceans at a rate of about 1.3 percent more per decade. Water vapor increases rainfall intensity, and a dramatic example of this relationship was shown in Hurricane Katrina (see p. 290) during which rainfall exceeded 12 inches (30.4 cm) near New Orleans.

Because, then, more heat leads to more evaporation, more wind energy, and therefore more violence, a greater frequency of extreme events—not only storms but also droughts—can occur.

And events at the end of the 20th century have seemed to bear this out. Islands usually hit by a cyclone once in a century have been pounded four times in a decade. Rivers that used to dry up once every few decades are now failing to reach the sea on 100 or more days a year.

The hottest year of the 1990s, 1998, had a total of 80 separate natural disasters, caused, according to some meteorologists, by global warming and one of its products, El Niño. There were terrible floods in China, ice storms in the United States, a huge pall of smoke and flames over Indonesia, and some of the most destructive hurricanes ever to hit Central America.

The following year, 1999, which was the fifth hottest recorded for the planet up to that time, was almost as bad, with floods and windstorms and droughts that either repeated themselves before the victims could rebuild, or continued for years and years.

According to some scientists, global warming will only magnify these events. Rising sea levels, according to them, will multiply storm surges, and higher sea surface temperatures will guarantee more frequent hurricanes.

And what is probably worst of all, the drying up of rivers will desiccate hitherto fertile soil. Farmers in India and Pakistan who depend on spring floods face ruin. The Yellow River—which once flooded so often it was known as "China's sorrow"—failed to reach the sea on 226 days in 1997.

Texas has continued to lose irrigated land at a rate of 1 percent a year. Even more strikingly, the European heat wave during the summer of 2003 produced the warmest August on record, unleashed massive forest fires, and claimed an estimated 35,000 lives. Vegetation growth across Europe was reduced by about 30 percent, which meant that not only crop yields and forest growth descended more rapidly than normal but also ecosystems absorbed less carbon dioxide from the atmosphere and therefore released less into the atmosphere. All of this hindered plant growth in a way that had been unprecedented in Europe over the past century.

Another cause of drought and damaging ecosystems is the melting of sea ice and snow cover. Warm air obviously melts more ice, but this in turn warms the air further, thus setting a chain reaction in motion. Satellite pictures taken in 2005 showed that the extent of Arctic ice in September of that year dipped some 20 percent below the long term average for the month, thus melting an extra 500,000 square miles (1.25 million sq. km), or an area twice the size of Texas. This melting of the sea ice accelerates warming because dark-colored water absorbs heat from the sun that was previously reflected back into space by white ice. Extrapolated, predictions were that if the trend continued, the summertime Arctic Ocean would be completely ice-free before the end of the 21st century. The Greenland ice sheet is, in the same way, melting at an alarming rate.

To the population—both animal and human—of the land within the Arctic Circle, all of this has caused huge changes. Vorkuta, a Russian coal-mining city of 130,000, is an example. In 2005, the entire city began to crumble into nonexistence. Many of the city's homes and factories were built not on hard rock, but on permafrost, a layer of perpetually frozen earth. The permafrost began to melt steadily, turning the ground upon which the city was built into mush.

Farther north, the extinction of polar bears and other wildlife that exist in the Arctic Circle has seemed to scientists as a portent of things to come. By September 2004, the polar ice cap had retreated a record 160 miles north of the northern coast of Alaska. Researchers counted 10 polar bears swimming as far as 60 miles (96.6 km) offshore that year, and most of them had drowned. While it is true that polar bears can swim long distances, these distances are usually measured between sheets of ice, where these bears spend most of their time hunting and raising their young. As the ice sheets have begun to disappear, the polar bears have been pushed into the sea, where they are drowning in ever increasing numbers.

According to the IPCC report in 2007, this type of event will become widespread by the end of the 21st century if nothing is done to decrease the production of the greenhouse gases that are warming the planet. The report estimated that 10 million people from the Gulf Coast of the United States to the Netherlands to Bangladesh will be flooded out of their cities and there will be massive dislocations in populations, ecosystems, and species during that time.

According to the report, and a number of advocates for preserving the environment, there can be an arresting of this progression of destruction by the enforcement of clean air standards. There have been periodic international meetings—in Rio de Janeiro in 1992, Kyoto in 1997, and Montreal in 2005. The Kyoto meeting produced the Kyoto Protocol that established specific and binding greenhouse gas reduction and targets for all ratifying industrial countries. It set a timetable for the achievement of these targets—from 2008 to 2012—as well as a process to negotiate later commitments. Ratified by 163 states and regional economic integration organizations, with the exception of the major industrial countries of the United States and Australia, it was hailed as one of the rare examples of the establishment of a binding, international, multilateral regime, and the basis for the sustenance of the world environment.

The 2007 IPCC report therefore held out hope for amelioration of global warming, but only if acted upon *immediately* to curb greenhouse gases. As the report was released, senior fellow David Wheeler of the Center for Global Development warned that ". . . we confront a stark reality here. Millions of poor people will be displaced by sea-level rise that has been caused by the affluent West (which releases by far the greatest amount of greenhouse gases into the atmosphere). When this happens, current international turbulence may seem placid by comparison." Bert Melcher, a former director of the Colorado Regional Transportation District, also held out some hope but only ". . . if we scare the hell out of ourselves and begin to reverse the trends. There's no time left for self-deception."

And yet, if self-deception began to wane in 2007, denial in favor of continued profits from polluting industries still seemed to thrive. Between 1998 and 2005, the Exxon Mobil Corporation gave $16 million to 43 ideological groups in an effort to discredit the science behind global warming. In September 2007, Britain's Royal Society of scientists wrote the oil company asking it to halt support for groups that "misrepresented the science of climate change."

Exxon Mobil's reply was a public statement that the Royal Society letter and a report in 2007 by the Union of Concerned Scientists in the United States were ". . . yet another attempt [sic] to smear our name and confuse the discussion of the serious issue of CO_2 emissions and global climate change."

When the IPCC report was released in 2007, Oklahoma Senator James Inhofe, the former chairman of the Senate Energy Committee, who repeatedly called global warming "one of the greatest hoaxes ever perpetrated upon the American people," issued a press release that stated that the report ". . . is a political document, not a scientific report, and it is a shining example of the corruption of science for political gain."

The American Enterprise Institute meanwhile promised $10,000 payments, plus "travel expenses" and "additional payments" to any scientist, economist, or policy analyst willing to rip the IPCC report as "resistant to reasonable criticism . . . and prone to summary conclusions that are poorly supported by the analytical work."

However, all large corporations in the United States charged with extraordinary pollution of the atmosphere did not go on the offensive. On the contrary, a group of major firms—including Alcoa, BP America, Caterpillar Inc., Duke Energy, DuPont, General Electric, Lehman Brothers, and PG&E Corporation—joined with the environmental groups Environmental Defense, the Natural Resources Defense Council, the Pew Center on Global Climate Change, and the World Resources Institute to form the United States Climate Action Partnership (USCAP). The organization is a group of business and leading environmental organizations that have joined to call on the federal government to quickly enact strong national legislation that will require significant reductions of greenhouse gas emissions. Foremost among its proposals is a cap and trade program in which definite caps would be put in place for industrial pollution. Polluting industries could buy credits from less polluting firms in order to bring down the overall level of industrial climate contamination that has formed greenhouse gases.

The U.S. administration's reaction to the IPCC report was to list the amount of money it had spent in education about global warming. Other world leaders—with the exception of Australia (at that time the only other nation with the United States not to sign the Kyoto Treaty on global warming)—issued clarion calls. "We are on the historic threshold of the irreversible," former French president Jacques Chirac said. "In the face of this urgency, it is no longer the time for half-measures. It is time for a revolution."

"When climate changes run like a rabbit, worldwide politics move like a snail; either we accelerate or we risk disaster," opined Italy's environmental minister Alfonso Pecoraro Scanio. And Canadian environment minister John Baird called on Canadians ". . . to get ready for some tough decisions on reducing greenhouse gas emissions."

Still, as late as fall 2007, neither Australia nor the United States had signed the Kyoto Protocol, citing threats to the economic stability of each country as their reasons for withholding ratification. However, in November 2007, Labor Party candidate Kevin Rudd roundly defeated Prime Minister John Howard in Australia and, immediately upon taking office, signed the Kyoto Protocol.

Further, in December 2007, when more than 10,000 participants joined in the United Nations Climate Change Conference in Bali, the United States, the world's largest carbon polluter, remained the single most effective stumbling block to an international agreement to move forward tangibly and internationally to control greenhouse gas emissions.

Finally, after a series of stormy sessions, the United States acceded to the Bali roadmap, which will chart the course for a new negotiating process to be concluded by 2009 that will ultimately lead to a post-2012 international agreement on climate change. At the conclusion of the conference, United Nations Framework Convention on Climate Change (UNFCCC) Executive Secretary Yvo de Boer said that "Parties have recognized the urgency of action on climate change and have now provided the political response to what scientists have been telling us is needed."

Meanwhile, glaciers continue to melt, seas continue to rise, and natural disasters become more intense. Much of the future of the planet, it seems, is being written in boardrooms and government assemblies. It will be interesting, to say the least, to see if it remains this way.

Now, to the thinking that went into the criteria for inclusion in this compilation of natural disasters.

While it is absolutely and tragically true that any disaster that affects one human being is, for that human being and those around him or her, a horrible happening, any attempt at fairly representing all of these sorts of tragedies would be a hopeless and endless task for writer and reader alike.

Therefore, some attempt at constraint had to be made, and this, in turn, came about through a combination of scientific and subjective judgments. It would have been easy, perhaps, to be entirely scientific, and to include only natural disasters that were of a certain magnitude. But that would be inaccurate on a number of levels.

First of all, certain natural disasters affect more people than others. A volcanic eruption of enormous power occurring on a remote and sparsely inhabited island, for instance, will not kill as many people as a slightly lesser one near a populated city. In such a case I chose a lesser explosion that caused more human damage.

A plague that takes place before the development of modern medical science will have more human casualties than one that takes place after this development. Which to choose? I tried to balance casualties with contemporary medical discoveries.

The same considerations were made when casualty figures were measured before and after the development of modern meteorology, seismology, and the entirely new science of volcanology. Obviously, those who are warned about approaching cyclones, tornadoes, typhoons, and hurricanes will, in most cases, clear out, as will those who are warned of the possibility of earthquakes or volcanic explosions. And those who don't form a small enough minority to affect final statistics very little.

I picked no earthquake under 6.5 on the Richter scale, except when the human factor, the suffering caused to people, outweighed this. Major disasters that did not fit this criteria are included in the chronological listing at the beginning of each section. References covering these events appear in the bibliography.

Casualty figures for natural disasters seem to bear out Mark Twain's assertion that "There are three kinds of lies—lies, damned lies and statistics." No two sources for casualty counts seem to agree, and sometimes the differences are both wide and incredible. When there were huge discrepancies, the disparate figures were indicated in the text.

AVALANCHES AND LANDSLIDES

THE WORST RECORDED AVALANCHES AND LANDSLIDES

* Detailed in text

Afghanistan
* * Northern (1997)
* * Salang (1998)

* Alps
* * (218 B.C.E.)
* * Tyrol (1915–18)

Austria
* * Blons (1954)
* * Galteur (2/1999)
* Galteur/Vent/Graz (12/1999)
* Montafon Valley (1689)

Brazil
* * Rio de Janeiro (1966) (1967)

Canada
* Alberta (1903)

China
* Kansu (Gansu) (1920) (see EARTHQUAKES)
* Kwangtung (Guangdong) Province (1934)
* Northwest (1983)
* Szechuan (Sichuan) Province (1981)

Colombia
* * Medellín (1987)

Ecuador
* Quito (1983)
* Southern (1993)

France
* * Les Orres (1998)
* * Le Tour/Montroc (1999)

Haiti
* Berly (1954)
* * Grand Rivière du Nord (1963)

Honduras
* Choloma (1973)

Iceland
* * Sudavik (1995)

India
* Assam (1948)
* * Bihar, West Bengal and Assam (1968)
* Darjeeling (1899)
* * Darjeeling (1980)
* Himalayas (1880) (1995) (1998)
* Jammu (1998)
* Sikkim (1998)
* * Uttar Pradesh (1998)

Iran
* * Roudehen (1998)

Italy
* * Belluno (1963)
* Chiavenna (1618)

Japan
* * (1972)
* * Niigata (1964)

Malaysia
* Kuala Lumpur (1993)

Mexico
* * Minatitlan (1959)

Nepal
* * Katmandu (1995)
* Mount Everest (1996)

Pakistan
* * Kel (1996)

Peru
* * Chungar (1971)
* * Huascarán (1962)

* Lima (1971)

Philippines
* (1814) (see VOLCANIC ERUPTIONS)
* * Leyte Island (2006)

Russia
* Caucasus (1997)
* * Pamir Mountains (1990)
* Karmadon, North Ossatía (2002)

Southeast Asia
* Afghanistan, India, Pakistan (2005)

Switzerland
* * Elm (1881)
* * Goldau Valley (1806)
* Great St. Bernard's Pass (1499)
* * Leukerbad (1718)
* Marmolada (1916)
* Panixer Pass (1799)
* * St. Gervais (1892)
* * Vals (1951)

Tajikistan
* * Anzob Pass (1997)

Turkey
* Northern Anatolia (1929)

United States
* * California (1969)
* Washington
* * Wellington (1910)
* (1980) (see VOLCANIC ERUPTIONS)

Wales
* * Aberfan (1966)

CHRONOLOGY

* Detailed in text

218 B.C.E.
October
* * Alps

1499 C.E.
Great St. Bernard's Pass, Switzerland

1618
September 4
Chiavenna, Italy

1689
Montafon Valley, Austria

1718
January 17
* * Leukerbad, Switzerland

1799
October 5
Panixer Pass, Switzerland

1806
September 2
* * Goldau Valley, Switzerland

1814
Philippines (see VOLCANIC ERUPTIONS)

1880
September 18
Himalayas, India

1881
September 11
* Elm, Switzerland

1892
July 12
* St. Gervais, Switzerland

1899
September 23–24
Darjeeling, India

1903
April 29
Alberta, Canada

1910
March 1
* Wellington, Washington

1915–1918
* Alps, Tyrol

1916
December 12
Marmolada, Switzerland

1920
Kansu (Gansu), China (see EARTHQUAKES)

1929
July 22
Northern Anatolia, Turkey

1934
May 23–25
Kwangtung (Guangdong) Province, China

1948
September 18
Assam, India

1951
January 20
* Vals, Switzerland

1954
January 11
* Blons, Austria
October 22
Berly, Haiti

1959
October 29
* Minatitlán, Mexico

1962
January 10
* Huascarán, Peru

1963
March 8
Tempayaeta, Southern Andes
October 9
* Belluno, Italy
November 13–14
* Grand Rivière du Nord, Haiti

1964
July 18–19
* Niigata, Japan

1966
January 11–13
* Rio de Janeiro, Brazil
October 21
* Aberfan, Wales

1967
February 17–20
* Rio de Janeiro, Brazil

1968
October 1–4
* Bihar, West Bengal, and Assam, India

1969
January 18–26
* California

1971
February 19
Lima, Peru
March 19
* Chungar, Peru

1972
July 17
* Japan

1973
September 20
Choloma, Honduras

1980
May 18
* Mount St. Helens, Washington (see VOLCANIC ERUPTIONS)
September 7
* Darjeeling, India

1981
May 18–20
Szechuan (Sichuan) Province, China

1983
January 12–14
China (Northwest)
September 8
Quito, Ecuador

1987
September 27
* Medellín, Colombia

1990
July 15
* Pamir Mountains, Russia

1995
January 13
* Sudavik, Iceland
January 18
* Himalayas, Kashmir
November 11–12
* Katmandu, Nepal

1996
March 15
* Kel, Pakistan
September 25
Mount Everest, Nepal

1997
March 26
* Northern Afghanistan
October 23
* Anzob Pass, Tajikistan

1998
January 13
* Roudehen, Iran
January 23
* Les Orres, France
February 26
Jammu, India
March 7
* Salang, Afghanistan
March 31
Sikkim, India
August 17
* Uttar Pradesh, India
October 16
Himalayas, India

1999
February 9
* Le Tour, Montroc, France
February 23
* Galteur, Austria
December 28
Galteur/Vent/Graz, Austria

2002
September 20
Karmadon, North Ossatía

2005
January–February
Afghanistan, India, Pakistan

2006
February 17
* Leyte Island, Philippines

AVALANCHES AND LANDSLIDES

Avalanches and landslides are usually secondary disasters, caused by such primary natural occurrences as heavy snowfalls, monsoon rains, volcanic eruptions, or earthquakes.

To occur at all, an avalanche needs an insecure base. Snow that has accumulated upon a mountainside can be loosened by tremors, echoes, or uneven melting of the snow base. Secure landmasses can be turned to mud by weeks of unrelieved rain. The underpinnings of cities can be rattled loose by repeated natural or man-made Earth tremors or by the superheating of subterranean depths under the soil brought on by volcanic activity.

Whatever the cause, an avalanche is sudden, unanticipated, and violent. Mountains, lake, and seaside waterfronts and entire population areas can be abruptly uprooted. Helped by accumulated force, speed, and gravity itself, these avalanches generally grow in size and destructive power as they accumulate loose debris, rocks, soil, trees, water, and anything that happens to be unlucky enough to find itself in their paths.

The most spectacular and lethal avalanches occur in regions of heavy snow and ice. In these places, as much as 1 million cubic yards of snow have been known to give way and thunder downslope at a time—an amount that would fill the beds of 10,000 10-cubic-yard (7.6 cu. m) dump trucks, which in turn is equivalent to a line of vehicles that would reach for approximately 200 miles (321.9 km).

Avalanches of snow can be triggered by the most delicate vibration, which is why some Alpine farmers muffle the bells on their livestock in winter. A falling stone, the movement of animals, thunder, the sonic boom of a jet, a man slicing across a slope on skis can become the genesis of an avalanche of snow and ice.

The results are sometimes instantaneous; sometimes delayed. Several skiers may safely cross a slope before the cumulative effect of their passing sets off vibrations in the snow that catch the last members of the party in an avalanche. Or, a rupture may streak through compacted snow from a point of first impact to break the anchorage of nearby snow and start a second avalanche. Fractures can run through ice at speeds of 350 feet per second (106.7 m/s).

One final, lethal effect of dry snow avalanches should be noted: They travel quickly. On a 43-degree slope in the Swiss Alps, one of them was clocked at almost 300 MPH. At these speeds, dry snow becomes airborne. Progressing downslope, its powdery clouds disturb the air in front of it, causing hurricane force winds, and these winds in turn whirl still more loose snow into frenetic suspension. Wind speeds within the vanguard of these avalanches have been recorded at nearly 400 feet per second (121.9 m/s), and the wind, like an unseen plow, can blast trees and buildings, toppling them even before the snow arrives. Death has struck skiers and mountain climbers and mountain residents untouched by the racing snow. When autopsied, their lungs have shown lesions of the sort produced by explosions.

Avalanches will occur wherever the conditions exist for them, and there is no avoidance except absence from them. Those who ski, who love to live amidst some of Earth's most beautiful and breathtaking scenery, or to travel to it, may well find this an unacceptable impossibility. For them, the experience is worth the risk.

The criteria for inclusion in this section are based upon size, fatalities, and/or unusualness. If the avalanche was a unique one, casualty figures were ignored. If not, large numbers of casualties—exceeding 1,000—became the yardstick.

AFGHANISTAN
NORTH
March 26, 1997

More than 100 workers, forced by the fighting between the Taliban army and its enemies to walk the 10 miles (16.1 km) from their homes to a bus on the northern end of the Salang tunnel in Afghanistan, were killed when an avalanche buried them on March 26, 1997.

In September of 1997, the Taliban, a hard-line, basic religious movement, conquered the southern two-thirds of Afghanistan, including its capital, Kabul. The Taliban army then went about setting up barriers between its two-thirds of the country and the northern third, which remained in the hands of the Taliban's enemies, an alliance of four groups led by former Afghanistan president Burhanuddin Rabbini and northern warlord Rashid Dostum. Fearing fresh fighting along these barriers, the Taliban closed down the Salang Highway, the only road between Kabul and northern Afghanistan—a wise military move, perhaps, but the severing of a lifeline for Afghani citizens who lived on one side of the border and worked on the other.

Only automobiles, buses, and trucks were barred from using the highway, however, and so, in order to keep their jobs in Mazar-i Sharif, these workers rose early and trekked by foot the 10 miles (16.1 km) across foothills and through the Salang tunnel, where they caught a bus to Mazar-i Sharif.

The morning of March 26, 1997, was cold and windy. Snow had fallen the day and night before, and walking along the highway was treacherous. Nevertheless, nearly 200 workers from the south set out that morning for the 10-mile (16.1 km) hike to the northern end of the Salang tunnel.

A small percentage of them had entered the tunnel, but most of them were walking along the road when the warning roar of an avalanche froze them in their tracks.

There was no escape. Tons of snow, loosened by the wind and the overload from the storm of the day before, thundered down the mountain, directly at the workers walking along the highway.

In an instant, they were either buried where they stood, or swept off the road and into the moving wave of frozen matter. And then it was over. Less than 30 survivors struggled, some with broken limbs and gaping wounds, away from the path of the avalanche.

There, they waited in subfreezing cold for rescue. It came in the form of old buses and trucks into which they were loaded and taken to Jebul Siraj and Charikar, the nearest towns. Over 100 of those on the forced march to work in the north were left behind, buried and dead beneath the detritus of the avalanche.

AFGHANISTAN
SALANG AREA
March 7, 1998

A small village, in the path of an avalanche that descended from the Hindu Kush Mountains in the Salang area of Afghanistan, was nearly demolished on March 7, 1998. Seventy of its residents were killed by the wall of snow that raged through the settlement.

The opposition to the Taliban Islamic militias governed the northern third of Afghanistan (see previous entry), and in the winter, their position was made more risky by the assault of snow and subfreezing temperatures. This was particularly difficult for those who lived and worked in the mountainous terrain of the Salang area and the Hindu Kush Mountains, notorious for the multiple avalanches that roar down their slopes every winter.

On March 7, 1998, the string of small farming villages at the foot of the mountains was steeped in snow from the seemingly unceasing storms of that winter. One such village, 75 miles (120.7 km) north of Afghanistan's capital, Kabul, was the recipient of an especially ferocious and large avalanche that roared unexpectedly down upon the village, crushing houses, uprooting trees and moving structures, and then burying them.

It would be a day before local rescuers could dig their way to the isolated village, where over 70 residents were killed by the huge wall of snow that devastated everything in its path.

ALPS
218 B.C.E.

Impatience caused the deaths by avalanche of 18,000 of Hannibal's men, 2,000 of his horses, and several of his elephants as he ordered them across the Col de la Traversette pass of the Italian Alps following an early October snowstorm in 218 B.C.E.

Hannibal met his match in natural forces in the early days of October 218 B.C.E. when, commanding 38,000 soldiers, 8,000 horsemen, and 37 elephants, he unwisely attempted to cross the Col de la Traversette pass in the Italian Alps. At least, according to incomplete records, that is the approximate location he had reached, in a heavy snowstorm and bone-numbing cold.

For two days, he and his expedition camped at the top of the pass. Finally, growing impatient, they resumed their march. A blanket of fresh snow covered the crusted snow of an earlier storm—a prime condition for an avalanche—and, as they began to descend, the animals' feet perforated the top layer of snow. That layer gave way and, in the words of the poet Silius Italicus, ". . . [Hannibal] pierce[d] the resistant ice with his lance. Detached snow drag[ged] the men into the abyss and snow falling rapidly from the high summits engulf[ed] the living squadrons."

Eighteen thousand men, 2,000 horses and several elephants perished in this historic and horrendous tragedy.

ALPS
TYROL
1915–1918

An estimated 40,000 to 80,000 men lost their lives during World War I in the Tyrolean Alps, not from enemy gunfire, but from avalanches caused by the sounds of war.

World War I was fought on many fronts, but one of the most dramatic—and disastrous—was in the Tyrol, where Italian and Austrian troops battled each other on forbidding terrain for three years.

During that time, there were appreciable casualties from gunfire. But by far the most lethal enemies of both armies were mountains and snow. Rattled by explosions and the noises of war, avalanche after avalanche cascaded down Alpine slopes, burying such villages as Marmolada, where, in one day, 235 men were lost, buried in their barracks.

The final death toll from avalanches was estimated to be between 40,000 and 80,000 men—an accumulation of lost life that eclipsed that of similar battles fought on more level terrain.

AUSTRIA
BLONS
January 11, 1954

The worst mass burial, by percentage, in history occurred when two avalanches roared into the small Austrian village of Blons on the same day—January 11, 1954. One hundred eleven of the 376 residents of the village were killed; 29 of its 90 homes were destroyed; 300 of the approximately 600 miners in the Leduc mine were buried alive.

The all-time record for mass burial by avalanche belongs to the twin slides that hit the small Alpine village of Blons, near Arlberg Pass in Austria at 9:36 in the morning and at 7:00 in the evening of January 11, 1954. Half of the miners in the nearby Leduc mine camp were killed; 111 of the 376 people who lived in the village perished outright; 29 of 90 homes were crushed.

This catastrophic loss of life came despite the villagers' constant preparation for the avalanches they had come to expect each winter. Every December, in fact, the village councilmen ordered the removal of the crucifix that stood close to a certain ravine so that it could survive the next months without damage. When villagers crossed a certain bridge over that same ravine, they automatically walked in widely spaced single file and stopped talking. If their voices or some other vibration began an avalanche, they reasoned that with wide distances between them, fewer would perish.

But there was no defense when these two slides—the largest and most powerful that have ever visited Blons—roared down the mountains. Of those who were buried, 33 extricated themselves, 31 were dug out alive by rescuers, and 47 were found dead. Eight survivors later died, and two residents were never found.

One woman perished from burns, even though she was buried in snow. She had been baking when the avalanche hit her house and coals from the oven seared her as she was swept downslope. A man who was trapped for 17 hours emerged alive when a search party reached him, but, according to wire service reports, he died of shock when they told him how long he had been under the snow. Others survived after being trapped under the snow for up to 62 hours.

AUSTRIA
GALTEUR
February 23, 1999

The worst series of snowstorms to hit Europe in 50 years rendered avalanche protection useless as a 16-foot-high snowslide roared down upon the Austrian village of Galteur at 4 P.M. on the afternoon of February 23, 1999. Thirty-eight tourists and villagers were killed.

Galteur, a small Austrian town in the Paznaun Valley of the Austrian Alps, near the Swiss border, had lived a calm, nearly charmed existence. Twenty million dollars worth of avalanche protection structures assured the flow of yearly tourists and professional skiers that there would be no repetition of the snowslide of 1689, in which 250 people died.

However, the winter of 1999 was a record-setter. All over Europe, snowfalls and the threat of avalanches isolated villages, stranded travelers in train stations, created traffic jams on buried highways, and disrupted rail traffic (see France, p. 9). In the Austrian Tyrol, some 20,000 tourists were stranded in February, and thousands of others were holed up in Vorarlberg.

The resort and skiing village of Galteur was cut off by the middle of the month by unprecedented heavy snow. Its 700 inhabitants and 3,000 tourists were unable to leave, but, because of the multimillion-dollar concrete barriers, they felt secure.

But this was no ordinary winter. As Thomas Huber of the Tyrolean Avalanche Prevention Bureau in Innsbruck explained it later, "It [had been] snowing for the last three weeks during which many differing layers [of snow] . . . formed. There [was] a mixture of new snow and wet snow together with high humidity and this [formed] an upper layer that [was] heavier than the ones below."

Shortly after 4 P.M. on February 23, a wall of snow 16 feet (4.9 m) high tumbled down the mountain and roared through the center of Galteur, burying some houses, flattening others, and leaving a track of horrible devastation behind it.

"That was not snow. It was like concrete," a Dutch tourist told an Austrian television reporter days later. It would be two days from the time of the avalanche until rescue teams arrived in the village. The roads leading into the village were blocked by huge drifts; snow was falling heavily when the disaster struck, and continued through the night, dropping another 20 inches (50.9 cm) of powdered snow on the wreckage.

Finally, the next afternoon, the storm lessened enough to allow helicopters from the Pontlatz Austrian army base to fly to the scene. Local firefighters and police and dozens of volunteers had already been hard at work, digging through the rubble and snow, probing into it with long metal poles.

Cars were crushed and hurled across the village. The avalanche had sliced the top off one house as cleanly as if it had been a razor blade. The rescuers, now aided by avalanche dogs brought in by helicopter, dug in the vicinity of houses first, since those trapped in houses stood a better chance of surviving than those caught in the open, where the heavy weight of the snow could suffocate them.

By late afternoon of the 24th, a steady stream of helicopters was flying in pallets of fresh fruits, vegetables, and foodstuffs, and evacuating survivors and tourists on the return trip.

"This is a catastrophe such as we have not had for centuries," said Wendelin Weingartner, the governor of Tyrol province, to reporters.

The final death toll was 38, with scores more injured, some of them seriously. There were none missing; all who had been caught in the path of the frozen juggernaut were eventually accounted for.

BRAZIL
RIO DE JANEIRO
January 11–13, 1966; February 17–20, 1967

Inattention to maintaining the terrain of the hills above Rio de Janeiro, Brazil, led to the enormous loss of life and property caused by the twin avalanches of January 11–13, 1966, and February 17–20, 1967. The slides, which followed a month of heavy rain in each case, killed 259, injured hundreds more, and crippled the city's three power plants.

Prior to the punishing landslides of 1966 and 1967, not much attention was paid to the insecure terrain of the hills above Rio de Janeiro nor to the obvious fact that mountainsides deprived of the roots of vegetation are prone to erosion and resultant slides following heavy rains.

For generations, the poor of Rio had built their shanties in what was called the *favelados* district, chipped into the sides of the hills above the city. The shacks were ramshackle at best and clung precariously to the hillsides.

Torrential rains battered that part of Brazil in early January 1966, causing only inconvenience to the rest of the country. But near Rio, from the 11th through the 13th, mudslides began to rumble and then slide inexorably toward the city. Two hundred thirty-nine residents of the districts of Santa Teresa, Copacabana, and Ipanema were killed and hundreds injured as their houses were either swept away or crushed by the walls of mud that caromed down the mountainside, burying or sweeping away everything in their paths.

The reaction of both residents and officials was either naive or philosophical. "Rains like this happen only once in a century," stated Governor Francisco Negrão de Lima of Rio. And so, the shacks were reconstructed, and still more foliage was stripped from the hills.

In February 1967, a little more than a year later, an 11-inch (27.9-cm) rainfall soaked the city of Rio de Janeiro. Once again, mudslides cascaded from the hills, toppling and smashing the primitively constructed homes, crushing and burying the *favelados* beneath them.

But this time, the wealthy and the white-collar workers who had been unaffected by the previous slides felt the impact. Mudslides crept into Rio's three power plants, cutting the city's electrical supply to 40 percent, which in turn caused air conditioners to burn out and elevators to malfunction. A score of heart attacks felled businesspeople who were forced to descend countless stairs in skyscrapers, thus raising the total of persons killed because of the landslides to 259.

This time, Governor de Lima forbade more construction in and about the hills. He set up a geological "police force" to patrol the mountains, instituted new laws to prevent the destruction of vegetation and sent out military helicopters to see that his rules were enforced. In addition, concrete shorings were set into the slopes to prevent what geologists predicted could be a burial under mud of most of the city of Rio de Janeiro.

COLOMBIA
MEDELLÍN
September 27, 1987

Mudslides caused by torrential downpours wiped out whole sections of the Villa Tina area of Medellín, Colombia, on September 27, 1987. Five hundred residents disappeared and were presumed dead, the bodies of 183 were found, and 200 were injured. Only 117 residents of the district survived the catastrophe.

The Villa Tina area of Medellín, a city in the mountains 160 miles (257.5 km) northwest of Bogotá, Colombia, is an achingly impoverished one, despite the affluence of the drug cartel based in the rest of the city. The entire population of approximately 1,000 lives from day to day, sometimes 10 to a dwelling, in improvised shacks.

On Sunday morning, September 27, 1987, tons of red soil, turned to thin mud by days of torrential rains, loosened from Sugar Loaf Mountain, and, picking up enormous boulders in their plummeting fall downslope, thundered into the Villa Tina section, just as a group of young children were receiving their first Communion.

An Associated Press reporter spoke to survivors. "We heard the noise that sounded like an explosion and soon afterward a huge mass of rocks and mud descended upon us," said Mary Mosquera, who lost three daughters in the avalanche. They were three of 43 children (seven died in church during the Communion ceremony) who would perish in a matter of minutes that Sunday morning. One hundred eighty-three people would be crushed under the roaring juggernaut of rocks and mud; nearly 500 would be missing and presumed dead; 200 would be injured. That would leave a mere 117 survivors to try to rescue their neighbors or families.

The destruction of the shanties in which they had lived was complete. Rescue workers, laboring for two days, were guided to many bodies by dogs howling at the spot at which their owners were buried.

The mayor of Medellín, William Jaramillo Gomez, ordered that all victims, many of them unidentified, be buried immediately to prevent any outbreak of disease. At least 50 people were interred in a mass grave.

All of this was done while the soaking rains continued to fall, threatening new slides. Thousands of people took refuge in shelters more substantial than their homes as several small slides hurtled down the mountain the night after the avalanche.

FRANCE
LES ORRES
January 23, 1998

Nine students and two adult guides were killed and 20 other students were injured when they triggered an avalanche as they were snowshoeing on an unmarked trail in the French Alps near the ski resort of Les Orres on January 23, 1998.

For 28 years, the resort area near Les Orres, in the French Alps, had been relatively free of major avalanche disasters. All of that changed when the worst disaster of its kind to hit the French Alps since 1970 killed nine students and two adult guides, injured 20 other students—six of them seriously—and left two missing. And this wouldn't have happened if the group itself hadn't ignored avalanche warnings and set out, against expert advice, to trek in snowshoes across a remote mountain area close to the Italian border.

The group of 34 students, ranging in age from 13 to 15, came from the St. Francis of Assisi school in Montigny-le-Bretonneux, south of Paris. They were on a holiday outing that had been momentarily kept cabin-bound by a heavy snowstorm that had dropped over three feet of snow in two days. Finally, on the 23rd of January, their eight adult supervisors, which included four mountaineering experts, decided it was time to tie on their snowshoes and do a little exploring.

Michael Roussel, the manager of the cabin in which the group was staying, vigorously objected, citing weather forecasters who had warned of an avalanche risk in the area rated at 4 on a scale of 5. "I told them they shouldn't go; there are warning signs everywhere," Roussel told Reuters reporters after the tragedy.

"We knew we had to be careful, that it was dangerous, but we were not really afraid," a young and unidentified survivor told Radio Alpes. Later, another 13-year-old boy testified that the mountain guides laughed at the teenagers for being afraid to take the trip along an 8,000-foot-high trail.

Finally, opposition silenced, the 26 teenagers and six adults set out on the marked trails through the forested area. The serenity of newfallen snow was relaxing, and the guides set the party along unmarked trails. Confidence and euphoria apparently united to beget carelessness. The experienced mountaineers should have steered the group clear of a patch of thickly packed snow, which overlaid more unstable snow. But by the time the group was in the middle of the patch, it was too late.

Their weight and movement broke the snowpack loose, dissolving the ground beneath their feet; now encased in a 1,000-foot-wide (305 m-wide) avalanche, they plunged down the mountainside. Some were instantly buried under the cloud of snow. Others were smashed into the intervening trees.

"Many of the victims were stuck in trees," a military police dog handler said later. "The kids who survived were screaming in panic. Some were wearing nothing more than a T-shirt because they had climbed the slope and had not yet cooled down and put their jackets on."

Rescue was swift and expert. Over 150 volunteers combed the deathly still countryside after the avalanche, probing in the snow with long poles for the dead, using military-trained dogs to sniff out survivors. Their search was hampered by scores of uprooted trees that had been torn from the mountainside by the swift-moving wall of snow.

The next day, grief therapy experts accompanied approximately 50 shocked parents on a grim round of hospitals. "This catastrophe . . . has saddened the whole nation," France's Prime Minister Lionel Jospin stated to the press as he accompanied the mourners.

The grim total, after much misinformation, stood at 11 dead (which included nine students and two adults), two missing and believed dead, and 20 injured. Forty-two-year-old Daniel Forte, one of the surviving professional guides, was taken into custody by French police for an investigation of suspicion of manslaughter. He was later acquitted, but his career as a mountain guide was justifiably terminated.

FRANCE
LE TOUR, MONTROC (MONT BLANC REGION)
February 9, 1999

On February 9, 1999, over 20 residents and tourists were killed in the worst avalanche to strike the Chamonix Valley in southeastern France on the face of Mont Blanc in 91 years.

"This was more like a California earthquake than an avalanche," said American skier Nathan Wallace after he and his girlfriend Alicia Boice had been rescued from the Chamonix valley's worst avalanche in 91 years.

It had been an unusually stormy winter so far. Snow had blanketed Europe as far south as the French Riviera and Rome, which had not seen snow since 1986. But it was a paradise for skiers, and hundreds of them filled the ski resorts in the French Alps. Chamonix, in the shadow of Mont Blanc, near the Swiss border, was one of the most popular and well skied destinations for these international sportspeople.

The snowfall was particularly heavy in the first week of February. Up to 16 inches (41 cm) fell in the Swiss Alps; in the Tyrol of Austria, thousands were trapped at resorts piled high with five days of steady snowfall. In northwestern Romania, a major highway was blocked by a major avalanche. Rail service between France and Switzerland was cut off. And in the French Alps, warnings were released by the authorities from the end of January onward. However, no serious avalanches occurred, and by the weekend of February 5 and 6, every available space, including the outlying villages around Chamonix, was packed with tourists.

And then, slightly before dawn on Tuesday, February 9, a huge field of snow loosened itself from the face of Mont Blanc and rocketed downward at 120 miles an hour directly toward the villages of Le Tour and Montroc-le-Planet. Chalets in the path of the avalanche were ripped off their foundations and crumpled like matchboxes slammed with a fist. Some literally exploded from the impact. Others were pushed miles from their original foundations, where they either came up against barriers and broke apart, or were buried under the mountainous snow.

Rescuers were prevented from entering the area for a day because of the risk of further snowslides, which set smaller quantities of snow and ice tumbling into the area, covering the wreckage with layer after layer of powdered snow.

Finally, dogs and men were able to reach the remains of the two villages. "There were blocks of cement and gravel everywhere," Jean-Marie Pavy, who survived at the edge of the disaster told reporters later. "It was the apocalypse."

Twisted metal from smashed cars and splinters of wood from crushed chalets jutted from hills of snow. The frozen crust was so hardened by the end of the day on the 12th that rescue workers resorted to drills and heavy machinery to bore through it. The force of the avalanche, which had hurtled 3,000 feet (914.4 m) down the mountain, forced it 300 feet (91.4 m) up the opposite slope, where it smashed 17 chalets in an area that had been registered as safe for construction.

In the middle of the rescue effort, news arrived of a major avalanche that struck the nearby village of Les Bossons, about 3,600 feet (1,097 m) high and in a straight line under the 15,771-foot (4,807 m) summit of Mont Blanc. The village, however, was spared, thanks to avalanche-breaks, shields of concrete blocks that had been anchored on the slope above to slow down and split up the snowslide. Thus, there was no damage except to the nerves of the villagers. "We heard a loud bang and felt strong wind," said villager Gilbert Cumin. "Chairs on the terrace flew up in the air and the lights went out."

Meanwhile, back in the buried villages of Le Tour and Montroc, the grim task of unearthing bodies from the wreckage continued. Snowplows, bulldozers, and dogs dug through the 20-foot-deep (6-m-deep) snow and debris, often finding nothing but foundations and beams, with furniture and mattresses strewn yards away. Over 20 bodies were uncovered and 20 chalets were smashed to pieces.

It would be several weeks before the area returned to normal. In the meantime, the Chamonix town hall and various schools were turned into shelters for the hundreds of residents and skiers who had been either injured or dispossessed by one of the area's worst avalanches, ever.

HAITI
GRAND RIVIÈRE DU NORD
November 13–14, 1963

Mudslides brought about by long periods of tropical downpours killed 500 residents and tourists in Grand Rivière du Nord on November 13 and 14, 1963.

Five hundred persons were killed when landslides devastated Grand Rivière du Nord, Haiti's most prosperous area, on November 13 and 14, 1963.

The slides, brought about by days of relentless rain during and following a hurricane, covered villages with mud and rock, flattened dwellings and totally destroyed all of the crops in the region.

Scores were crushed in their homes; scores more drowned when rivers overflowed their banks after mudslides displaced their waters. A bus carrying 20 passengers was picked up and hurled from a highway, then buried beneath the landslide that had displaced it. All 20 passengers and the driver died.

Communications with Port-au-Prince and the outside world were down for days, thus hampering the arrival of adequate relief supplies.

ICELAND
SUDAVIK
January 13, 1995

Most of the village of Sudavik, Iceland, was either destroyed or dislodged by an early morning avalanche on January 13, 1995. Fourteen residents, most of them children, died in the incident.

There is an avalanche zone in the mountainous south of Iceland, and its inhabitants, used to the vagaries and power of nature, stay well away from it during avalanche season. But the assault of snow and ice that slid from the higher elevations in the *vestfirdi*, or western part of Iceland, and obliterated the small village of Sudavik, in the northwest, came without warning or precedent.

Dawn was a long way to the east when the first rumblings occurred just before 6 A.M. on January 13. It had been snowing long and steadily enough to fill the crevices and slopes of the entire country, and it was possibly this overload of snow that caused the slide to begin shortly after 6 A.M. that fatal morning.

The mixture of ice and snow, as wide as the length of two football fields, thundered into the precise middle of the small, sleeping village. Its force was that of a tsunami, as it split municipal buildings in two and shattered houses into multiple pieces. Dwellings were ripped from their foundations and slid ahead of the main, frozen body of the slide. One was moved, intact, 100 feet (30.5 m) from its foundation.

Most citizens escaped either safely or with minor injuries. But 14 residents of Sudavik, most of them children, were crushed under falling buildings or the rushing mountain of ice and snow.

Although this seems to be a small number, all fatalities are relative. Iceland's small population made this a national tragedy that garnered the instant gathering of a disaster relief fund. Within a week, it totaled 300 million kroner, or about $3,000,000.

INDIA
BIHAR, WEST BENGAL AND ASSAM
October 1–4, 1968

One thousand residents of Bihar, West Bengal and Assam were killed by a combination of landslides and floods and the oil pipeline linking Bihar and Assam was largely destroyed between October 1 and 4, 1968.

One thousand people were crushed or drowned when India's three northeastern states—Bihar, West Bengal and Assam—were inundated by floods and landslides between October 1 and 4, 1968.

The Tista River, just outside of Jalpaiguri, overflowed, submerging the town under 10 feet (3.05 m) of water and thus accounting for many of the deaths. But equally vicious and lethal were the landslides, brought on by a combination of floods and monsoon rains. These cascaded down from the Himalayas to the north and the Naga Hills to the east, crushing native homes and burying parts of the cities of Darjeeling and Jalpaiguri, which were left without electricity and drinking water for weeks.

The entire area, in fact, was cut off from the outside world and the 500-mile (804.6 km) oil pipeline between Bihar and Assam was largely destroyed by the multiple mudslides which washed out the roadways and railbeds that connected this part of India with the rest of the subcontinent.

INDIA
DARJEELING
September 7, 1980

Thirty thousand residents of the Darjeeling area of India were cut off from the outside world and 250 were killed when avalanches, caused by monsoon rains, descended from the Himalayas on September 7, 1980.

Tons of earth and enormous boulders swept down the Himalayas in the tea-growing area of Darjeeling on September 7, 1980, killing 250 people and trapping 30,000.

Monsoon rains had ravaged this part of West Bengal for days before the landslides, which were brought on by the loosening of the earth from the pounding of the rains. Sheered off from the mountainsides, these huge walls of mud uprooted trees and rolled enormous boulders ahead of them as if they were snowballs. Entire villages were instantly crushed and obliterated.

The landslides, coupled with floods, increased the 1980 fatality total from monsoons to nearly 1,500.

INDIA
KASHMIR
January 18, 1995

Over 200 died in the avalanche that buried hundreds of vehicles on the Srinagar-Jammu Highway in Kashmir on January 18, 1995. Over 5,000 were rescued from their trapped buses or cars, or from the 1.7-mile-long (2.7-km-long) Jawahar tunnel.

There is one road—a modern, 110-mile-long (177-km-long) paved highway—that connects the cities of Srinagar and Jammu, in the northern portion of the Himalayan region of Kashmir. In the winter of 1994–95, snow had fallen for three solid days upon the elevated region through which the highway passed. However, since it was situated in the foothills, the highway was considered to be a safe passage, and it was cleared of snow by the end of the last three-day storm. On January 18, 1995, it was clogged, as normal, with trucks, buses, and automobiles.

From early morning on the 18th, there were rumblings above the highway, small avalanches that produced larger avalanches, but there seemed to be no cause for alarm or warning.

And then it began. The accumulation of snowslides combined in a monumental mountain of snow and ice that roared down from above the highway at express train speed. It smashed into the road with the force of an earthquake, crushing cars, trucks and buses. Five buses were swept off the highway and down into the fields below the foothills.

Some of the riders in these buses were rescued, and some 400 motorists were either in, or managed to gain entrance to, the 1.7-mile-long (177-km-long) Jawahar Road tunnel, where they escaped the avalanche. Later, Indian air force helicopters dropped food and blankets on the snow outside the entrance to the tunnel.

The toll was terrible. Some 5,000 people were rescued from their vehicles, but over 200 died, suffocated beneath the accumulation of snow, ice, and debris.

INDIA
UTTAR PRADESH
August 17, 1998

Over 200 pilgrims, residents, and officials were killed when a square kilometer of land, loosened by monsoon rains, slid down a mountainside in the state of Uttar Pradesh, India, on August 17, 1998. The entire village of Malpa was swept into a gorge of the swollen Kali River.

The Indian state of Uttar Pradesh curves like a parenthesis on its back around Nepal. Its easternmost region is flat and solidly within India; its northwestern end, dominated by the high Himalayas, touches the border between India and Tibet.

Every year, from June through October, Hindu pilgrims, in groups of 60, are met by Chinese officials at the border and escorted to the holy lake at Kailash Mansarovar in Tibet. Mansarovar lake and Mount Kailash, which flanks it, are revered by Hindus as the abode of the god Shiva, and these pilgrims make the 27-day trek from the villages in Uttar Pradesh on foot each summer to give homage to Shiva.

The region through which they pass is highly landslide prone, because it has been glaciated over the past few centuries. "All such glaciated areas are prone to landslides because of the mixed soil," explained an Indian geologist to Reuters. To make matters worse and more dangerous, the landslide area is close to the Kali River, ". . . which can be ferocious when it is in spate," noted an Indian foreign ministry official.

In 1998, over 700 pilgrims gathered in Uttar Pradesh, and by the middle of August, hundreds of them had already made the trip, despite the blinding monsoon rains and a series of mudslides that made the trip treacherous. On August 17, over 150 of them were in Tibet, while over 200 were on the way in India, among them Protima Bedi, a noted Indian dancer and the mother of Pooja Bedi, one of India's most popular film actresses.

Torrential rains continued to pound the area, forcing the 12th batch of pilgrims for that year to erect tents around the village of Malpa, in the Pithoragarh district, which is 7,000 feet (2,133.6 m) up in the Himalayas, between the Kali River and a gorge. The village itself is tiny, consisting of 40 houses with an average of 10 people in each house, and isolated. Only four unpaved roads lead to it—one from Darchula, another from Gunji and Garbuan, and the last from the Nepal border.

Officials and environmentalists had warned for years that deforestation above the site in the high Himalayas had made the entire mountainside unstable. And yet, farmers and villagers continued to clear patches for growing crops and used the trees for firewood. Iqbal Malik, a noted environmentalist had written that the greenery in the area had been disappearing for a decade. "There is no root system to hold onto the soil and rocks," she had concluded.

In the summer of 1998, her words turned into prophecy. All through the month, the monsoon rains pummeled the mountainside. Early in the month, a slide killed 42 people, including 20 children, in the Garhwal section. Other landslides swept away electricity poles and killed hundreds of animals.

And then, at 12:30 A.M. on August 17, the entire side of a mountain let loose. An enormous, moving mass of mud and boulders, freed by the rains, detached itself and roared down on the village, sweeping everything ahead of it over the lip of the hillock and into the gorge, and from there into the swollen river. Buildings, telephone poles, shelters, animals, and human beings alike were immediately buried under the ooze or drowned in the river.

It would be hours before news of the landslide, relayed by radio from a border police squad camped nearby, would reach New Delhi, where rescue helicopters were stationed. And it would be days before the rain and fog would lift enough to allow the helicopters to fly to the site of the tragedy. In the meantime, rescue crews that attempted to reach the area by foot sank up to their knees in mud, and were forced to turn back. Two hundred of them had set out that day to clear road blockages, but the impossibility of making the 60-kilometer distance between the site and their camp set rescue efforts back another 24 hours.

It would be August 20 before military helicopters reached a scene of enormous devastation that ranged over one square kilometer. Bodies decomposed from resting in the mud and in trees in the ceaseless rain, were everywhere. The entire total of victims was difficult to assess, since many of them had been swept under the waters of the Kali River.

Official totals reached over 200, which included all 60 pilgrims of the 12th batch, including Protima Bedi, residents of the area, and some Tibetan border police. The large death toll and the enormity of the disaster ended the annual pilgrimage abruptly, but not the political fallout.

Environmentalist Iqbal Malik, interviewed in the Indian press, asserted that the government of India had not been careful enough in protecting the pilgrimage route. "The tourism activity and blasting of mountains to build roads have been weakening the rock system," she said. "There is no planning. The government only reacts to disasters. It never prevents them."

Her final assessment reverberated far beyond the borders of India.

IRAN
ROUDEHEN
January 13, 1998

Thirty-two people were killed and over 80 injured by an immense avalanche that roared across the mountain row that links Iran's capital of Tehran with the Caspian Sea. The avalanche, which occurred near Roudehen, happened on the evening of January 13, 1998, while the road was crowded with traffic.

The coming of 1998 brought some of the heaviest snowfalls in Iran in 10 years. Over 1,000 villages were reported isolated by the snow in western and northern sections of the country. For the first two weeks of January, a series of 200 avalanches intermittently blocked the main road connecting Tehran and Mashad, on the Caspian coast. They were small and were cleared after a time, and the normally heavy traffic continued, until the night of January 13.

At the beginning of that night, the usual flow of buses, trucks, and cars moved between the packs of snow on either side of the road, which intermittently disappeared into mountain tunnels, then emerged on vulnerable stretches of open road that clung to the mountainside.

"Our bus came out of [a tunnel near Roudehen, which is 20 miles east of Tehran] and then suddenly we heard a horrible sound," reported bus passenger and survivor Hassan Eqtiedaii to an AP correspondent. Eqtiedaii's bus, like several other vehicles, was caught directly in the path of a giant avalanche that had detached itself from the upper slopes of the mountainous area moments before. Struck broadside by the avalanche, the bus plunged 700 feet (213 m) into the valley and broke in two.

"All the passengers were scattered in the snow," continued Eqtiedaii. "All I could hear was the sound of crying and shouting. Except for a few people like me, all the passengers were buried."

Thirty-five other vehicles were flung from the road into the valley at the same moment that the bus was hit. All of them were buried, and 32 people were killed. Over 80 were injured. Only the bus, two vans, a truck, and three cars were unearthed. It would be spring before the remaining vehicles and the bodies trapped in them would emerge from the snow.

ITALY
BELLUNO
October 9, 1963

An earthquake caused a landslide, which in turn caused a flood in the Piave River Valley of Italy on October 9, 1963. Over 4,000 residents of the area drowned.

At 11:15 P.M. on October 9, 1963, the Piave River valley in northern Italy, which had been the scene of the Italian army's stand against the Austrians after the Italians' defeat at Caporetto during World War I, and which was familiar to millions as the setting of Ernest Hemingway's novel *A Farewell to Arms*, faced a far worse natural disaster: The entire valley was flooded and the 873-foot-high (266.1-m-high) Valmont dam was at first thought to have collapsed under the assault of a mountain landslide, caused by an earthquake. Over 4,000 people drowned in this multiple cataclysm.

The 72-foot-thick (21.9-m-thick) dam, the world's highest, had not collapsed. The roar that survivors recalled hearing before an immense wall of water thundered into the valley was the cracking apart of the mountain on either side of the dam. Captain Fred R. Michelson, a U.S. Army helicopter pilot who flew out residents of the village of Casso above the dam (the village was threatened with destruction from residual landslides), described the scene: "There was a mile-long lake behind the dam, but it doesn't exist any more," he reported. "The entire tops of both mountains on either side slid into the lake and completely filled it."

The displaced water from the lake overflowed the dam, cracking it at the top and cascading in an immense 1,500-foot (457-m) waterfall at right angles to the Piave River valley.

"You can see where the burst of water swirled down the [Piave] river on both sides for about 30 miles [48 km] downstream and about 500 yards [548.6 m] on either side of the river," continued Captain Michelson. "You can't find any buildings for about five miles. There aren't even any foundations. You can't find anything of the towns that once existed."

Longarone, a village directly in the water's path, disappeared entirely, along with 3,700 of its 4,700 residents. At Pirago, a few miles downstream, only a church bell tower, a chapel in a cemetery, and one house remained standing. Nobody in the village lived.

"At first, there was a loud, distant roar, and then the window panes started trembling," was the way Alessandro Bellumcini, a resident of Longarone who survived because he was watching a soccer match in a tavern in Fae, outside of his native village, put it. "I raced outside and saw flashes of light on the mountains in the direction of the Vaiont Valley. They were probably high-power electric wires snapping."

"All of us rushed toward the mountain and started running uphill. We had not covered more than 300 yards when we saw something like a whitish, rolling form engulfing the valley at fantastic speed. It was just visible in the light of the half moon. I saw the few homes of our little hamlet being wiped out in seconds. It was horrible."

Bodies were carried as far as 40 miles (64 km) downstream; others were buried on the spot by the rubble of the landslide and other parts of the valley loosened and were carried along by the flood waters.

Huddled in an army blanket, Mario Faini, a survivor who had been asleep in his home on the fringes of the landslide and flood, recalled that he and his two sons ". . . felt what seemed like an earthquake. I got up and started dressing. I heard a terrible wind blowing outside just like a tornado. Suddenly, the windows were smashed in and water poured into the house. We were thrown off our feet. . . ."

His son recalled that ". . . our pajamas were torn off our bodies, and after a few horrible moments . . . we found ourselves rushing out the back windows and up the mountain, shivering in the cold."

Mario Laveder, municipal secretary of Longarone, observed this sort of evacuation: "Some villagers rushed into the streets and tried to climb up the mountainside. A few of them succeeded. [Others] were engulfed by a wave of swirling waters and drowned. Others died under the debris of their homes."

American helicopters from Allied headquarters in Verona were successful in lifting out stranded survivors, among them a pregnant woman who, assisted by the helicopter crew and a nurse, gave birth to a girl minutes after the copter had deposited her in a safe place.

Devastation was everywhere the next morning, augmented by a further, lethal danger. Authorities appealed to everyone in the area by radio not to drink water from the Piave or to allow their cattle to drink it. Five tons of potassium cyanide had washed from a riverside factory into the Piave, turning the waters poisonous.

Americans supplied 6,000-gallon (22,800-l) tank trucks from Allied headquarters. The grim rescue work continued for weeks, but many of the villages that had been wiped from the face of the valley by this huge combination of earthquake, landslide, and flood would never be rebuilt.

JAPAN
July 17, 1972

Torrential rains caused landslides that precipitated floods throughout Japan during the week of July 10, 1972. On Monday, the 17th, the destruction reached its climax. Three hundred seventy persons were killed; over $472 million in property damage was caused by the combination of landslides and flooding.

Three hundred seventy persons were drowned or crushed by landslides set off by floods, which were in turn brought on by a week of relentless rain throughout most of Japan in mid-July of 1972. Scores of major rivers and minor streams overflowed their banks, eroding them into rushing mudslides which embraced and then consumed whole villages and numerous farms.

It was, in fact, in the agricultural sections of Japan that calamity struck most clearly and widely. More than $472 million in both building and crop loss was caused by this combination of natural disasters.

JAPAN
NIIGATA
July 18–19, 1964

An earthquake combined with heavy rains caused landslides near Niigata, Japan, on July 18 and 19, 1964. One hundred eight died; 223 were injured; 44,000 were rendered homeless.

Landslides that were caused by a combination of a minor earthquake and torrential rains in Niigata, Japan, caved in river banks, destroyed 150 bridges, and collapsed 50-year-old dikes during the two days of July 18 and 19, 1964. The five districts near Niigata and along the Sea of Japan were likewise devastated and swept clean of 295 dwellings, which were either buried under the landslides or carried away by the waters released by over 200 gaps that had been opened up in protecting dikes by the cascading mud.

Several villages were wiped out completely by landslides; in cities such as Ishikawa, Toyamma, Niigata, Tottori, and Shimane, entire sections crumbled.

The final death count was 108. Two hundred twenty-three residents of the area were injured, and 44,000 were made homeless.

MEXICO
MINATITLÁN
October 29, 1959

An earthquake, a tsunami, and a series of landslides combined to kill 5,000 people in Mexico on October 29, 1959. The village of Minatitlán was obliterated, and 800 residents died in their beds.

A massive combination of natural forces conspired to kill 5,000 residents of Mexico on October 29, 1959.

First, an earthquake that barely registered on the Richter scale brought on a giant tsunami that slammed into the Pacific Coast of Mexico, drowning thousands, sinking 10 small freighters, and sending the passenger-freighter *Sinola* to the bottom of the Pacific with all on board.

That night, massive mudslides virtually buried the village of Minatitlán, crushing people in their beds. Eight hundred died within minutes when the landslide roared with comparatively little warning over virtually every building in the town.

Nearby, another 1,000 were killed when rocks and mud from the same surrounding hills cascaded murderously into their villages. The town of Zacoalpan in northern Colima state was obliterated by landslides and then flooded to the tops of the surviving buildings' roofs. A pilot flying over the area noted that only the church steeple protruded from the waters covering the town.

In the grim aftermath, swarms of snakes, scorpions, and tarantulas, unearthed from their lairs when the landslides tore apart the hillsides, slithered and crawled into what was left of Minatitlán, killing another 200 residents before serum could be flown in from Mexico City.

NEPAL
KATMANDU
November 11–12, 1995

Ninety-one people died in one of Nepal's worst avalanche disasters on November 11–12, 1995, on or near the slopes of Mount Everest. Eight helicopters managed to rescue and evacuate 237 survivors.

The weather was atrocious in the Himalayas of Nepal in November of 1995. Early snowstorms and winds caught hundreds of climbers of Mount Everest by surprise, isolating and in some cases burying mountain climbers of a multitude of nationalities.

The first report of a cascading series of avalanches was received from a Japanese trekking group on Mount Everest. On the 11th of the month, rescuers arriving at the site of the group's encampment discovered 13 Japanese trekkers, 11 guides and porters, and two residents of the Pangka region buried and dead under the avalanche.

As more and more avalanches roared down the mountain, it became apparent that an estimated 500 foreign climbers were trapped in the mountains by the subfreezing temperatures, deep snow, and avalanches.

A particularly lethal series of snowslides roared down the slopes of Mount Everest near Katmandu, Nepal, on November 12. Fifty-nine people were killed in various locations in Nepal that day, 42 on the slopes of Mount Everest, 17 others in houses that were crushed by the force of the falling snow and ice in nearby locations.

Eight helicopters were dispatched to the area in the continuing, sometimes blinding storm by the Nepali government. Two hundred and thirty-seven people, including 111 foreigners, were airlifted to safety in a series of flights that were conducted during comparative lulls in the storm. But the total of 91 fatalities made this one of the worst avalanche disasters in Nepal's history.

PAKISTAN
KEL
March 15, 1996

Forty-four residents of the village of Kel, on the Kashmir-India border, were buried under an avalanche that despoiled the small military post on March 15, 1996.

The small Pakistani village of Kel is a military post, located just inside Pakistan-controlled Kashmir, on the disputed border with India. It was accustomed to tension and attacks, but none were more unexpected or devastating than the one that came from the mountain above Kel on March 15, 1996.

On that day, the peak divested itself of several tons of ice, snow, trees, and rocks and showered them upon the tiny village. Five houses were engulfed and two others were flattened.

A day later, rescue teams from Islamabad, the capital of Pakistan, counted the 44 dead who were dug from the debris of crushed houses and roof-high snow and ice that had overwhelmed the village.

PERU
CHUNGAR
March 19, 1971

Chungar, a mining camp 10,000 feet up in the Andes, was destroyed by an avalanche caused by an earthquake on March 19, 1971. Between 400 and 600 were killed; 50 were injured.

An avalanche rained tons of water, mud, and rocks down upon the isolated mining camp of Chungar, high in the Andes of Peru, on the morning of March 19, 1971. Between 400 and 600 people were buried in seconds by the thundering slide.

The avalanche was touched off by an earthquake that struck at 8:30 A.M., crumbling a nearby mountaintop and toppling it into a lake. Water spilled over the banks of the lake and swept the nearby terrain, picking up soil, trees, and immense boulders. The slide continued on, obliterating the main road to Chungar from Lima, wiping out bridges and demolishing living quarters.

The camp, 10,000 feet (3,048 m) up in the Andes and an eight-hour journey by foot from the nearest town, had been inhabited by about 1,000 people. But rescuers, after crossing a 12,000-foot (3,657.6-m) mountain range on their trip in from Lima, Cerro de Pasco, and Canta, found only a third of that population still alive. The others had been buried under tons of mud, rocks, and the remains of their homes. Fifty injured people were flown out and taken to hospitals.

It was the worst disaster in Peru since an earthquake the previous May, which had struck an area 180 miles (289 km) north of Lima, with a death toll estimated at 70,000 (see EARTHQUAKES). As in this event, many of those killed in that quake had died when an avalanche of mud and rocks thundered down the slopes of Mount Huascarán, Peru's highest mountain, and buried two towns.

PERU
HUASCARÁN
January 10, 1962

A glacier at the summit of Mt. Huascarán in the Andes shattered and triggered an avalanche on January 10, 1962. Thirty-five hundred people died in the valley below.

At exactly 6:13 P.M. on January 10, 1962, 3 million tons of ice from Glacier 511, located 21,834 feet (6,655 m) at the top of Mount Huascarán, cracked, loosened, and began to slide toward the valley below. Within seven minutes, 3,500 people were dead.

Fattened by freak snows and warmed by unseasonal sunshine, the glacier simply came apart and, in seconds, started its highspeed, cataclysmic journey. A man in Yungay, looking upward at the instant that the glacier let loose, first thought it was a cloud turning golden in the sunset. "But I saw," he later reported, "that the cloud was flying downhill."

The ice sped into the gorge of the Callejon de Huailas with a roar "like that of ten thousand wild beasts," according to another man.

"I could feel the rumble in the walls of the belly," added still another.

This lethal barrage of ice first shaved down an uninhabited slope, then bounced in an insane ricochet back and forth from gorge to gorge, carving and collecting topsoil, boulders, and flocks of sheep. Finally, it entered the village of Yanamachico and three others nearby, flattening them and killing all 800 residents. Sweeping the ruins of houses along with other debris, it continued on.

Now the avalanche flattened to 60 feet (18.28 m) in thickness, slowing to 60 MPH from hundreds of miles per hour moments before. Hills banked near a gorge diverted it from the town of Yungay, but the larger town of Ranrahirca, with a population of 2,700, still remained in its path.

Within an instant, the avalanche crashed into this town, toppling its church steeples and crushing houses as if they were cardboard. Ricardo Olivera, the chief engineer of the local power station, hearing the telltale roar growing closer, grabbed the hands of two young girls playing in a playground in an effort to get them to the church, a sturdily built shelter. But, as Olivera later described it, "The girls were torn from my hands—by the winds or by a wall of mud. Electric wires had fallen around me. Somehow, I came free. I regained my senses, and saw only a waste of mud and ice."

The girls, all of the buildings around him, and the church itself had all been crushed, and every one of their inhabitants were dead. "I was impressed by a profound silence," Olivera went on. "Realizing that my wife, my children, my parents were all buried under the debris, I suddenly found myself sobbing."

According to *National Geographic* writer Bart McDowell, who came upon the scene shortly after the avalanche struck, the scene ". . . resembled an Old Testament visitation. White rock and pale mud stretched a mile across the green Andean valley. No ice was visible on the surface. Boulders were mortared together by a crusting mud of granite dust, and streaked by small, disoriented brooks of melt. Following a team of stretcher-bearers to recover the dead," McDowell sadly concluded, "we sank thigh-deep in mire."

The few surviving villagers were mute in their sorrow. According to McDowell, "As the priest intoned the Latin words [of absolution over the dead], some women wept, quietly, without sobbing. Their faces seemed numb beyond the curing salt of tears."

PHILIPPINES
LEYTE ISLAND
February 17, 2006

A giant mudslide obliterated the farming village of Guinsaugon on Leyte Island, 420 miles (675.9 km) southeast of Manila, on February 17, 2006. More than 1,000 people were officially reported dead, buried under mud that was between 30 and 100 feet (9.1 and 30.5 m) deep and that totally destroyed the village. The casualty figure may have been underestimated, since the population of the village was 1,857 and only 57 of the remaining 560 survivors were accounted for.

For two weeks in early February 2006, La Niña-fueled nonstop rains that mercilessly pounded the unstable countryside of Southern Leyte in the Philippines, dumping 27 inches (68.6 cm) of rain on the area. During the week of the 14th, the Sun appeared momentarily and hundreds of residents of the valley villages returned. At 10:00 A.M. on February 17, the precariously perched, shallow-rooted coconut trees on the side of one mountain began to move. Although the muddy slopes of the mountains in the area had been heavily and illegally logged from the 1970s to the 1980s, vegetation had begun to return to the area by 2006. Not enough vegetation, apparently, for on that particular Friday morning, an entire muddy mountainside, with its trees intact and standing, slid into the valley and swallowed hundreds of houses and an elementary school that was in session in the village of Guinsaugon (see color insert on p. C-1).

"It sounded like the mountain exploded, and the whole thing crumbled," one survivor later told the BBC. "I could not see any house standing anymore." Those structures that stood after the landslide were buried under mud 30 feet (9.1 m) deep in some areas and 100 feet (30.5 m) over the elementary school.

"Our village is gone, everything was buried in mud," survivor Eugene Pilo, whose family had disappeared under the rush of mud and displaced trees, told a newspaper reporter. "All the people are gone."

Within hours, rescue workers arrived, ferrying mud-covered survivors on the blade of a bulldozer across a stream to waiting ambulances, which carried the survivors to a nearby clinic. The rains resumed shortly after the landslide, loosening, following landslides and further softening the mud, which prevented heavy rescue equipment from entering the devastated area.

Rescuers dug with their hands and felt with their feet for survivors, trapped under what now appeared to be a plowed field, with bits and pieces of roofing and debris from the 281 destroyed homes pushing through its surface. "It took only two minutes to do this [destruction]," survivor Alfred Guab told an AP reporter.

The concentration of the rescue squads became centered on the area containing the buried elementary school. Relatives of some of the children reported text messages sent from cell phones. "We're still in one room alive," Agence France-Presse quoted one message. Another read, "We are alive. Dig us out."

Seismic sensors and sound-detection gear, brought in by U.S. and Malaysian forces, detected sounds of scratching and a rhythmic tapping, which intensified the rescue efforts and raised the hope of parents.

On the following Wednesday, U.S. Marines brought in a two-ton drill. But their work proved to be fruitless and discouraging. Apparently, the detected sounds were merely of the mud settling. No survivors from the buried school were ever retrieved.

The rain continued, raising fears of other landslides, exacerbated by the movements of the rescuers and their equipment. Altogether, a total of 1,112 missing and presumed dead residents of the area were determined, and the landslide directly affected 2,981 people. The overall damage to the area's infrastructure and agriculture—mostly rice paddies—was estimated at $2.2 million. And the village of Guinsaugon became, at least for the time being, a massive cemetery (see color insert on p. C-1).

RUSSIA
PAMIR MOUNTAINS
July 15, 1990

Forty-three mountaineers were killed when an earthquake triggered an avalanche at 19,500 feet in the Pamir Mountains of Soviet Central Asia on July 15, 1990.

An earthquake rumbled through the Pamir Mountains on July 14, 1990. It was a small quake, hardly visible on the seismographs of the world, but it was enough to destabilize the boulders and outcroppings in this mountain range that traverses the Kirghizia-Tajikistan border.

It was midsummer, a time for mountain climbers from Europe and beyond to travel to this part of Soviet Central Asia to climb the 20,000 foot-plus (6,096 m-plus) peaks, and on July 15, there were hundreds of them at various points in the Pamir range.

Some were far enough away from the earthquake's center not to feel the earth's pitch and roll. But at the 19,500-foot-point (5,943.6-m-point), precisely at the border, 43 climbers and residents of a narrow part of the mountains felt the quake and, looking up, faced a fatal avalanche that tore through the area with express train speed.

Most were swept off the mountain and into the valley below; several Swiss mountaineers were later found buried under the debris that had been shaken loose by the earthquake and sent tumbling down the mountainside.

SWITZERLAND
ELM
September 11, 1881

The undermined peak of Plattenbergkopf, hovering over Elm, Switzerland, collapsed upon the village on September 11, 1881. The village was destroyed; 150 villagers were killed; 200 were injured.

One hundred fifty people were killed when the peak of Plattenbergkopf, undermined by years of slate mining, came loose and hurtled down on the village of Elm in the Sernf valley of Switzerland on Sunday afternoon, September 11, 1881.

Cracks had begun to appear in the mountain as early as 1876, and by 1881, one of them had opened to a width of five yards. The random and careless blasting away of parts of the peak and the random positioning of mine shafts had, by the middle of that year, produced cave-ins and minor rock slides which were enough to cause some alarm among the residents of the hamlet of Elm.

But all of this was merely a prelude to the rumbling that first began at 5:30 P.M. on September 11, 1881. At that moment, the roof of the largest slate quarry caved in with a roar. And then there was a pause, enough for the mountain residents to settle into a shortlived feeling of relief.

Seventeen minutes later, the top of the mountain came loose. Over 10 million cubic yards (9,144,000 cu. m) of rock broke loose and thundered downslope at hundreds of miles per hour, shoving boulders, livestock, houses, and human beings before it. "Trees were snapped like matches," a survivor recalled, "and houses were lifted through the air like feathers and thrown like cards against the hillside."

The village inn, located above the village itself, was crowded with drinkers, gathered for the usual Sunday afternoon "social watch," of the daily discharge of random boulders from the mountaintop. Some, more sober than others, left after the first rumblings.

One of these was Meinrad Rhyner, who, shortly after he quit the inn with a wheel of cheese under his arm,

turned to see the entire building and its occupants disappear under tons of earth. Rhyner managed to escape the avalanche, but his family was not so lucky. Breathlessly reaching his home after the last of three separate slides, he found it intact. But the hurtling rocks loosed by the slide had killed every member of his fleeing family—his wife, daughter, son, daughter-in-law, and two grandchildren. "The doors were open," he remembered later, "a fire burning in the kitchen, the table laid and coffee hot in the coffeepot, but no living soul was left."

Over half the village was flattened. Some residents escaped, like Kasper Zentner, the fastest runner in the village. He outpaced the slide for a while, then, when it threatened to engulf him, he saved himself by jumping over several stone walls and, with a broken leg, dropping into a deep gully while the slide passed over him.

Most others were far less lucky. One hundred fifty villagers lay dead, 200 more were seriously injured and the village was in ruins.

SWITZERLAND
GOLDAU VALLEY
September 2, 1806

An avalanche caused by the sudden erosion of the top of Rossberg Peak in the Swiss Alps on September 2, 1806, set forest fires and inundated four villages in the Goldau Valley. Eight hundred were killed.

Eight hundred people in four villages were killed in a matter of minutes on September 2, 1806, when the entire top of Rossberg Peak in the Swiss Alps crumbled, then plummeted into the valley below.

A thick forest covered the slopes almost to the top of the peak, and this forest remained intact, sliding downward, in one destructive slab, at ferocious speed. Rock ground against rock, shooting geysers of steam in the air, finally erupting in flames as the friction increased. Horrified onlookers witnessed a flame-orange forest fire that rocketed downward at hundreds of miles an hour, filling up the entire valley and consuming everything in its path.

SWITZERLAND
LEUKERBAD
January 17, 1718

Sixty-one people were killed and almost every building in the village of Leukerbad, Switzerland, was smashed when cyclonic winds caused a snow avalanche on January 17, 1718.

The monster avalanche of January 17, 1718, was only one of a series that had plagued this popular tourist resort and hot springs for centuries. But it was by far the most devastating. Practically every building in the village was destroyed, and 61 people died—over half of Leukerbad's population.

It had snowed unrelievedly for two weeks before the avalanche, according to Stephen Matter, a local scribe. At the end of the two weeks, a large avalanche roared through the outskirts of the village, killing three young men. A search party, using sounding rods to detect bodies buried in snow, located the corpses and returned to the village that evening with them, thankful that Leukerbad had escaped a more destructive slide.

The party had hardly entered the village when a second, cataclysmic avalanche of tons and tons of powdered snow, preceded by cyclonic winds, thundered down the mountainside from the top of the 10,000-foot (3,048-m) Balmhorn. The Church of St. Laurentius, where parishioners were gathered for evening vespers, was wrecked, killing all of those within. The three luxury baths were destroyed. Entire families were crushed and buried.

Even the stories of survivors were tempered by tragedy. One man, caught in his wine cellar as he was searching for a bottle of wine to accompany his evening meal, was trapped alive for eight days. Finally rescued, he died soon thereafter of malnutrition and frostbite.

SWITZERLAND
ST. GERVAIS
July 12, 1892

One hundred forty people were killed in the giant avalanche caused by the calving of the Tête Rousse glacier on Mont Blanc on July 12, 1892.

Only 10 of 150 persons in the 19th-century resort towns of St. Gervais and La Fayet survived the massive avalanche that occurred at 2:00 A.M. on July 12, 1892.

Mont Blanc, 14,318 feet (4,364 m) tall, towered above these two resorts. St. Gervais was particularly crowded with tourists enjoying its luxurious hotel and sulfur springs baths. It was the middle of summer, and even to the seasoned mountaineers who lived in St. Gervais, thoughts of snow avalanches were distant, if not entirely absent.

But halfway through that July night, the Tête Rousse glacier, suspended at the brink of a gorge above the Glacier de Bionnassay on the western side of Mont Blanc, broke off. Within an instant, it plummeted toward the two sleeping towns, sweeping rocks, trees, snow, debris, and water along with it.

Not one person in the hotels was able to leave his or her bed. Every building—from the simplest lean-to to the stone edifices of the church and the hotel—was crushed and mutilated. Only those on the very outskirts of the village survived.

SWITZERLAND
VALS
January 20, 1951

Two hundred forty people died, and over 45,000 were trapped on January 20, 1951, when a series of avalanches, caused by a combination of hurricane force winds and wet snow overlaying powder snow, thundered through the Swiss, Austrian, and Italian Alps.

A horrendous series of avalanches—the worst since the 1916 series of slides that had buried hundreds of Italian and Austrian troops (see AVALANCHES, Alps, Tyrol)—roared and raged through the Swiss, Austrian, and Italian Alps on January 20, 1951. Before tranquillity returned to the mountainsides, more than 45,000 people were trapped for weeks, 240 were dead, and dozens of villages were in ruins. Even the posh resorts of Davos, Zermatt, Arosa, and St. Moritz did not escape the tragedy. Perhaps the hardest hit of all was the Swiss village of Vals, located 4,100 feet (1,249.6 m) above sea level in the most picturesque region of the Swiss Alps. The village disappeared entirely, and 19 of its residents were crushed to death.

The conditions that caused the avalanche were classic. First, very little snow fell in December of 1950, disappointing hundreds of international skiers. Then, with the new year, extensive, powdery snow storms struck. Then rain. Then snow mixed with rain. And finally, rain entirely.

On January 20, the sides of Sustenhorn and Dammastock shuddered, and tons of wet snow overlaying the shifting and unstable mounds of powder snow broke loose at the higher elevations. Hurricane force winds ran ahead of the white wall of snow and rocks, helping to topple the trees that had been planted to slow just such an avalanche, and allowing the force of the slide to snap off or uproot every one of those that escaped the winds.

The St. Gotthard rail line between Switzerland and central Europe was rendered inoperable for a week, blocked by immense walls of snow, ice, and rocks. Communication lines and towers were toppled, cutting off all contact with the outside world.

Dozens of bittersweet tales of survival surfaced, as survivors wandered into the few standing structures. Typical of these was the story of Johann Lutz, who buried his family in a common grave in Vals. When the first of the series of avalanches struck, Lutz climbed to the roof of his home to sweep the snow off the roof, lest it cave in. While he was there, four more avalanches roared down in quick succession. It was all he could do to hold on to the roof and not be swept into the valley with the rest of the debris. When he finally climbed down from the roof, he found that both his wife and two-year-old son had been crushed to death by the snow that had poured in through the windows and doors of his home.

TAJIKISTAN
ANZOB MOUNTAIN PASS
October 23, 1997

Forty-six people were killed when a sudden avalanche buried 15 trucks and cars on the 11,000-foot-high (3,352.8-m-high) Anzob Mountain Pass in Central Tajikistan. Only four survivors were pulled from a 40-foot blanket of snow.

The Anzob Pass is one of the most difficult mountain passes in the Central Asian nation of Tajikistan. Poised 11,000 feet (3,352.8 m) high near the provincial dividing line with Vilyati Leninobod in the antimony mining area 60 miles (95.56 km) north of the capital, Dushambe, it is closed entirely to traffic in the winter.

October is considered a "shoulder" season: Sometimes it snows; more often, it does not. The third week of October of 1997 brought with it a major snowstorm, which was still not heavy enough to shut down the pass to the trucks that hauled antimony out of the area.

But this decision proved to be foolhardy. On October 23, a giant avalanche began above the pass, and roared down on a string of 15 cars and trucks that was directly in its path. In an instant, all of the cars and trucks were buried under a glacier of snow and ice 40 feet (12 m) high.

For two solid weeks, rescuers dug into the snow. Only four of the 50 men and women who were in

the 15 trucks and cars were pulled out alive. Thirteen bodies were found. It would be spring before the remaining 33 were finally unearthed by rising temperatures.

UNITED STATES
CALIFORNIA
January 18–26, 1969

Ninety-five persons died and over $138 million in damage was caused in southern California by a series of mudslides, brought about by nine days of torrential rain, and a subtropical storm, from January 18 to 26, 1969.

For nine straight days, rain fell along the coast of California, fueled by a subtropical storm that slammed into the southern coast on January 18, 1969. By January 26, 10 inches (25.4 cm) of rain had eaten into the hillsides, particularly those insubstantial ones in the San Gabriel and Santa Monica Mountains above Los Angeles.

Landslides began on January 22, as running yellow mud threatened the posh homes of movie stars and movie moguls on Rainbow Drive in Glendora and in Mandeville Canyon. The slides, some of them half mud and half water, oozed into estates, trapping some residents in their homes. Other houses tipped and fell, shoved by the mudslides into the valleys below.

On El Paso Drive, more and more homes were demolished, and more and more people perished as the force of the mudslides increased in size, speed, and scope.

Santiago Creek rapidly filled with mud, rain water, and floodwaters, and threatened Santa Ana. Five thousand volunteers arrived and for hours labored at creating dikes and levees to hold back the overflowing floodwaters. Finally, Marine helicopters appeared, carrying wrecked cars in the slings beneath them. The combination of levees and junk cars formed a barrier that thwarted a potentially disastrous flood.

Movie sets in the Santa Monica Mountains were swept away. Over 100 boats, large and small, were sunk at Ventura and Santa Barbara. The final total of dead came to 95, and over $138 million in damage was finally tallied when the rain and mudslides ceased. Federal relief funds of $3 million were immediately funneled into the devastated countryside, which was declared a disaster area by President Richard M. Nixon.

UNITED STATES
WASHINGTON
WELLINGTON
March 1, 1910

....

Snow loosened by rain broke away from a mountain in the Cascade range of Washington on March 1, 1910, and buried the village of Wellington and a stranded train at its railroad station. Ninety-six died in one of the worst avalanches in U.S. history.

The worst single avalanche in United States history took place in the small mountain rail stop of Wellington, Washington, located partway up the Cascade range, at 1:20 A.M. on March 1, 1910.

A blizzard of enormous proportions had buffetted the area for nine days in February. The snow that fell at the alarming rate of a foot an hour—on one particularly ferocious day, 11 feet (3.35 m) of snow accumulated—managed to close down the Great Northern Railroad completely.

Unfortunately, it also stranded a passenger train, loaded with over 100 travelers, which had stopped at the high outpost of Wellington. Plows brought in to free the train had become stuck in the mounting snow just outside of town. Several locomotives and a mail train, all equipped with special plows, were unable to break through the steadily increasing wall of heavy powdered snow surrounding this tiny rail stop.

Residents of the village surveyed the mountain range above them nervously. A huge forest fire had swept the slopes clean of the trees that could prevent or at least dissipate an avalanche. The conditions were ideal for a disaster.

Then, late on February 28, the snow stopped, and rain, accompanied by warm winds, began to fall. What had become a possibility rapidly turned into an inevitability, and shortly after 1 A.M., a slab of rain-heavy snow resting upon unstable powder broke loose from the side of the mountain. A 20-foot-high (6-m-high), half-mile-long (.80-km-long), quarter-mile-wide (.40-km-wide) wall of snow roared down the slope, headed directly for the village of Wellington.

It missed the local hotel, but plunged, at enormous speed, directly toward the railroad depot. With the force of a thousand battering rams, it slammed into the locomotives, boxcars, engine house, water tower, mail train, and passenger train, in which 100 unsuspecting people slept. Within seconds, it picked up the trains and buildings and plowed them into a 150-foot (45.7 m) gorge.

Another minute, and it was all over. Trainmen who lived in the hotel rushed to the gorge with shovels and picks and, listening for muffled screams and groans for help from beneath the displaced mountain of snow, immediately started to dig for survivors.

Twenty-six people were rescued. Ninety-six died, buried in the pieces of the passenger train that had been crushed and transported into the gorge by the avalanche. The rescue work had to be abandoned within days; it took a late spring thaw to unearth all of the bodies and cars.

UNITED STATES
WASHINGTON
May 18, 1980

....

The volcanic explosion of Mount St. Helens on May 18, 1980, triggered the largest recorded avalanche on the mountain's north slope. The velocity of the avalanche reached a record-making 250 MPH.

The largest recorded avalanche thundered down the north slope of Mount St. Helens on May 18, 1980, following the cataclysmic eruption of that mountain (see VOLCANOS). The landslide measured 2.8 cubic kilometers—enough to cover an area slightly larger than downtown Portland, Oregon, to a depth that would bury the city's 40-story First National Bank Tower.

Its velocity reached an astonishing 250 MPH, which is 125 MPH faster than the wind speed in a maximum force hurricane.

According to George Plafker, a survey geologist who studied landslides in Alaska and South America, the Mount St. Helens slide dwarfed even the cataclysmic 1963 Mount Huascarán avalanche in Peru (see p. 17).

WALES
ABERFAN
October 21, 1966

....

The collapse of a slag heap outside of Aberfan, Wales, on October 21, 1966, caused the worst landslide in Wales's history. One hundred forty-five persons—116 of them children—were killed.

"Buried alive by the National Coal Board" was the statement a score of ravaged and angry parents demanded be put on their children's death certificates in the wake of the worst landslide in the history of Wales. One hundred forty-five persons were killed in the collapse of an

800-foot-high (243.8-m-high) slag heap outside of the village of Aberfan on the morning of October 21, 1966. One hundred sixteen of those crushed by the landslide were children.

There was certainly culpability on the part of the owners of the colliery that had built this characteristic slag pile to such a mountainous height. But according to geologist Robert Price, the primary cause was a "geological freak," an underground spring that had established itself beneath the slag pile after the pile had already grown to immense size.

Be that as it may, the combination of forces caused a cataclysm, which began, ironically enough, at the same moment that a maintenance man, sent by the mine operators, who had in turn been goaded by complaints from local residents, climbed around the peak of this man-made mountain, hoping to pronounce it safe.

A heavy ground fog clung to the land at dawn on October 21, but the slag heap inspector, David John Evans, remembered that ". . . it was like a summer's day on top."

Everything seemed normal to him at first. And then, suddenly, as he sat on it, the heap began to shift. "We could see the tip moving 300 yards [274.3 m] away from us," he remembered, in disbelief even after it had happened. "Down it went, but we could not see the result because of the thick fog. . . . The movement was like thunder. We could hear the trees on the side of the tip being crushed. It was frightening."

Mr. Evans managed to scurry to safety in the mine office, which by now was a maelstrom of hysterical activity.

Two million tons (1,814,369.5 tonnes) of rock, coal, and mud was catapulting down the hill, directly toward the Pantglas school, which had just begun its day. Almost every school-age child in the village was there, directly in the path of the avalanche.

One of these children was, however, late for school, and described the next few moments: "I could not see at first, because it was very foggy," he recalled. "When I could, it looked like water pouring down the hillside. It uprooted a great tree on its way. [My two friends] ran the other way. It just sucked them away and they ran right into it. It hit the school like a big wave, spattering all over the place and crushing the building. It was like a dream . . ."

According to another survivor, who by this time was, with miners and parents, running toward the school building, "I saw scores of houses, some with a slime-like river running through them. Running down the mountainside was a cataract, working itself through the tip which [had] collapsed."

The quarter-mile-wide mass hit the school dead center, pitching it off its foundation and entering any open space, be it door, window, or cracked shingling. The building folded like a paper box, crushing children and teachers within it.

Some managed to escape. One teacher, his leg broken, hobbled to a free, open window and handed surviving children out through it.

Hundreds of sobbing townspeople descended upon what was left of the school house and clawed at the wreckage, following the sounds of moaning survivors. One mining boss who directed a party to one of the collapsed rooms in the school related his findings: "We found four children underneath a lot of brickwork which had slipped down on top of them," he said. "One small boy was still alive. He was standing against the heater in the schoolroom and was crying because his leg had been caught in something. By his seat were three other children. They were dead."

The loss of any life is tragic; the loss of half of a village's children in a few horrible moments is almost beyond comprehension. A mass grave was dug for the children, marked by a hundred-foot cross upon which garlands, sent by miners from all over the world, would be hung, for months to come. Just before the grave was closed, mothers threw their dead children's stuffed animals into it as hardened police sergeants, sent to control the crowds, broke into unmuffled sobs.

EARTHQUAKES

THE WORST RECORDED EARTHQUAKES

* Detailed in text

Afghanistan
* Hindu Kush Region (2002)
  North (1999)
* Northeast (May 1998)
  Pakistan Border (1991)
* Rustaq (February 1998)

Africa
* (217 B.C.E.)

Algeria
  (1716)
* Al Asnam (1980)
  Northwest (1994)
* Thenia (2003)

Armenia (Soviet)
* (1988)

Asia Minor
  Cieilia (1268)

Belleny Islands
  (1998)

Bolivia
  North (1994)

Caucasia
  Shemaka (1667)

Ceram Sea
  (1998)

Chile
* (1822)
* (1939)
* (1960)
  Northern Coast (1995)
* Valparaíso (1906)

China
* (1556—Largest death toll in history)
  Chihli (1290)
* Kansu (Gansu) (1920)
  Kansu (1927)
  Kansu (1932)
  Peking (Beijing) (1731)
  Shansi (Shanxi) (1038)
  Southern Sinkiang (Xinjiang)
    (2003)
* Southwest (1988)
  Talifu (1925)
* Tangshan (1976)
  Yunnan (1996)

Colombia
* (1875)
* Quindio (1999)

Cuba
  South (1992)

Ecuador
* (1949)
* Quito (1797)

Ecuador and Peru
  (1868)

Egypt
* Alexandria (365)
  Cairo (1754)
* Cairo (1992)

France and Italy
* French and Italian Rivieras (1887)

Greece
* Corinth (856)
* Sparta (464 B.C.E.)

Guam
  South of Mariana Islands (1993)

Guatemala
* (1902)
* (1976)
  Santiago (1773)

India
  (893)
* Assam (1950)
  Calcutta (1737)
* Gujarat (2001)
* Kangra (1905)
* Kashmir (1885)
* North (1991)
  Quetta (1935)
* South (1993)

Indonesia
* Flores Island (1992)
  Irian Jaya Region (1996)
* Sumatra (2004)
* Yogyakarta (2006)

Iran
  (1962)
* Bam (2003)
  Central (2005)
  North (1990)
* Northeast (May 1997)

* Northwest: Caspian Sea Area
    (1990)
* Northwest (February 1997)
  Qir Valley (1968)
  Qir Valley (1972)
* Tabas (1978)

Italy
  Avezzano (1915)
* Calabria (1783)
  Calabria (1797)
  Calabria and Naples (1805)
* Calabria (1857)
* Calabria (1905)
  Catania (1693)
  Genoa (1819)
* Messina (1908)
  Naples (1456)
  Naples (1626)
  Naples (1693)
  Naples (1706)
* Southern (1980)
  Umbria, Marche (1997)

Jamaica
* Kingston (1907)
* Port Royal (1692)

Japan
  (1605)
* (1737)
  (1876)
  (1891)
* (1896)
  Hokkaido (2003)
* Hokkaido/Okushiri Island (1993)
* Kobe (1995)
  Niigata (2004)
  Tokyo (1703)
* Tokyo (1857)
* Tokyo and Yokohama (1923)
* Unsen (1793)
  Zenkoji (1847)

Kuril Islands
  (1994)

Martinique
  (1767)

Mexico
* (1973)
  Jalisco (1995)

Manzanillo (1995)
* Mexico City (1985)
Mongolia
Inner (1996)
Morocco
* Agadir (1960)
Near North Coast (2004)
Northern (2004)
Nepal and India
* (1988)
New Guinea
New Britain, Papua (2005)
Nicaragua
* Managua (1972)
Pakistan
* Baluchistan Province (1997)
* Kashmir (2005)
Papua New Guinea
* North (1998)
Persia
Gansana (1139)
Northern Khorassan (1929)
Tabriz (1040)
Tabriz (1727)
Peru
(1892)
North (1990)
South (1996)
* Yungay (1970)

Philippines
* Luzon (1990)
* Manila (1863)
* Mindanao (1976)
Portugal
Lisbon (1531)
* Lisbon (1755)
Romania
* Bucharest (1977)
Bucharest (1990)
Russia
* Sakhalin Island (1995)
* Soviet Georgia (1991)
South and Central America
* (1868)
Sumatra
(1995)
Near North Coast (2004)
North/Indonesia (2004)
Syria
(19 C.E.)
(365)
(742)
Antioch (115)
Antioch (526)
Taiwan
* San Cha Keng (1999)
Tonga Islands
(1995)

Turkey
* Adano (1998)
Bingol (2003)
Deinar (1007)
* Duzce (1999)
* Erzincan, Sivas and Samsun (1939)
* North (1992)
* Northwest (August 1999)
* Scio (1881)
United States
Alaska (1964)
California
* Los Angeles (1971)
Los Angeles (1994)
San Fernando Valley (1994)
* San Francisco (1906)
San Francisco Bay Area (1989)
Missouri
* New Madrid (1811–12)
South Carolina
Charleston (1886)
Vanuatu Islands
(1999)
Venezuela
* Caracas (1812)
Yugoslavia
* Skopje (1963)

CHRONOLOGY

. . . .

* Detailed in text
464 B.C.E.
* Sparta, Greece
217 B.C.E.
June
* Africa
19 C.E.
Syria
115
Antioch, Syria
365
Syria
July 21
* Alexandria, Egypt
526
May 26
Antioch, Syria
742
Syria
856
* Corinth, Greece
893
India

1007
Deinar, Turkey
1038
Shansi (Shanxi), China
1040
Tabriz, Persia
1139
Gansana, Persia
1268
Cieilia, Asia Minor
1290
September 27
Chihli, China
1456
December 5
Naples, Italy
1531
January 26
* Lisbon, Portugal
1556
February 2
* China (Largest death toll in history)

1605
January 31
Japan
1626
July 30
Naples, Italy
1667
May–June
Shemaka, Caucasia
1692
June 7
* Port Royal, Jamaica
1693
Naples, Italy
September
Catania, Italy
1703
December 30
Tokyo, Japan
1706
November 3
Naples, Italy

1716
May–June
Algeria

1727
Tabriz, Persia

1731
November 30
Peking (Beijing), China

1737
* Japan
October 11
Calcutta, India

1755
November 1
* Lisbon, Portugal

1767
August
Martinique

1773
June 7
Santiago, Guatemala

1754
September
Cairo, Egypt

1783
February 5
* Calabria, Italy

1793
April 1
* Unsen, Japan

1797
February 4
* Quito, Ecuador
September
Calabria, Italy

1805
July 26
Calabria and Naples, Italy

1811
December 16–February 7, 1812
* New Madrid, Missouri

1812
March 12
* Caracas, Venezuela

1819
August–September
Genoa, Italy

1822
November 19
* Chile

1847
Zenkoji, Japan

1857
March 21
* Tokyo, Japan
December 16
* Calabria, Italy

1863
July 3
* Manila, Philippines

1868
August 13
* Ecuador and Peru
August 13–15
* South and Central America

1875
May 15
* Colombia

1876
June 15
Japan

1881
April 3
* Scio, Turkey

1885
June–July
* Kashmir, India

1886
August 31
* Charleston, South Carolina

1887
February 23
* France and Italy

1891
October 28
Japan

1892
Peru

1896
June 15
* Japan

1902
April 18
* Guatemala

1905
April 4
* Kangra, India
September 8
* Calabria, Italy

1906
August 16
* Valparaíso, Chile
April 18
* San Francisco, California

1907
January 14
* Kingston, Jamaica

1908
December 28
* Messina, Italy

1915
January 13
Avezzano, Italy

1920
December 16
* Kansu (Gansu), China

1923
September 1–3
* Tokyo and Yokohama

1925
March 16
Talifu (Dalifu), China

1927
May 22
Kansu (Gansu), China

1929
May 3
Northern Khorassan, Persia

1932
December 25
Kansu (Gansu), China

1935
May 31
Quetta, India

1939
January 24
* Chile
December 27
* Erzincan, Sivas, and Samsun,
Turkey

1949
August 5
* Ecuador

1950
August 15
* Assam, India

1960
February 29
* Agadir, Morocco
May 21–30
* Chile

1962
September 16
Iran

1963
July 16
* Skopje, Yugoslavia

1964
March 27
Alaska

1968
August 31
Qir Valley, Iran

1970
May 31
* Yungay, Peru

1971
February 9
* Los Angeles, California

1972
>
> April 10
>> Qir Valley, Iran
>
> December 21
>> * Managua, Nicaragua

1973
>
> September 29
>> * Mexico

1976
>
> February 4
>> * Guatemala
>
> July 28
>> * Tangshan, China
>
> August 17
>> * Mindanao, Philippines

1977
>
> March 4
>> * Bucharest, Romania

1978
>
> September 16
>> * Tabas, Iran

1980
>
> October 10
>> * Al Asnam, Algeria
>
> November 23
>> * Italy (Southern)

1985
>
> September 18–19
>> * Mexico City, Mexico

1988
>
> August 21
>> * Nepal and India
>
> November 6
>> * China (Southwest)
>
> December 7
>> * Armenia (Soviet)

1990
>
> June 21
>> * Iran (Northwest: Caspian Sea Area)
>
> May 31
>> Northern Peru
>
> May 31
>> Bucharest, Romania
>
> June 22
>> Northern Iran
>
> July 16
>> * Luzon, Philippines

1991
>
> April 23
>> Costa Rica
>
> April 30
>> * Soviet Georgia, Russia
>
> October 19
>> * Northern India

1992
>
> February 2
>> Tokyo, Japan
>
> March 14
>> * Northern Turkey
>
> May 26
>> Southern Cuba
>
> June 28
>> United States
>> Yucca Valley, California
>
> September 3
>> Pacific Coast, Nicaragua
>
> October 12
>> * Cairo, Egypt
>
> December 12
>> * Flores Island, Indonesia

1993
>
> January 15
>> Northern Japan
>
> July 14
>> * Hokkaido, Japan
>
> August 8
>> Guam
>
> September 29
>> * Southern India

1994
>
> January 17
>> United States
>> Los Angeles, California
>> San Fernando Valley, California
>
> February 9
>> Pereira, Colombia
>
> June 9
>> Northern Bolivia
>
> August 9
>> * Northwest Algeria
>
> October 4
>> Kuril Islands

1995
>
> January 17
>> * Kobe, Japan
>
> April 7
>> Tonga Islands
>
> May 27
>> * Sakhalin Island, Russia
>
> October 9
>> Jalisco, Mexico

1996
>
> February 4
>> Yunnan, China
>
> February 17
>> Irian Jaya Region, Indonesia
>
> May 5
>> Inner Mongolia

November 12
>> Southern Peru

1997
>
> February 27
>> Pakistan
>> * Baluchistan Province
>> * Northwestern Iran
>
> April 12
>> * Northeastern Iran
>
> September 26
>> Umbria, Marche, Italy

1998
>
> February 4
>> * Northern Afghanistan
>
> March 25
>> Belleny Islands
>
> May 30
>> * Northern Afghanistan
>
> June 27
>> * Adano, Turkey
>
> August 17
>> * Papua, New Guinea

1999
>
> January 25
>> * Western Colombia
>
> February 4
>> Northern Afghanistan
>
> August 17
>> * Northwest Turkey
>
> September 9
>> * San Cha Keng, Taiwan
>
> November 12
>> * Duzce, Northwest Turkey
>
> November 26
>> Vanuatu Islands
>
> November 29
>> Ceram Sea
>
> July 30
>> Northern Chile

2001
>
> January 26
>> * Gujarat state, India

2002
>
> March 25
>> * Hindu Kush Region, Afghanistan

2003
>
> February 24
>> Southern Sinkiang (Xinjang) China
>
> May 1
>> Bingol, Turkey
>
> May 21
>> * Thenia, Algeria
>
> September 25
>> Hokkaido, Japan

December 26
 * Bam, Iran

2004
 February 24
   Near North Coast,
   Morocco
 October 23
   Niigata, Japan

December 26
 * North Sumatra/Indonesia

2005
 February 22
   Central Iran
 September 29
   New Britain, Papua, New
   Guinea

October 8
 * Kashmir, Pakistan

2006
 May 27
   * Yogyakarta, Indonesia

EARTHQUAKES

The rock-solid character of the Earth beneath our feet is only a comforting illusion, if a now fairly universally accepted theory termed plate tectonics is accurate. According to this postulation, the seemingly solid surface of the Earth is actually composed of a series of drifting plates.

According to the theory, which is a distillation of findings in geology, oceanography, and geophysics, the lithosphere, or outer crust of the Earth, is divided into seven major plates and 12 minor ones. Each is about 60 miles thick, and each rests upon a less stable, softer layer called the asthenosphere. Imbedded in these plates and poking up irregularly above them are the continents, which are roughly 40 miles (64.4 km) thick. Scientists theorize that the transfer of heat energy within the Earth causes the plates to drift.

In the course of this process, both continents and plates collide, and some slip over and under each other. Others drift apart. Along these boundaries, earthquakes and volcanic activity occur.

There are two basic causes of earthquakes. One, the shallow earthquake influence, occurs when two plates slide past each other in a grinding, shearing manner along great transcurrent faults, such as the San Andreas Fault in California and the Alpine Fault in New Zealand. The other, deeper influence comes from subduction zones along the edges of the shifting plates, where the rims of these crustal masses dive steeply into the earth mantle and are reabsorbed at a depth of 400 miles (643.7 km).

The tremor of an earthquake results from vibrations set up when these masses collide or override each other, causing compressional or tensional stresses that eventually find their way to the surface.

Picture it this way: The skin of the Earth is like the surface of a sea, shifting with the tides. A solid object, once snagged in some subterranean place beneath this body of water, breaks free and shoots to the water's surface. Where the solid object emerges is exactly akin to the focus point of an earthquake. The water directly above the location of the erupting object is called the epicenter. The ripples cascading outward from this epicenter correspond precisely to the ripple effect of the shock waves of an earthquake.

The only improvement reality can make upon this analogy is this: Unlike the watery target of a pitched or erupting solid object, earthquakes usually have a multitude of epicenters occurring simultaneously along a fault line; thus, there is a certain irregularity and latitude in the shock waves and their effect.

An earthquake customarily begins with a slight tremor. Following this—sometimes with alarming rapidity—is a series of violent shocks, which can cause volcanic eruptions, rockfalls, and explosions of the Earth's surface. Land areas can rise and fall during this time, triggering landslides and tsunamis, the giant Pacific Ocean waves that frequently and suddenly engulf land areas in Asia and the Pacific. (Elsewhere, these frightening walls of water are called seismic waves.) And finally, the third stage of an earthquake's life is marked by vibrations of gradually diminishing force.

When earthquakes take place in urban areas, the devastation is generally acute and cataclysmic. Flexible structures built upon bedrock weather earthquakes better than rigid structures built on loose soil, and the greatest tragedies have occurred in urban centers built in this latter way. In the last 4,000 years, over 13 million deaths have been caused by earthquakes and their aftereffects of fire, landslide, flood, and flying debris.

The strength and destructive power of earthquakes are both measured by seismologists, who gather information, via a seismograph, of the velocity, depth, and longevity of the quake. There are two types of body waves that pass through the earth: P, or primary waves, which are compressionary in nature and are fast travelers; and S, or secondary waves, which are transverse—which means that they cause the Earth to vibrate perpendicularly to the direction of their motion. S waves cannot pass through water; thus, tsunamis are caused by a third type of wave, called the L, or long wave, which churns around the earthquake's epicenter.

Interestingly enough, both P and S waves are affected by changes in the density and rigidity of the materials through which they pass, thus allowing scientists to theorize with great certainty regarding the precise boundaries between the three layers of the Earth—its core, mantle, and crust. (Earthquakes originate in the mantle and crust; volcanic eruptions in the core.) Thus, the disappearance of S waves below depths of 1,800 miles (2,896.8 km) indicates that at least the outer part of the Earth's core is liquid.

The intensity of earthquakes is measured in two ways: by the Richter scale and the Mercalli scale.

The Richter scale, devised in 1935 by the American seismologist Charles F. Richter, is a measurement of magnitude of the energy released by an earthquake at its point of origin. Since this is based upon measurements arbitrarily set at 62 miles (99.7 km) from the epicenter, a combination of multiple seismographs and conversion tables too complex for other than a seismologist to unravel is used to arrive at a Richter number.

The scale is logarithmic—in other words, the energy release increases by powers of 10 in relation to the Richter magnitude numbers. Thus, the magnitude of an earthquake measuring 6 on the Richter scale has a force 10 times that of one measuring 5.

There is no fixed lower or upper limit on this scale. Small earthquakes are measured at figures near 0, and some are actually given negative numbers. An earthquake of magnitude 1 can normally be detected only by a seismograph. One of magnitude 2 is the weakest disturbance that can be noticed by people. A magnitude 5 earthquake releases energy equivalent to that released by 1,000 tons (907.2 tonnes) of TNT.

Any earthquake with a Richter value of 6 or more is commonly considered to be a major disturbance, and a magnitude of 7 on the scale releases energy equivalent to 1 million tons of TNT. There have been earthquakes—the Alaskan earthquake of March 27, 1964, was one of them—that have hit 8.5 on the Richter scale.

The scale developed by the Italian seismologist Giuseppe Mercalli mixes in subjective factors. It measures the severity of an earthquake in terms of its effect upon the inhabitants of an area—the damage to buildings and people, or in the case of slight earthquakes, whether or not sleeping persons were awakened by it.

On the Mercalli scale, a quake can climb in intensity from Degree I to Degree XII. A Degree I earthquake is not felt except by a few people under especially sensitive or favorable conditions. A Degree II earthquake is felt by most people, and it causes delicately suspended objects to swing. A Degree IV earthquake cracks walls and produces the sensation that a heavy vehicle has just struck the building.

A Degree IX earthquake is capable of shifting buildings off foundations and cracking them conspicuously. With a rating of XI, an earthquake leaves few if any masonry structures standing, pulverizes bridges and produces broad fissures in the ground. A Degree XII earthquake is one in which damage is total and waves are seen on ground surfaces.

All of the earthquakes in the following section registered at least 6.5 on the Richter scale and were classified as Degree IX or higher on the Mercalli scale.

....

AFGHANISTAN
HINDU KUSH REGION
March 25, 2002

A 6.1 magnitude earthquake struck the Baghlan Province of the Hindu Kush region of northern Afghanistan at 7:26 P.M. on March 25, 2002, killing 1,000, injuring 4,000, destroying 1,500 homes, and rendering 20,000 families homeless.

By March 2002, the significant drought that had plagued the Hindu Kush mountain region of northern Afghanistan had begun to abate, and families were returning to their homes, particularly in the thriving village of Nahrin, 96 miles (154.5 km) north of Kabul, and 112 miles (189.2 km) southwest of Feysabad.

No sooner had they settled in when at 7:26 P.M. on the night of March 25 a 6.1 magnitude earthquake struck almost directly under the village of Nahrin. The epicenter was only 5 miles (8.1 km) beneath the surface of the Earth, thus intensifying the effect of the quake, which collapsed 90 percent of the mud and brick buildings in the village, burying hundreds of the inhabitants under rubble that continued to accumulate during repeated aftershocks.

In an area spreading out from the epicenter, massive rock slides in the mountains east of Nahrin, triggered by the vibrations, threw huge clouds of dust into the sky and shut off access to not only this village but also dozens of other nearby smaller villages.

Afghanistan, in the throes of a war that was wresting by force the control of the government from the oppressively fundamentalist Taliban regime, responded with a great deal of confusion and misinformation (see the color insert on p. C-2). At first, 2,000 were reported dead, then 1,800, and finally, when the United Nations arrived with its more precise methods of counting, 1,000 more were dead (for a total of 2,800), most from the now totally demolished town of Nahrin.

The 4,000 injured were tended to by a variety of aid groups (see color insert on page C-2). Frederic Roussel, the head of the French aid agency ACTED (Agency for Technical Cooperation and Development), told the BBC: "Everyone is in a state of shock. No one says anything. That's what struck me the most, the silence. One is there in the old town of Nahrin, and it's completely destroyed. Flattened."

"My house was destroyed," said shopkeeper Ghulam Rabbani, whose four-year-old granddaughter died in the quake. "And now we are sleeping outside on the ground. We are afraid right now, but we have nowhere else to go."

The shortage of food became acute as days went by, despite the influx of food, medicine, and emergency shelters that began to arrive in the region after workers used explosives to clear roads leading into the area. The United Nations estimated that 1 ton [1,000 tonnes] of food, 17,000 tents, and 10,000 blankets were distributed to survivors.

As the days passed, aid was delivered in the worst hit areas by trucks and donkeys. "The supplies are taken home by donkey," one correspondent reported. "In fact, some roads are almost congested with donkeys and their loads."

With mobile medical units supplied by the International Security Assistance Force in Kabul and the Russian Emergency Situations Ministry, some of the 20,000 homeless survivors began to return to Nahrin to assist in the rebuilding. Relief efforts were further hampered by land mines and the fear of al-Qaeda attacks, after al-Qaeda released a statement blaming the quake on God's punishment for the war against the Taliban.

In a possible countermeasure, a former Green Beret from Fayetteville, North Carolina, who only gave his first name and current occupation—Jack Does House-Calls—moved between the tent camps, offering help, particularly for women and children.

"I thought I was dead and I thought the child was dead for sure," said eight-months pregnant Sharifa, who had crouched on her hands and knees under the rubble of her home to protect her newborn until they were rescued by Jack. In gratitude, she asked Jack, who also treated her back injury, to name the daughter. He named her Suzzana, meaning "new beginning" in the Dari language.

AFGHANISTAN
NORTHEASTERN REGION
May 30, 1998

The second earthquake in four months tore through the northeastern portion of Afghanistan near the Tajikistan border on the morning of May 30, 1998. Worse than the February quake by .8 on the Richter scale, this one produced far more casualties: 4,000 were killed and many thousands were injured and/or made homeless.

Rescue workers were still clearing the ruination of the February 4 earthquake that had rumbled through the Rustaq region of northeastern Afghanistan, near the Tajikistan border, when another, more powerful earthquake ripped through the region. Measured at 6.9 on the Richter scale, the quake struck at 10:52 A.M. on May 30. Its epicenter was located approximately 190 miles east of the country's capital, Kabul.

Though the quake hit in daylight, when most people were outside of their houses, its devastation would surpass that of the earlier quake. "I think this is going to be worse than February for three reasons," the United Nations aid director in Rustaq, Alfredo Witschi-Cestari said, after taking a helicopter tour of the area. "The magnitude of this earthquake is greater, a lot of houses were already badly damaged from the previous quake, and the affected area is larger."

Not only that. In imitation of the events following the first quake, the weather turned foul. Heavy rains blanketed the area, threatening land- and mudslides, and lowering the cloud ceiling enough to prevent rescue flights. Aid workers who traveled in a Russian-built MI–8 helicopter on a reconnaissance mission, reported that one village over which they flew was completely destroyed, with survivors huddling amid the rubble in makeshift tents. The area's airstrip, repaired after the February quake, now seemed to be buckled and useless.

Conflicting reports of casualties cascaded from various agencies. The consensus seemed to be that over 100 villages had been destroyed in the quake, nearly 2,000 residents had been killed, and 2,000 injured.

Finally, the weather lifted enough for emergency supplies of food, medicine, and tents, left over from the February disaster, to be brought to the disaster area by the United Nations Office for the Coordination of Humanitarian Assistance to Afghanistan and the World Food Program.

United Nations spokespeople returning from the area were worried. "Aftershocks that continue to shudder through the region, sometimes only minutes apart, keep people from moving back to their houses and remind a population traumatized by Saturday's quake that another could easily strike.

"In addition to fractures and bruising caused by falling buildings," they went on, "there are fears that alternating sunshine and rain could increase the risk of malaria."

Noting the number of survivors sleeping in the open, British nurse Valerie Powell added, "Their supplies of water and food have been disrupted and there is a serious risk of epidemics breaking out, like cholera and dysentery."

Nevertheless, relief efforts, profiting from experience, proceeded more smoothly than those in February. And it was well that they did. The carnage from this quake was considerably worse than that caused by the February one. By June 6, the weather had cleared, and the United Nations and the Red Cross were able to fly helicopters in and out of the city of Faizabad to remote villages, where many survivors had had no access to food or medicine for over a week.

In Angarain, located in the rubble of the Shari-Buzurg area, villagers struggled to reach the top of a hill to receive food from a Red Cross helicopter. "We have been eating these grasses over the past week," a trembling refugee said as she received food from the workers.

It would be the remainder of the summer before casualty figures could be rightfully assembled. The final count: 4,000 dead and many thousands injured and/or homeless.

AFGHANISTAN
RUSTAQ
February 4, 1998

One thousand, five hundred and sixty residents of the northeastern part of Afghanistan were killed in a 6.1 force earthquake that struck at 7:37 in the evening of February 4, 1998. Eight thousand ninety-four houses in these isolated mountain villages were destroyed, and the livelihood of their residents was eliminated with the deaths of 6,725 livestock.

Afghanistan's battles with natural disaster in 1998 were surpassed only by its civil war between the Islamic Fundamentalist Taliban, which controlled 80 percent of the country, and an alliance under former president Rabbini, who had fled the capital, Kabul, when it was seized by the Taliban.

The country is mountainous, thinly populated, and landlocked, wedged between Pakistan on the east, Iran on the west, and on the north by several former Soviet states, Tajikistan, Uzbekistan, and Turkmenistan. In the best of times and weather, communication is minimal between the scattered villages clinging to the sides of mountains and nestled within ranges of hills.

Into this mix of misery, the first of two major earthquakes struck the northeastern province of Takhar, and particularly the city of Rustaq, at 7:37 in the evening of February 4, 1998. As in other countries of the area, the houses in which subsistence farmers and their families lived were made of sun-baked bricks and mud, which collapsed cataclysmically during the first tremor, which measured 6.1 on the Richter scale.

This was terrible enough. But to add to it, the winter of 1997–98 had been particularly fierce. Snow, fog, and destroyed roads left by the civil war all conspired to isolate the area still further from rescue squads.

Immediate assessment of the damage was first made in the area around the city of Rustaq, which had a population of 10,000. The city itself contained hundreds of damaged or destroyed buildings; former roads were impassable because of rubble that clogged them. The nearby villages of Guzara Darra and Ganda Chashma had almost totally vanished. "The hills collapsed into each other, making a huge crater in the earth," a spokesman for the alliance told reporters.

United Nations relief agencies set out from Rustaq across snowy mountain passes; another agency left from Dushanbe, Tajikistan. The weather made airlifts impossible, and it worsened quickly. To add to the misery, more tremors rumbled through the disaster area, further blocking mountain passes and roadways.

"Now we may only be able to reach the survivors by helicopter," said Sayed Ali Javed, the leader of one rescue team. "The cold must be the major killer now," added Andrew Wilder, director of the Save the Children program in the Pakistani capital of Islamabad. Without blankets, fuel, and plastic sheeting, there was no way that refugees from the quake could endure, and snowstorms and fog continued to keep pace with the repeated aftershocks. The only positive development during the first week after the quake was the announcement by the Taliban that it would establish a three-day truce so that opposition soldiers could help in relief efforts.

On February 8, another full-sized earthquake hit the region, laying waste to three more villages. A week later, the weather worsened considerably, canceling flights of airplanes with relief supplies. Two United Nations planes based in Pakistan circled a northeastern airstrip, waiting in vain for a break in the clouds. When it failed to occur, they returned to their base. Two Russian-made cargo planes, scheduled to fly to Pakistan from Rustaq, were unable to take off.

Only helicopters and some foot patrols, containing members of Doctors Without Borders, managed to break through the ring of mountains and snow to tend to survivors and help in pulling the dead from ruins.

Finally, in desperation, the United Nations appealed to the international community for $2.5 million to drop supplies by parachute to the isolated, devastated villages. Two weeks after the initial earthquake, the weather lifted enough to allow a plane loaded with emergency aid to drop, by parachute, supplies to various villages in the area of the temblor.

Ultimately, as the winter finally lessened a little, it was possible to make a general assessment of the destruction. One thousand five hundred and sixty people had died, 818 were injured, 8,094 houses had been destroyed, and 6,725 head of livestock had been killed in this, only the first episode in a steadily worsening tragedy that would further devastate this isolated region of the world.

AFRICA
June 217 B.C.E.

One hundred cities were affected and between 50,000 and 75,000 people were killed by a giant earthquake in North Africa in June 217 B.C.E.

According to Roman historians, a gigantic earthquake tore through all of North Africa early in June of 217 B.C.E., a moment in history best known for Hannibal's conquering of the Roman legions at Trasimeno Lake, which was a momentary stop in the Roman conquest of all of central and southern Italy, and the continued building of the Via Appia from Capua to Tarentum.

The parts of the Appian Way that were in place by June 217 B.C.E., however, must have been dislodged by this quake, which, according to the historians, ". . . tumbled lakes and streams from their beds . . ." in several Italian cities.

It destroyed 100 cities in North Africa and killed 50,000 to 75,000 people.

ALGERIA
AL ASNAM
October 10, 1980

Two earthquakes, rated at 7.5 and 6.5 on the Richter scale, swept through Al Asnam, Algeria, on October 10, 1980. Six thousand died; 250,000 were rendered homeless and 80 percent of the city was destroyed.

Slightly after 12:30 P.M., while thousands of residents of the Algerian city of Al Asnam were at home celebrating the Muslim sabbath, a monstrous 7.5 earthquake raged through the city, followed closely by another 6.5 tremor that toppled most of the public buildings, including a hospital, a girls' high school, and the central mosque.

"You could hear the screams of the injured and the dying," a French resident said, describing the impact of the first tremor, which lasted a full two minutes. "But it was mostly destroyed in the first 30 seconds," was the way the resident summed up the initial impact of this quake, which destroyed 80 percent of the city, killed 6,000 and left 250,000 homeless.

The shocks ripped through the countryside, forcing apart fissures in the sweeping plains and abrupt hills that surround Al Asnam and run down to the Mediterranean. In Al Attaf, two pink apartment buildings tilted crazily in opposite directions, a pair of leaning towers in imminent danger of collapse.

Rescue began immediately, but in a disorganized fashion. Reflecting the hysteria of the populace, officials initially released wildly inflated casualty figures, estimating between 17,000 and 25,000 dead. The highway leading from Algiers to Al Asnam quickly became snarled and choked with truck and automobile traffic. Guards at military checkpoints let only essential vehicles travel to the worst hit areas. Dozens of ambulances hurtled down the road, which buckled in places, while civilian cars moved over and waited patiently in line.

In some parts of the city center, hands and arms protruded from beneath heavy concrete slabs. Reports from the scene said that doctors and sometimes nurses amputated arms and legs without anesthesia to free trapped people.

As the night wore on, aftershocks panicked people as rescue workers toiled on by floodlights powered by portable generators. The Algerian government appealed to other countries for aid, and several moved swiftly—notably Switzerland, which sent in Alpine rescue teams, and France, which sent firemen and dogs trained to sniff out survivors. Soviet nurses arrived upon the scene, and planeloads of blood plasma, medical equipment, blankets, and tents arrived from the Netherlands and West Germany. The United States sent $25,000 from its emergency fund.

In one bizarre footnote, rescue workers, digging through ruins two weeks after the quake hit, found six survivors who had been having a drink in a local cafe when the quake hit. They had managed to stay alive for 14 days, drinking lemonade.

ALGERIA
NORTHWEST
August 9, 1994

One hundred and fifty people, mostly in rural areas, died in an earthquake that hit northwestern Algeria on August 9, 1994. Over 10,000 were rendered homeless.

At 2:15 A.M. on August 9, 1994, a major earthquake roared across northwestern Algeria. The 5.6 force quake instantly made 10,000 residents of the area homeless, collapsing their straw and mud-brick homes as thoroughly as if a giant had slammed them with a fist.

The center of the quake area was in the vicinity of the city of Mascara, 250 miles (402 km) west of Algiers, but the greatest damage was in the outlying rural region, composed of farms and small villages. There were few paved roads to these villages, some of which disappeared overnight, as the multiple tremors flattened most of their structures.

Rescuers, who set out at daylight from Algiers, treated the wounded and survivors in the hot springs resort of Ain Fekan and nearby Bou Henni. Besides damage in Mascara, there were also collapsed buildings and fatalities in Oran, the regional capital, on the coast about 62 miles (99.7 km) north of Mascara.

But the greatest tragedy awaited rescuers later, after they cleared the rubble from the unpaved roads that wove through the farmland and small villages. Typical was the plight of an elderly man in the village of Hassine, which was almost entirely destroyed. "I lost my wife, my children, my house, my everything," he told reporters. "I have nothing left."

Most of the rescue work that turned up 150 bodies was done by Algerian teams. International and foreign organizations, mindful of repeated threats against foreigners by Islamic militants, weighed their entrance into the disaster carefully.

ALGERIA
THENIA
May 21, 2003

A 6.8 magnitude earthquake struck northern Algeria on the afternoon of May 21, 2003, killing 2,268, injuring 10,147, and rendering 200,000 people homeless.

At 5:44 P.M. on Wednesday, May 21, 2003, a 6.8 magnitude earthquake ravaged the countryside near the town of Thenia, in the Zenmount Region, 45 miles (72.4 km) east of the capital Algiers. Originating 6 miles (9.7 km) beneath the surface of the Earth, the quake devastated the heavily populated area that also included the towns of Bourmedes, Zenmount, Belouizdad, Ruiba, and Reghia. Terrified residents who escaped from collapsing houses were faced with more than 20 continual aftershocks, some measuring 5 on the Richter scale. It was the worst earthquake to strike the region since the twin quakes at al-Asnam in 1980 (see p. 37).

Water supply lines burst, electricity disappeared, roads and bridges buckled, and normal communication with the outside world was impossible for days. The Algerian government was slow to respond, and as a result looters descended from Algiers, determined to clean out abandoned houses. Local police and local residents fought off the thieves. "People in vans were seen looking around for things they could steal, but they saw we were well prepared," Samir Helli, a resident, later told a reporter for the Associated Press.

Meanwhile, other local rescuers dug through the rubble to rescue buried survivors. Japanese rescuers, among the first to arrive on the scene, pulled a 21-year-old waiter from the rubble of a hotel on the Mediterranean coast, and there were reports of young and old buried residents revived from spending days and nights under the wreckage of houses and office buildings.

The 86°F (30°C) heat worked against aid workers as bodies rotted beneath the rubble and a lack of clean running water and sanitary facilities added to the health hazard. "It's the lack of water and the heat," Dr. Malika Lamare, a doctor at the health center in Reghia, told a reporter. "Without water, they risk death."

By the end of May, the United Nations Children's Fund had flown in infant hygiene kits, oral dehydration salts, first aid kits, water purification units, cooking kits, and vaccines. The International Red Cross and Red Crescent brought in potable water, and other countries, including the United States, flew in blankets and tents to house the 200,000 displaced people made homeless by the quake.

Finally, Abdelaziz Bouteflika, the president of Algeria, arrived to survey the damage. Angry crowds pelted him with debris and insults.

ARMENIA (SOVIET)
December 7, 1988

An earthquake measuring 6.9 on the Richter scale struck Soviet Armenia on December 7, 1988, destroying two-thirds of Leninakan, Armenia's second largest city, and eliminating a score of villages and towns.

Officially, 28,854 were killed, 12,000 were injured, and 400,000 were made homeless. Unofficial reports stated there were 55,000 casualties.

The inefficiency of bureaucracy, the contrast between the Leonid Brezhnev and Mikhail Gorbachev years in the USSR, and the terrible capriciousness of natural disasters all collided in a catastrophic, 6.9 earthquake at 11:41 A.M. on December 7, 1988, in Soviet Armenia. Gorbachev, in New York City on a diplomatic mission to the United States, cut short his visit and rushed home when the news reached him.

It was the worst quake in the region in 80 years and one of the worst in Soviet history. "I have Chernobyl behind me," said Yevgen I. Chazov, the Soviet health minister, to the newspaper *Izvestia*, after surveying the region by air, "but I have never seen anything like this. The scope is just catastrophic."

A woman, in the wreckage of her home, surveys what is left of her belongings after the Armenian earthquake of December 7, 1988. (Rudolph von Bernuth, CARE)

Leninakan, the second largest city in Armenia, was two-thirds destroyed. Half of Korovakan, a city of 150,000, was toppled, and Spitak, a town of 30,000, disappeared entirely, collapsed to its foundations by the quake. Tremors were felt as far west as Baku, the capital of the Soviet republic of Azerbaijan, and north to Tbilisi, the capital of Georgia.

According to Dr. Robert Wallace of the U.S. Geological Survey's earthquake center in Menlo Park, California, the area is a "structural knot," made by the interactions of several rigid tectonic plates. A broad zone extending from the Mediterranean Sea to the Himalaya Mountains is a relic of Tethys, an ancient sea that once separated Eurasia from Africa and India. These continents converge over what once was a sea and press against each other. In turn, the pressure causes earthquakes, volcanic eruptions, and the creation of nascent mountain ranges. In 893 C.E., a quake in the area was reported to have caused 20,000 deaths, and in 1667, in nearby Shemaka, another consumed 80,000 lives.

What was particularly horrendous about the 1988 Armenian earthquake was the extensive loss of life caused by the faulty construction of many of the houses in the area. Built cheaply and carelessly during the Brezhnev era, they simply became unhinged and crumbled inward on their inhabitants. In Leninakan, the clock in the town square was eerily frozen at 11:42, one minute after the quake hit. All around it, slabs of concrete were piled in crazy confusion, twisted girders poked fingers at the sky where apartment buildings once stood, and hollow-eyed survivors wandered about, looking for familiar faces.

Trains were derailed, tracks were obliterated by landslides, cars were tossed off roads like toys, and hundreds of fires were ignited by broken gas and electric lines.

"It was like a slow-motion movie," said Ruzanna Grigoryan, who was working at the stocking factory in Leninakan when the quake struck. "There was a concrete panel slowly falling down . . . I used every piece of concrete, every rod, and I was pulling myself out centimeter by centimeter," she continued. "Then I found out that I had been doing this for four hours."

"There was a loud humming noise, then steam burst out of the ground, buildings began to rock like boats and it was as though the earth was boiling," said Leninakan resident Gevork Shakhnasaryan, as he described the first and strongest of three shocks.

Soviet television showed a distraught man standing in a pile of wood and concrete and pointing despairingly at a charred pile of rubble, which had once been his kitchen, and where his brother had been eating lunch when the quake struck. The body of his father

A couple searches through the wreckage of their home following the Armenian earthquake of December 7, 1988. (Rudolph von Bernuth, CARE)

had just been pulled from the wreckage. The brother had not been found. "No one could live there . . . The grief is horrible," said the man, in a choked and terrible voice.

Leninakan's mayor, Emile Kirakosyan, lost his entire family of 15. "They're all gone," he wept on television, "I must keep working."

Workers continued for days to dig through the wreckage. Four days after the quake, a crane operator, Anton Sukisikanyan, said on Soviet television, "We've brought out 23 people alive. I don't want to talk about how many bodies—but there were 280 people in this building."

One hundred twenty of Leninakan's apartment buildings disappeared entirely. What had been a nine-story structure close to the central square was reduced to a 40-foot mound of debris flecked with remnants of clothing, curtains, and mattresses. Ironically, one of the buildings in Leninakan that collapsed into dust was the Seismic Institute of the Armenian Academy of Sciences.

In addition to the shoddy construction of prefabricated houses, the backward state of Soviet medicine

affected local rescue efforts. However, Gorbachev's perestroika facilitated a quick response from the outside world, and an openness to international relief efforts that would have been impossible in earlier years. By December 10, a team of 160 French firefighters with 36 dogs was unloading two DC-8 airliners and a Hercules transport plane that contained tents, blankets, medical supplies, and special equipment at Moscow airport.

And then, later that same day, heavy fog closed in on much of the USSR, and both supplies and help were further delayed. On December 11, a Soviet military transport plane crashed on its approach to Leninakan, killing nine crew members and 69 military personnel. A day later, a Yugoslav Air Force transport plane plummeted to earth near the congested airport in Yerevan. Its crew of seven was killed.

Chaos now began to escalate, as *Pravda*, the Communist Party paper, blamed the unruliness on entrenched bureaucrats and the shoddy construction on the "period of stagnation" of the 18 years of power of Leonid Brezhnev. Still, 46 countries, in the greatest foreign aid effort since World War II, continued to work around the clock despite rain slickened roads, shortages of cranes, medical equipment, antibiotics, and blood plasma, and two new problems: looting and an outbreak of ethnic tensions.

On December 12, Soviet television reported an attempted break-in at a savings bank in Spitak, a town leveled by the quake. An army officer reported that soldiers had to restrain local crowds who wanted to kill the thief on the spot. "You can see that grief does not always unite people," chief Soviet spokesman Gennadi I. Gerasimov observed.

By the next day, trucks with loudspeakers were roaming the streets of Leninakan, advising the homeless to leave. But a virtual army of soot-covered survivors remained in the central city, hunting for relatives, belongings or, more insidiously, for mementos.

One hundred fifty people were arrested for looting, including one man in Korovakan caught stealing jewelry and watches from corpses. Desperation drove some survivors to huddle in dangerously unstable ruins or to sleep around campfires in the rubble. Driven by despair, a 46-year-old man in a Leninakan hospital who had lost his entire family grabbed a knife and stabbed himself to death.

Gradually, the international rescue teams began to withdraw. Tent cities were erected for refugees, particularly in Spitak, once known for its sugar factory and elevator plant, and now reduced to nothing.

Thirty-five days later, in Leninakan, a miracle was reported. Six men, led by Aikaz Akopyan, a Soviet carpenter, were said to have been discovered in the basement of a wrecked building. They had, Akopyan told the world press, stayed alive on "pickles, canned fruit salad, jams, apples and smoked ham." Akopyan said he had kept up the spirits of the other younger men by singing and telling them stories.

But the rescue turned out to be a hoax. "It is understandable [that we all believed it]," reported Tass, the government news agency. "One wants so much to believe in the miracle of six people being saved on the 35th day after the quake."

Hoaxes aside, there were lessons to be learned from the quake and its aftermath. Armenia and Azerbaijan, friendly for a few days after the quake, returned to armed conflict, as if nothing had happened. Staggering misery seemed to make no difference where ethnic rivalries were concerned.

The weaknesses of the old Soviet system and the strengths of the new were revealed as dramatically as the ribs of destroyed buildings. "These days of death and diplomacy have exposed backwardness and inflexibility in the Soviet system that are certain to provoke months of official recriminations," reported Bill Keller in the *New York Times*. "But they have also seen a society long secretive about domestic tragedies and ashamed of soliciting foreign help open itself to the world's pity—and defer to outside advice—as never before."

"It is a big transformation since Chernobyl," agreed Colonel Khludnev, the commander in the village of Stepanavan. "After Chernobyl, they allowed foreigners to participate in the investigation of why it happened, but that was it. This time, we are asking for concrete help."

In the rest of the world seismic experts applied lessons learned in Armenia in setting up a new network of highly sensitive seismic stations to monitor the Earth's tremors with greater accuracy. "A safer world awaits our resolve to act together," said Dr. Frank Press, president of the National Academy of Sciences.

Peter Wilson, a London fireman who was interviewed by the *Times* as he was packing up 11 days after the quake, concluded, "I wouldn't want to see anything like this again. It gives a very good impression, I suppose, of what the Second World War must have been like."

CHILE
November 19, 1822

An earthquake centered in Valparaíso, Chile, on November 19, 1822, killed 10,000 people and raised the elevation of that city permanently.

The land around the central Chilean port city of Valparaíso was raised four feet by the earthquake that struck on November 19, 1822. The quake hit the Chilean coast, collapsing a number of towns and wiping out entire villages. Its focal point, however, was the city of Valparaíso, where 10,000 people died in massive building collapses, fires, and open fissures.

According to observers, the quake raised the ocean bed so far that thousands of fish were stranded on dry land, where they died, sending a horrible stench through the city. Hundreds of boats, once used to gather in these fish, landed out of the water, where, left behind by their dead owners, they gradually deteriorated.

CHILE
January 24, 1939

A record 50,000 people were killed, 60,000 were injured, and 700,000 were made homeless by the Chilean earthquake of January 24, 1939.

No earthquake in South American history claimed more lives than the earthquake of January 24, 1939. Striking at 11:35 P.M., it caught an unwary population asleep and unprepared. Over 50,000 people lost their lives, 60,000 were injured, and 700,000 were rendered homeless by this cataclysmic quake, which devastated five of Chile's oldest towns.

Concepción lost 70 percent of its buildings, from the most historic churches to the lean-tos of some of its poorest inhabitants. Underground, hundreds of coal mine shafts collapsed, burying the miners working in them.

Coihueco, Coronal, and Angol suffered similar devastation. But the worst force of the earthquake occurred in Chillan. Located 250 miles (402.3 km) south of Santiago, Chillan was the city closest to the quake's epicenter. Of its hundreds of buildings, only three remained standing after the three minutes of tremors ceased.

Over 300 patrons of Chillan's National Theatre were crushed when the building collapsed on them. Bewildered survivors wandered the streets.

A night watchman working outside of the main power plant saw a score of live wires split, erupt in sparks, and then snake toward the ground, obviously in the path of running refugees. The watchman ran into the power plant and turned off the power switch, thus saving hundreds from electrocution. A moment later, the building collapsed, crushing him to death.

Other individual tales of tragedy and heroism abounded, but the most chilling statistic from this particular disaster is a resounding one: Seventy percent—almost an entire generation—of the 50,000 dead were children.

CHILE
May 21–30, 1960

Multiple earthquakes occurring between May 21 and 30, 1960, killed 5,700 people, rendered 100,000 homeless and destroyed 20 percent of the country's industrial complex.

In a seven-day trial by terror, a series of earthquakes very nearly destroyed the entire country of Chile during the period from May 21 to 30, 1960. Ninety thousand square miles of Andean countryside were ripped apart and reshaped by these multiple and, to the inhabitants, at least, seemingly unending series of major and minor quakes, volcanic eruptions, and gigantic tsunamis. To some, it was as if the end of the world had arrived. Fifty-seven hundred people died; 100,000 were

made homeless; 20 percent of the country's industrial complex was demolished. The cost of the wholesale destruction was estimated at $400 million.

There was some warning of the earthquake's approach: A series of minor tremors shook the ancient city of Concepción, rebuilt with wider streets after the disaster of 1939 (see previous entry). But nobody could possibly have predicted the hammer blows of five earthquakes occurring within 48 hours. The first began shortly after 6 A.M. on the morning of May 21. And once more, the population of Concepción ran into the streets to escape the slow-motion crumbling of ancient buildings, the tearing apart of the walls of their homes and the sickening heaving of the Earth as it opened up into fissures that swallowed homes, animals, and people indiscriminately.

The city of Concepción was almost entirely destroyed, as were Valdivia, Osormo, and Puerto Montt. A 25-mile (40.2-km) area of the Chilean countryside sank 1,000 feet (304.8 m).

Soon after the first quake had ravaged the country, the ocean waters began to fall alarmingly, a certain sign that a tsunami was in the making. It was. One of the largest and most terrifying tsunamis ever to roar into Chile, 24 feet (7.3 m) high and traveling at a speed of 520 MPH (836.8 km/h), crashed over the coastline, flattening the fishing villages of Anoud, Lebu, and Quelin and drowning everyone in its path. It flooded the entire coast from 1,500 feet (457 m) inland to the former coastline. Withdrawing, folding back on itself, it then traveled in the other direction with equally withering force, breaking on the shore of Japan and killing 150 people there.

As the earthquakes continued, six dormant volcanoes sprang to life and three more were born. The erupting volcanoes spewed ash 23,000 feet (7,010.4 m) into the air. The lava oozed into Lake Ranco, which burst its banks and created multiple landslides that drowned yet more people and livestock.

Thousands of homeless refugees were then subjected to a mixture of rain and sleet that fell on the ravaged countryside after the quakes finally subsided.

Millions of dollars in aid and supplies were flown into Chile from 35 countries. Eventually, this would allow Chile to rebuild, but not before 100,000 Chileans, some of them berserk and hungry enough to commit cannibalism and murder, rioted for crusts of bread and containers of milk. Men fought knife duels, and the police and the army fired into the hysterical mobs. It would be months before the doctors, nurses, and relief workers flown into Chile from other countries could restore something even resembling their former living standards to hundreds of thousands in this country that had literally been torn asunder.

A scene in the Waiakea area of Hilo, Hawaii, following a tsunami caused by the Chilean earthquake of May 21, 1960 (National Geophysical Data Center/U.S. Navy)

CHILE
VALPARAÍSO
August 16, 1906

A monster earthquake ripped apart Valparaíso, Chile, on August 16, 1906. Fifteen hundred died, over 100,000 were made homeless, and over $200 million in damage was caused by the quake.

The entire north-south axis of Chile was shaken by a major earthquake on August 16, 1906—a mere four months after the fabled San Francisco earthquake (see p. 98).

Far more were killed in Chile than in San Francisco, although the epicenters of both quakes were in the center of major cities. Perhaps because it was more populous, perhaps because its buildings were older and situated more closely together than those in San Francisco, or more probably because of the inequality and randomness of natural disasters, much of Valparaíso collapsed in upon itself. Fifteen hundred people were killed, over 100,000 were rendered homeless, and over $200 million in damage was sustained by this city that had been damaged extensively by earthquakes many times before and would be many times again.

CHINA
February 2, 1556

Eight hundred twenty thousand people perished in an earthquake that took the largest toll of human life of any earthquake in all of recorded history, in China, on February 2, 1556.

As observers of natural phenomena, the Chinese are without equal in the ancient world. Besides noting astronomical events such as the appearance of new stars, comets, and sunspots, they chronicled the statistics of every major earthquake since 1831 B.C.E. Their laborious and diligent recordkeeping thus left the world a virtually uninterrupted account of nearly 3,000 years of earthquakes.

So, although details are missing, and only a dispassionate list survives, it seems indisputable that the worst earthquake in terms of loss of life in all of recorded history took place on February 2, 1556, in an immense territory of Shensi (Shaanxi) Province in central China, along the valley of the Wei River, a tributary of the Huang River.

A staggering total of 820,000 people perished in this cataclysm. And that is all that is known about it.

CHINA
KANSU (GANSU)
December 16, 1920

One hundred eighty thousand people were killed outright by an 8.6-force earthquake in Kansu (Gansu), China, on December 16, 1920. Another 20,000 froze to death after their homes had been destroyed by the quake.

One of the most severe earthquakes of the 20th century—it registered 8.6 on the Richter scale—rolled through the remote province of Kansu, China, on December 16, 1920. Villages, constructed mostly of flimsy material insulated with skins, were quickly destroyed by the huge quake. Ten ancient cities collapsed in ruins in a matter of moments. One hundred eighty thousand occupants were killed by the quake. Another 20,000 froze to death when their homes and the villages in which they had lived only a day before were wiped from the face of the Earth.

Adding to the destruction from the immediate rolling and opening of the Earth were landslides that were touched off by the quake. The area of Kansu is not only mountainous but also composed of caves in which there are deep deposits of loess, a fine and unstable sand. Set in motion by the quake, these loess deposits cascaded down the sides of mountains like falling water, loosening huge rocks and gigantic clods of turf.

In one case, a gigantic landslide actually saved a village and several thousand lives. Located at the junction of two valleys, Swen Family Gap suffered enough damage from the quake's first shock to kill one-tenth of its population and destroy roughly that percentage of its buildings. A giant landslide, set in motion by the same tremors, roared down the mountainside, and

would have buried the remainder of the population of the village if two bodies of earth hadn't collided just above the village and formed a natural dam that contained the rest of the landslide.

Other natural dams of timber and rockslide had to be demolished to prevent flooding. Ten thousand survivors of this monster quake set out to do just that within days of its occurrence.

CHINA
SOUTHWEST
November 6, 1988

In the thinly populated southwest region of China, 938 people lost their lives on November 6, 1988. The quake measured 7.6 on the Richter scale.

An enormous earthquake that measured 7.6 on the Richter scale rumbled through Lancang and Menglian counties, a lightly populated area of China about 240 miles southwest of the Yunnan provincial capital of Kunming and slightly north of the Burmese border, at 9 P.M. on November 6, 1988. Oddly, for a quake of that magnitude, no damage or injuries were reported in the city of Kunming, which contains 1.5 million inhabitants, even though there were aftershocks that registered a startling 7.2.

The scene was quite different in the two counties of Lancang and Menglian. There, the worst quake in China since the 7.8 monster that killed over 650,000 in Tangshan (see following entry) left only a handful of buildings standing and destroyed thousands more in the 14 surrounding counties.

A death toll of 938 was reported by the New China News Agency, a small enough figure, considering the magnitude of the quake, but the fact that the two counties were fairly sparsely settled—for China—might have accounted for this. Eighty-one thousand people live in Menglian County and 237,000 in Lancang. The region is one of dense jungle, and the poor roads into the area were rendered unusable. Neither communication with the quake area or accessibility for relief supplies was possible for weeks after the disaster.

CHINA
TANGSHAN
July 28, 1976

Over 650,000 people were killed and more than 780,000 were injured in a monster earthquake that centered in the city of Tangshan, China, on July 28, 1976.

The year of killer earthquakes was 1976. Twenty-three thousand were killed in Guatemala on February 4; the Friuli, Italy, quake on May 6 caused over 900 deaths; a 7.1 shock on June 25 killed 6,000 people in Iran; an earthquake and tsunami killed over 4,000 in the Philippines on August 17; and over 4,000 lost their lives in a quake near Muradiye, Turkey, on November 24.

But all of these pale alongside the staggering death toll of over 650,000 from an earthquake that virtually razed the industrial city of Tangshan, located some 95 miles east of Peking (Beijing). The monster shock reverberated for nearly a hundred miles in all directions, collapsing mud walls and brick buildings and killing 100 people in Peking itself. Over 780,000 people were injured and hundreds of millions of dollars in damage was caused.

There were also political reverberations set off by the quake. The Chinese view that predates communist doctrine maintains that natural disasters are a mandate from heaven, and earthquakes have been considered a bad omen for governments as far back as the Sung dynasty. Local politicians credited their losses to the effects of this one.

COLOMBIA
May 15, 1875

The brief earthquake that struck Colombia on May 15, 1875, claimed 16,000 lives.

More than 16,000 people were killed when a 45-second earthquake rumbled through the South American country of Colombia on May 15, 1875. Most of these fatalities occurred along the Venezuelan border, but the major cities of Colombia also suffered heavy casualties.

The city of San Cayetano sustained a large number of fatalities and severe damage to its business district. In Cucuta, a nearby volcano erupted simultaneously with the earthquake. Enormous fireballs were catapulted from the volcano into the city, some of them landing on churches and businesses, and burning them to the ground.

COLOMBIA
QUINDO
January 25, 1999

One thousand one hundred and eighty-five residents of Quindio Department in western Colombia were killed by a 6.3-force earthquake that struck near the capital city of Armenia on January 25, 1999. Over 700 were missing and presumably dead, over 4,750 were injured, and 250,000 were made homeless.

Earthquakes are common in the Andean nation of Colombia. But not since 1875, when 1,000 people died in a tremor near Cucuta, a border city with Venezuela, had an earthquake caused such destruction and carnage as the one that struck in the heart of Colombia's west coast coffee-growing area on January 25, 1999.

The first tremors hit at 1:19 P.M. in the Cordillera Central, in western Colombia, near the Ruiz-Tolima volcanic complex in the department of Quindio. Quindio is fundamentally a coffee-raising region, encompassing the cities of Córdoba, Barcelona, Calarca, Pijao, and the capital, Armenia, which was very close to the quake's epicenter.

The coffee fields that surround these cities were filled with workers at the time the temblor hit, and, terrified, these workers avoided the fissures that opened in the fields. But their troubles were amplified manyfold in the cities, which were built upon soft soil and experienced instant damage, their buildings swaying and collapsing. Armenia and its surrounding villages were hardest hit, and the city's populace panicked, causing intense traffic jams, injuries, and death from falling debris. A six-story office building accordioned in upon itself; a theatre and the city jail crumbled and fell in upon patrons and prisoners; the police station was partially reduced to rubble, killing 18 policemen who were trapped in it when it collapsed. Armenia's hospital was destroyed, but a new wing, built of stronger materials, withstood most of the earthquake and the following aftershocks. Its emergency room became a center for the rush of injured who immediately descended upon it.

The only fire station in Armenia collapsed upon itself, crushing all of the fire trucks. Fortunately, there was only one earthquake-induced fire in the city, in a collapsed three-story building that contained chemical products.

The nearby city of Periera, the capital of neighboring Risaralda province, was not so lucky. Fires roared unchecked in the ruins of crumbled buildings, and rescue workers concentrated their first efforts on extracting the bodies of victims trapped by the fires.

"There's no way to measure the crisis," Armenia mayor Alvaro Pulido said to reporters. The city had an earthquake emergency plan, but with the absence of police and fire fighters, it was unable to put it into use. Survivors wandered aimlessly, digging at the wreckage with shovels and bare hands, crying out for help in finding their families.

Twenty-six-year-old Liliana Patricia Vega stumbled through the Brasilia Nueva district, crying out. She had been on the second floor of her building with her son and six-year-old daughter when the building collapsed around them, killing her 10-year-old instantly. She held her daughter close to her, covering a bloody gash in the child's forehead. "Does anyone have any medicine?" the woman cried. "Can I get some medicine?"

In the nearby town of Cajamarca, the quake unearthed a grisly sight: old skulls and bones tumbling from wrecked crypts into the graveyard.

Outside rescue squads were prevented from getting to the area by the hundreds of landslides that made the roads impassable. Health officials, noting the shortage of body bags, became concerned about a possible epidemic. A warehouse and a football field were converted into morgues.

By nightfall, improvised shelters made from tents and pieces of wreckage began to appear on the pitch-dark streets of Armenia. Sixty percent of the city had been destroyed.

The next day, the overflowing hospital in Armenia began the task of transporting the wounded it could not treat to other cities. Approximately 400 people were taken to Piajo, Medellín, Bogotá, and Cali.

Airlifts now began to arrive, with rescue workers and machinery, but without food or other supplies. Some of the citizens of the city of Armenia began to loot wrecked supermarkets. Soldiers and police officers who tried to control the pillage were met with barrages of stones.

"We are hungry and the Government has done nothing to feed us!" one man in the mob cried out.

As a night and another day passed, the intensity of the looting increased. Roving bands of looters struck not only grocery stores, supermarkets, and warehouses, but also jewelry, hardware, and furniture stores. Increasingly larger contingents of police and soldiers in trucks and armored personnel carriers attempted to control the rampage, but by the second night it had gotten beyond control, which slowed the delivery of food and supplies to the remainder of the residents of the gutted city.

Eventually, international reinforcements controlled the mobs, and the slow rebuilding of the devastated area, where most of the world's coffee was grown and sold, began. One thousand one hundred and eighty-five deaths were officially tallied, but over 700 remained missing and were also presumed dead. Over 4,750 people were injured, and 250,000 were rendered homeless.

ECUADOR
August 5, 1949

Over 6,000 died, 20,000 were injured, and 100,000 were rendered homeless by the worst earthquake in the history of Ecuador on August 5, 1949.

There was no celebration of Ecuador's 140th birthday on August 10, 1949. Just five days before, the worst earthquake in Ecuadoran history decimated a 1,500-square-mile area of the central highland plateau, cut mountains in half, and wiped out the entire city of Pelileo. Over 6,000 were killed, 20,000 were injured, and 100,000 were made homeless. Fifty-three cities and towns were hit by the quake and damaged; some were decimated. The total dollar value of the destruction was placed at $66 million.

Fortunately, the earthquake sent out warning tremors shortly before it hit with full force at 2:10 P.M. As a result, thousands who might have been crushed by collapsing buildings had time to evacuate them. Not so fortunate were some who sought the shelter of churches. In the Ambato Cathedral, 70 children were crushed to death when the roof fell in.

Others were sucked into crevices that closed over them. One Indian woman held her baby over her head as she fell into a fissure. The woman was crushed to death; her baby, still held aloft in her arms, lived.

Numbed survivors of the vanished city of Pelileo (only 1,000 of the population of 3,500 escaped death) wandered about, unable to respond to the cries of their fellow citizens buried under the city's rubble. Enrique Mejia, a seven-year-old boy, lay buried beneath 12 feet (3.65 m) of bricks for 117 hours, yelling for help. Red Cross workers finally dug him out, but it was too late. He died an hour later.

All that was left of Pelileo was a 20-foot-deep pit submerged under a river that had once meandered through the town.

Relief agencies arrived only hours after the quake stopped, but some rescuers from the United States and other South and Central American countries met with other disasters. A Shell Oil plane carrying 34 workers who were to help in the excavation of buried survivors of the city of Ambato crashed, killing all on board the plane. Relief crews climbing into remote towns on the upper reaches of the Andes were set upon and slaughtered by primitive tribes of Salasaca Indians.

ECUADOR
QUITO
February 4, 1797

Most of the population of Quito, Ecuador, died in the earthquake of February 4, 1797. Forty thousand people were killed, and much of the city was destroyed in the first of two cataclysmic quakes to strike that area.

Two major earthquakes, one in 1797 and one in 1859, battered the Ecuadoran city of Quito, located on a 10,000-foot-high (3,048-m-high) plateau in the Andes. Although the 1859 quake destroyed most of its historic monuments and buildings, including the Government Palace and the Archepiscopal Palace, the quake of February 4, 1797, took a far greater toll in human life, killing 40,000 people—most of the population.

Within a few seconds, this cataclysmic series of earth tremors buckled buildings, tore open the earth throughout the city, and then radiated out into the countryside, activating Cotopaxi and Chimborazo, two immense volcanoes, which then showered lethal rainstorms of lava and stones upon Ambato, a town in the foothills.

ECUADOR AND PERU
August 13, 1868

An earthquake centered on the border between Ecuador and Peru killed 25,000 people in both countries on August 13, 1868.

Roaring and rolling along the border of Ecuador and Peru, a giant earthquake killed 25,000 people late in the afternoon of August 13, 1868. Very little is known except the statistics and this bizarre account by Lieutenant L. G. Billings, who was stationed at the time on the U.S. Navy ship *Wateree,* anchored in the harbor at Arica, a port that had been repeatedly beaten down and churned up by earthquakes:

> The last rays of the sun were lighting up the Andes when we saw to our horror that the tombs in which the former inhabitants had buried their dead, in the slope of the mountain, had opened, and in concentric ranks, as in an amphitheater, the mummies of natives dead and forgotten for centuries appeared on the surface. They had been buried sitting up, facing the sea. The nitrate impregnated soil had preserved them astonishingly, and the violent shocks that had crumbled away the desert-dry earth now uncovered a horrifying city of the dead, buried long ago.

EGYPT
ALEXANDRIA
July 21, 365 C.E.

Fifty thousand people lost their lives in a giant quake that also toppled the Alexandria lighthouse—the fourth wonder of the ancient world—on July 21, 365 C.E.

On July 21, 365 C.E., the fourth wonder of the ancient world, the 600-foot-high (182.8-m-high) lighthouse in Alexandria, Egypt, whose beacon lights reached 30 miles (48.3 km) out to sea, was demolished in a huge earthquake that affected the greater part of the Roman Empire.

The gigantic tremors also toppled other, less monumental edifices and killed thousands. Enormous seismic waves formed in the waters surrounding Greece, Dalmatia, Sicily, and Egypt, crashing onto their shore and pulverizing the ships and structures on them.

But the greatest destruction and loss of life occurred in Alexandria. In that city, 50,000 citizens drowned or were crushed by the quake, which left only the magnificently carved base of the lighthouse intact. It would remain in just that condition for another 500 years.

EGYPT
CAIRO
October 12, 1992

Five hundred and fifty-two people were killed, 10,000 were injured, and 3,000 families were rendered homeless by an earthquake that struck Cairo, Egypt, and its environs on October 12, 1992. A disturbingly large number of the dead were children, who were trampled to death in the panic that accompanied the disaster.

It was enough to tear the heart out of the most steely-souled person. "What happened to our children?" a Cairo mother shouted at the teachers and administrators who were surging around the ruins of a school. It, and 3,238 other buildings, had either been damaged or collapsed in the entirely unexpected earthquake that struck the capital of Egypt at 3:10 P.M. on October 12, 1992. The quake was centered 20 miles southwest of Cairo, a few miles from the pyramids and the Sphinx on the Giza Plateau, and it lasted for a mere 20 seconds and registered only 5.9 on the Richter scale. But that was enough to kill 541 people, injure 6,512, leave thousands homeless, and destroy 3,239 buildings, most of them in poor neighborhoods.

The pyramids, the Sphinx, the posh tourist hotels on the Nile, and the upper-class neighborhoods, built to withstand disasters of several kinds, remained upright. Still, some 150 antiquities, including pharaonic monuments and Islamic, Jewish, and Coptic Christian antiquities, were damaged.

But the worst human disasters occurred to the children of Cairo. Many of them escaped the death of being crushed beneath collapsing buildings. But over 100 were trampled to death by panicked, unsupervised hordes who were given no instruction when their schools began to move and buckle.

"The kids threw me down the stairs as they were going out of the classrooms," said 13-year-old Amira Ahmed from her hospital bed in Heliopolis, an upper-class suburb. "I didn't feel the quake."

"When the building started to shake, the teacher, instead of maintaining order, told the students the school was collapsing," echoed Megda Anwar, one of the mothers gathered outside of Medina al-Omal primary school in central Cairo. Her 11-year-old son had fallen down the stairs trying to escape. "Everyone just panicked," he said.

Later, in a Cairo hospital, Farouk Nimad's 13-year-old daughter Nevine Farouk lay swathed in bandages. "I was on a trip playing hide-and-seek," she whispered, "and then I fell."

"She doesn't understand what happened," sobbed her mother. "She is O.K. for a few minutes, but then she does not know what is going on around her. She knows us for a few minutes, then she forgets who we are."

The disorder that led to so many young deaths was caused partly by a lack of training among the staff, and partly because of severe overcrowding, with as many as 100 students assigned to a classroom.

But if there was chaos in the schools during the earthquake, there was an equal amount of disorder after it ended. Sweating men, leaderless, shouted contradictory orders, dazed survivors wandered aimlessly through the streets, and an occasional ambulance picked its way through crowds of people.

Samia Ragab Khalil was pulled from the rubble of her apartment building in Heliopolis. "Don't dig me out!" she shouted to the men who were removing bricks and stones above her. "Save my son!"

When the rescuers finally unearthed her, she was clutching the body of her dead son, which workers wrested from her as they put her in an ambulance. In the hospital, she refused to understand what had happened. She lay with an intravenous tube running into her arm, weakly repeating her son's name, over and over.

"There have been many cases of fractured limbs, intercranial brain hemorrhages and one broken spine," said Dr. Mohammed Ibrahim of the hospital. "A lot of children had concussions. It's a disaster. We could never have anticipated something of this magnitude. We can't cope with so many victims in such little time."

In the slums of Cairo, row after row of shoddy tenements lay in ruins. Substandard housing, allowed by a failure to enforce building codes, became tombs for hundreds.

"The large number of casualties was due to the fact that there are many rickety old buildings as well as shoddily constructed newer ones," was the cold conclusion of Joseph S. Mikhail, director of the National Research Institute for Astronomy and Geophysics.

But on the streets, there was panic and hunger. The government seemed to be immobilized by the vastness of the calamity. It set up tables to hand out money to the homeless and panicked people. Three days after the earthquake, there had still been no distribution of food, water, or clothes.

That need, the Islamic Fundamentalists, seeing a vacancy in service, managed to fill. They erected cardboard lean-tos and tents for shelter, and circulated regularly with meat and bean sandwiches for the thousands shuddering on the streets.

The scene around Shahata Boutros was typical. In front of the wreckage of his home on a dirt alley next to a domed Coptic church in Cairo's Dahir slum, he sat, motionless, on a blue wooden chair. Alongside him was his six-year-old daughter, her ankle, which she had broken when they fled from their collapsing home, in a cast. On the ground, wrapped in a blanket, was his four-year-old son. The only other belongings the family had salvaged were a washing machine and a couch.

On the sixth day after the tragedy, still with no organized government intervention, a mob rioted outside government offices, throwing stones and attempting to enter the buildings. Armored trucks and other vehicles, filled with black-uniformed riot squads, followed police, who lobbed tear gas at youths throwing stones.

Egypt's President Mubarak pleaded for peace, but offered no solutions. Two days later, an expensively dressed government official stood on a street handing out money to earthquake victims. "I have helped them," said the official to a *New York Times* reporter. "I gave her—" and she pointed to a woman whose daughter had been crushed to death in the earthquake, "500 pounds and I will give her the rest of the money soon."

"We don't want your money!" shouted one of the disheveled victims. "We want a place to live!"

The discord and chaos continued, and the political aftershock, brought about by decades of government corruption and neglect, seemed to grow. The housing crisis, in which as many as a dozen people often squeezed into a narrow room lacking running water, had been exacerbated thousands of times by the natural disaster that had exposed it. "There is housing for those with money," said another person in the crowd. "But there is nowhere for the poor to go."

It would be months before an organized plan to rebuild the thousands of destroyed buildings the earthquake left behind was formulated, and it would be piecemeal. Various diplomats went to various countries to raise money for the project. In Washington, Hoda El-

Sayed, the wife of Egyptian ambassador Ahmed Maher El-Sayed, formed the Friends of Egypt, whose goal was to raise $3 million to help rebuild the more than 7,500 schools destroyed or damaged in the earthquake.

FRANCE AND ITALY
FRENCH AND ITALIAN RIVIERA
February 23, 1887

Two thousand people—almost all of them entertainers or workers—were killed when a huge earthquake struck the French and Italian Riviera at the height of Mardi Gras, February 23, 1887.

All along the French and Italian Riviera, Mardi Gras was drawing to an end. The climactic balls and parties roared on all night on Shrove Tuesday, February 22, 1887. When dawn broke on the morning of Ash Wednesday, February 23, revelers were still rollicking in the streets.

But at 6:02 A.M., a major earthquake tore apart the Riviera from Leghorn, Italy, to Lyon, France. Two enormous shocks followed each other in quick succession, followed by several steadily diminishing ones. Huge sections of buildings cracked, wavered, and then tumbled, crushing thousands of costumed celebrants on the avenues and streets of Nice. The Jardin Public and the Place de la Liberte, relatively free from threatening buildings, were choked with terrified refugees.

Two thousand people perished in this earthquake, but ironically enough, they were mainly local residents. Not one of the well-heeled American, British, Russian, or French tourists was even slightly injured by it. In fact, the *Illustrated London News* reported that

Refugee camp near Nice following the cataclysmic earthquake that devastated large sections of the Riviera on February 23, 1887 (Illustrated London News)

the earthquake had ". . . caused a certain joy, because [it has] sent back to Paris some hundreds of people of leisure whose presence is desirable in the interests of luxury and elegance of life."

Meanwhile, back on the Côte d'Azur, life was considerably less elegant and more grim. On the Italian Riviera di Ponente, entire towns collapsed in tragic ruins. Following the first tremors, the entire population of Bajardo—1,500 persons—crowded into the local church. In the midst of their terrified prayers, another quake struck, bringing the heavy walls of the church down onto the townspeople, and killing 300 of them.

In Diano Marino, 250 dancers still celebrating in a ballroom were crushed to death when the building in which they were celebrating was split asunder. In Savona, the railroad was ripped apart. In Oneglia, a penitentiary was obliterated, and its prisoners moved to Genoa, where the Ducal Palace was scarred by huge, meandering cracks. When great fissures opened in the earth in Cervo, 300 people were killed.

The shock waves were felt as far north as Switzerland, and seismographs as far away as Washington, D.C. leaped when shock waves from this horrendous quake, traveling at more than 500 MPH (804.6 km/h), reached the east coast of the United States.

GREECE
CORINTH
856 C.E.

A death toll of 45,000 was recorded in a cataclysmic earthquake that struck Corinth in 856 C.E.

The earliest accurate recording of an earthquake occurred in Corinth, Greece in 856 C.E. The city—which has been destroyed by earthquakes no less than nine times—was reduced to absolute rubble, as were several other Grecian cities. Statuary and temples alike collapsed; the countryside was reduced to torn, ruptured earth.

The death toll was terrible to consider: 45,000 residents of Corinth and the surrounding countryside were killed by what must have been a monstrous quake.

GREECE
SPARTA
464 B.C.E.

Twenty thousand people died in a historically important earthquake that struck Sparta in 464 B.C.E.

The Peloponnesian War may have had its beginnings in a gigantic earthquake that tore apart a large section of Ancient Greece in 464 B.C.E., during the reign of Archidamus.

Sparta was unusually hard hit. In one account, a mere five houses remained standing in the entire state. Twenty thousand were killed and many more injured. "The flower of Spartan youth was overwhelmed . . ." was the way one overwrought chronicler of the time put it.

In the Spartan city of Ithome, which was horrendously ripped apart, the Helots, slaves to the Cacedemonians, were freed by the quake and rioted through that city's rubbled streets, murdering former masters who had survived the disaster. They eventually took control of the city.

Depleted by the earthquake, the Spartans called upon the Athenians for aid in putting down the uprising. When this alliance succeeded, the Spartans peremptorily dismissed the helpful Athenians, thus setting in motion resentments that would eventually grow into the Peloponnesian War.

GUATEMALA
April 18, 1902

In a series of earthquakes that struck Guatemala on April 18, 1902, 12,200 people lost their lives.

The worst earthquake in 200 years of Guatemalan history devastated a large part of that country on the night of April 18, 1902. There were a score of earthquakes, actually, preceded by torrential rain, lightning, and thunder.

Guatemala City was immediately flooded, as huge fissures opened in the streets, water mains burst, and huts and cathedrals alike crumbled and fell, burying hundreds. Within an hour, 80,000 people were made homeless. Throughout the ravaged country, 12,200 were killed.

Fissures hundreds of feet deep opened in the countryside. In the town of Escuintla, 4,000 of the 10,000 inhabitants were killed as a shock lasting a full two minutes crisscrossed the village, ripping its center apart. Ocos, on the Pacific Ocean, was totally destroyed, first by fissures spouting lava, then by a giant tsunami which roared through the city. The banks of a nearby river were squeezed together, and the muddy bottom spit up a sailing vessel, intact. It had sunk in the river decades before.

The volcano Santa Ana, which had not erupted in so long it was thought to be extinct, suddenly leaped to fiery life, flinging lava and rocks into the nearby countryside and towns. San Marcus, San Pedro, San Juan Ostancalco, Champerico, Cuyoenango, Mazatenango, and Tucana were all reduced to rubble.

GUATEMALA
February 4, 1976

Twenty-two thousand were killed, 70,000 were injured, and over 1 million residents of Guatemala were made homeless on February 4, 1976, when a giant earthquake struck that country.

At two minutes after 3:00 A.M., on the morning of February 4, 1976, a huge earthquake rippled along the Motagua Fault, which is the boundary between the Caribbean and North American plates. The fault ruptured spectacularly and horrendously, opening up parallel fissures like bursting veins from Guatemala City eastward as far as the Gulf of Honduras and south to Puerto Barrios. An astounding 22,000 people lost their lives and another 70,000 were injured. Over 1 million people in a 3,400-square-mile (5,471.7 sq. km) area were made homeless.

The tragic loss of life and huge injury toll was undoubtedly caused by the timing of the quake. Families were sleeping at the time, and with no warning whatsoever, were buried beneath the exploding walls and falling roofs of their homes. In the towns of the Motagua River Valley and to the west of Guatemala City, most of the dwellings are constructed of adobe mud brick walls, which are notoriously unresistant to horizontal motion. Once the quake began, it was only a matter of seconds before the walls cracked and broke, allowing wood beams and tile roofs to thunder down upon the occupants.

In Guatemala City, which is built on an elevated plain deeply cut by ravines, there was no building code requiring earthquake resistant design. This was despite the fact that 40 percent of the city had been destroyed by a 1917 earthquake, and Antigua, the city that had previously stood on the same site, had been totally destroyed by earthquakes in 1586, 1717, 1773, and 1874. Thus, when the quake hit, causing huge landslides and enormous, gaping fissures in the streets, damage was extreme. Thirty percent of the adobe dwellings were annihilated; some of the hospitals, the city's largest cathedral, and its largest hotel were demolished, although its two steel, multi-story buildings remained unscathed.

In the countryside, there was extensive flooding when river banks closed in on each other, spilling the

Guatemala, February 4, 1976. These railroad tracks were twisted and offset 1.07 miles (1.7 km) by the Motagua Fault, which runs perpendicular to the tracks. (USGS)

rivers through the countryside. Landslides buried villages, livestock, and people, and dammed some rivers. In the resort settlements along Lake Atitlán, huge upthrusts in the sand, resembling giant boils, toppled and damaged hotels and holiday dwellings.

The immense devastation brought aid in the form of medical supplies and food from the rest of the world within a few days.

INDIA
ASSAM
August 15, 1950

One of the most severe quakes ever recorded killed over 1,538 people in Assam, India, on August 15, 1950.

51

More than 1,000 people perished in the most violent quake known to man, a quake so fierce that it sent needles skidding off seismographs. A reading of at least a 9 on the Richter scale was later assigned to it. The violence of this quake was so extreme it confused seismological extrapolations. American geologists thought it took place in Japan; Japanese geologists pinpointed it in America.

In the vicinity of Assam, India, the situation was far less benign, though no less confusing. The catastrophic quake opened and closed the earth for five full days, sending geysers of steam and superheated liquid skyward and swallowing whole villages at a time. It wrecked dams, flooded fields and towns, and drove residents into the trees for safety—one woman was reported to have given birth in one of these trees.

The natives of this agricultural part of India likened the sound of the approaching quake to the stampede of elephants. British planters said it sounded like an express train entering a tunnel.

Its ruination exceeded that of the second most violent quake ever recorded, which also took place in this same province in 1897, and which claimed the lives of 1,542 people. Landslides spawned by the repeated tremors tumbled from the mountains, blocking off the tributaries that flowed into the Brahmaputra River. After the next quake struck, the blockages dissolved, sending huge waves of water into the fields.

In addition to the 1,538-person death toll, 2,000 homes were destroyed. Damage estimates totaled $25 million.

INDIA
GUJARAT STATE
January 26, 2001

The worst earthquake to strike India in 50 years occurred on the morning of January 26, 2001, in Gujarat State, the northwest corner of India on the Pakistan border. The estimated death toll was 19,403, with 68,478 injured; 228,906 homes were completely destroyed and 397,615 were partly damaged. The estimated total loss of property totaled $3 billion.

January 26 was Republic Day in India, a national holiday of parades and festivities celebrating the adoption of the Indian Constitution. Businesses were closed; families with schoolchildren rose early, in anticipation of dozens of local processions.

In Bhachau, a village in the northwestern state of Gujarat, on the Pakistan border, a contingent of 30 girls from the local girls school lined up with their teachers, preparing for the parade. But they never marched. At 8:46 A.M., the earth heaved under Bhachau and throughout most of Gujarat, in an immense, 7.7 earthquake, the worst since 1950, when a 9 magnitude quake killed 1,538 people in Assam (see p. 51).

The girls were instantly buried under the collapse of their school. Only three of the 30 survived. When the earth stopped heaving, the smoke cleared, and the damage was made visible, an entire town lay in ruins. Its misfortune was that it was located only 12 miles from the epicenter of an earthquake strong enough to shake high-rise towers in New Delhi, 600 miles (965 km) away, shatter windows in Nepal, 1,000 miles (1,609 km) north, and be felt in Calcutta and coastal Bangladesh, 1,200 miles (1,931 km) south and east.

The immensity of the quake and its damage became immediately apparent. High-rise office buildings, hotels, and hospitals had collapsed into themselves. Homes were translated into dust and mounds of stone. Cars, wagons, and trucks were flattened as if by a cosmic fist. Marketplaces, only moments ago whirling vortexes of activity, their merchants setting out their wares for the holiday, were suddenly silent wastelands, barren of people and products.

Telephone and power lines crisscrossed the ground. In town after town, city after city, red tile roofs lay in fragments, and brick walls were cleaved as if by a massive hatchet. Personal belongings were strewn everywhere. Beds, toys, and clothes lay abandoned in the debris, lamp posts and electric pylons were twisted into surreal shapes, and the small, precariously leaning buildings tottered on the verge of collapse.

Bhuj, Anjar, Wankaner, and the major city of Ahmedabad were all in various state of ruin. Bhuj, nearest to the epicenter, suffered the most. Ninety-five percent of its buildings were totally wrecked and beyond saving. Wankaner, farthest from the center of the quake suffered least, but even it resembled a bombed out city after a major war. Its hospital teetered on the verge of collapse, and would eventually have to be razed, once the surviving patients had been removed.

Moments after the tremors ceased, dazed and dirt-smudged survivors began to slowly emerge from the wreckage. From within the rubble, agonized cries from the trapped and the dying rose in a discordant chorus. Some of the survivors began to claw with their bare hands at the rubble of their homes, frantically trying to free family members trapped and pleading to be rescued.

Reaction from the Indian government was swift. "The earthquake is a calamity of national magnitude," India's Prime Minister Atal Bihari Vajpayee announced. "We have decided to meet the emergency on a war footing. This is the time for people to rally around."

Within hours, the Indian army appeared with bulldozers and cranes. As family members wept, watching helplessly and hopefully, squads of soldiers and volunteer workers began the terrible task of digging out the trapped, moving tons of stone that had once been homes and shops, and trying to rescue those who still lived without collapsing the precariously poised wreckage on them.

Over 100 buildings of supposedly solid construction collapsed in Ahmadabad, including government buildings housing the offices of Gujarat state's political and commercial leaders.

There was no power anywhere. Field generators were brought in to run the digging equipment. And though there was no damage to the two 220-megawatt nuclear plants in Gujarat, gas pipelines, most power supply stations, and water service were totally disrupted.

Panic began to develop among the survivors, some of whom besieged the fire station in Ahmedabad, pleading with the firemen to dig out their relatives. "This is an emergency. We are facing a riotous crowd," fire chief Rajesh Bhat told reporters who arrived on the scene.

Nearby, at the city's hospital, which had been spared, corpses were arriving, to be piled up on the hospital's veranda. Patients overflowed into the hallways, wailing and screaming with broken limbs and open wounds. The Press Trust of India reported that night that 70 people died waiting for treatment.

In the town of Surat, three people were killed in a stampede at a diamond factory as workers crowded into a narrow stairwell and tried to push their way to the only exit.

But by far the greatest number of casualties came from crushed limbs and bodies when walls and roofs collapsed on the residents of apartment buildings and homes.

Using crowbars and hand tools, cranes and bulldozers, local residents joined with soldiers and volunteers to dig out those who could be heard or seen in the wreckage. And all the while, hundreds of aftershocks, some of them as great in magnitude as 5.0 on the Richter scale, moved wreckage, collapsed some buildings that had been left standing, buried some victims who might have survived, and freed others who were in danger of being buried out of sight or hearing of the rescuers.

As night fell, nearly the entire population of the state spread blankets or whatever pieces of cloth they could on open ground. The 55-degree weather was no deterrent against the fear of being trapped inside buildings that continued to collapse, throughout the night and into the next day.

By this time, international rescue crews, with body- and survivor-sniffing dogs, began to arrive. The Indian air force and the International Red Cross started to fly in medical supplies, doctors, tents, and food. Over 3,800 Indian troops moved into the area and immediately began the grim business of extracting the dead and the wounded from the rubble, which pervaded the landscape for as far as it was possible to see.

All day and night, the roar of earth movers, the groan of timbers as they were being lifted, the movement of concrete slabs, mixed with the cries of the trapped. Those who were brought from the wreckage often had limbs that were crushed; others were barely alive or entirely hysterical.

The town of Bhabhau and its 25,000 residents no longer existed. The hillsides were just expanses of demolished concrete, some of it hiding the bodies of those trapped beneath it. "Not a single house is left," Atanu Chakraborty, the state emergency coordinator for the area told international reporters, then amended himself. "Oh, maybe one is still O.K."

It was scarcely comforting. Those residents of the demolished village who lived began to board buses to take them to Ahmedabad, which had become the center of relief and rescue efforts.

And this was the situation in other neighboring villages. Most of the rescue efforts began, at least, in the larger cities such as Anjar, Bhuj, and Ahmadabad. Meanwhile, in the typical, 400-year-old village of Paddhar, east of Bhuj, hardly a house remained, and the 3,500 survivors of the quake were largely left on their own for three days, until Indian army and international rescue groups reached them.

Even then, these survivors were forced to live in the open under plastic sheets that kept out the rain but not the cold. Each survivor was given 10 rupees (20 cents) a day by the government, which was not enough to buy food to remain alive. Denied even enough cooking oil for their needs, most of the villagers, according to Panchabhai Velji, the village chief, were surviving on chapatis and buttermilk.

Throughout the affected parts of the state of Gujarat, the foul stench of putrefying bodies began to sharpen the air. Rescuers were forced to wear surgical masks or cover their faces with handkerchiefs. The women, weeping and watching the rescue and removal efforts, pressed their saris to their faces.

Makeshift cremation pyres began to rise next to the scenes of most destruction. Enormous truckloads of wood rolled from town to town, and gray smoke rose in identification of them.

Hope began to be replaced by the reality of the massiveness of the death toll, which would be estimated, two weeks after the quake, at over 30,000. And yet, here and there, miraculous rescues occurred. On February 5, in Bhuj, soldiers, clearing away the ruins of collapsed apartment buildings, were approached by a local policeman who reported hearing faint cries for help in the city's Karsana neighborhood.

The soldiers spotted a man waving through the grill in the second story window of a damaged apartment building. Rubble blocked the entrance to the building, so 20 of the soldiers formed a human pyramid that reached the window, and there, they found two teenagers, a brother and sister, who had survived on cereal and water—all that was left in the ruined kitchen in which they were trapped.

Nearby, in Bhachau, which was now a ghost town of smashed houses, pieces of concrete, and collapsed roofs left at crazy angles, a Russian team was conducting the grim task of extricating corpses from the wreckage when they heard faint cries from another collapsed apartment building.

A woman was pulled from a kitchen, where she had counted the days by watching the light that seeped through the cracks in the kitchen walls. Her husband was trapped under the wreckage of his den, where he had been watching television. He was pulled from the chaos and loaded into an ambulance with his wife.

Finally, on February 4, eight days after the quake, the last survivor was rescued from wreckage in the village of Sikaravadi, eight miles from Bhachau. A teenager had thought the quake was a bomb dropping and had run into a shed for shelter. A moment after he entered it, the shed fell into a well, and this protected him from further injury. He attracted attention to himself by throwing rocks against the walls of the well.

The aftershocks continued. Even as the stench of dead bodies was being replaced by disinfectant, spread by hand by soldiers wearing rubber gloves, over 300 aftershocks of varying intensity terrified the survivors all over again. On February 8, a 5.3 magnitude quake rattled buildings and panicked survivors who had reentered their precariously perched buildings. Over 40 people were injured, 15 of them in Ahmadabad when they jumped from the windows or balconies of their homes.

Armies of rescue teams from all over the world descended upon India. Relief supplies of food and medicines arrived even from India's enemy, Pakistan. Immediate attention to sanitation problems removed the necessity of administering anti-cholera vaccines, but tetanus shots were given to adults and measles vaccine was administered to children during the second week of February.

The final statistics were as staggering as they were inconclusive. Because of the extent of the damage, final totals had to be recorded as estimates: According to the Department of Agriculture and Cooperation of the Indian Government, 19,403 people died and 68,478 were injured, and approximately 18,600 head

of cattle were killed. Altogether, 15.7 million people were affected by the disaster, with overall losses reaching $5.5 billion.

The material damage was staggering. Some 228,906 houses were completely damaged, and therefore uninhabitable, and 397,615 other houses were partially damaged. The price of damage to public property totaled $220 million. In the city of Bhuj, 90 percent of the homes were destroyed, and several schools and the local hospital were obliterated. It would be years before the region would return to a semblance of normalcy.

INDIA
KANGRA
April 4, 1905

An immense 8.7 earthquake killed 19,000 residents of Kangra, India, on April 4, 1905.

At 6 A.M. on the morning of April 4, 1905, the sleeping city of Kangra, India, was almost entirely destroyed by a monster earthquake that registered 8.7 on the Richter scale. Nineteen thousand inhabitants of the city were crushed to death, most of them in their beds.

The ancient and revered Bhowan Temple was filled with 2,000 chanting pilgrims when the earthquake struck. Every one of them, including the guru, was killed instantly when the temple collapsed on them. It has remained a communal tomb for the pilgrims to this day.

In the countryside, and for an area of approximately 1.6 million square miles (2,574,950.4 sq. km), devastation was rampant. The neighboring towns of Dharmsala, Naggar, Sultenpur, Suket, and Mandi were heavily damaged. Landslides, dammed rivers, and flattened farms turned the landscape into one resembling that of a nation after a war. Entire tea plantations, lush and vital a day before, were now nothing but piles of blue slate, thatched roofs, and ravaged crops.

INDIA
KASHMIR
June–July 1885

In a series of earthquakes that struck the Kashmir province of India in June and July 1885, 3,100 were killed and over 5,000 were injured.

In June and July of 1885, a relentless series of earthquakes opened huge fissures in the earth of India that swallowed entire herds of cattle.

On and on the earth rippled, and towns and villages throughout Kashmir province were either swallowed up or flattened. The death toll in the city of Srinagar and the surrounding countryside rose to 3,100 before the earth ceased to tremble on July 8. Over 5,000 hapless residents of this normally tranquil part of India were reported injured.

INDIA
NORTH
October 19, 1991

Faulty construction, isolation, the time of year (October)—and a 6.1 strength earthquake that hit the Uttar Kashi region of northern India—conspired to produce a fatality figure of 1,600, and utter devastation of a formerly rich agricultural region.

Deforestation is rampant in the foothills of the Himalayas, particularly in the region of Uttar Kashi near the border of Nepal. Traditionally, the homes in this region have been constructed of wood and earth, but because of the deforestation, more and more homes have been built of stone and plaster.

In addition to this, envy by the poor of the rich, who can afford homes of brick and mortar, led some villagers to plaster cement over rock walls to give visitors the impression that they, too, lived in "proper" houses.

Both of these decisions contributed mightily to a huge fatality figure of 1,600 deaths in the earthquake that struck the region on October 19, 1991. Though the U.S. Geological Survey rated it at 7.1 on the Richter scale, Indian geologists insisted upon the lower figure of 6.1. Whatever the measurement, the 45-second tremor, followed by a series of aftershocks, was strong enough to set off a multitude of landslides and to obliterate 400 villages. Many of the dwellings, built as they were, became death traps. "The reason why so many died was not because of the magnitude of the earthquake," engineer A.S. Rawai said afterward, "but because the buildings were badly designed and badly built. Light,

Earthquakes are equal opportunity destroyers. The most solidly built structures of the wealthy were no match for the fury of the earthquake that blasted the Uttar Kashi district of India. (American Red Cross)

flexible structures, such as tin and wood, are among the best in earthquake-prone areas."

The area was not, in the classic sense, earthquake-prone, though it had had its share of winter avalanches. A tract of rugged mountains and valleys with widely dispersed villages, it is known for its temples and ashrams. The shrine of Kedar Nath in the tiny village of Chamoli, just east of Uttar Kashi and near the Indian-Tibetan border, is particularly popular, and pilgrims who were there during the earthquake perished.

The epicenter of the quake was in Almora, a hill station about 30 miles (48.3 km) west of the Nepalese border, and it, too, was devastated, its dwellings and temples flattened or ruined by the disaster.

At least three bridges and one dam were damaged, and rescuers, arriving overland and in helicopters, dynamited rocks that blocked the path of the Bhagirathi River, which flows into the Himalayas.

A month later, survivors were still living in tents in the open. In the village of Netala, clusters of tents sheltered the entire population of 1,000, who now faced a winter of further deprivation. The quake had occurred at the precise time that local farmers would have sown their wheat crop. Furthermore, buffaloes and other cattle were killed, depriving the villagers of draft animals. Tractors were not used at that height, and walls that kept the terraced fields in place had collapsed.

And finally, the homes that had crumbled to dust represented the disintegration of years of privation and toil. "Women sold their jewelry to help build the houses, and now entire families are paupers," said Reghubir Singh, a trader at Netala, to Western reporters. It would be years before the impact of this not very strong earthquake would begin to lessen.

INDIA
SOUTHERN
September 29, 1993

Official records state that 9,748 inhabitants of southern Maharashtra state in India were killed in an earthquake that struck the area in the early morning hours of September 29, 1993. Other sources locate the casualty figure at 11,000, with over 20,000 homeless.

For millions of years, a slow collision—if that term is valid—has been taking place between the Indian subcontinent and the Asian landmass. This impact has created the Himalaya Mountains and causes residual tensions that periodically erupt in bursts of underground energy in the Indian tectonic plate.

According to Dr. Lynn R. Sykes, a geologist at Columbia University's Lamont-Doherty Earth Observatory, "The area to the south is under high compression. Compared to similar areas it has one of the highest rates of seismic activity.

"What is going on in peninsular India," the professor continued, "is that it is attempting to move to the north and is being resisted." This resistance results in earthquakes when rocky plates break apart to relieve the stress.

The earthquake of September 29, 1993, was an extraordinarily deadly manifestation on this tension and release analysis, partly because of its location, partly because it struck at 3:56 A.M. At that moment, most of the residents of the Maharashtra section of southern India were asleep in their badly constructed mud and thatch huts, held together by boards and capped by corrugated tin roofs.

Though the most intense of the earthquake's five tremors achieved only 6.4 on the Richter scale, its overall effect was so severe that the data on seismic recorders in India emerged as a black smudge.

Fissures opened in the earth; the serenity of an area that is a checkerboard of sunflower fields, acres of grapevines, and dense rows of sugar cane, was decimated. Small villages that dotted the landscape simply disappeared. There was no hardship in the destruction of rail lines or airports; the area contains none. But the single macadam road that runs 420 miles (675.9 km) from Bombay to Hyderabad, and thus is the only normally navigable way into and out of the area, was buckled badly and made impassable.

As far away to the south as Bangladore and Madras, vibrations from the earthquake disturbed inhabitants of these cities and sent them scurrying into the streets in panic.

But in the circle bordering the epicenter in southeast Maharashtra, there was more silence than panic, as the mud huts simply collapsed upon their inhabitants, crushing them to death.

"The problem is in the construction of their houses," one official said afterward. "The people who live here construct them with what they have—mud bricks, thatch, and more and more, poor quality cement."

Survivors struggled to excavate the buried bodies and cremate them, according to religious custom. But even this was defeated by nature, as heavy rain arrived, which doused the cremation fires. Shelter for the living was fashioned out of bamboo poles and corrugated metal. And here these frightened survivors waited for aid, which was a long time coming.

Though world organizations flew in large quantities of supplies, the Indian government took days to deliver it to the homeless who huddled in groups in

Ascetically constructed stone and plaster dwellings crumbled under the pounding of the Maharashtra earthquake. (American Red Cross)

the countryside. Food, water jugs, plastic sheeting, and tents were eventually brought to the refugees, along with military teams to help in the digging out of bodies.

Ahead of the soldiers, social workers and volunteers set up field kitchens, medical clinics, and counseling services. In the village of Tawashigad, which originally possessed 500 dwellings and had none after the quake, retired newspaper editor Dwarkanath Lele led the efforts of a social relief organization set up by his former newspaper, *Sakal.* "It is better not to speak about the government," he told Western reporters. "We are doing it only. Our newspaper set up a relief fund in 1944 when there was a drought in Bengal. Our founder said why should we depend on the Government? We should be doing it ourselves. From then on, we have been doing this work."

Four days after the quake, in the flattened village of Killari, a miraculous rescue was nevertheless effected by soldiers. Eighteen-month-old Priyanka Javalge had been sleeping on the floor of her parents' home and, during the night, had apparently rolled under the cot upon which her parents were sleeping. Everyone else

in the house was injured, and they assumed that the child had been crushed to death. But the cot sheltered her, and when rescuers removed the huge stones that obscured the cot, they discovered the wide-eyed child, conscious but dehydrated.

Over 20,000 inhabitants of the area were made homeless by the disaster, the worst natural catastrophe to strike India since 1935, when 30,000 people perished in an earthquake in Quetta. The final official total of dead in 1993 was 10,000.

INDONESIA
FLORES ISLAND
December 12, 1992

The island of Flores was hammered by an earthquake and a consequent 80-foot-high (24.4-m-high) tidal wave on December 12, 1992. Two thousand five hundred inhabitants of this Indonesian island and two smaller ones near it were killed.

The island of Flores, a poor copra-growing region in the province of East Nusa Tengara, approximately 1,100 miles (1,770 km) east of Indonesia's capital of Jakarta, sits, as does all of Indonesia, on the Pacific Ocean's "rim of fire," a necklace of underwater volcanoes that regularly produce temblors ranging from 5.87 to as high as 8.0 on the Richter scale.

None of these, however, began to match the devastation and fatality count of the 7.5 (according to the U.S. Geological Survey) or 6.8 (according to Indonesian officials) earthquake that struck the island and its major city, Maumere, on December 12, 1992. Not only were 80 percent of the buildings in the city and hundreds more in the hinterland flattened by the quake, but also waves 80 feet (24.4 km) high, produced by the underwater tremors, crashed into the island and roared 1,000 feet (304.8 km) inland, smashing trees and dwellings and sucking people out to sea, where they drowned.

Entire fishing villages and all of their inhabitants were washed away; government buildings, schools, mosques, churches, and shops were totally destroyed in Maumere and Larantuka, another town on the eastern part of the island. "All our development efforts for the last 25 years have disappeared in just one day," Hendrik Fernandez, the governor of the province said, as he surveyed the damage.

It would be days before rescue ships arrived with basic supplies for the thousands of refugees from the earthquake, who, because of constant aftershocks, remained, tentless and without shelter, in open fields in a tropical rainstorm that followed the earthquake.

Over 1,120 people died in the city of Maumere, whose population was 40,000. The final death toll for Flores and two small islands near it was 2,500, a shocking figure for any earthquake anywhere, but particularly devastating for an isolated, normally tranquil island in the Pacific Ocean.

INDONESIA
SUMATRA
December 26, 2004

A colossal 9.3 earthquake, the second-largest earthquake since seismographs were invented in 1900, erupted under the Indian Ocean off the northwest coast of the Indonesian island of Sumatra at 7:59 A.M. on December 26, 2004. The deadliest tsunami in world history, triggered by the enormous quake, spread outward from its epicenter, washing away whole cities in Indonesia and bringing devastation to every country on the shore of the Indian Ocean, killing 287,000, injuring more than 500,000, and leaving 1.8 million homeless.

Tsunamis, the huge waves that are sometimes the offspring of only the most violent of underwater earthquakes or volcanic eruptions—usually in the Pacific Ocean—radiate out, like the ripples caused by the dropping of a stone into otherwise placid water, for hundreds of miles. Some, traveling as fast as 450 MPH (724.2 km/h), have been known to reach heights of 100 feet (30.5 m), and when these aquatic juggernauts reach land, the land and everything and everyone on it is swept away like dirt before a broom.

A minute before 8:00 A.M. on December 26, 2004, just such a colossal earthquake began to vibrate through the northern part of Sumatra and the Andaman Sea to its north. Seven minutes later, the full force of the gigantic earthquake erupted, a 9.3 event on the Richter scale, which made it, after the 1976 quake in Tangshan, China (see p. 44) the second largest earthquake since seismographs were invented in 1900. The eruption tore the ocean bottom apart in a gash 600 miles (956.6 km) long.

Its effects were monstrous. The earth tremors spread throughout all of southeastern Asia. Buildings and bridges collapsed, roads were ripped apart, forests were uprooted, and the earth came apart with particular fury in northern Sumatra and in the Nicobar and Andaman Islands (see color insert on p. C-2). The Sumatran capital of Banda Aceh was devastated, losing almost all of its buildings in the first earthquake wave of what became, in an instant, a double disaster.

With the force of an enormous gong being struck, the Sumatra earthquake made the entire Earth ring with free oscillations. At every spot on Earth, the ground was raised and lowered at least a full centimeter by seismic waves from Sumatra, and these waves sped around the world several times before dissipating.

But other waves, far more destructive and dangerous, also rushed out from the quake's underwater epicenter. An immense tsunami, the deadliest in world history, sent its 350-foot (106.7-m) waves, traveling at hundreds of miles an hour, out in a giant path of destruction, laying waste to the shorelines, the cities, and the populace of 11 countries bordering the Indian Ocean. The hardest hit were Indonesia (particularly the province of Aceh), Sri Lanka, India, Thailand, and the Maldives. But this was just the beginning. The power of the tsunami caused lives to be lost on the coast of Africa, and was even detected, without damage or loss of life, on the East Coast of the United States (see color insert on p. C-3).

The nations and the people of the countries bordering the Indian Ocean were all relatively poor and therefore most vulnerable to any unexpected natural disaster. There was no warning system whatsoever to allow the residents of the villages located on the shores of the Indian Ocean to know that a tsunami was approaching. Most of the structures in these villages were fragile and were simply swept, with their inhabitants, out to sea in the back draft after the tsunami hit. A resident of Chennai (formerly Madras) told reporters afterwards that in his district, the Tamil Nadu district, which was India's hardest hit area, he saw several of his neighbors swept, at a high rate of speed, out to sea.

But the fragile structures on the beaches and in the villages were not the only physical casualties of the thunderous waves. Some villages were totally submerged (see color insert on p. C-3). Some 2,900 schools were razed to the ground. Witnesses near Phuket in Thailand reported that guests drowned in their coastal hotel rooms as 30-foot (9.1-m) waves crashed into the structures. Phuket's famed Laguna Beach resort area, which provided 40 percent of Thailand's $10 billion annual income from tourism, was flattened and eradicated completely.

One story of escape in Thailand was a minority report, but an encouraging one: A tourist boat had set off from shore shortly before the quake and tsunami hit. The captain of the boat, feeling the underwater tremors, turned his boat around immediately and headed for a nearby shore, screaming at his passengers to abandon the boat as soon as he beached it. These passengers, mostly Spanish tourists, scrambled up the beach, turned, and saw the boat crushed into a formless mass by the arriving waves.

In Sumatra, Irwandi Yusuf, a 45-year-old senior official in the Free Aceh Movement was locked in a government prison. The first tremors of the earthquake shook the walls of his cell, and then the sound of waves breaking sent him on a panicked journey in which he punched a hole in the asbestos ceiling of his cell and clambered to the roof of the prison. He was among only 40 prisoners who survived; 238 others died.

In Sri Lanka, more than 4,500 died, most of them in the eastern district of Batticaola. Thousands more were missing and more than half a million lost their homes. A large portion of the dead drowned at sea. Thousands of fishermen, including 2,000 from the Chennai area alone, were at sea when the wave struck, and never returned (see color insert on p. C-4). Reports were similar from almost every area.

As the waters receded and the first of several aftershocks—two measuring over 7 on the Richter scale—continued to strike terror in survivors, it was apparent that the farmlands and forests between the villages had been contaminated, livestock had drowned, and even the fishing stocks in the sea had been destroyed. Dead fish littered the beaches of Indonesia. Even the islands of the Indonesian archipelago that had fewer deaths were devastated. Fourteen of them were left uninhabitable, requiring their inhabitants to be permanently evacuated, and another 79 islands lost their safe drinking water.

The effects of the disaster would remain for a long time, and the effects would sometimes be paradoxical. In Aceh province in Indonesia, a 30-year-old civil war was raging when the tsunami and earthquake struck. The ethnic separatist guerrillas, seeing their villages swallowed by the tragedy and losing their own families, turned in their weapons and retooled themselves into a political party. A shared disaster seemed to shrink political and ideological differences.

There was a record $13.6 billion of aid pledges received by the United Nations, and a year later, 75 percent of the $10.5 billion earmarked for reconstruction had been secured—a record that contrasted to the mere 10 percent of pledges that were honored after the 2003 earthquake in Bam, Iran (see p. 60). In addition, the International Federation of the Red Cross raised a record $2 billion for tsunami aid.

Swift intervention generally prevented major outbreaks of infectious diseases. In Indonesia and Sri Lanka, nearly all of the surviving tsunami-affected children were back in school a year after the disaster. In Sri Lanka, 70 to 85 percent of the adults who had lost their livelihoods had regained their main source of income, and 41 of the island's 52 damaged hotels were open for business.

In Sri Lanka, nearly all of the displaced refugees had been moved out of tents and schools and housed in shelters designed to last up to two years. Drainage pipes were laid in new villages, hospitals were built, and nearly 32,000 homes that had been destroyed were rebuilt. In Indonesia, nearly a quarter of the 100,000 homes that needed to be reconstructed had been.

A tsunami early-warning system in the Indian Ocean—which, had it been in place might have saved hundreds of thousands of lives—was built. The intent of the system is to prepare every country's weather service to receive updates and warnings on a range of climate and weather shifts within two minutes of their occurrence.

But problems still remain in this cruelly demolished part of the world. Of the 1.8 million people made homeless throughout the region, only one in five, as this is being written, had been able to move into a permanent home. According to the aid agency Oxfam, some 67,000 Achenes were still languishing in tents a year later, and most of the rest were in charity-financed shelters designed only to last another year or two.

The single largest impediment to putting up permanent homes is the question of whether to allow people to return to the edge of the sea, where the huge fishing industry and enormous tourist attractions have been historically located. As of this writing, Indonesia and Sri Lanka, including the territory controlled by the Tamil Tigers, have prohibited new housing on the seafront.

One effect of this major calamity is clear: Nothing will ever be quite the same as it was before the tsunami occurred. Jan Egelund, the emergency relief coordinator for arriving U.N. rescuers took in the damage and pinpointed the reason: "I think," he said, "we are seeing now one of the worst natural disasters ever."

INDONESIA
YOGYAKARTA
May 27, 2006

More than 5,700 people perished, more than 36,000 were injured, and more than 200,000 were rendered homeless in an early-morning 6.3 magnitude earthquake on May 27, 2006, that affected a 193-square-mile area (499.9 km²) south of Yogyakarta, Indonesia. The quake was the worst disaster in Indonesia since the December 26, 2004, earthquake that triggered a giant tsunami (see previous entry).

At 5:54 A.M. on Saturday, May 27, 2006, an earthquake measuring 6.3 on the Richter scale erupted six miles (9.7 km) beneath the surface of the Indian Ocean and approximately 15 miles (24.1 km) southwest of the city of Yogyakarta, Indonesia, 250 miles (402.3 km) east of Jakarta, the country's capital. The area is located on the so-called Pacific Ring of Fire, an arc of volcanoes and fault lines encircling the Pacific Basin that contains 76 volcanoes, the largest number in the world. The Ring of Fire is prone to earthquakes—there were four in the 17 months preceding this one.

Awakened by the huge tremors that collapsed buildings and buckled the landscape, panicked residents fled into the streets of the small villages of the Bantul plain, a densely populated agricultural area of villages separated by rice fields. Houses, hotels, a hospital, and government buildings imploded, burying their residents in the rubble.

Roads leading in and out of various populated areas were destroyed and collapsed bridges kept possible survivors in and hindered rescue vehicles from entering the worst struck areas. Hospitals were immediately overwhelmed and exhausted their supplies of bandages, drugs, and even anesthetic. Operations were performed without anesthesia, and without nurses to provide basic care, relatives saw to the injured under a baking sun in hospital parking lots.

Late that night, a driving rain arrived and made conditions nearly unbearable. Hospitals set up tents to try to protect the overflowing masses of people constantly arriving for help. At one hospital, many victims lay on the wet ground with only newspapers beneath them, while nurses ran in and out, administering to whom they could.

The death toll continued to rise as the rain continued, and two other threats terrorized survivors. First, there was the possibility of a tsunami from the underground quake and coastal residents fled inland in anticipation of it. It never happened.

The second threat was more tangible. The 9,800-foot (2,987-m) Mount Merapi, one of the world's most active volcanoes, is located 250 miles (402.3 km) east of Jakarta, near the site of the earthquake. Within days of the quake, it began to stir, belching superheated gases and spurting a small amount of lava flow. Government officials immediately began to evacuate 11,000 villagers from around the mountain, and Mbah Marijan, a 79-year-old mystic who lived in Kaliurang, directly below Merapi's lava dome, and who had been directed by the sultan of Yogyakarta to make annual offerings to the volcano, began a barefoot walk through these villages, where offerings of rice and other bits of food hung from the doorways of the houses.

Marijan noted the offerings, but concluded that their ultimate fate rested with God. "I am not a scientist and I am not psychic," he admitted. "We can only pray for Merapi not to harm us."

The volcano quieted, and no eruption occurred, but the destruction from the earthquake would last for longer than the fear of a volcanic addition. Ultimately, it would take $3 billion to reconstruct the area, and during that time upwards of 1.5 million people would live in tents or improvised shelters.

IRAN
BAM
December 26, 2003

On December 26, 2003, a 6.6 earthquake destroyed this historic quarter as well as 85 percent of the remainder of the buildings in the Silk Road city of Bam, located in the southeastern province of Kerman in Iran. More than 26,000 people were killed and 200,000 were either injured or rendered homeless.

At 4:00 A.M. on Friday, December 26, 2003, a small earthquake hit the city of Bam, in the southeastern province of Kerman in Iran. People left their houses and gathered in the streets, fearful of their houses collapsing. But the tremor passed, and most returned to their homes. At 5:27 A.M., a 6.6 earthquake ripped open the earth 6.2 miles (10 km) southwest of the city and homes did collapse, burying tens of thousands of unsuspecting residents beneath roofs that contained multiple layers of bricks designed to keep the houses cool in summer.

Bam is a city built upon the ancient Silk Road city of Arg-e-Bam, a thriving trading and commercial center on the Silk Road originally founded during the Sassanian period (224–632 C.E.). The tremors spread instantly to the city and flattened and effectively erased the historic portion of it, which was originally built from mud bricks, clay, and straw. The ancient, pre-Islamic Argi-I Bam, or Citadel of Bam, founded more than 2,000 years previously, and a United Nations world heritage site, simply disappeared.

Rescue teams, entering the city after the dust settled, found a scene of utter demolishment in which 85 percent of the houses had imploded or collapsed (see the color insert on p. C-4). A huge confusion of counting followed. At first, it was reported that 43,000 were killed. Two days later, the count was reduced to 26,000. The infrastructure within Bam totally collapsed and it would be two days before rescue teams from the government of Iran and 60 other countries began to arrive to tend to the injured and try to dig out survivors.

There were individual tales of astonishing survival. For example, on January 2, 2004, 97-year-old Shahrbanoo Mazandarani was pulled from the wreckage of his house after being buried for eight days. Thirteen days after the earthquake, a 56-year-old man was found alive but in poor health. Aid workers told Reuters the rescued man had traveled from a nearby village to Bam for medical treatment and was staying with his sister when the earthquake struck. These were, however, exceptions to the tragic loss of life and shelter for a large portion of the population of Bam.

IRAN
NORTHEAST
May 10, 1997

Iran, which is prone to earthquakes, particularly along its border with Afghanistan, suffered its second and more devastating temblor in 10 weeks on May 10, 1997. One thousand five hundred and sixty Iranians, most of them from small villages, died; 4,460 were injured, and 60,000 were rendered homeless. The cost of rebuilding after the quake was estimated at $100 million.

A mere 10 weeks after northwestern Iran was pummeled by a major earthquake (see p. 63), an even larger one roared into the remote mountains of the northeastern part of the country at 12:28 P.M. on May 10, 1997. Measured at 7.1 on the Richter scale, it was centered 65 miles north-northeast of Birjand, near Quain, about 70 miles west of the Afghan border.

It was bright sunlight when the quake first began to pitch the ground into waves that sometimes opened up and swallowed objects and animals. "I was outside when I heard the mountain roar like a dragon, and suddenly the air became dark as night from the thick cloud of dust," survivor Gholamreza Nowrouz-Zadeh told officials.

He had much to mourn. One hundred and ten children, including all six of his grandchildren, were killed when the schoolhouse collapsed in the village of Ardakul, about 60 miles (96.5 km) east of Quain. The tiny village of Abiz, five miles away, suffered terribly. One-third of its residents died in the quake.

Over 155 aftershocks shook the area, terrifying survivors and bringing what was left of standing huts crashing down, sometimes killing survivors who were on the brink of rescue.

Iranian aircraft flew in food, clothes, and medicine and thousands of volunteers descended upon the area in convoys of trucks and buses. There was little heavy equipment, and rescuers often dug with their bare hands through crumbled cement and broken bricks to search for the living and the dead.

Streets in the small villages were turned into streams of rubble, over which distraught men and women clambered, shouting out for missing relatives and friends.

For days, as the temperature dropped to 41 degrees overnight and climbed to 84 in the daytime, the Iranian Red Crescent sent 9,000 tents, more than 18,000 blankets, and canned food, rice, and dates into the area. The Iranian government flew 80 tons of aid to the region aboard four American-made C-130 cargo planes and six helicopters.

The city of Meshed was the rallying point for relief work. From there, supplies and workers were trucked for five hours over rough terrain and roads that were nearly nonexistent. When the workers reached the villages, which were inhabited mostly by the families of subsistence farmers who either tended camels or sheep or grew wheat or saffron, they found injured people who appeared to have been weak and malnourished even before the quake had struck. As the time lengthened, concern rose among officials and villagers that the hundreds of bodies still buried under the rubble might begin to rot and spread disease.

Eventually, hundreds of thousands of dollars worth of blankets, tents, food, and clothes arrived from outside of Iran, from Japan, France, Britain, Italy, Russia, and neighboring Persian Gulf countries. In the United States, $100,000 was given by the government to the Red Cross to defray its costs in Iran.

The statistics were appalling: 1,560 people died, 4,460 were injured, and 60,000 were left homeless by this, the third earthquake in Iran within one year.

IRAN
NORTHWEST: CASPIAN SEA AREA
June 21, 1990

An earthquake that registered 7.3 on the Richter scale at Teheran University and 7.7 at the U.S. Geological Survey at Golden, Colorado, ripped through northwestern Iran on June 21, 1990, killing an estimated 50,000, injuring 200,000, and rendering 500,000 homeless.

The worst earthquake to hit Iran in 12 years struck the historically quake-plagued northwestern part of that country at 12:30 A.M. on June 21, 1990. The epicenter was located in the Caspian Sea, but its destructive force cut a wide swath through Zanjan and Gilan Provinces, a region bordering the Caspian Sea 100 to 200 miles (160.9 to 321.8 km) northwest of Teheran. The massive tremor toppled mountains and buildings, and heavily damaged or destroyed over 100 towns and villages. Gilan's capital city of Rasht was particularly hard hit, and a dam south of it burst, flooding the countryside and causing even more casualties.

More than a dozen aftershocks—one of them measuring 6.5—terrorized the already hysterical survivors of the quake for four days after the initial impact, thus confirming the opinion of geologists that Iran and its surrounding region is an area prone to catastrophic quakes.

Ever since 700 C.E., when records of the area's earthquakes were begun, this part of Iran has undergone devastating earth tremors. Twelve earthquakes with a magnitude greater than 7 rumbled through the area from 1960 through 1990. Iran lies in an earthquake belt extending from Turkey through the Caucasus Mountains and farther east into the Himalayas. This belt straddles the area in which the Arabian tectonic plate is grinding gradually northeastward, at a rate of an inch (2.5 cm) to five inches (12.1 cm) a year.

As it works its way in this direction, it continually collides with the larger Eurasian plate, and energy, geologists at the National Earthquake Information Center in Golden, Colorado, reason, is transferred from the collision to fault lines running from it. These fault lines run through northwestern Iran.

Geological explanations aside, it was a night of continual, ascending terror for the 2.7 million residents of the area, most of whom were either in bed or watching World Cup soccer from Italy on television when the quake struck. "A rock as big as a building crushed my home," wailed a man in Rudbar, a mountainside village in Gilan province, when rescuers reached him. The rock, loosed from a mountain that collapsed onto the main road into this village and hampered the influx of rescue workers, killed four of his six children and buried his wife for hours before help came.

"There is not a single house in this area that has been left standing," added Ali Mohammadi, a farmer from Rudbar, where 6,000 died and 90 percent of that city's buildings were destroyed. In the nearby towns of Abn-Dar and Bouin, all in Zanjan province, every building was flattened and every resident killed.

And it was much the same throughout the entire region of cities and villages located in the midst of inaccessible peaks, lush plains, vineyards, and wheat fields. Particularly hard hit was the jewel-like city of Manjil, located in an abundant valley in the foothills of the Elburz Mountains. Rescuers, battling bad weather and blocked roads, descended by helicopter and found mile after mile of buildings that had been smashed flat, as if the boot of a giant had ground them into the soil.

"I saw how the earth was trembling, like nature kicking the cradle," said 71-year-old Baba Kamalyari to a reporter for the *New York Times*. He was standing outside of his crushed home in the nearby mountainside village of Hir. "I saw the mountain slide toward the village and said, 'Allah Akbar! I am ready to die!' But I lived, and all the younger ones were taken from me." His wife, his children, his brother, and his brother's family were all crushed by a rock as big as a three-story house.

"The earth was whistling," a veteran of Iran's war with Iraq told a Reuters reporter. "It reminded me of the sound of an incoming mortar bomb."

In the city of Rasht, frantic young men clawed at the rubble with bare hands, digging through smashed concrete buildings in a search for survivors. Victims were extricated by hand, then taken out on stretchers and flown by helicopter to hospitals in Teheran, while the dead were hurriedly buried in mass graves. The haste of the operation and the efforts to bury the dead before disease spread made an accurate casualty count impossible.

Within days, international aid began to arrive, sparking a furious debate in the Iranian parliament. Hardliners wanted no interference from the west, and, in some places, Western doctors and relief workers were turned away. In other places, like Rasht, French rescue workers used trained dogs to search for victims and aided in the search for survivors while cranes lifted concrete slabs and bulldozers cleared roads for ambulances and other rescue vehicles. Tons of supplies were funneled to Iran from the outside world through the Red Crescent.

In Rudbar, more than 50 hours after the quake struck, workers uncovered a woman who emerged with a dead child in her arms. She joined thousands of wailing refugees who were housed in tent cities. Others continued to wander from place to place, searching for missing family members.

On Sunday, June 24, three days after the original earthquake hit, a 6.5 aftershock roared through the area, burying some victims who were near rescue. Claude Sins, a French fireman who had come to Iran as a relief worker, told the *Times* that he was preparing to dig through to a woman who was only a few feet below the surface. She had been desperately pounding on a doorframe in answer to his knocks when the aftershock hit, and nearly buried him, too. "[The aftershock came,] I knocked again, and there was no answer this time," he said. The woman was later discovered, dead, by other workers.

In Hir, rescue workers broke through the rubble of a home into what once was a bedroom and found a woman wrapped in bedsheets and a blanket. She was still breathing, and workers dashed to her aid, pulled her from the wreckage and laid her gently on the ground. Moments later, she died.

All through the area, the stench of death filled the air, as, day by day, mass graves overflowed with victims, and confusion among the rescuers began to grow. "The scope of the damage is so massive that much may be going undone because of simple confusion and a lack of coordination," said a Western diplomat. The general consensus was that the official report from the Islamic Republic News Agency of at least 50,000 dead, more than 200,000 injured, and 500,000 homeless was modest, by far.

There was even a discrepancy in the seismograph readings. Teheran University recorded 7.3 on the Richter scale; the United States Geological Survey at Golden, Colorado, recorded 7.7. Either way, it was a major earthquake in a region that has been periodically ripped asunder by this natural disaster for 1,300 recorded years.

IRAN
NORTHWEST
February 27, 1997

. . . .

Nine hundred sixty-five people died and 40,000 were rendered homeless in a major earthquake that hit northwestern Iran 16 hours after another earthquake struck neighboring Pakistan. There was no relationship between the two earthquakes.

Sixteen hours after a 7.3 force earthquake roared through Baluchistan province in neighboring Pakistan (see p. 79), a 6.1 force earthquake ripped through northwest Iran near the border with Azerbaijan. For 30 seconds, the territory that included the city of Ardebil and the town of Meshkinshahr rocked and opened gaping fissures that swallowed trees and animals.

Fifty-two snow- and icebound villages near these centers of population were devastated. In one, 720 residents—nearly the entire population of the village—were laid out in the village square. A teacher in the town of Sarab reported that he counted 2,000 dead from the surrounding villages, whose bodies were taken to the cemetery in Ardebil.

Later that same afternoon, two large aftershocks terrified survivors, who panicked in the streets of Ardebil and Meshkinshahr, but by that night, 4,000 relief workers had entered the area and set up temporary settlements in which survivors were given tents, blankets, heaters, bread, and canned fruit. By that night, snow was falling.

Three days after the quake, an eerie calm, muffled by new snow, settled over the area. Crying children broke the silence from time to time, and an acrid, biting smoke from burning tires and timber that survivors had lit to keep warm hung heavily on the air.

"Those who have survived are dying inside because they have lost everything." wailed Gafur Lutfi, a farmer in the village of Villadaragh. "All we had in the village was our animals and they are all dead now."

As daylight arrived, mud began to flow down the steep streets of the village as mechanical diggers sent by the Iranian government shifted the mounds of debris in the search for corpses.

Nearby, in the village of Saraien, a hot spring was converted into a morgue, where bodies were washed according to Islamic ritual before they were buried. In Ardebil, the municipal soccer stadium housed some of the homeless. In some of the streets, villagers clutching chits gathered around Iranian health officials, pleading for food. Trucks, piled high with food, clothing, and plastic sheeting thundered through the city streets, heading for the worst-hit areas.

The eventual official total of dead was 965, with over 40,000 homeless. Unofficial totals were considerably higher.

IRAN
TABAS
September 16, 1978

Twenty-five thousand people died in a sudden earthquake centered in the city of Tabas, in eastern Iran, on September 16, 1978. Fifteen thousand of the 17,000 residents of Tabas were killed in the quake.

The city of Tabas, an agricultural center in eastern Iran, was almost completely demolished by a huge earthquake whose epicenter was located directly in the thickly settled city on September 16, 1978. Twenty-five thousand perished in this 7.7-force quake, 15,000 of them in the city of Tabas itself, whose total population was only 17,000.

It could be said that because the quake struck in a relatively unpopulated section of Iran many lives were saved. But to the populace of this city on the edge of the central Iranian desert, 400 miles (643.7 km) southeast of Teheran and close to Iran's border with Afghanistan, it was a devastating loss.

As in the case of the lesser, though still major quake of 10 years before, there was no warning. Thousands were crushed to death when the mud walls of their dwellings caved in on them. Others were swallowed alive when their fields opened up beneath them.

Although the greatest and most dramatic devastation occurred in Tabas and the surrounding countryside, the quake was strong enough to be felt throughout two-thirds of the area of Iran. More than 40 villages within a radius of 60 miles of the epicenter were completely leveled, and buildings in Teheran, 400 miles (643.7 km) from Tabas, swayed.

ITALY
CALABRIA
February 5, 1783

One hundred eighty-one towns were destroyed and 60,000 residents of the Calabria region of Italy were killed during two months of earthquakes that began on February 5, 1783. During this period 949 separate earthquakes hit the region.

For two long, terrible months, beginning on February 5, 1783, a continuous series of earthquakes shook the entire western portion of southern Italy. Nearly 60,000 people perished in the 949 recorded shocks that occurred during the period, most from the quakes, but 10,000 from epidemics that resulted from the rupturing of waterlines; massive injuries; the crowding of survivors into small, unsanitary shelters; and exposure to the elements.

The epicenter of the quake was the earthquake-prone region of Calabria, located in the instep of the boot of Italy, and bordering on the Gulf of Taranto and the Ionian Sea. Thirty thousand people died in this region during the first, most serious and disastrous quake, which was particularly terrifying to its inhabitants because of the immensely powerful thermal springs that were unleashed by the rupturing of the earth. Some of the fissures were described by eyewitnesses and scientists as 150 feet (45.7 m) wide and 225 feet (68.6 m) deep—enormous by any standards. Like the quake itself, these cracks in the earth opened without warning, swallowing hundreds of humans and animals that were almost instantaneously regurgitated on geysers of boiling lava. Those few who survived this terrible ordeal were so badly burned they spent the remainder of their lives hopelessly crippled.

In total, 181 towns in Calabria were leveled. Landslides occurred everywhere, but were most acutely felt by residents along the seacoast, who were set upon by landslides on one side and giant 20-foot (6.1 m) tsunamis on the other. Sixteen hundred coast dwellers perished from one or the other of these two murderous side effects of the earthquakes.

ITALY
CALABRIA
December 16, 1857

Over 10,000 people perished in the earthquake of December 16, 1857, in the Calabria region of Italy. The quake was the last in a series that rippled through the region for 74 years and killed 110,000 people.

According to the Italian historian Lacaita, 110,000 inhabitants of the region of Calabria lost their lives during the period between 1783 (see previous entry) and 1857. The 74-year period contained the most violent earthquakes to ever hit this often shaken region.

The final earthquake of this series struck on December 16, 1857, splitting the peninsula open from the Adriatic to the Mediterranean. Entire villages were either flattened, engulfed in landslides or swallowed up by gigantic earth fissures. Over 10,000 perished in this quake alone, the last of a series, but by no means the last earthquake to devastate southern Italy.

ITALY
CALABRIA
September 8, 1905

Twenty-five villages were demolished and 5,000 people lost their lives when an earthquake once again tore through the Calabria region of Italy on September 8, 1905.

The quake that struck in the middle of the night of September 8, 1905, throughout southern Italy killed 5,000 people, most of them in their sleep. Over 25 villages were obliterated by the quake, including Conidoni, San

Distressed refugees huddle in a public square in the town of Martirano, in the Calabria region of Italy, after the September 8, 1905, earthquake that killed 5,000 and destroyed more than 25 villages. (Illustrated London News)

Constantino, Sanfeo, Bratico, Zammaro, Tripapni, and Piscopio. Martirano alone lost 2,000 of its populace to the series of 18-second shocks that traversed the countryside and rocked pantry shelves as far north as Naples and Florence.

On the island of Stromboli, the volcano that gives the island its name was stimulated to erupt by the quake, strewing the night air with flaming rocks and lava and killing several dozen residents of the sparsely populated island.

Twenty thousand were made homeless by the series of shocks that forced the survivors of collapsed dwellings into the streets. Mobs of terrified refugees ran insanely through city streets, where they smashed storefronts, wrecked cafes, and generally added to the unbridled turmoil brought about by the natural disaster.

A supremely ironic drama unfolded in the local jail in the town of Monte Leone. Terrified prisoners staged a riot in which several inmates and guards were killed. Free of their cells, scores of rampaging prisoners perished when a secondary quake loosened the already perilously pitched walls of the prison, sending them crashing in on the newly liberated prisoners and burying them under the debris.

ITALY
MESSINA
December 28, 1908

Depending upon the source of statistics, either 160,000 or 250,000 people were killed in the cataclysmic quake that demolished the city of Messina and 25 surrounding towns on December 28, 1908.

One of the most vicious quakes in all of recorded history began under the waters of the Straits of Messina at 5:25 A.M. on December 28, 1908. The official death toll was set at 160,000, but Professor Antonio Rioco of the Mount Etna Observatory reported 250,000 dead.

This unimaginable loss of life was matched by the fact that the entire Sicilian city of Messina was wiped out. Only one structure, an iron-reinforced home built by an eccentric merchant, remained. Cities and towns within a 120-mile radius of the quake's epicenter were destroyed. Among them Reggio de Calabria (in which 25,000 of the city's 34,000 people were killed), Caltanissetta, Patti, Augusta, Mineo, Naro, Marianopoli, Terranova, Paterno, Vittoria, Chiaramonto, Noto, Floridia, Cannitella, Lazzaro, Scylla, San Giovanni, Seminaria, Riposta, Bagbara, Cosenza, Casano, Palmi, Catania, and Castroreale.

The quake struck with the fury of an apocalypse, combining earth tremors, a 50-foot-high (15.2-m-high), 500-MPH (804.6 km/h) tsunami, hurricane force winds, and driving, relentless rain—and all of this occurring in that deep darkness that precedes the dawn. Small wonder that so many of the few survivors went hopelessly insane. It was more than the most stable of minds could imagine, much less endure.

The preliminary shocks roused people from their beds at precisely 5:25 A.M. The second, rending ones lasted roughly 10 seconds, and roused still more. And then, the shocks hit like an artillery barrage, each one lasting from 35 to 45 seconds, and each bringing down a series of buildings, a cathedral, a hotel, an entire city block.

The great Norman Cathedral of Annunziata dei Catalani which had sheltered ancient art treasures for centuries collapsed, burying the treasures under tons of granite. The architectural wonders of the Munizone and Victor Emmanuele Theatres crumbled to dust. Plush and thickly populated tourist hotels, the Victoria, Metropole, Trinacria, and the France wavered and then toppled, crushing their occupants inside. The Duomo, another storehouse of centuries of priceless art, imploded, leaving only the colossal figure of Christ, in mosaic in the dome of the apse at the east end standing. In the words of survivor Alexander Hood:

> . . . with serene countenance and hand uplifted in the act of blessing as for five hundred years or more it has remained, gazing benignly on [while] the monster monoliths of granite with gilded capitals, which were the columns of Neptune's Temple at Faro, lie half or wholly covered by the painted woodwork and debris of the roof, among which are fragments of marble tombs and inlaid altars, golden figures of angels and sculptured saints . . . with the mosaic and frescoes, with the arches and cornices which had made the Duomo so rich a treasure house of art.

Convents that had sheltered centuries of other art crumbled and destroyed their contents; the Castle Durante collapsed, burying its prehistoric artifact collection.

But that was only a small part of the tragic story. Homes buried families; complexes buried citizens. The Santelia army barracks collapsed, killing almost all of the soldiers sleeping there and creating tragedy. With the chief of police dead in his home and most civilian authorities also killed, 750 prisoners liberated by the quake from the Cappuccini prison were left free to roam the streets, creating a residual menace as they looted and murdered at will.

When the tremors finally quieted, 65,000 had survived out of the previous population of 147,000. The rest had been killed, some outright, some by the fires set by broken gas mains, some by the flooding caused by the splitting apart of the city's reservoirs; others along the port were drowned by the force of the tsunami.

Captain Owen of the Welsh steamer *Afonwen* described conditions in the port:

> . . . It was a cyclone from all points of the compass. The wind howled and the waves battered and swept the decks. Amazing and terrifying things were happening all around us. Great holes opened in the sea itself and seemed to reach down twenty to thirty feet . . . the water at first appeared to grow livid and then became white with foam. . . .

Sailors from these ships rescued many and eventually restored order after the unbridled chaos that followed the quake. At the Trinacria Hotel, some survivors found themselves trapped in parts of rooms left after the rest of the structure had fallen away. Precariously perched, up to 10 stories above the street, some of them were rescued by the aforementioned Captain Owen and his first mate, Read.

One family, huddled on a balcony of the remains of the hotel called to the two sailors. Read and Smith, another sailor, climbed up a small ladder to a lower balcony and shouted to the trapped children to lower a string with a rock attached. Grabbing the rock, the two sailors attached an eighth-inch manila rope to it and told the terrified youngsters to secure it to the balcony railing. While the balcony still swayed from aftershocks, Smith shinnied up the rope and, with Read at his side, lowered 10 children and several adults to safety.

Meanwhile, at ground level, chaos turned to madness. Looters hacked fingers from corpses to retrieve jewelry. Famished crowds fought each other to death over food in semi-collapsed warehouses.

Finally, at dawn, Russian battleships, informed of the tragedy, steamed into the Straits of Messina from Agoata. Six hundred Russian sailors went ashore, organizing rescue parties and restoring some semblance of law and order. They rounded up looters, executed some of them on the spot, and ultimately set up an open air hospital that treated over 1,000 wounded survivors during its first hour of operation.

By afternoon, Britain's Royal Navy arrived, erecting soup kitchens and aiding the Russians in maintaining order.

The American navy did not acquit itself nearly so well. Stung by its rejection the previous year by the British governor of the earthquake-struck island of Jamaica (see p. 68), the American fleet in the Red Sea refused to respond to distress calls unless they came personally from King Victor Emmanuel. However, within weeks, the civilian American government, apparently rethinking the situation, dispatched millions of dollars in supplies to the city that had been virtually wiped from the face of the Earth by the forces of the Earth.

ITALY
SOUTHERN
November 23, 1980

A seven-part earthquake, measuring 6.8 on the Richter scale, devastated parts of southern Italy on November 23, 1980, killing over 3,000, injuring 3,000, and rendering 200,000 homeless.

The earthquake that roared through the south of Italy early on the evening of November 23, 1980, killing over 3,000, injuring 3,000 more, and leaving 200,000 homeless, was the strongest to hit that quake-prone part of the world in 65 years. Measuring 6.8 on the Richter scale, the quake's epicenter was located at Eboli, a town near the bay of Salerno and 30 miles southeast of Naples.

The first of seven shocks hit at 7:34 P.M., just as residents of some cities were sitting down to their Sunday meal, and just as others were beginning the ceremony of the Mass.

The parish priest at Balvano, the Reverend Salvatore Pagliuca, his vestments ripped and covered with dust from his efforts to free victims from his collapsed church, told a reporter, "There were at least 300 people at the Mass tonight, including many children. The front wall collapsed as people were trying to get out."

Thousands of people in Naples spilled hysterically out of their homes and into the streets, as their evening meals accumulated dust and debris from rocking walls and collapsing ceilings. Most roamed the streets, afraid to return to their homes.

In Potema, 90 miles (144.8 km) east of Naples and near Balvano, virtually all of that city's 50,000 residents fled to the hills nearby, to spend the night either in their cars or in the open.

There and elsewhere, fires began to ignite from stoves left burning when apartment and home dwellers abruptly left their kitchens. Water mains were severed, hampering fire fighting efforts, but leaving firemen free to aid in the rescue work, which began as soon as the nation's doctors, who had scheduled a strike for the next day, decided to cancel their work stoppage.

As far away as Rome, the Leonardo da Vinci airport closed for 40 minutes when air traffic controllers fled their swaying 195-foot-high (59.4-m-high) control tower.

At the Poggioreale prison in Naples, prisoners rioted as their cell walls began to crumble around them and fissures yawned in the prison yard. Afraid of an escape, guards hurled gas grenades and fired submachine guns into the air.

A heavy fog rolled in over Naples, making it slow going for teams with bulldozers, tents, and medical supplies. Trains were halted south of Naples, and traffic was blocked on highways.

When rescuers finally reached isolated towns, they found heartrending desolation. Tiny Pescaopagano, the closest town to the center of the quake, was virtually razed. New tremors rumbled through the area, collapsing half-demolished buildings and creating panic among the survivors.

But it was neither the aftershocks nor the terrain that ultimately hampered rescuers. It was the poverty of the people. The area of the Apennine Mountains along the spine of the Italian Peninsula is one of the poorest in Europe. There was virtually no modern heavy equipment, such as bulldozers, tractors, or cranes, in the afflicted areas.

This, plus a bureaucracy that sent well equipped rescue teams into the larger towns like Potenza and Avellino and ignored the more inaccessible, smaller places in the mountains undoubtedly contributed to an increase in both suffering and casualties.

Still, by any standards, the rescue effort was massive. Not counting volunteers, 9,000 men joined in the search for bodies and survivors. Thirty army helicopters flew mercy missions, and regular airlifts were established to ferry in food and transport survivors out to hospitals. Six hundred army tents were brought in, five field hospitals, and 28 field kitchens were erected in various sites.

And finally, belatedly, these rescuers climbed their way to remote villages, bereft of food, water, money, or young men—most of their ablebodied young men had long since left, to work in northern Italy, Germany, Switzerland, and Scandinavia. Conza Della Compagna, formerly a town of 2,000 on a mountaintop, was transformed into a shapeless heap of stones, collapsed roofs, beams, and debris. Firemen, their faces masked to avoid the pervasive stench of death, dug for bodies with bulldozers. Coffins were placed singly or in neat stacks in front of wrecked homesites.

Some of those who survived were housed in prefabricated barracks once used by construction workers at a nearby dam site. Others were put into tents. Some went to relatives in other towns. The small children, some of them with aged faces that reflected the holocaust they had experienced, were housed in a hotel in a nearby village, to await—perhaps fruitlessly—the return of their parents.

JAMAICA
KINGSTON
January 14, 1907

The major earthquake that struck Kingston, Jamaica, on January 14, 1907, destroyed most of the city and killed over 1,400 of its residents.

Over 1,400 inhabitants of Kingston, Jamaica, were killed by a grueling earthquake that ripped that city apart on the afternoon of January 14, 1907. Not since the 1692 earthquake in nearby Port Royal (see next entry) had such devastation struck the island of Jamaica in such a concentrated space and time.

According to reports, the earth actually thrust upward for 36 seconds, toppling buildings, snapping water and electric lines, opening huge longitudinal cracks in roads, sidewalks, and buildings, and setting off a tsunami which slammed, with gigantic force, into the dwellings and inhabitants rimming Anotta Bay. Within moments, hundreds of houses, with their inhabitants inside, were swept out to sea.

The Kingston powerhouse was one of the first buildings to be destroyed, but not before live wires split and broke, spewing sparks, electrocuting scores of people, and setting a huge number of fires. With no available water, since all of the city's water mains were broken, the fires raged out of control, destroying, along with the quake, more than 25 square blocks of dwellings and businesses. A cigar factory collapsed, killing all 125 workers trapped inside of it.

With no usable treatment facilities available on land, the wounded were taken to vessels in the harbor, which were quickly turned into hospital ships. On the *Arno,* a doctor, said to have gone crazy halfway through the quake, performed 79 amputations—not all of them, according to observers, necessary.

JAMAICA
PORT ROYAL
June 7, 1692

One-third of the population of Port Royal, Jamaica— 1,600 persons—was killed when the worst Western Hemisphere earthquake recorded to that time ravaged the island on June 7, 1692.

When, at 11:40 A.M., the most powerful earthquake to yet rock the Western Hemisphere made matchsticks out of the roaring port of Port Royal, Jamaica, it was regarded by many pastors as the just desserts of sin. A kind of combination Sodom and Gomorrah to some, a

jolly good time to others, Port Royal consisted of 2,000 dwellings, some of which were fleshpots, some ship refitting centers, some homes to its natives. Within two minutes, 1,600 people—one-third of the population— would lie dead, the victims of a cataclysm that struck some distance north of the port and literally moved mountains—two of them—a mile to the south.

The first of three shocks was a major one, dislodging crockery, upsetting furniture, and cracking walls. The second, longer and more violent (it lasted a full minute, which is an eternity during an earthquake), started the real damage, crumbling the wavering walls, sending them crashing down upon occupants of disappearing buildings, opening up huge crevices in the earth that would run the entire length of streets and consume vehicles, animals, and human beings.

Simultaneously with the last two shocks, a tsunami was formed, which, combined with landslides, caused the entire northern portion of Port Royal to slide into the sea, drowning every inhabitant in that section of the city.

For some others, the tsunami was a lifesaver. The frigate *Swan,* resting on its side for repairs, was caught up in the tsunami as it roared inland. Skimming along over the submerged part of the city at what must have seemed unimaginable speed, it dragged its lifelines, which were grabbed by survivors thrashing about in the turbulent waters. Yanked along by the wildly racing frigate, they were able to emerge from the shallow water, shaken but safe, when the ship finally beached on the roof of a partially submerged building.

Others were far less lucky. Many were crushed by the closing of fissures into which they had slipped, only their heads protruding from the ground, and these heads, according to Reverend Emmanuel Heath, a surviving minister, turned into food for roving dogs.

One venerable city resident, merchant Lewis Galdy, had better luck when he fell into a fissure. Trapped under water and sand, he concluded that his time had come, only to feel, a moment later, a geyser erupting beneath him, which forced the fissure open again, and popped him, like a human cork, into the air. Mr. Galdy lived on to be one of the island's leading repositories of first-person anecdotes about the great earthquake of 1692.

This sort of selective salvation contributed to the legend that this particular earthquake had been the product of divine intervention.

JAPAN
1737

An enormous earthquake struck Japan sometime in 1737 and launched what is believed to be the largest tsunami in recorded history.

Information is decidedly sketchy, but if that information is even remotely correct, possibly the greatest tsunami of all time was set in destructive motion by an earthquake that rocked Japan sometime in 1737. According to records that border on rumor, this tsunami reached a height of 210 feet (64 m) as it curled away from, then rushed ashore from Yezo Island to Kamchatka. Thousands were killed by the earthquake; thousands more perished from the tsunami, which completely wiped out the coastal city of Kamaishi.

JAPAN
June 15, 1896

Tsunamis set in motion by an earthquake in the Tuscarora Deep, off the coast of Japan, drowned nearly 28,000 people on June 15, 1896.

Sometime during the late morning of June 15, 1896, a subsea crater seismologists call the Tuscarora Deep collapsed. The shock waves hit the shore of the Sanriku District of Japan, where an ancient ceremony known as the "Boy's Festival" was at its height. Traditionally conducted at this time every year along Japan's northeast coast, it attracted thousands of people.

Although the shock seemed mild, many of the celebrants retreated to the hills, away from the coast, in case the quake spawned a tsunami. Rain fell, but no wave appeared, and by sunset, when the skies had cleared, so had the memories of the multitudes that had evacuated the beaches. They resumed their places on the sand for the conclusion of the ceremonies.

And then it happened. Not one, but a series of tsunamis, each traveling at 500 MPH (804.6 km/h), varying in height from 30 to 110 feet (9.1 to 33.5 m), slammed into 100 miles (160.9 km) of shore with a horrendous roar. They broke over the beach and then continued to blunder inland for almost 100 miles (160.9 km).

Nearly 28,000 people were consumed by these raging waters, which pounded and then pulverized scores of seaside towns. Kamaishi disappeared; when the waters receded, only 143 of its 4,223 buildings were left standing, and a mere, 1,857 of its 6,557 people escaped drowning.

The story was the same in other nearby villages. In the obliterated village of Toni, only 97 of the village's 1,200 people escaped drowning. In the Kissen District, 6,000 were drowned or crushed. At Yamada, 1,200 survivors remained of a village that had once housed 4,000. Every person in the hamlet of Hongo died except for a group of old residents who were wiling away the time playing a game of tiles in a hilltop temple.

Some people were swept out to sea and then back again. Others were transported from one shore to islands offshore, and survived. But they were few, and many of these wandered listlessly among the mounds of rotting, dismembered corpses piled high in villages and along the beaches. Some villages would never be rebuilt; others would be years in reconstruction.

JAPAN
HOKKAIDO/OKUSHIRI
July 14, 1993

A 7.8-force earthquake followed by a series of 10 tsunamis created havoc and death on the coast of Hokkaido and the fishing and resort island of Okushiri in northern Japan on July 14, 1993. One hundred eighty-five people were killed and 57 were reported missing and presumably drowned as a result of the quake and the waves.

In May of 1960, a series of earthquakes that killed 5,700 people on and off the coast of Chile (see Chile, p. 42), spawned a giant tsunami, 24 feet (7.3 m) high, which, after devastating the Chilean coast, folded back on itself and traveled, at speeds estimated at 540 MPH, across the Pacific and crashed against the coastline of Hokkaido, Japan's northernmost territory, and particularly the nearby island of Okushiri. One hundred forty-two residents of Hokkaido and Okushiri were killed in that tragedy.

Thirty-three years later, on July 14, 1993, an equally devastating earthquake, which measured exactly the same as the Chilean quake—7.8 on the Richter scale—erupted in Hokkaido itself, and once again a tsunami was set in motion. This one was measured at 30 feet (9.1 m), and its spoliation of the Hokkaido coast and the island of Okushiri was similarly widespread and horrendous. This time, 185 people were killed by the quake or the wave, and 57 were reported missing, presumably sucked out to sea by the backwash and drowned.

The epicenter of the quake was centered about 50 miles off the southwest coast of Hokkaido, in the Sea of Japan. Striking with only minimal warning at 10:17 P.M., it collapsed houses, opened up fissures, and swallowed trucks and cars on the coast. Homes collapsed and gas lines ruptured, starting fires that burned out of control and leveled entire villages.

But the worst damage was not revealed until 24 hours after the earthquake struck. The island of Okushiri, 40 miles (64.4 km) southwest of Hokkaido, and thus only some 10 miles (16.1 km) from the epicenter, was—silently to the world, because of downed communication lines—suffering far worse damage. Not only had gaps in the earth swallowed buildings on the sparsely populated island, but a series of giant tsunamis had pounded its shores repeatedly, inundating and collapsing buildings, drowning fleeing residents, and dragging others into the Sea of Japan, where they drowned.

The wreckage was appalling. Dwellings and Buddhist temples were shattered and sunk in story-high mud; boats were flung through the walls of houses or deposited in unlikely, inland places. Gas lines were ruptured in greater profusion than on Hokkaido, and by the next day, the landscape was pockmarked by smoldering ruins where clusters of dwellings once stood.

The Yoyoso Hotel, standing below a hillside, was demolished by a huge landslide. The fishing docks that allowed the livelihood of the island's fishermen were universally smashed into useless sticks.

But this palled next to the human toll. Divested of their homes, thousands sought shelter where they could. Schools and community shelters were packed. One television news program showed pictures of refugees sitting near corpses in a gymnasium, where incense was burning in a Buddhist ceremony of mourning for the dead.

The most severely injured were airlifted by helicopter to Hakodate or Sapporo, on Hokkaido. Okushiri's airport was useless, its runways buckled and pitted by the heaving earth of the quake.

Relatives of the inhabitants of Okushiri attempted to reach them by taking the ferry from the coastal town of Esashi, on Hokkaido. But ferry service was severely disrupted. Over 20 cars had been swept into the sea around Esashi, preventing the ferry from docking. Eventually, tugs were put into service to carry passengers around the wreckage to the ferry, which then made the 40-mile journey to Okushiri.

Once there, relatives of the island's residents sought out their families and friends. Michihiko Inagaki, a printer who lived in Tokyo, told the story of the tragedy to reporters afterward. "My father was taking a bath [when he felt the rumble of the earthquake], and he ran

Pools of standing water and a shattered landscape remain after a giant tsunami inundated the coast of Japan in 1896. A staggering 28,000 were drowned. (Frank Leslie's Illustrated Newspaper)

out in his underwear and bare feet," said Mr. Inagaki. "[My mother and he] escaped with nothing but their lives."

A small quake and wave had struck the island 10 years before, with no fatalities, and some of the residents therefore knew the signs. Others either had no history of experience or ignored it; they took time to gather the valuables in their houses and were drowned by the tsunami, which smashed through a retaining wall built 10 years ago.

Relatives of Akiko Morikawa, the owner of the restaurant in the Yoyoso Hotel, were not as lucky as Mr. Inagaki. She was crushed, along with two of her three children, by the landslide that collapsed the hotel. Her sister, who managed to escape from the restaurant, was swept up in the waters of the series of tsunamis, and clung to some grass. It saved her from being pulled out in the backwash.

Ironically, the tsunami warning system that had been developed in Japan functioned, but too late. It reached Okushiri at the precise moment that the first of 10 waves, traveling at 300 miles an hour (482.8 km/h), reached the island.

JAPAN
KOBE
January 16, 1995

Five thousand five hundred and two people were killed in Japan's second largest earthquake and the most expensive disaster in the world on January 16, 1995. Tens of thousands were injured, 310,000 were made homeless, and 46,000 buildings were destroyed. The port, which had been the second largest port in Japan and the sixth largest in the world, was rendered inoperable for nearly a year. It would cost over $120 billion to rebuild Kobe.

The 310,000 people left homeless by Japan's second largest earthquake and most expensive disaster of all time welcomed all of the ways in which life slowly inched toward normality after the January 16, 1995, cataclysm. But nothing equaled the return of public baths, set up in tents by the army nine days after the quake. To hundreds of displaced residents of this ruined city, it was like the opening of a door upon darkness.

"I'm so excited I can have a real bath now," exulted 66-year-old Toshiko Hakenaka. "I've been cleaning myself with wet tissues, but now I can have a real bath. I'm going to dash home and get my friends and get some clean underwear, and then I'll be back."

To a nation that places cleanliness at the top of its list of personal priorities, the loss of public baths was as terrible as the loss of over 46,000 buildings, and the return of the baths marked the beginning of the return of normalcy.

It would be, experts agreed, a decade before Kobe would be rebuilt, at a cost of over $120 billion, a reexamination of Japanese building design, and the erasure of hundreds of thousands of terrible memories.

The earthquake, which would kill 5,502 residents of the city, injure 26,800, and render 310,000 homeless, struck at 5:46 A.M. on January 16, 1995, as most of the city slept. Registering 7.2 on the Richter scale, it was, later examination showed, a shallow quake, occurring at a mere 6.2 miles (10 km) beneath the surface of the earth. Therefore, its shocks were intensified and multiplied.

The city had a population of 1.47 million people and occupied an area of 208 square miles (334.7 sq. km). The sixth busiest port in the world, and Japan's second largest port after Yokohama, it had been a lifeline to the outside world for more than a thousand years. But in the 20 seconds that the quake lasted, that lifeline was instantly severed. The port would be closed for nearly a year while the immense wreckage caused by it was cleared and the businesses that ringed the port were rebuilt.

But that was only part of the story. The bullet train that normally roared past Kobe on its way to Osaka would be out of service for months, as would all rail service in and out of the city. Two cargo trains that tried to make the trip the day after the earthquake derailed and ended up on their sides beside the unaligned tracks.

The elevated Hanshin Expressway, which connected Nishinomya with Kobe, collapsed like a suddenly freed ribbon, its pillars crumbled, its roadway on its side, its load of automobiles scattered over the countryside.

Buildings in the city either tipped over, were sliced in two, or telescoped within themselves. "I was on the fourth floor of a five story hotel," said one television journalist. "The quake continued for 20 seconds or so, and I just lay on the floor and couldn't move. I tried the door, but it wouldn't open, so I kicked it down and barely escaped. The lower section of the hotel under the third floor had completely collapsed."

Story after story of cataclysmic horror emerged as Japanese journalists fanned out and captured dazed inhabitants of the city, many of them still in their underclothes or pajamas, trying frantically to get to buried relatives or friends. A 47-year-old businessman complained that his business was in ruins, then suddenly realized that his brother was missing, buried beneath mounds of debris after the second floor of his house had sunk into the first.

His body and hundreds of others were transferred to a nearby school that had become a combination shelter and morgue. The room reserved for the dead was labeled, via a paper sign, "The Room of Peaceful Spirits."

Some 70 fires were begun by the quake, and as they burned themselves out, their flames were high enough and intense enough to be seen in Osaka. Fire companies had no water pressure. The city's mains had burst.

But unlike the terrible 1923 earthquake that had killed 143,000 residents of Tokyo and Yokohama, mostly in fires (see p. 73), the 1995 temblor killed most of its victims by crushing them. Narrow streets in Kobe became impassable because of houses that had toppled over onto sidewalks, caved in storefronts, and downed telephone poles.

The sides of buildings were peeled away, so that the buildings resembled ill-kept dollhouses. Glass windows from storefronts had shattered, spraying broken glass before them. A Nissan agency's front window had disappeared, leaving half a dozen shiny new cars exposed. But there was no looting, nor would there be much. The only looting that was reported was one man drinking another's sake in a shelter. Reflecting long-held prejudices, the small amount of vandalism that eventually did occur was blamed on Koreans. Japanese, it was reported, waited in lines to buy food, get water, and make telephone calls. Though the schools were closed, teachers made house calls to check on their students.

The Itami railroad station collapsed, buckling under the weight of two trains that had been parked on its second level. And so, Kobe was effectively cut off from the rest of the world, except for some small secondary roads that rapidly became clogged with the cars of relatives who lived outside of the city, frantically trying to get to their families or friends. Fifty-four-year-old Masayoshi Ogawa took his first day off from work in 27 years. The manager of a steel company on the island of Kyushu in western Japan, he traveled to Kobe to help out his mother and relatives. The usually short trip would take him 14 hours.

Although most Japanese are taught from an early age to prepare for earthquakes (some elementary schools require pupils to sit on fireproof cushions that can be worn as a hat to shield the head from falling debris), the terrible actuality of this quake far exceeded the limits of their preparedness. Plastic water buckets that were given to each home by the fire department to either contain a week's worth of drinking water or enough water to douse a fire, were empty when the quake struck.

And so, afterward, a major goal for many residents was to find drinking water. Some scooped water in plastic soda bottles out of street puddles. Others ladled water from swimming pools. When these ran out, they stood in line to receive water distributed by army troops. As night fell, fires made from the remains of houses appeared on streets, where groups of survivors cooked whatever food they could find in whatever containers they had. "It reminds me of the time after the war," said Masakazu Koga, throwing a kitchen cabinet door on the fire. "Everything is a wreck, and I don't see how we're going to rebuild it all. But the earthquake also drew us together. I've seen generosity in people that I thought had disappeared from Japan."

Houses that had survived the quake now began to crumble, and their former residents moved into their cars, fearful that the houses would fall on them. Hospitals became part hospitals and part shelters, though some of them also became disaster areas. The sixth floor of a seven-story hospital in the center of Kobe collapsed, trapping 51 patients, all of whom were rescued.

By the third day after the quake, avalanche dogs were brought in to sniff through wreckage for survivors, and heavy bucket-loaders were used to lift slabs of concrete and heavy beams.

As the number of dead mounted, mass funerals were held. "I think some people feel that the people who died in the same disaster should be buried together because they shared a destiny," said Dr. Kaneatsu Miyamoto, the director general of Kobe's Public Health Bureau. The bureau had other worries: The city was already in the grip of a plenitude of flu victims. There was concern that it would become an epidemic because of increasingly cold nights and great numbers of people living outdoors.

An interesting social phenomenon developed. The Yamaguchi-gumi, Japan's largest crime syndicate, which was headquartered in Kobe, began to distribute free food and water, which the public eagerly lined up to receive.

And still the total of fatalities mounted, as more and more bodies were unearthed. A woman, digging to find her family, came upon the wife and son of her landlord. The mother's body was spread over the boy's, as if protecting him from the falling house. Both were dead.

Some were like 75-year-old Chieko Inohara, who had lain for 75 hours in the wreckage of her home with only a non-functioning electric blanket over her, but then was pulled out alive and taken by ambulance to a hospital, where she recovered.

One of the most touching stories of survival was that of Shizuko Hirajima, who communicated with the veterinarian who lived around the corner from her. Trapped in his collapsed house, he shouted that he was all right. Mrs. Hirajima shouted back, telling him that he would be okay, while she and some neighbors dug in other houses, looking for trapped neighbors.

And then, suddenly, a fire erupted in the veterinarian's house. Mrs. Hirajima and the others rushed back, but they were unable to get through the blistering flames. "We could hear him call out," Mrs. Hirajima later told reporters. "'Help me!' he shouted. 'Help me!' But we couldn't do a thing. We just had to stand there while he burned to death."

For a while, Mrs. Hirajima heard the screams in her sleep, and then, she abruptly lost her hearing. She could no longer hear the screams, but she could hear nothing else either.

Gradually, as the weeks dragged on, some stores opened, and the government set about building 25,000 temporary homes for the homeless. Organized crime, in the person of the Yamaguchi-gumi gang, used motor scooters, boats, and a helicopter to move in goods for them. By the end of January, the syndicate was handing out 8,000 meals a day from a parking lot next to its headquarters.

On January 21, heavy rains arrived, adding to the misery of the survivors. Some areas became too dangerous to inhabit and were evacuated. Others were coated in blue plastic sheets that were distributed by the government.

Depression coated the city along with the rain. "In the first day and the second day and the third day after the earthquake you felt lucky to be alive," said 30-year-old Manabu Takai to a reporter. "Now, the depression of it all is setting in. We are all very tired."

But gradually, the city began to rebuild. It would take nearly a year for the port to return to its former business, and by then, some of its trade had moved permanently to ports in South Korea and Hong Kong. By May, the bullet train was running again.

"It seemed as if every bulldozer in Japan was in Kobe," one resident observed in the spring. Twenty-five thousand huts filled parks, tennis courts, and parking lots, but they did not shelter all of the homeless. Tents, some of them as near as people could get to the homes in which they once lived but which were now piles of rubble, dotted every neighborhood.

But there was one happy ending. Mrs. Hirajima recovered her hearing. "Recently my hearing got better," she confessed to a reporter. "And now it's almost back to normal. There's no point in dwelling on the past. I've got to keep struggling ahead."

JAPAN
TOKYO
March 21, 1857

Fires caused by an earthquake on March 21, 1857 swelled the total fatalities in Tokyo, Japan, to 107,000.

An incredible 107,000 people died fiery deaths in the horrifying aftermath of an earthquake that rocked the city of Tokyo on March 21, 1857. Thousands perished in the first shocks that devastated large sections of the city, collapsing thousands of structures that had not been built to withstand earthquakes—nor would they be rebuilt to withstand them, either.

But the greatest and grimmest toll came from fires, fanned by 60-MPH (96.6 km) cyclonic winds that swept, uncontrolled, through the city. With its water mains split apart, the firefighters of Tokyo could only look on helplessly as huge sections of the city were eaten up by towering flames that spread with all the rapidity and power of a tsunami.

JAPAN
TOKYO AND YOKOHAMA
September 1–3, 1923

The massive earthquake of September 1, 1923, combined with two days of heavy winds, which in turn fanned out-of-control fires, killed 143,000 residents of Tokyo and Yokohama, injured 200,000, and rendered 500,000 homeless.

There was horror enough for a hundred earthquakes in the great Kwanto earthquake of September 1, 1923. Preceded by a rainless typhoon and followed by a tornado, this underwater quake that consumed Tokyo and Yokohama and its inhabitants in a fiery fury claimed a staggering 143,000 lives, injured another 200,000 and made a half-million people homeless.

The first tremors hit at noon, just as the charcoal burners were being lit for the midday meal—a circumstance that would add to the holocaust of flames that would rage unchecked through both cities for two full days.

Tokyo had already begun modern construction in 1923. Chief among these structures was the Imperial Hotel, designed by Frank Lloyd Wright, who had learned from the great San Francisco quake (see p. 98) and had specified that not only the hotel but the ornamental pool at its entrance be constructed of steel frames that would expand and contract, be sunk into bedrock, and be equipped with diagonal supports to protect the structure from an earthquake's lateral movements. His design was sound; the Imperial Hotel was one of the few structures to survive the quake, and its ornamental pool became the only source of water for the hard-pressed firefighters.

Fanned by high winds, the flames sent fiery fingers 20 and 30 feet (6.1 to 9.1 m) into the air. Nothing beneath them, buildings or people, survived. On the Tokyo waterfront, thousands leaped into the water to escape the flames. Clinging to the sides of boats and to floating pieces of docks ripped apart by the earthquake, they at least felt safe from the fires. But their safety was abruptly shattered when the gigantic Standard Oil Building and its oil storage tanks suddenly exploded with an ear-shattering roar, flinging 100,000 tons of oil into Yokohama Bay, where it immediately caught fire around the swimming survivors, immolating thousands of them.

Yokohama suffered the same fate. Within hours, there were no identifiable landmarks. The city was one vast plain of fire that roared on unchecked, since there were neither water supplies nor firemen to fight the flames.

In Yokohama harbor, the liner *The Empress of Australia* was preparing to set sail. The wharf collapsed around those seeing their friends off on the festive trip, but most of these well-wishers managed to scramble to safety. The ship, tossed about sickeningly by swells in the harbor, was rendered inoperable when its propeller tangled with the anchor chain of another nearby ship. It was fortunate that it could not set sail; over 2,000 frantic swimmers, escaping the fires ashore, were pulled to safety aboard the *Empress*.

Others near the Yokohama waterfront were less lucky. In one public park, hundreds slid gratefully into pools, hoping to shield themselves from the flames. But they soon found that they were only partially protected. Flying, flaming debris began to ignite their hair. Only by slapping wet mud on each other were some of them able to extinguish the flames. In nearby Yokohama Park, 24,000 people were surrounded by flames. When the fire closed in on them, some dove into the park's lagoon, which had already been vaporized by flaming bits of buildings, and those in the water were literally boiled to death.

In the surrounding countryside, more chaos accumulated. The earth actually capsized upon itself in some places. A 750-year-old bridge near the village of Chigasaki was first buried under tons of mud, then flipped, intact, to the surface.

The ground turned to liquid and swallowed huge trees, so that only their tops were visible. Potatoes were pulled from the ground and flung about like baseballs.

Landslides roared through the countryside. An entire forest detached itself from the upper slopes of Mount Tanzawa and slid at 60 MPH (96.5 km/h) down the slopes, consuming a village and a railway and flinging the entire mass into Sagami Bay, which turned the color of blood for miles around the site.

A passenger train that had stopped at the station at Nebukawa with 200 passengers aboard was hit broadside by another landslide that plowed the train and the entire village into Sagami Bay, killing everyone on the train and in its path.

At 4:00 P.M., when it seemed that no horror could exceed the present devastation, a new one occurred. A tornado appeared over the Sumida River, slamming into and crushing boats filled with wounded survivors of the fire. It then scooped up fireballs and flung them across the river and into a huge military clothing depot that had become a sanctuary for 40,000 survivors of the fire. In moments, this haven was turned into an inferno, burning to death all but a dozen of the 40,000 who were gathered there.

Hirohito, a regent at the time, shrugged off the traditional blame that would normally befall an emperor for a holocaust of these dimensions. (According to legend and religious lore, something of this vastness could only be caused by the sun goddess's displeasure with the reigning monarch.) Instead, he placed blame on Koreans and socialists who, according to his official dispatch, ". . . had offended the spirits before the earthquake and were taking advantage of the disaster by setting fires and pillaging shops."

Four thousand Koreans were publicly beheaded by the emperor's Black Dragon Society, adding still more casualties to the tragic event.

JAPAN
UNSEN
April 1, 1793

The island of Unsen and its 53,000 inhabitants disappeared beneath Japan's Satsuma Sea during the earthquake of April 1, 1793.

Although the size and effect of disasters sometimes become distorted with time, there seems no doubt that the statistics recorded by the series of earthquakes and volcanic eruptions that played back and forth across Japan between 1780 and 1800 remain staggering, almost beyond belief.

The worst disaster took place on April 1, 1793, when the island of Unsen and all of its 53,000 inhabitants were swallowed up by the earth, as an earthquake split the island apart while simultaneously igniting its volcano, which in turn blew asunder.

The fallout from this cataclysm formed dozens of small, new islands in the Satsuma Sea. Some said that the Asama volcano unleashed a lava stream 425 miles long and catapulted a steady arc of stone blocks, one of which was reported to be 42 feet (12.8 m) in diameter, into the sky.

MEXICO
September 29, 1973

Seven hundred residents of southeastern Mexico were killed and tens of thousands were made homeless by the earthquake of September 29, 1973.

Striking at 3:51 A.M. on the morning of September 29, 1973, a 6.5-force earthquake ripped apart 300 miles of southeastern Mexico, centering somewhere near the cities of Ciudad Serdan, Tehuacán, and Orizaba. Seven hundred people, most of them asleep in their beds, were killed by the sudden quake. Twenty-one other villages, towns, or cities in the provinces of Puebla and Veracruz, located in the Sierra Madre, were severely damaged by the tremors, leaving tens of thousands homeless.

Centuries-old churches were swallowed up by the huge fissures that opened in these villages. Particularly wide and deep crevices extended outward from Pico de Orizaba, the extinct volcano that rests in the midst of these provinces.

MEXICO
MEXICO CITY
September 18–19, 1985

Two stupendous earthquakes, the first registering 8.1, the second registering 7.5 on the Richter scale, tore through Mexico City on two successive days, September 18 and 19, 1985, killing 5,526, injuring 40,000, and leaving 31,000 citizens of that city homeless.

The epicenters of the two monster earthquakes that roared through Mexico City on September 18 and 19, 1985, were located on a front along which the Pacific Ocean floor drives under the Mexican coast. Pushing under the land at this point, the sea floor is dragged down to form the Middle America Trench, which parallels the coast offshore. According to officials analyzing it afterward, the relentless pressure along these two colliding plates "... has caused scores of earthquakes ... much like firecrackers on a string. It was," according to

Between 800 and 900 people are believed to have died in the collapse of the Nueva León building during the 1985 Mexico City earthquake. (U.N. photo/Jean-Claude Constant)

one seismologist, "just a matter of time before the next one struck."

That next time fell on the morning of September 18, while many citizens were either at work, at school, or watching television, which at first flickered, then, as live announcers expressed disbelief and finally panic, went blank. Hundreds of buildings in Mexico City collapsed immediately. Others continued to crumble during the day as their weakened structures slowly gave way.

Many residents felt as if the buildings in which they found themselves were swinging back and forth, then heaving up and down. Furniture shot across rooms, windows shattered, refrigerators toppled, books and pictures went flying, doors and shutters flapped crazily on their hinges. For some, it was as if they were at sea in a terrible storm.

Others said they heard walls cracking with a powerful, horrifying force of a gun shot.

The 12-story Hotel Principiado collapsed, burying its 140 guests. Only 39 survived. Witnesses told of schoolchildren standing on street corners sobbing for their parents as the city rumbled beneath their feet and its customary smog was slowly replaced by clouds of dust and smoke rising from the rubble. On the Avenida Juarez, rescue workers, troops, and citizens used any available tool to dig out survivors.

In the city's suburbs, similar tragedies occurred. In Guzmán, in Jalisco state, the cathedral collapsed during morning Mass, killing at least 26 people.

Then, at 7:37 the next night, just as rescue work had begun to reach an organized state, the second quake struck, collapsing buildings, sometimes on top of rescue crews. The Hotel del Carlo resembled a disassembling accordion as it cracked and then collapsed upon itself.

There were multiple, individual scenes of suffering. Arturo Lara Rivas wandered through streets looking for his family. As he picked his way past broken windows and fallen bricks, he occasionally whispered words of comfort to his canary, which he carried in its dented cage.

Mrs. María de los Angeles Lara Rivera talked expressively to reporters. "Oh, I was scared," she said to anyone who would listen. "It swayed and swayed and then people shouted, 'Run, run, the building will fall.' But oh, God, we didn't know where to run." As she spoke to newsmen, the Valentin Zamora Orozco public school collapsed behind her in a noisy geyser of green brick and broken glass.

Jose Costello, resting on a bench in a government shelter, his fedora still powdered with dust, said he had clung to a bedroom wall "like a drunk." His sister hung onto a doorknob and a chair. "When the thunder came," he said to reporters, "we scrambled out with only the clothes on our backs and our white dove, Linda."

Temporary morgues were set up at government offices in the 18 districts of the capital, and when they overflowed, bodies were brought to the city baseball stadium, where they were laid end to end on the playing field. Workers and pedestrians in the street donned blue surgical masks or held scarves or handkerchiefs to their faces as a defense against the rising odor of death.

As the rescue attempts began to reorganize themselves, it became apparent that the quake had limited itself to a small geographical area of the city. But, tragically, that particular area was also the most thickly populated one. The six square miles of absolute destruction were centered along Avenida Juarez, the Plaza de la Republica, and the neighborhoods off the southern part of the Avenida Insurgentes. The giant Nueva León apartment building in Tiateloico had once housed 200 families. It had disappeared. The nearby General Hospital collapsed, burying 600 patients and staff members.

Specially trained dogs and relief experts from France, Switzerland, West Germany, and the United States flew into Mexico City and immediately went to work seeking signs of life under tons of wreckage. When the dogs barked, indicating the possible presence of a body, rescue workers rushed to the site to begin digging.

In one international incident, on Sunday morning, September 22, two days after the last quake, Mexican soldiers told French rescuers they were moving too slowly. The Mexicans proposed, none too politely, that they were about to tear holes in a main wall with heavy cranes. The French answered that it would collapse their tunnel and kill two survivors who were in it. The argument rapidly escalated into a shouting match, and the French, their pride injured, walked away, only to be entreated by the survivors' relatives to resume their work. A truce was arranged, and the Mexican soldiers dug elsewhere.

Another incident involved an infant who had been born just before the first quake. A premature child, she had been placed in an incubator. Just as the door on the incubator was shut, the hospital floors above it tumbled down, burying child and incubator under tons of steel and concrete.

Fifty-five hours later, rescuers reached the site. A steel beam had wedged itself above the incubator, protecting it from the cascading concrete, and the baby was alive and unharmed, safe within the environment of its only slightly dented incubator.

MOROCCO
AGADIR
February 29, 1960

An earthquake and tsunami devastated the port city of Agadir, Morocco, on February 29, 1960, destroying 70 percent of its buildings and killing 12,000 of its residents.

At 11:45 P.M. on February 29, 1960, the third day of the Muslim observance of Ramadan, one of the most destructive earthquakes of modern times swept through the international resort city of Agadir, in French Morocco. A few seconds later, after the earth ceased to tremble and the giant tsunami had receded, 70 percent of this once shining city would lie in ruins, and 12,000 residents would lie crushed to death beneath the remains of the grandly designed buildings.

Some earthquakes strike without warning; others announce their imminence. In this case, a slight tremor ran throughout the city on February 28, 24 hours before the major jolt. It was a warning, heeded by most, misinterpreted by others as a loud knock at the door or the backfiring of a car.

Two major shock waves passed through the city the next night. The first, at 10:50, was severe enough to cause some consternation. It was the third tremor that sealed the fate of thousands. It struck with the force of a cosmic fist, which instantaneously clawed the earth apart a full four feet, and then, within six seconds, slammed it back together again. Luxury hotels packed with American and European tourists, apartment buildings, office buildings, markets, the Casbah that had stood in squalor for centuries on the side of a hill all toppled, then disintegrated.

Simultaneously, water mains burst, sending up geysers of water that might have been used to extinguish the hundreds of fires that ignited. They would never be put out, since every fire station was destroyed, and most firemen were dead. Sewers exploded, releasing tens of thousands of marauding rats.

The Mediterranean, its floor buckled, gathered itself into a tsunami that rose, broke, and rushed inland for 300 yards, drowning everything in its path.

In the harbor, boats bobbed like toy vessels, slamming into each other. Steam rose in gusts as underwater crevices opened and closed. A pilot flying above the city later commented that "it looked like a giant foot had stepped on the city and squashed it flat."

There was no possibility of rescue and relief operations from within the city. By dawn, French army personnel and sailors from the American Sixth Fleet anchored nearby descended upon the city, shooting rats that were already spreading plague among the survivors and taking part in a gigantic, often touchingly successful digging out of survivors who cried out from beneath the mountains of wreckage.

One rescue had a storybook quality about it. Mrs. Margaret Sue Martin, the wife of a U.S. Navy lieutenant, would lie pinned in the wreckage of the Saada Hotel for 40 hours. A French lieutenant, digging close to her, buoyed her spirits by noting her beauty. "How do you know that when all you can see is my feet?" Mrs. Martin is reported to have asked the lieutenant.

With classic French gallantry, he answered, "To have your courage, you *must* be beautiful."

It would take hours of painstaking digging to release her finally, long after a doctor examining her legs had predicted that she would be dead within an hour.

But she survived, remembering, for the press, that ". . . at the top of the hole I felt the sunlight on my face. People were crowding around. They were all cheering. They seemed to be crying. I think I was crying, too, with sheer happiness."

NEPAL AND INDIA
August 21, 1988

Northern India and the adjoining district of Nepal were raked by a 6.5 earthquake on August 21, 1988. Impassable terrain made rescue and the tallying of casualty figures difficult, if not impossible. Between 1,000 and 1,500 people were thought to be killed, tens of thousands injured, and hundreds of thousands rendered homeless.

A 6.5 earthquake whose epicenter was located east of Darbhanga in northern India wreaked its greatest havoc across the border in neighboring Nepal. The worst quake in 54 years in either India's Bihar state or Nepal, it claimed over 700 deaths in Nepal and between 300 and 800 in India. Tens of thousands were injured and hundreds of thousands were made homeless in both countries.

Precise fatality figures were difficult to determine because of the area in which the quake hit.

The causes of the large loss of life were multiple. The third straight year of monsoon rains, which flooded India's Darbhanga district and washed away the summer rice crop, weakened buildings in cities and villages and accounted for the instantaneous collapse of thousands of structures in the city of Darbhanga. This, plus the fact that the first tremors were felt shortly before dawn, when most families were indoors, accounted in part for the immense casualty figures in his fairly unpopulated part of the world.

"It was a rumbling sound that seemed evil, demonic," was the way Ishrar Jha, an automobile mechanic living in the Indian city of Patna described the coming of the quake. He and others said that tens of thousands of people rushed into the streets and fields of this sprawling city when the first tremors occurred. "It was like the end of the Earth," remarked Beshan Jha, a farmer who brought his badly hurt wife and young son to the Darbhanga Medical College Hospital, which was so overcrowded that mattresses had to be spread on its floors to accommodate the overflow.

In Nepal, one of the world's least developed countries whose population is strung along the foothills of the Himalayas, the death toll was higher, and the damage more severe. Three days after the quake, 700 bodies had been dug from the rubble of villages in the quake zone. Thousands more were injured and far more were believed to be buried under the numerous landslides the earthquake had spawned. Unlike India, which has a widespread irrigation system based on ground water, Nepal is primarily dependent upon rain-fed irrigation for its farms, and these irrigation ditches added to the instability of the hillsides near many of the villages.

Nearly 18,000 buildings were either destroyed or severely damaged, and tens of millions of dollars in property loss was estimated. Over $1 million in funds was immediately funneled into the stricken country from the United States, the European community, Australia, and Britain. The United States flew in 650,000 square feet of heavy duty plastic sheeting to be used to roof temporary shelters, since most families were afraid to reenter their former homes. "This heavy rain may be the next one, and it will all fall in," commented one farmer in Nepal, as his family huddled under a makeshift tent that was being pelted by yet another fierce rainstorm.

NICARAGUA
MANAGUA
December 21, 1972

Seventy-five percent of Managua, Nicaragua, was destroyed on December 21–22, 1972, by earthquakes that killed 7,000 and made 200,000 homeless.

The first of a series of quakes struck the capital city of Managua, Nicaragua, at 11:10 P.M. on December 21, 1972. Ten minutes later, a stronger shock hit, produc-

ing consternation but nothing more. At 1:30 A.M. on December 22, the third, heaviest, most terrifying tremors rumbled through the city, hitting 6.25 on the Richter scale and triggering an entire series of aftershocks that would tumble buildings as if they were children's blocks.

Managua had been destroyed twice before by earthquakes and fires, once in 1885 and again in 1931. This time, the city would again be leveled and turned into a roaring inferno that would kill 7,000 of its people and render 200,000 homeless. Seventy-five percent of its buildings would be destroyed, another 15 percent to 20 percent would be left uninhabitable. Its population of 325,000 was diminished to 118,000.

The first tremors on December 22 took their toll. Collapsing buildings were accompanied by explosions, the rending apart of the earth into huge fissures, the release of sulphurous fumes, and multiple fires which could not be put out because of the mass rupture of the city's water mains.

Survivors fled in panic, clogging the roads out of the city in the eerie predawn darkness. "Hundreds of mutilated bodies were strewn along the streets, some still wrapped in bedsheets, some missing heads, hands or feet," said one witness who was fleeing the city.

All communication with the outside world was cut off when transmitting towers toppled. One man, however, known only as "Enrique," managed to transmit news of the catastrophe over his ham radio. For hours, he was the only link between Managua and the rest of humanity. "People run through the streets like zombies, with terror," he said at one point. "Big buildings are cracked. There is blood on people's faces, legs, arms as they leave their houses. We have never seen a catastrophe like this."

Fires burned through the night. From the air, downtown Managua looked as if it had been under incendiary attack. At dawn, the commander of the Army Corps of Engineers, Lieutenant Colonel Jose Alagret, led his men into the city to control rampant looting and to aid in rescue attempts. He was appalled. "This is a city that was but is no more," he said, simply but eloquently.

The only solution would be to destroy what was precariously standing and begin again. Some survivors remained in the city, picking through the smoking ruins. In an attempt to clear the city of survivors before the decaying bodies beneath the rubble created an epidemic, the Nicaraguan government cut off food supplies, but distributed bottled water and drinks free.

An American demolition team stood by, ready to blow up the heavily damaged United States Embassy, among other buildings. The fire department declared part of the city a "contaminated area" and proceeded to level it, covering it with lime to serve as a mass grave for the unknown number of people who had died and were buried beneath the ruins.

Looting continued, then grew, as troops abandoned any attempt to contain it. Hundreds of people raided a market center and its warehouses, carrying away everything transportable, including washing machines and clothing.

Outside of the city, 2,000 people were buried in a mass grave. Other bodies were burned where they were found in the city.

The United States sent $3 million in aid, and transport planes began to arrive with food and supplies from all over the world.

There was almost no damage a mere few miles from Managua, thus giving credence to the repeated warnings of geologists, who had stated over and over that the city was built upon a fault and that, had it stood even 25 miles (40.2 km) from its present location, this and the previous two tragedies might never have occurred.

The lesson would not, however, be learned. The city would be rebuilt on the same fault and suffer more destructive earthquakes.

PAKISTAN
BALUCHISTAN PROVINCE
February 27, 1997

A major 7.3-force earthquake struck Baluchistan province in northwestern Pakistan on February 27, 1997. Mountain villages were totally destroyed. Over 100 residents were killed, all of them in remote locations near the Afghan border.

Almost simultaneously with an earthquake in Iran (see p. 63), a major, 7.3 temblor struck Baluchistan province in northwestern Pakistan on February 27, 1997. Quetta, the provincial capital, was shaken, but the greatest damage occurred in villages in mountainous terrain near the capital.

"I thought doomsday had arrived and the world was about to end," said a teacher, Feroza Begum, in Quetta. The greatest damage and the most casualties occurred in the district of Sibi, where small villages composed of sunbaked mud dwellings were totally destroyed. Seventy-five people, a large percentage of the population, died in Harnai, a village 30 miles (48.3 km) from the epicenter of the quake.

Mountain roads, normally a challenge, became impassable for rescue vehicles. The main road and the rail line to Quetta collapsed. It would be long hours before rescue teams arrived, and even then, there was no heavy equipment to dig out the dead and wounded from the rubble of flattened houses. The ultimate total of those killed by the quake rose to over 100; hundreds more were injured, and scores were unaccounted for.

PAKISTAN, NORTHERN
KASHMIR
October 8, 2005

A 7.6 magnitude earthquake roared through Kashmir, part of both Pakistan and India, on October 8, 2005, killing 75,000, injuring 130,000, and rendering 3.5 million people homeless, in the greatest disaster to strike the region in recorded history.

"I thought it was doomsday, that the earth would open and swallow me up. The houses on the ridge—they were exploding, one by one." So related Ihsanullah Khan, a former Washington, D.C. cab driver who had become mayor of his hometown of Batagram on October 5, 2005. Three days later, at 8:50 A.M. Pakistan Standard Time (9:20 A.M. India Standard Time), a 7.6 magnitude earthquake, the worst natural disaster in Pakistan's history, smashed his village and dozens more, into oblivion, killing 75,000 residents of this densely mountainous region of the Himalayas and injuring 130,000 more (see color insert on p. C-5). For two years, 3.5 million people were displaced in an area the size of Maryland.

Ironically, the earthquake was triggered by the same forces that had created the Himalayas, the mountain range in which this quake occurred. In Pakistan, the country worst hit by the disaster, the Indian continental plate to the south has for hundreds of years tried to subduct, or dive beneath, the Eurasian plate to the north, without success. This ongoing collision has forced the Earth's crust to buckle, producing the Himalaya, Karakorum, Pamirs, and Hindu Kush mountain ranges—a long range and fairly benign process.

However, the compression of these two plates also creates a sinuous array of smaller faults in the upper layers of the Earth's crust. Movements in these shallow faults (known as thrust faults) are responsible for devastating earthquakes, the worst among them the October 8 cataclysm, which originated only about 6.2 miles (10 km) deep. Because it was so shallow, the shaking forces were much greater than similar magnitude earthquakes that occur deeper in the crust.

There were other, human-made influences that conspired to compound the force of the earthquake and its consequent vast human tragedy. To build on the Himalaya terrain, villagers dug into the mountainous territory and piled the dug-up dirt and rocks below to create a flat spot. "What that does," said Wayne Pennington, a geologist at Michigan Technological University in Houghton, "is [to] steepen the mountain even more than it originally was. A little shake and the uphill side of the road will likely have rocks that come loose and fall, and the downhill side just slides away completely."

And this is precisely what happened to the villages and dwellings that clung to the side of the Himalayas on that fateful Sunday morning. The earth came apart, swallowing entire villages and great portions of their populations, and buried more residents in the rubble of collapsing buildings. In addition, this ripping apart of the earth triggered scores of landslides that buried villages and buildings and people even more deeply. The once boisterous town of Muzaffarabad, Pakistan, with its 150,000 residents, resembled a refuse-strewn plain, masking its role as a gigantic graveyard containing 12,000 of its residents. In India, farther along the fault, it was the same story. And in Kashmir, the transcendentally beautiful land of rivers, lakes, and valleys that is controlled, under continuing disputation, by both Pakistan and India, the devastation was at its worst (see color insert on p. C-5).

The Kashmir location added a further political aspect to the tragedy. Kashmir is the site of the world's largest and most militarized territorial dispute. India, Pakistan, and China all have staked claims to parts of the territory, and the region is bisected by the Line of Control that separates Indian and Pakistani forces, who have operated under a cease-fire since 2003.

So, there was already ample presence of both Pakistani and Indian military units in the area when the natural disaster struck. But the first five days after the original quake were marked by 900 aftershocks, some of which were six points on the Richter scale in severity. The injured, dying, and homeless had to fend for themselves, as parts of the mountains continued to slide away (see color insert on p. C-5). Helicopters, some from U.S. Army bases in Afghanistan, some from India, and some from Pakistan, flew over the wreckage. But in the Indian Kashmirian mountain village of Skee, residents received no drops of food, water, or medicine, despite the fact that the village overlooked a base for thousands of Indian troops. All too often, eyewitnesses in both countries told reporters, military troops took care of their own casualties before climbing over the rubble and into the towns and villages.

In Balakot, Pakistan, a town of 20,000 people was reduced to a mud hole. It took the Pakistani army three days to arrive, even though its base was only 20 miles

(32.2 km) away. When the troops finally converged upon a collapsed school building to help dig out some 200 students trapped inside, enraged parents hurled stones at the soldiers.

Damage to schools was extensive and terrible. Since the quake struck while children were in their morning classes, many in shabbily built schools that crumbled under the first shock wave, thousands were crushed within their schools. Four days after the quake, a teacher named Said Rasool traveled down from his village to seek help in Balakot. His clothes were still covered in the blood of his dead students, and he wandered from one cluster of soldiers to another, pleading for help to try to dig out the children. The reply was the same from each cluster. There was too much work still to be done in Balakot for the soldiers to climb up into the mountains and begin rescue work there.

As helicopters touched down in wrecked mountain hamlets, survivors mobbed the crews and fought one another for blankets and biscuits. Some Pakistani officials reported that several times, stranded earthquake victims clung to a chopper as it lifted off, nearly causing it to crash.

Desperation was at the heart of the chaos. Roads that had once linked small mountain villages were atomized, and confused refugees found themselves trapped within the collapsed buildings of their hamlets. Giving up on the armies, thousands of volunteers headed into the mountains, carrying shovels, pickaxes and iron rods to dig for survivors. Others dug with only their hands.

Rain set in, making the lot of survivors even more miserable. It would be several days before tents were delivered by the tens of thousands of rescuers who finally descended upon the area. What they found was sometimes terrifying. The dead and dying were everywhere. Rescue workers found the bodies of 60 road workers in a bus crushed by a landslide on a highway. Hundreds of the most badly injured were flown by helicopter to Islamabad. And as the days went by, the rain increased, grounding rescue teams from Jordan, Malaysia, and Russia.

The United Nations appealed for $550 million in humanitarian aid before the winter set in, but the money was slow in coming because, as some governments avowed, there was "donor fatigue" from the massive amounts needed to help the victims and destroyed infrastructures left by Hurricane Katrina (see p. 290) and the Asia tsunami (see p. 58), both of which had occurred only a few months prior.

Medical care was a continuing problem, as more of the injured were freed from the rubble and there was only a 50 percent chance of receiving institutional care. Half of the 564 hospitals in the area had been either damaged or destroyed. Dr. Ali Shehada, who treated Hurricane Katrina victims in Houston (see p. 290), told reporters for the *New York Times* that he was overwhelmed. "What we're seeing [in this earthquake] is that it's like you've got a bleeding wound and you're putting a band aid on it," he said. In Islamabad, in the first two weeks after the quake struck, and where the most seriously injured were taken, doctors performed 165 amputations.

There was a race against time, as December 1 approached, when winter would prevent helicopter flights. Tent cities appeared, and makeshift shelters of plastic and wood were built for families who huddled over meager supplies of rice and turnips. Conflicts erupted over whether families should be evacuated or remain in the place they knew and trusted. Aid workers ferried everything from tarps and corrugated tin to hammers and nails to help people cobble together the remains of their houses from the rubble.

The winter that set in was, fortunately, a mild one, and when the spring of 2006 arrived, the United Nations launched a recovery plan, setting up pre-fabricated basic health units and more than 32 schools. Monsoon rains in the summer of 2006 destabilized rebuilt villages and washed away agricultural areas. By the winter of 2007, approximately 100,000 people were still living in camps.

During all of this, Ihsanullah Khan, the former taxi driver turned mayor, spearheaded a home-supported rebuilding and relief program. Using $200,000 of his own money, he bought all the medicine and bandages he could find, set up tent hospitals, and arranged makeshift ambulances to ferry the injured over the mountains. He established a fund to help villagers rebuild their own demolished homes. And he fought Pakistan's bureaucracy to send in bulldozers and start clearing out fallen buildings. Interviewed by Western journalists, he said he was convinced that God had something special in mind for him. "I just didn't know what it was until the earthquake happened," he concluded.

PAPUA NEW GUINEA
NORTHERN COAST
July 17, 1998

Two thousand one hundred and eighty-three inhabitants of the north coast of Papua New Guinea were either killed or drowned by an offshore earthquake that spawned three giant tsunamis on July 17, 1998. Five hundred were never found, and 9,500 were made homeless.

On Friday night, July 17, 1998, in the Pacific Ocean just off the north coast of Papua New Guinea, and almost facing the village of Sissano, a 7.1-force earthquake ripped apart the ocean floor. At the instant of occurrence, some rumbling and pitching was felt on the island; some of the thatched and homemade huts groaned and leaned. Lagoons grew waves; the earth split in various places, there were some minor mud slides. Though terrifying, these land manifestations were manageable.

But the worst fury of the earthquake had just begun. An underwater landslides was touched off by the quake at the point at which the sea floor fell off sharply from the northern beaches of Papua New Guinea. More than two cubic miles of material was suddenly displaced, and the energy of this roiled the Pacific into three enormous waves. The first, over 40 feet (12.2 m) high and stretching back two and a half miles from its crest, rushed forward, breaking on the beach and surging into the villages that were arranged along the coast in a nearly unbroken string. Within seconds of the first wave's breaking, two more tsunamis, each over 20 feet (6.1 m) high, slammed into the beach, piling water and force upon the first wave's destructive swath, and forcing the reinforced wall of water higher and farther inland.

The multiple tsunamis destroyed everything in their paths, uprooting palm trees and turning them into lethal lances, rushing against buildings and smashing them to pieces, surrounding people and animals, floating them on the churning waters until they were smashed into trees or torn apart by the wreckage that spun around them. Those who survived were swept violently out to sea by the rush of backwaters that followed the inland path of the three monster waves.

And then it was over. The normally tranquil island, devoted to fishing and agriculture, located 375 miles (603.5 km) north of Australia's northeast tip, was momentarily silent again. "The waves went slowly back into the sea and it looked like nothing happened," said Fabian Nakisony of Warapu, a village where nearly all the homes were destroyed. "The beach was new sand. Except no houses."

There would be an aftershock, 20 minutes after the original quake, but it would produce no more tsunamis. The first three caused enough devastation.

Lucien Romme was standing in his village near Sissano lagoon when he felt the earth lurch and then saw "the sea rising up and coming toward me."

He turned and ran as fast as he could, shouting to the others who lived in the small village of Arop. He looked frantically for his wife and six-year-old daughter, but suddenly the wave hit him in the back and hurled him forward, into the lagoon. Debris slammed into him, breaking two of his fingers and one of his ribs.

But he survived, holding onto a coconut tree when the backsweep began. "There was nothing left except for coconut trees. No houses. No houses at all," he recalled later to a reporter. After the interview, he received word that his daughter was safe, but his wife had drowned.

"We heard a large bang, then saw the sea rising up," Paul Saroya, who lived in Nimas Village, told Australian television. "We had no choice but to run for our lives."

"The wave crashed into the house," Raymond Nimis, another survivor said. "People were dying everywhere. Some died under the house. Others got rolled in by the wave."

Fabian Nakisony, who lost his year-old son, and who himself was immobilized when a mangrove root pierced his leg, shook his head sadly. "We have nothing," he said. "We can get timber from the bush, but we have no hammer, no nails, no saw."

The lagoons and mangrove swamps were rife with floating bodies, dead fish, and the debris of coconut palms and houses. "We are finding more dead bodies now than yesterday—every minute and every hour we find more dead bodies," Bill Skate, the prime minister of Papua New Guinea stated. "I will always remember this as long as I live."

"There were so many bodies together I had to move the boat slowly to pass through them," Jerry Apuan, a fisherman, added. "I was afraid. It was the first time I had seen so many bodies."

These bodies appeared in such profusion and in such a state of advanced decay that they could not be given proper funerals. On one beach, soldiers, missionaries, aid workers, and the healthiest survivors worked together, using spades to dig shallow graves, wrap the bodies in black plastic, then stamp the dirt down on the grave. Some of the bodies were in such poor condition, they were doused with gasoline and cremated. The thousands displaced by the tragedy were warned to stay away from the beaches. Officials sent hunting parties out to shoot crocodiles, dogs, and pigs that were eating bodies that had not yet been reclaimed.

The most heartrending aspect of the disaster was the age of the victims. Most of the dead were children. "What chance would a two-year-old or a three-year-old child have?" asked Father Austin Crapp, the administrator of the Catholic Diocese of Altape. "It wipes out everything, destroys everything, bounces people off trees, bowls them into the lagoon. The children may be hiding somewhere. We hope so. But the current fear is that they have drowned."

Father Crapp's fears were realized. Seventy percent of the survivors were adults. Most of the 2,183 dead and 500 missing were children.

Australia and New Zealand sent 6 C-130 Hercules transport planes full of personnel, food, and building supplies to the stricken island. One hundred relief workers, including doctors and nurses, set up a field hospital in Vanimo, about 50 miles (15.2 km) away from the earthquake site, to protect it from further aftershocks. Many survivors underwent amputations because bacteria-filled coral sand had infected wounds, causing gangrene. Over a hundred died in the hospital of their injuries.

Elsewhere on the island, several helicopters hired by local mining companies picked up the wounded and ferried them to the hospital. Seventy grave diggers were airlifted to the Sissano lagoon, but they found that many of the bodies were too decomposed to be buried without machinery.

"For three nights I was crying for my lost wife and three daughters," said Fabian Tombre, one of the 9,500 homeless on Papua New Guinea after the final count had been made. "The people will go back, but to a better place," he added. "We will build new homes away from the sea."

PERU
YUNGAY
May 31, 1970

Only 2,500 of the 20,000 residents of Yungay, Peru, survived the 7.75 earthquake that devastated much of that country on May 31, 1970. A total of 70,000 were killed, 50,000 were injured, and 200,000 were rendered homeless.

An awesome total of 70,000 people died in the cataclysmic 7.75-force earthquake that struck Peru at 3:24 P.M. on May 31, 1970. Fifty thousand people were injured, and 200,000 were made homeless. Government figures released immediately after the quake listed a mere 200 dead "to prevent panic" as they later explained it. But just 24 hours later, that figure had increased 150 times, and reports of horrendous devastation began to filter through to the outside world.

There were reasons for the inaccuracy of casualty figures. First, a combination of fog and cold would hamper rescue efforts for days. Also, much of the area affected was in the high Andes and therefore difficult to reach. And although the epicenter was located in the Pacific Ocean 210 miles (338 km) northwest of Lima,

Stone houses decimated in the Yungay, Peru, earthquake of May 31, 1970 (CARE)

much of the damage was outside the capital city. In fact, the quake was felt as far south as the village of Nazca, about 300 miles (482.8 km) south of Lima, as far north as the southern part of Ecuador and as far east as the Amazon jungle city of Iquitos, where some damage was reported.

The immediate damage visible to observers who flew over Peru were scores of raging rivers that had not been there before. Landslides and burst dams turned formerly habitable valleys into river basins in a matter of minutes.

The city of Chimbote, a naturally protected harbor and the object of intense investment in the two years prior to the earthquake, was located a mere 12 miles (19.3 km) away from the quake's epicenter and suffered extreme and dramatic damage. Three-fourths of its flimsy housing was totally destroyed; over 200

people lost their lives. Nearby, the 10,000-foot-high (3,048-m-high) resort city of Huaraz—in which 6,000 people were killed in an avalanche in 1941—was totally leveled. Iquitos, with a population of 100,000, located in the middle of an area that produces gold, oil, iron, rubber, quinine, and palm oil, was devastated.

But it was in the city of Yungay, located in the valley of Huaylas, that the most dramatic disaster took place. Only 2,500 of its 20,000 inhabitants survived the quake. Roaring flood waters, gigantic fissures in the ground that swallowed buildings and human beings alike, avalanches of rocks, ice, and snow all conspired to virtually wipe out the city. The final blow was a wall of icy water that roared into the city from lakes high in the Andes. Dams had burst and lake shores had folded upward, spilling these lakes beyond their banks and into the nearest valleys.

A young boy surveys the devastation left by the earthquake in Yungay, Peru, on May 31, 1970 (CARE)

The terrified survivors—all 2,500 of them—scrambled upward to a hillside cemetery, where they clung to the sides of the Andes and waited for help from the circling aircraft and helicopters that soon began to arrive from the United States and neighboring countries.

From the beginning, however, these paratroops and helicopter units were hampered by the seasonal fog that rolls in off the cold waters of the Pacific and becomes trapped on the eastern slopes of the Andes by the warm moist air from the Amazon. Pilots reported fog rising to 18,000 feet (5,486.4 m) over parts of the affected region, and this made the dropping of supplies by parachute impossible. Thus the misery of the 200,000 made homeless by the quake intensified as the cold deepened and food and water began to run out.

Three days after the quake, the fog lifted, and supplies and rescue teams began to parachute in, meeting those on the ground who were frantically opening blocked roads in order to get medical supplies to the injured.

More and more died as too little effort and too few supplies finally determined the ending to the story. In Huaraz, a month later, there were still too few tents to protect homeless survivors from the cold. Two hundred thousand people in the area were without any kind of shelter, and many were forced to subsist on contaminated water and very little food.

According to one observer, the bulldozers that were seen in a few towns clearing away mountains of adobe rubble looked as if they were tidying up a cemetery rather than rebuilding a community.

Most cynically, the tragedy ultimately turned into a political parade ground. Lieutenant General Juan Velasco Alvarado, mindful of the fact that Peru's treasury was nearly empty, hoped that relief moneys would go toward replenishing it, and the magnitude of the disaster would unite the people of Peru behind, as his government put it, ". . . the programs of their revolutionary government."

Internationally, foreign governments went to extraordinary lengths to obtain maximum publicity for their donations to Peru. Premier Fidel Castro of Cuba personally donated a pint of blood for transfusion into an injured Peruvian child. Mrs. Richard Nixon toured Huaraz and other devastated towns as a compassionate representative of the United States government.

Both gestures were dutifully recorded by an army of television crews and newspaper reporters.

In Huaraz, ragged survivors were heard to grumble that the immaculately uniformed Peruvian generals who accompanied Mrs. Nixon had until then never set foot in the quake zone.

PHILIPPINES
LUZON
July 16, 1990

One thousand six hundred and fifty people lost their lives in an earthquake measuring 7.7 on the Richter scale, which struck the main island of Luzon in the Philippines on the afternoon of July 16, 1990.

Cabanatuan, the provincial capital of Nueva Ecija Province, which is located 55 miles (88.5 km) north of Manila on the main island of Luzon in the Philippines, boasted a major educational institution in 1990. Called the Christian College of the Philippines, its six-story structure housed classes that ranged from elementary school through college, and it was the pride of that part of the world.

At 4:20 P.M. on July 16, classes were in session, as were the many businesses nearby. At precisely that hour, faint tremors rippled through the city and the college building, then increased. Within seconds, a major earthquake that measured 7.7 on the Richter scale was in full progress, opening fissures in streets, precipitating landslides in the nearby hills, and collapsing buildings—among them the six-story Christian College of the Philippines.

Two hundred and fifty teachers and students, trapped in the collapsing building, panicked, as walls caved in on them and upper stories tumbled downward, crushing the floors beneath them.

When the 45 seconds of the earthquake ended, the building's six stories had telescoped into one flattened story, filled with debris, dead children and adults, and screaming survivors. "I could hear the boy next to me muttering the 'Our Father,'" said 14-year-old survivor Fernando Memphis to a television reporter later. "He was gasping for breath, and he could barely speak."

Elsewhere in the area, chaos erupted. In Manila, occupiers of schools, shopping malls, and movie houses stampeded, and a college professor died of a heart attack in one of these near riots. Patients scurried out of hospitals carrying bottles of intravenous fluid.

Makati, Manila's financial district, emptied itself of hundreds of workers, as tall buildings swayed and cars bounced out of control on nearby highways.

In Baguio, a mountain resort north of the epicenter, 10 people died when the main market collapsed. Nearly 1,000 tourists and workers were trapped in the rubble of four luxury hotels that collapsed around them. The Hyatt Terrace and the Nevada Inn, which were hosting an international conference, were nearly entirely destroyed, if not by the initial quake, by the four aftershocks that followed within minutes of the first one.

On the outskirts of Baguio, in the factories of the export-processing zone, hundreds of workers were trapped. Forty people were killed in the collapse of a gold and copper mine in the neighboring town of Tuba. The airport control tower of Baguio was destroyed, and the airport's runways buckled, rendering them useless; power lines were down; telephone communication and the radio stations were silenced.

And now the rescues began. Junjun Merosa, a 19-year-old vendor in the market at Cabanatuan, was the first to pull survivors from the wreckage of the Christian College of the Philippines. He personally retrieved 25 bodies and 12 living people by dousing them with motor oil and pulling them gently from the collapsed building with ropes.

Roads to Baguio had been rendered impassable, leaving the town's surviving inhabitants without drinking water, fresh food, and electricity. Frightened of entering the remaining buildings, they moved into tents to sleep and live until rescue teams could fly in by helicopter.

U.S. Marines arrived from the Subic Bay naval base. On July 19, an OV-2 observation plane running rescue flights crashed into a mountain near Baguio, killing its entire crew.

The rescue effort continued for several weeks, while a chorus of discontent slowly rose. Most of the criticism was leveled at President Corazon Aquino, who was accused of moving too slowly in organizing relief efforts. "Let us not blame others," she said in a tele-vised address, "What we need to do now is help each other."

President Aquino asked lawmakers in Manila for a half-billion dollars for earthquake recovery. As the recovery continued, miraculously, 11 days after the quake, two people were rescued from the wreckage of the Hyatt Hotel. Miners working near the wreckage heard voices, and dug out two hotel workers, Arnel Calabia and Luisa Mallorca, from a room that was protected by supporting but collapsed beams. The bodies of four other employees were discovered near the two survivors. Mr. Calabia said that he, Miss Mallorca, and a male employee had been on the third floor of the hotel when the quake struck. All three dove under tables before the ceiling caved in on them. "We recovered consciousness later," he told reporters, "and we called out to each other." The third person, a man, died from his injuries on the seventh day; the other two spent the rest of the time until their rescue praying and talking.

Two days later, the final survivor of the catastrophe, a 27-year-old employee, was discovered, dehydrated and near death, in another part of the wreckage of the Hyatt Hotel. No more rescues were possible; over 1,650 people had perished in this, one of the Philippines' worst recorded earthquakes.

PHILIPPINES
MANILA
July 3, 1863

One thousand people were killed and Manila was almost destroyed by a huge earthquake that struck that city on July 3, 1863.

Manila was nearly leveled by an enormous earthquake that struck at 7:30 P.M. on July 3, 1863. Only a few of the historic and magnificently carved churches and government buildings were left standing, but severely damaged. One thousand people were crushed to death, buried under tons of stone and rafters that trapped worshipers at vespers and shoppers in the main market.

Two shocks, each of a minute's duration, toppled the Binondo Cathedral and the huge Church of St. Domingo. Three Dominican convents collapsed, burying scores of nuns and friars, most of whom did not survive. At the Binondo Cathedral, seven priests remained unscathed by huddling underneath a gothic arch, which held; hundreds of others, worshiping beneath its dome, were crushed when the dome fell.

The nearby Pampasinga River overflowed its banks, causing huge mudslides that carried away warehouses, buildings, and $2 million worth of tobacco, the mainstay of Manila's economy. The nearby volcanos of Taal, Albay, and Arayat were nudged into borderline activity, but fortunately soon quieted.

The cataclysm of the earthquake was missed by Manila's governor, who was, at the moment of its arrival, riding with his son in the hills. His residence collapsed in his absence, narrowly missing his wife and his daughter, who, hours later, greeted the returning riders with plates laden with their dinners. The governor complained. His food, he said, was unpardonably cold.

PHILIPPINES
MINDANAO
August 17, 1976

A huge earthquake struck the island of Mindanao in the Philippines on August 17, 1976, killing 5,000 and making 150,000 homeless.

Five thousand people were killed, 3,000 were reported missing, and 150,000 were rendered homeless when a huge, 7.8 earthquake struck the Philippines at dawn on August 17, 1976. Mindanao, the largest island in the southern Philippines, was hardest hit, sustaining huge damage and enormous loss of life and property. The epicenter of the quake was located in the Celebes Sea, about 500 miles (804.6 km) south of Manila and 100 miles (160.9 km) east of Zamboanga City.

This epicenter, in the Moro Gulf, is about 250 miles (402.3 km) west of the Philippine trench, where the Pacific Ocean floor is currently sliding into the Earth's interior, under the Philippine Islands. This overbiting of moving plates is scientifically explainable and almost predictable. But to the human beings caught in the quake, there was very little explanation and far less comfort in facts.

The fierce nature of the quake, which also spawned an 18-foot (5.5-m) tsunami, plus the ramshackle nature of the dwellings of the fishermen and others who lived along the shoreline, were no match for these twin forces of nature. Entire villages were wiped from the face of the Earth, either swallowed up by the giant fissures the quake produced, or drowned by the giant wave that reached in and gathered up entire settlements and dragged them into the Pacific Ocean.

Hardest hit were the towns of Malabang in Lanao Province, where more than 300 were reported killed; Pagadian City, with 180 dead; and Margosatubig and Zamboanga City, where more than 200 were killed.

In Malabang, two squads consisting of 16 men of the 33rd Infantry Battalion were swept out to sea with all their equipment. The headquarters of the 41st Infantry Battalion in Curuan Zamboanga was likewise washed away.

Ralph Consing, a newsman from Colabato, said the tremor continued ". . . for what seemed an interminable time, the road cracked, fires broke out here and there and the flames cast a ghastly glow on the crumbling city."

"For a time people panicked," said a young teacher from Harvardian College, which, along with the bigger Notre Dame University, suffered extensive damage. "There was shouting. Soldiers began shooting in the air and common folk, recalling superstition, beat on drums and gongs to frighten away evil spirits."

Days later, hundreds were still without shelter. The 21st Infantry Battalion, which had a camp in Sangali, took in the homeless temporarily, doling out food rations and inoculations against cholera.

Syd Wigen of the International Tsunami Information Center, which assists nations in organizing local warning and evacuation procedures, voiced his frustration and that of his fellow scientists as they helped in the cleanup:

> Five thousand people died. You can talk about statistics and they don't mean a hoot, but it's different when you know the people. Every day I'd meet someone I knew who had lost relatives or friends, and I'd go cry my eyes out. It gave me an awareness of the human situation, an awareness that reading never would.
>
> But what really gets me, is that most of those deaths were needless—the people just didn't know how to react. They don't know that when they feel an earthquake they should take to the hills as fast as possible. They're fishermen, living on the beach or in huts just above the water, totally exposed to the waves. Today, they've rebuilt their homes, all in the same areas. Those places got whacked fifty years ago, too. Looks like the next time will be just as bad. But it doesn't have to be that way.

PORTUGAL
LISBON
November 1, 1755

Fifty thousand people died in the Great Lisbon earthquake—actually a series of over 500 shocks—on November 1, 1755. The entire city was destroyed, and along with it, priceless works of art and much of the momentum of the Age of Enlightenment.

The Great Lisbon earthquake of 1755—the most severe in modern times—was cataclysmic on a number of levels. First it destroyed the entire city. What was not leveled by the tremors was consumed in fire, so that where once a proud city stood, only a smoldering, stench-ridden wasteland remained. On a deeper level, the human toll was staggering. The official estimate of loss of life was set at 50,000; other, equally reliable estimates ran as high as 100,000.

The range of the quake was awesome: Its destruction was felt over an area of 1.5 million square miles (2,414,016 sq. km), and in North Africa, specifically in Fez and Mequinez, in Morocco, 10,000 died from the quake's ripple effects.

Finally, enormous libraries and irreplaceable works of art—all of Lisbon's museums and libraries, both public and private—burned to the ground. The past was forever destroyed in the minds of some of the most influential thinkers of the Great Enlightenment. There was nothing left but the present, and it was at best uncertain. The Inquisition was still thriving in Lisbon, creating the same tremor as this natural disaster. But its opposite, the unbridled optimism of the Enlightenment, died that day in Lisbon. Reality rushed in with the first shock waves the morning of All Saints' Day, Saturday, November 1, 1755, and the world would never be the same for it.

The first tremors rumbled through the city at 9:30 A.M., while thousands crowded Lisbon's many cathedrals. Walls swayed, chandeliers gyrated wildly. Sacramental objects, including the candles that would soon kindle the wildfires that would sweep through Lisbon, tumbled from altars and pulpits. In the harbor, a sea captain watched transfixed as, with a slow, almost stately grandeur, the stone structures of Lisbon, erected on terraced hillsides overlooking the River Tagus began to rock back and forth—"like a wheat field in a breeze," he would later recall.

It would be 40 minutes before the second of some 500 shocks and aftershocks (A Spanish nobleman would one day ask a Lisbon dignitary, "Will your earth never be quiet?") would strike the city, killing 50,000 in two minutes, opening a huge fissure 15 feet (4.57 m) wide in the middle of the city, and collapsing 18,000 buildings.

After the first shock, hundreds were buried under masonry and marble as churches—Santa Catarina and the Church of Sao Paulo among them—fell. The square before the Basilica de Santa Maria, Lisbon's ancient cathedral, became jammed with screaming, praying refugees. When the second, larger shock hit, the basilica and surrounding buildings collapsed inward with a horrendous roar onto the square, burying all of the refugees beneath them.

A contemporary engraving captures the horror of the great Lisbon earthquake of November 1, 1755 (New York Public Library)

Hundreds of other refugees sought safety on a new marble quay recently built on the banks of the River Tagus. After the initial shock, the water in the Tagus receded, exposing the bottom all the way out to a sandbar at the river's mouth; none of the refugees noticed what seismologists would regard as a sure sign that a seismic wave would soon appear. It did—measuring from 50 to 70 feet (15.2 to 21.3 m) in height, and it washed horribly over the quay, carrying off every single person on it. Two more seismic waves followed and consumed people and boats in the bay in deadly whirlpools.

And then the fires began. Whipped by fierce northeast winds, flames ignited by altar candles toppling onto tapestries and timbers collapsing onto kitchen hearths became one uncontrollable inferno. The city burned for three days and nights, and when the flames finally died, the destruction was complete. Even the structures that had managed to survive the quake were consumed by fire.

Piled high upon the human loss was the loss of history. Lisbon's two convents burned to the ground, a new Opera House was leveled, and the royal palace was destroyed. The palace of the Marques de Lourical, containing over 200 paintings by Rubens, Correggio and Titian, was obliterated. Also included in its priceless library were 18,000 books, among them a history, written in his own hand, by Charles V; maps and charts of the world made over centuries by Portuguese seamen; and an especially valuable niche of Incunabula, those first-person views of the world published before 1500. Illuminated medieval manuscripts that had been stored within the Dominican convent were all consumed by the raging flames.

The tremors spread outside Lisbon and were so powerful that they were felt through all of Europe and North Africa. In addition to the calamity in Morocco, 500 soldiers died when a barracks collapsed in Luxembourg. As far north as Scandinavia, rivers and lakes overflowed their banks. In the English county of Derbyshire, nearly 1,000 miles (1,609.3 km) from the quake's center, plaster fell off walls and a fissure opened in the ground.

In Lisbon, revenge for real or fancied responsibility for the disasters of the earthquake came swiftly. Gallows were erected by Dom Joseph, the young king, and hundreds of prisoners who had escaped when the walls of the prison collapsed, were rounded up and publicly hanged. Some confessed, before their deaths, to looting and setting some of the fires in the city.

Armies of priests in the black hoods of the Inquisition roamed the city looking for heretics to burn, rounding up and forcing some Protestant ministers to be baptized in penance over their sinful instigation of this natural calamity.

But fortunately for the 200,000 survivors, the level thinking of the secretary of state, the Marquês de Pombal, finally prevailed. Asked by the king for his rehabilitation plan, the marques uttered words that would echo for centuries: "Sire, we must bury the dead and feed the living."

That was what was done; tons of food were trucked in from the provinces, and the city would, for the next 15 years, be slowly rebuilt, this time with new streets that were 40 feet in width, bordered by wide pavements.

But the impact on the Age of Reason was lasting. Voltaire would immortalize the earthquake by letting Candide and Doctor Pangloss arrive in Lisbon in the midst of it. Jean-Jacques Rousseau would regard it as a vindication of his theory of natural man. If more people had lived outdoors, he wrote, more would have survived.

But for the other thinkers of the Enlightenment, it was a cold shower, a shocking, chastening immersion in pragmatism.

ROMANIA
BUCHAREST
March 4, 1977

Fifteen thousand died, 10,500 were injured, and tens of thousands were made homeless in Romania's worst earthquake, which was centered in Bucharest on March 4, 1977, and measured 7.5 on the Richter scale.

The worst earthquake in the history of Romania struck its capital city, Bucharest, on March 4, 1977. Although the greatest damage was done to the city itself, the effects of the 7.5 quake were felt throughout the Romanian countryside. Fifteen thousand people were killed, 10,500 were injured, and tens of thousands were left homeless.

Seventy miles south of Bucharest, the Danube River port of Zimnicea lay in almost total ruin. Dumitru Sandu, its mayor, who had taken office only minutes before the quake hit, was faced with a horrendous disaster. "The earth swayed round and round. . . ." he said afterwards. "Eighty percent of my town vanished. We had no electricity, no gas, no water. Everywhere, I saw people crying. Some had to be pulled from the rubble; we didn't know if worse tremors would follow, so we took to the streets, where we lit fires and shared blankets to keep warm. Nobody slept that night for fear of aftershocks."

Thirty miles to the north of Bucharest, tremors crippled factories in the oil-rich Ploieşti. And in the capital itself, dozens of brick and cement buildings crumbled, burying hundreds. Some of the buildings were sliced in half and, poised in the early spring sky that followed the quake, resembled nothing so much as giant doll houses, their contents tumbling to the streets behind them.

Rescue work continued for weeks. The air hung heavy with plaster dust and chlorine, which was sprayed on the debris to avert epidemic. Some survivors were trapped under the wreckage for as many as eight days.

RUSSIA
SAKHALIN ISLAND
May 27, 1995

A 7.5-force earthquake struck Sakhalin Island, a Russian possession off the east coast of Russia in the Sea of Okhotsk, north of Hokkaido, Japan, on May 27, 1995. The town of Neftegorsk was completely destroyed and 1,989 people were killed.

Sakhalin Island, once a home to 750,000 people, is located in the Sea of Okhotsk, north of Japan and some 4,000 miles (6,437.4 km) and eight time zones east of Moscow. Its value to Moscow has been constant and multiple: Rich in oil, gas, coal, timber, and fish, it was also an easternmost location for Soviet military bases, and so has been historically closed to foreigners. In 1983, as a defense against spying, Soviet authorities authorized the shooting down of a Korean Air Lines plane when it strayed over the region. All 269 people aboard, including 61 Americans, were killed when the jet plunged into the sea off the island's southwestern coast.

The workers in its oil pumping stations and oil fields were given high salaries and bonuses, and they and their families were treated to three-month vacations at Black Sea resorts. Their supply of food was generally considered to be better than that of most other citizens of the Soviet Union.

In 1995 the island was, therefore, an anachronism, a throwback to the previous regime. Its statue of Lenin remained in the public square of Neftegorsk, one of its major towns, and much of its population lived in badly and hastily constructed Soviet-era apartment houses, designed to give basic shelter for workers, and little else.

All of this figured into the enormous toll of 1,989 fatalities when a 7.5-force earthquake struck the island at 1:03 A.M. on May 27, 1995. It was a Sunday, and the island's populace was asleep and therefore taken unawares by the major temblor, which completely destroyed the town of Neftegorsk. The town, with a population of 3,500, was 40 miles (64.4 km) northwest of the epicenter. Three thousand people were buried in the collapse of 19 five-story apartment houses made of prefabricated blocks.

Nearby, the town of Okha, with a population of 35,000, sustained only minor damage. Balconies fell from two five-story buildings and cracks appeared in the walls of others. But the concentration of damage in Neftegorsk, which, translated, means "oil town," spread to its chief source of work: An oil pipeline running north from the town was ruptured by the quake, and a number of oil wells were knocked down.

Even before dawn, under floodlights powered by portable generators, rescuers dug at the rubble of the collapsed apartment houses. Sheets of twisted metal made the rescue effort slow going. In many places, mangled bodies protruded ghoulishly from the wreckage. In others, toys and household belongings peppered the rubble, mute evidence of the carnage caused by the quake.

Seventeen-year-old survivor Irya Golovchinka told workers on the scene, "I did not hear anything. I just felt air beneath my feet and then I fell. I managed to drag myself out, but my mother and father were buried. My mother was unrecognizable. She was all burned."

"There weren't any tremors or shaking or any of that," recalled 19-year-old Yelena Tischenbko. "It was Boom! and it was over. And everything was quiet again until the screaming started."

By afternoon of the first day after the tragedy, planes and helicopters were ferrying in more rescue crews, food, clothing, and medical supplies. Thick ice lying off the island prevented a hospital ship from getting near enough to accept patients.

Four days later, the town of Neftegorsk resembled a gigantic cemetery. Most of the survivors had been pulled from the wreckage in the early hours of the rescue. The dead were collected under blankets on the scene until they could be placed in coffins and taken to a proper cemetery.

Dazed survivors sat in clusters, without shelter or direction. "We were asleep when the quake struck," said Olga Bespalenko, "and I don't remember much after that until little Vova started to cry and cry. It was only then that I realized we were buried alive."

Mrs. Bespalenko's husband had been crushed when three floors of concrete fell on him, and his body had been recovered and buried before Mrs. Bespalenko and her child were rescued.

Of the 3,500 residents of Neftegorsk, only 875 were not injured. On June 1, the Russian minister for emergency situations, Sergei K. Shbigu, announced, "We will continue to look for survivors for the next five days. but then the water runs out." The official assessment was that by then most of those still alive would have died of thirst, if not from their wounds or the freezing night-time temperatures.

Meanwhile, the pipeline that ran 17 miles (27.3 km) across the Sakhalin landscape from offshore fields in the Sea of Okhotsk to terminals in mainland Russia was spewing oil from 18 different rupture sites. Blackish green petrochemical pools had formed in declivities, and had begun to seep into the Sakhalin natural drainage system, which ran into the rich salmon and crab grounds that surrounded the island.

Serafina Varlamova, the head of a section on oceanographic research in the Russian Ecology Ministry surveyed the damage. "The problem is that all the people who ran the pumping stations and knew the pipeline are dead, missing or so distressed that they can't possibly work," she said. "Neftegorsk was not just an oil town. It was a town created exactly to house those who worked in the industry. Now it—and they—are gone, and no one else seems to know what is where or what to do."

The symbol of this to Mrs. Varlamova was an abandoned bulldozer. "Drunk," she said, in exasperation. "We are looking at an ecological disaster in the wake of this human tragedy, and all anyone can think about is drinking vodka."

Only three of the 19 prefabricated apartment houses were remotely habitable, and finally, three weeks after the quake struck, authorities gave up the search. The decision was made not to rebuild the town. Survivors were moved out and the ruins were sealed, like the concrete sarcophagus placed over the nuclear reactor at Chernobyl. The devastation would be transformed into a monument to carelessness.

RUSSIA
SOVIET GEORGIA
April 29, 1991

Over 200 people were killed in a powerful earthquake that hit the northern part of Soviet Georgia on April 29, 1991. Seventeen thousand houses and over 80 percent of the hospitals, schools, and day care centers in the area were demolished.

The South Ossetia region, in the north of Soviet Georgia, was beset by political and physical battles in 1991, as were many of the newly created entities spawned by the new Russia. Ossetians, who wanted independence from the Georgian authorities, and the Georgians who governed them were locked in ethnic feuding.

On April 29, a cause for a different sort of battle presented itself: Just after noon, a powerful earthquake, which was rated at 7.0 on the Richter scale, struck in South Ossetia, near the village of Dzhava. This, the worst earthquake to strike the former Soviet Union since the Armenian earthquake of 1988 (see ARMENIA, p. 38), struck when schools and businesses were full, and thus the tragedy was deepened by the number of adults and children trapped within collapsing schools, apartment buildings, and factories.

The area's only hospital was reduced to rubble, and its surviving patients were brought in stretchers to the public square of Dzhava, where doctors and nurses tended them in the open.

Two lesser aftershocks crumbled additional structures, brought down power lines, and made roads impassable. Extensive damage occurred in the mining center of Chiatura, and tremors were felt in Yerevan, the Armenian capital; Tbilisi, the Georgian capital; Sochi, a Black Sea resort town; and villages in neighboring Turkey.

Rescue teams responded swiftly. In contrast to the past, when the Soviet government refused international aid, European teams arrived within 24 hours, and began the grisly business of extricating bodies and survivors from the ruins. Their work was further hampered by a shortage of heavy equipment, and the refugees suffered from a lack of medicine, clothing, food, and housing.

The fact that the earthquake occurred in a relatively thinly populated area suppressed the death toll to slightly over 200. But the physical devastation was appalling—over 17,000 houses were destroyed and 80 percent of the schools, day care centers, and hospitals in the region were totally demolished. It would be years before the area returned to a life that began to resemble that which had existed before the earthquake.

SOUTH AND CENTRAL AMERICA
August 13–15, 1868

Multiple earthquakes ripped through South and Central America from August 13–15, 1868. Over 25,000 were killed, and 30,000 were made homeless.

Over 25,000 people died in a series of earthquakes that shook much of South and Central America for three days, August 13 through 15, 1868. Thirty thousand more were made homeless, and the damage totaled $300 million.

Damage was most severe and casualties most numerous in cities and villages, since population was dense, buildings were built in close proximity to each other, and—in the villages particularly—building materials were largely timber and mud. The cities of Arica, Arequipa, Iquique, Tacna, and Chencha were wiped away totally.

Huge seismic waves slammed into coasts, engulfing villages and cities, and drowning thousands. Entire fishing fleets were demolished.

Suffering from depressed economies, the countries affected by the quake appealed for international relief. Most of this came from England, in the 1860s a thriving mercantile nation.

TAIWAN
NANTOU AND TAICHUNG COUNTIES
September 21, 1999

The 7.6-force earthquake that struck Taiwan at 1:47 A.M. on September 21, 1999, killed 2,405 people and injured 10,718. Over 100,000 were rendered homeless when 31,534 residences were destroyed and 25,506 were damaged. The ultimate cost of rebuilding would reach $40 billion.

The island of Taiwan, located 100 miles (160.9 km) east of mainland China, is equal in size to a combination of the states of Massachusetts and Connecticut and situated on the earthquake-prone Pacific Rim, where it is crisscrossed by no less than 51 fault lines. As a result, its 22 million people are used to being rattled by small quakes.

However, no amount of experience could have prepared them for the 7.6-force earthquake that struck at 1:47 A.M. on September 21, 1999. Most of the country's population was at home and asleep in bed at that moment, which accounted for the fact that a great majority of the 2,405 deaths and 10,718 injuries occurred in the collapse of residential structures.

Though all of the country felt some tremors, the worst damage and the most casualties occurred near the quake's epicenter, which was located near the town of Chi-Chi in Nantou County, approximately 90 miles south of the capital, Taipei. Nantou and neighboring Taichung counties are noted for their mountains and tourist resorts, a category of buildings almost never identified by deep-seated, solid construction. Like their momentary inhabitants, they tend to have a transitory quality about them, and thus, the destruction in these resorts was particularly appalling. The casualty statistics bore this out: 85 percent of all injuries and 90 percent of the earthquake's deaths occurred in three places: Taichung County, Nantou County, and Taichung City.

In addition to dwellings and office buildings, all telecommunication connections were damaged in an area that manufactures a third of the world's computer chips and 10 percent of its memory chips. Roads were twisted into impassability, power lines were severed, hospitals and morgues were without power and therefore not functioning, and two major hydroelectric dams in Taipei were damaged and put out of service.

In the immediate aftermath, residents tried to dig family members out of the rubble by hand. Ironically, because of stated threats from mainland China to overwhelm Taiwan, there was an enormous military buildup, and thus thousands of soldiers were immediately available to assist fire and local groups in search and rescue operations. Twenty-four hours later, international search and rescue teams from 21 countries were also on the scene.

Red-shirted rescue workers with flashlights taped to one side of their helmets gave an aspect of mining to the rescue efforts. "It's really like a war," marveled Kao Chen-ting, a taxi driver who observed the frantic work, the digging, the fetching of ladders and pails and rope. "It's big and terrible. You have to fight it."

And fight it they did, against dangerous odds. One strong aftershock registering 6.8 on the Richter scale and nearly 4,000 smaller ones rocked the island, forcing workers to step back from collapsed or nearly collapsed structures. These fresh jolts cracked Sun Moon Lake Reservoir, one of Taiwan's largest dams and a prime tourist attraction. But most disastrously, they also triggered massive mudslides that swept homes, restaurants, hotels, and temples from the hillsides as thoroughly as if some giant hand had swiped at them.

The village of Chifenerhshan in Nantou County was totally obliterated by one of these slides, killing 36 villagers. In the nearby village of Kuoshin, a massive flow of mud and rock that descended from three hillsides simultaneously killed 60 villagers and buried 40 others, which was two-thirds of the village's population. One woman told reporters that she and her two daughters were carried along on the slide for half a mile "as if sitting on a magic carpet."

In the village of Tali, a few miles from the epicenter, a 13-story apartment building collapsed, burying one-quarter of the 200 residents of the building. Three days later, dazed survivors still hovered around the site, as soldiers cautiously picked through the destruction, concrete piece by concrete piece. Asked why she didn't leave the search to the soldiers, a 62-year-old survivor answered, "I don't know what else to do."

Nearby, on Hsientai Street, a section of sidewalk opened up into a dark infinity of space. Nearby, a building leaned like the tower of Pisa against another, its foundation crumbled, but its upper stories intact. One survivor, a Mr. Tsai, told reporters, "I was lucky. I don't know why [I] lived. My good friends did not."

Later, the Taichung prosecutor would freeze the passports of 16 local builders and architects pending an investigation into possible faulty construction of some residential buildings that had toppled while other buildings around them had remained virtually unscathed.

The pattern of destruction seemed haphazard, and was later explained as a product of selective adherence to building codes that were patterned upon those of California. "I suspect that we will see that some of the buildings which have been destroyed have been built on land which according to the planning rules should not have been constructed upon," noted Portsmouth University earthquake expert David Pertley afterward. "There's quite a lot of illegal construction work which goes on in Taiwan as a result of the very rapid development. In Taipei we will find that some of the older buildings, which have come down, in one way or another would have broken the building code."

All of this scarcely touched the bewildered refugees from the quake, who were gathered in schools, meeting halls, and stadiums. Water shortages loomed in Taichung after the quake ruptured a section of the Shihkang Reservoir. The Health Department warned of possible epidemics because of the lack of water and the improper care of corpses. The morgues were overflowing and bodies were lying on the floors of hospitals and community centers.

Tents began to appear, the shelter of choice or necessity for those whose homes had been reduced to tangled piles of concrete and mesh. In Taichung, approximately 1,100 people were sheltered in tents arranged around the grounds and athletic fields of the Vocational Technical High School. Nearby restaurants worked with the city government to feed these survivors. Elsewhere, improvised shelters began to appear in parks and vacant lots.

Over 3,000 medical volunteers staffed hospitals, clinics, and shelter locations, volunteers worked side by side with organized search and rescue teams, religious organizations provided both material and spiritual support for victims and their families, and corporations donated products and labor.

Six days after the quake struck, two brothers, trapped in the collapse of a residential high rise in Taipei, were rescued by a city fire rescue team. It would be the last rescue of living survivors.

All in all, 31,534 housing units were destroyed and 25,506 were damaged. Over 100,000 people were therefore made homeless. The ultimate cost of rebuilding the destruction left by the earthquake was estimated at $40 billion.

TURKEY
ADANA
June 27, 1998

A 6.2-force earthquake struck the city of Adana, in the south of Turkey, on June 27, 1998, killing 112 people and injuring hundreds more.

Adana, a city of 1 million in the south of Turkey, absorbed the brunt of a 6.2-magnitude earthquake on June 27, 1998.

The quake, which occurred in daylight, tore through the city, which emerged relatively unscathed except for its slum area, where older buildings and badly constructed dwellings collapsed.

The nearby town of Ceyhan, the site of an oil terminal, was also heavily hit, but, miraculously, the terminal and its pipeline remained intact. An American military base on the outskirts of Ceyhan sustained considerable damage, and some servicemen were injured by falling walls and collapsing runways.

One hundred and twelve people were killed and hundreds were injured in this relatively moderate temblor in a country that over the ages has experienced two catastrophic earthquakes (see p. 94) and would, in 14 months, be ravaged by its worst (see p. 95).

TURKEY
DUZCE
November 12, 1999

Just three months after the monumental earthquake that devastated northwestern Turkey on August 17, 1999, another 7.2-force quake struck on the eastern fringe of the same area on November 12, 1999. The city of Duzce was almost totally destroyed. The death toll was 700, with thousands injured and some still missing.

"Oh God, What Pain!" screamed a headline in the Turkish newspaper *Sabah* on November 13, 1999. And under it, the corresponding article began: "Just as we were bandaging the wounds of August 17, a new blow pierced our hearts."

It was a repeat tragedy: At 6:57 P.M. on November 12, 1999, just three months after one of Turkey's worst earthquakes (see p. 95), and just two days before President Bill Clinton was to arrive in Turkey for a summit meeting of the Organization for Security and Cooperation in Europe, a 7.2-force quake split the earth beneath the village of Duzce, on the main highway between Istanbul and Ankara, and only a few kilometers from the site of the August temblor.

As before, buildings crumbled, fissures opened in the earth and swallowed cars, roads, trees, and parts of buildings. Duzce, a prosperous town of 80,000, set amid tobacco and hazelnut plantations, was at the epicenter. It had largely escaped August's devastation. Now, it was flattened, and its stunned residents were divested of their belongings, their homes, and in hundreds of cases, their lives.

The terrible aspect of this quake was that its harbingers were ignored and mistaken for aftershocks from the August temblor. Leman Ongor, one of the survivors of the November disaster, recalled it vividly. "The earth [had] been shaking and shaking for weeks here," Ms. Ongor told a *New York Times* reporter. "Everyone in Duzce was asking if we should move out of our houses, but we were told, 'Don't worry, it's just aftershocks, nothing bad will happen.'"

But the trembling had become so intense that Ms. Ongor had sent her mother to Istanbul. "We should have all gone," she said, as she surveyed the ruins of her home. "It started getting worse," she sobbed, "[Then] the ground opened up and there was a huge crash. Suddenly we weren't on the second floor anymore. The street was right outside our window. So we broke the window and jumped out."

Equally but differently tragic was the state of 27-year-old Celil Akbal, a buyer for a food processing plant. His home had been destroyed in the August quake, and for 10 weeks, he and his two-year-old son had lived in a tent supplied by the company. But at the end of October, authorities, attempting to clear out the clusters of tents in the area, told Mr. Akbal he would have to find some other shelter.

He moved into a nearby house of friends, and the second quake destroyed it. "When the house I was in started to shake," Mr. Akbal said, "I grabbed my boy, ran out onto the balcony and waited to die. I thought, 'This is it. This is your last moment on earth.'" The building fell, but he and his son survived.

As in August, people tore frantically at the rubble with their bare hands, trying to dig out relatives and friends. Ambulance sirens, cries for help, and screams of the living pierced the steady rumble of earthmovers and the punctuation of jackhammers. Turkish authorities, criticized for their slow reaction to the August tragedy, responded swiftly and efficiently. "No Mistakes This Time!" shouted the Istanbul paper *Milliyet*.

Another paper summed it up: "This time the state reacted with agility."

However, to those trapped in buildings, it seemed that their rescuers were a long time coming. The injured, taken to the local hospital, were treated and put in beds arranged in the hospital's garden, in freezing temperatures. The hospital itself was considered too unstable to house those it treated.

One of the survivors being administered to in the hospital's courtyard was Turkan Buyuk, who had lived in a building slightly damaged by the August quake, and was ordered to move. She had spent four months looking for an apartment in a secure building in Duzce. A week before the quake, she had signed a lease, and on the day of the temblor, she was unpacking her belongings with her daughter and mother-in-law.

"I threw myself over them, to protect them," she told reporters. They were in a fourth floor apartment, which fell to the top of a mound of rubble just above ground level. She and her child and mother-in-law were rescued through a hole in the heap of cement blocks.

Overturned trucks lined the road into Duzce, making it difficult for rescue crews and firefighters to get to the multiple blazes that the earthquake kindled. In Bolu, a neighboring town, the earthquake caused explosions that triggered fires. The road to Istanbul that skirted both Bolu and Duzce was torn apart and unnavigable, and so local officials called for medical aid from Ankara, 160 miles (257.5 km) to the east.

As night fell, electricity was restored in hospitals and clinics, but not in private residences. Terrified refugees rode out powerful aftershocks that brought down electric poles and more buildings.

The next morning, rescue teams from the United States, Italy, France, Israel, Germany, and Greece—whose aid after the August quake had produced a new and warmer Greek-Turkish relationship—began to arrive. By nightfall, over 4,000 Turkish soldiers were in place. They would be put to work building a tent complex (designed to house 2,000 people) in Duzce's central park. It would be the first step in the construction of shelter for the majority of the populace of this nearly totally destroyed city of 76,000. Add to this the 46,000 already displaced by the August 17 quake, and a monumental crisis in housing in sub-freezing temperatures faced Turkish authorities.

Three days after the quake, hundreds of people were still camping in the open, huddled around bonfires of burning tires and broken furniture. The Turkish Red Crescent and the American Red Cross provided 1,300 winter tents, 6,500 mattress/blanket sets, and 2,000 field beds. President Clinton pledged aid, and promised that the U.S. military would promptly deliver 500 winterized tents capable of housing 10,000 people.

And still rescuers dug in the wreckage, following directions from survivors and cries from beneath the rubble. But hope faded as the temperature plummeted and a freezing rain hampered rescuers' efforts.

Finally, on November 15, rescuers discontinued their efforts. "It's finished," said Belgian fireman Jean Paul Dezutter, "You can't find live people after 72 hours."

And so, rescue turned to relief for the thousands of displaced people, some of whom settled down in tents for the winter, while others found trucks and piled them high with furniture, rugs, and clothing, and joined the caravans creeping along the cracked roads out of Duzce.

The final toll as of this writing stood at 700 killed, thousands injured, and more than 700 buildings destroyed. However, hundreds were still missing and the exact number of dead may never be known, since many families undoubtedly burned the bodies of loved ones immediately, according to Muslim tradition.

TURKEY
ERZINCAN, SIVAS, AND SAMSUN
December 27, 1939

Over 50,000 were killed in the seven shocks of an earthquake that struck the Erzincan, Sivas, and Samsun provinces of Turkey on December 27, 1939. Hundreds of thousands were injured.

More than 50,000 inhabitants of the provinces of Erzincan, Sivas, and Samsun, Turkey, perished in the most violent earthquake ever to hit that country, at 2 A.M. on December 27, 1939. Hundreds of thousands were injured or made homeless, and even the homeless continued to die after the tremors stopped, as a huge blizzard tore into the countryside directly on the heels of the earthquake.

The worst hit city was Erzincan, which was totally leveled. But it was the severity of the winter weather that accounted for a large number of the deaths from the quake. As a primitive kind of insulation against the cold that often reaches 30 degrees below zero, citizens pile mounds of dirt and rock on their rooftops. When the first shock waves hit, the weight of these piles instantly collapsed roofs, caving in ceilings and walls on top of the sleeping inhabitants.

All in all, seven major shocks rocked the city, turning it into a tumultuous cemetery. Every building except the prison fell. All of its doctors and nurses were killed. Most of its government officials were buried under the rubble.

In an unusual act of heroism, dozens of prisoners, some of them convicted murderers, helped dig out over 1,000 buried victims of the quake. Finishing this task, they built shelters, provided warm clothing, often from their own backs, built fires against the raging blizzard and fought off hundreds of wild dogs that were roaming the city, feeding on the dead and the injured.

Although seismologists insisted that there was no inter-relationship, a series of earthquakes traveled the globe for the next three weeks, while multiple aftershocks rumbled through Erzincan and part of the rest of Turkey. Minor quakes hit Los Angeles, Bolsena (on the outskirts of Rome), Nicaragua, El Salvador, and Honduras. In South Africa's Rand, no less than 25 tremors struck during this period.

TURKEY
NORTH
March 13, 1992

One hundred and four residents of Erzincan, in northern Turkey, died in the 6.8 earthquake whose epicenter was 10 miles (16.09 km) from the city. Hundreds were injured; thousands were rendered homeless, and a quarter of the city was reduced to ruins.

"It is a tragedy for Erzincan to suffer a second disaster when the scars of the first have not healed," said Turkey's Prime Minister Suleyman Demirel after the 6.8 earthquake that erupted in that northern city, the prime stopping place on Turkey's east-west highway, at the precise hour that its inhabitants were either sitting down at the meal called *iftar* that marks the end of the daylight fast in the Muslim holy months of Ramadan, or at worship.

It could not have been totally unexpected. Erzincan, a thriving metropolis of 175,000 in the part of Turkey in which the Euphrates River has its headwaters, and in which snow-capped mountain peaks rise 11,000 feet into the sky, lies directly upon the Anatolian fault, a faultline that has been compared to the San Andreas fault in California. The area has had a history of earthquakes; the one to which Prime Minister Demirel was referring was the cataclysmic 1939 quake, but the area's earthquake history stretches back at least to 1043 C.E., when the first recorded earth tremors destroyed the city. In the past 1,000 years, it has been hit by no less than 11 earthquakes.

Still, predictable or not, the event of a disaster is always an unwelcome and terrifying surprise, and the earthquake that struck Erzincan at dusk on March 13, 1992, was no exception.

In the business center of the city a building collapsed upon 150 worshipers in the midst of their prayers. A six-story school, a hospital, an orphanage, and a sugar factory were decimated. Seventy-five prisoners who thought they might benefit from the tragedy when the quake rocked the foundations of the jail in which they were incarcerated tried, but failed, to escape.

"It was like Hiroshima," said Mustafa Ates, a Turkish journalist who saw the quake from a distance as he was driving toward the city. "A dust cloud rose above the whole city. The road lurched under my car."

The injured and dying cried out from beneath the wreckage, and survivors clawed with their bare hands at the piles of twisted girders and crumbled plaster. In the early evening, a cruel aftershock, measured at 6.0 on the Richter scale, sent vibrations through the ravaged city and shifted its rubble.

The inhabitants of Erzincan slept in the open that night, frightened to enter any of their former homes. By morning, Swiss and Turkish rescue teams were on the scene with avalanche dogs to sniff out any sign of life in the ruins. At 10:15 A.M., in what was left of a nursery school that had been four stories high, dogs sniffed positively and rescuers thought they heard a voice calling for help.

The team rushed into excavation mode, using drills and power saws to sculpt a cavity in the wreckage. A crowd gathered expectantly. But as the workers dug, the voice faded to a whisper and then to silence. At 11:20, they reached a teenage girl. All they could do was to wrap her entirely in a blanket, as they did the others who had died in the quake.

It had been a year of natural disasters for Turkey. In January, a series of avalanches had struck the region. In the beginning of March, a coal mine disaster on the Black Sea coast had claimed 300 lives. This earthquake destroyed a large portion of the city of Erzincan and miraculously killed only 104 residents. But a death toll is often not indicative of the suffering and the injury—both physical and psychological—to hundreds and hundreds of residents of a region that has undergone such a sudden disaster.

TURKEY
NORTHWEST
August 17, 1999

In one of the worst disasters of modern times, 15,657 residents of cities and villages in the northwestern region of Turkey, at the eastern end of the inland Marmara Sea, were killed in an early morning earthquake that struck the area on August 17, 1999; 24,941 were injured and 250,000 were rendered homeless.

A ring of apartment buildings once looked out on the Gulf of Izmit, at the eastern end of the inland Marmara Sea, some 50 miles (80.5 km) east of Istanbul. No more. A gigantic earthquake, rated 7.4 on the Richter scale, roared through the city of Izmit and its high-income environs on August 17, 1999, destroying thousands of buildings and killing a staggering 15,697 people, injuring 24,941, and turning 600,000 into homeless refugees.

The enormously high casualty figures were explainable: Whereas many earthquakes in the region in the past few years had taken place in isolated, agricultural, or mountain areas, this one struck in the midst of highly populated cities. There were more people and more buildings to feel the impact of the temblor, more human beings packed into smaller spaces, and thus more people vulnerable to the sort of destruction earthquakes cause. For instance, Tekirdad, one of the cities affected, contained over 42,000 people. This was multiplied many times by both cities and summer resorts along the Marmara Sea. It was August, and these resorts were packed with families. And finally, the quake occurred in the middle of the night, when most people were in their houses, sleeping.

The weather preceding the earthquake had been unusually hot. Windows were open to the night as the quake hit at 3:15 in the morning. Pinar Onuk, a young girl who was asked to write her recollections of the quake, captured those moments vividly:

> Without understanding what was happening I was staring at the window from where I lay . . . It was as if something had grabbed hold of us from underneath, turned us upside down and was shaking us. Then the house was moving from one side to another without stopping. While this was going on there were terrible deep noises coming from the ground. Just as it was finishing there was a loud noise of buildings collapsing. Screams, the noise of breaking glass. Our house was buried in a deep silence. Then I heard my aunt's voice say that the top floor of our house had collapsed. In the inky black darkness I couldn't feel my own feet . . . When my mum called me, my feet revived and I ran out into the pitch dark night. But the minaret of the mosque had fallen down and split the block opposite us in two. There were voices coming from the upper floors . . . For two days we slept in the street.

The devastation was widespread and horrific. Golcuk, Eskisehir, Yalova, Cinarcik, Izmit, Adapazari, and Avcilar, all bustling hubs of activity, were nearly silent fields of rubble. In Izmit, the giant TUPRAS refinery—Turkey's largest—was burning, its flames huge orange pyres in the sky. As far away as Istanbul, a score of buildings collapsed. Everywhere, there was disbelief and horror.

As daylight arrived, the people commandeered picks, shovels, sledgehammers, and scoops in their search for survivors in the debris. Apartment buildings filled with sleeping families were now mountains of tangled concrete and metal. Trapped victims cried out for water and were given it through holes in the collapsed walls.

Paul Adams, a BBC correspondent, arriving upon the scene in Adapazari, tried to describe it:

> It may not help much, but the best comparison I can think of is parts of wartime Sarajevo—a similar and equally hideous architecture, cracking and peeling in the summer sun.
>
> Buildings slumped down on their foundations, balconies concertinaed, walls lurching at insane angles. Rubble, broken glass, suffering and perhaps a measure of resignation too.
>
> In Sarajevo it took two years and the Bosnian Serb army to achieve this dismal effect. In Adapazari, Golcuk, Izmit and parts of Istanbul it took an act of God just 45 seconds long.

Rescue teams with dogs to sniff out the dead or the living arrived later on the first day, followed by bulldozers.

Outside of the cities, an equal chaos ruled. There was a solid logjam of vehicles, inching its way toward and away from the cities. Again, according to Mr. Adams: "The job of keeping lanes open for the rescue services had been entrusted to angry young vigilantes who wielded sticks and in some cases guns."

The Gulf of Izmit had invaded the land near it, while across the bay, the flames and smoke from the rapidly shrinking refinery dominated the scene. "A poisonous mix of oil and sewage slopped noisily over the redefined shoreline," Mr. Adams concluded. "The town was bracing itself for disease."

By the second day, international rescue teams began to arrive. They found unbelievable mayhem. Turkish authorities, inexperienced, disorganized, and bereft of supplies, were often helpless, and foreigners frequently took charge, guiding armies of volunteers.

Survivors were dug out from the ruins. Fourteen-year-old Onur Umit, trapped for 27 hours in the ruins of a five-story apartment building in Golcuk, remembered hearing other people trapped in the wreckage screaming and calling for help. As the hours passed, the cries died away. "Around me I heard people screaming," he told rescuers, "I said 'Don't shout; you need to conserve your energy.' But after maybe 10 hours there was silence."

Conserving his energy and reciting a comedy sketch to keep his spirits up, he finally heard rescuers coming near, and cried out. The men of his family used drills to bore into the cement and steel to get to him.

Siena Bulet, a young married woman, remained trapped for 80 hours in the remains of her Izmit apartment block. The last thing she remembered before passing out was being in bed with her husband. When she woke up, she was alone, pinned under a ceiling. Her husband had escaped, and he was calling to her. She screamed answers and workers toiled for seven hours before they could free her.

More than 1,000 relief workers from 19 countries dug through the ruins, pulling survivors from it days after the quake. One of the last to be extracted from the wreckage of Cinarcik was four-year-old Ismail Cimen. For 140 hours, the boy, who had been playing with his truck when the quake hit, had remained conscious but trapped beneath the crumbled remains of his family's apartment block. Finally, a hole just 18 inches high was knocked through the wreckage, and a worker saw Ismail squinting back at him. Within minutes, Bulgarian and Turkish rescuers pulled the dehydrated and emaciated boy from his sculpted place of safety and sent him to a hospital.

The remainder of his family did not fare as well: His mother had been pulled alive from the wreckage; his father and three sisters, aged 8 to 13, one of whom was crushed just inches away from Ismail, all had died.

Refugees took up places where they could. "We had lots of aftershocks," wrote Deniz Aydin, a young survivor. "We lived in the park for a week." Fourteen-year-old Kemal Murat Olur recalled that "The first night after the earthquake we just slept on the ground. On the fourth night we got the beds out of the house, cut down branches and made ourselves a shelter . . . On 30 August a tent came for us. A relative had sent it. We set it up and started sleeping in it. Because people weren't in their houses, there was stealing. They raided one village and three people died."

Izmit was described by a *New York Times* reporter as ". . . wildly out of control. Soldiers and relief workers are all but invisible at many of the worst sites," he continued, "A few minutes' walk from . . . tent[s] of scrap wood and carpets, Turkey's biggest oil refinery continues to burn uncontrollably and spew vast clouds of black smoke."

In Golcuk, the Turkish naval base was in ruins. Two hundred sailors and officers had been killed there. The surviving military men set up a civilian crisis center to coordinate relief. But because it was in the center of town, most of the relief trucks could not reach it.

As the days went by, a grim, checkerboard pattern began to emerge: Some buildings seemed unscathed, while those next to them had crumbled into nonexistence.

Other collapsed face forward onto the sidewalks. Still others tilted like sinking ships onto neighboring walls.

The pattern, survivors began to notice, was not random or selective. The difference between survival and nonsurvival was the difference between the construction of the buildings. Contractors had cut corners by mixing too much sand into cement and using cheap iron. In fact, it had been an open secret that many apartment blocks, built to house the area's exploding population, had been in violation of local housing codes. They had been constructed by contractors who used shoddy and inferior materials, added extra stories, avoided soil tests, and ignored earthquake-proof requirements.

Ninety-eight percent of the Turkish population lived in earthquake-prone areas, but over half the buildings in the nation failed to meet construction requirements. "The inevitable happened, despite years and years of repeated warnings," said Ahmet Ercan, a professor of geophysics at Istanbul Technical University, to a *New York Times* reporter. "Officials refused to face facts. They never insisted that contractors survey the risks and build earthquake-resistant structures. Maybe after this bitter experience, we will update our regulations along the lines of Japan, the United States and Mexico."

Turkish citizens took matters into their own hands in various areas. In Yalova, a seaside resort near Istanbul, relatives and neighbors of victims burned the car and stoned the house of a local contractor, seven of whose 16 buildings collapsed when the earthquake hit.

All in all, it was a tragedy of immense proportions, and the bill for it would be equally immense. The cost of rebuilding would, experts agreed, be 16 times higher than that for the reconstruction of Kosovo—$20 billion compared to $1.23 billion. It would impact mightily upon a Turkish economy just recovering from a recession.

But the greatest tragedy continued to be the human one. Young Ersu Berkcan Kosem concluded his memories of the disaster with a prediction: "Hopefully there won't be another earthquake," he wrote. "If it happens, I hope to God that it won't hit Istanbul. Because no-one could help Istanbul, no-one could save us . . . Please God, I worship you, if you kill me I won't be able to worship you."

TURKEY
SCIO
April 3, 1881

Seven thousand died and 20,000 were injured in an earthquake that struck Scio, in the Aegean Sea off Greece, on April 3, 1881.

The tiny island of Scio (Chios), located in the Aegean Sea but controlled, in the 19th century, by Turkey, suffered three damaging tremors on April 3, 1881. Of a population of 80,000, 7,000 persons were reported killed and 20,000 were injured.

Scio is dotted with tiny villages, composed of earthquake-prone stone structures bereft of the support of either beams or adequate mortar. Thus, when the first of the three shocks hit, many of these structures immediately collapsed, crushing their inhabitants before they were able to escape.

Forty-four villages were virtually obliterated; 10 in Kempos, a district in the south, were absolutely leveled. In Kalimasia, the district's largest village, only 330 of 1,000 residents survived, and half of these were seriously injured.

Many of the fatalities were caused by the peculiar behavior of the residents. Panicked to the verge of insanity by the multiple quakes, many ran as fast as they could for the seashore, ignoring the cries of the injured, some of whom were buried only under light layers of debris. Hundreds of friends and family members were thus crushed to death by the hordes of refugees running helter-skelter over the wreckage.

UNITED STATES
CALIFORNIA
LOS ANGELES
February 9, 1971

Fifty-nine were killed and several hundred injured in a 6.6 earthquake that struck Los Angeles on February 9, 1971.

At a few seconds after 6 A.M., on February 9, 1971, a 6.6 earthquake rattled Los Angeles and its surrounding environs. In most buildings in Los Angeles proper, this meant only minor inconvenience and a loss of sleep. To 59 people who died in this quake and the several hundred who were injured, it proved the adage that there is no such thing as a mild disaster.

The cause of the quake was determined to be a sudden earth movement along a minor fault in the San Fernando Valley, north of Los Angeles. This "minor movement" along a "minor fault" resulted in a quake that released more energy than the Hiroshima atom bomb and caused $1 billion in damage.

At the Sylmar Veterans Hospital, walls that had been built before earthquake-proof construction was legislated simply collapsed inward, killing 45 patients.

Spectacular damage to a freeway overpass that was under construction at the time of the February 9, 1971, earthquake in the San Fernando Valley (National Geophysical Data Center/J. R. Evans)

There was a supposedly earthquake-proof hospital nearby, but it too was damaged, and two of its patients died when a power failure shut down their respirators.

Near the hospital, a freeway overpass collapsed, crushing a car and its occupants, and homes near the overpass were damaged, killing several more people. Nine more died from heart attacks. More than a thousand landslides touched off by the quake caused no injuries.

A great potential tragedy was averted by seconds. The old dam of the Lower Van Norman Reservoir, only partially filled at the time, suffered immense damage. Experts later affirmed that a few more seconds of tremors would have brought down the entire dam, releasing a huge wall of water that would have swamped hundreds of houses built in the shadow of the dam. Tens of thousands of lives would have been lost.

UNITED STATES
CALIFORNIA
SAN FRANCISCO
April 18, 1906

Close to 700 people perished, and most of San Francisco was destroyed, in the great earthquake and fire of April 18, 1906.

No earthquake in the history of the world has been better recorded than the famous San Francisco quake and fire of April 18, 1906, in which an estimated 700 people died, 500 city blocks were obliterated, and nearly $500 million of damage was done to what was once—and would be again—one of the most fabulous cities in the United States. On that day, and for two days afterward, San Francisco was practically burned to the ground.

The quake, which measured an awesome 8.3 on the Richter scale (later downgraded to a 7.8), struck at 5:13 A.M. The epicenter was created by a section of rock snapping along the San Andreas Fault, an inevitable (and, seismologists say, repeatable) circumstance caused by the inexorable movement of land masses on either side of this fault. Today, Los Angeles is grinding northward at a rate of two centimeters a year, which means that in 30 million years, it will be a suburb of San Francisco.

But back to 1906. The quake came in two shocks: one 40 seconds in duration, the other 75. The fire—set off by not only the quake but also the bumbling efforts of troops trying to stop it by dynamite—lasted for three days.

The quake itself was, by all accounts, apocalyptic, tumbling huge buildings in seconds, caving in walls, splitting water and gas mains, and igniting thousands of fires instantaneously. Cavernous cracks in the earth opened—and swallowed people and vehicles and then snapped shut, crushing them.

The product of several economic booms, the city had been built haphazardly in many sections, including the mansions of Nob Hill, the huge commercial establishments and hotels of Market Street, the subterranean catacombs of Chinatown, and the honkytonk structures of the Barbary Coast.

It was here, on the Coast, that some of the most dramatic destruction from the earthquake took place. Sitting on sandy landfill, this area was a prime target for the first tremors. Almost instantaneously, block after block of shoddily constructed, thin wood-frame dwellings resting on the shifting sands of the landfill splintered into unrecognizable ruins, burying the occupants within them. On the Barbary Coast, hundreds of shanties collapsed upon themselves. The wholesale district was an enormous 15-foot-high (4.57-m-high) pudding of bricks, beams, dying produce, and men and dead horses.

The rooming houses at Ninth and Brannon streets were leveled. Writer William Bronson described the crazily perched buildings along the roller-coaster landscape of buckled Dore Street as ". . . a row of tottering drunks."

Nearby, the $6 million, seemingly indestructible city hall, located at McAllister and Larkin streets, crumbled into a mess of stone and cast iron. Its steel-framed dome resembled a cracked egg, its columns splayed into the surrounding streets, resting atop the unlucky pedestrians they had crushed in their precipitous fall.

In the Mission District, the Valencia Hotel simply slid into the street, then folded up like an accordion, the topmost of its four floors ending up where the first had been. All 80 of its occupants were crushed within the ruins.

The spires of churches—St. Patrick's, St. James's, St. Bridget's, St. Dominick's—buckled and became lethal spears falling at those fleeing in the streets beneath them. These streets filled with hysterical people soon after the first tremor hit. A survivor, Sam Wolfe, described them vividly:

> The street seemed to move like waves of water. On my way down Market Street the whole side of a building fell out and come [sic] so near me that I was covered and blinded by the dust. Then I saw the first dead come by. They were piled up in an automobile like carcasses in a butcher's wagon, all over blood, with crushed skulls, and broken limbs, and bloody faces. A man cried out to me, "Look out for that live wire!" I had just time to sidestep certain death.

The ornate and grand hotels that had once given San Francisco part of its patrician identity went down like dominos. The Denver, the Cosmopolitan, the Brunswick, the Palace, the St. Francis were total losses; the Fairmont was heavily damaged.

World celebrities and millionaires were sleeping in some of these hotels when the quake struck, and their adventures added another layer of experience to the bizarre happenings of that April morning. Enrico Caruso, who had opened in *Carmen* at the San Francisco Opera House the night before, was in residence at the Palace Hotel. Ironically, he had heard the previous day that Vesuvius had erupted over his native Naples. Wiping his brow, he proclaimed, "Maybe it was God's will that, after all, I should come this far."

Caruso emerged from the quake shaken but uninjured, and one story has it that, although terrified that his voice would be ruined forever by the dust and the disturbance, he reared back and sang, loudly and clearly, thus bringing comfort to other survivors nearby.

Whether or not this or Caruso's supposed later meeting with an inebriated John Barrymore actually happened is highly questionable, but the lore and facts of this earthquake are so interwoven by now that perhaps it makes little difference.

Barrymore, the great actor and great consumer of spirits, was, at the time, appearing in *The Dictator,* a play by Richard Harding Davis, at the Columbia Theatre on Powell Street. Barrymore had little interest in the play. He apparently had great interest in the fiancée of another man, and he was pursuing this interest in his suite at the St. Francis Hotel when the quake struck. Attired in the evening dress of the night

The Cell Building, the heart of San Francisco's communication network, burns out of control during the San Francisco earthquake of April 18, 1906. (New York Public Library)

before, he emerged from his suite unscathed (the story neglects to note the whereabouts or condition of the lady involved).

Failing to find the St. Francis bar open, he worked his way over the rubble to the street, where he found a sobbing Enrico Caruso atop a broken down peddler's cart. Caruso was clutching a picture of Theodore Roosevelt to his bosom—all he had rescued from his huge wardrobe bursting with possessions.

The dapper Barrymore, in his tails and diamond shirt studs, surveyed Caruso with an appraising eye. "Hello, old boy," he is supposed to have said. "Rather dumpy about the whole thing, eh?"

Caruso looked at his sartorial colleague for a moment, then reportedly smiled, his depression disintegrating in the absurdity of the moment. "Mr. Barrymore," Caruso said, "You know, you're the only man in San Francisco—the only man in the world who would dress for an earthquake!"

This was the legend. The very real conflagration roared on for three days, fed not only by the overturned coal burners and ruptured gas mains of the city but also by foolhardiness brought on by criminal corruption.

Dennis T. Sullivan, the farsighted and inventive chief of the San Francisco fire department, had evolved a logical plan of dynamiting to control a fire of this magnitude. But he was killed in the quake's first moments.

The mayor of San Francisco was a corrupt embezzler (he was later removed from office and jailed) named Eugene Schmitz, who was in turn elevated to office by a grafter named Abe Ruef. Both of these scoundrels had so looted the city treasury that no funds had been appropriated for Sullivan's safety measures. Moreover, Schmitz had no idea of how to run a city, much less save it from a fiery demise.

That task fell to a self-appointed military dictator named Brigadier General Frederick Funston, who, without consulting civil authorities, declared martial law and ordered troops from the nearby Presidio garrison mobilized. The troops descended upon San Francisco with fixed bayonets at 7 A.M., with orders to shoot any looters on sight.

Undoubtedly, they were needed to stem the tide of wholesale mayhem that immediately followed the quake. Homeless toughs from the Barbary Coast broke into bars, consumed the stock, and thus fortified, rampaged through the city, looting stores, breaking down the doors of banks and rifling their safes.

At the height of this chaos, a ragtag mob tried to loot the United States Mint building. They were fortunately met by a phalanx of policemen, armed clerks, vigilantes, and troops who turned them back in an exchange that left 34 of the mob dead and the Mint's $39 million in gold, silver, and currency intact.

While Mayor Schmitz did what only the president of the United States had a right to do—authorize General Funston's unconstitutional action after the fact—troops, often joined by vigilante groups, stalked looters and set up instant firing squads and makeshift gallows. Lawbreakers—and some who did not break the law—were executed on the spot. Told of these executions, Mayor Schmitz gave them his full endorsement. So, while legitimate arrests were made, the swiftness and indescriminateness of the punishment grew. In one instance, three men were dragged from a basement of a building on Stockton Street where they were rummaging through belongings. They were lined up against a nearby wall and shot to death.

Even more disturbing were the soldiers' attempts to stop the fire. Laying down dynamite charges without design, they managed to set more fires than they put out. Most of the time, they used too much powder and instead of creating a backfire, leveled whole blocks, sending flaming mattresses into the air in one instance, thus setting fire to the whole of Chinatown.

This "unmasking," as one writer of the period put it, of Chinatown unearthed more than opium caverns. Thousands of rats—a large number of them infected with bubonic plague carried across the Pacific from the Orient—swarmed out of their lairs, driven streetward by the red hot coals descending into the catacombs of Chinatown. They fanned out throughout the city, and within a year, more than 150 cases of bubonic plague would be reported by victims of rat bites.

Half the city went up in flames. The next day, the survivors boarded crowded ferries to Oakland or climbed to higher ground to escape the flames. All of the water mains in the city had burst, and on Telegraph Hill, the Italian community improvised successfully, clearing their wine cellars of 1,000 gallons of wine in order to put out the flames. According to one observer, "... barrel heads were smashed in, and the bucket brigade turned from water to wine. Sacks were dipped in the wine and used for beating out the fire. Beds were stripped of their blankets, and these were soaked in the wine and hung over the exposed portions of the cottages, and men on the roofs drenched the shingles and sides of the house with wine."

All night and into the next day, explosions rocked the city and the fire burned on. Gougers began to charge $1 for a loaf of bread. Drivers extracted as much as $1,000 to transport a load of family belongings 10 or 12 blocks. A glass of water went for 50 cents.

In retaliation, troops broke open standing warehouses and distributed food to the starving. More than 75,000 residents of the burning city took ferries to Oakland and went from there to Berkeley, Alameda, and Benicia.

Finally, on the third day after the quake, the fire was halted at Van Ness Avenue, where troops and firemen were able to create successful backfires.

The devastation was horrible to behold. Five hundred million dollars of damage in 1906 translates into hundreds of billions of dollars today. Funds that might have begun the restoration were burned up in the banks that housed them. The small, privately owned Bank of Italy, headed by Amadeo Giannini, managed to salvage $80,000 of its deposits and to lend this to all who wished to rebuild. This one altruistic stroke marked the beginning of the Bank of America.

The city would rise like a stately phoenix. But to those who were there that day, it would be all they would need to know about hell. An old man who had heeded the pleas of another man crushed beneath a collapsed building and committed a mercy killing turned himself in to an exhausted police sergeant. The overworked policeman looked at him for a moment and then said: "Go home, old man. All of San Francisco is dying this day and it no longer matters how."

UNITED STATES
MISSOURI
NEW MADRID
December 16, 1811–February 7, 1812

A series of earthquakes that began in the tiny town of New Madrid, Missouri, on December 16, 1811, with a pre-Richter scale estimated force of between 7 and 7.5, eventually affected 50,000 square miles (80,467 sq. km) of the United States and parts of Canada. New Madrid was destroyed and six known deaths occurred. It was the greatest recorded natural disaster in the United States to that date.

At 2 A.M. on the morning of December 16, 1811, the worst natural calamity to hit the United States to that date struck the small town of New Madrid (population 800), on the Mississippi River in what was then the Louisiana Territory. In a place that on the surface seemed an unlikely location for an earthquake to strike, a force estimated between 7 and 7.5 pounded the countryside for a full minute.

Trees were torn apart, chimneys buckled and fell, furniture overturned, and the terrified populace fled outdoors into the night.

It turned out to be just the beginning. At dawn, a stronger tremor ran through the settlement, and this time, as one eyewitness wrote, "... the whole land trembled like the flesh of a beef just killed." In nearby Kentucky, naturalist John James Audubon wrote that "... the ground weaved like a field of corn before the breeze."

Long fissures split the earth, swallowing homes and stores, and sulfurous fumes, escaping from these upheavals, sent an odor of brimstone through the settlement.

The normally placid Mississippi went wild. During the repeated series of quakes, pressures deep underground exploded through the riverbed from bank to bank; this in turn sent a wall of water upstream, thus allowing a steamboat captain accurately to observe that, for a moment at least, the Mississippi actually reversed its flow.

As the tremors continued for an unconscionable two months, land masses began to rise, fall, and shift. The tremors radiated south, north, and east. Chimneys fell in Tennessee, Georgia, South Carolina, and Virginia. Church bells began to ring in Charleston, South Carolina, pendulum clocks stopped in Washington, D.C., and windows rattled in New York City. Lesser effects were felt as far away as Boston and Canada. All in all, nearly 50,000 square miles (80,467 sq. km) of land area were affected by this endless string of quakes.

But the greatest damage was caused nearest the epicenter, which was located directly under the village of New Madrid. A nearby stretch of land measuring approximately 25 square miles (40.2 sq. km) was raised 20 feet (6.09 m) and earned the name of Tiptonville Dome.

In northwest Tennessee, a region of swampy area fed by creeks sank, while an adjacent area was thrust upward, cutting off the creeks' outlet. The water accumulated itself into what is now known as Reelfoot Lake—a highly descriptive title for a body of water formed by an earthquake.

In some locations, the shocks and upheavals created new channels of the Mississippi, and by the middle of January, what was left of New Madrid was largely under water.

Fortunately, no more than 3,000 people lived around the settlement at the time, and this undoubtedly kept down the loss of life. Only half a dozen residents are confirmed to have been killed. Still, as many as 100 deaths may have gone unrecorded, mostly of travelers trapped on the Mississippi. To this day, farming in the region suffers from the sandy soil spewed out of the earth in those terrifying months.

New Madrid became newer still. The survivors rebuilt it on the banks of the Mississippi redefined by the quake. Today, the 3,000 descendants of these survivors seem to consider the possibility of a future earthquake. A popular T-shirt sold in the local novelty store reads: VISIT NEW MADRID (WHILE IT'S STILL THERE).

A contemporary newspaper lithograph depicts the fate of some of the citizens of Charleston, South Carolina, during the earthquake of August 31, 1886. Most of the 100 who died were killed as they ran through the streets shouting that the end of the world had arrived. In the inset: refugees huddling on Battery Place, Charleston. (Frank Leslie's Illustrated Weekly)

UNITED STATES
SOUTH CAROLINA
CHARLESTON
August 31, 1886

Ancillary shocks from the Charleston, South Carolina, earthquake of August 31, 1886, were felt over 2 million square miles (3,218,688 sq. km). One hundred were killed.

According to contemporary accounts, most of the 100 persons who died in the Great Charleston Earthquake of 1886—one of the most widely felt earthquakes in United States history—were killed while running hysterically through the streets, some of them shouting that the end of the world had arrived. Secondary shocks sent buildings toppling on them; fissures spouting sulfuric fumes swallowed them.

The damage to historic buildings in Charleston itself was extensive, surpassing the 1687 quake, which had likewise been regarded by the citizenry as the Day of Judgment.

What was particularly unique about this quake, however, was its extensive ancillary effect. The shock was felt over 2 million square miles (3,218,688 sq. km)—as far north as New York and Chicago, as far south as Mobile, Alabama, and as far west as Omaha, Nebraska.

Gas flames flickered, walls cracked, and office workers were knocked from their chairs in New York City. In Richmond, Virginia, prisoners in a federal penitentiary rioted, prompting officials to call in troops to surround the prison walls to prevent an escape. In Cincinnati, Ohio, panicked office workers leaped from second story windows when buildings began to sway.

One of the most curious stories surviving from this quake deals with a Professor Capen, a New York City weather forecaster who had eerily predicted the date and time of the quake. Despite turmoil in the offices around him, he failed to notice any of the effects of the cataclysm he had predicted, and had to be told about it after the fact.

VENEZUELA
CARACAS
March 12, 1812

Twenty thousand people died in an earthquake that destroyed nine-tenths of the city of Caracas, Venezuela, on March 12, 1812.

The course of history was changed for Venezuela on Holy Thursday, March 12, 1812, when the worst quake in that country's history destroyed nine-tenths of Caracas and laid waste to the countryside surrounding it. The viciousness of the quake, the enormity of its destruction—20,000 people perished in Caracas and its environs—and its timing on a holy day when the cathedrals were filled with worshipers, convinced multitudes of people that God was displeased with their efforts to overthrow Spanish rule and the rebels' leader, Simón Bolívar. Wholesale defections took place in the armies of Miranda, the revolutionary leader, who had managed, to make large inroads into Spanish landholdings and military installations. Terrified peasants deserted the cause, and efforts to establish the United States of Colombia were crushed in one terrible day. Buoyed by this unexpected turn of events, the Royalists returned to power, claiming divine intervention.

The magnitude of the quake and the resulting terror must have seemed like an omen. Striking after dark, its epicenter erupted in the northern part of the city of Caracas, a hilly promontory thickly populated with both people and churches. The first tremors toppled the churches of Alta Gracia and Trinity, both built of stone. The splendor of Alta Gracia was made particularly impressive by the presence of 15-foot-thick (4.6-m-thick) pillars supporting its massive, 150-foot (45.7-m) height. Not one worshiper emerged alive from either church.

In the streets, chaos abounded. Gigantic fissures swallowed people, livestock, vehicles—even entire houses. In one horrible moment, the entire complex of San Carlos barracks was consumed, along with every soldier sleeping within it.

All of this had taken a mere minute, but the rest of the night and days afterward would be spent extricating the 2,000 survivors trapped in the wreckage of what once was a city. Ten thousand were killed in Caracas alone. In the surrounding villages of San Felipe, La Guaya, Mérida, Antimano, La Vega, Baruta, and Mayquetia, 5,000 more perished.

The Caracas authorities, fearful of an epidemic of plague resulting from unburied bodies and rats attracted to the carnage, ordered gigantic funeral pyres of hundreds of bodies apiece to be erected throughout the ravaged city. Their flames could be seen for miles throughout the night.

What the authorities had not planned for was famine and dysentery spread by a polluted water supply caused by the rupturing of fresh water conduits. Another 5,000 would die as a result, lending further credence to the exhortations of priests that this was God's revenge for straying from the cruel, inhuman, vicious but nevertheless devout Spanish rulers.

YUGOSLAVIA
SKOPJE
July 16, 1963

Two thousand people died, 3,000 were injured, and 170,000 were made homeless by a monster 9-point earthquake that leveled Skopje, Yugoslavia, on July 16, 1963.

A cataclysmic 9-point earthquake destroyed the city of Skopje, Yugoslavia, when it struck at 5:15 A.M. on the morning of July 16, 1963. Two thousand were killed outright; more were buried under the rubble and never found. Three thousand were injured and more than 170,000 were made homeless.

A tourist transfer point at a road and rail junction 125 miles (201 km) north of Salonika, Greece, and 110 miles (177 km) west of the Bulgarian capital of Sofia, the city lies on a direct fault line with the Italian and French Rivieras, which had experienced tremors just days before the devastating quake hit Skopje.

"I thought it was a hydrogen bomb," a Skopje man said. "There was a terrible roar. I woke up, looked out the window and saw the Hotel Macedonia swaying from side to side."

The Macedonia, a popular hotel, was full, its 180 beds occupied by tourists. It collapsed into a 20-foot-high (6.09-m-high) rubble heap of stone and steel, killing all but a few of its guests and staff.

The Macedonia had company in destruction. Nearby, the Skopje Hotel disintegrated. The main post office collapsed, leaving only one wall remaining upright. The center of the railroad terminal fell in on hundreds of travelers waiting to catch the early morning train for Belgrade.

The Kaarpus, a five-story office building, suddenly shrank to three stories as the earth swallowed up its bottom two floors. An ancient mosque on the left bank of the Vardar River was completely shattered.

From the moment the tremors ceased, a haze of brick and mortar dust hung like a curtain over the city. From the air, it seemed as if a giant foot had stamped it out. The scene up close was appalling. Silent groups of homeless men and women dug wordlessly through the ruins, trying to save whatever they could.

There was nothing left of the city. Its two movie houses, library, museums, the Yugoslav National and Investment Bank, many schools, the City Council, radio station, and 85 percent of its dwellings were simply no more.

Ironically, the Emperor Duahan Bridge, originally built in 520 C.E., in the reign of Justinian and now a main thoroughfare that connected the old city of Skopje on the left bank of the Vardar River with the newer part on the right bank, remained unscathed.

Thirteen survivors of the railway station underwent a similar, miraculous escape. The 13, trapped in a tunnel beneath the wreckage of the railway station, were detected by a supersensitive French listening device that heard them through 26 feet (7.9 m) of rubble. They had spent more than 72 hours under the tangle of steel and stone.

The rebuilders of Skopje consulted seismologists before undertaking the task of reconstruction of their demolished city. Fearful of a typhus epidemic, they were eager to begin construction.

But one epidemic almost certain to strike any disaster site did not roar through Skopje. Hundreds of stores lay with their display windows shattered. Goods were there for the taking. But there was no looting, none whatsoever, despite the citizens' desperation and homelessness.

FAMINES AND DROUGHTS

THE WORST RECORDED FAMINES AND DROUGHTS

* Detailed in text

Africa
* Eastern and sub-Sahara regions (1983–present)
Uganda, Kenya, Somalia, Ethiopia, Darfur (1997–present) (see PLAGUES AND EPIDEMICS)

Caucasus: Georgia, Tajikistan, Armenia
* (2000)

China
(1333)
(1810–11)
(1846)
(1849)
* (1876–78)
(1920)
(1928)
* (1939)
* (1942–43)

Denmark
(1087)

Egypt
* (Earliest recorded famine) (3500 B.C.E.)
* (1708)
* (1064 C.E.)
* (1199)
Fustat (968 C.E.)

England
(680 C.E.)
(695–700)
(822)
* (1069)

(1235)
(1563)
* (1976)

Europe
(2003)

France
(987–1059)

Greece
(1942)

Holland
* (1944–45)

India
(1669)
(1745)
* (1769)
(1782)
* (1790)
(1812)
* (1833)
(1837)
* (1866)
* (1876–77)
* (1898)
* (1943)
* (1972)
Hyderabad (1677)

Ireland
(1316)
(1816)
* (1845–50)

Italy
(450 C.E.)

Jamaica
* (1788)

Micronesia
* (1998)

Nigeria
* (1968)

North Korea
* (1996–97)
(1997–06)

Roman Empire
(79–88 C.E.)

Russia/USSR
(1650)
(1906)
(1914)
* (1921–23)
(1932)
* (1975)

Scotland
(856 C.E.)
(936)
(1047)

Sudan (see AFRICA)

United States
* East (1998–99)
Midwest
* (1909–14)
* (1934–41)
* Southwest (2005–06)
* Texas (1996–present)
Virginia
* Jamestown (1607)

Yemen
(1970)

CHRONOLOGY

(There are no specific dates for famines. Thus, only the year of the famine's origin is noted.)
* Detailed in text

3500 B.C.E.
* Egypt (Earliest recorded famine)

1708 B.C.E.
* Egypt

79–88 C.E.
Roman Empire

450
Italy

680
England

695–700
England

822
England

856
Scotland

936
Scotland

968
Fustat, Egypt

987–1059
France
1047
Scotland
1064
* Egypt
1069
* England
1087
Denmark
1199
* Egypt
1235
England
1316
Ireland
1333
China
1563
England
1607
* Jamestown, Virginia
1650
Russia
1669
India
1677
Hyderabad, India
1745
India
1769
* India
1782
India
1788
* Jamaica
1790
* India
1810–11
China

1812
India
1816
Ireland
1833
* India
1837
India
1845–50
* Ireland
1846
China
1849
China
1866
* India
1876
* China (1876–78)
* India (1876–77)
1898
* India
1906
Russia
1909–14
* Midwestern United States
1914
Russia
1920
China
1921–23
* USSR
1928
China
1932
USSR
1934–41
* Midwestern United States
1939
* China

1942
* China (1942–43)
Greece
1943
* India
1944–45
* Holland
1968
* Nigeria
1970
Yemen
1972
* India
1975
* USSR
1976
* England
1983
Africa
* Eastern and sub-Sahara regions
1996
* North Korea
United States
* Texas
1997–present
Uganda, Kenya, Somalia, Ethiopia, Darfur, Africa
1997–2006
* North Korea
1998
* Micronesia
United States
* East
2000
* Caucasus: Georgia, Tajikistan, Armenia
2003
Europe
2005–06
* Southwestern United States

FAMINES AND DROUGHTS

....

Most natural disasters are mercifully brief. An earthquake usually lasts under a minute. A tornado careens through a midwestern town in something under five minutes. Cyclones and hurricanes lay waste to cities in hours. Even the duration of floods can usually be measured in days.

But not so famine and drought. These are long-lasting disasters that can go on for decades, and their effects continue for generations. There is a Fourth World, some social scientists assert, and it consists of the half-billion people on this Earth today who live their lives under conditions of famine.

Droughts tend to do their worst on land that is already arid, and where societies are on the margin. In a world where rain is rare, people can cope with a dry year or two, but then, quite quickly, wells fail and livestock begin to die. The first sign of famine is usually a sudden series of grain price rises in local markets. Within weeks, this translates into the beginnings of a deprivation of food in human populations.

The reasons for drought and famine are likewise more complex than those for all other natural disasters. Along with nature and the changing state of the earth, there are distinctly human, economic, cultural, and political causes for both drought and famine. And although drought is the most common cause of famine, it is by no means the only one. In fact, famine can be brought about by just the opposite phenomenon—flood. Other natural causes abound: heavy rains; unseasonably cold, hot, or dry weather; typhoons; pest infestation; and plant disease.

All of these are reasonably beyond the control of human beings, although much has been invented and put to use to counteract pests and disease.

Similarly, drought, which occurs naturally when evaporation and transpiration (the movement of water in the soil through plants and into the air) exceed precipitation for a considerable period, is also affected by human behavior.

There are four basic kinds of drought: permanent drought, which occurs in the driest climates, where nothing could possibly grow without constant irrigation; seasonal drought, which occurs in climates that have well defined dry and rainy seasons; unpredictable drought, which occurs as a result of a sudden reduction in rainfall; and invisible drought, which is a borderline situation in which high temperatures induce abnormal evaporation and transpiration, so that even regular rainshowers fail to irrigate crops, and they die.

All of these are typical conditions in nature. Yet human beings often choose to live on and farm land that hasn't a prayer of supporting them. Or, human beings take good land in a benign climate and turn it into worthless dust by misfarming it.

The natural causes of famine that frequently elude human correction sometimes occur outside of the region that is affected. Drought, for instance, may occur in the headwaters of a major river used for irrigation, thus causing famine in an irrigated region hundreds of miles downstream—possibly across a national border.

And so, the famines of Egypt and the Middle East, regions whose environments are naturally hostile to intense sedentary agriculture. Because of this, sources of irrigation often occur miles away across national boundaries.

Asia is another example. It has land that is resistant to predictable planting. Alternately drought- and flood-prone, this land has traditionally been incapable of supporting its population; thus the reputation—unfortunately not well documented—of China as the most drought- and famine-prone country in the world, with India closely behind. Each has farmland that is irrigated from rivers with their headwaters in the other's country, and their needs—often conflicting—have caused some of their most terrible droughts and famines.

Also entering into the famine picture of both of these countries are political and cultural factors. And these make natural causes pale by comparison.

Consider the severe food shortages of Roman times, in which citizens purportedly flung themselves into the Tiber rather than starve. The shortages were caused by Roman emperors who hoarded valuable grain.

Or consider another common human cause of famine: warfare. One way of conquering a country is to starve it to death, and the blockade of food supplies was widely employed in Europe between 1500 and

1700. In 1812, the "scorched earth" policy of the Russians not only deprived Napoleon's armies of needed food but also inflicted the same deprivation on the Russian people. During World War II, the German policy of starvation in occupied Holland was a particularly infamous use of famine as a weapon of war. In the late 1970s, the genocidal policies of the Khmer Rouge regime in Kampuchea, in which there were massive deportations of urban populations into the countryside without food or shelter, caused over 1,000,000 deaths from starvation. And the civil wars that have torn the sub-Sahara region of Africa asunder in the 1980s and 1990s have killed millions more, with no end in sight.

As the 21st century began, genocide—the worst of all possible causes of famine and drought—grew and spread like a deadly plague through Darfur, in the Sudan. In 2003, a violent conflict began between rebels protesting the neglect of farming tribes and the government of Sudan in Darfur. And as the years went on, the conflict escalated, fueled exponentially by the Sudanese government's employment of the Janjaweed, a group of Arab herders on horse and camelback who turned a hunt for rebels into a wholesale slaughter. Villages were burned, inhabitants murdered, and women raped: Refugee camps were routinely raided and razed. What the Janjaweed failed to accomplish, government air attacks finished. And when worldwide relief efforts and U.N. intervention tried to stem the tide of destruction, they were met with regulation and visa barriers, the cutting off of fuel by the government, military threats and the killing of U.N. and humanitarian workers.

By 2007, hundreds of thousands of inhabitants of the region were dead and there were more than 2.5 million refugees spilling over the border into neighboring Chad. Jan Eliasson, the U.N. envoy to Darfur, summed up the horror succinctly: "It's a critical situation . . ." he told reporters. "We have a crisis of humanitarian operations. We have harassment of U.N. personnel and [aid] workers. I saw so much suffering . . . we have to find a solution now."

The fight between the government forces and the African populace was seen on the surface to be one between an Islamist government and Africans, but as fighting progessed and increased, it became far more complex and diffuse, pitting Arab against Arab, African against African. Underlying this tragedy was the continuing drought in Africa. By 2007, 5 to 6 miles (8 to 10 km) of former farmland was turning to desert by the year and fights for land added yet another layer to the crisis.

Cultural influences were responsible for the famines of medieval Europe, as well as those of Asia. The feudal social system combined with enormous overpopulation resulted in extended food shortages, which begat malnutrition, widespread disease such as the Black Death, and famine. From the birth of Christ to 1800, there are records of famine occurring in Europe in 350 different years, while in England during the same period there was a food shortage in one year out of 10. According to historian A. Porter in his *Diseases of the Madras Famine of 1877–79,* it was estimated that famine occurred somewhere in France every six years between 1000 C.E. and the 19th century. Throughout history, the food supply of humankind has been, at best, highly precarious.

The effects of overpopulation continue to compound a burgeoning problem worldwide. Add to this the exploitation of the environment, atmospheric pollution by industrial and nuclear waste, and, in the 21st century, intertribal and intergovernmental violence that has burst into genocide. Short of a grim, Malthusian housecleaning, famine can only grow in this world in which we live.

There is hope, in the presence of relief agencies battling, against overwhelming odds, both the causes and the effects of famines. The UNRRA (United Nations Relief and Rehabilitation Administration), WFP (World Food Program) and FAO (Food and Agriculture Organization) all represent the United Nations's war upon famine. Voluntary agencies joining in the anti-famine fight include CARE (Co-operation for American Relief Everywhere), Catholic Relief Societies, Save the Children Fund, War on Want, Christian Aid, and the League of Red Cross Societies.

Their enemy is as formidable as nature, the universe, human cruelty, and stupidity. But they have been successful against such famines as the one in the Bihar section of India in 1967. Their cause is correct, and is the only hope of reducing, even fractionally, the growing Fourth World of starvation on this planet.

· ·

AFRICA
EASTERN AND SUB-SAHARA
1983–Present

Millions of people have died so far in a famine that peaked first between 1984 and 1986, and which arrived *with a greater vengeance in 2003, fueled by natural factors, civil wars in Ethiopia and the Sudan, and ethnic cleansing tending toward genocide in Darfur.*

Most of Africa is marginally less prone to drought and famine than Asia, but intertribal warfare, generations of poor farming practices, and complex civil wars have increased the impact of famine upon its populace. Today, in the 21st century, while most of the world gets wealthier, 150 million Africans are, in the words of the United Nations Food and Agricultural Organization's former director general, Eduard Saoums, "... in the most serious economic distress and shortage of food, which may reach proportions of hunger and malnourishment on a massive scale." By 2006, this prediction had already come true.

Even in the best of times, Africa is, by Western standards, a poor continent. It depends upon farm products for the survival of the 12 billion people who live in its countries, and most of the agricultural methods it uses are ancient and sometimes counter-productive. For instance, the nations of the sub-Saharan region—Chad, Niger, Mauritania, Mali, Upper Volta (Burkina Faso from 1984), Gambia, and the Cape Verde Islands—lose valuable agricultural land each year as the Sahara advances southward at an average of five miles per year, while rainfall has decreased 25 percent in the last 20 years.

The lack of rainfall is a natural phenomenon. The encroachment of the Sahara, however, is worsened by unwise overfarming and overgrazing by sheep and cattle. South of the Sahara, where trees once stood and greenery abounded, there is now nothing but barren, eroded land.

In countries like Zimbabwe, rebuilding after an eight-year war, with its northwest portion suffering from year after year of drought, there is little hope for self-sufficiency in the near future. Governmental corruption in Ghana has created food shortages for 10 million people.

So, Africa is a land that has been and will continue to be, for the foreseeable future, a region in which drought and famine hold the upper hand over the populace, at least while that populace is governed by warring tribal and governmental factions.

No two nations in Africa exemplify this situation more dramatically than the neighboring states of Ethiopia and the Sudan. Since 1983, both countries have been wracked by drought, famine, and civil war. Their governments have been accused of genocide through starvation. Both countries have, to a certain extent, been used as pawns between the East and the West—the governments of the United States and the U.S.S.R.—supporting those in power (or guerrilla groups as the case may be) and adding to the general turmoil and privation.

A pivotal year was 1983. As the U.S.S.R.-supported Marxist government seemed to have the upper hand in Ethiopia, a truce was requested by the United Nations to end the fighting. But 1983 was also the year that the civil war between the north and south of Sudan began. In this case, the U.S.-supported government of the country advertised itself as a democracy, though its Islamic Fundamentalists were responsible for Islamic law being proclaimed throughout the entire country, thus pitching a battle between Arab Muslims in the north and Christians and others in the south. The battles, drought, and famine—both natural and forced by the government—would, between 1983 and 1988, claim 1 million lives in the Sudan, and the dying, the fighting, and the famine are still continuing, in varying degrees.

The Darfur conflict in western Sudan between the Janjaweed, an armed militia group of Baggara herders first recruited by the government of Sudan in 1996 and a rebel group, was both bloody and one sided. From the time this rebel force of farmers attacked Sudanese government outposts in 2003 to the end of 2006, as many as 10,000 people died monthly in the region, mainly, according to World Vision, a relief organization, from disease and hunger. As of this writing, over 2 million people have been driven into homelessness by the conflict, their huts and villages pillaged, their homes burned and destroyed. These refugees have ended up living in ramshackle huts in numerous camps along the edge of the Sahara, with meager access to food, water, clothing and shelter. Health care is nearly nonexistent, and killings and sexual assaults are rampant.

To sort out this tangle of troubles, it is necessary to go back nearly 30 years:

In 1973 and 1974, several hundred thousand people in West and East Africa died of famine and attendant malnutrition, while the Western world was absorbed in the economic crisis precipitated by OPEC's dramatic oil price rise. As a consequence, little aid came to the starving of Africa, and, according to some analysts, this made the truly terrible famine which peaked in 1984–86 even worse than it might have been. In fact, the Marxist government of Ethiopia seized upon this assumption based upon some fact as a smoke screen to cover its heartlessly extravagant expenditure of $200 million on a celebration to mark the 10th anniversary of its coming to power, while millions of poor Ethiopians were threatened with starvation, and hundreds were dying of it every day.

It could be that the problems of famine might once have been successfully addressed, for, beginning in 1984, various international relief organizations had begun to make inroads against famine and disease. But the problem of constant population shifts caused by refugees being driven from one part of the country to the other has made prior planning almost impossible. Supplies may be abundant in one area and pitifully inadequate in others, with virtually no chance of swapping supplies, since even U.N.-marked supply convoys were attacked and destroyed by Somali guerrillas.

This sort of activity goes back to 1980. At this time, 1.8 million of the 5 million people in Ethiopia affected by famine were from the Ogaden region of the country, where ethnic Somalis conducted frequent guerrilla raids against governmental outposts and villages. Frequently, these raids occurred in Gamu-Gofa, in the southwest, where the drought had hit the hardest, where virtually no rain fell in 1980 and the U.N. officials visiting Gamu-Gofa, Baje, Harar, and Wallow reported that 50 percent of the 600,000 cattle in the land had died of starvation, too.

As the years 1981 and 1982 dragged on, even irrigation became pointless. Rivers dried up, while tides in the Indian Ocean forced themselves inland, leaving the water brackish. Even if the marketing policies of some of the African countries had been perfect, there was little food to market. More and more of the population turned nomadic, wandering from place to place and talking of 1968, the last year of good rains in the sub-Sahara region.

In Ghana, in 1983, the hot wind that usually comes in January lasted twice as long as it normally does,

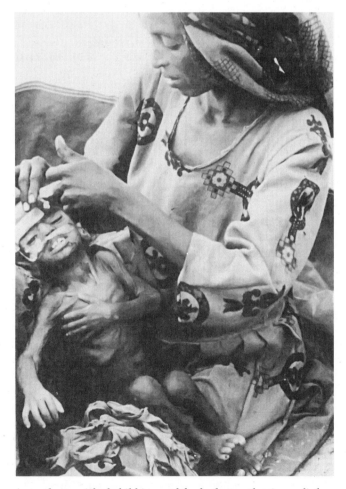

An undernourished child is cared for by her mother in a relief camp in Bati, Ethiopia. (U.N. photo/John Isaac)

fanning brush fires that destroyed both fields and food storehouses. Along with the drought, the fires cost Ghana a third of its annual food production.

Finally, in mid-1983, most of the world woke up and began massive efforts to stem the tide of starvation in Africa. In September of that year, the U.N. urged a truce between the Ethiopian government and the guerrillas. The U.S. administration under Ronald Reagan, reluctant to send foodstuffs through Ethiopia's Marxist government, finally came down on the side of humanity and upped its expenditures for relief to something over $10 million.

By the end of 1984 the U.N. reported in a *New York Times* story that "up to 7,000,000 Ethiopians [were] said to be 'at risk of starvation,'" Many others were dying of attendant diseases.

U.N. teams discovered horrifying conditions under which the populace had lived for a decade. In the 10 years since the government had taken over in a military coup, virtually no land had been irrigated and little had been done to correct environmentally destructive agricultural practice. Farmers in the north were farming in ways that eroded the land completely. Forests were cut down, overgrazing by livestock grew, and the government did nothing—or worse. It dropped the prices it would pay for grain, through its government-owned Agricultural Marketing Corporation, thus discouraging farmers from producing a surplus, or selling whatever surplus they were able to produce.

"The fact is," one U.N. worker noted, "many farmers hoard their excess grain, not to sell later but because they'd rather have the food than the little money the government would pay them for it."

Meanwhile, as the drought dragged on, hundreds of Ethiopians died daily. Projections put the number that would starve to death between May 1984 and May 1985 at half a million.

"Many others, particularly children, will suffer problems for the rest of their lives, including impaired physical and mental growth," said High Goyder, field representative of the British-based relief organization Oxfam.

The workers who toured the camps which were feeding tens of thousands daily described conditions ranging from grim to hellish. "At Korem, for example, things have improved," noted William Day of Save the Children, another independent relief agency. "[In the feeding center 225 miles (362.1 km) north of Addis Ababa,] three weeks ago, 150 were dying ... every day. Three days ago [November 1984], that figure was down to 40."

It was a comparative improvement; but the deaths from famine continued. At camps in the Ethiopian highlands, severe cold and the resulting hypothermia

Dead cattle and devastation are mute evidence of the gigantic drought that has ravaged Ethiopia and the Sudan since 1984. (U.N. photo)

killed many. Lacking other shelter, thousands of people dug holes in the ground and surrounded the rims with rocks as their only protection against winds and frost. Disease invaded even these camps. Typhus, pneumonia, dysentery, meningitis, and measles claimed the lives of hundreds.

Meanwhile, various governments blamed each other for the problem, while the dying went on, and nomads filtered into feeding centers, where they set up their way of life in *tukuls*—low gumdrop-shaped huts fashioned from grass mats and sticks. One such woman, in Harerge, in eastern Ethiopia, said to a *Times* reporter in December 1985, "All the animals died because of the drought. For three years' duration there was nothing. We don't have any sheep or goats, and we can't do anything now even if the rains come."

That sort of hopelessness extended to many of the 1.2 million people in Harerge who were affected by this prolonged dry spell, described by a representative of another relief organization, Interact, as a "green famine. There's sorghum in the fields, but there's no seed on it. The maize is withered," he concluded. Thus, the

starving would not, could not be fed in the foreseeable future.

By January 1985, the relief supply to Ethiopia from the United States had increased to $40 million. But the Ethiopian government was misusing much of this, utilizing blankets and supplies as bait to supposedly resettle hundreds of thousands of Ethiopians from the poor and overcrowded North to the fertile and under-populated South. The hidden agenda may have been genocidal. Weakened by starvation and disease, hundreds of thousands of these refugees died by the sides of roads already littered by the corpses of humans and animals.

Finally, in 1986, the rains came, and the drought dissipated. But all of the problems did not miraculously cure themselves with a change of weather. A "normal" situation in Ethiopia means importing 15 percent of food needs and feeding 2.5 million people in residual pockets of famine. (During the peak famine years of 1984–86, 6.5 million people were being fed.)

In January 1987, Ethiopia's leader, Colonel Mengistu Haile Mariam, began a three-year drive for food and self-sufficiency, saying that "hunger has left its indelible scar on the history of our country, the honor and morale of our people."

Noble words. But the underpayment for farmers' produce, plus the resettlement horror of January 1985, which was to be resumed, made U.N. agencies look with some trepidation upon the Ethiopian government's ability to handle poverty, drought, and famine. Not only that. In late 1987, U.N. convoys bearing food were attacked in the drought stricken provinces of Eritrea and Tigre by the Eritrean People's Liberation Front, an anti-government guerrilla group. Thirty-five trucks were set afire, incinerating food supplies that were on their way to thousands of starving Ethiopians.

And more problems arose: Now, the civil war in Sudan, Ethiopia's neighbor to the west, accelerated to full fury, and hundreds of thousands of refugees from Sudan began to pour over the borders into Ethiopia. Tens of thousands of these refugees walked hundreds of miles, and the roads in Sudan, as they had been in Ethiopia a few years earlier, became littered with the corpses of those who had died of starvation along the way, or from the blows and bullets of Sudanese militiamen.

Because the Sudanese government represented a semblance of democracy, the Reagan administration sent $1.7 billion in aid to it. Over $1 million a day was to be spent on distributing food. But independent relief agencies were barred from the country, and the army was given the responsibility of distributing the food. As a result, much of it went to the army rather than to the starving populace.

To complicate matters, massive floods paralyzed the country in August 1988. When these subsided, various U.N. representatives and relief agencies that began to filter into the country saw evidence of genocide through neglect. As in Ethiopia, masses of people were herded together by militiamen and moved from place to place. Many of them died along the way. But, whereas in Ethiopia, there was some evidence of positive sense to the moves, there was none in Sudan, except to kill off the populace. Country dwellers were moved to cities, urban dwellers to the country. Neither could live in these strange surroundings, and so they died.

Disease rose like a wraith. Tuberculosis swept the ravaged countryside.

In May 1989, a truce was declared between the north and the south of Sudan. For the first time, the International Red Cross was allowed in. Food supplies were in the country already, but the government had not distributed them. A crisis prevailed. A massive international airlift was begun, and for a moment, at least, these two poor, drought- and famine-plagued countries looked forward to some hope of reducing the devastation of death by starvation.

But their euphoria was short-lived. Religious wars do not die easily. The Muslim, Arabic-speaking north continued to wage war against the black African, non-Muslim south, first in small ways, then in larger ones. By the mid-1990s, the country and its neighbors in eastern Africa were once more plunged into civil strife, and shipments of food from the remainder of the world diminished.

By 1999, 1.6 million people in Somalia were cut off from food supplies that continued to feed the military and the politically connected. Drought returned in the 1990s with a vengeance, and by the end of the 1990s, 15 countries in the sub-Saharan region faced exceptional food emergencies. The worst affected countries were Angola, Burundi, Sierra Leone, Somalia, and Sudan—all during not only droughts but also civil conflict.

In Somalia, the constant civil war displaced large numbers of farm families. The United Nations Food and Agricultural Organization (FAO) revealed in a report that "The escalation of violence has reduced the distribution of humanitarian relief assistance and a number of starvation-related deaths have been reported."

The estimation of the starving in Sudan was set at 2,000,000 people. But it was not the only country in eastern Africa ravaged by dry spells, erratic rains, and violence. Kenya, Uganda, Tanzania, and Ethiopia faced depleted food supplies and the danger of their people moving from their own neighborhoods to find food. In Eritrea, 500,000 people displaced by the war with Ethiopia were starving.

International relief organizations tried to fill in the gaps. More than a quarter of a billion dollars in food relief from the outside world was spent each year in the 1990s in Sudan alone.

In 1999, many people in the Republic of the Congo were uncertain of their next meal. In Burundi, even the 821,000 people in camps provided by international groups were threatened. "Living conditions in these camps are reported to be extremely poor, with no clean water and sanitary facilities," the U.N. reported. "The overall crop prospects are also unfavorable, due to dry weather and reduced planting. A reduced harvest this season will follow a below normal harvest last season."

As the century ended, Somalia became a prime target of humanitarian aid from the U.N. "The effect of the drought is compounded by the upsurge in civil strife that led to large numbers of farm families being displaced," its December 1999 report stated. "The conflict has also disrupted farming activities and the delivery of humanitarian assistance to the people who need it most."

The violence against citizens escalated to violence against humanitarian workers. In late July of 2000, a French and a British fieldworker in Somalia for the international group Action Against Hunger (AAH) were kidnapped in south Mogadishu. The group immediately suspended all but minimal life-saving activities in the country and entered into high-level negotiations through Somalia's new president, Abdulkassim Salat Hassan. Finally, on September 18, the two hostages were released, but Action Against Hunger and other international organizations were forced to cut back even more in their activities because of security concerns.

And meanwhile, two threats escalated the famine and despair in the area. The spring planting season in the Ogaden region of Ethiopia in 2000 was plagued by continued drought. Some rain fell in May, but, as AAH reported that month, "Rains will not make up for dead livestock. People who used to herd cattle will probably remain destitute and may not have other alternatives than the sale of their remaining shoats."

Most of the farm families in the area had herds, before the drought, of 40 shoats and between 80 and 100 cows. By May of 2000, the average herd consisted of 20 to 30 shoats and two cows.

Another threat presented itself in 2000. Though 2,000,000 Sudanese had died in 17 years of fighting and famine, and 4.4 million in southern Sudan had been driven from their homes—which made it the largest displaced population in the world—the Sudanese government led by General Omar Hassan al-Bashir, whose military coup seized power in 1989, worsened the condition by engaging in a ruthless campaign of depopulation of the areas around oilfields and a pipeline in southern Sudan.

Amnesty International reported that civilians in the oil region were suffering violent abuses. Sudanese air force planes were attacking hospitals, schools, and airstrips where relief flights usually landed. In 1999 and 2000 alone, tens of thousands of people were displaced in the name of oil to pay for the country's military campaigns.

Chaos is too gentle and inadequate a word to describe the situation in Darfur. From the beginning of the recruiting of the Janjaweed by the Sudanese government, there was never a straightforward line of demarcation between the two sides, though, for the sake of history, it is possible to state that the conflict began when two rebel groups opened attacks on the government in early 2003, accusing it of neglecting the African farming tribes of Darfur. The Islamist government struck back, enlisting the Janjaweed, composed of Arab herders, and the Janjaweed proved to be a terrifying force, destroying hundreds of villages, raping and pillaging as they supposedly sought out rebels and their sympathizers.

But this was just the beginning. In a short period of time, the conflict transformed itself from a clash between Arabs and Africans to part Arab versus African, part government versus rebel, part nomad versus farmer. In the beginning, there were two rebel forces, which eventually split into five. And on the other side, the Janjaweed raids were aided by government aircraft attacks.

In November 2003, Jan Egeland, the U.N. humanitarian chief, pronounced the conflict a genocide, prompting both a heated denial from the Sudanese government and attacks on U.N. relief workers. By spring 2004, several thousand non-Arabs had been killed and as many as a million more had been driven from their homes. Some 100,000 refugees poured across the border into neighboring Chad.

To add to this, drought increased in the region. From November 2004–November 2005, no rains came to relieve it. A scant corn harvest was exhausted in six months. In Malawi, more than 4.6 million of its 12 million citizens became dependent upon international aid to survive.

The entire horn of Africa was being beset by drought and recurring food shortages. Zambia, Mozambique, Lesotho, and Swaziland were swept with drought and resultant malnutrition. Villagers were reduced to boiling bark from the surviving trees for food.

And in Darfur, the horror increased. The rest of the world began to pay attention, but it was slow in sending aid to the starving refugees. Desperate members of some refugee camps kidnapped Sudanese aid workers and held them hostage in their camps to gain attention. International aid workers trying to feed more than 2 million refugees told reporters that roads in Darfur were so crowded with bandits and killers that they had to deliver food by air. But the Sudanese government suddenly cut off supplies of jet fuel.

Antigovernment rebels in Darfur carried out attacks in vehicles painted as if they were carrying aid workers. Government forces attacked and strafed civilians and fleeing refugees from aircraft, while promising the United Nations that they would ground their air force in Darfur.

A 7,000-member African Union force was dispatched to the area, but its soldiers were overwhelmed. The U.N. Security Council passed resolutions and at a U.N. summit in September 2005, world leaders pledged to protect civilians caught in armed conflict from genocide, war crimes, and ethnic cleansing. In 2006, the Security Council adopted a resolution pushed by England and the United States to transfer peacekeeping in Darfur from the African Union force to a larger, better-equipped U.N. force. But the Sudanese government rejected this, claiming it would violate the country's sovereignty and was in fact a thinly veiled attempt at recolonization.

And so the dilemma of Africa seems to sustain itself eternally. Drought, famine, and conflict continue to unite in killing millions of its inhabitants. As 2006 came to an end, the government backed Janjaweed invaded both Chad and the Central African Republic in pursuit of escaping rebel groups. Nicholas Kristoff, in the *New York Times,* described it chillingly: "[They've] unleashed their fury on villages in Chad," he wrote, "riding in and killing and raping, accompanied by their standard shouting of racial epithets like 'black slaves.'"

As he stepped down from his post as U.N. humanitarian chief in December 2006, Egeland pronounced Darfur in "free fall" with 6 million people facing the prospect of going without food or protection. In his farewell speech, he said that one of his greatest regrets was that key global leaders had not come together to offer the sticks and carrots to settle the conflict in 2004, when it had only involved one million people

As he spoke, the United Nations, because of the intensifying violence and insecurity, was evacuating its international staff. "We're not protecting the lives of the vulnerable women and children, and there are four times more of them now than when we started in 2004," he added.

The situation in this part of the world, ravaged by drought, pillaged and plundered by armed conflicts, continues to deteriorate. And the human toll inexorably rises, as part of a poem by an aid worker in Sudan has vividly noted:

Baat Wol is my starving black kid,
My very own because I've seen him and held his hand.
He's about six years old
Naked with a bloated stomach, a name tag
And a very dry sense of humour.
He peers out from behind a long face,
Only slightly drawn by hunger,
Waiting for you to entertain him . . .
We saw all the women waiting for their hand-outs,
Men with red eyes and spears looking more confused
   than fierce,
UN people bustling by the airplane
And rebel officials, the politicians of aid,
Mentally rubbing their hands with glee,
As they counted the sacks of maize,
   the new harvest . . .
The thing is, Baat Wol's life hangs on such a thread.
He's not from the town where we met,
And he didn't have any family around.
He's OK right now, today,
But if he twists an ankle tomorrow, playing,
It might not heal right, he could limp,
The food might be flown to another town,
He might not be able to get there
And he could die . . .

CAUCASUS: GEORGIA, TAJIKISTAN, ARMENIA
2000

Political instability, small wars causing large refugee populations, and a record drought conspired to cause famine and a bleak outlook for families in the former Soviet republics of Georgia, Tajikistan, and Armenia in the summer and winter of 2000.

The former Soviet republics suffered in many ways in the 1990s. Poverty, disorientation, civil unrest, and inflation all made the transition from communism to democracy a rocky path. It seemed quantitatively unfair, then, that in the summer of 2000, drought and a resultant famine added to an already substantial storehouse of woes for the average citizen of Georgia, Tajikistan, and Armenia.

The winter of 1999–2000 was notably low in snow and rainfall, and this meant that winter crops did not flourish. Still, spring planting took place as usual. But beginning in May, an extended drought set in, with scorching temperatures and hardly any rain. And the summer crop also wilted and died.

It was catastrophic for the population of the three countries, which is made up in large percentages of subsistence farmers. A third of Armenia's population lives in rural areas and approximately 70 percent exist by farming. In Georgia, agriculture is the main source of income and employment for more than 50 percent of the population.

Tajikistan has a greater problem, since it is a land-locked country, and the only way to transport relief supplies into it is from one of the Baltic ports, then by rail through Uzbekistan to Khatlon, in the southern province of Tajikistan—a distance overland of approximately 2,700 miles (4,345.2 km).

In Armenia, the estimated damage to its agricultural sector was over $100 million. Georgia was far more ravaged; its entire harvest in the eastern part of the country failed to materialize.

This meant a loss of food for subsistence farmers who could not afford to buy food from the village markets. The same markets were usually sources of survival, places to which they normally took their produce and sold it on barter terms. Thus, over 695,000 people needed to be fed in order to live through the 2001 harvest. And this would be only a temporary reprieve. Twenty-six thousand tons (23,586.8 tonnes) of seed would have to be supplied by the rest of the world for them to continue to survive.

Georgia, once a thriving part of the world, had become one of the world's poorest nations. In 1999, the Georgian government was able to finance only 37 percent of its budget, and its transition to a market economy was a painful one.

And so, when the low level of rain and snow in the winter and spring of 2000 reduced the water level for irrigation, the farmers' usual fallback upon secondary and tertiary canals became impossible, since these canals had fallen into disuse and thus had not been maintained since independence.

The cereal and maize harvest of 2000 was half the average of the past five years. Barley production was one-third of the 1999 harvest. Besides this, potatoes, vegetables, fruits, vineyards, oilseed, fodder, and livestock production fell dramatically because of the drought.

The shortage of all of this in the markets of Georgia increased dramatically—far beyond what the average Georgian could pay. Tomato and onion prices increased by 100 percent, and the price of cheese rose by 55 percent.

Besides subsistence farmers, pensioners, widows, and the disabled suffered acutely. Beneficiaries of pensions received erratic payments from the government. Some pensioners waited as long as 15 months for their payments. By the end of 2000, it was estimated that 58.6 percent of all citizens of Georgia were living below the poverty line.

And then, to add to the misery, an energy crisis beset the country. The government promised the entire country light and heat, but was unable to deliver it. In the city of Tbilisi, 2,000 demonstrators took to the streets, protesting that they had electricity, water, and telephone communication for only five hours a day. "My children became sick as it grew colder," one protester told Western newsmen, "and I have no money to pay either for medicine or for fuel."

Finally, Russia agreed to supply Georgia with electricity for 16 to 17 hours a day for the winter of 2000, but Georgia would have to pay back the debt in the summer of 2001.

In Armenia, the summer's drought had burned away range vegetation for livestock grazing and had dried up grains, potatoes, and other edible crops. Livestock, the livelihood for one-third of Armenia's families, were particularly affected by the shortage of feed. Undernourished cattle could not be expected to survive the winter, and so were slaughtered by farmers. The estimation in the fall of 2000 was that 20 percent of the existing animal herds would be eliminated throughout the winter.

The uplands in the north and center of Armenia, where a mix of potato, livestock, and cereal production predominated, were particularly devastated, and the families of the region, already economically vulnerable, were flung abruptly into poverty. UN teams estimated that 77 percent of the populace would suffer food shortages in the winter. In Shirak, Lori, Tavush, Aragatsotn, and Gegharkunik, some 258,000 farms were affected by the drought, and 63 percent of them had a loss of 66 percent of their normal production. Pastures in this semi-mountainous area were, by early autumn, mostly dry and unfit for grazing. By September, a noticeable drop in milk production was evident.

Whatever was left of the potato crop nationally was of poor quality, and there was a widespread shortage of potato seed. The U.N. estimated a loss of 81 percent in most areas. This was particularly important, since the winter wheat plantings had been reduced by 50 percent. Without potatoes, widespread starvation into 2001 was inevitable.

The drought began to ameliorate at the beginning of 2002. But it would be a long time before families in this part of the world who had become refugees from the fighting and victims of both the instability of the area's new countries and an extended drought would find anything resembling a normal life.

CHINA
1876–1878

The worst recorded famine in the history of the world was the China famine of 1876–78. Thirteen million people (according to some missionary sources) or 9.5 million people (according to Chinese government sources) died as a result of this famine, caused by a drought in northern and central China.

The great and terrible famine of 1876–78, which killed between 9.5 and 13 million people and affected another 70 million, began simply, as the majority of famines do, with a drought.

For three endless years, from 1876 to 1878, not a drop of rain fell upon northern and central China, in the area bordered by the Yangtze River on the south and west, Peking (Beijing) on the north, and the Korean border on the east. Normally, this region is beset by yearly monsoons, thus ensuring a plenitude of rice and other staple crops of the Chinese diet. But from 1876 through 1878, the monsoons did not come.

The neighboring provinces of Kwangtung (Guangdong) and Fukien (Fujian) to the south suffered massive crop damage by monsoon-induced floods. The British Crown colony of Hong Kong, in the southernmost section of Kwangtung, was deluged with rain.

But not so the immense agricultural area to the north, which was transformed into an immense hell of starvation, murder, slavery, and cannibalism.

A normally benign countryside, it was transformed into a savage jungle in which night travel became suicidal. Roving bands of starving men set upon travelers, killing their mules, oxen, camels, or horses out from under them. By the second year of the famine, it became unsafe to even rest by the side of the road.

According to the chairman of the Foreign Relief Committee set up in the coastal city of Tientsin:

> In November, 1877, the aspect of affairs was simply terrible. The autumn crops over the whole of Shanzi and the greater part of Chihli and Honan had failed. . . . During the winter and spring of 1877–78, the most frightful disorder reigned supreme along the route to Shansi. Hwailuhien, the starting point, was filled with officials and traders all intent on getting their convoys over the pass. Fugitives, beggars and thieves swarmed. The officials were powerless to create any sort of order among the mountains. The track was completely worn out and until a new one was made, a dead block ensued. Camels, oxen, mules and donkeys were hurried along in the wildest confusion, and so many perished or were killed by the desperate people in the hills, for the sake of their flesh, that transport could only be carried on by the banded vigilance of the interested owners of the grain, assisted by the trained bands, or militia, which had been hastily got together. . . . Night traveling was out of the question. The way was marked by the carcasses or skeletons of men and beasts, and the wolves, dogs and foxes soon

put an end to the sufferings of any wretch who lay down to recover from or die of his sickness in these terrible defiles.... Broken carts, scattered grain bags, dying men and animals so frequently stopped the way, that it was often necessary to prevent for days together the entry of convoys from the one side, in order to let the convoys from the other come over....

The Ch'ing (Qing) dynasty of Manchu rulers may have contributed to the huge death toll by not revealing the extent of the famine to the outside world. Perhaps fearing that a revelation of any internal weakness might bring about a collapse like the one that had befallen its predecessor, the Manchus forbade foreign travel through the afflicted area for two entire years.

Still, fragmentary stories began to leak out, and piecemeal descriptions only heightened the horror to those who were told of the catastrophe. Stories (later verified) of "ten thousand man holes"—enormous pits into which the dead were thrown—leaked beyond China's borders.

A Shanghai resident, Frederick H. Balfour, reported in detail his first-hand findings:

> The people's faces are black with hunger; they are dying by thousands upon thousands. Women and girls and boys are openly offered for sale to any chance wayfarer.
>
> When I left the country, a respectable married woman could be easily bought for six dollars and a little girl for two. In cases, however, where it was found impossible to dispose of their children, parents have been known to kill them sooner than witness their prolonged sufferings, in many instances throwing themselves afterwards down wells, or committing suicide by arsenic.

Finally, by January 28, 1878, the Manchus could hide the facts no longer, and a British investigating envoy opened the drought and famine area to the remainder of the world by dispatching a telegram to his home office: "Appalling famine raging throughout four provinces North China. Nine million people reported destitute. Children daily sold in markets for food. Foreign Relief Committee appeal to England and America for assistance."

The Great Chinese Famine of 1876–78 remains, to this day, the worst recorded famine in the history of the world.

CHINA
September–November 1939

Destruction by the government of a record abundance of crops in the Hunan, Anhwei (Anhui), and Kiangsi

(Jiangxi) Provinces of China and flooded crops in the adjacent province of Hopei (Hebei) caused a famine that killed 200,000 by starvation in three months, from September through November 1939.

The enormity of China produces grim ironies. In the fall of 1939, the provinces of Hunan, Anhwei, and Kiangsi had bumper crops of rice—25 million bushels, by local government count. It was too much for either the populace or the economy to accommodate, and, to avert a depression, officials and farmers in the region destroyed hundreds of tons of rice.

But at the same time in an adjacent province, 25 million people were being made destitute and 200,000 were perishing from a famine caused by flooding in the Yellow River basin. Ninety of Hopei's 130 districts remained under 10 feet (3 m) of water for several months, and this in turn wiped out the entire grain and rice crop in this area. Over 500,000 bushels of grain that could have kept the region's inhabitants alive were destroyed by the flood waters.

To add to the problem of starvation, Japan and China were at war, and Japanese troops consistently cut off food and aid supplies that were sent to the area by the International Red Cross. None of these supplies ever reached the starving people they might have saved.

CHINA
1942–1943

A combination of drought, internal political infighting, and World War II combined to cause a famine in Honan (Henan) Province in China during 1942 and 1943. Nearly 3 million people died of starvation.

One of the most profound famines in the history of the world took place in the Honan Province of China during World War II. It was a natural disaster compounded by two human factors.

First, there was the war with Japan that had been raging since 1936. By 1943, Japan had occupied much of Honan province.

Second, there was the increasingly uneasy alliance between Chiang Kai-shek, heading the Nationalist government, and Mao Tse-tung (Zedong), heading the Communist government. United to fight the Japanese in 1937, they fought each other with almost as much vigor as they pursued the war against the Japanese.

Thus, when the great famine of 1942 and 1943 struck this southern province of China, the combination of forces was overwhelming for the people of the province. Of the 30 million people who lived in Honan, close to 3 million died of hunger in one year.

The immediate causes of the famine can be traced to 1940. Until that time, Honan was noted as a fertile province with richer-than-ordinary soil. The cash crop was spring wheat, which the peasants sowed in late autumn and harvested in mid-May. This was followed by the secondary crops of millet and corn, sown after the harvesting of the wheat and gathered in by late autumn. In 1940 and 1941, all crops were poor. In 1942, a full-scale drought hit, destroying all three crops.

No provision had been made by either the government or the populace to deal with the effects of the drought; each assumed that it would not last, and when it did, it was the peasants who suffered.

The central government was slow to react. Crops failed in the summer of 1942. In November, the government sent in observers, and then, instead of sending food to the starving populace, it sent money—$200 million in famine relief funds, most of which disappeared on the way to Honan. By March of 1943, only $80 million of the original $200 million had reached the provincial government. It was put into local banks, collecting interest for the government, while the populace was reduced to eating bark and dying in the streets.

When the money was finally distributed, taxes were deducted up front, and after that, the local banks skimmed off operating costs. Peasants who were subsisting on dried leaves and elm bark had to haul their last sack of seed grain to the tax collector's office. To add to the charade, the government distributed the money in $100 denominations. Wheat could only be bought from the hoarders in small denominations, and the banks charged the peasants for changing the large bills for small ones.

In the midst of this, there was some tiny, almost Machiavellian gesture of food distribution made by the government. Ten thousand sacks of rice and 20,000 sacks of mixed grain were meted out between January and March. This came to slightly less than a pound apiece for 10 million people who had been starving since autumn.

Ironically, just across the border from Honan, in Shensi province, grain was plentiful, as it was in Hupeh, on the other side. But ancient provincial distrust and the delicate balance of power and conflict in the central government prevented the shipping of food across the borders of these provinces.

Meanwhile, the extent of the famine was being recorded by Protestant missionaries. Early in the autumn of 1943, mobs of hungry peasants stormed wealthy homes and farms that had, through irrigation, survived the drought, rifling the homes and seizing the standing crops.

Horror stories abounded. The parents of two small children near one mission murdered them rather than hear them beg for food. Some families sold all they could for one last meal and then committed suicide.

Authors and correspondents Theodore H. White and Annalee Jacoby, gathering material for their book *Thunder Out of China*, toured the province in the early spring of 1943. Their written accounts were graphic and terrifying.

> A great stink suffused [everything]. Dry sweat, urine, common human filth, scented the morning. The peasants shivered in pulsing reaction to the cold, and their gray and blue rags fluttered and quivered in the wind.... When we walked down the street, children followed crying, "K'o lien, k'o lien (mercy, mercy)." If we pulled peanuts or dried dates from our pockets, tiny ragamuffins whipped by to snatch them from our fingers. The tear-stained faces, smudgy and forlorn in the cold, shamed us. Chinese children are beautiful in health; their hair glows then with the gloss of fine natural oil, and their almond eyes sparkle. But these shrunken scarecrows had pusfilled slits where eyes should be; malnutrition had made their hair dry and brittle; hunger had bloated their bellies; weather had chapped their skins. Their voices had withered into a thin whine that called only for food....
>
> There were corpses on the road. A girl no more than seventeen, slim and pretty, lay on the damp earth, her lips blue with death; her eyes were open, and the rain fell on them. People chipped at bark, pounded it by the roadside for food; vendors sold leaves at a dollar a bundle. A dog digging at a mound was exposing a human body. Ghostlike men were skimming the stagnant pools to eat the green slime of the waters....

The city of Chengchow (Chengzhou), before the war a thriving center of 120,000 people, contained less than 40,000 by late 1943. In the countryside, it was even worse. White and Jacoby went there, too, and recorded it. "The people were slicing bark from elm trees, grinding it to eat as food," they wrote. "Some were tearing up the roots of the new wheat; in other villages people were living on pounded peanut husks or refuse. Refugees on the road had been seen madly cramming soil into their mouths to fill their bellies, and the missionary hospitals were stuffed with people suffering from terrible intestinal obstructions due to the filth they were eating."

And, as in many famines, cannibalism followed. A missionary doctor told White and Jacoby about a woman who was caught boiling her baby. The authorities let her go when she convinced them that the baby had died before she had begun to cook it. Another woman was caught cutting off the legs of her dead husband. She was also released when it was ascertained that he had died first.

Because of the war, no relief came from the West; because of the Nationalist/Communist face-off, no relief came from within China. Millions died, waiting for nature to right both its wrongs and the ones committed by those in power.

In the winter of 1943, snow did fall, and this irrigated the fields. Passing a field of green wheat, White and Jacoby encountered an emaciated old man. Gesturing toward the field, they told him that there was hope. The man nodded hollowly, and told the writers, "It is fine, yes, but who knows whether we will be alive to eat it?"

By 1944, the Japanese had decided to wipe the province clean of Chinese soldiers and government officials. They did it in three weeks, thanks to armed uprisings by peasants who had been taxed and starved, literally, to death by their own government, and were all too eager to help in rooting that government out, even if it meant aligning themselves with an enemy of their country. They must have thought this enemy might, at least, feed them.

EGYPT
3500 B.C.E.

An inscription on a tomb on the island of Sihel, off the coast of Egypt in the Mediterranean, dating back to the Third Dynasty, leaves a fragmentary but revealing picture of the first recorded famine, in Egypt, 3500 B.C.E.

The earliest recorded famine, which occurred in Egypt in the third millennium B.C.E., was preserved visually in a relief that survives on the causeway of the Fifth Dynasty Pyramid of Unas in Sakkara. The relief shows emaciated famine victims apparently consoling each other. Their faces and posture in which they are frozen reflect, as if in a long-distance mirror, the twisted and wasted forms of Biafran children in the 1980s.

The written account was found in an inscription on a tomb on the Mediterranean island of Sihel, created in the reign of Djeser during the Third Dynasty. The inscription was restored and rewritten centuries later, during the time of the Ptolemies, and offers a remarkable insight not only into the physical effects of famine—which are, not surprisingly, the same as those

experienced by famine victims today—but also into the very attitude of those in power 5,000 years ago:

> I am mourning on my high throne for this vast misfortune [says the ancient, anonymous chronicler of this first famine] because the Nile flood in my time has not come for seven years. Light is the grain; there is lack of crops and of all kinds of food. Each man has become a thief to his neighbor. They desire to hasten and cannot walk. The child cries, the youth creeps along, and the heads of the old men are bowed down; their legs are bent together and drag along the ground, and their hands rest in their bosoms. The counsel of the great ones in the Court is but emptiness. Torn open are the chests of provisions, but instead of contents there is air. Everything is exhausted.

EGYPT
1708 B.C.E.

The beginning of the biblical seven-year famine in Egypt has been set by scholars in 1708 B.C.E. Located in both Egypt and Palestine, it killed tens of thousands.

The Great Famine of the Book of Genesis ("And the famine was over all the face of the earth . . . and the famine waxed sore in Egypt . . .") took place, by consensus of historians and biblical scholars, in 1708 B.C.E.

The passage of time and intervening conflict have erased most of the firsthand information concerning the event. What is known is that it was at the time of Egypt's Middle Kingdom, when its pharaohs had been replaced by rulers from Syria and Canaan, and strict control over agriculture and the use of the land was, like much of the country, in disarray.

Thus, crops failed, corruption in the distribution system kept food from the populace, and for seven years, from 1708 to 1701 B.C.E., famines and plague ranged back and forth across Egypt and Palestine. Tens of thousands died of starvation in both countries.

EGYPT
1064 C.E.

Forty thousand people died in the eight-year famine that gripped Egypt from 1064 to 1072 C.E. Drought caused the famine; crime and cannibalism exacerbated its effects.

An eight-year famine ravaged Egypt from 1064 to 1072. During that time, approximately 40,000 Egyptians died from starvation, and crime and cannibalism were rampant. W. R. Aykroyd quotes survivors in his study, *The Conquest of Famine:* "Organized bands kidnapped the unwary passenger in the desolate streets," noted one observer, describing a creatively cruel twist upon this ancient custom: ". . . principally by means of ropes furnished with hooks let down from latticed windows."

The famine was the result of a drought brought on by the unusually low state of the Nile in 1064, which made irrigation impossible. At that time, Egyptian farmers customarily depended upon the regular overflowing of the Nile's banks to irrigate their crops. For eight years, this natural phenomenon did not occur, and the resultant drought caused famine and then pestilence.

EGYPT
1199

One hundred thousand people died in Egypt in the famine of 1199, which was caused by a failure of the Nile River to provide natural irrigation to the crops bordering it. Increased food prices were secondary culprits. Medical science advanced when Islamic law was softened, allowing the first autopsies to be performed.

The cause of medical science was tangentially advanced and crime and cannibalism reached new lows in a recordbreaking famine once again caused by a drought that was brought on by the failure of the Nile to overflow its banks (see previous entry). This time, the low state of the river occurred in 1199. Over 100,000 Egyptians lost their lives, either through starvation or at the hands of their fellow human beings. In the city of Maks alone, 20,000 perished.

In no other recorded famine until that time were such extremes of human desperation reached. Murder and cannibalism were rampant. The drought, caused by the drying up of the fields neighboring the Nile, was extraordinarily widespread and continued for three years, through 1202.

Ironically, the human-induced consequence of drought, economic deprivation and starvation because of decreased food supplies and increased prices, produced the only positive outcome of this otherwise grisly period in Egyptian history.

Sometime in the early part of 1200, Abdul Latif, a physician from Baghdad, observed an increased number of people emigrating from Egypt to Arabia, Yemen, and Syria. Questioned about their reasons, they answered that they were moving on because of spiraling food prices brought on by the drought-caused famine in the Nile River valley.

More and more refugees poured into the cities of Egypt and its neighboring countries. The cities could not accommodate them, and so, instead of starving at home, they starved in strange surroundings, on city streets, or were murdered by roaming bands of brigands who sometimes lived, according to Latif, on the excrement of animals.

The bodies of the starved, mutilated, or murdered were everywhere, and local law enforcement officials gradually began to ignore the Islamic law forbidding the dissection of bodies for medical experimentation. Thus, anatomists like Abdul Latif and his colleagues were able to perform the first crude autopsies on record.

This event, however, was overshadowed by the horror. Perhaps in no other famine on record did cannibalism reach such vast proportions. Children particularly were caught, slaughtered, and roasted as if they were lambs or pigs. And, most difficult to believe of all, these children were often eaten by their own mothers and fathers.

Latif recorded this nightmare of nightmares. "I myself saw a small roasted child in a basket," he wrote. "They carried it to the Emir and led in at the same time the mother and father of the child. The Emir sentenced both of them to be burnt alive."

Once these cannibalistic offenders were publicly and municipally roasted, they were considered legal food and distributed to the masses for consumption.

Other examples of cannibalism are part of the public record of the time. Friends were invited for an evening and slaughtered and eaten by the host and hostess. Repairmen were called to homes and butchered as they worked. It was proof beyond contradiction of the theory that hunger is a pervasive motivation for many of mankind's most heinous crimes.

ENGLAND
1069

In 1069, following the Norman Conquest (1066), a famine raged over the northern counties of England. Over 50,000 died.

Thousands sold themselves into slavery and more than 50,000 others perished in a deep and pervasive famine that overran the northern counties of England in 1069, shortly after the Norman Conquest of 1066.

Records of the time give no specific reason for the famine. They only describe it, in some detail: ". . .

between Durham and Yorke lay waste, without inhabitants or people to till the ground, for the space of nine years . . ." states the *Harleian Miscellany*. ". . . Many were forced to eat horses, dogs, cats, rats and other loathsome and vile vermin," the record continues, "yea, some abstained not from the flesh of men. . . ."

ENGLAND
1976

A freak drought in South Wales brought on a summer famine, of sorts, in England in 1976.

By the last quarter of the 20th century, the more primitive and criminal tendencies of famine-stricken human beings had been all but tamed, and in the summer of 1976, the British even managed to treat their plight, in the worst drought in 500 years, with a certain degree of lightness.

Centered in South Wales, which normally is treated to heavy downpours and then misty stretches of unrelieved precipitation, the drought brought on a dry spring in 1976. By the end of May, crops were withering all over the British Isles.

Marion and Arthur Boyars, two London publishers writing to a friend, chronicled the summer:

> June: . . . tankers hauling water to rural areas, heath fires in Somerset. Greatest shortage of potatoes in memory. They cost more than Mediterranean oranges.
>
> July: Passersby cheer lady in Chiswick for walking the street cool and naked. Pleasure boats ordered off the canals. The Thames looks sick, low, muddy. . . .
>
> August: Water cut off overnight in South Wales. . . . Authorities with power to fine water-wasters accused of sending spy planes to spot green gardens. . . . Queen decides to let royal gardens die. London's fountains turned off. . . . Clouds of smoke from burning forests and heaths. Drought grips all western Europe. Shortage of beef, wine, milk saves Common Market huge price subsidies. Rainy London day makes news on the telly. St. James's and Hyde parks are like deserts. . . . Water-diviners do a roaring business.
>
> September: Wales . . . worries about cutback in factories of big water users—American industrial chemicals, plastics. . . . Water for homes cut off in parts of southwest England.
>
> Mid-September: Finally! Torrential rains! Six months later: Rain still falling. English again living under umbrellas.

HOLLAND
1944–1945

The most extreme example yet of human manipulation of natural forces to bring about a natural disaster was the 1944–45 famine in western Holland. Nazi occupiers brought on a two-year famine that killed 10,000.

Although the death toll from the famine of 1944–45 in western Holland was 10,000—relatively small compared to the millions that have perished in other famines—its circumstances and placement in history make it important. Unlike other famines caused entirely by natural circumstances, this famine was purposefully created by the occupying Germans during World War II and specifically designed to bring about mass starvation. The natural resources that supplied food were intentionally manipulated.

Prior to the beginning of World War II, the Dutch lived on an excellent diet containing an abundance of animal products, with wheat and rye as the staple cereals. However, most of their food was imported, including fodder for livestock. And to add to a precarious internal balance, much of the butter, cheese, and eggs produced in Holland was exported.

The war broke out in 1939, and it became immediately apparent to the Dutch that they would have to stockpile imports against a possible blockade. This they did, but their facilities were inadequate, and by the beginning of 1940, rationing of all food, including fodder, was instituted.

It was only an interim measure that soon became outdated. The Germans invaded Holland in May 1940 and not only confiscated the stockpiles but also commandeered, for their own armies, 60 percent of everything that was produced agriculturally in Holland.

The Dutch made do with the remaining 40 percent by self-imposed rationing, by cutting down on the production of pigs and poultry, and by increasing the planting of potatoes. The eastern part of the country, where most of the farms are located, naturally fared better than the low-land, industrial west, but overall, the entire populace managed to survive.

Then, in September 1944, the Dutch government in exile ordered a nationwide railroad strike. At the same moment, General Bernand Law Montgomery launched an airborne attack on Arnheim, on the border between eastern Holland and Germany, as a prelude to an advance on the Ruhr. There were huge casualties on both sides, and the attack was judged a failure. The strike went on.

But the worst was to come. Retaliation was the rule in occupied countries, and Seyss-Inquart, the Nazi

Reichskommissar, immediately cut off all movement of food from the north and east into western Holland. The Reichskommissar warned that unless the strike was terminated, famine would result.

And it did, rapidly. By October, stocks of food began to be depleted. Ships with cargoes of food that were already in harbor were boarded by German soldiers, and their cargoes were confiscated and destroyed. Factories and warehouses of food were emptied and their contents carried off. Eastern Holland became isolated.

Those who could went out into the country and foraged for food on bicycles with handcarts. After a while, even the few potatoes or sugar beets that were lying untended in abandoned fields had been gathered up.

People began to die. The elderly perished first, then those who lived alone, and then, as the famine deepened, families started to starve.

In certain towns, "starvation hospitals" were set up—waystations, where people could check in for a prescribed period to be fed and then discharged to make room for others. The diet in the hospital, shared alike by doctors, nurses, and patients, was anything but grand:

Breakfast: 1 slice of bread; 1 cup of tea.
Lunch: 2 potatoes, a small portion of "vegetables," some watery sauce.
Dinner: 1 or 2 slices of bread, 1 plate of soup, 1 cup of "coffee substitute."

In desperation, Dutch doctors sent an open letter to the Reichskommissar, Seyss-Inquart:

We hold your Administration responsible for the dire shortage of even the most necessary foodstuffs. The want and distress of the Dutch people living in the most densely populated parts of the occupied territory increase day by day. The ration allotted to adults has a nutritive value of only 600 to 800 calories. That is even less than half what is needed for an adult to survive, even when resting; it is less than a third of what is required for work. The small stocks of food which many families had been able to put aside are disappearing or are already exhausted . . . extra rations for the sick and aged have been withdrawn altogether.

As a result of serious malnutrition, insufficient clothing and the great shortage of fuel, endurance has been seriously undermined, so that there is an increase of grave illness. These evil consequences are made even more serious by the shortage of means for carrying on medical work, cleansing and disinfection. Tuberculosis, dysentery, enteric fever and infantile paralysis are rapidly increasing in severity, while epidemics of diphtheria and scarlet fever have already reached proportions formerly unknown in Holland. The danger of typhus must be seriously faced.

The Occupying Authorities are to blame for these conditions. In the first place, because they broke International Law by transporting to Germany the large reserve supplies available in 1940, and in the years following 1940, by carrying off a considerable portion of the livestock and food produced in our country. Secondly because, now, in 1944, they are, by confiscation and abduction of nearly all transport material, preventing the Dutch people from distributing the remaining food satisfactorily over the whole country. . . .

The letter was, needless to say, never answered. But by this time, the tide of war had turned, and British and Americans, aware of the famine, were organizing teams of doctors and trained personnel to fan out over western Holland immediately after the liberation.

As it turned out, the relief teams entered ahead of the Allied armies. Seyss-Inquart, increasingly repentant as he received an avalanche of hysterical memos from Berlin, advising him to break down the sea-dikes and flood western Holland, met secretly with Allied and German officers and arranged for a nonmilitary medical mission, under a flag of truce, to pass through the lines and treat the starving.

The rescue team was greeted by the cheerful faces of the thin people on the street. Their ruddiness was in stark contrast to the emaciated bodies and faces of those in emergency hospitals and in bed at home, suffering from famine edema and lingering on the verge of death.

The official report, issued after the liberation of Holland, noted that the Dutch famine was within days of becoming ". . . a very terrible catastrophe. Had the German occupying forces held out another two or three weeks against the Allied attack, nothing could have saved hundreds of thousands in the towns of the western Netherlands from death from starvation."

INDIA
1769

Eighteen months without rain in the northern part of India from 1769 through 1770 produced a drought which created a monster famine that killed 3 million people.

Three million people perished in the three-year famine that followed an 18-month drought in the Ganges Plain of Northern India, known as Hindustan, from 1769 to 1770.

No rain fell throughout the entire area, from the Himalayas on the north to the Decca Plain on the south, and from Punjab to Assam, for a year and a half. When the monsoons finally returned, and crops began to grow again, the farmers who might have harvested them had already died of starvation. Thus, the crops that might have fed and saved thousands also died, unpicked and unprocessed.

The devastation was widespread and dramatic. Entire villages became devoid of human beings. According to one written record, in many of these villages ". . . the air was so infected by the noxious effluvia of dead bodies, that it was scarcely possible to stir abroad without perceiving it. . . ."

According to others, the air was alive with the cries of those who were in various stages of starvation, a sound and a sight that those who witnessed it were never wholly able to forget.

INDIA
1790

More than 1 million residents of the western province of Baroda in India died in a famine caused by a two-year drought, from 1790 through 1791.

An extensive famine climaxed in cannibalism among families when crop ruination and an absence of monsoons in the western province of Baroda, India, dried the countryside to dust from 1790 to 1791.

For almost two solid years, no crops were able to grow or be harvested in this province of India that had heretofore been regarded as a rich agricultural area. When the remnants of the scorched crops were exhausted, desperate residents of the area turned to cannibalism—even among families. Over 1 million perished, either from starvation or murder.

INDIA
1833

Over 200,000 died in the Guntoor Famine, caused by a drought in the Madras Presidency of southeastern India in 1833.

Portions of the Madras Presidency, in the southeast of India, were scenes of widespread human suffering dur-ing 1833. Over 200,000 persons died of starvation in what later became known as the Guntoor Famine, since the greatest concentration of drought and consequent famine occurred in this district.

According to some British authorities, the official death figures were actually skewed on the conservative side, to save face for the British colonial government. In actuality, it was the most serious famine to occur in India since the beginning of the British occupation in 1757.

INDIA
1866

Off-schedule monsoons caused a famine in the south-ern Indian provinces of Bengal, Orissa, and Behar in 1866, killing 1.5 million people.

In an odd twist of fate, an abundance of rain was respon-sible for the immense famine that caused the deaths of more than 1.5 million residents of the southern prov-inces of Bengal, Orissa, and Behar in India in 1866.

The regularity and abundance of monsoons have always been the basis of India's agricultural economy. Irrigation of crops is dependent on the monsoons. With-out them, famine results.

The monsoons arrived in 1866, but not on schedule. The first rains arrived two months before the beginning of planting season and ceased before the sowing began. The later rains, upon which farmers depended for sus-tenance of the crops and which regularly arrived at the end of September and October, did not arrive at all.

Thus, a million and a half people died of starvation and its related diseases of cholera and scurvy.

INDIA
1876–1877

India's largest recorded famine, in which 6 million peo-ple died, occurred in the Madras Presidency and the Bombay district of Poona from 1876 to 1877. Brought on by an extensive drought, its devastation was deep-ened by the spread of cholera, which was responsible for half of the recorded deaths.

The most extensive famine in India occurred in the two years of 1876 and 1877. More than 6 million people starved or succumbed to cholera within the Madras Presidency and the Bombay district of Poona.

The onset of the so-called Great Famine began two years earlier when the customary monsoons did not materialize. To make matters considerably worse, no appreciable rainfall occurred between monsoon seasons. There were a few short-lived rainstorms during the two years—enough to raise hopes—but these hopes were short-lived, as an unrelieved drought consumed whatever crops managed to break through the dry earth.

By October 1876, nine districts of the Bombay Deccan had been hit by famine, and malnutrition, coupled with polluted drinking water, conspired to produce 3 million deaths of cholera. It was one of the most tragic instances of drought and famine on record.

INDIA
1898

One million residents of a 300,000-square-mile area of southern and western India died in a drought-induced famine in 1898.

Over 61 million people were affected and 1 million died of starvation and related diseases in the immense famine of 1898 which ranged over 300,000 square miles (482,803.2 sq. km) of southern and western India.

Once again (see preceding two entries), a solid two years of insufficient rainfall and the failure of crops came together to produce a famine of major proportions that resulted in enormous loss of life, with its ancillary horrors of scurvy, cholera, leprosy, murder, cannibalism, poverty, and mass misery.

As in any drought, the scarcity of even the most basic staples of wheat and rice spawned an alarming rise in prices. Women and men, paid two and a half cents a day for hard physical labor, could not afford a minimal amount of grain to sustain their families. As a result, families were decimated. Fathers and mothers died, and children joined the vast, destitute armies of people who wandered the countryside in search of work.

These starving nomads were in turn exploited by entrepreneurs who sometimes gave them three cents a day in salary for several days' labor. A precious few survived the trek and the first few days of hard labor. Others, less fortunate and more desperate, were forced to survive by eating berries, roots, thorny cactus, and grass seed. Some resorted to cannibalism.

An ancillary result of this particular famine was the proliferation of so-called "poorhouses." These were nothing more than exploitation centers which traded marginal amounts of food in exchange for near-slave labor, prostitution, and crime.

There were good people including missionaries and government officials who offered aid during the famine. Reverend J. Sinclair Stevenson was one of these who recorded some of his efforts to stem the tide of mortality in Parantij, Gujarat: "My chief work was to take care of orphans . . ." he wrote, "but often you get them . . . just in time to fill your cemetery. . . . To go out every morning and whenever we see a child lying beside its dead mother, we, of course, take it back with us. Yesterday morning, within two hundred yards of our house, I saw sixteen corpses; today, within the same distance, ten. Must people really see ribs and skeletons to make them give?"

INDIA
1943

The immense Bengal famine of 1943, in which 1.5 million died, was caused by a confluence of factors, natural and man-made. World War II, wrong decisions by the Indian government, and human greed were the man-made factors. Heavy rains, a typhoon, and river flooding were the natural ones.

For 40 years before the Bengal famine of 1943, there had been no famines in India. In fact, the British pointed with pride to the apparent fact that one of the triumphs of the Raj was the elimination of famine in India.

But in 1943, a combination of natural and man-made factors converged and changed the records forever.

The man-made causes, so pervasive in this scenario, began on January 15, 1942, when the Japanese captured Singapore, in one of their first major victories of World War II. Following this, Burma was overrun and captured within two months. Thousands of refugees from Burma poured into Bengal through Assam and Chittagong, carrying with them not only horrible stories of atrocities but also a virulent form of malaria.

The invasion of Bengal by the Japanese was expected at any moment. That it never occurred, and that the Japanese remained merely a menacing presence on the borders of Northeast India from then until the end of the war in August 1945 did nothing to decrease the tension of expectation or the preparations for it.

Two of these preparations, in early 1942, paved the way for famine. First, stocks of paddy and rice in excess of local requirements were removed from coastal areas in the delta that were perceived to be vulnerable to invasion. Instead of being earmarked for the future needs of the populace of Bengal, this rice was distributed in Calcutta.

Secondly, in order to further frustrate the awaited Japanese, all boats capable of carrying 10 passengers or more were removed from the province. This not only cut the transport lines for the populace, it also removed the only means local fishermen had of getting their catch and distributing it to the area.

These were devastating decisions on the part of the Indian government, and they were not helped by natural conditions in Bengal. In 1941, there was a poor harvest of rice. During 1942, a combination of heavy rains and increased exports weakened the supply of rice still further. On October 11, 1942, a typhoon, accompanied by three tidal waves, struck the western districts in the delta, inundating 3,200 square miles (5,149.9 sq. km) of it, causing a sizable loss of life and the destruction of all of the standing crops and stocks of rice. Even in the areas away from the coast, the level of rivers was raised, which caused the flooding of approximately another 400 square miles (643.7 sq. km).

All of these influences, plus the factor of human greed, conspired to raise the minimum price of rice, which is the staple of the Indian diet and accounts for 80 percent of the caloric intake of the average inhabitant of Bengal.

From a price of approximately six rupees per maund (82 pounds) in January 1942, the price of rice skyrocketed to 40 rupees per maund in June of that year. By the end of 1943, it was reckoned that the unprecedented profits made in the buying and selling of rice had reached Rs 150 crores, which translates into approximately $24 million.

The toll of humanity was far more grim. By June 1943, the price of rice was fantastically out of reach for the poor. Landless laborers, small tradesmen, weavers, and potters who had managed marginal livings were forced, in order to buy food, to sell at cutthroat prices their domestic utensils, ornaments, tools, clothes, even the doors and windows of their dwellings to more fortunate neighbors. Eventually, their short rations began to dwindle, while the residents on the other side of the economic fence, those who could buy and store rice, began to reap monumental profits.

By May, the death rate from starvation began to rise in six districts in the Delta, particularly in Chittagong. In June, it was double the quinquennial average and three to four times the normal July average.

The government in Calcutta remained unaffected by this situation until August, when thousands of people began to drift into the city, propelled there by rumors that food was to be had. They were a placid army, begging for food, often dying in the street just outside locked storehouses that were filled with the very rice that could have prevented their deaths. The dying were the poor, used to accepting their misfortune, and by the time they reached Calcutta, most of them had sunk into the apathy that precedes death by starvation.

Author W. R. Aykroyd, who was a member of the National Inquiry Commission appointed in July 1944 by the government of India to investigate the causes of the famine, recalled, in his book *The Conquest of Famine*, the scene in the outlying districts:

> In August I was traveling by rail from Madras to Calcutta. . . . It was customary for the Madras-Calcutta mail to pick up a dining car at Khargpur Junction, some thirty miles outside Calcutta, to provide first and second-class passengers with breakfast. I stepped cheerfully down from my compartment en route for a hearty meal. The whole platform was thronged with emaciated and ragged people, of all ages and sexes, many half-dead, hoping to board a train for Calcutta. What I remember is a loud, bleating, wailing noise which the starving crowd made, a combination of begging and misery. . . . I could not eat breakfast in the dining car and went back to my compartment.

The death rate from the famine reached a peak in December 1943. In its early days, most deaths were caused by starvation, but later, hundreds of thousands of deaths were attributed to smallpox, cholera, and malaria.

The government of Bengal tried to stem the tide of deaths. But their methods were too little, too late—distracted, at that, by the war and the perceived necessity of keeping the factory workers adequately fed and thus capable of turning out materiel for the army to defend India against a Japanese invasion which never occurred. Thus, out of 260,000 tons (235,868 tonnes) of rice that the government bought at exhorbitant prices from profiteering dealers, 140,000 tons (127,005.8 tonnes) were used to feed Calcutta, while only 65,000 tons (58,967 tonnes) were sent to rural Bengal, where the famine began and was raging most furiously.

It would take the end of the war to end the famine, and the Commission would estimate that one and a half million victims would be officially counted. But this figure was, at best, conservative, according to Aykroyd, who noted, ". . . whenever I come across [the historical figure of 1.5 million dead] I remember the process by which it was reached. I now think it was an underestimate, especially in that it took too little account of roadside deaths, but not as gross an underestimate as some critics of the Commission's report, who preferred 3 to 4 million, declared it to be. The lower figure is tragic enough."

INDIA
1972

Drought in the Ganges basin caused the 1972 famine that affected most of India and killed 800 people.

Over 800 persons died of heat exposure or starvation, 50 million residents of India were affected, and the Indian economy was $400 million poorer as a result of a combined heat wave and drought that struck 14 Indian states in May 1972.

May is generally a benign month in India in terms of hot weather. In this particular year, however, the annual monsoons that enter the Bay of Bengal and soar up the Ganges basin, generally watering the crops and lowering the temperature, failed to arrive. As a result, thousands upon thousands of acres of sugar cane and jute broiled on the vines or failed to mature, while the sun baked the land and kept the thermometer at an even 110 degrees for nearly a month.

When the rains finally came, the soil had been so devastated, it washed away in gigantic mud flows, making the resowing of crops extraordinary difficult.

IRELAND
1845–1850

One-fourth of the entire population of Ireland— 2,209,961 people—either died of starvation or emigrated as a result of the Great Potato Famine of 1845–50.

If the sources of a national, binding emotion can be traced to one event, the Irish hatred of all things British would have to be traced to the Great Potato Famine that raged across this small and verdant country for five years, from 1845 to 1850. During that time, fully one-quarter of the country's population was eliminated, either from death or emigration—1,029,552 died of starvation, typhoid, typhus, or scurvy and 1,180,409 emigrated, mostly to America—and the seeds would be sown for a deep-seated enmity that would erupt in a bloody revolt whose echoes and actuality continue to reverberate into the present.

A short 45 years before the famine, Ireland was, if not rolling in wealth, at least touched by it. But with the departure of Wellington's forces in 1815, its Irish contingent was abruptly thrown upon the Irish labor market, glutting it.

That might have been absorbed, since Ireland has an abundance of rich soil and the crops to prove it. But the British withdrew more than Wellington. They were, the Irish soon found, experts in extraction through tariff and taxation. The Corn Laws imposed impossible tariffs on small landowners and funneled herds of cattle, boatloads of oats, rye, and corn away from Ireland's increasing population and into England. "Immense herds of cattle, sheep and hogs," reported Irish writer John Mitchel, "floating off on every tide, out of every one of our thirteen seaports . . ."

And so, while the populace of Ireland continued to multiply (in 1800 it reached 5 million—a population exceeding that of America at the time), its breadbasket continued to empty, until the Irish, along with the Belgians, earned the name of "potato-eaters," after the new staple of their diet.

And so Ireland was not a pretty sight in the early 1800s. Poverty was everywhere, and, with most of its agricultural output being exported to the ruling British, hunger soon followed. Thomas Carlyle, touring the land, wrote, "Never saw such begging in the world. . . . Often get in a rage at it . . . [beggars] storming round you, like ravenous dogs round carrion . . . human pity dies away into stony misery and disgust at the excess of such scenes."

Worse was to come when British landlords evicted tens of thousands of starving peasants when they were unable to pay their rent. The Earl of Lucan in County Mayo, eulogized by Alfred, Lord Tennyson in his "Charge of the Light Brigade," evicted 40,000 peasants from the hovels upon which he was collecting rent.

And then the potato blight hit. It was a chance for the British to save both their honor and millions of lives. Instead, the Earl of Lucan stepped up his evictions and his rents. The *London Times* Irish correspondent, the Reverend Sidney Godolphin Osborne, called these actions "philanthropic," claiming that they helped stabilize populations. Gentle Tennyson added his voice, noting that "Kelts are all made furious fools. They live in a horrible island and have no history of their own worth the least notice. . . . Could not anyone blow up that horrible island with dynamite and carry it off in pieces—a long way off?"

So much for disaster relief in the 19th century.

In Ireland itself, misery spread like the black plague. Large parts of the population, swelled to 8.2 million in 1845, were existing on "lumpers," a gray tuber used for pig fodder in all other parts of the world, and water.

When even this was reduced by blight, starvation began to increase. Tens of thousands died in the privacy of their homes; others perished along the sides of public roads. Along with cholera and scurvy, there was death by exposure as a result of mass evictions when peasants could no longer harvest wheat crops with which to pay their rent. There were incidents of cannibalism.

Shallow graves were dug along the sides of roads, but the graves were frequently raided by hungry dogs, who dismembered the corpses and scattered them about the countryside.

In *The Black Prophet*, William Carleton wrote, "The roads were literally black with funerals, and as you passed from parish to parish, the deathbells were pealing forth in slow but gloomy tones, the triumph which pestilence was achieving over the face of our devoted country—a country that was every day filled with darker desolation and deeper mourning."

Still, the ships loaded with grain and produce continued to set out from Ireland's 13 ports, while other ships loaded with emigrants began to fill with the nearly 2 million who would flee Ireland and give rise to its continuing reputation as a country whose greatest export was its young men. But in the 1840s, these emigrants found life scarcely better elsewhere. In England, they were treated as pariahs and forced into hovels and cellars. In America, they met similar resistance, the infamous "No Irish Need Apply" signs everywhere, forcing them to gather in Irish ghettos in Boston, New York, and Baltimore, where a half-million of them would perish, miles away from home.

England did make some halfhearted attempts to establish work projects in Ireland—hard labor at low wages, building roads to nowhere. The large public projects that might have turned the death rate around were lost in parliamentary squabbles. Benjamin Disraeli remarked, "One day the Pope, the next day potatoes." Lord Salisbury, as prime minister, compared the Irish to "Hottentots," incapable of both self-rule and self-survival.

Only Lord John Russell, speaking in the House of Lords on March 23, 1846, voiced some concern and responsibility. "We have made Ireland, I speak deliberately—we have made it the most degraded and most miserable country in the world. . . . All the world is crying shame upon us; but we are equally callous to our ignominy and the results of our misgovernment."

But his remarks would be buried beneath indifference and other matters, like the beginning of the Anglo-Sikh War in India. Time would be the curer of Ireland's famine and the sustainer of its rage.

JAMAICA
1788

A series of major hurricanes left much of Jamaica unfarmable in 1777. By 1788, a resultant famine killed 15,000 of the island's 25,000 slaves.

Both the rulers and the ruled, the privileged and the underprivileged, were affected by a huge famine that swept through the island of Jamaica in 1788. Landowners and slaveholders lost money. Fifteen thousand of the 25,000 slaves on the island lost their lives.

The famine had its roots in a series of violent hurricanes that swept through the island in the fall of 1777. The devastation from these hurricanes was particularly felt by the 775 sugar plantations, owned by whites, worked by slaves. Over 200 of these estates lay in utter ruin after the hurricane left, unfarmable until more financing could be brought from England.

In the ensuing six months, before rebuilding funds could arrive and the moderately damaged plantations could be replanted, work disappeared for the island's slaves. And their masters refused to feed them if they did not work. Riots erupted throughout the colony and dozens of whites were killed. But the riots proved ultimately ineffectual, and by the end of the six months, over two-thirds of Jamaica's slave population had perished from starvation.

MICRONESIA
1998

A crippling shortage of water produced an encompassing drought that afflicted 60 of the 65 inhabited islands of Micronesia for the entire year of 1998. Only the intervention of U.S. and international relief agencies averted a national tragedy.

Micronesia, a grouping of the westernmost islands of the Pacific, was formed from volcanic mountain ranges that rose abruptly from the deep ocean. It comprises the former island groups of Kiribati (formerly the Gilbert Islands), Guam, Nauru, the Commonwealth of the Northern Marianas, the Federated States of Micronesia (Yap, Pohnpei, Chuuk, and Kosrae; formerly, the Caroline Islands), the Republic of the Marshall Islands, and the Republic of Palau.

Mangrove and tropical forests and scrubland constitute most of the ground cover, while the reefs and lagoons abound in lobsters, shrimp, clams, oysters, octopuses, turtles, and innumerable fish species.

As a result, the economies of most of the Micronesian islands are based on subsistence farming and fishing. The principal cash crop is coconuts, though Guam and some other large islands rear some cattle and water buffalo. Pigs and chickens constitute the livestock on most farms.

At the beginning of 1998, the El Niño factor caused a major drought on 60 of the 65 inhabited islands of Micronesia. Streams and rivers dried up, and sources of irrigation were halted because of a lack of water supply.

By the beginning of March, many areas were either totally without water, or on water rationing systems. Schools and public facilities ran dry; wells began to fall to dangerously low levels, and the salinity in them increased, making them unfit for drinking or feeding livestock.

Lack of irrigation caused crops to wither and die, but the threat to human beings was of even greater concern. A health crisis seemed imminent in April when an increase in bacteria and contaminants in the remaining water supplies was reported. Three cases of cholera surfaced, but were immediately treated.

And then, for unknown reasons that probably had to do with variations in the water temperature, fish, the staple of the diet of Micronesians and a major portion of their economy, began to become less and less available. Fish catches dropped precipitously.

Nearly 29,000 people, particularly on the outer islands of Pohnpei, Chuuk, and Yap, were in dire need of water, and it began to be shipped in by international relief organizations. Water rationing became universal throughout Micronesia.

The Clinton administration declared the entire collection of islands a disaster area, and began to fly in water supplies to all of the afflicted islands except Kosrae, which was able to tap into newly drilled water wells, rich enough in water to sustain the island's population of 7,317.

On Pohnpei and Chuuk, all normally operated shipments of cargo and passengers to the outer atolls were canceled and the ships were used to transport water.

From April through January of 1999, emergency food was distributed in huge quantities to the afflicted islands. Approximately 57,000 people at 58 locations received assistance in the form of tons of dairy products, fruits, vegetables, flour, and other edible commodities.

By February of 1999, the crops that had been destroyed by the drought were cleared, and new ones planted. Normal rainfall resumed, and the native vegetation and food crops responded. William Carvile, the coordinating officer for FEMA (Federal Emergency Management Agency), which oversaw, with local governments, the distribution of relief supplies, stated in his final report, "We're gratified that this effort has greatly relieved a dreadful situation for thousands of people."

NIGERIA
1968

One hundred thousand died—most from ancillary reasons—in the drought-caused famine in Nigeria from 1968 through 1973.

Much of Nigeria lies in the Sahel region of Africa, one of the most drought-prone areas of the world. Over and over, thousands die when rain refuses to fall, rivers dry up, and crops wither.

The six-year drought that began in 1968 and ended in 1973 claimed 100,000 Nigerians. The causes of their deaths were recorded in a startling report by medical missionary Dr. John A. Dreisbach, a 30-year resident of the area who noted, in the National Geographic Society's 1978 survey of natural disasters, *Powers of Nature:* "I saw thousands of cattle carcasses. . . . Nomads who once owned them were in relief camps eating gift food sent by the rest of the world. We saw much malnutrition and sickness. In their weakened condition, many people died of pneumonia, measles, and whooping cough. I saw little starvation as such."

It was there. But it accounted for only a fraction of the overall death rate. More than three-quarters of the deaths were from disease—typhus, pneumonia, measles, and whooping cough.

NORTH KOREA
1996–1997

Famine and drought brought about a small rapport between North and South Korea, as lines of aid were established to counteract a severe drought and its resultant famine in North Korea in late 1996 and the summer and winter of 1997.

Droughts at the end of the 20th century seem to have resulted from a multiplicity of factors, only some of which are natural. In North Korea, two years of floods in 1995 and 1996 were followed in 1997 by a summer of extreme drought, which was brought about not only by scorching heat that dried out millions of acres of crops but also by useless irrigation pumps disabled by an absence of spare parts, wholesale deforestations and eroded hillsides planted with the wrong crops, and cutworms devouring fields because there were no pesticides to fight them.

Coupled with this was a refusal by the North Korean government to move forward economically, to adopt modern agricultural methods, and more importantly to accept aid from outside of its borders. Particular distrust of South Korea prevented the import of offered aid, until it became apparent that vast numbers of North Korea's 23 million people might starve to death.

Han S. Park, a political scientist from the University of Georgia, described the situation in North Korea in graphic terms. "The entire population is in the process of slow death," he reported, and a UN official described it as "famine in slow motion."

Finally, in the last week of May 1997, an agreement was signed with South Korea in Beijing. Aid was also accepted from the United States, but more importantly for peace in the area, 50,000 tons (45,359.2 tonnes) of corn, noodles, and other food covered by the agreement began its circuitous trip from South to North Korea.

Reportedly concerned that it would be demoralizing for its hungry soldiers to see truckloads of food rolling in from enemy territory across the border at Panmunjom, the government of North Korea routed the trucks through China, and thence transferred the supplies to trains, which carried them over three rail lines to two seaports.

It was a revealing occurrence: One of the most closed societies in the world, North Korea was particularly suspicious of anything from the South. And yet the reality of drought and its resultant famine finally pierced a half-century of enmity.

It was a start, which grew as the enormous, El Niño–induced floods hit all of Korea and much of Asia in 1998 (see pp. 153, 158, 168).

NORTH KOREA
1997–2006

A combination of natural and political forces joined to both establish and prolong one of the worst famines of modern times in North Korea. From 1997 through the present, reports from U.N. aid agencies report that 57 percent of the 23 million people in North Korea are without enough food to keep healthy; 36 percent are undernourished; and of those under six years old, 37 percent suffer from chronic malnutrition. Because of the secretive nature of the North Korean government, there is a wide disparity of the numbers of casualties in this disaster: The Democratic People's Republic of Korea's government sources quote 220,000 deaths, the U.S. Agency for International Development quotes 2.5 million, and the Institute for International Economics quotes from 600,000 to 1 million dead.

When the forces of nature and the forces of a careless government unite, the effect can be catastrophic. Such is the continuing case in North Korea, where the suffering of millions of people, as a result of an extended drought, has been exacerbated by the government of Kim Jong-il. This government, for reasons that are its own, ignored a gathering disaster, hid its effects from the rest of the world for too many years, and continues to be seemingly detached from the gathering calamity.

In 1995 and 1996, a series of floods caused a shortfall of 2,500,000 tons (1.8 million tonnes) of food, and flooded North Korea's most productive coal mines, which reduced the power production in the country by 30 to 40 percent. This, in turn, fostered a huge rash of deforestation, brought about by the people who, deprived of heat and fuel, began to burn forests woods for cooking and heating. This desperation led to flooding and the loss of top soil in the enormous areas denuded of trees. Then, a mild El Niño event caused a drought in 1997 that ruined 700,000 tons (635,030 tonnes) of maize, a staple in the Korean diet. In short order, the inefficient collective farming system was pushed to the brink of collapse. All the while, the dikes and dams used for irrigation in North Korea were deteriorating because of insufficient maintenance and because government funds were redirected toward other projects, such as the military.

Finally, in the summer of 1997, word began to filter out as United Nations and other relief agencies were allowed into the country. "The people of North Korea appear to be suffering from hunger on the level of the notorious Somalia and Ethiopia famines," Ted Yamamori, president of Food for the Hungry told western reporters. "Only in North Korea, they are suffering in silence out of the view of the world's media. In kindergartens and nurseries I saw cases of severe malnutrition."

By 1998, the famine left North Korea's 23 million people largely dependent upon international aid. In three years, this famine had already killed an estimated 1 million people—nearly five percent of the population, largely from famine-related illnesses like pneumonia, tuberculosis, and diarrhea.

"Hospitals are really hospices," Mark Kirk, a member of a U.S. congressional delegation to the country remarked on his return. "There is little or nothing in North Korea's entire health care system. The hospitals have no food, no X-ray film, no aspirin."

By 2001, the world finally began to respond with huge amounts of foreign food aid for a country whose agricultural system was in a state of total collapse, compounded by, with the fall of the Soviet Union, the disappearance of trading partners and sanctions imposed because of foreign missile sales.

The combination of famine and economic collapse had cut the life expectancy of North Koreans by more than six years, and the mortality rate for children under age 5 had risen from 27 deaths per 1,000 to 48 per 1,000. The infant mortality rate rose from 14 to 22.5 per 1,000 births. Norbert Vollertsen, a German doctor who toured the county, reported a grim scene: "They have no running water," he said. "No electricity . . . They do not have any medicine, no bandage material, no drugs, no nothing. Some of the children are in such bad condition, they've no emotional reaction anymore. They can't even scream."

In the late 1990s, North Korea had set up "alternative food" factories, making small bricks from bark and leaves. But the food had hardly any nutritional value, and it also caused internal bleeding, dysentery, and diarrhea, and the program, over time, disappeared into the fog of privacy thrown up by the North Korean regime.

By 2002, according to the U.N.'s World Food Programme (WFP), people were reduced to eating grass and acorns to survive. Hundreds of thousands of them had abandoned work and school to forage for these last stands against starvation. "They're going up into the mountains in search of edible grasses," WFP spokesperson Gerald Bourke reported. "They're on the beaches collecting seaweed. Teachers say attendance at school is down because children are out collecting wild foods. Teachers themselves and so-called caregivers at kindergartens, nurseries, and the like are having to take time off from work for the same reason."

In 2003, reports from some of the thousands of refugees escaping the famine through China stated that children had been killed and corpses cut up by people desperate for food. When the WFP requested permission to be allowed to access farmers' markets where human meat was said to be traded, they were turned down, citing "security reasons."

The WFP has had its own troubles with the government in the capital P'yongyang. From the late 1990s to 2003, the WFP fed 1.9 million of North Korea's 23 million people with donations from the United States, Japan, South Korea, the European Union, Australia, Italy, Germany, Canada, Sweden, Russia, Ireland, and Norway. But as soon as the aid began to arrive, the North Korean government took the curious step of reducing its commercial imports, diverting the money it saved to other priorities including the military. In 1999, at the same time it was cutting grain imports, it spent scarce foreign exchange on military equipment.

In January 2003, North Korea withdrew from the Nuclear Non-Proliferation Treaty, telling the United States it had nuclear weapons and might test them or transfer them to other countries. At the same time, P'yongyang modified its rationed food program to cut cereal rations to 7–9 ounces (200–250 g) a day—half the recommended amount. In October 2005, after international negotiations, it raised its cereal ration to 16 ounces (500 g) a day, but at the same time, the government closed down 19 food factories established by the WFP to produce fortified foods and oversee food-for-work programs and shut down the monitoring program installed by the WFP.

In addition, more horror stories were related by defectors from the country. Labor camps, they said, had been set up for anyone challenging North Korea's system of governing. They described forms of torture, public executions, forced abortions and campaigns to kill disabled babies. Former prisoners told how they survived by catching and eating rats.

In December 2005, the WFP pulled out of North Korea, citing disagreements over the monitoring of future aid, but in May, after months of negotiation, it signed a new agreement, for a two-year program that would focus on providing vitamin- and mineral-enriched foods processed at local factories to young children and pregnant and nursing women.

And so the forces of nature and the forces of politics continue to define the terrible famine conditions in North Korea. In July 2006, its government defied international warnings and test fired seven missiles. At almost the same time, the country was hit by three major storms that caused floods, killed hundreds of people, and displaced tens of thousands more. The government initially turned down offers of aid, but later changed its mind and accepted humanitarian assistance from the South Korean Red Cross and the WFP.

And so the grim seesaw continues in its abrupt interchanges. The United States has withdrawn its contribution of food for North Korea through the WFP. Some nations have asked whether the world should be giving aid at all if it means extending the life of a despotic regime and whether aid is just prolonging the very policies that contributed to famine in the first place.

"We have to pose the question whether through giving humanitarian aid we are at the same time reinforcing perhaps the worst political regime on the planet," suggested former Czech president Vaclav Hável in the introduction to a Hunger and Human Rights report to the United Nations.

RUSSIA
1921–1923

The Russian famine of 1921–23, in which over 3 million people died, was caused by a combination of drought, the depletion of granaries from World War I, a civil war, and an international blockade.

Although the Volga River basin in Russia is one of the richest sources of agriculture in the world, it is also one of the most vulnerable to drought. Since 900 C.E., Russia has suffered over 100 famines, and one of the worst of these occurred in 1921. More than 30 million people were affected by it; over 3 million died from it.

The causes of the famine were twofold. First was a lack of rain in 1920, followed by a complete drought in 1921 that left one of Europe's most productive cornlands a blackened waste.

That in itself would not have been enough of a natural setback to cause a famine that covered an area of a million square miles and affected so many millions of people. It was the economic effect combined with this that produced the cataclysm of 1921–23.

During World War I, granaries in Russia were depleted, and most of the able-bodied farmers were conscripted into the army.

Immediately following World War I, Russia became embroiled in civil war, which did not conclude until 1920. During that time, fully half of the arable land lay fallow. And if this were not enough, Russia was blockaded by other nations until 1921, thus preventing any slack in domestic supplies from being made up through imports.

The new Bolshevik government found itself in great trouble. Seven provinces on the Volga were in drought and in need. In the province of Samara alone, 70 percent of the cornfields had failed completely, and without help, most of the peasants would not survive the winter. By September 1921, 1.2 million people were already starving.

Several international relief organizations were formed, and the most prominent of them was the English and American Society of Friends, which faithfully recorded the event. Michael Asquith, a member of the first Quaker unit to be headquartered in the town of Buzuluk, in Samara, documented the famine in his book *Quaker Work in Russia 1921–23*. By September, when Asquith's party arrived, the populace was already subsisting on grass and acorns, a diet that was rapidly killing most of the young children, who could not digest it. Asquith wrote:

> I saw in practically every home benches covered with birch or lime leaves. These are dried, pounded, mixed with acorns, some dirt and water, and then baked into a substance which they call bread, but which looks and smells like baked manure. . . . There are practically no babies and those that survive look ghastly. The mothers have no milk and pray that death may come quickly. . . . All the children have distended stomachs, many are rachitic and have enlarged heads. . . . According to Government figures, ninety percent of the children between the ages of one and three have already died from the famine. . . .

The first snowfall took an enormous toll; one Quaker reported seeing 40 people die in one morning at the beginning of November.

The government attempted to stop the death of children by setting up receiving homes, distributing homes, and permanent homes. These became graphic exercises in futility, shelters where the young went to die. One receiving home in Buzuluk was described in detail by a Quaker worker:

> As a home it was intended for 50 children, but yesterday 654 children were crammed within its walls. On such days as many as 80 are brought in. The stench inside was indescribable . . . as we entered we became aware of the continual wailing sound that goes on day and night. In each room . . . there were at least a hundred children packed like sardines in canvas beds—six on a bed intended for one, and underneath the beds as well. The typhus cases, some of them completely naked, lay on straw in a separate room. They had neither bedding, medicines nor disinfectants, though we had been able to give them a little soap and clothing. They had no doctor.
>
> Each morning the attendants picked out the dead from the living and put them in a shed to await the dead cart which every day makes its round of the Children's Homes.

One home's record summarizes the situation: "Admitted, 1,300; Died 731."

As the famine increased, cannibalism occurred, but it was not as widespread as in other famines of this extent. Still, the Friends Relief Committee met and decided to give up buying cheap sausages when some were found to contain human flesh.

It was not until the middle of 1922 that the International Russian Relief Commission was able to surmount various bureaucracies and begin to set up soup kitchens—a process that had succeeded in the waning days of the Great Irish Potato Famine (see p. 127)—to feed up to 2 million people.

Finally, in the autumn of 1923, a sizable Russian harvest was reaped, and the International Relief Commission turned its attention to clothing the poor, replacing farm equipment and harnesses that had been boiled and eaten, importing farm horses from Siberia, and tending to some 13,000 abandoned children. It was a baptism by fire and famine for the new Soviet government.

USSR
April–December 1975

Few died in the Soviet famine of 1975, but the economic impact of the loss of a wheat crop caused the starvation of tens of thousands in Third World countries.

Although there was little or no loss of life from the extreme drought that swept across the USSR from early spring through the end of 1975, the worldwide economic impact was immense. Deprived of most of its wheat crop, the Soviet Union bought millions of tons of wheat from the United States, Canada, and Argentina. As a result, American farmers reaped a bonanza, and the price of bread doubled worldwide—a situation that caused starvation and deprivation in Third World countries thousands of miles from the drought itself.

Statistics of the 1975 drought in the USSR were characteristically withheld from the rest of the world, but what is known is that the absence of rain in a small spot near Kuibyshev, between the Volga River and the Ural Mountains, in April of that year was the harbinger of far worse conditions to come.

Ninety percent of the USSR's wheat is grown in a fertile triangle formed by the Baltic, Black, and Caspian Seas. Ordinarily, this is an ideal spot for wheat: rolling plains, rich soil, a steady but not overly abundant supply of rain, which irrigates the soil without leaching it out. In 1975, Russian farmers relied exclusively upon natural irrigation to keep their huge wheat crop healthy, despite the fact that the Soviet Union's farmlands are the most drought-plagued in the world.

In 1975, the rains did not come. The great plains east of the Urals baked under unrelieved skies all summer. By autumn, the rich topsoil blew off farms in enormous, dun-colored clouds.

Autumn passed; winter set in with no appreciable precipitation but with widespread drought and poverty. It finally took the onset of the deepest part of the Russian winter, in January 1976, for snow to reirrigate and enrich the soil enough to prepare it for a successful early spring planting.

UNITED STATES
EAST
1998–1999

The worst drought to hit America since the dust bowl drought of 1934 beset the eastern United States in 1999. Thousands of farms went bankrupt and most of the states east of the Mississippi and north of South Carolina were declared a national disaster area by President Clinton.

The second worst drought of the century (the worst was the great dust bowl calamity of 1934 [see p. 134]) struck the United States east of the Mississippi River at the end of August of 1998 and continued through the summer of 1999. West Virginia, Kentucky, Ohio, Maryland, Pennsylvania, New Jersey, New York, North Carolina, and the New England states were all victims of scorching temperatures and less than an inch of rain during the summer of 1999. Of these states, Maryland, Ohio, Virginia, New Jersey, and Rhode Island suffered the heaviest crop losses.

Wells and rivers ran dry, pastureland turned brown and disappeared, corn that normally rose to a height of eight or 10 feet (2.4 or 3.1 m) barely reached farmers' knees, the apple orchards in upstate New York were flogged by sun; apples that should have been three inches in diameter were only half that at the beginning of August, and those that grew were splotched with sunburn.

Sixty-three-year-old Mark Roe, a New York farmer who lost 60 percent of his corn crop, told a government inspector, "This is the driest and hottest I've ever seen. And that's a tough combination for the crops."

In West Virginia, there was a 50 percent failure of hay crops, a 60 percent loss of corn crops, and vegetable growers faced a wipeout. The drought, the worst ever in West Virginia history, drained aquifers. The south branch of the Potomac River was at less than half its record low from the 1930s. Thirty-four percent of wells across the state went dry and 35 percent of farmers began to haul water for their livestock from distances as far as 25 miles (40.2 km).

With a shortage of hay and feed for the winter, farmers were faced either with economic disaster or selling off their cattle. The problem in August was that the price of beef and milk was severely depressed, and selling off their herds became a losing alternative.

The drought was a cruel blow to residents of the area, who had just recovered from three major floods that had swept through West Virginia in 1997. Heavily in debt from repairing fencing, homes, roads, and crops, many farmers were left with little or no financial reserve, and it was estimated that at least 10 percent of West Virginia's 21,000 farmers would have to give up and sell their farms.

Farmers in southeastern Wisconsin reported that they had lost 50 percent of their corn crop; in New Jersey, more than 7,000 family farms reported crop losses.

In Ohio, where most farms relied upon three hay harvests a year to feed their animals, dry conditions limited it to one crop cutting. Farmers there, too, were forced to sell off their livestock.

Corn and soybean losses were up to 80 percent in Ohio, and at least 67 counties, more than two-thirds of the state, were declared agricultural disasters. Nationally, drought-related losses totaled over $1 billion.

One ancillary problem visited Pennsylvania. In the western part of the state, tourist-based businesses that offered white-water rafting and other river recreation were forced to shut down, which meant the loss of about 1,000,000 visitors—a huge blow to the local economy.

President Clinton declared the entire, multi-state expanse a national disaster area, and eventually, in the winter of 1999, the rains returned. But they fell on a dramatically decreased number of farms.

UNITED STATES
MIDWEST
1909–1914

Several thousand died in the Great Plains drought of 1909–14, in which unique and ineffectual methods of inducing rain were employed.

Although mild in comparison to the great dust bowl of the 1930s (see following entry), the Great Plains drought of 1909–14 caused widespread deprivation. For five years, the winter winds brought nothing but cold and the spring breezes only isolated showers. Summers were unbearably, hot, and crops either did not germinate or withered on the vine.

This was the age of rainmaking folktales. In the 1870s, settlers had been lured to the southwest desert by folklore that stated that "rain follows the plow." In the 1890s, perhaps taking their cues from Indian rain dances, farmers paid as much as $500 to itinerant and ineffectual rainmakers.

In the drought of 1909–14, famous cereal magnate C. W. Post, hearing from Civil War veterans that rain always followed heavy bombardments, employed armies of men to set off scores of dynamite explosions. The hope was that the first explosions would cause clouds and the later ones rattle rain loose from these clouds. It didn't work. Only time and the cycle of nature returned rain to the parched land.

UNITED STATES
MIDWEST
1934–1941

Thousands of farmers-turned-migrants died when a combination of drought, unwise use of land, and bank foreclosures converged to cause the great dust bowl of 1934–41.

A combination of natural and man-made circumstances brought on the famous "dust bowl" drought of the 1930s, in which 25,000 square miles (40,233 sq. km) of the midwestern United States became one vast wasteland of dried soil and unproductive land. More than 300,000 farmers and their families, driven off their ancestral land by a combination of a hostile nature, the economics of the Great Depression, and their own folly, became a vast, exploited army of nomads, chronicled vividly in John Steinbeck's powerful novel, *The Grapes of Wrath.*

The troubles of this area began during World War I, when the high price of wheat and the needs of the Allied troops encouraged farmers to expand their wheat fields into pasture land. By cramming their livestock into abbreviated grazing pastures and plowing under every square foot of available land and planting it with wheat, these farmers made small fortunes in a short period of time.

But when the war ended, demand shrank to its prewar levels, and these same farmers returned the plowed fields to their livestock without first properly seeding and preparing them. Within a few years, the cattle's hooves pulverized the unprotected soil, and when strong winter winds blew in from the northwest, this topsoil was borne eastward on the wind.

Now, a combination of forces gathered to produce drought and famine. The topsoil that was blown away was gone forever. Spring and summer rains became more infrequent. Beginning in 1934, the northwest winds blew relentlessly from December to May, picking up not only the soil but also seeds that had lain in it for six months without sprouting. Tumbleweeds whirled across the landscape, fetching up against fences and houses. Stock tanks were emptied of water and filled up with dust. The dry soil lifted itself in thousand-foot clouds that obscured the sun, engulfing farmhouses and forcing itself inside, into cook pots and beds. Traces of dust and wisps of winter wheat were driven by these winds as far east as New York and Washington, D.C.

In July and August, the bitter northwest winds were replaced by furnace blasts from the deserts of the southwest. The bare land baked under horrendous heat which often reached and sustained, for days on end, temperatures of 110 degrees—enough to kill rattlesnakes.

The wind-driven dustclouds resembled horizontal tornadoes, according to Kansas farmer and author of the memoir *Empire of Dust,* Lawrence Svobida, who

The effects of the great drought and famine of the 1930s in the American Midwest are etched into the faces of this mother and her child, two of thousands made homeless and forced to wander from shelter to shelter, temporary job to temporary job. (Library of Congress)

described conditions in his state to the *National Geographic*. When a duststorm came, he said,

> You got inside the house quick, watched the cloud coming, and felt it envelop you. I've known storms to last 12 hours. Dirt clicked like sleet against the window glass.
>
> You'd hear almost continuous thunder, and the crackle of lightning—the friction of dust particles throws off a lot of static electricity. Streaks of it ran back and forth around metal structures—gasoline pumps, for instance. Radio aerials lit up like fiery crosses. That was scary. Static electricity made the ignition systems on cars fail. That could be fatal if you got caught on the road in a storm.

Lives were claimed by these storms. Winds of 30 MPH (48.3 km/h) with gusts up to 60 MPH (96.6 km/h) blew enough dirt into the nostrils of cattle to suffocate them.

"Dust did kill some people I knew," Svobida remembered. "One young fellow about my age spent his last day alive plowing in a dust storm, trying to roughen the topsoil to keep the wind from blowing it away. He stumbled into the doctor's office that night to be told he was dying on his feet, filled up with dust."

These assaults of man upon nature and nature upon man were fatal, but not altogether insurmountable. A drought cannot last forever.

But this occurred during the time of the Great Depression, which formed the final third of the equation. The depression brought on an overload of poverty, starvation, and hopelessness, which, piled upon the other misfortunes, collapsed the strongest will. Farms that contained some planting room could count on small subsidies—usually under $1,000 a year—for, ironically, *not* raising wheat they could not raise anyway.

Even with the dust bowl tragedy, there was a surplus of wheat in the United States in the 1930s. But those farmers who had only grazing land which would grow nothing starved, and were eventually run off their land by the banks holding mortgages the farm families could no longer pay.

And so that great, nomadic migration began. The disparagingly named "Okies" (even though most of them came from Kansas), set out for California with all of their possessions strapped to broken-down cars and trucks, lured by promises of jobs and prosperity. Once there, they were turned away, or, if they were employed, were paid not in money but in company scrip, worthless any place but at the exorbitantly priced company store.

Finally, in 1940, the rains came, and a decade of normal-to-wet weather, combined with regrassing and erosion-preventing procedures, revitalized the Central Plains of the United States and restored its reputation as the "Breadbasket of the Nation."

But that was too late for many of the migrant workers who had lost both their land and their belongings in the 1930s. Thousands of them perished miles from their roots.

UNITED STATES
SOUTHWEST
2005–2006

From autumn 2005 through summer 2006, the entire southwestern United States was in the grip of a record-breaking drought, caused by two forces: a weak La Niña pattern along the equator in the Eastern Pacific and an active storm track that brought storms into the northern Rockies while leaving areas to the South untouched. Wildfires burned 3.7 million drought-dried acres (1.2 million hectares) in Texas alone, killing livestock and residents, and rendering an entire crop of winter wheat unharvestable.

The La Niña phenomenon—the cooling of ocean waters in the east-central equatorial Pacific—seemed to be hard at work in the southwestern United States from 2005 to 2006. Drought conditions began in parts of the south and southwestern United States from Arizona to Arkansas and Louisiana in early 2005. By January 2006, the La Niña events began to assert themselves. As La Niña simultaneously caused above-normal precipitation in the Northwest and Tennessee Valley areas, the drought conditions intensified.

As winter turned to spring, the drought grew in central and eastern Texas, central and eastern Oklahoma, and western Arkansas, all of which were declared by the National Oceanic and Atmospheric Administration (NOAA) to be "exceptional" drought areas in which grass and other vegetation had been starved of rain and snow for over a year.

Phoenix, Arizona, had no rain from November through February for the first time in history. Sante Fe, New Mexico, had its driest winter since 1890, with only 0.27 inch (0.6 cm) of rain from November through the following summer. "What can you say?" Joe Garcia from the 1.1 million-acre (445,154 ha) Lincoln National Forest in New Mexico told a reporter. "It's been a historical winter that wasn't." *The National Drought Monitor* reported that Tucson, Arizona, was so dry, some homeowners were watering cactus and other desert plants to keep them alive.

A grim side effect of this widespread and long lasting drought was the threat of wildfires. Texas and Oklahoma suffered fierce winter grass fires that scorched thousands of acres, destroyed about 500 homes, and killed at least five people. In Texas, four oilfield workers died in a roaring wildfire and the Texas Animal Health Commission stated that about 10,000 cows and horses were believed killed in fires there. Altogether, from December 2005 through summer 2006, fires in Texas consumed about 3.7 million acres (1.5 million ha), nearly 400 homes, and killed 11 people. In February, the largest fire ever in the Arizona area burned

more than 4,000 acres (1,618 ha) in the Tonto National Forest, and in New Mexico, a grass fire of over 26,000 acres (10,521 ha) forced the evacuation of a farming and ranching community.

The uncertainty of water supply, a continuing problem in the Southwest, was exacerbated by the drought. A confluence of weather forces in the late spring 2006 conspired to drop extraordinary amounts of precipitation in the northern Rockies, while leaving the Southwest untouched. And then, on March 12, as many as 100 tornadoes touched down in five states from Illinois to Oklahoma, killing 10 people, but dropping no rain.

There was another effect of the drought that was not confined to the Southwest. A scarcity of rain in fall 2005 parched hard red winter wheat and dried up stock ponds and pastures in the Southern plains. And with below-normal precipitation predicted for the remainder of 2006 in the same area, the prospect of this huge wheat crop—as well as the rest of the livestock and agricultural picture in the entire Southwest—looked dismal.

UNITED STATES
VIRGINIA
JAMESTOWN
1608

A consuming fire caused the death by starvation of 62 of the original 100 settlers of the Jamestown Colony in 1608.

One hundred settlers established the first English settlement in Jamestown, Virginia, in 1607. By the end of that year, 38 of the original 100 remained alive. The other 62 died of starvation following, not a drought, but a fire.

On January 7, 1608, a devastating fire of unknown origin swept through the colony, consuming lodgings, clothing—and, what was most important—provisions. Most appalling was the fact that the fire occurred in the midst of a particularly bleak and bitter winter. With neither shelter nor the means to grow food, homeless and threadbare colonists froze to death in various stages of starvation.

FLOODS

THE WORST RECORDED FLOODS

* Detailed in text

Afghanistan
Amu Dary River (2005)

Africa
* Southern (2000)
Uganda, Kenya, Somalia, Ethiopia (2003)

Asia
* Southeast (2000)
* (and China, India, Nepal, Bangladesh) (2002)
* South and Central (Bangladesh, India, Pakistan, Nepal) (2004)

Bangladesh
* (1970) (see CYCLONES)
(1972)
* (1974)
(1987)
* (1988)
* (1998)

Brazil
* Tubarao (1974)

China
(1851–66)
* (1887)
* (1911)
(1931)
* (1939) (see FAMINES AND DROUGHTS)
(1949)
* (1950)
* (1954)
Canton (1915)
Eastern Shensi (Shaanxi) Province (1942)
Foochow (Fuzhou) (1948)
* Fukien (Fujian) Province (1948)
Hankow (Hankou) (1908)
* Hankow (Hankou) (1935)
Hong Kong (1906)
Hupei (Hubei) (1926)
Kaifong (Kaifeng) (1642)
* Kwangtung (Guangdong) (1982)
Kweichow (Guizhou) (1938)
Senchow (Senzhou) (1933)
Shantung (Shandong) Province (1957)

Szechuan (Sichuan) Province (2004)
* South, Northwest (2005)
* Szechuan (Sichuan) Province (1981)
* Yangtze and Shoshun River Basins (1998)
* Yellow and Yangtze River Basins (1996)

Colombia
(1970)

Congo, Democratic Republic of the
Kinshasa (1999)

Denmark
Zealand (1717)

Dominican Republic/Haiti
* Solie River, Hispaniola, Southeast Haiti (2004)

England
(48)
(1099)
Cheshire (353)
* Lynmouth (1952)

Europe
Central and Eastern (2002)

France
(1208)
* Frejus (1959)

Germany
* North Sea Coast (1962)

Greece
Attica (1760 B.C.E.) (The "Second Deluge")
Deucalion (1504 B.C.E.) (The "Third Deluge")

Guatemala
(1949)

El Salvador
(1982)

Haiti
* (1935)

Holland
* (1530)
* (1570)
(1916)
* (1953)

* Dort (1421)
Friesland (1228)
Friesland (1646)
* Leyden (1574)

India
(1864)
(1955)
Bengal (1876)
* Bengal (West) (1978)
Bihar (1961)
* Gujarat (1968)
Gujarat (1970)
Lahore (1947)
* Maharashtra, Gujarat (2005)
* Morvi (1979)
* North (1998)
Punjab (1787–88)
Rajputana (1943)
Surat (1959)

Indonesia
Bohorok (2003)
Lomblem Islands (1979)

Iran
* Farahzad (1954)

Italy
* Belluno, Pirago, Villanova, and Rivatta (1963) (see AVALANCHES AND LANDSLIDES)
* Florence (1966)
* Po Valley (1951)
* Stava (1985)

Japan
* (1896)
* (1964)
* (1972)
Kyushu (1953)
Tokyo (1947)

Java (and Sumatra)
(1883) (see VOLCANIC ERUPTIONS, Krakatoa)

Manchuria
Harbin (1932)

Mozambique
Quelimane (1971)

New Guinea
Papua (1998)

New Zealand
Queenstown (1999)
North Vietnam
(1971)
Norway
(1219)
Pakistan/Afghanistan
Punjab, Sind (1972)
* Shadikor, Gaggo, and Chelvi
(2005)
Peru
Huaraz (1941)
Philippines
Central, South (2003)
Poland
(1813)
Portugal
Lisbon (1967)
Russia
* St. Petersburg (1824)
Salandria
(1287)
Scotland
Glasgow (758)
* Inverness (1829)
Sicily
(1161)
South Korea
* (1972)

Spain
(1617)
* Barcelona, Sabadell, Tarrasa
(1962)
Consuegra (1891)
Sri Lanka
* Ratnapura (2003)
Sudan
* Khartoum (1988)
Thailand
* (1988)
Tibet
Himalayas (1970)
* Shigatse (1954)
Tunisia
* (1969)
Turkey
(1946)
United States
* California (1928)
* Colorado (and Montana, Kansas,
Wyoming, and New Mexico)
(1965)
* Colorado (1976)
Connecticut
* Putnam (1955)
Florida (1928) (see HURRICANES)
* Kansas (1951)
* Midwest (1993)
* Midwest (2007)

* Mississippi River (1874)
* Mississippi River (1890)
* Mississippi River (1912)
* Mississippi River (1927)
* Mississippi River (1973)
North Dakota (1999)
* Ohio (and Indiana and Illinois)
(1913)
* Ohio (1937)
* Ohio River Valley (1997)
* Oregon (1903)
* Oregon (and Washington, Califor-
nia, Idaho, and Nevada) (1964)
* Pacific Northwest (1997–98)
Pennsylvania
* Johnstown (1889)
South Atlantic Coast (1893)
South Dakota
* Rapid City (1972)
Southeast (1994)
Texas
* Galveston (1990) (see
HURRICANES)
Venezuela
* North (1999)
The World
* (2400 B.C.E.) (The Great Deluge
of the Book of Genesis)
Yugoslavia (and Romania)
(1970)

CHRONOLOGY

. . . .

* Detailed in text
2400 B.C.E.
* The World (The Great Deluge
of the Book of Genesis)
1760 B.C.E.
Attica, Greece (The "Second
Deluge")
1504 B.C.E.
Deucalion, Greece
(The "Third Deluge")
48 C.E.
England
353
Cheshire, England
758
Glasgow, Scotland
1099
England
1161
Sicily
1208
France

1219
Norway
1228
Friesland, Holland
1287
Salandria
1421
April 17
* Dort, Holland
1530
November 1
* Holland
1570
November 1
* Holland
1574
October 1–2
* Leyden, Holland
1617
Catalonia, Spain
1642
Kaifong, China

1646
Friesland, Holland
1717
Zealand, Denmark
1787
Punjab, India
1813
Poland
1824
November 19
* St. Petersburg, Russia
1829
July
* Inverness, Scotland
1851
China
1864
October 5
India
1874
April
* Mississippi River

1876
Bengal, India
1883
August 26
Java (and Sumatra) (see VOLCA-
NIC ERUPTIONS, Krakatoa)
1887
* China
1889
May 31
* Johnstown, Pennsylvania
1890
January–April
* Mississippi River
1891
June 20
Consuegra, Spain
1893
August
United States, South Atlantic
Coast
1896
* Japan
1900
August
* Galveston, Texas (see
HURRICANES)
1903
June 14
* Oregon
1906
Hong Kong, China
1908
April 14
Hankow (Hankou), China
1911
September
* China
1912
April
* Mississippi River
1913
March 25
Ohio (and Indiana and Illinois)
1915
June 12
Ohio (and Indiana and Illinois)
1915
June 12
Canton, China
1916
January 14
Holland
1926
August 4
Hupei (Hubei), China

1927
April
* Mississippi River
1928
March 13
* California
September 10
Florida (see HURRICANES)
1931
August
China
1932
August 3
Harbin, Manchuria
1933
September 1
Senchow (Senzhou), China
1935
July 4
* Hankow (Hankou), China
October 22
* Haiti
1937
January
* Ohio
1938
June 11
Kweichow (Guizhou), China
1939
China (see FAMINES AND
DROUGHTS)
1941
December 14
Huaraz, Peru
1942
September 28
Eastern Shensi (Shaanxi) Prov-
ince, China
1943
August 4
Rajputana, India
1946
May 12
Turkey
1947
September 17
Tokyo, Japan
September 29
Lahore, India
1948
July 6
Foochow (Fuzhou), China
August 7
* Fukien (Fujian) Province,
China

1949
July 17
China
October 14
Guatemala
1950
August
* China
1951
July 12–31
* Kansas
November
* Po Valley, Italy
1952
August 15
* Lynmouth, England
1953
February 1
* Holland
June 27
Kyushi, Japan
1954
August
* China
August 10
* Shigatse, Tibet
September 17
* Farahzad, Iran
1955
August 19
* Putnam, Connecticut
1957
July 21
Shantung (Shandong) Province,
China
1959
September 18
Surat, India
December 3
* Frejus, France
1961
October 9
Bihar, India
1962
February 17
* North Sea Coast, Germany
September 26
* Barcelona, Sabadell, Tarrasa,
Spain
1963
October 9
* Belluno, Pirago, Villanova,
and Rivalta Italy (see AVA-
LANCHES AND LANDSLIDES)
1964
July 18–19
* Japan

December
* Oregon

1965
June 16–26
* Colorado (and Montana, Kansas, Wyoming, and New Mexico)

1966
November 4–6
* Florence, Italy

1967
November 26
Lisbon, Portugal

1968
August 7–14
* Gujarat, India

1969
September–October
* Tunisia

1970
June 1
Yugoslavia (and Romania)
July 22
Himalayas, Tibet
September
Gujarat, India
November 12
* Bangladesh (see CYCLONES)
November 12
Colombia

1971
January 31
Quelimane, Mozambique
August 30
North Vietnam

1972
March 13
Bangladesh
June 9
* Rapid City, South Dakota
July 17
* Japan
August 12
Punjab, Pakistan
August 19
* South Korea

1973
April–May
* Mississippi River

1974
March 24
* Tubarao, Brazil
July–August
* Bangladesh

1976
July 31
* Colorado

1978
September
* Bengal, India

1979
July 17
Lomblem Islands, Indonesia
August 9
* Morvi, India

1981
July
* Szechuan (Sichuan) Province, China

1982
May
* Kwangtung (Guangdong), China
September 17
El Salvador

1985
July
Bangladesh
July 19
* Stava, Italy

1988
August 4–5
* Khartoum, Sudan
September–November
* Bangladesh
November
* Thailand

1993
June–August
* Midwest United States

1994
February
Southeast United States

1996
July–August
* Yellow and Yangtze River Basins, China

1997
March
* Ohio River Valley, United States
December–January 1998
* Pacific Northwest, United States

1998
July
* Bangladesh
July–September
* India
July
Papua New Guinea
July–August
* Yangtze and Shoshun River Basins, China

1999
July
North Dakota, United States
September
* Mexico (and Central America)
November
Queenstown, New Zealand
December
Kinshasa, Democratic Republic of the Congo
* Northern Venezuela

2000
February–April
* Southern Africa
July–October
* Southeast Asia

2002
June–August
* Asia (and China, India, Nepal, Bangladesh)
August
Central and Eastern Europe

2003
April–May
Uganda, Kenya, Somalia, Ethiopia, Africa
May 17–18
* Ratnapura, Sri Lanka
November 2
Bohorok, Indonesia
December 17–23
Central, South, Philippines

2004
May 18–26
* Dominican Republic/Haiti
June–August
* South and Central (Bangladesh, India, Pakistan, Nepal)
September 7–8
Szechuan (Sinchuan) Province, China

2005
February 6–13
* Shadikor, Gaggo, and Chelvi, Pakistan
March 18–22
Amu Dary River, Afghanistan
June
* South, Northwest China
July 26
* Maharashtra, Gujarat, India

2007
August 20–29
* United States, Midwest

FLOODS

At first glance, the causes of floods seem easy to discover and simple to define: melting snow, frequent storms, heavy rainfall.

But these obvious factors form only a small part of the story. This single most catastrophic natural disaster known to humankind—one study found that between the years 1947 and 1967, 173,170 persons died as a direct result of riverine floods, whereas the grand total of 18 other categories of catastrophe, including hurricanes, tornadoes, earthquakes, and volcanoes, came to just 269,635—springs from a multiplicity of subtle, complex factors.

One of these is inevitability, manifested in time, tides, and the hydrologic cycle—an endless, natural circle of stabilization, in which water passes from the oceans into the atmosphere, onto the lands, through and under the lands, and back to the oceans. As certainly as the moon will rise and set, rivers will rise and fall. And as certainly as the seasons change, the hydrologic cycle will continue. For 3 billion years, the total amount of water on the Earth and in its atmosphere has been almost exactly the same. And if it has remained this way for 3 billion years, it is fairly safe to say that it will probably stay that way for the next 3 billion years—unless, in our reckless disregard for the world's ecosystems, we upset that balance as we have in other natural phenomena.

This water balance and its cycle is achieved by a combination of the sun's heat and the pull of gravity, both of which constantly recycle moisture, which evaporates and enters the air as vapor, then condenses and falls back to Earth as rain or snow. Thus, though it may boggle the mind, it is entirely possible that the glass of water you drink today could very well have been Cleopatra's bathwater. Or, if that image is a little distasteful, consider that the water in your swimming pool could very well have fallen as snow on Hannibal's troops.

There is also an interesting and perhaps controversial side to the scientific theory of the hydrologic cycle, and that is that if all of the water in the atmosphere at any given time were suddenly to be loosed upon the Earth, it would cover the planet to a depth of only a fraction of an inch. Believe this—as most scientists do—and the great deluge of biblical times, which must have been thousands of feet deep, and for which there is much archaeological and historical evidence (see p. 200), could not have occurred.

Naturally, the hydrologic cycle does not manifest itself regularly in one place. If it did, there would be neither deluge nor drought. It constantly shifts, from time to time and place to place, taking from here and overfeeding elsewhere.

Thus a logical inference to draw from this is the constant warning that if you live near a river or a sea, sooner or later, you're liable to experience a flood.

Why, then, do humans build some of their finest cities near rivers? The answer is twofold: commerce and food. Since Mesopotamia, rivers have been arteries of commerce. Even Barotseland tribesmen in the northwestern floodplains of Zambia float goods on the Zambesi, and when the flood season arrives, they simply move to higher ground and wait it out.

As far as food is concerned, humankind learned early that alluvial soils (soils deposited by moving water) were the richest for growing crops. According to one estimate, as many as 1.5 billion people, or one-third of the world's population, still depend upon alluvial soils for food. Small wonder, then, that cities, towns, villages, and farms are commonly located on the floodplains of rivers, or on the seashore. In the United States alone, nearly 3,800 settlements, each one of them containing 2,500 people or more, are located in spots that are prone to flooding.

What is worse, the farming and construction methods employed in these locations often contribute to flooding. Vegetation captures precipitation, often before it hits the ground, and returns it to the atmosphere. Denuding the landscape by grazing, tilling, building, or the feckless felling of trees removes that process.

Certain soil characteristics also contribute to flood potential. If the ground is coarse and composed of sand or gravel, rain water is absorbed quickly. But if the ground is fine, composed of clay, for instance, less water gets through, and runoff is inevitable. Sooner or later, of course, no matter what the character of the soil, the water level above subterranean, less permeable rock—known as the water table—will be reached, and will back up to the surface, also causing runoff.

Not surprisingly, the most resistant surface is caused by human construction. The present, wholesale concretizing of the landscape allows for no absorption, and so cities create their own threats of flood through runoff.

This man-made flood potential has forced humankind to invent methods to divert and/or contain waters swollen by precipitation, tidal waves, tsunami, or melting snow. These include dikes, diversions and dams. But none of these have been effective against a major catastrophe. In fact, flash floods and bursting dams are the most calamitous of all flood disasters, because they are unpredictable, and because those who live in the path of potential danger frequently develop a tragically false sense of limitless security.

This egregious mindset might have been at least partially responsible for the monumental loss of life in New Orleans from Hurricane Katrina in August of 2005 (see pp. 290–293). The only city in the United States ever to be so completely devastated by a natural disaster, it is also the only U.S. city that is 70 percent below sea level, surrounded by water on three sides and protected by levees lower than 25 feet (7.6 m). In addition, wetlands that had formed natural barriers to flood had been drained, excavated, and built upon long before the arrival of Katrina.

The flood, which occurred after the bulk of the hurricane had sideswiped the city, came not from the Mississippi River, but from the backwash of Lake Pontchartrain, which breached levees in multiple locations and sent waters roaring into the streets of the city, literally drowning most of it. Even the Army Corps of Engineers, responsible for maintaining the levees, did not foresee the fragility or the failure of these levees.

In retrospect, after the devastation and death that were certainly compounded by the bungling of the relief efforts of local and national governments, it seemed that New Orleans had, all along, been a prime candidate for a disastrous flood. Some 1,277 lives were lost that August, all but a very few by drowning. The total number of homeless was 374,000, and over a million people were evacuated from a city that had been built upon the belief that it was possible and even safe to live below sea level in a place surrounded on three sides by water.

There is simply no defense against unleashed water. A gallon of it weighs about 8.5 pounds; a cubic foot of it weighs approximately 62 pounds; and a bathtubfull—one cubic yard—weighs approximately three-quarters of a ton.

Now, try to contain an astronomically large amount of this heavy substance behind concrete or earthen walls, and you get a rough idea of the pressures that dams and dikes must withstand. The reservoirs of dams impound colossal amounts of water. Lake Mead, behind Hoover Dam on the Colorado River, is 115 miles (185 km) long and has a capacity of 1.4 trillion cubic feet of water, or 10.5 trillion gallons (40 trillion l), or, as scientists measure it in acre feet (one acre foot is the equivalent of one acre of land covered by water to a depth of one foot), 32 million acre feet.

This is all well and good if the dam holds. But consider the other characteristic of water besides its volume and its weight: its force. An inch of rainfall descending 1,000 feet (304.8 m) and draining one square mile has the energy potential of 60,000 tons of TNT, or three times the force of the Hiroshima nuclear bomb. Once this energy enters a stream or a conduit, it concentrates itself into a battering ram that can destroy stone buildings and bridges and lift tons of debris and fling them around like discarded toys.

The velocity of a river is largely determined by gravity. The greater the volume of water and the steeper the grade, the faster it goes—within certain limits. The friction caused by the combination of the Earth's surface beneath, the air above and even within the water itself usually keeps it at a maximum of around 20 MPH. Still, given all of those factors, the potential for disaster is staggering.

So, compound this with the natural forces of extreme rainfall, seasonal storms, the natural inclination of the hydrologic cycle to concentrate itself in selective areas, and other human factors, and you have the makings of a cataclysmic flood. And why, if all of this is known, is the casualty rate from floods so sorrowfully high?

Perhaps it is because human beings make compromises, with life and with nature. Like our perception of death, we feel that maybe it might not happen to us. And the price of these compromises and illusions is often measured in the casualties caused by floods.

AFRICA
SOUTHERN
February–April 2000

A gigantic flood ripped through Southern Africa for three months, from February through April 2000. Fed by monsoons and two cyclones, its greatest wrath was visited upon Mozambique, whose infrastructure was nearly erased. Nine hundred thousand people were affected and 400,000 lost their homes in Mozambique. In Zimbabwe, 80,000 were rendered homeless, and thousands more were dispossessed in Madagascar and South Africa. Over 700 people in the region died, the majority of them children.

Of all the poor countries in the world, Mozambique is one of the poorest. Ninety percent of its population of 16,000,000 live on less than a dollar a day. Their life expectancy, during the best of times, is between 45 and 50 years. Most wrenching of all, Mozambique has the highest child mortality rate of any country in the world. Twenty-seven percent of its children die before they reach their fifth birthday.

And yet, in 1995, Mozambique joined the Commonwealth of Nations, and by 1999 it had begun to emerge from the worst of its poverty.

And then, on February 9, 2000, heavy rainfall began across all of Southern Africa, and Mozambique bore the full impact of it. The Save, Limpopo, and Incomati rivers swelled to overflowing in less than two days. As the Incomati bore down on the capital of Maputo, tens of thousands of people were forced to evacuate their houses. Those with substantial homes were eventually able to return to them. But the poor, living in makeshift shacks in the slums surrounding the capital, lost everything. In an instant, it seemed, whatever shelter or material wealth they had was swept away by floodwaters.

Since agriculture employs 83 percent of Mozambique's labor force, these workers lived in the most fertile areas of the country, along the Save and the Limpopo rivers, which now turned from suppliers of life to destroyers of it.

In district after district, homes were flooded and their inhabitants were forced to flee. In March, after the waters subsided, one woman walked into an international relief shelter in Chaqualane and told the workers how she had lifted one of her children into a tree and balanced the other on her head while she clung for two days to the tree's trunk in chest-high water. Aid workers in a helicopter rescued the children, but the mother refused help, staying in the tree for a week until the waters receded and she was able to walk to Chaqualane, where she found her children.

In the north of the country, in Gaza province, hundreds of thousands were deprived of their homes as floodwaters and mudslides devastated entire villages.

The downpours continued, unabated, and more roads, bridges, and crops were inundated and washed away by the floods. Electricity supplies were disrupted. Pumping stations in towns were swept away by the torrents, leaving the towns without clean water. The main north-south road was rendered impassable, isolating the capital and the second city of Beira from each other.

As the week unfolded, reports of the dead began to emerge from much of Southern Africa. In Swaziland, the capital of Mbabane lost its drinking water. Southern Botswana received 75 percent of its average annual rainfall in three days. In the Limpopo Valley north of Maputo in Mozambique, the Limpopo River burst its banks.

But all of this was a mere prelude. On February 22, Cyclone Eline hit the Mozambique coast near the central city of Neira with 160 MPH (257.5 km/h) winds and torrential rains. At the same time, swollen rivers in the rest of Southern Africa, which empty into Mozambique, increased the force and amount of water rushing into the country. Throughout the region, 23,000 people were flooded out of all they possessed, and large portions of the country disappeared under water.

At the end of February, flash floods erupted throughout Mozambique, and floodwater levels rose an astonishing 26 feet (7.9 m) in five days, inundating low farmlands around Chikwe and Xai-Xai, isolating 100,000 people, and trapping 7,000 who had climbed into trees to escape the floodwaters.

Mozambique was not the only area ravaged by the sudden appearance of Cyclone Eline. In Zimbabwe, crops and village granaries were washed away, taking with them food supplies for the populace. Roads, bridges, and dams were destroyed; livestock and people alike drowned in the torrents spawned by the deluge.

Survivors, part of a population staggering under record rates of unemployment and inflation, criticized government inaction in relief, and this inaction, in turn, became an immediate problem for President Robert Mugabe, who had deployed a third of Zimbabwe's army and much of its equipment in the war in the Democratic Republic of the Congo.

Supply ships were unable to dock at Mozambique's port of Beira, and food and gas supplies rapidly dropped throughout the entire region, until they were in critically short supply.

To the south, South Africa braced for a massive influx of refugees from both Zimbabwe and Mozambique. Better situated economically, South Africa was able to send 12 helicopters to rescue people from trees and rooftops in Mozambique. But South Africa, too, had its share of disasters. More than 90 people died in its Northern Province and in neighboring Mpumalanga in the February floods. Families in the area were forced to keep the dead in their homes for several days because access routes to mortuaries and hospitals had been severed by floodwaters.

Madagascar was hit directly by Cyclone Eline. It happened without warning; later, meteorologists on Madagascar would explain that their equipment was too obsolete to allow them to follow the trajectory of the storm. Trees were torn up by the roots, roofs were blown off, waters rose in the lower areas and in the ricefields around the capital, Antananarivo.

And then, three days later, Cyclone Gloria hit the country, ripping through Madagascar from north to south. Landslides, floods, collapsed bridges, and torn up and flooded roads prevented travel from one part of the country to the other. The towns of Sambava, Antalaha, Vohemar, and Dndapa, in the Sava region, were most cruelly hit by the storm and by its subsequent floodwaters. Houses were destroyed and rivers flooded onto fields, killing livestock and people alike.

Unsanitary conditions had already created a cholera epidemic that had claimed 1,000 lives since March 1999; now contaminated waters created the threat of more deaths from the disease. Over 130 people died in Madagascar from drowning, but 384 died from cholera in February alone; 560,000 were affected by the flood; 40,000 were rendered homeless.

In Zambia, thousands faced starvation. Their crops were inundated as the overspill gates of Zimbabwe's giant Kariba Dam were opened when authorities feared it might burst under the pressure of floodwaters. Further trouble lay downstream, since this spill flowed directly toward Mozambique's largest dam at Cabora Bassa. Fortunately this dam held.

In Botswana, over 10,000 houses collapsed or were washed away and 34,000 people were dispossessed.

But Mozambique remained the country most savagely demolished by the three-month flood, which finally began to recede in March. The country's infrastructure was devastated. Hundreds of roads and bridges were unusable. And worse: Land mines left over from its recent civil war were dislodged by the settling floodwaters, thus forming new minefields.

There was, as there always is after a flood, the threat of disease. But because of Mozambique's location, tropical diseases were also present. Dysentery and *E. coli* food poisoning, water-borne diseases, were caused by the contaminated water supply from pit latrines washed out by the floods. But there were added natural dangers: Mosquitoes, thriving in stagnant pools left behind when the waters receded, carried malaria and dengue fever. Finally, Mozambique, already beset by a cholera epidemic, now was beset by a

further, huge increase in the disease, carried by wreckage and corpses from the floods.

Thousands of orphaned children, their parents drowned in the flood, were cared for in UNICEF shelters. Aid workers from Save the Children counted up to 250 lost children in just two camps in the Save River area.

In Xai-Xai, a young boy spoke to Portuguese radio about his plight, which was reflective of that of many survivors. "Aid has arrived from Chibuto," he said. "There's no money, and corn for the poor which was meant to be given out is being sold. We have food because people are gathering together and each one contributes something from what they have brought. But soon we will run out because a bag which cost 120,000 meticals [$8] now costs 700,000 [$49]."

The boy had no shelter, and no directions about where to find one. "At my school—which is a brand new school and has just been built," he continued, "the teachers and the director say they don't want to let the refugees stay, because they will damage the school."

Most transportation between points was still done by boat, and the water was rapidly becoming polluted by fuel. People slept in the open air, which meant they were being bitten by mosquitoes, some of which carried malaria. "There are no medicines, so they just die," the boy said, "and many people have died here in Xai-Xai."

It was a scene of utter confusion to most refugees. "Children are running about crying," the boy told the radio reporter. "Women don't know what to do with themselves. People are selling their things. And then there are the thieves. Even with the water they want to go into the city to steal people's goods. Up here there are many thieves. It is not safe. They try to steal our rucksacks, so they can sell what's inside to survive."

Nearly 1,000,000 people became destitute in Mozambique as a result of the flood. The infrastructure of the country was completely shattered. Roads, railways, and bridges were destroyed, as were clinics and hospitals. "Ironically," one aid worker said, "in a country that has been under water for weeks, dehydration is one of the biggest problems and possibly the biggest killer, particularly for the young and old."

As rescue operations continued, Cyclone Gloria broke up near the coast, dumping still more rain and fears of more rising waters on the battered country, its survivors, and its benefactors.

Mozambique's damage figures were staggering: Besides the irrigation systems, sugar refineries, sugar plantations, railroad lines, roads, and sanitation facilities that were destroyed, two hospitals, 37 health centers, and 508 schools had to be replaced. Over 100,000 acres of crops were washed away and over 40,000 head of cattle were drowned. The total bill for reconstruction would run to nearly $500,000,000, some of it paid for by forgiveness of debt by other nations.

In South Africa, 340 schools were damaged or destroyed, mostly in the Mpumalanga and Northern provinces, and the country's prime tourist attraction, Kruger Park and its animals, was totally annihilated. Houses, farms, dams, electricity, water, and sewer lines were demolished.

More than 800 people died in Mozambique, 900,000 people were injured, and more than 300,000 lost their homes. Eighty thousand people were rendered homeless in Zimbabwe.

ASIA (CHINA, INDIA, NEPAL, BANGLADESH)
June–August 2002

In the worst flooding since the floods of 1998 (see p. 153), China, India, Nepal, and Bangladesh were engulfed by floodwaters from June to August 2002. Some 2,000 people were killed and nearly 2 million more were made homeless by a series of river overflows and mudslides caused by three months of torrential monsoon rains.

The El Niño phenomenon was furiously active in the monsoon season of 2002 in Asia. Four neighboring nations—China, India, Nepal, and Bangladesh—were inundated by raging floodwaters that spread from the interior of China to the Bangladesh villages on the Bay of Bengal.

In early June, incessant deluges swelled rivers that had been drained nearly dry from a prolonged drought. Farmers had populated the riverbanks, tilling them in the erroneous belief that if water did return to the rivers, it would be a gradual homecoming. Caught unaware, these farmers and their families were swept into the rivers of the provinces of Shansi (Shanxi), Fukien (Fujian), Kiangsi (Jiangxi), Hunan Kwangsi, (Guangxi), Ch'ung/Ch'ing (Chongqing), Szechuan (Sichuan), and Kweichow (Guizhou). "Most of the deaths were caused by torrents of water, mud, and rocks tumbling down from hills in the remote areas," one of the survivors told a Western reporter after the flood.

All of China's seven major rivers overflowed, and when September finally arrived, China tallied the damage:

- 1,543 people had died of drowning or waterborne diseases
- 1.05 million homes were totally destroyed
- 2.7 million homes were damaged
- 32.1 million acres (13 million hectares) of cropland were rendered useless
- a large number of infrastructure facilities—roads, bridges and railroads—were destroyed.

In the mountainous country of Nepal, landslides compounded the terror brought on by rivers whose overflow gouged huge chunks of the countryside loose and turned parts of that countryside into rolling juggernauts. In the village of Thapra (a remote mountain retreat typical of the construction of Nepal), one such rain-fed landslide struck in the middle of the night, sweeping away 40 houses and killing 65 people in an instant.

The only way to reach these villages was by helicopters, and helicopters were ready in the Nepal capital of Katmandu. But the relentless rain grounded most of them, and particularly in the eastern state of Bihar, survivors were succumbing to waterborne diseases.

In Bangladesh, which suffers annual monsoon flooding because it is largely below sea level, more than 4 million residents fell victim to the roaring overflow of its 250 rivers, constantly refilled by the driving rain. The worst such flooding in four years submerged or washed away thousands of tin, bamboo, or straw houses, sweeping them away as if they were made of cardboard. Flood barriers, roads, and bridges disappeared, drowning tens of thousands of hectares of crops.

Chandpur, a central district on the confluence of Bangladesh's two main rivers, the Padma and the Meghna, was largely swept away by roaring tidal surges from the merged rivers. Eventually, a third of this delta nation of 130 million people disappeared underwater.

In India, the teeming overflow of its two biggest rivers, the Brahmaputra and the Ganges, literally swallowed up thousands of villages. Over half a million people were made homeless in the northern states of Bihar and Assam, where major highways buckled, then collapsed, preventing supplies from arriving in the areas.

Even in the western state of Maharashtra, cases of cholera were reported, and throughout the four flooded nations, the waterborne diseases of typhoid and dysentery spread with the waters. And as the monsoon season wore on to its final conclusion in the fall, mosquitoes bred in the stagnant floodwaters, spreading malaria and encephalitis. The final tally was crushing: Some 2,000 died and over 2 million were displaced by the fury of the 2002 floods.

ASIA
SOUTH AND CENTRAL (BANGLADESH, INDIA, PAKISTAN, NEPAL)
June–August 2004

Some 5 million residents of Bangladesh, India, Pakistan, and Nepal were made homeless and 1,800 died in the worst floods in the region in a century, caused by record monsoon rains during the summer of 2004.

The combination of yearly monsoon rains and melting snow in the Himalayas has become an expected danger in South and Central Asia from mid-June to mid-October. In early spring 2004, the added dimension of increasing global warming induced a higher-than-average snow melt in the high Himalayas, and, during the same time, the intensity of monsoon rains was the highest in a century.

As a result, floods and landslides in the mountain kingdom of Nepal killed scores and left thousands stranded (see color insert on p. C-6). Throughout the region, rivers, fed by the monsoon rains and melted snow runoff swelled to flood stage, then overflowed their banks, inundating rice paddies, villages, and Bangladesh's capital Dhaka, and its port city Chittagong, with some four million inhabitants. Assam, the northeastern province of India (through which the Brahmaputra River flows before joining with the Jamuna and Ganges Rivers in the flood plains of Bangladesh) was hard hit with entire villages swept away in the surging overflow of the Brahmaputra.

By mid-July, the entire region resembled the Indian Ocean. Lightning played across the region from the storms, causing some deaths, while drowning caused others. Houses, sometimes with their residents inside, collapsed or were swept along in the vicious currents caused by winds and surging waters. And in practically all regions, poisonous snakes were swept along by the tides, and caused a plenitude of deaths by snakebite.

Underground water reservoirs and gas outlets were inundated, causing shortages of clean water and cooking fuel, and waterborne diseases began to spread through July and August. Schools were turned into shelters. A school housing 3,500 evacuees in Dhaka's Mugdapara district was plagued by a lack of government-distributed rations that were cooked around small stoves on the verandahs. Kanakfool Begun, huddled with her two children and her husband, a rickshaw puller, spoke to Western journalists. "We have been here for seven days," she said. "But we only got some rice and lentils from relief workers."

As some floodwaters began to recede at the end of July, bodies began to be found. Eventually, the death toll, from drowning, landslides, electrocution, and waterborne diseases such as diarrhea, dysentery, and typhoid would total 1,800, mostly in the impoverished villages of eastern India and Bangladesh. More than 5 million people would be rendered homeless throughout central and south Asia.

ASIA
SOUTHEAST
July–October 2000

An epic flood consumed most of Southeast Asia during a sharply extended monsoon season from July to October of 2000. Rivers in India, Bangladesh, Cambodia, Vietnam, Laos, and Thailand, fed by constant torrential downpours, killed at least 1,634 and rendered 20,000,000 homeless.

The year 2000 is the year of the dragon in Asia. In the Mekong Delta in Vietnam, those who lived through the immense flood of 2000 referred to it as the Dragon Flood. "I am seventy years old," one survivor told reporters. "In all my life, I have not seen floods worse than what we are experiencing. I have lived through a series of wars. Fleeing from these wars was easier than fleeing from this flood."

Heavy rain began to fall in July 2000—45 days ahead of the beginning of the normal monsoon season, from September through October. And it would not truly cease for another three months, enough to inundate and cause untold devastation throughout Vietnam, Cambodia, Laos, India, Pakistan, and Bangladesh.

India and Bangladesh, barely recovering from the great floods of 1998 (see p. 153), were once again plunged into chaos and misery. Rivers overflowed across eastern India and western Bangladesh. Thousands died of drowning, snake bites, and murder by pirates looting abandoned villages.

In both countries, 18 million people were marooned, some in trees, some on the roofs of their houses. Millions more lost their homes entirely. Those who survived battled the cholera, diarrhea, and typhoid that ran as rapidly as the rivers through the crowded relief camps in Bangladesh.

In India's West Bengal state, villagers fled their inundated homes, submerged by the Hooghly River, on rafts made of tied-together banana trees. Others waded through waist-high flood waters carrying children and the elderly on their shoulders. Over 40,000 mud and straw huts were demolished in the state, leaving 200,000 people homeless.

Once in safe havens, most of the survivors were desperate to find the food they had not been able to consume for days. In the town of Debagram, police were forced to fire in the air to disperse people who were stealing food and other relief supplies from a relief station. Nearby, in Guptipara, mobs beat railway officials and snatched goods from a relief train.

As in 1998, most of India's Kaziranga National Park, the home to 1,600 one-horned rhinos and hundreds of elephants, was flooded. But this time, most of the elephants and rhinos were moved to higher ground.

By the beginning of October, the floods had forced 600,000 people from their homes in Cambodia, Vietnam, and Laos. Cambodia's Prime Minister Hun Sen told Associated Press reporters that the level of the Mekong River, one of three that meet in Cambodia's capital, Phnom Penh, was higher than at any other time in the past 70 years. "I have participated to stop the killing fields, the genocide of Pol Pot . . . but it is impossible for me to stop the natural disaster," Hun Sen concluded.

The water level at the meeting point of the Mekong, Tonle Sap, and Bassac rivers in Phnom Penh lapped continually at its flood level of 38 feet (11.6 m). By the middle of October, it had exceeded this level and flooded the streets of Phnom Penh.

The Mekong also overflowed its banks in Dong Thap and Long An provinces on the border between Vietnam and Cambodia, turning rice fields to lakes. In Laos, the major rice-producing areas of the central and southern parts of the country were destroyed for seasons to come by enormous flooding.

This was the situation throughout most of Southeast Asia. In Cambodia, the flooding affected 2.2 million Cambodians—one-fifth of the country's population. Many of those affected were rice farmers who lost all or most of the year's primary rice crop.

In Vietnam, 35,000 families were evacuated from 700,000 submerged homes. Nearly 25,000 soldiers and health workers were dispatched to flood areas to face the worst flooding in nearly 40 years in Vietnam. There was a growing fear in the Mekong Delta that the pollution from the ground water had reached dangerous levels.

In the third week of September, the Mekong River reached a record 36.7 feet (59.1 m) outside of Phnom Penh, and would have flooded the capital and its population of 1 million if hundreds of thousands of sandbags had not been laboriously placed at the tops of its dikes.

The landscape of Bangladesh becomes a seascape during the crippling floods that have lately become yearly occurrences. The 2000 flood was unusually severe and far-reaching. (U.N. photo/DPI)

The flood continued through the entire months of September and October. By the beginning of October, over 2,000,000 people across Vietnam, Cambodia, and Thailand were homeless, flooded out of their dwellings and either at the mercy of the elements or crushed into makeshift shelters. In Vietnam alone, official estimates put the number of those without shelter at 1,750,000. Thousands of families in that country faced hunger and epidemics as they camped out on narrow dikes that themselves threatened to crumble under the constant assault of fresh rain and increasing floods.

In Cambodia, 200,000 people abandoned their homes for higher ground, and in Thailand, the flooding caused a massive outbreak of leptospirosis, a disease caused by a bacteria spread through rat urine. At least 224 people were killed by the disease, according to public health ministry officials.

All over the area, pagodas, schools, and elevated areas became more and more crowded, as more and more refugees arrived. By the beginning of October, all low-lying areas in Vietnam, including cities and towns, were submerged. Roads were either underwater or ripped asunder. And the river dikes were showing signs of giving way.

The monsoon season still had six weeks to wreak its havoc, and that is precisely what it continued to do. At the beginning of October, new storms hit India and Bangladesh from a depression in the Bay of Bengal. The storm moved over the western coast of Myanmar, which borders Bangladesh and India, then lashed northeast India with torrential rains, which in turn further inflated the flood level of India's rivers.

The constant rain had an additional effect, preventing rescue workers from reaching even moderately remote villages. However, in the northern portion of India, there was some subsiding of floodwaters during the first week of October. But the momentary cease did not stop the rising death toll. Relief workers found as many as 142 bloated and decomposing corpses floating in floodwaters in the northern India district of Murshidabad.

And, as the month deepened and the torrents increased, additional flooding took place in the northern India districts of Cooch Behar, Jalpaiguri, Darjeeling, Malda, and Murshidabad.

Farther south, Cambodia's upstream rivers began to fall, but not Vietnam's. In this country, vast tracts of land were now vast seas, up to 16 feet (4.87 m) deep, except where some 700,000 houses abruptly shallowed the water. In Vietnam's northern mountains, meanwhile, landslides claimed more lives, particularly in the mountain district of Sin Ho in Lai Chau province, close to the Laotian border.

Downstream, Vietnamese faced further horrors. Cholera began to break out in the crowded refugee shelters. And crocodiles began to appear, swimming downriver from Cambodia. One fisherman caught a 55-pound crocodile in the Hau River, a tributary of the Mekong in An Giang province.

"It was raining and the wind was very strong," a survivor told international rescuers. "People could not distinguish where the river and the rice fields were located. They all resembled the ocean in a typhoon."

The particularly tragic aspect of the fatalities now began to make itself known: In the southern zone of Vietnam, most of the fatalities were children. In September alone, in the eight provinces of this area, 225 of the 296 deaths were those of children, who slipped off makeshift islands or succumbed to disease. "We're looking at 20 to 30 kids dying a day—it's crazy," said John Geoghegan, the chief delegate to the area of the International Red Cross and Red Crescent societies.

Now even the two major cities in Vietnam, Ho Chi Minh City, just north of the Mekong Delta, and Danang, on the central coast, were threatened. Sandbags and makeshift dikes were swiftly put up around each city, and some residents evacuated to higher ground. As October waned, the total of people in the Mekong Delta affected by the floods rose to 5,000,000.

Finally, at the beginning of November, the floods began to recede, and assessments began. Children in Vietnam had not attended school in three months, and some would be out of school far longer—2,673 schools were destroyed by the floodwaters.

Shabby tents sheltering thousands of flood victims lined the roads from which the water had receded. Families and their livestock were crowded into these tents poised between roads and lakes that were once rice paddies.

In Tan Lap, in the midst of the village, rescuers found a typical sight: Of the 993 households in the village, 100 percent had been forced to seek refuge from the flood on earthen dikes. "We still face starvation," one of the villagers told a representative of the International Red Cross. "The rice and other material aid that was given to us only lasted for a few days, and then we go for stretches without food. There were times when we were so desperate we sought water hyacinths to eat at the risk of drowning."

Dysentery was everywhere. The water survivors used for drinking was the same water that contained human waste and corpses. Boiling the water was no option, since there was hardly any wood for burning. The low absorptive capacity of the soil in the region, as well as the salinity intrusion from the Bay of Bengal, added to the floodwater's contamination.

Agriculture was devastated. Orchards of mango, guava, longan, and banana were universally destroyed. Rice crops, once the water receded, could be revived within four months. But it would be four years before the citrus, mango, and longan orchards would again produce crops.

Nearly 8 million people were affected by the floods. The official death toll was 2,000. Some 338 people died in Cambodia, 224 in Vietnam, 47 in Thailand, 15 in Laos, and more than 1,000 in India and Bangladesh. More than 200,000 people were left homeless.

Certain conclusions have been drawn, and some of them are disconcerting for both the present and the future. The terrible loss of life and property in the Mekong Delta was blamed to a great extent, by the United Nations Economic and Social Commission for Asia and the Pacific (ESCAP), upon deforestation. "Forests have been reduced in most countries from 70 percent of land area in 1945 to about 25 percent in 1995," the report stated. "Partly because of the widespread felling . . . the intensity of flood disasters appears to have increased in the region in the past decade, especially during the past few years," the report concluded.

U.N. weather scientists noted that the combination of heavy rainfall in Southeast Asia and devastating drought across Central Asia in 2000 was at the extremes of what had been experienced over the last 100 years. Thus, they concluded, while conceding that there is still no direct link, heavy rainfall in the region was consistent with the climate change expected from global warming. "This is a taste of things to come," prophesied Peter Walker, head of the delegation for Southeast Asia to the International Federation of Red Cross Societies.

BANGLADESH
1970

(See CYCLONES.)

BANGLADESH
July–August 1974

Over 2,000 died, more than a million were injured, and millions were made homeless by a monsoon-created flood that covered Bangladesh throughout July and August of 1974.

Bangladesh, the world's most densely populated country and one of the world's 10 most heavily populated countries, is, in terms of natural disasters, an accident waiting to happen. And, sadly enough, that accident occurs practically every year. Every year, the monsoon rains come, and every year, the land, which is, at its highest point, a mere 300 feet (91.4 m) above sea level, is flooded, killing anywhere from scores to thousands, rendering millions homeless, causing crop loss and disease and starvation.

Both the Ganges and the Brahmaputra Rivers empty into Bangladesh, and even in normal times, the country is composed almost entirely of tiny islands surrounded by rivers, canals, and other waterways. This allows for easy irrigation, and the alluvial soil is rich enough to support abundant harvests of rice, cashews, and spices, upon which—along with textiles and paper products—the country's economy is based.

In the past, this natural state of affairs was controllable. But in recent years, the snows have increased in the Himalayas, where the two rivers originate, and the deforestation of Nepal continues. And so, as the inevitable monsoons return, the richness of the soil is depleted, as, year after year, the crops are washed away by floodwaters, composed of melting snow, mudslides, and monsoon rains.

The monsoons were unusually severe in July 1974—more so, residents noted, than in 20 years. Seven of the 20 states of India were affected by serious flooding. By the beginning of August, the floodwaters had reached Bangladesh, and the delta region of that country, already soaked by monsoon rains, disappeared under water. By the middle of August, two-thirds of the country was flooded, and 80 percent of the annual summer crop was destroyed, along with the seedlings planted for the main winter crop. Officials estimated that at least 40 percent of the annual food output of 12 million tons (10,886,217 tonnes) was lost in the floods.

But these statistics pale in contrast to the human suffering, borne stoically by the inhabitants of hundreds of villages that were completely inundated with such regularity that the more farsighted families had built bamboo living platforms in nearby trees.

Cut off from the rest of the world, their food supply under water, those who did not drown in August 1974 frequently succumbed to starvation or disease. In total, more than 2,000 died, more than a million were injured, and millions were made homeless.

In the village of Sunamganj, 160 miles (257 km) northeast of the capital city of Dacca, a 45-year-old farmer, Mohammed Ahmadullah, spoke in despair to a reporter for the *New York Times*. His village of 4,000 remained under water for over a month, while its inhabitants huddled either on their tree perches or in the local school building. The cattle they owned were left standing in four feet of water, where they died, one by one.

"We can eat their flesh but we cannot cook," Mr. Ahmadullah told the Western reporter. "We have no fuel. Our stock of rice is exhausted. We have eaten the seeds meant for the next crop."

Their resistance to infection lowered by starvation, the people of Bangladesh then became victims of typhoid and cholera, spread by the stagnation and pollution of the sewage systems. In the 2,000 relief camps established by the government and the U.N. relief organizations, the death toll from these two diseases reached 100 a day in the month of August. By October, 15 million starving citizens of this stricken country were without homes, food, and jobs. Thirty-five million had been affected by the disaster.

The World Health Organization established a headquarters in Dacca to combat the cholera epidemic that established itself in September, and the government of Bangladesh appealed to the rest of the world to help them through a period of starvation. Help came over the next few months, but it came slowly, and people continued to starve to death through the winter of 1974–75.

BANGLADESH
September–November 1988

Indiscriminate deforestation in the Himalayas was the major cause of the devastating flood in Bangladesh at the end of 1988. Independent estimates reported 5,000 dead; official figures placed the death toll at 2,100. Three-quarters of Bangladesh disappeared under water, and half of its people were rendered homeless.

Deemed "man-made" by Tom Elhaut, director for projects in Bangladesh for the International Fund for Agricultural Development, the horrendous floods that engulfed three-quarters of that country from September through November of 1988 were also called by U.N. relief agency officials, "one of the worst natural disasters of the century."

Casualty figures varied according to the source. John Hammock, director of Oxfam America, estimated that 5,000 died; Bangladesh's foreign minister, Hamayun Rasheed Choudhury, set the figure at 2,100. Both reported that three-quarters of the country had disappeared under water, and half of its people were rendered homeless by overflowing rivers caused by heavy rains in June that were compounded by unusually severe monsoons in August, September, and October.

Yearly monsoons are both a bane and a blessing to Bangladesh (see STORMS). They provide the natural flooding that sustains the rice crop which forms the staple of the diet of Bangladesh. But the man-made destruction of the Earth's ecosystem in the 20th century, accounted for the severity of this disaster, and caused Elhaut to look at the floods of 1987 (see STORMS: MONSOONS) and 1988 and admit, sadly, that "This phenomenon is bound to recur. These are the first two years of a sustained series of catastrophic floods."

The man-made causes of the 1988 flood took place in the Himalayas, upriver from Bangladesh. Indiscriminate deforestation and soil erosion in the Himalayas removed natural barriers to the monsoon rains, and dikes built to shield development projects only exacerbated the problem. The Nepalese forests were subjected to the same wanton and wholesale destruction as the rain forests of the Amazon. A population increase in Nepal heightened the demand for wood faster than the trees could be replaced, and growing herds eroded the remaining grasslands. Thus, the flood-swollen rivers, carrying tons of silt, rushed annually into Bangladesh's rivers and streams, which in turn became clogged with silt and overflowed onto dry land. Downriver, the water finally drained into the Bay of Bengal in India, which created barrages to slow the rivers long enough for the silt to drop to the bottom before it flowed into Calcutta Harbor. But this construction also caused the rivers to back up into Bangladesh, thus exacerbating the floods.

Added to this was the difficulty of achieving regional cooperation among Bangladesh, India, Nepal, Bhutan, Tibet, and China, because of India's reluctance to share power with China in the region. It was an international recipe for disaster.

This fear was voiced by Thomas Drahman, the Asia regional manager for CARE, after surveying the area by helicopter: "I hate to sound hyperbolic, but this could be one of the largest natural disasters of the century."

To those villagers huddled on tin rooftops or in trees, fighting off poisonous snakes who climbed into the same trees for shelter, it must have been that, and more. In the village of Nichkasia, in the Bhola district, 550 drowned, and the rest went without food for days. An 11-year-old girl named Fajilatunnesa told *New York Times* reporters that she had survived for four days on a small package of biscuits.

No deaths were attributed to starvation because of the country's extensive stores of food, collected for just such purposes, but thousands were made homeless and thousands more died of disease. In Bhola alone, 70,000 people were rendered homeless. An estimated 3 million people in the entire country contracted diarrhea or dysentery.

More than 85 percent of the population of the country was thrown out of work as a result of the floods, and its economic activity was brought to a virtual standstill because of destroyed factories and equipment. Most of the livestock and crops were totally destroyed.

Ironically, the contemporaneous Hurricane Gilbert in Jamaica occupied the attention of many of the relief agencies of the world, thus reducing aid and attention to Bangladesh. The United States sent $125 million to Jamaica for hurricane relief and only $3.6 million to Bangladesh. It fell to Japan to be the greatest donor, sending $13 million and rescue workers to the stricken region, where they joined private and U.N. relief agencies in the nearly hopeless task of rebuilding and rehabilitation.

BANGLADESH
July–September 1998

Bangladesh was devastated in the summer of 1998 by a flood that lasted three times longer than any other in its history. Two-thirds of the country was under water for three months. Thirty million people were affected by the flood; 25 million were rendered homeless; 1,050 died.

The longest-lasting flood in the history of flood-prone Bangladesh gripped the country during a summer of unusually strong monsoons. Nineteen ninety-eight was a year of extreme weather throughout Asia. India (see p. 168), China (see p. 158), and Bangladesh were pummeled by winds and rain that seemed ceaseless. Bangladesh, besides being one of the world's most poverty-stricken nations, has the added deficit of being a nation of newly formed islands, many of which are at or slightly below sea level. Thus, every monsoon season produces floods throughout this country. But none quite so widespread or terrible as those of 1998.

Most of the flooding in Bangladesh comes from the great rivers like the Ganges that rise in the Himalayas, course through India, and finally empty into Bangladesh. In June and July of 1998, there were torrential downpours in Uttar Pradesh in northern India, which in turn swelled the Ganges River to flood heights in India and eventually in Bangladesh.

Ordinarily, the floods from the Ganges last in this country for only a few weeks, then empty into the Bay of Bengal. But in the summer of 1998, the sea level was abnormally high, which trapped the waters in Bangladesh.

It was a long summer of horror for 20 million Bangladeshi. For two months, residents of entire villages were marooned on the tin roofs of their shacks by the flooding of the Padma, Jamuna, and Brahmaputra Rivers. Two-thirds of the country remained underwater for two months. Ten thousand miles of roads, 14,000 schools, and a million homes were heavily damaged. Twenty-five million people were left homeless; 1,050 people died from drowning, cobra bites, or dysentery.

These cold figures, while staggering, tell only part of the story. In terms of human misery, this particular flood outstrips practically all others in this century. In Chor Shibola, on the Jamuna River, the family of Mohammad Harunuddin Sheik huddled on a wooden platform. The mother and father kept all-night vigils to keep their daughters from rolling off the platform and drowning in the tea-colored water that lapped at them day and night.

In Dacca, Bangladesh's capital, half of its 9 million residents were dispossessed by the floods. Even its landfills were flooded, and so with no place to dump garbage, the populace threw their refuse into the already putrid floodwaters, which included the human waste of the poor, who had no toilets.

These poor were particular prisoners of the flood. In their neighborhoods the water was so contaminated it was black, but residents still waded through it, washed their dishes in it, and even drank it.

Finally, in the beginning of October, the floodwaters receded. Mud, disease, and hunger were the three hallmarks of the aftermath. A heavy gray muck covered most of the country. Millions were affected by life-threatening and often life-taking diarrhea as they continued to live in the filthy sludge that nourished disease.

Worst of all, over 2 million tons (1,814,369.5 tonnes) of rice that would have been harvested that year were either drowned out or never planted. In a country in which two-thirds of its children are normally seriously undernourished, this became a staggering tragedy.

BRAZIL
TUBARAO
March 24, 1974

Two hundred people drowned and over 100,000 were made homeless when the rain-swollen Tubarao River overflowed its banks and spilled into the city of Tubarao, Brazil, on March 24, 1974.

Tubarao is a city of 65,000 nestled in one of the final curves of Brazil's Tubarao River before it snakes into the Atlantic Ocean. Under ordinary circumstances, the river flows peacefully past Tubarao toward the river's nearby outlet.

In March 1974, torrential downpours swelled the river and accelerated its flow. Days and nights of rain caused streams and rivers to begin to overflow all over Brazil, and on March 24, at high tide, the swollen Tubarao River flowed over its banks. Once freed of its customary path, the river's water became a juggernaut, smashing houses flat in the suburb of Sao Joao—50 people were crushed to death in one house near the flooded river bank. Within moments, every street in Tubarao was flooded, and buildings disappeared up to their first stories. Two hundred people drowned, and more than 100,000 fled to higher ground.

The surrounding countryside was wrecked. Over 8,000 head of cattle drowned, and growing crops of rice, potatoes, corn, and cassava were washed from the earth. The final damage to property would total $250 million.

CHINA
1887

Depending upon the record-keeping source, between 1.5 million and 7 million people perished, and over 2,000 towns, villages, and cities were inundated and destroyed when the Yellow River overflowed its banks in China's northern provinces in late spring of 1887.

The worst flood disaster in modern history, which claimed somewhere between 1.5 million and 7 million lives, occurred in the spring of 1887 in the northern provinces bordering upon China's Yellow River.

Beginning in 2297 B.C.E., when the first flood was recorded by Chinese historians, the Yellow River (or Huang Ho), a 3,000-mile-long (4,828-km-long) carrier of the silt which gives it its proper name, has consistently, in the rainy summer season, raged over its banks, causing floods and famines and earning it its second name, "China's Sorrow."

On the other hand, China's agricultural plain gained its richness from the Yellow River. In fact, the great North China Plain, which extends over much of Honan (Henan), Hopei (Hebei), and Shantung (Shandong) Provinces, was formed alluvially over many millennia from the waters of the Yellow River. This is China's agricultural heartland, producing the corn, kaolin, winter wheat, vegetables, and cotton that form the basis of its economy.

The river has its origins in the Kunlun Mountains of Northwest Tsinghai (Qinghai) Province. From here, it flows generally east through a series of gorges to the fertile Lan-chou valley. Here, it develops into a slow-moving stream as it flows around the "great northern bend" that skirts the huge Ordos Desert. The west end of this bend passes through the heavily populated oasis of the Ningshia (Ningxia) agricultural district, a source of cereals and fruits. At the northwest corner of the bend, it divides into small fingers of tributaries which were helped by ancient constructions to form natural irrigation sources. At the northeast corner, more fertile fields, farmed without irrigation until 1929, receive the Yellow River's largess of moisture.

From here, the Yellow River plunges south, through the loess region, where it picks up most of the yellow silt that has earned it its name. As the river picks up speed and power and bridges the Great Wall, it turns east, through the San-men gorge and onto the Great Delta it has formed at its mouth and which is constantly meandering toward the Po Hai (Bohai), a portion of the Yellow Sea.

During the winter, the Yellow River shrinks to a trickle, barely visible from the banks it nurtures; in the spring it overflows its banks and has caused major floods more than 1,500 times since 2297 B.C.E. The spring of 1887 was the worst instance.

Heavy rains throughout the entire province of Honan had swollen the river that spring, and the first incursion occurred at a sharp bend near the town of Chengchou (Chengzhou). Generations ago, the town had erected a protective wall, and the population attempted to shore it up against the torrent that battered it. The wall failed, and within seconds, both populace and town were engulfed by 20 feet (6.09 m) of swirling water.

Day after day, the surging river waters slammed into towns, inundating or washing them away—a total of 600 located near its banks, including the walled city of Chungmow. It continued to consume crops, animals, towns, and people in a 40-mile-wide (64.4 km-wide) swath, with floodwaters that reached 50 feet in depth.

By late November, eyewitness reports had started to filter through to the English-language *North China Herald*. "Every night," a correspondent in Anhwei Province related, "the sound of the winds and waters, and the weeping and crying, and cries for help, make a scene of unspeakable and cruel distress. It is slow work going from terrace to terrace against often both wind and tide; on these terraces from a dozen to 100 families were often congregated. Of houses, not more than one or two in 10 are left with walls in ruins and half under water. Men rest on tops of these houses, and those of the old who do not die of hunger, do of cold."

Straw ricks became rafts that refugees clung to, "... which," the correspondent continued, "in a high wind are driven along the water, each with its weeping load of men and women. The tops of poplars which lined the roads now float like weeds on the water, but here and there an old tree with thick strong branches has strong men clinging to it crying for help. In one place a dead child floated to shore on the top of a chest where it had been placed for safety by its parents, with food and name attached. In another place a family, all dead, were found with the child placed on the highest spot ... well covered with cloths."

A missionary later told *Goldthwaite's Geographical Magazine:*

> In Cho-chia-kow itself fifty streets are swept away, leaving only three business streets, on the north side, which are all flooded. The west and south parts of the city are on opposite sides of the stream. The whole area is one raging sea, ten to thirty feet deep, where there was, only a month ago, a densely populated, rich plain. The newly gathered crops, houses and trees, are all swept away, involving a fearful loss of life and complete destruction of next year's harvest. The river is all coming this way now, and a racing, mad river it is. The mass of the people is still being increased by continual arrivals, even more wretched than the last. There they sit, stunned, hungry, stupid and dejected, without a rag to wear or a morsel of food.

The devastation left behind when the flood waters eventually receded was awful to behold, and the final statistics would never be accurately computed. By 1889, when the Yellow River was at last back to normal, pestilence added its afflictions to flood and famine, and an estimated half a million people died of cholera.

CHINA
September 1911

Over 200,000 died and a half-million were made homeless when the Yangtze River flooded the Chinese provinces of Anhwei (Anhui), Hupei (Hubei), and Hunan in September 1911.

Over 100,000 people died of drowning, another 100,000 perished from starvation, and hundreds more were murdered by roving bands of starving marauders when China's Yangtze River burst its banks in September 1911.

The so-called water basin of the 700-square-mile (1,126.5 sq. km) area that includes the provinces of Anhwei, Hupei, Hunan, and the city of Shanghai is a rich and lush area, housing 2 million people and feeding much of the surrounding area. It was turned into an inland sea by the floodwaters of the Yangtze in 1911. Nothing survived—no crops, no cattle, and very few human beings.

Typical of the devastation was the condition of the city of Suchow (Suzhou), situated on the Grand Canal near Tai Lake. This proud population center of over 1 million was noted for the exquisite silks its artisans and mills produced for royalty and wealthy Westerners. In September, the floodwaters of the Yangtze, swollen by repeated rains and funneled through the canal, completely submerged the town within the first hour of the flood. Thousands were drowned instantly; others fled to the drier countryside. As days passed and the floodwaters failed to recede, robber bands formed, and one of these bands from Suchow sacked the American Baptist Chapel near Ch'uisan (Quisan), murdering the missionaries and leaving the chapel a scarred wreck.

Other Western missionaries wrote home of the macabre sight of thousands of wooden coffins floating down the Yangtze, as one of the largest cemeteries in the district was flooded.

Ultimately, over a half-million refugees from country-side and city alike fled to Manchuria and Mongolia, never to return.

CHINA
1939

(See FAMINES AND DROUGHTS.)

CHINA
August 1950

Official figures record that 489 persons drowned and 10 million were made homeless by flooding from the Hwai and Yangtze Rivers in eastern China in August 1950.

The Communist government of China refused to release final statistics on the flood caused when the Hwai and Yangtze Rivers overflowed their banks in August 1950, inundating much of the provinces of Anhwei (Anhui), Kiangsu (Jiangsu), Honan (Henan), Hopei (Hebei), Hupei (Hubei), Hunan, Kiangsi (Jiangxi), and Kwangtung (Guangdong).

What is known is that the hardest hit province was Anhwei, a low-lying section of eastern China that has traditionally suffered enormous losses of life and property from flooding. The government admitted to the death by drowning of 489 persons, but Western observers concluded that this was a gross underestimate.

Official figures were a little more accurate regarding property damage. More than 890,000 dwellings were destroyed, leaving 10 million people homeless; 5 million acres of cultivated land were left under floodwaters for a long period of time, rendering 3.5 million acres of it untillable and unworkable for the entire planting season.

In terms of effect upon the food supply and material devastation, the flood of 1950 must emerge as a major catastrophe.

CHINA
June–July 2005

In the worst flooding since 1998 (see p. 158), more than 1,000 people died and hundreds were reported missing during summer 2005. More than 2 million people were relocated, particularly in the worst-hit provinces of Kwangsi (Guangxi), Fukien (Fujian), and Shensi (Shanxi), the city of Tachou (Dazhou), and the areas surrounding the Yellow River, Yangtze River, Liao He (Liaohe), and Hai He (Haihe).

Every summer in China is a flooding season. The rainy season begins at the beginning of June and lasts through the end of July. During that time, torrential downpours swell China's rivers beyond their banks, inundating rice paddies, fields, villages, and cities. But in 2005, the torrents were heavier, the rain lasted longer, and the flooding was more severe than in any year since 1998 (see p. 158). In parts of the southern province of Kwangsi (Guangxi), the flooding was the worst in a century, with Hsiangchow (Xiangzhou) county receiving one third of its annual rainfall in less than three days.

In Tachou, in Szechuan (Sichuan) Province, floodwaters reached the third stories of some buildings and most of the roads both within and leading into and out of the city were cut. During the first week of July, some 150,000 people were evacuated.

In Shensi Province, mountain torrents, landslides, and mud-rock flows in 40 townships of 12 counties, affected more than 300,000 people. More than 35,000 houses were hit either by the landslides or floodwaters, and 862 acres (348 ha) of farmland, valued at $26.6 million, were destroyed. Near the Min River in Fukien, a mudslide swept a bus and a car off a highway and into a river near the city of Jian'ou, leaving 23 people missing.

As the downpours continued into late July, the Yellow, Yangtze, Lia He, and Hai He rivers began to overflow, washing more farms and houses away. These houses were built of poor quality construction materials because of the poverty of the populace; in these flood-prone areas, many farmers were unable to afford cement and other proper materials, so a great many houses were built with a mud-based sealant that was unable to stand against the onrush of the floodwaters.

The causes of the widespread destruction were multiple, but the 25 years of "market reforms" and cutting of social spending by the national government, along with pervasive official corruption, led millions of poor peasants to cut down forests for wood and fuel, thus exacerbating the destruction from the floodwaters. Declines in natural reservoirs such as forests and lakes and the increased silting of rivers and lakes from deforested land in the Yangtze basin contributed to the tragedy, as did the encroachment on riverbeds by farmers. And, to add to this, dams that were built to help control the flooding were too small. After the floodwaters finally receded in August, over 1,000 people had died, and hundreds remained missing.

CHINA
August 1954

Over 40,000 drowned and over 1 million were made homeless by floods caused by the overflowing of the Yangtze and Hwai Rivers in China in August 1954.

Old records as well as ancient structures swept away by the monumental flood of August 1954, caused by torrential rains and the overflow from the swollen Yangtze and Hwai Rivers. Over 40,000 people drowned, hundreds of villages and towns disappeared and the floodwaters rose to a level of 96.06 feet (29.3 m).

According to Peking Radio, the "heaviest rainfall in a hundred years" preceded the flood. It was enough to swell the Yangtze, the sixth largest river in the world, into a raging torrent that ultimately burst its embankments, exploding into the rich rice fields of the Peking-Shanghai-Hankow triangle.

For millennia, this sunken valley had been ideal for rice growing but dangerous for settlers. Prior to the Communist takeover of China in 1948, the United States had attempted to reverse the flow of the river by proposing a TVA dam—it would have been the largest in the world—and a reservoir system.

Instead of the American plan, the leading Chinese Communist architect, a woman named Chien Chen-ying, designed, in the Soviet mold, a dam that was constructed completely on the surface of the soil. Three million workers completed this monumental but ultimately flawed structure. Like similarly constructed dams before it, the large dam crumbled like a sand castle against the advancing floodwaters.

At one break point in the dam, 200 soldiers and 10,000 peasants with mats strapped to their backs linked arms to form a human wall. For three hours, they remained shoulder to shoulder, while the floodwaters battered at this concrete and human barrier. Finally, the waters won, crumbling the wall and sweeping several thousand soldiers and peasants to their deaths. More than the dam crumbled that day; Mao Tse-tung (Zedong)'s first Five Year Plan was also turned to ruin.

CHINA
FUKIEN (FUJIAN) PROVINCE
August 7, 1948

More than 80 percent of the Chinese province of Fukien was destroyed, 1,000 drowned, and 1 million were made homeless by a flood caused by the overflowing of the rain-swollen Min River on August 7, 1948.

Two solid months of torrential rains preceded the flood of August 7, 1948, which drowned 1,000 people, sent 1 million refugees onto the higher ground of the surrounding hills, and destroyed more than 80 percent of the Chinese province of Fukien.

The Min River, Fukien's principal source of irrigation for its plentiful rice crop, is ordinarily a benign resource. But the cloudbursts of June and July of 1948 filled the river to overflowing, and on the night of August 7, squads of gong beaters shouting "Chiu ming! Chiu Ming!" ("Save life!") preceded the floodwaters. Practically every shelter within a few miles of the river bank was destroyed, and for weeks, floating debris and bodies drifted with the currents in the Min.

A curious sidelight of this flood involved a political duel of words between the Communists and the Nationalists, both of whom were vying for the power that the Communists would claim within a year. The Communists, via Peking Radio, blamed the Nationalists for the tragedy: "It is impossible to complete dike repair work because of constant Nationalists raids. . . . We request Nationalist troops and air forces to cease their obstruction."

Almost simultaneously, the Nationalists, through Nanking Radio, were firing their salvos of accusation: "Since their occupation of this area, Communists have methodically destroyed dikes. With floods coming they are wildly firing accusations against the Central Government. . . . We hope they will show a sense of humanity and withdraw. . . ."

CHINA
HANKOW (HANKOU)
July 4, 1935

The Yellow River flood of July 4, 1935, killed 30,000 and rendered 5 million homeless in the North China Plain.

When the Yellow River burst its banks on July 4, 1935, it inundated 6,000 square miles (9,656 sq. km) of the North China Plain. Thirty thousand died, 5 million were left homeless, and the damage was estimated at $300 million.

The flood was caused by torrential downpours that swelled the Yellow River until it broke through the levee in western Shantung (Shandong) Province. In an alarmingly short amount of time, the breach extended until it measured 1.5 miles (2.4 km) in length. River water poured through this breach at the rate of 3.79 million gallons per second, and all buildings and life, both animal and human, were swept before it.

While rescue attempts—most of them futile—were launched, an army of 35,000 peasants was organized to repair the levee. First, 15,000 of the workers constructed four dikes into the river a short distance above the break to deflect the tumultuous flow of the water.

Simultaneously, on the far bank, 20,000 more workers excavated a canal around the levee, the break, and the main channel of the Yellow River. For two solid months, while the dead and the dying were collected, and medical facilities were readied for a possible plague, this work continued.

It took another month to rebuild the broken levee and reduce the gap to 130 feet (39.6 m). At this point, the canal on the opposite shore was opened, and the river, still swollen, was diverted to it. With pressure off the levee, workers dumped bales of kaolin stalks into the breach, added bags of earth and sand, and waited several days for the silting to close in minor leaks. At this point, the river returned to its normal flow, and the breached levees held. It was a method that worked well after the fact. But it did little to save lives.

157

CHINA
KWANGTUNG (GUANGDONG)
May 1982

Four hundred thirty people drowned and over 450,000 were rendered homeless when the North River flooded the Chinese province of Kwangtung in May 1982.

The populous province of Kwangtung, in the southeast of China, was pelted with nearly 24 inches (61 cm) of rain in 13 hours early on May 12, 1982. As a direct result, the North River rose to its highest level since 1949, burying roads under six feet of water, inundating 286,000 acres of farmland, and collapsing 46,000 homes. Forty bridges spanning the river collapsed under the rising floodwaters; dikes meant to contain the river were leveled. Before two days passed, Guingyuan and Yingde counties were entirely under water.

The floodwaters spread with alarming speed and force for several days, sweeping away more than 20 miles (32.2 km) of the main rail line between Canton and Peking (Beijing). Some travelers arriving in Hong Kong reported being stranded in trains for 30 hours as the waters rose and landslides blocked the tracks.

After the first five days, Chinese television began to show army units rescuing stranded peasants, some huddled on rooftops. Military aircraft dropped tons of biscuits, clothing, and candles. Sailors, piloting landing craft and rubber dinghies, rescued others.

In all, over 100,000 troops were pressed into service to rescue the more than 450,000 marooned and displaced people. Less fortunate were the 430 who died in this tragic and unexpected disaster.

CHINA
SZECHUAN (SICHUAN) PROVINCE
July 1981

Ecologically unsound practices produced the flood in China's Szechuan Province in July 1981. Seven hundred fifty-three drowned, 28,140 were missing, and 1.5 million were made homeless.

In July 1981, a combination of natural overabundance and human foolishness combined to provide Szechuan Province, China's most populous (100 million people) province, with one of the worst flood disasters of this century.

For three decades preceding the disaster, the province's watershed area, in the upper reaches of the Yangtze River, had been virtually swept clean of trees. This

wholesale and wanton deforestation had allowed the water from normally heavy summer rains to fill the river's tributaries, building up force and sweeping tons of valuable topsoil downriver. This not only reduced the amount of land available for crops; it created a potential hazard, since the force of flooding water is increased many times by the addition of silt and soil.

Ironically, the reason for this denuding of the landscape was agricultural reform. Under the edicts of Mao Tse-tung, farmers were authorized to strip the land of trees in order to plant wheat and corn, and to plow up grasslands in the north that had held back the deserts.

The resultant ecological chaos came to a head in early July 1981, when the Yangtze River, swollen from a three-day rain, overflowed its banks. Within hours, bridges, roads, and over 400,000 structures were leveled or swept before the raging floodwaters. The highest tide of the century swept unchecked through the Yangtze gorges and then, in an 18-foot-high (5.5-m-high) wave, smashed against the pride of China, the $2.2 million Gezhou Dam.

The dam held, but this did little for the land upstream of it, which contained Szechuan's two largest cities, Ch'eng-tu (Chengdu) and Chungking (Chongqing). Backwaters swept the record-breaking tide even higher, and more and more land disappeared under water.

Two hundred thousand workers were formed into an army that shored up dikes along the riverfront, hauling huge rocks with their bare hands or on shoulder poles.

The effort tamed the torrents and saved millions of lives, particularly in Hupei (Hubei) Province. But when the waters finally receded, the death toll would still be appalling: 753 were dead, 558 were missing, 28,140 were injured, and 1.5 million were made homeless.

CHINA
YANGTZE/SHOSHUN RIVER BASIN
July–August 1998

The worst flood in 44 years raged through the basins of the Yangtze and Shoshun Rivers in northeast and south-central China in the summer and fall of 1998. One hundred and eighty million people were directly affected by the flood; 18.3 million were evacuated and 4,150 died. Millions of acres of agricultural land were made untillable; 13.3 million houses were damaged and 6.9 million were totally destroyed. The ultimate cost of the disaster was $26 billion.

It is not unusual for China's Yangtze River to flood. But the inundation along the Yangtze in the summer of

1998 was the worst in 44 years, affecting over 180 million people and killing 4,150. And that, even in one of the most thickly populated countries in the world, was unusually tragic.

A convergence of forces—natural, human, and governmental—caused this cataclysmic flooding. First, there was relentless rainfall combined with the melting of a deep snow cover on the Tsingha (Qinghai)-Tibet Plateau. The runoff into the valleys was enormous.

Then there was rampant deforestation on the upper reaches of the Yangtze, which allowed a record amount of silt from the eroded soil to filter into the Yangtze basin.

There were also the flood plains of the Yangtze, which for thousands of years had contained the runoff of the yearly floods that poured from the river and irrigated thousands of acres of rice paddies. In recent years, China's burgeoning population had begun to inhabit these flood plains, and so became waiting victims, directly in the path of disaster.

And finally, there was the government-directed destruction of points in the dike system by Chinese troops in order to prevent breaching of the main dikes that protected the huge concentrations of the people surrounding Wuhan, one of China's largest metropolitan areas.

It was in Hupei (Hubei), the south-central province that contains Wuhan, that the worst flooding occurred from August into October. More than half a million people were evacuated from the province in the first weeks of August by the Chinese army. A state of emergency was declared as the river, cresting higher and higher, broke through embankments at more than 100 places.

In Jiayu County, 200 people were washed away and drowned when a dike suddenly burst on August 8. Some 40,000 people were stranded near Chiuchiang (Jiujiang), in neighboring Chiangshi (Jiangxi) Province, when another dike collapsed, surrounding them with water.

Nature seemed to be hurling its fury at the Yangtze and its shores. Downstream, in Anhwei (Anhui) Province, a deluge of water poured from two sides as the river swelled in the west and a typhoon dumped torrents of rain in the east.

By the middle of August, 2,000 people had been killed by landslides and mud flows that consumed houses and villages. And as refugees fled the waters and evacuated their homes, thieves moved in to take advantage of flooded homes and unprotected evacuees. The police and the army organized night patrols to prevent theft from submerged houses and refugees.

As the intensity of the flood increased, floodwaters poured into the city of Jiujiang. It was apparent that a crisis situation had arrived, and officials and the army ordered tens of thousands of residents of the areas along the river above Wuhan to abandon their homes. Engineers then dynamited secondary levees to allow the swollen river to drain off some of its bulk and power before it reached the populated urban areas.

At Shashi, in central Hupei, the Yangtze crested at 47 feet, just eight inches short of the level that would have required levees to be blasted and the flooding of the homes of as many as 250,000 people.

At the end of August, the huge Taching (Daqing) oil field was inundated when the Nen River drove through the dikes guarding it. One thousand, eight hundred of the field's 25,000 oil wells were inundated and rendered useless.

At the same time, Harbin, the capital of Heilongjiang Province, was threatened by the Shoshun River, which, like the Yangtze and Nen, was swollen from runoff and rain. On its north bank, opposite the city, the river swallowed dikes and covered fields for as far as the eye could see. Families who had lived along dikes they thought would protect them were forced to take shelter in trains that were parked atop raised rail beds on sidings.

They were some of the 13.8 million people the government of China evacuated to higher ground, while soldiers struggled in chest-high water, hauling sandbags to dikes that were in danger of crumbling, and factories worked around the clock to produce the sandbags that were the last line of defense against the ceaselessly advancing waters.

Some families lived on top of the larger dikes; others were housed in schools and factories that had been turned into temporary shelters. The conditions in the shelters were safe but unhealthy. Some families took furniture, chickens, pigs, and even tractors with them, and a combination of poor sanitation (many people were drinking water that was contaminated by human waste) and human beings and animals living in close proximity made the refugees vulnerable to disease. Outbreaks of skin infections and infectious diarrhea greeted the Red Cross when it arrived, distributing water purification tablets, medicines, and chlorine spray for disinfecting sanitation pits and the submerged land, after the waters receded.

As dike after dike gave way, critics of the policy of using dams to control the river's flow, and particularly the huge and then unfinished Three Gorges Dam Project in Hubei, faulted the government for not maintaining the involved dike system along the Yangtze and Shoshun rivers.

Village after village simply disappeared, like the village of Shinkankow (Xinkankou) on the Yangtze flood plain. The only evidence that it had existed at all consisted of three metal rooftops, barely visible above the surface of the water.

It was a result—perhaps a consequence—of the policy elicited by President Jiang Zemin earlier in the summer: First, protect the river's main dikes; second, protect major cities; third, protect human lives.

It would be another two months before the waters would finally recede. And China would, after a half-century of rampant clear-cutting, ban cutting of the old-growth forests that had once carpeted the mountains in Szechuan (Sichuan), Yunnan, and Kansu (Gansu) Provinces, at the upper reaches of the Yangtze River.

All told, more than 180 million people were affected by the flooding; more than 18.39 million people were evacuated, and 4,150 were killed. Close to 13.3 million homes were flooded and 6.9 million were destroyed. The total bill for the devastation was $26 billion.

CHINA
YELLOW AND YANGTZE RIVER BASINS
July–August 1996

At least 2,775 people were killed and 234,000 were injured in the flooding caused by relentless rain and typhoons in the summer of 1996 in 21 provinces of China through which the Yellow and Yangtze Rivers flowed. Eight million people were evacuated, 4.4 million were left homeless, and 8 million acres of cropland were turned useless. The damage totaled $20.5 billion.

Torrential rainstorms pelted northern and northwestern parts of China along the Yellow River during early July of 1996. It was a sensitive time in the grain belt of China. July is the time of the summer grain harvest, and August is the sowing season. Neither would take place in the summer of 1996, as an assault of rain and typhoons, and the floods that both caused, would kill both humans and crops, covering the countryside for as far as the eye could see with an encasement of floodwater turned a sickening brown by the mud that laced it. Rotting crops, pieces of dwellings, and the corpses of humans and animals were everywhere in the provinces of Hopei (Hebei), Shansi (Shanxi), Heilongjiang, Shangtung (Shandong), and Honan (Henan).

By August 5, over 400,000 people were evacuated from Shandong and Henan, and the monitoring post in Henan showed that the Yellow River had reached its highest flood peak in recorded history. This spelled enormous danger for the 1 million to 3 million residents of the North China Plain, who were protected by a series of dikes. Fortunately, these dikes held, but only barely, and only because they were constantly shored up by ultimately exhausted army troops and young farmers.

Those who lived along the Yangtze fared less well. On July 23, its main dike gave way near the village of Kaohuang (Gaohuang). A 30-foot (9.14-m) wall of water crashed through the barrier and continued across the countryside, wiping out, in an instant, half of the houses of the village and killing half of its inhabitants.

The breach was patched but not before thousands of acres of grain were reduced to rotting stalks, protruding here and there above the water.

Both rivers continued at flood stage, as more natural catastrophes struck. The eighth and most powerful typhoon of the season struck Fukien (Fujian) Province full force, and continued on to damage Hunan, Hupei, Chekiang (Zhejiang), and Honan Provinces. In the flooding caused by this one storm alone, 250 people were killed, 300 were reported missing, and hundreds of thousands were left homeless.

A week later, as the Yangtze surged dangerously close to the top of its dikes, officials opened a dike at Longku, wiping away the livelihoods of hundreds of farmers. Seventy-year-old Liu Yueyi remained, afterward, sorting through the pile of bricks that had once been her home. "All of our income came from our fish farm," she told a reporter, "but all the fish swam away in the flood. Now we are living one day to the next."

The causes for the flood and its particular devastation, as would be the causes of the China flood of 1998 (see p. 158), were multiple and similar: The Tibetan-Tsinghai (Qinghai) Plateau, where the Yellow and Yangtze Rivers have their sources, is 20,000 feet (6,096 m) higher than the alluvial plains toward which the waters of the rivers flow. Earthquakes, movements of tectonic plates, unwise clear-cutting of forests, and summers in Asia that seem to be increasing in their rainfall, storms, and general severity, all collide with a huge population, many millions of whom are forced to live and farm in the shadow of the high earthen dikes that hold back the country's major rivers.

More than 2,775 people died in the China floods of 1996; more than 4.4 million were rendered homeless, 234,000 were injured, and 200 million people in 21 provinces were directly affected by the floodwaters. Eight million acres of cropland were made unproductive for at least two seasons.

DOMINICAN REPUBLIC, HAITI
May–June 2004

In one of the worst natural disasters in Caribbean history, a flood caused by two weeks of torrential rains in May 2004 claimed 3,300 lives, most of them in Haiti.

With no tree roots left in the surrounding mountains to slow the overflowing rivers, entire areas of Hispaniola, which the Dominican Republic and Haiti share, were inundated.

Haiti, one of the poorest nations in the world in the 21st century, has a population whose average income is $400 a year. The island country's next door neighbor, the Dominican Republic, on the island of Hispaniola, has a similarly poor populace, though, with an average yearly income of $2,000 a year, it is not nearly as poor as Haiti's.

Haiti was once lush with tropical trees. By 2004, only one percent of the land had tree cover, and in the Dominican Republic, only 15 percent of its land was forested. The trees had all been felled by poor residents in need of charcoal for their own cooking, or to sell as their only means of survival.

Thus, when two weeks of torrential downpours relentlessly pounded both countries in May 2004, there were no trees on the surrounding hills to hold back a cascade of rainwater. This, augmented by overflowing rivers which collapsed the banks that had contained them, resulted in a giant and terrifying flood which swept away houses, killed livestock, destroyed crops, inundated villages, and drowned a total of 3,300 people in both countries (see the color insert on p. C-6).

The town of Mapou, in the southeast part of Haiti, simply disappeared under 10 feet (3 m) of water rimmed by mud and rubble. In and around Fond-Verrettes, just north of Mapou, hundreds of village residents and those in the hills near the town disappeared in a moment, as rivers and streams burst their banks.

In the town of Jimaní, the Dominican Republic, those not buried under mud were swept downstream for 19 miles (30.6 km) to a lake called Lago Enriquillo. In the La Cuarenta neighborhood of Jimaní, where the poorest residents—mostly Haitians—lived, the fatalities were staggering. The neighborhood had been built on a riverbed that had been dry for years. In May 2004 it simply filled itself to overflowing. "We have nothing left," Socorro Moquete, a 67-year-old grandmother from Jimaní told rescuers who arrived within a day. "The river took everything, even the dead in the cemetery."

No roads were passable in the region along the Haiti–Dominican Republic border on the island of Hispaniola. Aid workers could only ferry supplies in by helicopters from the Dominican Republic and from a multinational force stationed in Haiti after the revolt in the winter of 2004. (The revolt had overturned the government of President Jean-Bertrand Aristide, leaving a bankrupt, cobbled together government without many functioning agencies and only three operative helicopters.)

The rescue operation was a frantic and feverish effort, exacerbated by the uninterrupted rain, a shortage of supplies, and the fact that Haiti, even before the flood, was a country in a deep crisis.

ENGLAND
LYNMOUTH
August 15, 1952

A flash flood totally destroyed the town of Lynmouth, England, on August 15, 1952. Thirty-four people died; the remaining 1,200 residents were made homeless.

The quaint seaside town of Lynmouth derives its name from an old Anglo-Saxon term meaning "the town on the torrent." For centuries it contradicted its name, and with its cobbled streets and thatched houses and benign climate served as a haven for such 19th-century poets as Percy Bysshe Shelley, and, later, as a seaside resort for vacationing Londoners.

The flood that destroyed the village in August 1952 was of the flash variety, caused by torrentially intense and sustained rains, which swelled the West and East Lyn Rivers in north Devon. Lynmouth itself sits in a narrow Y-shaped valley at the confluence of the two rivers. Originating in 39 square miles (62.8 sq. km) of the Chains of Exmoor, which is a heather-coated plain, the two normally placid and streamlike rivers funnel through gorges punctuated by oak forests and plunge some 1,500 feet (457.2 m) in four miles before joining together for a short, final fall to the sea.

For the entire first two weeks of August 1952, rain fell intermittently and heavily over the entire area. August 15, a Friday, was a particularly dismal day. Three inches of rain had fallen, and by evening, the East and West Lyns were swollen and roiling.

At dinner time, the rain stopped, but only momentarily. From 6:30 until 11:30 P.M., over six inches of rain fell on Lynmouth.

At 7:30 P.M., a canal carrying water to Lynmouth's hydroelectric stations roared over its banks, crashing into the generators, and the plant went dead. A backup emergency diesel system was brought on line, but by 9 P.M. this too shut down, leaving Lynmouth in darkness, which made the invasion of the debris-laden flood all that more terrifying.

By 9:30, the floodwaters had roared down the gorges above Lynmouth and had begun to take down trees and brick walls of homes that had heretofore rested a comfortable 30 feet (9.1 m) above the ordinary flow of the twin rivers. Thomas Floyd noticed the garden wall of his property starting to crumble. Before he could get back to his home to warn his family, the floodwaters roared into his yard. He managed to seize a piece of brickwork that withstood the flood; the eight other members of his family, trapped in the house, drowned.

The next morning, as is often the case after a natural disaster, was brilliant and calm. Lynmouth had been demolished; 93 houses were either swept away or irreparably damaged, 132 vehicles were swept out to sea, and at least 34 people were dead. Those who survived did so through luck or fate. The 60 guests in the Lyndale Hotel, for instance, managed to scramble to safety on the upper floor of the hotel. The next morning, they were able to step out of their upperstory sanctuary directly onto a rocky flooring composed completely of huge boulders. Beyond the village, on the beach, the detritus of the flood was strewn for miles. Splintered lumber, broken telegraph poles, pretzelized steel girders, ruined automobiles, and thousands of saplings that had been uprooted and peeled of their bark by the grinding action of the flood, lay everywhere. They would continue, in some fraction, to remain as reminders of the terrible dark night of August 15, 1952.

FRANCE
FRÉJUS
December 3, 1959

The collapse of the Malpasset Dam above the small town of Fréjus, France, caused the destruction of most of the village and the death of 419 residents on December 3, 1959.

Four hundred nineteen residents of the small town of Fréjus, located near Cannes on the French Riviera, were swept to their deaths in a few terrible moments on the night of December 3, 1959.

For two years, the Malpasset Dam, a supposed miracle of modern engineering designed by world famous engineer Andre Coyne, had doled out water from a five-mile-long by two-and-a-half-mile-wide Alpine lake, supplying water for a number of Riviera towns and irrigation for the Reyran valley, the major source of Europe's peaches. The dam, begun in 1952 and completed to much fanfare in 1957, had been accepted as a monumental example—much like the local Roman ruins of a gladiator's arena—of humankind's ability to build to last.

Not so. Without warning and with a thunderous roar, the dam gave way, splitting in half and unleashing a 15-foot-high (4.57-m-high) juggernaut of water that roared downhill and into the sleeping village. Trees, bridges, and buildings were ripped apart; livestock and human beings were tossed, doll-like, downstream. In

a few moments it was all over. Most of the village was destroyed, except for one landmark, which remained unscathed: the Roman ruins.

GERMANY
NORTH SEA COAST
February 17, 1962

Germany's most serious flood of the 20th century occurred on February 17, 1962, when a storm-agitated North Sea breached seawalls and flooded most of the coastline. Three hundred forty-three people drowned; 500,000 were made homeless.

Three hundred forty-three residents of the North Sea coast of Germany were drowned on February 17, 1962, when the waters of the sea, churned by a hurricane, breached seawalls and roared inland, swelling rivers and streams in an echo effect that exceeded the first inundations along the seacoast. Thousands of homes and buildings were swept from their foundations or collapsed by the floodwaters, thus rendering 500,000 homeless. More than $6 million in damage was caused by this, the most serious flood along Germany's coast in modern times.

In the first few hours of the flood, the Elbe River rose to particularly extended heights, engulfing the city of Hamburg and killing 281, inundating a large part of the city of Bremen, and isolating the island of Krautsand for days.

A distinctive quality of this flood was its suddenness and its longevity. Though the storm that was its origin blew itself out in hours, the flood continued for days, as its floodwaters pushed inland, decimating ancient breakwaters and buildings. In Schleswig-Holstein and Lower Saxony, rich agricultural land was rendered temporarily useless, as crops washed away and livestock drowned. With little opportunity for runoff, the fields were turned into marshes for weeks, and did not return to their true character for a multitude of months.

HAITI
October 22, 1935

Over 2,000 drowned in the floods caused on the island of Haiti by the so-called hairpin hurricane of October 22, 1935.

More than 2,000 people lost their lives to floodwaters spawned by a hurricane that swept across Haiti on October 22, 1935 (see HURRICANES, p. 264). The "hairpin hurricane," so called because of the pattern of its path, hit Jamaica head on, pouring huge amounts of rain on the island in a very short time, swelling rivers and streams, and catching farmers unaware in their fields.

These workers formed the bulk of the casualties from floodwaters, and the plantations in which they worked suffered $2 million in economic damage. For days, livestock and human bodies floated on the increasingly stagnant waters that covered plantations like some murky inland sea.

HOLLAND
November 1, 1530

The North Sea, powered by gales, breached Holland's dikes on November 1, 1530, causing the most catastrophic flood in Holland's history. Four hundred thousand people drowned; entire villages disappeared.

Forty percent of the land area of Holland lies beneath sea level, and most of the country is actually land reclaimed from the sea. Bounded by the North Sea on the north and west, by Belgium on the south and by Germany on the east, it is crossed by drainage canals, and the main rivers, the Scheldt, the Maas, Ijssel, Waal, and lower Rhine are canalized and interconnected by artificial waterways that are linked with the river and canal systems of Belgium and Germany.

The country has always been thickly populated, and, from its earliest recorded history, has served as a battleground for many of Europe's fiercest wars. In the 1500s, the so-called low countries were under the rule of the House of Habsburg, represented on the throne by Charles V—who would, in the 1550s, give them to his son, Philip II of Spain.

In 1530, Holland was a prosperous country, protected from the rest of Europe by the Hapsburgs and from the sea by a system of dikes. On November 1, 1530, after a furious gale which whipped the North Sea into a frenzy, the dikes gave way. With no high ground available for escape, 400,000 hapless people drowned. Hundreds of homes and entire villages were inundated and collapsed, and the property damage, in today's figures, amounted to the billions.

HOLLAND
November 1, 1570

Gale-driven waves from the North Sea breached Holland's northwestern dikes on November 1, 1570, *drowning 50,000 people and destroying the northern provincial capital of Friesland.*

Forty years to the day following the most destructive flood in Holland's history (see previous entry), its rebuilt dikes gave way once more under the assault of gale-driven waves from the North Sea.

For days before, these newer, stronger dikes were battered by heavy seas. Finally, at high tide on November 1, the dikes in the northwestern part of the country crumbled, once more cutting wide swaths of destruction throughout the countryside, wiping out dozens of villages and drowning more than 50,000 people. Friesland, the capital city of this northern province, was totally destroyed. Twenty thousand residents of the city perished in the first assault of the floodwaters.

At the time of the flood, Holland was at war with Spain, and the Spanish blamed the flood, which occurred on All Saint's Day, on the Calvinism of the Dutch, ". . . the vengeance of God upon the heresy of the land . . ." according to the account of the period.

HOLLAND
February 1, 1953

Fifty of Holland's primary dikes melted away under the assault of the North Sea on February 1, 1953. One thousand eight hundred thirty-five people drowned; 72,000 were evacuated; 43,000 homes were either damaged or demolished.

A horrendous death toll of 1,835 followed the breaching of 50 of Holland's primary dikes in the pre-dawn hours of February 1, 1953. It was one of the worst floods in modern history and in all of Holland's history. Over 72,000 people were evacuated, 3,000 homes were totally destroyed and 40,000 were damaged in a tragedy that was intensified by a false sense of security.

Until then, the 10-foot (3.04-m) dikes of Holland had held for nearly 400 years against enormous assaults. Thus, when hurricane winds of over 100 MPH moved 15 billion cubic feet of water from the Atlantic Ocean into the North Sea and continued to drive high tides and waves at the dikes from January 29 through January 31, the Dutch felt no particular sense of alarm. But a few minutes before dawn on the next day, February 1, the combination of battering wind and increased water volume finally combined to crush 50 dikes simultaneously. Within minutes, 133 towns and villages were buried beneath the onrushing water. It took virtually no time at all for the floodwaters to rise to the tops of town towers.

The populace, caught unaware, improvised survival procedures. Residents of Burghsluis, Stellendam,

Ouddorp, and Kortgene escaped the waters by climbing through upper windows and floating or swimming to the roofs of houses. Many of those who hesitated to pack bags or rescue valuables were drowned instantly. One man chose to stretch out on his dining room table. Water gushed into his home and turned the table into a raft. The next day, rescuers found him alive, floating just beneath the ceiling beams of his dining room.

In one of these villages, a couple who were to be married the next day used their bodies to hold an embankment together for 36 hours at the height of the punishing assault of rain and wind and water. The embankment held; the human beings who had morticed it did not. When rescuers reached them, she was dead, and he had gone mad.

In Kortgene, in a move that would be duplicated almost exactly a year later in China by Chinese soldiers and peasants (see FLOODS, China, August 1954, p. 156), a hundred fishermen linked arms and pressed their backs against one crumbling dike. The town constable, strolled by, looked at them curiously, refusing to heed their cries to him to ring the alarm bell. His reasoning: The water at his feet was merely the overflow from a few high waves and the dike was impregnable. In the next few minutes he was proved wrong. The dike gave way and soldiers and constable barely escaped with their lives.

Elsewhere, terror rippled through villages awash with the floating carcasses of drowned people and livestock. When the waters receded, as they would in a few days, the stench was enormous and overwhelming. Over half a million acres were under salt water, 625 square miles (1,005.8 sq. km) of agricultural lands would be rendered temporarily useless. By the end of the first day, 12,000 soldiers of the Dutch army and helicopters of the Royal Netherlands Air Force joined vessels of the Royal Netherlands Navy in rescuing clumps of survivors clinging to chimneys and bell towers.

Within a week, 25 of the world's nations had rallied to the support of a reeling Holland. Food, medical supplies, and personnel, including battalions of engineers, super amphibious ships, helicopters, and planes came from the United States. Even England and Belgium, ravaged by the same storm, sent rescue teams and money. The cleanup was swift and efficient; the price tag would run into the hundreds of millions.

HOLLAND
DORT
April 17, 1421

The second worst flood in Holland's history, on April 17, 1421, was caused by heavy winds and prolonged rain. One hundred thousand people drowned and the city of Dort was leveled.

The second-worst flood disaster in Holland's history (the worst was in 1530 [p. 163]; the third-worst [see previous entry] was in 1953) took place in and around the city of Dort, in the southern part of Holland. A deluge that had lasted for days, whipped by strong winds, swelled both the North Sea and the Waal River that had flowed from it and past Dort and the nearby city of Dordrecht.

On April 17, 1421, the dikes surrounding the city burst, and battering rams of water rushed onto the thickly populated countryside. Even in 1421, Holland contained an average of 500 people per square mile, and within a day, more than 100,000 drowned. The city of Dort was not merely inundated; it was totally destroyed. Seventy-two villages surrounding it were also obliterated, never to rise again. The city of Dordrecht, prior to this April flood a solid portion of the mainland of Holland, was permanently detached by the gushing waters of the Waal River. To this day, Dordrecht remains surrounded by water.

HOLLAND
LEYDEN
October 1–2, 1574

The tide of war was turned, literally, when a storm-driven flood drowned 20,000 occupying Spanish troops near Leyden, Holland, on October 1, 1574.

Holland was at war with Spain in October 1574, and things were going badly for William the Silent and the Dutch populace. The Spanish Duke of Alba, engaged in subjecting the Godless, Calvinist Netherlanders in the name of the Inquisition, had successfully surrounded the walled city of Leyden. While they were unable to storm the city because of the fierce determination of the inhabitants, who had chosen to fight to the last grain of powder, the Spanish were having some success in cutting off all supply lines, thus starving the city's defenders and inhabitants alike.

Starvation was and still remains a weapon of war (see FAMINES AND DROUGHTS), and, according to historian A. H. Godbey, the dwellers of the besieged city "... were digging up every green thing, devouring roots of grass, old leather, offal, anything that could in the least aid to sustain life ... so long as a dog barked in the city the Spaniards might know they held out."

The situation was desperate indeed, and a miracle was clearly needed to save the absentee William the Silent and the defenders of Leyden. The miracle arrived, ironically, in the person of a natural disaster. A fierce storm struck the coast on the night of October 1, 1574, and North Sea waves crumbled dikes miles away from the encamped Spaniards. By October 2, the roaring floodwaters reached the Lowlands surrounding Leyden, and an estimated 20,000 Spanish troops drowned.

The tide literally turned that day, and, with the reinforcement of floodwaters firmly in place, the defenders of Leyden joyously declared the siege at an end.

INDIA
BENGAL (WEST)
September 1978

Monsoon rains caused extensive river flooding in India's Bengal state during September 1978. Thirteen hundred drowned and 15 million were made homeless.

In four months of torrential monsoon rains and repeated flooding, 1,300 people lost their lives, 15 million of Bengal's 44 million people were displaced, 26,687 head of cattle drowned, and 1.3 million dwellings were destroyed. The economic loss was put at $11.3 million. Unofficial estimates were two and sometimes three times that amount.

The cataclysmic flooding that took place in the summer and fall of 1978 in northern India and profoundly affected one in 18 Indians, began with the expected, annual monsoon rains. By September 5, rivers were regularly overflowing their banks in seven northern states, forcing hundreds of thousands to the flee their inundated villages. Reports began to filter into New Delhi of survivors in West Bengal perched in housetops and in trees. In the Midnapore district, west of Calcutta, the United News of India reported that the first patients of the inevitable cholera outbreak were being treated on the roof of the area hospital because the ground floor was under water.

By the next day, which was the Day of Id, the Muslim festival marking the end of a month of fasting,

Improvised transportation is the only alternative for residents of a village in India's Bengal State during the monsoon flood of September 1978. (CARE)

the monsoon-fed Yamuna River overflowed its sandbag dikes and flooded the low-lying suburbs of New Delhi. More than 200,000 residents were evacuated, but not all of them were saved. According to Lt. Gov. D. R. Kholi, speaking to the Associated Press, at least 20 persons, including some women and children, drowned when an army boat rescuing flood victims capsized in the river.

Water overflowed New Delhi's main crematorium and flooded the nearby cremation site of the late prime minister Jawaharlal Nehru, which had become a national monument. Two thousand flood victims taken to a stadium north of New Delhi were forced to flee when water surrounded the building. Road and rail traffic to and from New Delhi was in chaos. Four main bridges across the Yamuna were closed, and the trunk road to the northwest was submerged in knee-deep water for 12 miles.

The floodwaters traveled southward, rising to record levels and overrunning low-lying areas of Agra, the site of the Taj Mahal, the 17th-century tombs of Itimaduddaula, and the holy cities of Benares and Mathura. Workers toiled feverishly to protect the Taj Mahal by piling sandbags around it.

Southeast of the Taj Mahal, the Mathura region was swamped. The famed Dwaradhesh Temple, a landmark in what is considered to be one of the seven sacred Hindu cities, disappeared under the rising floodwaters. Two-story houses in nearby Sudamapauri and Mathura's Masai railway station also vanished under the rising, silt-stained, and disease-carrying floodwaters.

By mid-September, the waters peaked. The Taj Mahal was not damaged, and the Krishna Janma Bhoomi temple complex at Mathura, said to be the birthplace of Lord Krishna, the Hindu deity, was also spared, but thousands of pilgrims, housed in hostels nearby, were trapped in their quarters.

Slowly, the waters receded, as army units ranged over the countryside, delivering food and medical supplies. But toward the end of September, heavy rains again began to fall near Calcutta, forcing the Ganges River to overflow its banks, trapping army troops and residents alike in seven districts in West Bengal. The city of Calcutta was under five feet of water, and in Durgapur, west of Calcutta, the power plant was flooded, plunging the city into a 48-hour period of absolute darkness.

Within a day, more than 20,000 persons were rescued from flood-lit districts. Nearly 300 villages in the already flood-ravaged district of Midnapore were again inundated.

By September 29, Calcutta was entirely cut off from the rest of India. Low-lying districts of the city were under 10 feet (3.1 m) of water, and some 500 houses collapsed, while tens of thousands of Calcutta's pavement dwellers were forced to seek shelter in school buildings. Railway tracks leading out of Calcutta were entirely washed away and all of the major roads were under at least a foot of water.

But the worst was yet to come. For days, the 7 million residents of Calcutta would be without fresh food supplies. Gougers who had food raised their prices by 100 percent, making the acquisition of life-sustaining staples by the poor absolutely impossible.

And then, in October, more rain fell, and more flooding hit western Bengal. Some areas received 21 inches (53.4 cm) of rain in three days, inundating some recently dried out villages for as much as three weeks. Fewer than 10 percent of the 30,000 mud-walled dwellings were left standing in the Midnapore district 30 miles (48.3 km) west of Calcutta. "The disaster is so total that books on relief management never envisaged any natural disaster of this magnitude," said state spokesman Dr. Nitish Sangupta to the Associated Press.

With this new onslaught of floodwaters, food shortages increased dramatically. In some districts bands of refugees raided relief trucks, including some bearing CARE supplies. Armed police were pressed into service to accompany the relief caravan. Karen Kandeth, a relief worker, lamented, "It is the worst flood I've seen in this country and I've been with CARE for 17 years and I've seen a lot of floods."

It would be weeks before order and disease would be brought under control by Indian and World Relief agencies, and months before the devastation would be cleared.

INDIA
GUJARAT
August 7–14, 1968

Monsoon rains caused the Tāpi River to flood Gujarat state in India on August 7, 1968. Over 2,000 people died, either from drowning or cholera.

Over 1,000 people were drowned when the Tāpi River in the Indian state of Gujarat overflowed its banks on August 7, 1968. For a solid week, the monsoon-swollen river waters swirled across the countryside, inundating and uprooting crops, smashing farm structures, sweeping other, smaller ones ahead of it like so many bobbing corks.

The city of Surat was one of the first major population centers to feel the floodwaters. Within hours it was submerged under 10 feet of water, and remained that way for a full seven days.

But the aftereffects of this flood were, in many ways far worse. In Gujarat and the neighboring state of Rajasthan, over 80,000 head of cattle drowned. The corpses of these cattle were considered sacred, and so rotted, untouched in the streets, and this, coupled with the widespread contamination of the drinking water throughout the state, ignited a cholera epidemic that killed at least a thousand people.

INDIA
MAHARASHTRA, GUJARAT
July 26, 2005

More than 1,000 died from drowning, electrocution, entrapment in collapsing structures or burial under landslides, and over 10,000 homes were destroyed when an unprecedented rainfall in the monsoon season of 2005 struck the western India states of Gujarat and Maharashtra, and particularly the major city of Mumbai.

Each year, the monsoon season arrives in India, bringing with it months of unrelieved, sometimes torrential rain, which swells rivers over their banks and inundates much of the largely below-sea-level neighboring country of Bangladesh. But no monsoon season on record matched that of 2005.

The rains began as usual at the beginning of July, but they were particularly insistent, strong, and, what was worse, relentless. Mumbai, the city formerly known as Bombay, and still the economic and financial heart of western India, experienced gigantic deluges. On July 26, 37 inches (94 cm) of rain—the most any Indian city had ever received prior to that date—fell within 24 hours. By the end of the month, the city had been immobilized. Rail traffic had stopped, all public transportation had been halted, and floodwaters had reached a level of 7 feet (2.1m) in some parts of the city.

Meanwhile, in the rest of Maharashtra, the state in which Mumbai rests, and in the neighboring Gujarat state, village after village was being swept away or swallowed up by floodwaters. South of Mumbai, several hundred people died in landslides brought on by the combination of rain and overflowing rivers. In Mumbai itself, fatalities resulted not only from drowning and burial beneath the wreckage of tin and wooden houses, but by electrocution, when power lines were severed and fell into the floodwaters.

As the rain began to lessen enough for rescue workers to begin the monumental task of trying to clear away the debris and rescue survivors, a further, unique danger began to assert itself. Rumors began to circulate that a super cyclone, and then tsunami, were on their way. On July 28, 22 people, including several children, were killed in a stampede that was prompted by rumors of a collapsed dam.

Outside of Mumbai, 354 people were trapped for 30 hours in a stalled train in which waters climbed to neck level. Fortunately, military forces finally evacuated them, and all survived.

Help came too late for hundreds of the dead and those who survived the flood in the city of Mumbai. A furious Hafeez Irani, his face covered with a handkerchief against the overwhelming stench of dead and decaying bodies, complained to Western reporters. "For so many days we have been lifting the bodies of the dead and now we are clearing animals from the roads," he shouted. "Is this our work?"

The monsoon continued into August, but not with the same fury it had possessed in July. When the skies cleared enough for officials, rescuers and international aid agencies to survey the damage, the estimate was that more than 1,000 died, and at least 10,000 homes had been either rendered unlivable or totally destroyed. The cost of rebuilding was in the hundreds of millions of dollars.

INDIA
MORVI
August 9, 1979

The Machu Dam 2 near the city of Morvi, India, burst on August 9, 1979, allowing the Machu River to flood the city. The official death toll was 1,000; the unofficial, 5,000. Seventeen thousand people were evacuated from the area.

A spectacular flash flood was unleashed on August 9, 1979, when a rain-weakened dam burst in Gujarat state, four miles (6.43 km) above the city of Morvi, which is 300 miles (482.8 km) northwest of Bombay.

Prior to the collapse of the Machu Dam 2, 25 inches (63.5 cm) of rain fell on the region in a 24-hour period, which is the usual amount for a whole year in this arid region of Saurashtra, bordering the Rann of Cutch. The Machu River, which the dam had heretofore comfortably contained, was turned into a raging monster. Even the dam's 52-foot (15.85-m) height and its ability to withstand pressures of 200,000 cubic feet (60,960 cu. m) per second was not enough, and it burst with a gigantic roar, sending tons of water in a 20-foot-high (6.09-m-high) wave crashing forward, sweeping dozens of villages into obliteration and smashing into the city of Morvi with its population of 60,000 unprepared citizens.

Fully 60 percent of the dwellings in the city were leveled in minutes. Mud, debris, and water engulfed streets and buildings, suffocating and drowning thousands. (The official death toll would be placed at slightly more than 1,000. Unofficial sources claimed that it was nearer 5,000.)

Within minutes, the floodwaters receded, leaving silt covered chaos behind. A blanket of mud, which was 19 feet (5.8 m) high in some outlying areas of Morvi, reached well up into the second stories of buildings within the city. Roadways were littered with strewn bodies.

The city's central telephone exchange was crushed by the floodwaters; thus, word of the tragedy was slow in reaching the outside world, and relief teams were not dispatched until the following morning. High winds and a steady rain made these efforts even more difficult. Still, the army was able to evacuate 17,000 people within the first week following the disaster.

Local officials, faced with this sudden disaster, seemed unable to cope, and by the end of the week, a state investigation into the reasons for the fumbling first relief efforts was ordered. The findings would be inconclusive.

International relief agencies would soon appear; the army would recede as quickly as the flood waters had; and the state would grant relatives of each victim cash compensation equivalent to $125.

INDIA
NORTH
June–September 1998

The extreme flooding that immobilized much of Asia in the summer of 1998 was particularly devastating in India, where several rivers inundated 22,000 villages and nearly 5 million acres of farmland. Eight million people were rendered homeless. Over 1,000 died.

The extreme and prolonged convergence of monsoons in Asia, believe to be caused by the La Niña phenomenon in the latter part of June 1998, created giant floods throughout China, Bangladesh, Korea, and India. In India, these rain-induced torrents swelled the Ganges, Jamuna, and Brahmaputra Rivers. While the worst ravagement from the months of torrential rains and consequent floods came in Bangladesh (see p. 153), India, particularly in its northern states of Uttar Pradesh,

Bihar, Assam, and West Bengal, suffered enormously from the raging torrents of the rivers that flowed through them.

August was the worst month. From the beginning of the month forward, the rain fell unendingly. Rivers, already swollen by the snow melt in the Himalayas, overflowed their banks, forcing the evacuation of over 150,000 people and cutting off all communication with the eastern portion of Uttar Pradesh. Railroad lines were ripped up and highways destroyed by the torrents that washed out millions of acres of farmland.

In hilly terrain, landslides buried villages and stranded millions, some on rooftops, some on improvised barges, some in tree tops. Airdrops of food by the army sustained some of these refugees; others were sent provisions by a flotilla of 6,000 boats commandeered by the army.

By the end of August, 223 people had died from the floods in the state of Bihar alone. Throughout the entire area, fresh bodies appeared daily, retrieved from under landslides or washed ashore from the raging rivers.

In the tea- and oil-rich state of Assam, through which the Brahmaputra River wends its way for over 440 miles, over 200 people were drowned or killed in landslides that continued to grow as August gave way to September. Though the state had 27,870 miles (44,852 km) of embankment, 629 protection works, and 406 miles (653.4 km) of drainage channels designed to protect its land and people from monsoon flooding, these measures failed to safeguard the populace. The damage to the flood control mechanisms was estimated at nearly $50 billion. Over 3.6 million people were forced to leave their homes and sought shelter in nearly 400 government relief camps.

One of the natural wonders of Assam state is the Kaziranga National Park, a 175-square-mile (281.7 sq. km) tract of open land that borders on the Brahmaputra River. By August, the Brahmaputra had overflowed into the park, submerging 90 percent of it and drowning 17 rhinos, six elephants, and 100 deer.

The Holy River of the Ganges, a place where the devout regularly bathed, was a roaring torrent, covering the entire township of Melda in West Bengal.

In September, India's Defense Minister George Fernandes described the flood situation as "a human tragedy beyond imagination." Not only had thousands of lives been lost, hundreds of thousands of acres of agricultural land rendered useless, and 3.6 million people evacuated from their homes, but the danger of contagious disease was now rising. Assam state deployed hundreds of doctors throughout its territory to inoculate and treat refugees.

In northern Uttar Pradesh, the government remained poised to use newly installed pumps to rid some of its worst-flooded cities of water. But the floodwaters stubbornly refused to recede.

In Naresh Dayal, drinking water and chlorine tablets were distributed in the largest relief operation the state had ever undertaken.

In the western city of Surat, a textile and diamond polishing center of 2.5 million people near the Tapi River, water was released from the Ukai dam to prevent it from bursting. But the effect was enormous flooding in the city itself. Fifty thousand people were evacuated to higher ground.

When the waters finally receded in October, the grim task of reassembling lives and livelihoods began. Many of the flood victims lost not only their crops to the floods but also the use of their land for a long period to come. Sand and silt were everywhere. Twenty-two thousand villages had been inundated and nearly 5 million acres of agricultural land had been underwater. Over 8 million people were homeless. Over 1,000 died.

IRAN
FARAHZAD
September 17, 1954

A sudden storm washed a mountain shrine in Farahzad, Iran, into a gorge on September 17, 1954. Two thousand pilgrims worshiping in the shrine were swept to their deaths.

In the middle of September 1954, several thousand Iranian pilgrims arrived in Farahzad, the location of a mountain shrine called Imam Zadeh David. The shrine clung to the side of a slope giving onto a gorge that plummeted thousands of feet to a valley floor. Next to it were dozens of cottages designed to house pilgrims to the shrine.

On the night of September 17, 1954, a torrential rainstorm hit the area, uprooting trees, unleashing mountain streams, sending torrents of water cascading down the mountainside. Three thousand pilgrims were worshiping in the shrine when it suddenly let go of the side of the mountain and plunged downward, taking the cottages with it. It fell to the bottom of the gorge, where it was rapidly blanketed by rain water, rivers, and streams, which combined to flood the gorge. Over 2,000 pilgrims drowned within minutes; most of the bodies were carried miles downstream, where they remained, unrecovered, for days.

ITALY
BELLUNO, PIRAGO, VILLANOVA, AND RIVALTA
October 9, 1963

(See AVALANCHES AND LANDSLIDES, Italy, Belluno)

ITALY
FLORENCE
November 4–6, 1966

One hundred forty-nine people drowned, over 100,000 were trapped in their homes for days, and millions of dollars worth of priceless art works were either destroyed or damaged when the River Arno overflowed its banks and flooded Florence, Italy on November 4, 1966. A combination of heavy rains and human neglect—some of it centuries old—caused the flooding.

In the 19th century, poet Elizabeth Barrett Browning called the River Arno where it flowed into Florence, "This crystal arrow in the gentle sunset."

On the rain-soaked morning of November 5, 1966, novelist Katherine Taylor, observing the Arno from the fortunate safety of her upper-story apartment, described it as ". . . a snarling brown torrent of terrific velocity, spiraling in whirlpools and countercurrents that send waves running backward."

For centuries, the Arno had been both of these and more, but on the night of November 4, 1966, it would once again prove that Florence is a city built in the wrong place. In a few hours, it would be transformed into a dung and mud heap. Thousands of priceless art treasures would be destroyed forever and thousands more damaged. One hundred fourteen people would drown in the surrounding countryside; 35 would perish in the city, and 500,000 tons of mud would be deposited within its environs—one ton for each man, woman, and child living in Florence. Over 100,000 citizens of this Tuscan paradise would be trapped in their upper-story dwellings.

For over 900 years the Arno had overflowed its banks regularly and relentlessly. In the years between 1177 (the year of its first recorded flood) and 1761, the Arno crested its banks 54 times, with a major flood on the average of every 26 years and a severe, catastrophic flood every hundred years.

In 1545, Leonardo da Vinci drew up plans for an intricate set of dams, lakes, and locks that would prevent further flooding. They were never utilized. In

August 1547, Bernardo Segni recorded a huge flooding of the Arno, writing, ". . . because very great numbers of trees had been cut down for timber . . . the soil was more easily loosened by water and carried down to silt up the bottoms of the rivers. In these ways, man had contributed to the disaster. . . ." Late on the third of November 1966, modern men compounded the unwise farming habits of their ancestors.

They had help from nature. The entire month of October 1966 was unrelievedly wet and gloomy. But the first two days of November were crisp and clear, and November 4, the anniversary of the end of World War I in Italy and a national holiday, was anticipated to be an auspicious and fair one. But a blinding downpour occurred on November 3. In fact, during the 48 hours of November 3 and 4, Florence and the Arno would receive 19 inches of precipitation, which was more than one-third of the area's average annual rainfall.

It seems, however, that the operators of the Penna hydroelectric dam some 29 miles (46.7 km) above Florence were not paying attention during the deluge of the previous weeks, and did not release water from the dam in gradual, small doses. Rather, they let it go all at once, in a huge mass, which put an impossible strain on the Levane hydroelectric dam four miles downstream. In turn, the operators here were faced with a disturbing decision: whether to open the gates and flood the valley. Signora Ida Raffaelli, who lived below the Levane dam, remembered for the world press later, "I was appalled to see the gates slowly opening, and immediately an enormous wall of water started coming down the Arno toward us. I screamed to my sister and we ran for our lives."

Well that they did, for the Arno roared on, straight for Florence. Between 9 P.M. and 11 P.M. on November 4, the river rose 18 feet (5.5 m) in the city. Romildo Cesaroni, an elderly watchman, was patrolling the famed Ponte Vecchio to warn shopkeepers of any danger. By 11 P.M., the water was roaring by at 40 MPH, a mere three feet below the bridge. Huge tree trunks were slamming into the bridge. Cesaroni began to telephone shopkeepers, who came running to reclaim their valuables.

By 3 A.M., the waters had engulfed the bridge, and floating automobiles were smashing into it, causing it to shudder and tug at its moorings, which had been dynamited by the Germans in World War II. But miraculously, the bridge held.

Now, the lower parts of the city began to disappear under water. Underground transformers started to short out. Furnaces exploded. The antiquated sewer system, built 300 years before, gave way under the tremendous force of the water backing into it, and human waste shot like a geyser out of manholes, blanketing the city with a horrendous stench. Fuel oil from the furnace rooms floated out on the water, coating walls and buildings.

The old part of the city, on the river bank, also housed the city's poorest residents, and many of them drowned that night in their apartments. In the Santa Teresa prison, the floodwaters climbed to 13 feet (3.9 m), and the prisoners were herded to the top floor, where they summarily overpowered their guards. Eighty of them then clambered to the roof, where, to the cheers of other stranded Florentines in their apartments, some of them tried to escape by leaping onto tree trunks and other debris that sped by. Not many were successful.

In another part of the city, at the Cascine Park race track, handlers frantically struggled to load 270 horses into vans to evacuate them ahead of the flood waters. They only managed 200; 70 horses drowned, and two days later, when the waters receded, their corpses were incinerated by rescuers with flame throwers to prevent the spread of disease.

At exactly 7:26 A.M., every electric clock in Florence stopped, and power was not restored for 24 hours. Outer bridges leading from the city were washed out, roads were blocked, railroad tracks were covered. Florentines were isolated from the world.

The worst destruction took place while the guardians of the richest collection in the world of Renaissance art were busy elsewhere. Slowly, this universal treasure was dissolved, torn apart, or forever stained with the oil-laden waters. Before dawn, this destruction spread into the low-lying Piazza Santa Croce. In the ancient church of Santa Croce, the water climbed to a height of 20 feet (4 m), covering the tombs of Galileo, Michelangelo, Niccolo Machiavelli, and Antonio Rossini. Donatello's relief of the Annunciation was smeared with the oil and sludge-stained water. Next door, in the church museum, the famous 15-foot (4.6 m) painted crucifix by Giovanni Cimabue, the father of western art, was completely destroyed as the waters continued to climb and consume it. At the baptistry in the Piazza del Duomo, five of Lorenzo Ghiberti's 10 monumental bronze panels of Old Testament scenes were ripped from the portals that Michelangelo had named the "Gates of Paradise." At the Institute and Museum of the History of Science, the director, Maria Luisa Righini-Bonelli, edged along an inches-wide third floor ledge 28 times, saving priceless precious objects, including several of Galileo's telescopes from the Uffizi Gallery. In the basement of the Biblioteca Nazionale Centrale, a million books and manuscripts were turned to pulp by the flood waters.

Meanwhile, elsewhere in the city, rescue efforts began. Newspaper editor Franco Nencini witnessed helicopters lifting stranded people from housetops. One old woman, clinging to a rescue rope from the helicopter, tired midway to safety, and dropped to her death in the roaring floodwaters.

At the hospital of San Giovanni di Dio, patients were carried to upper floors when the generators flooded and the elevators failed. The food supply was rapidly covered with oily water, and all that workers were able to salvage were 10 chickens and 20 bottles of mineral water, which provided sustenance for the entire hospital until emergency food could be brought.

All day, the rain continued, without a sign of letting up. Finally, at 6 P.M., a delegation from the national government in Rome arrived. *Nazione* editor Enrico Mattei, accompanying the group, described the scene:

> Behind a curtain of driving rain, pierced here and there by dim, mysterious lights, the Piazza San Marco was a storm-tossed lake. This lake was fed by a violent torrent which poured down from the Piazza dell'Annunciata, lapped at the church and went swirling off down the Via Cavour in the direction of the Cathedral. Beneath the grim rumbling of the water, we could hear a subdued murmur of human voices.

The voices belonged to the survivors. Five thousand families would be left without a place to live; 6,000 of the 10,000 shops in the city would be destroyed.

Still, Florence belonged to the world, and the world responded almost immediately. The skies had scarcely cleared when the first contingents of rescue workers from 10 European countries, America, and Brazil arrived. Sustaining gifts from England, Germany, Austria, and the Soviet Union began to arrive. Scotland sent blankets, water pumps, and vaccines. The United States dispatched food, clothing, generators, and prefabricated houses. The Dutch sent engineers who brought water-decontamination equipment, and Israel provided a Christmas vacation on kibbutzim for more than 100 homeless children.

But the worst task, the cleanup, lay ahead. Students from the Florence branches of Stanford, Syracuse, and Florida State universities plunged in immediately. Because they wore blue jeans, they were lovingly referred to by Florentines as "Blue Angels." For two weeks, the students formed human chains from the basements of the Biblioteca Nazionale, passing up water-logged manuscripts and books to upper floors, where they were carefully blotted with a special paper to absorb the moisture. From here, the books were loaded into U.S. Army trucks and sent to tobacco-drying kilns in central Italy, overseen by some of the planeloads of scholars and restoration experts who had arrived upon the scene from all over the world.

Ghiberti's priceless panels were discovered, buried in the muck. They were returned to their places on the Gates of Paradise. Michelangelo's statues, coated thickly with talcum, then scrubbed with powerful detergents, seemed hardly the worse for wear. Donatello's 500-year-old wooden sculpture of Mary Magdalene, restored with solvents and surgical lancets, seemed in better condition than it had been before the flood. But Cimabue's crucifix was irreparable, as were invaluable frescoes.

More than half of the million books damaged in the flood have, as of this writing, been restored and rebound. More are continuing to be worked on, as are the frescoes, the paintings, the statuary. But, millions of lire later, projects designed to build more dams upriver on the Arno to put into practice some of Leonardo's plans have been shelved for lack of funds. And so, at any time, Florence may be—as it has for much of its history—flooded again.

ITALY
PO VALLEY
November 1951

Over 100 drowned when dikes and canals were inundated in the Po Valley in November 1951. Crop loss was placed in the millions of lire.

From ancient times, the Po Valley has undergone constant and regular inundation, thus making it a rich source of agriculture. Grain, sugar beets, livestock, and fruits feed not only the principal cities of Turin, Asti, Milan, Brescia, and Verona but also much of the rest of Italy and parts of Europe. But it was not until the 1930s that attempts were made to control the floodwaters of the River Po and its tributaries, the Dora Baltea, Tanaro, Ticino, Adda, and Oglio. An elaborate system of hydraulically controlled dikes and canals were installed, designed to drain off excess waters, refocus the tidal flow, and protect the populace, livestock, and crops of the valley.

And for 20 years and a World War, it held. But then, in November 1951, torrential rains fell on the valley for weeks. The force of the river, swollen by the unceasing downpours, fed into the tributaries faster and more vigorously than the hydraulic controls could handle. Dikes began to crumble, then dissolve completely. Levees were swept clean as floodwaters crested, then crashed through them.

Tens of thousands of people were trapped by the sudden and unexpected flooding. Over 100 people were drowned, 30,000 head of cattle were destroyed, and the damage to the valley's crops climbed into the millions of lire.

ITALY
STAVA
July 19, 1985

The collapse of an earthen dam caused a flash flood in the resort of Stava, Italy, on July 19, 1985, that killed 250 and injured nearly 1,000.

Stava is a small resort community in the Dolomite Alps, about 40 miles (64.4 km) south of the Austrian border. Its unusual countryside, made up of pine forest and broad, green meadows, its Alpine houses overlooking winding, cobble-stone streets that rise and fall along the hillsides and its spotless villages, make it a favorite tourist attraction and vacation spot for Italians and Austrians.

July 1985 was the height of a particularly busy tourist season. The area's four hotels, the Erika, the Stava, the Miramonti, and the Dolomiti, were filled; several civic groups had organized excursions, and children's camps were thickly populated in this particularly placid and sylvan part of the Flemme Valley.

About a half-mile above Stava was an earthen dam, built to hold back two artificial lakes which were used to purify minerals from a nearby fluorite mine. Fluorite is used in glass-making, and the Prealpi Mining Company, which owned the dam, was one of Italy's most successful producers of this mineral.

Over the years, the earthen dam had built itself up, and large amounts of chemical sediment had accumulated on the bottom of the lakes, thus increasing the weight and pressure on the dam. Over the months preceding July 1985 trees had been cut in preparation for the expansion of the pools.

In the days before July 19, heavy rains fell.

At 12:20 P.M. on Friday, July 19, with a roar that sounded to one local resident like an earthquake, the dam gave way, and an immense white wall of water thundered toward the crowded resort. "I saw the end of the world," a survivor told reporters. "I saw a white wall coming toward me. I couldn't tell if it was fire or what."

The enormous wave, made more lethal by mud and debris, smashed into Stava, collapsing three of its four hotels, the Stava, the Erika, and the Miramonti, and partially collapsing the fourth, the Dolomiti. Following the Stava River, the floodwaters roared into the nearby town of Tesero, damaging bungalows and a bridge over the Avisio River, which ultimately absorbed the rest of the wave of mud, houses, trees, and bodies.

The floodwaters had ripped a brownish-gray swath two to three miles long through pastures and forests of the Flemme Valley. Trees were stripped and uprooted,

houses were cut in two, and cars were upended and buried, their wheels sticking out of the mud.

Almost immediately, rescue workers began to dig through the overwhelming effulgence of mud and slime. Alma Bernard, a hotel owner in Tesero, said to reporters, "Many families were wiped out with their houses. Earth and mud covered [everything]." Another survivor, identified only as Pietro, said he had seen his 48-year-old brother, Lucio, climb a tree to escape the mud-laden tidal wave. "But then a second wave carried him away," he added.

It was a grisly sight for rescuers manning bulldozers, rescue trucks, and ambulances, or merely digging with anything that was at hand. One young soldier, doing his compulsory military service nearby and pressed into rescue duties, spoke of his horror at digging through the mud with a shovel and finding a child, its head crushed. Giuseppe Zamberletti, the minister of civil protection, said in a news conference that in the case of 15 bodies, disfigurement was so severe that it was impossible even to determine the victims' sexes.

A rescue worker from Bolzano wept when asked for a description of his work. "I felt helpless," he said. "I knew that under that moving sand, there were so many people who were suffocating. And I could not do anything to save them."

However, there were hopeful stories, as well. Maria Assunta Cara, a Sardinian, was pulled alive from the wreckage of a hotel around dawn. She had been buried up to her mouth in mud for 18 hours.

For days, traffic jams of rescue vehicles clogged the mountain roads around Stava. Bodies were taken to a hospital in Cavalese, three miles southwest of Stava, and to a school in Tesero, which was converted into a morgue.

The countryside looked as if some indiscriminate war had been waged back and forth across it. Commenting on the view from a helicopter, Zamberletti related, "The sites of the hotels and houses had to be pointed out to me. It's as if they never existed."

All in all, 250 people died and nearly a thousand were injured. It would be the worst tragedy from a collapsed dam since 1963, when a dam burst at Longarone, 35 miles (56.32 km) away, and killed 1,800 (see AVALANCHES AND LANDSLIDES).

The Italian parliament, acting upon multiple complaints and petitions, investigated criminal culpability and negligence on the part of the Prealpi Mining Company and local officials responsible for supervision of the dam. Because of a complex history and multiple responsibilities, coupled with inadequate proof of any single act of negligence, the charges were ultimately dropped.

JAPAN
1896

Twenty-seven thousand people drowned when a giant tsunami inundated the Japanese coastline on the Sea of Japan in 1896. No specific date is recorded.

Tsunamis (see EARTHQUAKES, Introduction) have worked particularly destructive havoc in the Sea of Japan. Entire villages have been obliterated or displaced by these gigantic, seismically-caused waves that appear without warning, create their havoc in minutes or seconds, and then disappear as suddenly and unceremoniously as they have appeared.

The second most catastrophic tsunami-caused flood in Japan occurred in 1896, 13 years after the horrendous explosion of Krakatoa, which created 100-foot waves that drowned more than 36,000 people. (See VOLCANIC ERUPTIONS AND NATURAL EXPLOSIONS, Krakatoa.)

This particular tsunami was the product of a gigantic undersea earthquake 700 miles (1,126.5 km) from Japan's coastline. A series of tsunamis were set in motion by this earthquake; they rolled over the coast of Japan in rapid-fire progression, wiping out villages, changing the coastline, and drowning 27,000 people (See EARTHQUAKES, Japan 1896.)

JAPAN
July 18–19, 1964

A combination of an earthquake and heavy rains conspired to flood the coast of Japan bordering on the Sea of Japan on July 18, 1964. One hundred eight people died; 233 were injured; 44,000 were made homeless.

A series of multiple natural disasters set the stage for a fatal flood along the Sea of Japan on July 18 and 19, 1964.

First, a minor earthquake rumbled through Niigata, laying waste to much property along the seacoast, but causing very few deaths. Immediately after the earthquake, torrential rains followed, ceaselessly.

Swollen rivers began to crumble structures that had been loosened and weakened by the earthquake. The sides of hills began to break away and slide toward the villages nestled at the bottom of these hills.

Before the end of the day on July 18, 150 bridges had collapsed, and nearby dikes which had also been weakened by the repeated shocks of the earthquake, cracked and fell in more than 200 separate places.

The onslaught of water, rain, and landslides collected together to destroy several villages and 295 homes in the small cities of Ishikawa, Toyama, Niigata, Tottori, and Shimane.

One hundred eight people either drowned or were crushed by mudslides and collapsing buildings, 233 were injured, and an astonishing 44,000 were made homeless.

JAPAN
July 17, 1972

Extremely heavy rains caused river flooding throughout Japan on July 17, 1972. Three hundred seventy people drowned; 70 were reported missing.

A week of heavy rains swelled rivers and streams and caused a series of floods throughout an area that covered almost the entirety of Japan in the middle of July 1972. Landslides added to the destructive force of rain and flood, particularly in the farming districts, which were flooded beyond recognition.

By the time the waters receded, and the rain ceased, 370 people had drowned, 70 were missing, and the crop loss was estimated at $472 million.

MEXICO
PUEBLA, TABASCO, VERACRUZ, HIDALGO
October 4, 1999

A monster rainstorm, fueled by a stalled tropical depression off the coast of Mexico from October 4 through 6, 1999, caused floods and mudslides that killed 341 people and rendered 300,000 others homeless in the Mexican states of Puebla, Tabasco, Veracruz, and Hidalgo. Over 1,300 schools and hospitals were severely damaged.

"We never dreamed that rain could hit so hard," said Gildardo Castaneda Dominguez, the mayor of Zacapoaxtla, a city of 27,000 in Puebla State, 95 miles (152.9 km) east of Mexico City. He was standing in the ruins of a former colony of riverside shanties, which had been swept away by flash floods and a monster mudslide. On October 4, 1999, a tropical depression stalled over the Gulf of Mexico and initiated Mexico's worst flood in 40 years, and one of the country's worst natural disasters. Thirty inches of torrential rain fell in three days over the states of Puebla, Veracruz, Hidalgo, and Tabasco, and streams, rivers, and still bodies of water became raging torrents as they overflowed their banks and inundated fields, barns, homes, and entire villages.

A 600-mile (965.6-km) stretch of the Gulf Coast was decimated. In Teziutlán, 20 miles east of Zacapoaxtla, an entire clifftop cemetery was uprooted and slid, at express train speed, down a hillside, sweeping away dozens of homes. In the coastal town of Gutiérrez Zamora, in Veracruz State, a 12-foot (3.6-m) wave at the front of a flash flood, swept away an entire riverside neighborhood, drowning everyone in it.

The village of Tapayula, in a valley 10 miles (16.09 km) north of Zacapoaxtla, suffered twin mudslides, from two sides. Every dwelling, every business, every road in the village simply disappeared under the mud. Only the bell tower of the church was visible above it.

"Tapayula doesn't exist anymore," Zeferino Ramos Peralta, a schoolteacher who escaped the mudslide reported as she walked into Zacapoaxtla, where refugees were lined up for bean broth at a sidewalk soup kitchen.

The combination of the deluge, mudslides, and flooding from rivers, streams, and dams that could not stem the torrents of water claimed the lives of 341 people. More than 300,000 in 179 municipalities lost their homes; over 500,000 acres of cropland were destroyed, and more than 1,300 schools and hospitals were severely damaged.

PAKISTAN, AFGHANISTAN
February–March 2005

A lethal combination of heavy snow and encompassing floods devastated residents of southwest and northern Pakistan and the bordering province of Paktia in Afghanistan in February 2005. In Pakistan, 2,000 people were reported missing and presumed dead, tens of thousands were made homeless, and 450 were known dead, most of them victims of the bursting of the Shadi Kor, Gaggo, and Chelvi dams.

A meteorological convergence of heavy snow in the north and torrential rains south of Pakistan combined to cause a natural disaster of huge proportions throughout all of February and the first week of March 2005. The torrential rains began at the beginning of the month, flooding farmlands, swelling rivers, and causing mudslides that carried houses before them, burying the houses that they did not demolish.

The rain and the destruction continued day and night, and on February 9, the Shadi Kor dam, near the town of Pasni Tehsil in Baluchistan, on Pakistan's southwestern coast, burst. Pasni, located about 25 miles (40.2 km) from the dam was inundated in seconds as tsunami-force waves rushed through the town, drowning it, and washing out a coastal highway and bridges linking the town with Karachi and Gwadar.

Shortly after this, two other small nearby dams, the Gaggo and Chelvi, burst and their waters, combined with the onrush from the Shadi Kor dam collapse, swept five villages and some of their inhabitants into the Arabian Sea. Some 215 residents of Pasni and its surrounding villages died in the flash floods. Between 25,000 and 30,000 people were affected, in one way or another, by the dam bursts.

Meanwhile, the heavy snow in Paktia province (the portion of Afghanistan near the border with northwestern Pakistan), in northern Pakistan, and in the Pakistani-held section of Kashmir, was creating havoc and death, triggering avalanches and, along the Afghanistan–Pakistan border, where the snow-rain line was crossed, and the melting snow-swollen Dasht and Koja Rivers overflowed, flood waters that entirely consumed the land. At least 4,000 families were affected; 2,364 homes were destroyed, 1,600 acres of land were laid to waste, and more than 4,000 animals, needed for the livelihood of families, died. Some 104 people died in Afghanistan; 209 perished in Kashmir and northern Pakistan.

Relief efforts began almost immediately, particularly in Baluchistan province in Pakistan, where the major casualties occurred. "Seven thousand people were affected (in this immediate area)" United Nations Children's Fund (UNICEF) country representative Omar Abdi told reporters. "We estimate that about one third of the affected (were) children . . . The main need now is shelter and clean water." Both arrived, but not until the end of the month.

RUSSIA
ST. PETERSBURG
November 19, 1824

The River Neva overflowed its banks in and near St. Petersburg, Russia, on November 19, 1824. Over 10,000 people drowned.

More than 10,000 people drowned in the swirling floodwaters of the River Neva on November 19, 1824, near St. Petersburg, Russia. It would remain the worst flood in the history of the Neva.

In 1824, St. Petersburg was the winter home of the czar, and his troops and consorts were everywhere that November. All were threatened by the roaring floodwaters. According to one report, ". . . a regiment of Carabineers, who had climbed to the roof of their barracks, were drowned."

Carriages and horses were swept away; the homes of the privileged as well as peasant shacks were inundated and sometimes swept from their foundations. Practically every dwelling in the city, including the Winter Palace, was flooded to the top of its first story.

Nearby, the coastal city of Kronstadt was also inundated. Refuse swept into dwellings and open places not only from the river banks but also from the sea. A 100-gun navy ship was floated from its berth and ended up in the middle of Kronstadt's marketplace.

By the time the waters had receded, 10,000 people were dead, the property damage amounted to millions of rubles, and there were too many homeless to count.

SCOTLAND
INVERNESS
July 1829

Hundreds drowned when floods caused by giant storms inundated Inverness, Scotland, in July 1829.

Enormous thunderstorms raged over the Inverness area of Scotland throughout the month of July 1829. Drops of rain powerful enough to kill small animals and birds pounded. All of the streams and rivers running northward from the mountains to the sea were filled to overflowing. By the middle of the month, tremendous torrents of water began to roar down mountainsides into the hapless valleys below. The Findhorn Gorge filled to overflowing so that only the tops of great trees were visible above the water.

Cottages in these gorges were picked up and flung ahead of the raging waters. Farms, bridges, mills, and factories were sucked into the flood and disappeared. One huge wave lifted a 65-foot (19.8-m) stone arch from a bridge into the air and carried it, floating like a raft on the raging waters, for miles before it finally sank.

The entire plain of Forres was flooded to the Moray Firth. Findhorn Village became an island, its people marooned on the upper stories of their homes. For weeks, broken furniture and vehicles dammed the streams. Fertile soil was replaced by gravel in the fields. In the rivers, hundreds of salmon were dashed to pieces or choked by the sand. Nearly all the bridges in the area

were swept away, and four sawmills were completely destroyed. The dead numbered in the hundreds.

SOUTH KOREA
August 19, 1972

Six hundred thirty-eight people died and 144,000 were made homeless in river and stream flooding caused by heavy rains in South Korea on August 19, 1972.

The rain that fell relentlessly in Seoul, South Korea, on August 19, 1972, measured 17.8 inches, which was a record.

No terrain can withstand this sort of attack from the heavens for long, and within hours, swollen rivers and streams, some of them redirected by massive landslides brought on by the downpours, flooded Kyonggi, Kangwon, and North Chungchong Provinces. Overnight, 463 would drown or be crushed by collapsing houses inundated by falling mud or pounding floodwaters.

Thirteen separate landslides occurred within the city of Seoul itself, accounting for 175 deaths. Here, 127,000 inhabitants were made instantly homeless, 158 were injured and 34 were reported missing.

Areas outside Seoul were similarly affected but because of smaller concentrations of population, the casualty figures were smaller. The city of Yongwol, 90 miles southeast of Seoul, was completely submerged, forcing its entire population of 17,000 to flee in panic. In Suwon, an ancient city wall collapsed under the assault of floodwaters, demolishing hundreds of homes. The agricultural losses were staggering; the loss of buildings was likewise enormous. But what was most appalling was the small amount of time it took to wreak such widespread havoc.

SPAIN
BARCELONA, SABADELL, TARRASA
September 26, 1962

Floods following catastrophic amounts of rain killed 445 and rendered 10,000 homeless in Barcelona, Sabadell, and Tarrasa, Spain, on September 26, 1962.

Famed painters Salvador Dalí and Pablo Picasso auctioned off some of their paintings to aid the 10,000 homeless residents of Barcelona following the enormous floods of September 26, 1962. When the worst flood disaster of modern times hit this premiere city of the Costa Brava, 445 were killed.

Torrential rains followed by heavy floods totally submerged the nearby villages of Sabadell and Tarrasa, and damage to agriculture and the closely packed dwellings of not only Barcelona but also the stuccoed towns that border it resulted in a high economic toll.

SRI LANKA
May 2003

An unprecedented flash flood caused by torrential rains hit southern Sri Lanka throughout May 2003, causing an inordinate number of landslides, killing 264, and displacing nearly a million residents of the area.

A torrential downpour throughout the month of May 2003 triggered the worst floods since 1947 in the five southern districts of Sri Lanka. The landscape, plagued by a lack of government control over deforestation, gem mining, and quarrying, simply collapsed under the assault of the rain, and flash floods and landslides washed away farms and entire villages.

The singularity of the event caught hundreds of its victims unaware. Although floods were a common occurrence in low-lying coastal areas, they were rarities in the highlands, and almost unknown at this scale and level in the mountains. But all precedent was overturned as accumulated rainwater from huge mountainous catchments roared into the narrow valleys, flooding homes up to the second story.

The worst hit district was Ratnapura, where 137 died, most of them members of families who were buried alive beneath repeated landslides. At the Abeypura housing scheme in Palewela, also in the Ratnapura district, more than 70 people disappeared beneath a landslide. Lali, a young mother, was buried up to her neck, but her seven-year-old son—who was dug out not by government assistance (which never arrived) but by surviving villagers—succumbed to his injuries in a local hospital.

Roads were washed away as if they had never existed in Matara, Hambantota, Galle, and Kalutara. Power and phone lines were cut, nearly 3,000 acres (1,214 ha) of paddy fields and other crops were destroyed, and the tea bushes on multiple tea estates—a major industry in Sri Lanka—were damaged beyond repair.

Most of the victims in the Ratnapura District were heartbreakingly poor. Those who did not farm were workers in the gem mines for the equivalent of $1 a week. And so, when the irrigation system was rendered useless, clean water also disappeared, and diseases proliferated. Diarrhea, viral flu, and typhoid and isolated cases of dengue and Japanese encephalitis appeared. A social worker from Ratnapura told reporters, "This disaster hits people and they don't know what is going to happen to them. Some are on the road. Some are with friends or relatives. Some have gone to refugee camps in schools. And with no clean water, fever is spreading in the camps." It would be two years before the region would return to a semblance of normalcy.

SUDAN
KHARTOUM
August 4–5, 1988

The city of Khartoum, in Sudan, was flooded when the Nile overflowed its banks on August 4, 1988. Over 100 drowned, hundreds were injured, and over 1 million were made homeless.

Khartoum, located where the Blue and White Niles join to make the Nile River that flows into Egypt, is the capital of Sudan, Africa's largest country. Ordinarily, Khartoum's population totals 334,000, but in August 1988, that figure had been swollen by an additional 1.5 million refugees, fleeing the civil war in Sudan's south. Most of these refugees lived in makeshift shelters of mud thatch and cardboard in shantytowns on the fringes of Khartoum, and it was this ancillary city of displaced persons that was most catastrophically hit when the floods of August 1988 poured through Sudan's capital.

Torrential rains soaked the city and its surrounding countryside on August 4 and 5, 1988. In the space of 48 hours, over eight inches of rain fell, compared with a total of less than two inches of rain in the previous year.

The Nile boiled furiously, overflowing its banks, rising higher, according to the Sudanese Irrigation Ministry, than at any other time in this century. Within a very short time, all of the roads leading in and out of the city were washed away, including the one paved road between the city and Port Sudan on the Red Sea. This would make later rescue and relief attempts hazardous to the point of impossibility.

Electricity was cut off early on when two thermal electricity plants were damaged by flood waters and five of the six turbines serving the national electrical supply were knocked out.

Streets of Khartoum were turned into high-speed streams; houses were flooded to the upper stories;

A residential section of Khartoum is totally inundated during the Sudan flood of August 4, 1988. (CARE)

automobiles and buses were overturned and smashed into stores and buildings; livestock drowned on the outlying farms. Without electricity—it would not return for four days—the life of the city virtually ground to a stop.

But the true carnage took place in the shantytowns. According to Sudanese ambassador to Kenya, Omar el-Sheik, "The shantytowns were completely washed away, leaving 1.5 million completely homeless. The rains made a bad situation worse. Houses of unbaked clay just washed away."

Water supplies were rapidly contaminated, and outbreaks of malaria, cholera, and typhoid began to erupt shortly after the flood waters eased. Access to the city was restricted to helicopters, which flew in food, tents, medical help, mobile generators, water tanks, and water purification tablets. The treatment of the injured was hampered by the loss of electricity, and when power was finally restored four days later, scores of people, including a worker for the International Committee of the Red Cross, were electrocuted as eroded pylons fell, bringing power lines down with them.

Final casualty figures were hard to come by. Over a hundred were known dead, hundreds were injured, and more than 1.5 million were totally without shelter. It would be months before relief efforts by Egypt, Britain, the United States, Italy, Saudi Arabia, and the United Nations Disaster Relief Organization would restore some semblance of normalcy to the city and its displaced hordes.

THAILAND
November 1988

Torrential rains caused a huge flood in Thailand in November 1988. Over 1,000 drowned; over 100,000 were rendered homeless.

A combination of natural and man-made conditions joined to produce one of Thailand's greatest natural calamities. Over 1,000 drowned, hundreds were reported missing, and at least 100,000 people were made homeless by enormous floods that raged through Thailand's southern regions and throughout northeastern Malaysia.

Five days of heavy rainfall beginning on November 19, 1988, swelled streams and rivers throughout this largely agricultural region. The situation was compounded by illegal logging activity. In addition to the indiscriminate elimination of forests that had hitherto contained flooded land, logs were left piled along river banks and in areas that, because of the absence of trees, turned instantly into moving quagmires.

When the rain reached torrential proportions, land that had been bound tightly by the roots of trees and protected by enormous umbrellas of leaves became malleable, and huge mudslides developed. Uncollected logs, left where they had fallen and often hidden in the mass of mud and water and other detritus, were swept up in the mudslides.

In the province of Nakhon Sri Thammarat, about 360 miles south of Bangkok, entire villages were mowed down by this murderous combination, and entire settlements were leveled within minutes, leaving nothing alive or standing.

Roads were blocked by flood waters, rail lines and telephone service were disrupted. Gangs of looters navigated the flooded streets of deserted villages in large trucks, scooping up whatever they could carry.

In the countryside, 700,000 acres of orchards and rice paddies were inundated, 1,000 shrimp farms were destroyed, and nearly 300 bridges were either damaged or swept away. The total cost of the carnage exceeded $400 million.

TIBET
SHIGATSE
August 10, 1954

The overflow of Lake Takri Tsoma, above Shigatse, Tibet, on August 10, 1954, caused a flood that killed between 500 and 1,000 people.

A flash flood drowned between 500 and 1,000 people in Shigatse, the second largest city in Tibet, on August 10, 1954.

For months, the Nyang Chu River, which snakes through the high Himalayas (Shigatse itself is located at an elevation of 12,800 feet [3,901.4 m]) had been swollen by pounding, daily rainfall. Its overflow normally ran into Lake Takri Tsoma, where it was safely absorbed. The lake, which hovers near and slightly above Shigatse, had been a natural safety valve for as long as anyone alive could remember—enough so that the Palace of the Western Paradise, housing the Panchen Lama, the religious leader of three million Tibetans who regarded him as the reincarnation of Buddha, had been built below it.

But on August 10, 1954, Lake Takri Tsoma failed in its function, and with a thunderous roar overflowed its banks, sending driving cataracts of water roaring downhill and into the palace. Towers and wings that had stood for centuries crumbled under the onslaught of the water, filling the palace's courtyards and drowning scores of Buddhist monks at prayer.

The Lama escaped unscathed. The Communist household troops who simultaneously served him and held him prisoner in his palace were all killed when their barracks collapsed on them, and the Providential differentiation was not lost upon devout Tibetans.

TUNISIA
September–October 1969

An extended period of rain resulted in widespread flooding in Tunisia in September and October of 1969. Five hundred forty-two people drowned; hundreds of thousands were made homeless.

For 38 days in September and October of 1969—just two days short of the biblical record—it rained over the entirety of Tunisia. Every river, stream, pond, and lake first filled, then overflowed its basin. The Zeroud and Margeulil Rivers became roaring torrents, carving out the alluvial banks that formerly bordered them, inundating fields and villages. Eighty percent of Tunisia was under water by mid-October, and by the end of the month, the death toll would reach 542.

The rivers crested at an astonishing 36 feet (10.9 m) above normal, and the force of their overflow actually moved 100-ton concrete anchoring slabs on the scores of bridges that spanned their normally tranquil waters. Thirty-five major spans were washed away completely; near one, a $7 million irrigation project was obliterated.

The rich soil that supported Tunisia's agriculture was shoveled up by the floodwaters and washed into the Mediterranean. Farmers lost everything—buildings, crops, animals, and in many cases their own lives. One hundred thousand head of livestock were drowned, their bloated and rotting carcasses causing health hazards for months. This, plus the total destruction of crops in the country, caused widespread sickness and death from disease and starvation.

Help came swiftly from the rest of the world: The United States offered $4 million in loans; West Germany provided $2.5 million in loans. France, Belgium, the Netherlands, and Spain sent contingents of engineers to help in the reconstruction of the destroyed bridges. Russia sent blankets and food. All in all, 80 nations responded with aid, and when it arrived, it was consumed hysterically by wild-eyed, starving refugees.

The door of the helicopter carrying food and piloted by Major Robert McDougal, USAF, was ripped completely off by starving villagers. Once they had pulled the food parcels from the helicopter, these same villagers clawed at each other, in feeble fights over food they had not seen in weeks.

UNITED STATES
CALIFORNIA
March 13, 1928

The collapse of the St. Francis dam caused heavy flooding in Southern California on March 13, 1928. Four hundred twenty people drowned.

Four hundred twenty people drowned in the early morning hours of March 13, 1928, when the St. Francis dam burst and sent a 120-foot-high (36.6-m-high) wall of water roaring past Los Angeles and into the Pacific Ocean, 44 miles (70.81 km) from the dam site.

The dam was a mere two years old when it was totally destroyed by the water it was to have contained, but it was probably doomed from its inception.

The creation of William Mulholland, the man who conceived and built the 233-mile-long (374.97-km-long) aqueduct through the Sierra Nevada Mountains that carried water and provided hydroelectric power to the city of Los Angeles, this monumental system included 52 miles (83.68 km) of tunnels blasted through the rock faces of the eastern slope of the Sierra Nevadas.

Forty-five miles north of Los Angeles, the mouth of one five-mile-long tunnel emptied over a precipitous cliff that plunged 1,000 feet (304.8 km) into the San Francisquito Canyon. Here, Mulholland constructed two concrete power stations that harnessed the plunging water into electricity for the city.

Shortly after this, Mulholland made the most unwise decision of his life, mapping out a huge dam that would be built across the canyon midway between the two power stations. Its reservoir, supplied by the aqueduct's surpluses, would store enough water to supply Los Angeles for one entire year, in case of an emergency.

Mulholland refused to heed the warnings of geologists who knew that the ground upon which he was

building the dam was composed of a combination of conglomerate (which ultimately dissolves into mud in water) and mica schist (which has a habit of breaking into thin, flat, slate-like pieces). To add to the dilemma, the dam was built along a geologic fault.

But the grand dam went up, 205 feet (62.4 m) high, 700 feet (213.3 m) wide, and 175 feet (53.3 m) thick at its base. Impounded in the reservoir behind it were 38,000 acre-feet of water.

The dam began to show cracks almost from the very beginning. Water dribbled continuously from its base. On the morning of March 12, 1928, a large crack appeared on the conglomerate side, and inspections were made, but the leak was judged by Mulholland himself to be insignificant.

At 11:58 P.M. on March 12, the dam gave way. Lights in Los Angeles flickered, a house above power station Number 2 began to shudder and the lights went out. With a tremendous roar, several 3,000-ton (2,721.55-tonne) sections of the dam were swept ahead of a roaring, thundering wave comprised of 12 billion gallons of water, cascading down toward the Santa Clara Valley, nine miles away.

In a construction camp, 150 men slept in their tents. Miraculously, 66 of them rode the wave to safety; the others were drowned. On California Highway 126 in the Santa Clara Valley, 50 automobiles, carrying 125 people, were swept off the road by the raging floodwaters. Later, some of these cars were found buried in mud more than 20 miles (32.2 km) from the highway.

Broadening out, the waters slowed into a two-mile-wide (3.2-km-wide) collection of sludge, pieces of houses, and fencing and other detritus which ultimately crept into the Pacific Ocean.

The final inquest and investigation laid the blame on the one man who had acquired not only fame but also the ability to make unilateral decisions about dam sites. The 73-year-old William Mulholland accepted the blame and the responsibility for 420 deaths.

UNITED STATES
COLORADO (AND MONTANA, KANSAS, WYOMING, AND NEW MEXICO)
June 16–26, 1965

Twenty-three people drowned in Colorado, Montana, Kansas, Wyoming, and New Mexico when the Arkansas River flooded the territory through which it ran from June 16 to 26, 1965.

The Arkansas River spilled over its banks in the middle of June 1965, bringing woe to the citizens of Colorado,

Trailer homes and cars were tossed about the streets of Denver, Colorado, in the flood of June 16, 1965. (American Red Cross)

Montana, Kansas, Wyoming, and New Mexico. Of all of the states, Colorado was the hardest hit. Fourteen of the 23 dead were drowned in Colorado; the property damage in the state exceeded $120 million.

Entire villages disappeared under the flood waters in the northeastern part of the state, among them, Fort Morgan, Brush, and Sterling. In the normally placid city of Denver, murky floodwaters covered a mile-wide part of it, immobilizing transit and rendering commercial establishments uninviting, to say the least. It would be the end of the summer before the villages and cities would return to normal, and longer than that for the agricultural sections of the state to begin to produce again.

UNITED STATES
COLORADO
July 31, 1976

One of the most tragic flash floods in U.S. history, caused by astonishing rains, occurred in Colorado's Big Thompson River on July 31, 1976. One hundred thirty-nine people drowned; 600 were missing.

Looking at Colorado's Big Thompson River in normal times would make the average person scratch his or her head in disbelief at its name. More a stream than a river, it measures roughly 18 inches (45.72 cm) in depth and several feet across from the time it leaves its headwaters in the Rocky Mountains until it finds its junction with the South Platte River, 78 miles (125.5 km) downstream. On the night of July 31, 1976, it measured 32 feet (9.7 m) in depth, demolished motels and automobiles, washed out bridges, floated concrete slabs and rocks weighing hundreds of tons for miles, and drowned 139 people. Six hundred more would remain missing in one of the most tragic flash floods in history.

Through the first 21 miles (33.8 km) of its run, the Big Thompson descends 5,000 feet (1,524 m) to the town of Estes Park; two miles (3.2 km) east of there, it sluices into the 25-mile-long (40.2-km-long) gorge of Big Thompson Canyon, a narrow alleyway in which it drops another 2,500 feet (762 m) before wandering for 30 miles (48.3 km) along flat land to the South Platte.

Big Thompson Canyon is a wilderness area that usually contains around 600 people. But Saturday, July 31, 1976, was the beginning of a three-day celebration weekend of the centennial of Colorado. The motels and campgrounds of the canyon were filled to overflowing, and the estimate of its population that day was around 3,500.

Late in the afternoon of the 31st, a line of thunderstorms formed from central Kansas to eastern Colorado, eventually deteriorating into a 12-mile (19.3-km) band of intense activity that spread from Estes Park to the town of Drake. At approximately 6:30 P.M., an intense rainstorm hit. It would continue for four and a half hours without letting up and would dump more than 12 inches (30.4 cm) of water into the upper one-third of Big Thompson Canyon—exceeding the average yearly rainfall in the region.

Despite advanced meteorological technology used to track the storm, the occupants of the canyon received no warning of the potential danger. Two satellite specialists in Kansas City were preoccupied with two larger storm systems developing in the Southern Mississippi River Basin over Missouri and Arkansas. Backup video equipment used to flash radar findings to Denver was down for repairs. Thus, from the time of the 7:35 P.M. weather forecast that mentioned "the possibility of flooding in low-lying areas," until the first frantic reports of a river gone instantly wild by state troopers at 8 o'clock, the occupants had no way of knowing the seriousness of the situation.

Tragically enough, many of those who were told to evacuate by the state police did not believe that they were in mortal danger, and by then, events had speeded up enough so that the troopers could not spend time trying to convince the unconvinced.

"Raindrops a half-inch in diameter were coming straight down," reported one trooper. "My slicker pockets filled with water almost instantly." Shortly after 8 P.M., the Colorado Highway Patrol Office at Greeley, 43 miles (69.2 km) east of Estes Park, received a report that part of U.S. Highway 34, a scenic route that passes through Big Thomp son Canyon and parallels the river, had a washout. Two troopers were dispatched—William Miller and Hugh Purdy. Miller sent a frantic message before he abandoned his floating car and swam to safety: "We've got to start taking people out. . . . My car's gonna be washed away. . . . I've got a real emergency here!" And then, seconds later: "The whole mountainside is gone. There is no way. I'm trying to get out of here before I drown."

Purdy was less fortunate. His message, "I'm stuck. I'm right in the middle of it. I can't get out," turned into his epitaph. His battered body was recovered later eight miles downstream.

By now, the incredibly engorged river was racing forward at a velocity of more than 20 feet (32.2 m) per second. Bursting out of the canyon's mouth, the floodwaters roared on at a rate of 233,000 gallons per second. "Campers were being washed away and big propane

tanks were coming downstream, spinning like crazy, starting to explode," recalled one survivor. Another saw Highway 34 ripping apart, "sending 10 to 12 foot [3.04 to 3.6 m] chunks of asphalt high into the air."

"You could hear the people inside . . . cars screaming for help," said Mrs. Dorothy Venrick, who lived about 250 feet (76.2 m) from the river in the town of Drake. "Above the roar of the river and the sound of homes smashing into each other, you could hear [them]. . . . It made your heart sick to stand there and know that there was nothing you could do."

"While I was outside on the porch," recalled Mrs. Barbara Nicholson, of Loveland, "I heard this sound echoing in the canyon. It sounded like three freight trains off in the distance . . . then I saw the water rising on our property."

Mrs. Nicholson's daughter, Christine, and her husband, Howard, attempted to rescue the Baileys, two neighbors. "The water knocked all of us down," related Christine Nicholson. "Suddenly we were in the flood—drifting past houses and getting struck by a lot of debris."

"You've got to realize that by the time the flood hit us, it wasn't just water," continued Mr. Nicholson. "There was a lot of solid material—dirt, rocks, buildings, cars and concrete—that was carried along with us."

Farther downstream, emergency medical technicians John McMaster and George Woodson, of Loveland, set out on an emergency call in their ambulance on Highway 34 on the night of the 31st, and were calmed by the sight of the river downstream. "The river was nice and low and moving along slowly," said McMaster afterward. "It was so peaceful you could have gone out fishing on it."

But a few miles upstream, they found the swollen river, debris, and roadblocks. Since they were driving an ambulance, they were waved through. A few miles of growing debris, and they realized they were in the wrong place. As they were turning the ambulance around, they heard the roar of the main floodwaters approaching them. "There was a huge, choking dust cloud ahead of the water," remember McMaster. "Then the water hit us like a big freight train. It picked up our ambulance about fifteen feet into the air and slammed it into a V-shaped wedge on one of the canyon walls."

The two jumped clear as the water picked up the ambulance a second time and flung it against the opposite canyon wall, smashing it to pieces. The two men climbed the canyon wall until they found a perch 50 feet (15.2 m) above the highway. The perch was insecure. McMaster lost his footing and plunged downward toward the water. "I fell about twenty-five feet [7.6 m] straight down, but as I fell I put out my arm and caught a rock," he recalled.

The water climbed up to his waist, but fortunately receded before it could rise higher. The two remained in the canyon until daylight, when they were lifted out by one of the many helicopters that plucked hundreds of survivors from the horrendously scarred and debris-scattered canyon.

The toll was tremendous. A motel log was found, with 28 registered names. Neither the motel nor any of the registered guests was ever found.

It would take months to clear the wreckage. Clarence Johnson, of Drake, remembered, for the national press, the Army Corps of Engineers coming into his yard to clean it of rocks. "They used the biggest tractors I ever saw," he said. "One rock they moved from in front of our house weighed 10 tons [9.07 tonnes], they said. There were three huge rocks a little way from our house that couldn't be moved. An engineer estimated that each weighed a hundred tons."

UNITED STATES
CONNECTICUT
PUTNAM
August 19, 1955

Swollen by Hurricane Diane, the Quinebaug River destroyed its dams and flooded Putnam, Connecticut, causing no deaths but spectacular property damage.

On the night of August 18, 1955, Hurricane Diane, the second of two back-to-back hurricanes—the other was named Connie (see HURRICANES, United States, East)—pelted Putnam, Connecticut, with eight inches of rain. Prior to that, Connie had dumped four inches on this small (pop. 8,200), riverfront colonial town.

On the morning of August 19, the Quinebaug River, customarily contained by a series of old stone and earth dams 20 to 50 miles (32.2 to 80.5 km) north of Putnam, erupted. One by one, the dams crumbled under the onslaught of the swollen river, and within moments, waves five feet high sent plumes of spray cascading over river banks, wiped away roads and took out bridges, carved up railroad embankments, and crashed headlong into the town.

Fortunately, the rising of the river and the falling of huge amounts of rain were duly recorded by weather bureaus throughout the East, and state police and civil defense personnel preceded the flood by a safe margin, evacuating as many people as they possibly could. The residents of Connecticut cooperated fully and by the time the roaring floodwaters, clocked at 25 MPH, roared into the town, it was vacant. Not one person was lost in this flood.

There was, however, considerable and spectacular property damage. Just above the town, water poured into a warehouse stocked with 20 tons (18.1 tonnes) of magnesium. Magnesium is noted for, among other things, its volatility when placed in contact with water. The warehouse went up in a gigantic eruption of flame and water, and hundreds of barrels of burning magnesium were launched into the raging floodwaters. They, too, exploded along the way, sending plumes of water and flame 250 feet (76.2 m) in the air.

Some holdouts were plucked from rooftops by National Guard helicopters after the first passage of the river. Others returned a week later to assess the $13 million worth of property damage caused by the flood.

UNITED STATES
KANSAS
July 12–31, 1951

Fifty-three days of rainfall preceded the flood of the Kansas River on July 12, 1951. Fifty were killed; hundreds of thousands were evacuated.

The costliest flood in U.S. history up to that point (over $1 billion) and the worst ever in the Central Plains claimed 50 lives and came after a solid 53 days of rainfall that totaled 16 inches (40.6 cm) by the middle of July.

Most of this relentless deluge fell in the basin of the Kansas River; the rest found its way into the Arkansas and Neosho Rivers.

There was warning of the impending flood. By late May most of the smaller rivers in Kansas and northern Oklahoma had burst their banks. On May 22, the small city of Hays, Kansas, was inundated by water from nearby Big Creek.

As June wore on, low-lying farmlands and cities throughout the state were on constant alert for floods. By the end of the month, all of the area drained by the Kansas River had absorbed all of the water it could. Scientists estimated that the soil in the area could safely hold only two more inches (5.08 cm) of rain. Much more than that would fall.

By the afternoon of July 11, all of the larger rivers in Kansas were over their banks, and the highest flood crest ever seen in the state moved down the length of the Kansas River.

Manhattan, Kansas, located on the Kansas River at the point at which it joins the Big Blue River, was hit by floodwaters on the night of July 12, 1951. In minutes, more than half the town was under five feet of water, and thousands of people had been made instantly homeless.

At dawn of the 13th, the floodwaters reached Topeka, cresting three and a half feet (1.06 m) higher than the previous record height, set in 1903. The city was cut in two as water topped levees and flood walls. Twenty-five thousand people were evacuated; railyards,

The wreckage of a home and boathouse following the Kansas flood of July 1951 (American Red Cross)

factories, and businesses were gutted or destroyed, and a large section of the city became an unpopulated lake.

Many miles downstream, the floodwaters poured across the Santa Fe railroad tracks for 55 hours, trapping 337 people aboard the passenger train *El Capitan*. From here, they sped on to the thickly populated twin cities of Kansas City, Kansas, and Kansas City, Missouri, where 35-foot-high (10.6-m-high) dirt-and-rock dikes had been built 40 years ago to shield the stockyards, packing plants, railyards, flour mills, and grain elevators from flooding. On the Missouri side of the river, an eight-foot-high by three-foot-thick (0.9-m thick) concrete wall had been added as reinforcement.

But by Friday, July 13, the Army Corps of Engineers estimated that the debris-laiden river would broach these dikes easily. Workers toiled into the night, adding sandbags to the dikes. Their efforts were fruitless, and in the middle of the night, the dikes gave way. Roaring floodwaters rammed through the railyards, blanketing $50 million worth of locomotives, freight cars, and buildings.

Factory whistles and police sirens set up a wail, warning people to evacuate the path ahead of the flood. Twelve thousand people scurried for the bluffs beyond the town.

In moments, most of the buildings in Kansas City, Kansas, were covered up to their second stories. Outside of town, six square miles of homes, grain elevators, and manufacturing plants were under as much as 30 feet of water.

On the Missouri side of the river, the reinforced dike held, but by midday on the 13th, it, too, began to waver under the assault of 500,000 cubic feet (14,158.4 cu. m) of water per second. Refugees from Kansas were threatened for the second time. Within two hours, the Kansas City stockyards were under 10 feet (3 m) of water. Six thousand hogs and sheep that were not moved to higher overhead shoots drowned.

Dead animals, smashed homes, uprooted trees, splintered telephone poles plummeted through the city. An enormous steel tank loaded with diesel oil floated into an area where two other, building-size tanks loaded with gasoline floated. The diesel tank became entangled with a live electric wire, exploded, and as the flame spread across the water to the other two tanks, a chain reaction of multiple explosions flung fire in all directions, igniting a naphtha tank, which went up like a volcano. Within minutes a quarter-mile square sea of fire ignited, and despite efforts of firemen in boats to put it out, it burned for five days.

The wreckage was enormous. Twenty-five hundred homes were totally destroyed. Streets of the two Kansas Cities were filled with 16 million tons (14,514,955.9 tonnes) of silt and sand. Only one of the 13 rail lines servicing the area was operable. Only two out of the 17 highways running into it were open. Seventeen major bridges had been washed away. Two million acres (809,371 ha) lay under polluted water. And only 10 percent of the area's sewers, water, and power systems were left in working order.

The floodwaters crashed on into the Missouri River, but workers in Jefferson City, downstream, saved their main bridge by piling nearly 100 tons of scrap metal on it.

Within days, President Harry Truman, a native of nearby Independence, Missouri, made $25 million of Federal Emergency Aid available, and sent in the Red Cross and other relief agencies. It would be nearly a year before a semblance of normalcy would return to the area.

UNITED STATES
MIDWEST
June, July, August 1993

Fed by ceaseless rains, the costliest flood in U.S. history inundated huge stretches of the Missouri River from Pipe Stem Reservation, North Dakota, to St. Louis, and the upper Mississippi River from Minneapolis to St. Louis, in the summer of 1993. The total cost of the damage was nearly $18 billion, 48 people were killed, 15 million acres (6,070,284 ha) were drowned and their crops destroyed, and 54,000 people were forced to evacuate their homes, businesses, and farms.

It all began on June 11, 1993. A foot of rain fell that day on southern Minnesota and lesser amounts on northern Iowa. If that were all that had happened, the weather that June day would have been a mere inconvenience. But four days later, 11 more inches (28 cm) fell in the same area, and a great deal of that water found its way to the Minnesota River, a Mississippi River tributary southwest of St. Paul, Minnesota. Simultane-ously, the Black and Wisconsin Rivers, filled to over-flowing by the intense rainfall, began to rise.

A series of 300 dams and reservoirs exist along this river system that includes the Des Moines and Iowa Rivers and eventually empties into the Missis-sippi. Ordinarily, they would drain off excess rainwater and prevent flooding farther south. But this storm sys-tem was an unusual one. It now stalled over Minnesota and Iowa, and continued to pour more water into the reservoirs than they could hold.

By mid-June, the Mississippi crested at Davenport, Iowa, at 18.5 feet (5.6 m), four feet (1.2 m) below the record of 1965. But that, too, was just the beginning. More rain fell, and June became the wettest in that region since record-keeping began in 1878.

The Mississippi began to move downstream at 70 miles (112.6 km) a day, and now it was the job of the 500 miles (804.7 km) of levees north of St. Louis to contain the torrent. They were not up to the job. By the end of June, the Mississippi from St. Paul to St. Louis was closed to most commercial traffic—a great loss for farmers along the river. Grain, farm chemicals, and fertilizer are traditionally shipped by barge along the Mississippi, and by the beginning of July, barge traffic had nearly disappeared from the river.

In Minnesota, the downpours carried high winds that, in concert with waters that began to overflow the banks of the Minnesota and Mississippi, damaged 800 homes, situated on 700,000 acres (283,279 ha), which were themselves inundated and ruined. Parts of 12 highways in the state were closed, and the flood's first fatality occurred. An 11-year-old girl, wading in a small lake near the Minnesota River, was sucked under and drowned in its fast-moving waters.

Davenport, Iowa, was the first major urban center hit by the flood. The Mississippi rose to 4.6 feet (1.4 m) above flood stage at the end of June, filling River Drive and Front Street, two blocks from the river's edge, with water, which lapped at businesses and homes in the neighborhood.

And as June gave way to July, the rains continued. "It's the most rain we've had in over 100 years," one farmer observed to President Clinton, who made a visit to the area.

The Mississippi continued to rise, reaching 22.3 feet (6.8 m), just shy of Davenport's record level of 22.5 feet (6.9 m). Six hundred people were evacuated from low-level dwellings in the city and 200 National Guard troops piled sandbags on the levee and patrolled the city. One thousand barges, the equivalent of 40,000 trucks, remained, fully loaded, motionless, and tied up.

Southward the torrent moved, converging on the little town of Grafton, Illinois, 45 miles (72.42 km) north of St. Louis and at the juncture of the Illinois and Mississippi Rivers.

In Missouri, at the town of La Grange, 30 miles (48.28 km) north of Hannibal, two of the village's three levees burst, flooding the small town and forcing 400 residents to abandon their homes. In Jefferson City, over 400 women were evacuated from a low-lying prison and taken to a drier one in central Missouri.

Meanwhile, Davenport, which had never built any flood control systems, continued to flood and reflood. By the morning of July 4, the Mississippi had risen to 21.9 feet (6.4 m), 6.1 feet (1.8 m) above flood level. That evening, it rose to 22.2 feet (6.76 m), and the next day, the rains, which had lightened, resumed heavily. One hundred and fifty bridges across the river were closed.

By July 6, it was apparent to engineers that the Mississippi had not crested yet. The rivers that fed into it were increasing in volume. The Iowa River was threatening to flood the intake pipes of Iowa City's water treatment plant, a situation that would contaminate the city's drinking water.

At a dam at Coralville Lake, water overflowed an emergency spillway for the first time in its 35-year history, and left the dam at three times its normal rate.

Flash floods appeared on tributaries of the larger rivers. In southwest Missouri, three people drowned when one of these flash floods swept two cars off a bridge over Flat Creek, near Cassville. The ground on either side of the river was saturated; small green frogs leaped where crops had once grown.

Entire towns began to be evacuated, and the Red Cross opened 19 shelters along the Mississippi for refugees from the flood. In West Alton, Missouri, river waters swept into town and mixed with wells and septic systems, contaminating drinking water with raw sewage.

Nearly 16,000 people had been evacuated by the middle of July, and the rain continued relentlessly; from being a flood the likes of which had not been seen in a couple of decades, this inundation became the worst in a century. "We've never seen anything like this before," a spokesman for the Army Corps of Engineers in Rock Island, Illinois, confessed, after noting that rainfall amounts were breaking records that had been kept since the early 1800s.

Dams were breached; levees crumbled by the dozens. "It's like water rising over the sides of a bathtub," another army engineer said. "The rain will just spill over the sides. There's not a dry spot in Iowa."

It was as if nature had turned endlessly vicious. Giant thunderstorms moved across Nebraska, with 100 MPH winds, damaging homes and downing power lines. National Guard personnel and volunteers worked feverishly, piling sandbags on top of threatened levees. But most of the time, their work was in vain. In South St. Louis County, a levee broke after 72 hours of sandbagging, and a town of 100 had to be evacuated.

In various locations, between 7.5 and 4.5 inches (19.05 and 11.43 cm) of rain fell within 24 hours, washing out roads and bridges and forcing further evacuations.

Farmers crossed their flooded fields and lawns in boats. Almost half of Missouri's 144 counties were inundated by rivers, streams, and creeks. For 120 miles (193.1 km), from Glasgow in west-central Missouri to St. Louis, there was hardly a levee that had not given way to the Missouri River. In Herman, Missouri, homes and businesses were rendered uninhabitable. A restraining wall collapsed into the town's only factory and chief employer, Steven's Toys. Flotillas of plastic toys washed through a ruined wall, and 200 employees were thrown out of work.

"There was no warning whatever," one farmer explained to a reporter, as he piloted his brother's boat into town across a cornfield that was eight feet underwater. "The water started coming. It just kept coming."

By July 10, the Mississippi had flooded 920 square miles (1,480.6 sq. km) along a 250-mile (402.33-km) stretch of the river. In places, there was water as far as the eye could see. Near Mark Twain's boyhood town of Hannibal, Missouri, the muddy waters of the Mississippi moved tons of uprooted trees and rusted automobiles. Children caught catfish in the streets of the town.

Kansas City was flooded with seven inches of rain in five hours. In Des Moines, Iowa, the Des Moines and Raccoon Rivers overflowed, ruining the fresh water supply, cutting off electricity to thousands of homes and businesses, and forcing hundreds of people to flee. Downtown bridges that linked two sides of the city were closed down, preventing thousands more from going to work. Emergency arrangements were made to truck millions of gallons of water into the city, and water rationing was instituted. Water was distributed for drinking purposes only. Thirty-five hundred people were evacuated by boat from various parts of Des Moines, as the Des Moines River crested at 45 feet (13.7 m).

At Des Moines General, the city's 100-bed hospital, barrels of filtered water were rolled in from a truck for hand-washing for surgery and to sustain the hospital's patients. With its electrical supply provided by one generator, and no way to sterilize equipment, the hospital was forced to shuttle surgical instruments to rural hospitals for sterilization.

Now, human waste began to wash up on floodwaters. Iowa health officials issued alerts to citizens to be vaccinated for tetanus if they had open cuts and were working in the water. In Missouri, officials warned residents to bathe after coming in contact with floodwaters.

The rains continued to fall, as manhole covers burst from streets in downtown Des Moines under the pressure of runoff from as much as two inches of rain falling in 20 minutes. Residents set out pots and garbage cans to catch rainwater to flush their toilets.

In Glen Haven, a town of 544 at the base of four hollows along the Mississippi in Wisconsin, a flash storm dumped four inches of rain in 20 minutes, unleashing a six-foot-high wall of water that rushed down Main Street and swept away 17 cars, washing five of them into the Mississippi.

It seemed like a war along the banks of the river. National Guard helicopters hovered overhead as bulldozers dumped piles of sandbags that had been filled by armies of volunteers. "One day we put down 40,000 sandbags," one of the workers near Des Moines said, "It was 95 degrees. There wasn't a breath of air."

In Iowa City, volunteers and National Guard troops stacked 10,000 sandbags around the city's water treatment plant to prevent the contamination that had occurred in Des Moines, where residents traveled 10 miles to the Altoona YWCA to take showers. At one point, 2,500 people were lined up outside the Y.

By July 16, the Mississippi and Missouri Rivers were threatening St. Louis. In Quincy, Illinois, the Bayview Bridge, the only crossing of the Mississippi for 200 miles (321.8 km), became unsafe and was closed. Nearby, a collapsed levee led to an oil explosion from tanks near the river. The resulting oil slick burned for two hours before it could be extinguished.

As the crest of the flood moved southward, there was some relief upstream. But near St. Louis, over 7,000 people were evacuated from rampaging floodwaters. The river was expected to crest at 46 feet on July 17, which set off a mad scramble to top off levees with sandbags. Most of these bulwarks against the flood were only 45 feet (13.7 m) high.

The river continued to rise, gorged with runoff from about 20 percent of the nation's continental land mass, propelled by water from the Missouri River. On July 18, it rose to a crest of 47 feet (14.3 m) as it rolled southward past the Gateway Arch in St. Louis. But the system held despite the assault by some of the worst flooding in the nation's history. Muddy, wide as five football fields, choked with flotsam, the water roared southward at 7 million gallons a second.

By the end of July, although there were some locations at which the floodwaters were receding, the Mississippi was still a menace downstream. A 480-mile stretch from Dubuque, Iowa, to the mouth of the Ohio River remained at flood stage. In St. Louis, the river rose and fell, hovering around a record 47 feet (14.3 m) at its crest.

And that would be the way it would be for another month for most locations on the Mississippi and its nearby rivers. Rain continued to fall, in lesser and lesser amounts, but the rivers, swollen far beyond their banks at depths that no one could remember experiencing before, would remain above flood stage through October.

At the end, it would be the most expensive flood in U.S. history, costing nearly $18 billion in damage and claiming to that date 48 lives. Over 15 million acres (6,070,284 ha) in nine states were inundated, displacing 54,000 people. Ninety-five points along the Missouri and upper Mississippi Rivers exceeded previous flood records, and on the tributaries along the rivers nearly 500 points exceeded their flood stage at some time during that remarkable and tragic summer.

UNITED STATES
MIDWEST
August 20–29, 2007

The double-barreled assault of a stalled weather front and the remnants of Tropical Storm Erin conspired to produce record-breaking floods in America's Midwest on August 20–29, 2007. Twenty-two people died in the week of torrents and flooding.

Ordinarily, the effects of tropical storms from the Gulf of Mexico do not reach up into Ohio. But Tropical Storm Erin, born in the Gulf on August 15, 2007, was an exception. When it first made landfall on August 16 near Lamar, Texas, it was entering a region of the country that was already experiencing its largest amounts of rainfall in years.

At the same time that Erin began its journey northward from Texas, the upper Midwest was being pummeled by a series of thunderstorms playing along a stalled cold front. Now, Erin added its potential for large amounts of torrential rains. By August 22, the situation was grim for Texas, Oklahoma, Wisconsin, Illinois, and Minnesota.

By August 22, more than 9 inches (228 mm) of rain had soaked parts of central Oklahoma, swelling rivers beyond their banks and washing away roads and houses. In Kingfisher, Oklahoma, 100 people were evacuated from their homes after the nearby Cimarron River burst its banks.

In Missouri, the combination of storms, which spawned tornadoes, had dropped 11.9 inches (303 mm) of rain on land already saturated with weeks of heavy thunderstorm activity. In Minnesota, nearly 11 inches (279 mm) of rain caused violent flooding that washed away bridges and roads and killed six people. In Wisconsin, up to 12 inches (304 mm) of rain fell, triggering mudslides that buried entire villages. Chicago's O'Hare International Airport was forced to cancel more than

200 flights. Ohio was one of the hardest hit states, with many sections of the Blanchard River rising to more than 7 feet (2.1 m) above flood stage level, matching a record set in 1913. More than 500 people were forced to evacuate from villages along its banks and officials declared a state of emergency across 9 of the 21 affected counties in the flood zone. Seven rescuer boats hunted for stranded residents and 130 inmates were moved from the threatened county jail to a regional prison away from the river.

In Humboldt, Iowa, where the flood cancelled the opening of schools, 33 residents were evacuated from a care center, and the city administrator told reporters "We're still swimming. There is no part of town that's unaffected. It just poured, and poured."

In Madison, Wisconsin, lightning from one of the storms struck a utility pole near a bus stop. A wire, loosed by the impact, landed in a flooded portion of the curb where a woman and her young child were waiting to board a bus. They were electrocuted on the spot. A person on the bus got off and tried to help them, but he was also electrocuted. Then, the bus driver tried to get off, but was shocked by the current and fell back into the bus. He and a child injured on the scene were taken to a nearby hospital, where they recovered from their injuries.

In Findlay, Ohio, Dick and Peggy McAllister and their fellow tourists who were stuck in a motel, were evacuated by dump truck to higher ground, and then transferred to a school bus, which took them to a shelter where 150 others were already staying. "It's raining and raining," McAllister told reporters. "I just need to get the hell out of here."

In Iowa, north of Fort Dodge, levees overflowed and forced 90 people from their homes, and State Highway 7 and U.S. Highway 20 were washed away. And in Hoka, Minnesota, more than 15 inches (381 mm) of rain fell in 24 hours. "Every 24 hours," National Weather Service meteorologist Todd Shea in La Crosse, Wisconsin, observed, "there's a resurgence of warm air that fires up more thunderstorms."

Finally, the skies began to clear on the afternoon of August 22 and temperatures began to rise. It would be days before the flood waters would recede. The Fox River feeds into Piskatee Lake in Illinois, and during the flooding, the lake had overflowed and inundated most of the homes in the village of 11,000 bordering on it. In the aftermath, picnic tables and debris from flooded homes floated on the now stagnant waters. "You take the good with the bad when you live on a lake," said John Paladino, as he shoveled mud from around his house. "Things could have been worse," he added. "As long as you don't need a life jacket I guess it's really not that bad." It had been worse for many. Twenty-two people had died in the floods before the weather finally changed.

UNITED STATES
MISSISSIPPI RIVER
April 1874

Two hundred to 300 people drowned in Mississippi and Louisiana in the first recorded flooding of the Mississippi River, during April 1874.

A graphic depiction of the first recorded Mississippi flood, in April 1874, caused by melting snow and heavy spring rains (Frank Leslie's Illustrated Newspaper)

The first reported flooding of the Mississippi River was sketchily told, largely because the areas hit were rural ones, where, in 1874, communication ranged from non-existent to feeble.

What is known is that an estimated 200 to 300 people were drowned in the floodwaters of the Mississippi that had been swollen from melting snow and heavy rains in the early spring of 1874. Tens of thousands of acres of farmland and grazing land in Louisiana and Mississippi were inundated, and over a thousand head of livestock were killed in hundreds of locations at which early levees had been erected.

UNITED STATES
MISSISSIPPI RIVER
January–April 1890

At least 100 drowned and 50,000 were made homeless by the seemingly endless flooding by the Mississippi River of its entire valley between January and April of 1890.

The spectacular Mississippi River flood of 1890 was a long time coming. Beginning soon after New Year's, with melting snows swelling its headwaters at Cairo, Illinois, the Mississippi kept on rolling over its banks and its levees, right up until the end of April, when it inundated three thousand square miles of Louisiana, rendered 50,000 persons homeless, and drowned at least 100.

A former steamboat pilot sadly concluded: "I have lived on the river for thirty years, and I have studied it, for it was my business to do so. I have been steamboating all that time. I am now certain that I don't know anything about it, or about what ought to be done to it."

What had been done since the flood of 1874 (see previous entry) and the flash floods of 1882, was a strengthening and lengthening of the levee system. Perhaps anything would have been too little and too late. The river washed away the dikes, the levees, the extra sandbags, and, in some cases, the men and women who were frantically piling on the sandbags in one last effort to stem the roaring tide.

First warnings of the impending flood were issued by the United States Signal Service in December. Massive snowfalls in Ohio would, they said, cause flooding. And certainly enough, on January 1, 1890, the Mississippi rose to 18 feet (5.5 m) in Cairo, Illinois, and continued to rise after that, at a rate of one and three-eighths feet a day.

By the end of February, Memphis, Shreveport, Vicksburg, and New Orleans were threatened, and by the end of March, the river was breaching its banks at Arkansas City and Memphis, drowning dozens and covering thousands of acres.

By April 3, Greenville, Mississippi, was inundated and most of its 10,000 residents had removed themselves to nearby high ground.

The Morganza Bend, south of this, suffered greatly in the middle of April when 400 feet (121.9 m) of dikes and levees collapsed simultaneously, drenching the countryside and collapsing many nearby homes.

Finally, by April 24, the entire Mississippi Valley was all or partially under water, from Illinois to Baton Rouge, Louisiana. Three thousand square miles of land in Louisiana alone were under water, and it would be months before the destruction would be cleared.

UNITED STATES
MISSISSIPPI RIVER
April 1912

Two hundred fifty people drowned and 30,000 were rendered homeless by a Mississippi River flood, caused by melting snows, in April 1912.

April has always, it seems, been the cruelest month for those who live along the banks of the Mississippi River, particularly in those years when the winters to the north have been severe and full of snow. In 1912, 250 drowned, 30,000 were made homeless, and at least $10 million in damages were caused by floodwaters that were first fed by melting snows in tributaries of the Mississippi and then swollen by falling rains.

On April 2, troops that had been brought into Cairo, Illinois, to build up and hold the dikes, ran for their lives as the river shot up like a geyser over the barriers and roared into the Mobile & Ohio train yards.

During the next week, river towns in Kentucky began to fill up with water, turning them into muddied lakes.

As usual, the worst flooding occurred in Mississippi, in Bolivar County, where great loss of life occurred and thousands of acres were inundated. The cities of Beulah and Benoit were turned into Southern Venices, and it would be weeks before they would return to some semblance of normality.

UNITED STATES
MISSISSIPPI RIVER
April–July 1927

....

A long rainy period culminated in a long flood in the Mississippi River Valley, from April to July 1927. Local and federal statistics conflict; from 246 to 500 were killed and 650,000 were rendered homeless in the flood.

Between 246 and 500 people drowned, 650,000 were made homeless, 2.3 million acres (809,613.8 ha) of land were inundated, and a staggering $230 million in damages were caused by a series of monumental Mississippi River floods from Tennessee to New Orleans between April and July of 1927.

The floods originated in August 1926, almost a year before the recession of floodwaters. Heavy, nearly ceaseless rains began to fall throughout much of the Mississippi's drainage basin. The deluge continued into the winter. On New Year's Day, 1927, the Cumberland River, a tributary of the Mississippi, stood at a depth of 56.2 feet (17.1 m), which was 41 feet (12.5 m) above its level before the rains began in August. In Cairo, the Ohio River stood at 44.9 feet (13.7 m). In August, it had measured 18.1 (5.5 m).

The rains let up in February and resumed in March, and on April 19, the Mississippi broke loose, riddling the levees at Mound Landing, above Greenville, Tennessee, with small drill holes of rushing water that erupted into boiling bubbles on the outside wall of the levees. All through the night of April 20, small farmers and townspeople worked in a downpour, attempting to shore up the levee. But, according to flood historian William Percy, ". . . about daylight, while the distraught engineers and labor bosses hurried and consulted and bawled commands, while the 5,000 Negroes with 100-pound sandbags on their shoulders trotted in long converging lines to the threatened [weak] point, the river pushed, and the great dike dissolved under their feet."

Scores were swept to their deaths there and in Greenville, where General Alexander G. Paxton of the Engineers recalled ". . . three words that I shall remember as long as I live—'There she goes!'"

One planter remembered the crest of the river as a "tan-colored wall seven feet high, and with a roar as of a mighty wind." Over two million acres of farmland were submerged, under water which swirled around eastern hills and flowed into the basin of the Yazoo River, and then back to the main stream of the Mississippi near Vicksburg.

Workers sandbagging levees on the Mississippi River during the enormous floods of April through July 1927; more than 650,000 were made homeless by this natural catastrophe. (Library of Congress)

Across from Mound Landing, on April 21, the levee at Pendleton, Arkansas, burst. "You could hear it roaring a long ways off," noted Robert Murphy, who lived below Pendleton. "Kind of like, they say, a tornado. Near about that loud." In Arkansas City, 32 miles (51.5 km) downstream, streets that had been dry at noon were flooded enough to drown people, and did, by 2 P.M.

A secondary danger, backwater flows from main streams into tributaries, threatened those living along the Arkansas River all the way upstream to Little Rock, 100 miles (160.9 km) from the Mississippi. Near Pine Bluff, which was 60 miles (96.5 km) from the mouth of the Arkansas, 500 people crowded together on Free Bridge, a mile-long steel span across the Arkansas.

They didn't realize the danger of being stranded when access roads flooded. For three days and nights, in a cold rain, they remained on the span, isolated by waters that were too turbulent to allow rescuers to reach them from the river. When the survivors were finally taken off by boat, two babies had been born on the bridge.

By the middle of May, the floodcrest had traveled southward to Louisiana. On May 17, the supposedly impregnable Atchafalaya levee at Melville crumbled. "The water leaped through the crevasse with such fury that it spread into three distinct currents," reported Turner Catledge in the Memphis *Commercial Appeal*, "One force shot straight west.... A second current raced north, quickly eating out 50-foot sections of the Texas & Pacific Railroad embankment . . . a third current struck out from the south . . . washtubs, work benches, household furniture, chickens and domestic animals were floating away."

One more city would be jeopardized by the rushing floodwaters, New Orleans. A difficult solution was decided upon: the dynamiting of the Caernarvon levee just below the city. This would drain the water off to the southeast, but it would mean that the thinly populated parishes of St. Bernard and Plaquemines would be inundated.

The advantages of saving New Orleans and draining off the upstream floodwaters won out over the protests of the inhabitants of the two parishes. State authorities trucked out the refugees, and the National Guard moved in to oversee the demolition.

The first attempt was a spectacular failure. Fifteen hundred pounds of dynamite, buried in the top of the dike, lifted tons of earth into the air and dropped them back in exactly the spot from which they had been loosened. Finally, diver Ted Herbert volunteered to place the charges at the levee's base. Twice, he nearly drowned in the strong current which twisted his diving

helmet askew, nearly strangling him. The third attempt succeeded, and New Orleans was saved.

Seven states had been inundated by the flood, and now 33,000 rescue and relief workers poured in under the joint direction of Secretary of Commerce Herbert Hoover and the American Red Cross's James Fieser. Hoover's success in the effort was widely credited with helping him in his later, successful campaign for the presidency.

His influence and a radio appeal immediately brought in $15 million for rescue efforts. The railroads provided trains and 6,000 riverboats of all sizes, from 700-ton (635.03-tonnes) river steamers to tiny skiffs. Sawmills on the river went to work and made 1,000 boats in 10 days. Navy planes circled the area, as the Coast Guard manned the larger boats, and engineers studied possible future courses of floodwaters.

Interestingly enough, some of the most effective and personal rescue work was achieved by bootleggers from the swamps. Their high-speed escape boats proved particularly valuable. Historian Percy reported, "No one had sent for them, no one was paying them, no one had a good word for them—but they came. They scoured the back areas, the forgotten places, across fences, over railroad embankments, through woods and brush, and never rested."

UNITED STATES
MISSISSIPPI RIVER
April–May 1973

Twenty-five drowned and 35,000 were made homeless by the April–May 1973 Mississippi River floods, caused by melting snow.

The work of Herbert Hoover and his commission in 1927 (see previous entry) prevented the enormous floods of 1973 from reaching the catastrophic proportions of the 1927 event. Still, this region suffered the inundation of 13 million acres of land, the displacing from their homes of 35,000 people, and 25 deaths.

The winter of 1973 was a severe one, with heavy snowstorms that killed hundreds of head of livestock in the northern portion of the Mississippi drainage system. And spring was scarcely any better; heavy rains began in March and continued into early April, preventing the sowing of spring crops in Wisconsin. Rain melted the snow quickly, and pressure began to build in the 340 storage basins built at a cost of 10 billion dollars as a result of the 1927 catastrophe.

But the basins held. By early April, more than 7 million acres in the upper floodplain of the Missis-

Floodwaters from the Mississippi and Illinois Rivers totally inundated the main street of Grafton, Illinois, in April 1973, forcing the evacuation of 900 of the town's 1,100 residents. (American Red Cross)

sippi were under water, and the storage capacity had been reached. Floodgates were opened, and the excess drained into Lake Pontchartrain, away from New Orleans. The dikes held, and the floodwaters dropped.

But the rain continued, and by early May, the drained water had been replaced and much more added to it. At Keithburg, Illinois, a levee broke and three feet of water gushed into the main street of the town. Fifteen city blocks disappeared underwater, and downstream, Hannibal, Missouri, was flooded under six feet of water. In Quincy, Illinois, 4,000 people had to be evacuated from their riverside homes, and now, more and more levees toppled and more and more campsites and resort communities disappeared under the Mississippi's muddy torrents. In Missouri, as dike after dike disintegrated, 30,000 acres of farmland were drowned in water, mud, and silt.

Finally, at the beginning of June, the rains ceased, and the waters receded. Despite the damage, most of the levees had held, and it was generally estimated that without the construction that had resulted from the 1927 floods, property damage would have risen another $7 billion, and human casualties would have soared into the hundreds.

UNITED STATES
OHIO (AND INDIANA AND ILLINOIS)
March 25, 1913

Five hundred people drowned and over 175,000 were rendered homeless in a flood in lands adjacent to the

191

Scioto, Mad, Miami, and Muskingum Rivers in Ohio, Indiana, and Illinois, on March 25, 1913.

The second-worst flood of the 20th century in America (see FLOODS, United States, Pennsylvania, p. 196, and FLOODS, United States, Mississippi River, 1927, p. 189) took place on March 25, 1913. Five hundred people drowned, most of them in Dayton, Ohio; over 175,000 people were made homeless; and over $147 million in damage was recorded from the several days that the Scioto, Mad, Miami, and Muskingum Rivers ran amok, inundating cities and farmland alike.

For decades, the population of Ohio, Indiana, and Illinois rested secure in the assumption that the levees they had constructed would contain the swelling of these rivers. But by the middle of March, levees in southern Indiana, Illinois, and Ohio had already been breached by river waters swollen from weeks of steady rain. In three days, according to a reporter in the *Cleveland Leader,* 18 billion tons (16,329,325,320 tonnes), or 575 million cubic feet (16,282,187 cu. m) of water, fell on Ohio. It was enough to make the Miami River smash through its dikes above Dayton, Ohio. Almost simultaneously, the Mad River broached its banks and roared toward Dayton.

Within minutes, the city was under seven to 12 feet (3.6 m) of water, and its 125,000 residents climbed to the upper stories of buildings, some of which collapsed under them. Seven thousand refugees sought shelter in landmark office towers such as the Bretheren Publishing Company, the Conover Building, the Calahan Bank, and the National Cash Register Building, whose 70-year-old president, John H. Patterson, immediately set out by rowboat to rescue drowning swimmers in the city's streets.

In those first few moments of the flood, all communication in and out of Dayton was severed as the city's sole telephone line snapped under the pressure of falling poles.

An 18-year-old Ben Hecht, then a cub reporter for the *Chicago Journal,* was one of the first newspapermen to arrive and report on the scene. He was nothing if not persevering, taking the train as far as he could from Chicago, walking through heavy snows, fast talking two trainmen into loaning him their handcar, which he pumped into town, while ducking the shots of militiamen and deputies ordered to shoot potential looters and strangers on sight.

In his book of recollections, *Child of the Century,* Hecht recalled spending the day in a canoe, ". . . paddling through flooded Dayton, interviewing other boatmen and gathering data on the catastrophe. . . ."

He had plenty to report, from the grotesque to the grim. In downtown Dayton, a dead horse floated through the broken doors of the First National Bank and was later found standing behind the bars of a teller's cage. A sow managed to make it to the second floor lingerie department of a large clothing store before drowning.

Martin Ellis and his wife found themselves marooned on top of the Algonquin Hotel in Dayton, when the hotel's front doors became blocked by a team of horses. "The water traveled like a sheet to the east," recalled Ellis, "passing over the city. A panic followed. People ran to the tops of buildings and were brushed off like flies. My four children were home. We lived on North Main Street. We saw the top of our house burn. . . . My wife couldn't stand it. . . ." Mrs. Ellis leaped into the raging waters and drowned.

The age of the horse-drawn vehicle saved some, who clambered atop swimming animals and thus saved themselves.

Meanwhile, the rest of Ohio was also disappearing under torrents of wild water. Train passengers, passing Toledo, witnessed wholesale destruction in that city as it filled with water and refugees stopped the train and clambered aboard. According to passenger Perry Hillister, "The wind was raw, too, and some of them were nearly frozen when we took them aboard. . . . Toledo was struck badly. The lower part of the city was under water."

Individual acts of heroism abounded. Ben Hecht described the telegraph operators in Miami Junction, who manned their posts for 30 hours: "Some have dropped from exhaustion," he reported. "The Western Union men, who were the first to break into the city, haven't slept since Tuesday. 'Safe.' 'Safe.' the monotonous words of rescue and death have jammed the wires since the first one was opened. . . ."

Students at Wesleyan University in Delaware, Ohio, were responsible for saving scores of lives. One unnamed student saved 30 people by swimming back and forth in the raging current. A boatload of other students saved a woman and her three children who were dangling from a railroad bridge by maneuvering under them and catching them as they dropped, one by one, into the boat.

In Columbus, Ohio, Midwestern ingenuity accounted for the survival of the George Roller family. As the flood approached their home, they somehow managed to maneuver the family cow through the kitchen door, and upstairs, where it was accorded its own private bedroom in return for providing themselves and their neighbors with milk for five days, until the waters receded.

As the Scioto River also rose, the towns of Troy and Tadmore were submerged entirely under water. Portsmouth, Columbus, Circleville, and Chillicothe were flooded. In Indiana, the Wabash River rampaged through Terre Haute and Lafayette, and the White River slammed into the western suburbs of Indianapolis. But it was in Ohio that the greatest destruction and loss of life took place.

There was a huge outpouring of relief supplies and Red Cross help from across the country, but it would be well into 1914 before the damage to dikes, buildings, and farmland would be remedied.

UNITED STATES
OHIO
January 1937

Twenty-five days of steady rain caused the Ohio and Mississippi Rivers to bridge their levees in Ohio in January 1937. One hundred thirty-seven people died primarily from fires and explosions; 13,000 were made homeless.

A flood of fearsome proportions was generated by 25 straight days of torrential downpours which dumped an estimated 15 trillion gallons of water on the Ohio River Basin during the month of January 1937. This in turn swelled the Ohio all the way from West Virginia to its mouth in Cairo, where it emptied an unprecedented tide of water into the Mississippi.

From Pittsburgh, Pennsylvania, to Cairo, levees crumbled and cities were inundated. The waters rose to a depth of 10 feet in the streets of Cincinnati, and Louisville, Kentucky, was likewise inundated. Over 204,000 square miles (328,306 sq. km)—more than 8 million acres (3,237,485 ha)—of floodplain and farmland, town and city were affected by the Ohio's raging floodwaters, and 13,000 homes were swept away.

Ironically, practically all of the 137 deaths that occurred were not from drowning, but from fires and explosions. On January 16, the Louisville Varnish Company went up in flames. Gas explosions caused by the displacing of mains in the flood erupted in Louisville for two weeks, as they did in Ironton, Ohio.

At Cairo, the Army Corps of Engineers was forced to dynamite the left-bank levee when the Ohio crested at 63 feet (19.20 m). Downstream on the Mississippi, thousands of workers manned the levees, and each of them held, thus preventing the overflow that might have recreated the 1927 flood disaster (see p. xx).

The Red Cross commandeered 5,400 boats to conduct rescue work on the Ohio. Hundreds of people were scooped from the roofs of crumbling houses, 300

emergency hospitals were set up, and 3,600 nurses arrived to staff them. It would take more than $25 million in relief funds to administer to the injured and displaced alone.

UNITED STATES
OHIO RIVER VALLEY
March 1997

Torrential rainstorms and tornadoes caused the flooding of the Ohio River in March of 1997. Hundreds of communities in the Ohio River Basin were partly or totally destroyed. Thirty-nine people died; thousands were made homeless. The cost of the flood was estimated at over $3 billion.

The 981-mile-long (1,569.6 km-long) Ohio River is born at the confluence of the Allegheny and Monongahela Rivers in Pittsburgh, Pennsylvania. From here, it flows generally northwest, then southwest and west through eight states until it ultimately enters the Mississippi River at Cairo, Illinois. Prone to spring flooding in ordinary times, it was turned, by a late winter storm in early March of 1997, into a raging, demolishing torrent.

From Pennsylvania west, through Ohio, then south through West Virginia, Kentucky, Tennessee, Arkansas, Texas, and Mississippi, floodwaters from the Ohio roared over embankments, inundating towns and villages to their rooftops, and threatening the cities of Cincinnati, Louisville, and eventually, as the Mississippi swelled from the torrent of the Ohio, New Orleans.

Nearly 3,000 people were evacuated near Louisville, where city workers slid concrete flood walls into place in an effort to hold back the Ohio. On the outskirts of Cincinnati, suburban villages were buried under lakes that surrounded and sometimes filled the structures under them.

High winds and rains pounded the towns and cities in the Ohio basin, and in Arkansas, tornadoes grew and migrated through the area, devastating farms and towns. On March 1, downtown Arkadelphia, Arkansas, was completely destroyed by a tornado, which killed six people.

Below Louisville, the village of Lebanon Junction was consumed by the Rolling Fork River, which washed over its banks at a rate of two feet an hour. It was the area's worst flooding in three decades, and a large percentage of the village's residents were forced into a Red Cross shelter. Mrs. Gregory Kelly described her plight to a reporter. "A fella came back just awhile ago and said all you could see of our house was about a foot of the roof," she said. "There were 7 copperheads lying on top of the water there, so he left."

In Brown County, Ohio, two days of relentless rain turned the Ohio into furious brown waves that ate away at the roads around the town of Greenfield. A police officer nearly drowned when his car was swept away from a collapsing road, but he broke a window and escaped. Fellow police officers found his car later, two miles downstream.

By March 6, 2,000 people were in shelters in 16 counties in Kentucky, and the Ohio was cresting in Cincinnati. Streets in the city became fast-moving streams as the river overflowed its banks, 12.6 feet (3.84 m) above flood stage. Industrial areas near the river were inundated and low-lying residential areas were evacuated.

South of Cincinnati, Louisville had begun building a flood wall in 1948, and had been extending it and adding various flood control systems ever since. The river crested on March 7 at nearly 16 feet (4.87 m) above flood stage, the highest in 33 years. But the flood walls, bolstered by 90,000 sandbags, held, for the most part. Only a few streets near the waterfront were submerged up to the tops of their streetlights.

As the flood crest moved west and south, deer and other wildlife were seen swimming in large groups to escape the flood. "Every kind of critter you can think of is trying to get out of here right now," Duke County executive James O. McCord told a reporter.

Tributaries of the Ohio flooded as rapidly and caused as much damage as the main river. Falmouth, on the Licking River in western Kentucky, was particularly hard hit. When the flood first visited it, filling up the village faster than anyone had thought possible, the entire town was evacuated, and remained so for six days because of gas leaks and downed power lines.

When the residents of Falmouth returned, they found incredible devastation. The village looked as if a tornado or a bomb had ripped through its houses, gasoline stations, and restaurants. Railroad tracks were torn up; cars and tractor-trailers were flung around like toys; houses were pried off their foundations and dumped hundreds of yards from their original sites. One house ended up on top of two ruined cars in a football field.

"We knew how nasty the Licking River could get," Lark O'Hara, the county sheriff noted, "but nobody had expected it to turn into a cold blooded killer. I don't ever need to see another disaster movie," the sheriff concluded. "I've just lived through one."

Water treatment plants in scores of towns and cities were overcome, and floating raw sewage became a health hazard. Returning residents, cleaning up the muck in their houses, were warned to boil drinking water, care for any wounds, and get diphtheria and tetanus shots.

The cleanup began a week after the first rains fell. In West Point, Kentucky, residents hosed down their bedrooms and threw out ruined furniture. Red Cross volunteers drove around the village, handing out brooms, mops, and bleach. The Salvation Army offered breakfast, lunch, and dinner.

By that time, the Ohio had emptied into the Mississippi, and it in turn was bearing down on New Orleans. But there, the Army Corps of Engineers, for the first time in 14 years, was able to divert the floodwaters away from the city and into Lake Pontchartrain. The opening of the spillway, built as a result of the 1927 flood (see p. 189), saved New Orleans from being flooded, but it distressed shrimpers and other fishermen who feared that the intrusion would drive brown shrimp and crabs out of the lake and destroy oyster beds in Lake Borgne. It did not.

The flood killed 39 people and caused over $3 billion in damage. Thousands of homes were destroyed and their inhabitants were rendered homeless.

UNITED STATES
OREGON
June 14, 1903

Three hundred twenty-five drowned and one-third of the town of Heppner, Oregon, was destroyed when a flash flood, the product of a hail- and rainstorm, transformed Willow Creek into a killer torrent on June 14, 1903.

A lightning-fast hail- and rainstorm that lasted a mere half hour triggered a flash flood at 4 P.M. on June 14, 1903, in the foothills of the Blue Mountains of Oregon, near Willow Creek. An enormous amount of rain fell in those 30 minutes, enough to turn Willow Creek into a raging, churning dynamo that generated 25-foot-high waves and sent them crashing into nearby Heppner.

For almost an hour, the floodwaters raged through the town, sweeping away 200 buildings—one-third of the town—and drowning 325 people. In the short space of an hour and a half, nature changed the face of an entire community, caused $250 million in damage, and dramatically altered the population of this formerly peaceful place.

UNITED STATES
OREGON (AND WASHINGTON, CALIFORNIA, IDAHO, AND NEVADA)
December 1964

Fifty were killed and over 12,000 homes were destroyed in the "Christmas Week" floods that roared through Oregon in December 1964 and, to a lesser degree, affected Washington, California, Idaho, and Nevada.

The Christmas Week floods of 1964 affected five states—Oregon, Washington, California, Idaho, and Nevada, destroying over 12,000 homes and killing 50 people. But by far the greatest damage and loss of life occurred in Oregon, where the cause of the flood originated.

A cold front from Alaska brought plummeting temperatures and a record snowfall in the Cascade Mountains and the surrounding countryside during the middle of December. Shortly before Christmas, meteorological conditions reversed, and sharply rising temperatures produced a sudden thaw. More than a foot of mountain snow melted each day for three straight days, and this, compounded by steady rains, swelled the Willamette, John Day, and Deschutes Rivers considerably beyond their floodstages.

During Christmas week, paper mills were wiped away, and tides 75 feet (22.86 m) above normal overflowed dams, crumbled roads, and uprooted bridges. The newly constructed 160-foot (48.768-m) bridge across the John Day River near Estacada was hit while holiday travelers were crossing it. With one horrendous roar, it split in the middle, hurtling one car and its driver into the floodwaters below and thoroughly terrifying other motorists and their passengers who narrowly escaped the same fate.

Civil Defense crews and Red Cross personnel were flown into the stricken areas of the state by helicopters on Christmas Day. It would be after New Year's before major cities like Portland and Corvallis would dry out.

UNITED STATES
PACIFIC NORTHWEST
December 1997—January 1998

The holiday floods of December 1997 and January 1998 in the Pacific Northwest, brought on by a combination of heavy rains and a resultant snow melt in the mountains, killed 23 people, destroyed up to 2,000 homes, and rendered 125,000 homeless.

It began with a one-two punch of snowstorms that dumped two feet of snow in western Washington State during the third week of December 1997. Almost immediately, temperatures rose and both Washington and Oregon were hammered by giant rainstorms, which unleashed mudslides and avalanches that blocked roads, stranding holiday travelers. And then the worst occurred: The rain began to flush away the snow, which caused pervasive flooding.

The rain continued, fed by a low pressure system that stretched practically from Guam to the northern Pacific Coast of America and packed winds of 40 to 50 MPH with gusts up to 100 MPH. Nicknamed by meteorologists the Pineapple Express because it had originated near Hawaii, it was really a series of disturbances imbedded in the jet stream.

"We finally got our natural disaster," said Seattle resident Wayne Rawley. "The whole Midwest flooded, there were earthquakes in California and people froze to their floors in New York, and we just sat back and watched it—until now."

The temperature continued to rise, exacerbating the situation. On New Year's Day in Seattle it was 54 degrees; in Walla Walla it was a balmy 77, with torrential rain. In Seattle, giant sinkholes began to open up in several roads, and more than 90,000 people were suddenly plunged into darkness when the power was knocked out by gale winds collapsing power lines.

In Northern California, rivers and streams began to overflow their banks, flooding homes and businesses. Flood warnings were issued for every river north of San Francisco. In Tehana, 300 people were forced to leave their homes when the Russian River reached a record crest of 49 feet (14.9 m).

Fifty thousand residents of Tuba City and Marysville, which face each other on opposite sides of the Feather River, were evacuated. The river blasted its way through a levee, submerging surrounding orchards and farmland.

In Del Norte County, in the northwest part of the state, the Klamath River carried off the Indian-run Golden Bears Casino and dumped it onto a nearby campground.

On January 2, the melting snow in the Sierra Nevadas dumped tons of water into Yosemite National Park, trapping 2,500 tourists and workers. Reno, whose casinos boast that they never close, closed. Up to seven inches of rain sent the Truckee River raging through downtown, inundating motels, restaurants, and wedding chapels. Eventually, the Truckee crested at 14.7 feet (4.48 m), a new record. The Reno-Tahoe airport was closed down, as was the Mustang Ranch brothel, two other firsts.

In Idaho, mud slides washed away huge sections of roads, including a 1,000-foot (304.8-m) stretch of Route 95, the state's only north-south highway.

As the rains and the floods continued, over 100,000 people were forced to leave their homes in California alone. Helicopters plucked stranded farmers from rooftops and sunken pickup trucks.

Major highways and rail lines were blocked in California, Idaho, Nevada, Oregon, and Washington. In the Sierra Nevada, boulders the size of a suburban home plummeted down on one major highway, and Highway 1, the scenic north-south route along the California coast, was cut in four places.

The Diablo Canyon nuclear power plant declared an "unusual event" (the lowest alert) when a mud slide blocked the main road leading to the plant.

Throughout the area, muddy floodwaters covered fields. The identity of communities was established by the rooftops visible above the surface of the flood. On the north fork of the Mokelumne River in San Joaquin County, California, floodwaters smashed through a levee and swept a marina and 230 boats downstream, where they were caught up on the pylons of a bridge.

By January 4, the weather finally cleared, raising hopes that the intensity of the floods would subside. But it would take some time. The rivers in California continued to surge as upstream dams were opened to make room in reservoirs for still more runoff water, which in turn continued to roar down the mountains.

There was thick muck everywhere. From 1,500 to 2,000 homes were destroyed completely. Over 125,000 people were forced from their homes. Twenty-three people died. Union Pacific railroad tracks were washed out in 40 places in a 30-mile stretch between Lakehead and Dunsmuir in California. Newly planted vines in the vineyards of the Napa and Sonoma Valleys were washed away.

But the vines would recover, and the older growths were unharmed. It was in the human arena that the damage was most severe, and the worst of it was in trailer parks and poor neighborhoods. People with little appeared to have lost the most.

"Seems like that's usually the way, doesn't it?" one of the survivors in one of the trailer parks observed to a reporter.

UNITED STATES
PENNSYLVANIA
JOHNSTOWN
May 31, 1889

Over 2,500 died and thousands were made homeless when the South Fork Dam collapsed, and the Conemaugh River roared into Johnstown, Pennsylvania, on May 31, 1889.

"Run for your lives! The dam has burst!" was a line in a joke in Johnstown, Pennsylvania, before May 31, 1889. After that date, it became a grim reminder of a monumentally cataclysmic day in which, in the space of one hour, over 2,500 people lost their lives as 4.5 billion gallons of water weighing 20 million tons (18.1 million tonnes) burst through a sorrowfully neglected and obsolete earthen dam above that Pennsylvania steel city. The water laid waste to everything and everybody in its thundering descent.

Johnstown is located in the flat land where the Little Conemaugh River and Stony Creek come together to form the Conemaugh River. In 1824, 14 miles (22.5 km) upstream of Johnstown, engineers hired by the state of Pennsylvania began a 12-year project that would create a system of canals, between Pittsburgh and Johnstown. The system would allow barges to move between the cities and be the last link in a chain of canal and rail conduits between the coal mines, iron deposits, and limestone quarries near Jamestown and the commerce hub of Pittsburgh.

Thus, work began on the South Fork Dam, in its time, the largest earthen dam in the world, holding back the largest man-made lake in the world. Looming 100 feet (30.4 m) above the old creek bed, its dirt atop an arched stone culvert rose a mere 72 feet (21.9 m) in height. However, it measured 272 feet (82.9 m) thick at its base, 10 feet (304 m) at its top, and stretched 931 feet (283.8 m) from rock wall to rock wall. The culvert housed five cast-iron pipes, each two feet in diameter, which would, through the regulation of valves in a nearby wooden tower, discharge water from a two-mile-long, mile-wide 70-foot-deep (21-m-deep) reservoir that contained approximately 5 billion gallons of water. In addition, a spillway 72 feet (22 m) wide was cut nine feet deep into the rock abutting the eastern end of the dam.

From the time the dam was finished, it was obsolete. By this time, canal-rail links had been replaced by purely rail links through the Allegheny Mountains, and so, in 1879, Pennsylvania sold the dam and its reservoir for $2,000 to promoter Benjamin F. Ruff. Ruff then incorporated the South Fork Fishing and Hunting Club, and sold $2,000 memberships to, among others, Andrew Carnegie, Andrew Mellon, Henry Clay Frick, and Philander C. Knox. Cottages and a clubhouse were built, and makeshift repairs were made to the deteriorating dam. Hemlock branches, tree stumps, and straw were used to patch up holes. The discharge pipes were removed, and the outlet was filled in, thus leaving only the spillway to prevent floods from topping the dam, whose crest was lowered to accommodate a 20-foot (6.09-m) roadway. The spillway itself was blocked by

The wreckage left behind after the South Fork dam burst above Johnstown, Pennsylvania, on May 31, 1889 (Library of Congress)

debris when a fish-catching, iron mesh net was strung across it.

The residents of Johnstown were not altogether unaware of the dangers from flooding that these modifications posed. Every year there was flooding of some sort, including the famous "pumpkin flood" of 1820, when thousands of pumpkins floated into the city from farms upstream.

Thus, in 1880, Daniel J. Morrell, the president of the Cambria Iron Company, retained an engineer to inspect the dam. His findings were disquieting, but nevertheless Ruff stated publicly, "You and your people are in no danger from our enterprise."

Nine years later, at the end of May 1889, an unusually heavy storm struck the mountain region of western Pennsylvania, saturating 12,000 square miles (19,312 sq. km) of earth that was already soaked by the sudden melting of 14 inches of snow the month before. Rivers were full to their banks, and the lake above Johnstown had filled to the point at which tons of water were already hurtling through the partially blocked overflow channel. (It was later calculated that up to 75,000 gallons of water per second were entering the lake by way of swollen feeder streams and the rainfall. The spill was equipped to handle only 45,000 gallons per second. And that was when it was entirely clear of debris.)

On the morning of May 31, 1889, a crew of 30 men, overseen by South Fork Fishing and Hunting Club president Colonel E. J. Unger and accompanied by engineer John G. Parke, attempted to shore up the dam. "Rip-rap"—the small stone wedges that had supported the dam—were beginning to fall away from the top, and no amount of shoveling of more earth onto the dam or attempts to clear away the obstructions in the fish gratings was going to forestall the inevitable.

At 11 A.M., the 23-year-old engineer mounted a horse and began a solitary ride through the valley to warn residents to evacuate.

At 3:10 P.M., the dam gave way. According to one witness, "When the dam broke, the water seemed to leap, scarcely touching the ground. It bounded down the valley, crashing and roaring, carrying everything before it." A wall of water 150 feet (45.71 m) high rushed toward the village of South Fork at an estimated 50 MPH. The village was wisely built on high ground, and only 24 houses were swept away.

But now, as the waters plunged forward, they picked up flotsam—houses, trees, the stones of a railroad bridge—a few miles downstream from South Fork. A freight train waiting to cross the bridge was abruptly shifted into reverse by its engineer, John Hess, when he received a telegraphed warning of the water's arrival. He backed the train six miles (9.61 km) downstream to East Conemaugh, just ahead of the flood.

While the train had been backing up, the floodwaters had smashed into the village of Mineral Point, wiping it out, killing 16 persons, and sending its Methodist Church spinning and hurtling downstream.

In the Pennsylvania Railroad yard and roundhouse at East Conemaugh, 33 Consolidation-model locomotives, weighing, with their tenders, 170,625 pounds apiece, were suddenly picked up—along with 315 freight and 18 passenger cars—and flung about as if they were toys. One engine was later found buried in rock and sand nearly a mile from its original location.

After gutting East Conemaugh, the now lethal debris-carrying wave careened through the town of Woodvale, uprooting 800 buildings, killing close to 1,000 of its residents, and smashing 255 houses to tinder. Hundreds of people were seen hanging from pilings and rooftops as their improvised rafts were driven furiously downstream.

Moments later, the water hit the steel mills of the Gautier Wireworks, bursting its boilers and sending a towering plume of steam hundreds of feet into the air.

At a little after 4 P.M., the enormous wave, containing the remains of houses and factories, railroad cars and engines, trees, telegraph poles, steel rails, rocks, and the bodies of hundreds of people and animals, reached the outskirts of Johnstown. It would take 10 minutes to span the city and in those 10 minutes more than 1,000 people would die. "In an instant," according to one reporter, "the deserted streets became black with people running for their lives. The flood came and licked them up with one eager and ferocious lap."

Within seconds, enormous stone buildings collapsed under the assault of the wave, trapping hundreds who doubtless thought these buildings would offer them safe haven. The German Lutheran Church and YMCA, both stone monoliths, crumbled and fell in on themselves. All of the municipal buildings were flattened. The Hulbert House, an elegant brick hotel, fell, drowning all 60 of its guests, who were trapped on a third floor staircase leading to the roof and supposed safety.

Families were decimated. Parents saw their children slip, one by one, from roofs into the water. Others were miraculously saved when they transferred themselves—or were transferred by the swirling currents and the persistence of fate—from one floating cluster of wreckage to another.

Finally, just below Johnstown, the floodwaters encountered a stone railroad bridge standing 32 feet above the normal river level. It remained standing, and the mountain of debris and river piled up against it. One huge, heaving mass of ruined homes, parts of the landscape and human beings, both alive and dead, rose to a height of 30 feet (9.14 m) and extended back into Johnstown for 30 blocks.

Coal stoves containing live coals had been picked up by the flood; a railroad car loaded with lime caught fire when its contents reacted chemically with the water. By 6 P.M., the mountain piled against the bridge had become a massive pyre of flame. People caught in the mass struggled to free themselves.

"It reminded me of a lot of flies on flypaper, struggling to get away, with no hope," said one observer. Another witness reported, "As the fire licked up house after house and pile after pile, I could see men and women kiss each other goodbye and fathers and mothers kiss their children. The flames swallowed them up and hid them from my view, but I could hear their shrieks as they roasted alive."

Two hundred people were pulled from the debris; countless others perished. The next day, more survivors were dug from under the rubble; others were found in the wreckage of their homes. One five-month-old baby was discovered nearly a hundred miles downstream in Pittsburgh, floating on the floor of a ravaged house. For weeks, bodies floated downstream and surfaced, threatening a spread of disease.

Help came quickly, and lasted through a long, horrible summer. Among the hundreds who arrived was Clara Barton, one of the founders of the American Red Cross. "The Angel of the Civil War Battlefields," then 68 years old, worked tirelessly for five months, distributing $500,000 worth of money and materials and supervising the building of three large structures to provide shelter to the thousands of homeless refugees.

All summer long, the city rehabilitated itself. Over $3 million in aid and workers filtered in, and Johnstown gradually took shape again.

UNITED STATES
SOUTH DAKOTA
RAPID CITY
June 9, 1972

Torrential rain and the collapse of two dams brought about the flooding of Rapid City, South Dakota, on June 9, 1972. Two hundred sixty-three people died; hundreds were injured; 1,200 homes were destroyed.

A flash flood that devastated Rapid City, South Dakota, and collapsed two nearby dams began with an unexpected 10-inch (25.4-cm) rainfall on the night of June 9, 1972. By 9 P.M., Rapid Creek, which ran directly through the heart of the city, had swollen from a creek into something resembling a large river. Twenty feet wide in the afternoon, it was 400 feet (121.9 m) wide after 9 P.M.

Within minutes, undermined buildings began to crumble and wave-driven waters started to smack against telephone poles and wooden dwellings, which collapsed like piles of matchsticks. Most residents of the city were either in bed or lounging before their television sets, which had reported no warnings.

Harold Higgins, a local reporter, later told the *New York Times*, "I was standing in the middle of [a] road when a four-foot bank of water came down the creek." Trailers shot by him like racing speedboats. "It [was] like a war zone," he went on, ". . . fires all over the place, and nothing [could] be done about it because the city [was] cut in half by the flooding of Rapid Creek."

Meanwhile, the dams at Canyon Lake and Deerfield burst, and the nearby village of Keystone was swept entirely away by the rushing waters. Not one building remained standing, and the death toll was high.

While residents were still huddled on rooftops or clung to trees and high promontories of land, 1,800 South Dakota National Guardsmen, quartered in Camp Rapid on the western edge of the city, moved in with rescue vehicles.

For days afterward, bodies floated through the streets and were unearthed when debris was sifted. Refrigerator trucks were used to pick up and store the bodies to prevent the spread of disease.

Shortly thereafter, President Richard Nixon declared Rapid City and its surroundings a national disaster area, which brought federal funds into the stunned area. When the casualties were counted, 236 people were reported dead, 1,200 homes were destroyed, another 2,500 were damaged, and 5,000 automobiles were wrecked beyond use. Hundreds of people were injured and the damage was reckoned at $100 million.

UNITED STATES
TEXAS
GALVESTON
August 1990

(See HURRICANES.)

VENEZUELA
NORTH
December 1999

The worst natural calamity to strike Latin America in the 20th century thundered down from the Ávila mountain range in northern Venezuela on December 15, 1999. Monster mudslides, loosed by three days of intense rain, buried villages and invaded city streets. Slum sections of cities and entire villages disappeared under the mud, which buried or swept out to sea at least 30,000 people. Final figures would never be ascertained, since thousands of bodies remained buried or missing. Over 400,000 people were rendered homeless.

Torrential rains, attributed to the La Niña phenomenon, struck 10 states in the northern part of Venezuela on December 13 and 14, 1999. It was the beginning of a cataclysmic disaster, the worst of the century in Latin America.

On December 15, most Venezuelans ignored the continuing rain and went to the polls to vote on a new constitution that would add to the powers of President Hugo Chávez. But throughout the entire Caribbean coastal area, destruction was beginning. The rain pelted the country mercilessly, flooding low-lying areas, washing out roads and bridges, collapsing buildings, but, most dangerously, turning solid ground into malleable mud. And here is where the worst tragedy took place.

Early in the morning of the 16th of December, the mudslides began. They started in the Ávila Mountains, north of Caracas. "The mountain [El Ávila] began to shake and tremble," a survivor told a *New York Times* reporter, "There was this terrible noise, and then all of a sudden a gigantic wave of mud and debris was upon us and we could do nothing but scramble up the side of the mountain and pray that we would be spared."

The mudslide was described by those who witnessed it as being 20 feet (6.09 m) high as it thundered down mountains and into valleys, picking up trees, livestock, houses, and human beings and either ripping them asunder or hurtling them, at express train speed, down the mountains.

"The mountain came down and the walls of our house broke. We lost everything. The house opened in two parts," another survivor told a Reuters reporter. The survivor and his sister escaped from their home and watched in horror as their neighbors' homes were lifted on a wave of mud and swept away.

The worst hit areas, as usual, were those occupied by the poor in shanties that barely withstood the normal assault of the elements but which folded into matchstick fragments when the mudslides crashed into them. One of the worst struck areas was Blandin, on the outskirts of Caracas.

In La Guaira and Tanaguarena on the coast and in hundreds of villages in between, it was the same sad story. Where there had been streets there were now rivers of mud, with tree trunks, refrigerators, automobiles, and buses embedded in the fetid mess. Streams containing clothing, furniture, household appliances, and Christmas decorations roared downstream.

Teams of rescuers began work immediately, but it was a grim task. The stench of death became their guide, but what they found was not what they had hoped for. "We are recovering hardly any whole corpses," Captain Vicente Campos Ron, the head of the search squad, reported. "What we are finding is a head here, an arm or a leg there, which makes it hard to identify the bodies or calculate how many people have actually died in this catastrophe."

A sports stadium in Caracas was turned into a shelter, and all over the countryside, improvised shelters received hundreds, then thousands of refugees, each bearing a story. Joanna Saavedra, a 25-year-old mother, related one with a hopeful ending to an aid worker for Disaster Relief. When the floodwaters started to rise outside her home, she grabbed her son and tried to escape. But she was able to take only one step from her door. Instantly, the floodwaters swept her off her feet and carried her past her home. Holding her three-year-old son in one hand, she grabbed for a tree, where she hung for the next 30 minutes. "The massive water was so strong," she said. And then, "almost out of nowhere" a firefighter appeared to rescue them. "The fireman made miracles happen," Mrs. Saavedra concluded.

The total of homeless climbed rapidly to 150,000, and the government immediately made plans to resettle these survivors elsewhere. A large percentage of them refused. They had built their shacks in unwise areas only because they were too poor to build them elsewhere, and habit was now guiding them.

For the remainder of December, troops and relief workers plowed through head-high mud and rubble to dig up bodies and pieces of bodies. Paratroopers patrolled ghost towns in the state of Vargas, pinning down looters emptying homes that their owners had evacuated when the mudslides began. In some places, troops warded off gangs of youths ransacking unoccupied apartments, but still, some escaped with television sets and other goods strapped to their backs.

It was clear, as December ended and the sun came out and baked the mud, that a 60-mile (96.56-km) stretch of Vargas state on the Caribbean and Caracas on the other side of the Ávila mountain range, were becoming graveyards for what was estimated as 30,000 victims of the mudslides. No firm fatality figures would ever be reached, since it was assumed that most of the bodies were either washed out to sea or lay buried in the mud, never to be found.

The number of homeless climbed to 400,000 by the beginning of 2000. In Vargas state, 51 percent of the population had lost their homes. In the Federal District in Caracas, 27.7 percent of the population had been rendered homeless. The total of dispossessed in the country reached 140,000.

Blame for faulty construction in the slum areas, which bore the greatest brunt of the tragedy, was placed on previous governments, but that was academic. Now, the long and arduous and expensive task of recovery had to begin. It progressed slowly. A year later, promises of new living quarters in safe agricultural areas for refugees had yet to be fulfilled. And the $20 billion that was estimated as the cost of the cleanup and recovery was a long way from being found.

THE WORLD
2400 B.C.E.

Scientific evidence indicates that the Great Deluge of the Old Testament probably occurred in approximately 2400 B.C.E., and covered the known world at that time.

Of course, the date of the Great Deluge and Flood of the Book of Genesis must be approximate. There are no written records of the precise day and hour, nor is there universal agreement about the year or the location. But scientific evidence gathered within the past 50 years has confirmed that such a flood actually did take place.

Consider the synchronicity of myths from various civilizations. The Chinese have a story of a universal deluge that, according to some historians, must have occurred at the time of the biblical deluge. Although there is no evidence of widespread disaster in this account, there is evidence that there was flooding on an enormous scale.

Peruvian records speak of a huge inundation occurring before the time of the Incas. The story parallels that of Noah, except that in it, six people are saved from the flood, and they owe their lives to a float. Geological evidence likewise proves that there have been many cataclysmic inroads of the sea into this region.

The coastal area of Chile, near Concepcion, has a long history of flood disaster, and the native Araucanian Indians have a flood legend. Earthquakes and sea waves permeate this story, and, in recorded history, earthquakes and sea waves repeatedly struck the area, including the devastating one of 1751. What is particularly fascinating is that the tribal legend parallels the Mosaic Deluge in that only a few people were saved by taking refuge on a mountain-top, although no ark is mentioned. Even today, when earthquakes occur in Chile, people of Indian descent flee to the mountains in fear that the sea will again cover the world.

There is also archaeological evidence. In 1872, George Smith, a young official in the British Museum, discovered a portion of the story of the Great Deluge and Flood on a broken tablet taken from the library at Nineveh. It spoke of a ship docking on the mountain, and a dove being set forth.

Other excavations at Nippur in the lower Euphrates valley yielded a library of over 20,000 tablets, made before 2000 B.C.E., and in one of the Sumerian texts, an Assyrian version of the Flood was discovered. In it, there is a hero who is a priest and king, a godly man very much like Noah. In it, God warns the hero, and a ship is mentioned.

The most dramatic archaeological find, however, was uncovered in 1929 near the pre-2000 B.C.E. dead city of Ur on the Euphrates River in what is now Iraq. A 62-foot (19-m) section was cut at Ur that showed some evidence of having been occupied by humans. In one part of the section, there is a 10-foot stratum of clay or sand in which there is no evidence of human life. Below this, there are remains of a human civilization that are consistent with remains found above the clay or sand interruption. Normally, it would take a very long time to accumulate 10 feet of silt, but the comparative thicknesses and the evidence below the belt of sand and clay show that this took place in a very short time.

Add to this the evidence from the Sumerian civilization, which occupied Mesopotamia from 5000 B.C.E. forward, in which records indicate that the period around 2400 B.C.E. was one of abnormally heavy rainfall in West and Central Asia.

PLAGUES AND EPIDEMICS

....

THE WORST RECORDED PLAGUES AND EPIDEMICS

* Detailed in text

Africa
(1799) plague
Angola (2005)
Ebola

Asia
(2005) Japanese encephalitis

Brazil
(1560) smallpox epidemic

China (and India)
(1910–13) plague

Constantinople
(746) plague

Egypt
(1792) plague

Europe
(558) bubonic plague
* (1348–1666) "Black Death"—
"second pandemic"
* (1493) syphilis epidemic
(1826–37) cholera epidemic

France
* (1518) St. Vitus's dance epidemic
(1632) plague

Great Britain
(430) plague
(1235) plague
(1625) plague
* London (1665) plague
* London (1832) cholera epidemic

Greece
* (431 B.C.E.) "Plague of
Thucydides"

Hungary
Budapest (1544) typhus epidemic

India
(1812) plague
(1876) (see FAMINES AND
DROUGHTS)
* (1896–1907) plague
(1921–23) plague
(1926–30) smallpox epidemic
(1958) smallpox, cholera epidemic
(1974) smallpox epidemic
(2005) Japanese encephalitis

Ireland
(1172) plague
(1204) plague

Italy
(1340) plague
* (1505–30) typhus epidemic
(1656) plague
Naples (1672) plague

Palestine (and Egypt)
(1097) plague

Panama
(1903) yellow fever epidemic

Rome
* (452 B.C.E.) epidemic pestilence
(169 C.E.) plague
(250–265) plague

(452) plague
* (541–590) plague—"first
pandemic"

Russia
* (1812) typhus epidemic
(1917–21) typhus epidemic
Smolensk (1386) plague

Syria
(1760) plague

United Kingdom (and western Europe)
* 1986–Present (mad cow disease)

United States
* California (1899) plague—"third
pandemic"
Massachusetts (1721) smallpox
epidemic
* Pennsylvania (1793) yellow fever
epidemic

West Indies
(1507) smallpox epidemic

The World
* (1200 B.C.E.) plague
(767 B.C.E.) plague
* (1846–60) cholera epidemic
* (1855–Present) plague
1889 (influenza epidemic)
* (1918–19) influenza epidemic
* (1959–Present) Avian (Bird) Flu
* (1980–Present) AIDS epidemic
* (2002–05) SARS

CHRONOLOGY

* Detailed in text

1200 B.C.E.
* The World (plague)
767 B.C.E.
The World (plague)
452 B.C.E.
* Rome (epidemic pestilence)
431 B.C.E.
* Greece ("Plague of Thucydides")
169 C.E.
Rome (plague)

250–265
Rome (plague)
430
Great Britain (plague)
452
Rome (plague)
541–590
* Rome (plague—"first
pandemic")
558
Europe (bubonic plague)

746
Constantinople (plague)
1097
Palestine (and Egypt) (plague)
1172
Ireland (plague)
1204
Ireland (plague)
1235
Great Britain (plague)

1340
Italy (plague)

1348–1666
* Europe ("Black Death"—"second pandemic")

1386
Smolensk, Russia (plague)

1493
* Europe (syphilis epidemic)

1505–30
* Italy (typhus epidemic)

1507
West Indies (smallpox epidemic)

1518
* France (St. Vitus's dance epidemic)

1544
Budapest, Hungary (typhus epidemic)

1560
Brazil (smallpox epidemic)

1625
Great Britain (plague)

1632
France (plague)

1656
Italy (plague)

1665
* Great Britain (plague)

1672
Naples, Italy (plague)

1721
Massachusetts (smallpox epidemic)

1760
Syria (plague)

1792
Egypt (plague)

1793
* Pennsylvania (yellow fever epidemic)

1799
Africa (plague)

1812
India (plague)
* Russia (typhus epidemic)

1826–37
Europe (cholera epidemic)

1832
* London, Great Britain (cholera epidemic)

1846–60
* The World (cholera epidemic)

1855–Present
* The World (plague)

1876
India (see FAMINES AND DROUGHTS)

1889
The World (influenza epidemic)

1896–1907
* India (plague)

1899
* California (plague—"third pandemic")

1903
Panama (yellow fever epidemic)

1910–13
China (and India) (plague)

1917–21
Russia (typhus epidemic)

1918–19
* The World (influenza epidemic)

1921–23
India (plague)

1926–30
India (smallpox epidemic)

1958
India (smallpox, cholera epidemic)

1959–Present
* The World (Avian [Bird] Flu)

1974
India (smallpox epidemic)

1980–Present
* The World (AIDS epidemic)

1986–Present
* United Kingdom and Western Europe (mad cow disease)

2002–05
* The World (SARS)

2005
Angola, Africa (Ebola)
Asia, India (Japanese encephalitis)

PLAGUES AND EPIDEMICS

Of all natural calamities, plagues and epidemics affect the human population of the Earth most directly and most devastatingly. And although other natural disasters may be more spectacular and abusive to buildings and landscapes, plagues and epidemics bear a disquieting resemblance to the neutron bombs that scientists dangled tantalizingly before the public in the early days of the atomic age, bombs which destroyed people but left cities standing. Plagues and epidemics also destroy people. Only people.

And they have a disturbing longevity about them. Tornadoes are over in an instant; volcanic eruptions in minutes; floods, earthquakes, and hurricanes in hours. Only famine and droughts, in some circumstances, challenge plagues and epidemics in longevity.

Strictly speaking, there is only one transmitted disease to which the term plague can be rightly applied: bubonic plague, carried by the *Rattus rattus* of antiquity. This disease spread by rodents is graphically chronicled in the cuneiform writings of Babylonia.

The plague bacillus, *Pasteurella pestis*, is spread among rodents by *Xenopsylla cheopis*, a flea that inhabits the fur of these rodents. These fleas also like to feed on the blood of humans, and thus, most plagues involve mass movements of rats, which invade the environs of man. When the host rat dies, its fleas, in search of food, attack humans.

When this vector bites, and infection occurs, there is a swelling of human lymph nodes, causing buboes (which is derived from the Greek word *boubon,* meaning groin or swelling in the groin). If the victim does not succumb to the primary infection, the organism often invades the lungs causing pneumonic plague, a secondary infection. Each results in a high fever, which is followed by delirium, vomiting, bleeding, and eventually death.

For centuries, bubonic plague roared through the inhabited world with mysterious velocity. In fact, the connection between fleas, rats, and plague was not made until the end of the 19th century. Even though Babylonian writing reported large quantities of rats, it did not link them to the pestilence that occurred at the same time. Nor did medieval historians writing of the Black Death make this connection.

It is only through informed hindsight, through connecting large migrations of rats and fleas, that medical historians have been able, in the 20th century, to put together the pattern, history and magnitude of plagues.

Epidemics can arise from a number of sources and are transmitted in a variety of ways. Again, before modern medicine it was universally accepted that plagues and epidemics alike were transmitted either through the intervention of the gods or through miasmas, or vapors. Thus, medieval families would often seal their homes from the inside, while the infection raged unchecked within their tightly packaged dwellings, killing them all.

One of the most persistent epidemics is cholera. A water-borne disease that is spread through sewage-contaminated rivers or urban water supplies, it is characterized by profuse diarrhea, vomiting, muscular cramps, dehydration and ultimate collapse. Rampant in the 19th century, it is still with us today.

Other epidemics such as yellow fever, syphilis, poliomyelitis, influenza, encephalitis, and puerperal fever (a hospital-related epidemic disease) have been largely brought under control through the efforts of modern medicine. Still, wherever starvation, privation or unsanitary conditions prevail, epidemics can occur.

A word, however, about what we term "epidemic." Each year, local television and radio stations broadcast illustrated maps and lectures about certain epidemics—of influenza, for instance. These localized epidemics are certainly worth recording, and will, alas, be worthy of note for the foreseeable future. The influenza virus, to cite one, has a vicious habit of adapting itself to new vaccines. As soon as a vaccine is developed to stem the tide of one strain, another develops. The only bright note in this frustrating picture is that in modern times, at least, these epidemics tend to limit and localize themselves.

Two current epidemics, however, do not limit and localize themselves, and they have sinister possibilities. The first, Lyme disease, often evades detection in its early stages by disguising itself behind symptoms of lesser afflictions—flu, allergies, etc. It is also a young epidemic, whose impact has yet to be felt in a wide geographical sense. AIDS, the second pervasive epidemic of today, which has entered its adolescence, has the potential, unless a cure is soon found, of escalating into one of the most devastating pandemics in all of recorded history.

And that final term, "pandemic" is worth a close examination, for it has given rise to a certain amount of confusion and conflict among medical historians and scientists. A pandemic is generally defined as an epidemic or plague that affects an enormous number of people across a wide geographic area.

The first of the three great plague pandemics, is generally thought to have begun in the 15th year of the Roman emperor Justinian I (thus its sobriquet, "The Plague of Justinian"). It raged from C.E. 531 until roughly C.E. 650, approximately 100 years.

The second pandemic, also agreed upon by most historians, is that of the Black Death. It began in 1348 and lasted over 300 years, until 1666 and the Great Fire of London—although some few historians seem to think that its pandemic stage lasted only four years.

The third pandemic began in China in 1892, and, according to some historians, ended 15 years later. To others, however, it continued until 1959. And to others, notably Charles T. Gregg in his compelling case study, *Plague!,* published in 1978, it is still with us. "Plague is now neighbor to us all," he states unequivocally, "for many, only a few hours away, at most a day's journey distant. Plague is a willing handmaiden to famine and war; these threaten us still—perhaps more so than ever. The plague bacillus and its hosts show increasing signs of resistance to antibiotics and pesticides. This raises the specter of our most potent weapons against the pestilence splintering in our hands at that moment when they are most needed."

And now the fourth pandemic has undeniably established itself. At the conclusion of the 20th century, AIDS had become a pandemic spread, at the beginning of the 21st century, primarily through heterosexual sex and the sharing of used needles by drug addicts. As the Middle East and Asia entered the contamination column, it became apparent that this scourge had entered a new, and, alas, more powerful phase.

By 2006, the 25th anniversary of the inception of the AIDS pandemic, 24.7 million people infected by the disease lived in sub-Saharan Africa, and the number of newly infected African adults and children who died of AIDS that year represented 72 percent of worldwide AIDS deaths.

The incidence of HIV infection in the United States dropped somewhat in 2006, except in the African-American population. Nevertheless, approximately 4.3 million people worldwide contracted HIV that year and an estimated 2.9 million died of the disease.

There were no signs in the rest of the world of a lessening of this, the most far reaching and horrific of all pandemics so far, despite steady if somewhat inhibited research for a vaccine, the only real cure for the disease. The inhibition has been the result of the following two major barriers:

- funding for research and treatment, while increasing worldwide, has not kept up with the growth of the disease
- ideological and political barriers in too many countries, including the United States, have prevented proper education, limited the distribution of condoms, and diminished the ability of researchers to apply their research to inhibit the growth of HIV infection

Scientists state that if the AIDS pandemic continues at its present pace, it will, by 2010, have become the deadliest epidemic in human history, eclipsing both the 14th-century Black Death and the 1918 flu pandemic. Hope, based upon increasing mainstream awareness of the seriousness of the situation, does exist for the discovery of a vaccine before then. But that hope is neither endless nor unassailable.

. . . .

EUROPE
THE BLACK DEATH
1348–1666

In the 300 years of its existence in Europe, from 1348 to 1666, the bubonic plague known as the Black Death killed 25 million people. Its demise was due to three possible causes: the fire of London, the change of seasons, and an improvement in sanitation.

The immense, horrific bubonic plague that roamed and ravaged Europe for 300 years killed 25 million people, or one-third of the population of Europe at that time. The Black Death, as it was known, is considered to be the most deadly natural catastrophe in recorded time.

Worse than a war, since it recognized no national boundaries, more cruel and painful in its attack upon its victims than an earthquake, and more terrifying, because of its hidden nature, than the eruption of a volcano or the coming of a hurricane, the Black Death—

A still verdant countryside near the village of Guinsaugon on the island of Leyte, Philippines, is slashed and disfigured by the giant landslide of February 17, 2006. The entire village was obliterated and more than 1,000 of its residents were buried beneath the mud. (Department of Defense)

An aerial view showing the giant swath of destruction made by the February 17, 2006, landslide on the island of Leyte, Philippines. Two weeks of nonstop, torrential rains loosened the side of a mountain, which ripped away and buried villages and villagers under tons of mud. (U.S. Navy/Photographer's Mate 1st Class Michael D. Kennedy)

(above) *One of the 20,000 displaced families who lost their homes huddles near a temporary tent shelter following the March 25, 2002, earthquake that struck the Hindu Kush region of Afghanistan. A battery-powered radio is the only contact with the outside world.* (IOM/USAID)

(right) *Displacement is the inevitable consequence of a natural disaster. After receiving temporary shelter from a relief agency, this man goes in search of a spot following the 2002 Afghanistan earthquake.* (USAID)

(left) *The deadliest tsunami in world history, triggered by a gigantic 9.3 earthquake off the island of Sumatra washed away entire cities and sank boats like this one that were the livelihood of their residents—278,000 died in this December 26, 2004, catastrophe and 1.8 million were made homeless.* (National Information Center of Earthquake Engineering) (right) *Not only was the Aceh province in Indonesia particularly hard hit by the tsunami of December 26, 2004, but it disrupted the lives of residents of the coasts of 10 other countries bordering the Indian Ocean. Lives were even lost as far away as the coast of Africa, and high tides were caused on the East Coast of the United States by this world-shaking natural cataclysm.* (EC/ECHO)

(below) *Some of the fortunate survivors of the explosive force of the 350-foot (106-m) wave of the world's deadliest tsunami stand near their homes, crushed into matchsticks by the water's awesome, destructive power. As far as the eye could see, in any direction, the entire coastline of Sumatra was reduced to just such uninhabitable ruination.* (EC/ECHO)

C-3

(left) *The coastlines of 11 countries were turned into turmoil by the monstrous tsunami unleashed by the giant earthquake in the Indian Ocean on December 26, 2004. This tranquil seaside town in India now became a repository for fishing boats from miles away in the aftermath of the disaster.* (EC/ECHO)

(below) *A rescue squad from Fairfax County, Virginia, inspects the devastation left by the 6.6 earthquake that totally destroyed the historic quarter and 85 percent of the rest of the Silk Road city of Bam, Iran, on December 26, 2003. The early morning timing of the arrival of the tremor was largely responsible for the deaths of 26,000 residents of Bam.*
(FEMA photo/Marty Bahamonde)

(right) *Material and human depredation greet rescue workers attempting to restore a semblance of normalcy to the refugees of the October 8, 2005, earthquake that ripped through Kashmir, part of both Pakistan and India, killing 75,000 unwary people.* (The Salvation Army)

(above) *Even the sturdiest of homes can be reduced to rubble in an instant in an earthquake. A simple structure like this home in the hills of Pakistan collapsed during the first tremors of the tri-country earthquake of October 8, 2005.* (The Salvation Army)

(right) *A young Pakistani girl searches through the wreckage of her home for links to a more peaceful past following the 2005 earthquake that rendered 3.5 million people homeless.* (The Salvation Army)

(left) *Record monsoon rains in the summer of 2004 caused the worst floods in a century in Bangladesh, India, Pakistan, and Nepal. Bangladesh, persistently threatened by flooding, resembled, in much of the country, an extension of the Indian Ocean.* (EC/ECHO)

(bottom) *Survivors wash their clothes in the polluted waters left after the flood from two weeks of torrential rains in May 2004. The water inundated scores of villages in Hispaniola, shared by the Dominican Republic and Haiti. It was the worst natural disaster ever to strike the region. Most of the 3,300 dead came from Haiti.* (Leslie Scott)

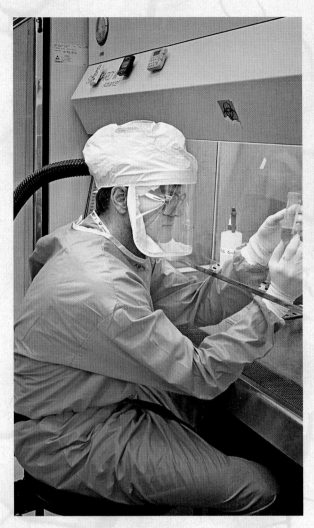

(above) *Chickens infected with bird flu await vaccination in the back of a truck in Romania.* (Calvin College)

(left) *Microbiologist Dr. Terrence Tumpey of the Centers for Disease Control and Prevention and a member of the National Center for Infectious Diseases examines a specimen vial containing a reconstructed 1918 pandemic influenza virus. In an effort to develop a vaccine and treatment for future influenza viruses, Dr. Tumpey recreated the 1918 virus in 2005.* (Centers for Disease Control and Prevention/James Gathany)

(right) *A photo taken by an Air Force cameraman at the height of Katrina's fury and flooding. The cars belonged to some of the 6,000 inhabitants of Keesler Air Force Base in Mississippi, which suffered catastrophic damage from the storm, but no casualties.* (U.S. Air Force)

(left) *The only possible transport after levee breaches flooded New Orleans during Hurricane Katrina was boats and light-armored vehicles of the Counterdrug National Guard. Guardsmen in their amphibious vehicles searched for and rescued scores of survivors stranded in their homes.* (U.S. Air Force)

(right) *Pets were handled with care after undergoing the assault of Hurricane Katrina and its floodwaters. A volunteer working with the Massachusetts Society for the Prevention of Cruelty to Animals cleans and grooms a bedraggled and undoubtedly thankful miniature poodle.* (U.S. Air Force/Tech. Sgt. Sandra Niedzwiecki)

(bottom) *The dramatic rescue of a young boy lifted to safety in the arms of Tech. Sgt. Lem Torres to a hovering helicopter after the boy was discovered clinging to the roof of his flooded home in New Orleans a day after the terrifying passage of Katrina* (U.S. Air Force/Tech. Sgt. Manuel J. Martinez)

so called because it was first carried by black rats—held all of western civilization in its thrall for generations.

The first evidence of *Pasteurella pestis,* the parasite responsible for the disease, as nearly as historians and scientists can ascertain, was in 1334, in the vicinity of Lake Issyk-Kul in the province of Semiryeschensk, in what is now Kyrgyzstan and called by Charles T. Gregg in his study, *Plague!,* "... part of the primordial plague reservoir of central Asia."

Carried by black rats transporting fleas which were infested with *Pasteurella pestis,* the plague spread south and east to China, India, and then west to Turkistan.

In 1347, in the Crimean trading port of Kaffa (which is now the southern Russian city of Feodoiya) a group of enterprising Genoese traders became trapped in a years-long siege by the Janibeg Kipchak Khan. Kaffa was the major port for goods from Genoa, but that made little difference to the Khan, who held Kaffa hostage, fighting off any and all invaders, except one: the Black Death. The disease arrived early in 1348 and mowed down his long-limbed Tartar troops as surely as if they had been defeated by overwhelming military forces.

According to the contemporary account of Gabriel de Mussis, a Piacenza notary who is thought to be an eyewitness: "Infinite numbers of Tartars and Saracens suddenly fell dead of an inexplicable disease ... and behold the disease invaded all the army of the Tartars ... every day ... thousands were killed ... the humors coagulated in the groins, they developed a subsequent putrid fever and died, all council and aid of the doctors failing...."

The Khan, ever resourceful and barbaric, decided to use the corpses of his fallen army as ammunition and thus launched the first known biological warfare attack. "The Tartars," de Mussis continued, "fatigued by such a plague and pestiferous disease, stupefied and amazed, observing themselves dying without hope of health ordered cadavers placed on their hurling machines and thrown into the city of Caffa [sic], so that by means of these intolerable passengers the defenders died widely. Thus there were projected mountains of dead, nor could the Christians hide or flee, or be freed from such disaster ... they allowed the dead to be consigned to the waves. And soon all the air was infected and the water poisoned, corrupt and putrefied, and such a great odor increased...."

The contaminated Genoan sailors now boarded their boats and sailed back to Italy, infested with the fleas themselves, and unleashing, down the anchor ropes and into the Italian port, hordes of black rats that were similarly contaminated. The Genoans, however, were not the only ones spreading the Black Death into Europe. Sixteen galleys brought plague to Italy, and only four of them came from Kaffa. Twelve oth-

ers, bearing returning Crusaders from Constantinople, docked in Messina, Sicily, at roughly the same time. The Crusaders were already infected.

By the end of 1348, all of Italy was blanketed by plague, and most of France was beginning to feel its grim effects. Switzerland was infected by August, as was England, via a ship from Calais that put into the Dorsetshire port of Melcombe. By the end of 1349, Scotland, Ireland, Denmark, and most of Germany were under its influence. Norwegian ships carried the plague as far as Iceland, where the entire population would be wiped out, and to Poland and Russia, both of which would be contaminated by 1351.

The number of fatalities was astronomical. Half of the population of Italy died. Nine out of every 10 residents of London succumbed. In the year 1348, 1,244,434 inhabitants of what is now Germany were killed by the plague. In the Russian city of Smolensk, only five persons remained alive by 1386.

It was not a pretty death. According to Michael Plateinsis of Piaza, quoted in Johannes Nohl's study, *The Black Death:* "Those infected felt themselves penetrated by a pain throughout their whole bodies and, so to say, undermined. Then there developed in their thighs or on their upper arms a boil.... This infected the whole body and penetrated it so far that the patient violently vomited blood. This ... continued without intermission for three days, there being no means of healing it, and then the patient expired."

Thoroughly terrified and helpless, the potential victims of the plague began to act in decidedly inhuman ways. "Not only those who had intercourse with them died, but also those who had touched or used any of their things ..." continued Platiensis. "Soon men hated each other so much that, if a son was attacked by the disease, his father would not tend him. If, in spite of all, he dared to approach him, he was immediately infected and ... was bound to expire within three days...."

In Florence, the plague was at its worst, so much so that the Black Death was sometimes referred to as the Plague of Florence. Here, according to Giovanni Boccaccio in *The Decameron,* "[some] went about, carrying in their hands, some flowers, some odoriferous herbs and other some divers kind of spiceries, which they set often to their noses, accounting it an excellent thing to fortify the brain with such odours, more by token that the air seemed all heavy and attainted with the stench of the dead bodies and that of the sick and of the remedies used."

In Parma, Francesco Petrarch echoed this: "Alas, my loving brother, what shall I say? How can I begin? Where shall I turn? Everything is woe, terror everywhere. You may see in me what you have read in Virgil of the great city, with 'everywhere tearing pain, everywhere fear and the manifold images of death.' Oh, brother, would that I had never been born or that I had already met my death!"

In France, the papal city of Avignon, the residence of Pope Clement VI, was horrendously invaded by the plague. An unidentified canon, writing to his family in Belgium and quoted in George Deaux's book, *The Black Death 1347,* chronicled the sad scene: ". . . one-half, or more than a half of the people at Avignon are already dead. Within the walls of the city there are now more than 7,000 houses shut up; in these no one is living, and all who have inhabited them are departed; the suburbs hardly contain any people at all. A field near 'Our Lady of Miracles' has been bought by the Pope and consecrated as a cemetery. In this, from the 13th of March, 11,000 corpses have been buried. . . ."

Later, Pope Clement would consecrate the Rhone River so that surplus corpses might be dumped into it. The Pope himself would survive, protected by two huge fires kept burning night and day on either side of him.

In England, William Dene, a monk of Rochester, recorded the scene: "To our great grief, the plague carried off so vast a multitude of people of both sexes that nobody could be found who would bear the corpses to the grave. Men and women carried their own children on their shoulders to the church and threw them into a common pit. From these pits such an appalling stench was given off that scarcely anyone dared even to walk beside the cemeteries."

And so it went throughout all of Europe. Frantic to escape the pain and horror and inevitable death of the plague, victims turned to physicians who knew no more than they how to cure or prevent the scourging sickness, but who continued to experiment with all sorts of palliatives.

Some doctors prescribed human feces to be worn around the neck in a bag. Others prescribed bathing in and drinking urine. Leeches and dried toads and lizards were applied to boils to suck out the poison. Butter and lard were lathered into open wounds. Testicles were run through with needles. The blood of freshly slain pigeons and puppies was smeared on feverish foreheads. A French physician, Guy de Chauliac, sliced open the boils and inserted red hot pokers into the open wounds. This primitive purification method did meet with some success with those who didn't die from a heart attack, go into irreversible shock, or become insane from the pain.

There was, too, the problem of the "poisoned rooms" of those who died from the plague. Fresh milk was placed on a platter in the middle of a sick room to absorb the poisoned air. An anonymous London physician offered the following prescription for decontaminating a home in which a plague victim died: ". . . take large Oynions pale them and lay three or foure of them upon the ground, let hem l ie ten daies, those picled Oynions will gather all the infection into them that is in one of those Roomes, but burie those Oynions afterwards deep in the ground."

Bewildered that neither physicians nor priests could stop the Black Death, many poor souls turned either to extreme piousness, or, despairing of a God whom they felt had turned from them, turned to hunting down scapegoats, to unregenerate licentiousness and voluptuousness, to amulets and charms, or even to devil worship. In many respects, the Black Death set civilization back centuries.

There were isolated pockets of positiveness; some of the very pious set in place some traditions that have endured. The villagers of Oberammergau, for instance, vowed to present a religious performance at intervals forever if the dread hand of the pestilence was lifted from them. Their vow was given toward the end of the plague, in 1634, and the plague left them. They still present their Passion Play.

But these shafts of sunlight in a dark time were perhaps brighter for their rarity. The zealous and feverish in their religious fervor, did monumental ill. The flagellants, the so-called "Brethren of the Cross," roamed countryside, arriving in village squares, where they performed their rituals of self-flagellation in atonement for the sins that had supposedly brought on the plague, while at the same time probably spreading the plague itself by their travels.

The search for scapegoats fueled anti-Semitism. In May 1348, the Jewish communities in three cities of southern France were exterminated. Aged and newborn, healthy and firm, men, women, and children were all savagely murdered. In September of that same year, a Jewish physician in Chillon, Switzerland, confessed, after having been brutally tortured, that he and other members of the Jewish community had poisoned the wells. The news spread fast, and all over Europe, some 60 large and 150 small Jewish communities were wiped out in 350 separate massacres.

A nursery rhyme and a fairy tale found their roots in the plague.

> Ring around the Rosies,
> Pocket full of posies,
> Achoo, Achoo,
> All fall down.

This was a description, supposedly, of the habit of wearing garlands of flowers in plague time to ward off miasma. However, the failure of the remedy is evidenced by the last two lines, in which the wearers gasp and fall dead on the spot.

The more cheerful fairy tale involves the Pied Piper of Hamelin, the German town afflicted by the plague and a rising rat population in 1348 and again in 1361. Historically, the facts pretty much parallel the story and Robert Browning's poem: An army of rats invaded Hamelin. The town corporation hired an itinerant ratcatcher. When the rats were cleaned out, and the ratcatcher demanded his fee, the corporation offered him a pittance. The ratcatcher stormed out of town, swearing revenge. Meanwhile, the children of Hamelin gathered up the rat corpses that were littering the streets and tossed them into the swift-flowing River Weser. But, having been exposed to the plague, the children died, and they were buried in a new cemetery dug on the side of Koppelberg Hill, the place in the tale in which the earth opened up and the Pied Piper and the children disappeared forever.

Unaffected by the measures of God or man, the plague persisted, as did its manifestations in the consequent habits of the populace. All sorts of remedies and penances continued to be tried—some effective, some not. Dancing was one of the more macabre ones that did not. In this fanatical act of penance, thousands of victims would dance wildly on platforms erected for them in public squares, in a goulish, aptly named Dance of Death, until they dropped from fatigue or the effects of the plague. Others trampled on them, thus assuring their deaths.

In a more effective method of containing the plague, the tradition of quarantine was put into practice for ships in port. Forty days (probably because of its religious significance) was the time assigned to ships kept at anchor in various ports, and, although this probably prevented the spread of the plague into these ports from ships at sea, it also set some ships adrift, as every man aboard succumbed to the Black Death.

All in all, it was a dolorous time of the world, a Veblenian way of trimming down a population that had strained the natural resources of Europe. The one weapon that could have conquered the plague—sanitation—would be implemented in 1666, when the plague would suddenly disappear. Some credited its end to the London fire; others to the change of seasons. Few realized at the time that it was probably soap and water that were responsible.

For details concerning the end of the Black Death, see PLAGUES AND EPIDEMICS, Great Britain, London, 1665, p. 212.

EUROPE
SYPHILIS EPIDEMIC
1493

Millions not afflicted by the Black Death died in a 50-year epidemic of syphilis that ravaged Europe from 1493 to 1543.

In the midst of the Black Death, parts of both Europe and the New World were riven by a 50-year epidemic of syphilis. Its origins are contradictory. According to one school, the so-called Columbian School quoted by J. F. D. Shrewbury in an article, "The Saints and Epidemic Disease" in the *Birmingham Medical Review,* the disease was endemic among the natives of the West Indies, and was transmitted sexually by the Carib women to the sailors of Columbus, who then imported it into Europe.

A much more persuasive case than the Columbian one can be made for the *Morbus Gallicus* or "French Disease" theory, in which the disease originated in France in 1493, and then spread from there to Spain, the islands of the Mediterranean and Italy via the French army of King Charles VIII. Giovanni di Vigo, surgeon to Pope Julius II, blamed its entrance into Italy on the siege of Naples, in 1494, and Rodrigo Ruiz de Isla, in his *Treatise on the Serpentine Malady,* reaffirmed this: "In . . . 1494 the most Christian King Charles of France who was then reigning having gathered a great army passed into Italy. And at the time he entered the country with his host many Spaniards infected with this disease were in it and at once the camp began to be infected with the aforesaid malady and the French, as they did not know what it was, thought it came from the atmosphere of the region. The French called it the disease of Naples."

Whatever the source, it was a new disease, which the poet, physician, and philosopher Girolamo Fracastoro named syphilis in his essay, "Syphilis or the French Disease." Sipylus was the son of Niobe, who claimed that she should be worshiped equally with Leto the mother of Apollo because she had borne seven times as many children as Leto. In retaliation for this effrontery, Apollo destroyed all her sons.

It apparently took little time for physicians to discover that the disease was transmitted sexually. Armies left the then fatal disease behind as they invaded, and subsequent armies contracted it from female citizens with whom they had sexual intercourse. Fracastoro described the progress of the disease on an "illustrious YOUTH [sic] of Verona":

> . . . The wretched man, too confident to fear such dire mischance, was attacked by this disease. . . . By slow degrees that shining springtide, that flower of young manhood, fell to ruin as his vital force declined; then, dreadful to relate, foul corruption fastened on his wretched limbs, and deep within, the larger bones began to swell with loathsome abscesses. Foul ulcers (shame on the mercy of heaven!) began to devour the fair eyes which the light of day loved to look upon; to feed, too, on nostrils gnawed away by bitter wounds. And then, doomed too soon by unkind fate, the luckless youth left behind him the now hateful air of heaven and the light of day. . . .

Treatment evolved swiftly; mercury was used to dissolve the scabs, and this treatment was continued into the mid-19th century. Still, millions would die in this epidemic of the end of the 15th and beginning of the 16th century, but according to British researcher R. S. Morton, "With the spread of the Renaissance influences of education, sophistication in dress, and improved personal hygiene, the complex known so widely as *Morbus Gallicus* faded from central Europe. Residuals were found in the most backward areas, causing sporadic outbreaks."

In the 20th century, first arsenic-bismuth and then penicillin were introduced to treat syphilis and thus prevent further epidemics. Still, the disease is reported to be on the increase.

FRANCE
ST. VITUS'S DANCE
1518

Hundreds danced themselves to death in an outbreak of an unexplained affliction termed "St. Vitus's dance" in France in 1518.

The curious outbreak of St. Vitus's dance in Strasbourg, France, in 1518 was, conceivably, a side effect of the religious and fanatic dancing that accompanied the Black Death (see p. 208). According to medical historian George Rosen of Yale University, this epidemic of continued dancing occurred

> . . . eight days before the feast of Mary Magdalene [when] a woman began to dance, and after this went on for some four to six days she was sent to the chapel of St. Vitus at Hohlenstein, near Zabern. Soon thereafter, more dancers appeared and the number grew until more than a hundred danced at a time. Eventually, the municipal council forbade all public gatherings and music, restricted the dancers to two guild halls, and then sent them off to the chapel of St. Vitus. According to one account, more than four hundred people were affected within four weeks. . . .

Hundreds literally danced themselves to death in this, one of the strangest epidemics ever.

GREAT BRITAIN
LONDON
PLAGUE
1665

The end of the Black Death occurred in England in 1666. Some historians and writers of the time, among them Samuel Pepys, insisted that it was a self-contained "English Plague" that lasted for a mere two years.

Although most historians consider the Great Plague of London actually to mark the end of the Black Death, which began in 1348 (see pp. 208–211), others consider it to be a separate event, confined solely to London and the surrounding countryside.

It was Samuel Pepys, in his diaries, who graphically chronicled the Plague of London:

> 20th [June] . . . Walked to Redriffe, where I hear the sickness is, and indeed it is scattered almost every where, there dying 1,089 of the plague this week. . . . Lord! to see how the plague spreads! it now being all over King's Streete, at the Ace, and next door to it, and in other places. . . . 12th [August] . . . The people die so, that now it seems they are fain to carry the dead to be buried by daylight, the nights not sufficing to do it in. . . . 31st . . . the plague having a great encrease this week, beyond all expectation, or almost 2,000, making the general [Mortality] Bill 7,000, odd 100; and the plague above 6,000. Thus the month ends with great sadness upon the publick, through the greatness of the plague everywhere throughout the kingdom almost. Every day sadder and sadder news of its encrease. In the City died this week 7,496, and of them 6,102 of the plague. But it is feared that the true number of the dead this week is near 10,000. . . . 16th October . . . I walked to the Tower; but, Lord! how empty the streets are, and melancholoy, so many poor, sick people in the streets full of sores; and so many sad stories overheard as I walk, everybody talking of this dead, and that man sick, and so many in this place, and so many in that. And they tell me that, in Westminster, there is never a physician and but one apothecary left, all being dead. . . .

The disease was spread to the surrounding countryside. A village tailor in the tiny village of Eyam in Derbyshire, received a box of patterns and some articles of discarded clothing from London. He would be the first of 85 members of the tiny village to die of the plague.

Finally, November and its frosts seemed to stop the growth of the plague. The Great Fire of September 2–6, 1666, is also thought to have ended it.

GREAT BRITAIN
LONDON
CHOLERA EPIDEMIC
1832

Sixty-eight hundred Londoners died from a cholera epidemic, introduced from Germany in 1831. The epidemic would last for one year.

Cholera came to the British Isles via a ship from Hamburg which docked at Sunderland in the north of England in October 1831. It would be February 1832 before the first case would be reported in London, but the terrified populace would soon realize its full impact.

According to Francis Sheppard, in his book, *London 1808–1870: the Infernal Wen,* "[the disease] with its usual unpredictability [leaped] to outlying districts such as St. Marylebone and Hoxton. By the autumn, when few new cases were occurring, some 5,300 Londoners had died, and during the second short attack in 1833 there were another 1,500 victims."

London hospitals were ill equipped to handle the victims of the disease. Sanitary conditions were being improved in St. Bartholomew's Hospital as a result of new discoveries regarding the spreading of germs. However, modifications were not completed when the epidemic struck. Thus, hastily improvised houses were utilized, often over the extreme objections of neighbors. They were actually dying houses. Fully half or more of the patients admitted never left.

As devastating as this first exposure to cholera was for Londoners, it would pale before the arrival of the next invasion, termed the "third pandemic," which would dock in England within a short 16 years (see p. 219).

GREECE
"PLAGUE OF THUCYDIDES"
431 B.C.E.

Hundreds of thousands of citizens of Athens died from the so-called Plague of Thucydides, a combination of typhus and measles, from 431 to 427 B.C.E.

The so-called Plague of Thucydides took place in 431 B.C.E., one year into the Peloponnesian War. It lasted for two years, then reoccurred for a year at the end of 427 B.C.E. Actually an epidemic of several mysterious diseases (post analysis by medical historians identify it as a combination of typhus and measles), it affected hundreds of thousands of soldiers and citizens of Athens alike. Ironically, the largest fatality rate at first took place among the physicians themselves, since the epidemic proved to be wildly contagious.

According to Thucydides, in his chronicle of the Peloponnesian War, ". . . the disease is said to have begun south of Egypt in Aethiopia; thence it descended into Egypt and Libya, and after spreading over the greater part of the Persian empire, suddenly fell upon Athens."

The first incidents of the epidemic came after the Peloponnesian army, under the command of the Lacedaemonian king Archidamus, had invaded and established itself in Attica, where it set about ravaging the country and presumably spreading the disease. "It first attacked the inhabitants of the Piraeus," continues Thucydides, "and it was supposed that the Peleponnesians had poisoned the cisterns, no conduits having as yet been made there. It afterwards reached the upper city, and then the mortality became far greater."

During the summer of 431 B.C.E., Thucydides reported that his army and the population of Athens was in remarkably good health. And then, ". . . all in a moment, and without apparent reason, were seized with violent heats in the head and with redness and inflammation of the eyes. Internally the throat and the tongue were quickly suffused with blood, and the breath became unnatural and fetid. There followed sneezing and a hoarseness; in a short time the disorder, accompanied by a violent cough, reached the chest; then fastening lower down, it would move the stomach and bring on all the vomits of bile to which physicians have ever given names. . . ."

Convulsions followed, then

> . . . the body externally was not so very hot to the touch, nor yet pale; it was of a livid colour inclining to red, and breaking out in pustules and ulcers. But the internal fever was intense; the sufferers could not bear to have on them even the finest linen garment; they insisted on being naked, and there was nothing which they longed for more eagerly than to throw themselves into cold water. And many of those who had no one to look after them actually plunged into the cisterns, for they were tormented by unceasing thirst, which was not in the least assuaged whether they drank little or much. They could not sleep; a restlessness which was intolerable never left them. While the disease was at its height the body, instead of wasting away, held out amid these sufferings in a marvelous manner, and either they died on the seventh or ninth day, not of weakness, for their strength was not exhausted, but of internal fever, which was the end of most; or, if they survived, then the disease descended into the bowels and there produced violent ulceration; severe diarrhea at the same time set in, and at a later stage caused exhaustion, which finally with few exceptions carried them off. For the disorder which had originally settled in the head passed gradually through the whole body, and, if a person got over the worst, would often seize the extremities and leave its mark, attacking the genitals and the fingers and the toes; and some escaped with the loss of these, some with the loss of their eyes. Some again had no sooner recovered than they were seized with a forgetfulness of all things and knew neither themselves nor their friends.

When neither physicians nor priests could offer cures or even some relief from the suffering, the populace lapsed into lawlessness and anti-religious fever. With the end of the scourge, Athens once again pulled itself together into an orderly and dedicated city-state.

INDIA
PLAGUE
1896–1907

The third pandemic had its origins in China in 1892, but its first devastation occurred in India from 1896 to 1907. During that time, nearly 5 million Indians died of plague.

The third pandemic of bubonic plague affected a broad geographic area. It is, however, generally agreed that it had its beginnings in the Yunnan province of China in 1892, but did not acquire world attention until it reached Canton in the spring of 1894. From here it was a short distance to Hong Kong, with its trade connections to the ports of the Pacific in India, Australia, Japan, and America.

India, at least in the first stages of this pandemic, seems to have been most mortally affected, despite the fact that the plague bacillus had been discovered in 1894 and vaccine inoculation was introduced into India in 1897. Perhaps because of faulty communication, the immensity of the population or ancient religious beliefs and practices, the average annual death toll was 548,427 in India. For the period from 1889 to 1908, plague deaths in India would account for 47.88 percent of these deaths.

As in the Black Death, the second pandemic (see pp. 208–211), this plague was exported by ship, to ports in Asia, South America, Australia, Africa, Europe, and even San Francisco, on the west coast of the United States (see p. 218 for the development of this pandemic).

ITALY
TYPHUS EPIDEMIC
1505–1530

A typhoid epidemic raged through Italy from 1505 to 1530, killing hundreds of thousands.

Typhus (a rickettsial disease transmitted by human body lice) and typhoid (a Salmonella bacillus found in human waste which is transmitted by contaminated food and water) have long been known as the basis for epidemics. Even in the second century B.C.E., a 56-mile aqueduct was built to supply Rome with fresh water, since the Romans recognized the importance of clean household water and sanitation in the combating of waterborne disease.

Still, in the 16th century C.E., epidemics of typhus and typhoid were common throughout Italy, particularly between 1505 and 1530, when hundreds of thousands succumbed to them. Girolamo Fracastoro chronicled the disease and its fevers, which he said were ". . . vulgarly called 'lenticulae' (small lentils) or 'puncticulae' (small pricks) because they produce spots which look like lentils or fleabites. . . ."

Fracastoro described the onset of this form of typhus as gentle, in fact so mild that most victims waited until too late to call a physician. A general kind of lassitude was reported until the seventh day, when ". . . the mind would wander; the eyes became red, and the patient was garrulous; the urine was usually observed to be at first pale, but consistent, but presently red and clouded, or like pomegranate wine; the pulse was small and slow . . . about the fourth or seventh day, red, or often purplish red spots broke out on the arms, back and chest, looking like flea-bites . . . though they were often larger and in the shape of lentils, whence arose the name of the fever. . . ."

ROME
EPIDEMIC PESTILENCE
452 B.C.E.

Epidemic pestilence swept through Rome in 452 B.C.E. killing thousands, among them, many members of the Roman Senate.

An epidemic hit Rome in 452 B.C.E. that not only killed thousands of citizens but also reached into the high councils, killing Sextus Quintilius, one of the consuls, Spurius Furius (who had been appointed to succeed him), four of the tribunes, and an untold number of senators.

Despite the planning engineers who already realized the importance of uncontaminated water (see previous entry), the general populace seemed to ignore such good advice, and thus this epidemic (most likely typhoid) spread quickly and viciously. Entire households perished and then, because of the disposal methods of the bodies of rich and poor alike, the infection was spread through both the air and the water supply.

According to Greek scholar and historian Dionysius of Halicarnassus:

> . . . they burned the bodies and committed them to the earth, at the last, either through a disregard of decency or from a lack of the necessary equipment, they threw many of the dead into the sewers under the streets and cast far more of them into the river; and from these they received the most harm. For when the bodies were cast up by the waves upon the banks and beaches, a grievous and terrible stench, carried by the wind, smote those who were still in health and produced a quick change in their bodies; and the water brought from the river was no longer fit to drink, partly because of its vile odour and partly by causing indigestion. . . .

As in the case of most early epidemics and plagues, the populace called upon physicians and priests, and when neither could help them, gave up all pretense of morality. As German classical historian Barthold G. Niebuhr pointed out in his *History of Rome,* "Pestilences, like inhuman military devastations, corrupt whom they ruin. . . ."

However, in the case of this particular pestilence, reform in some Roman laws resulted. As Niebuhr also pointed out, "Very calamitous times however serve to awaken a sense of the defects of existing institutions; many cheer themselves with the belief that the correction of these would restore their past prosperity; and this motive unquestionably seconded the proposals made at Rome, after the pestilence and the military reverses, for the reformation of the laws."

ROME
PLAGUE OF JUSTINIAN
541–590 C.E.

Millions died during the first pandemic, called the "Plague of Justinian," which spread over the Roman Empire from 541 to 590 C.E.

The first of the three great pandemics (the second was the Black Death, the third began in China, spread to San Francisco at the end of the 19th century, and continues today—see introduction to this section for details) occurred in the year 541, during the 15th year of the reign of Emperor Justinian I. Thus, its title, the "Plague of Justinian."

The coming of the plague was reportedly heralded by a 20-day, spectacular nighttime display by Halley's comet in 531. Its departure was marked by Pope Gregory the Great crossing the Tiber on the Bridge of Hadrian, heading a procession to Saint Peter's to pray for relief from the pestilence. There, the pope saw the Archangel Michael standing on Hadrian's mausoleum, sheathing a bloody sword. At that moment, the plague ceased, and thereafter a chapel was erected atop Hadrian's tomb, called Saint Michael's in the Clouds.

Procopius of Caesarea chronicled the first appearance of the plague in Constantinople, where it was brought, apparently by Egyptians living in Pelusium. From there, it proceeded to Alexandria, then to Palestine, and from there, it spread over the entire Roman Empire.

Procopius goes on to write what has come to be known as the first detailed description of bubonic plague: ". . . a bubonic swelling developed," he wrote, "and this took place not only in the particular part of the body which is called 'boubon' [groin], that is below the abdomen, but also inside the armpit, and in some cases also behind the ears, and at different points on the thighs. . . ."

He then proceeds to describe the classic spreading of bubonic plague, from the victims to the attendants of the victims, of the enormous anguish of the patients, who rushed from their houses, attempting to immerse themselves in the sea or to throw themselves from great heights.

"Death came in some cases immediately, in others after many days, and with some the body broke out with black pustules about as large as a lentil and those did not survive even one day, but all succumbed immediately. With many also a vomiting of blood ensued without visible cause and straightway brought death. . . ."

However, the vigilant Procopius did notice that:

> . . . in those cases where the swelling rose to an unusual size and a discharge of pus had set in, it came about that they escaped from the disease and survived, for clearly the acute condition of the carbuncle had found relief in this direction, and this proved to be in general an indication of returning health; but in cases where the swelling preserved its former appearance there ensued those troubles which I have just mentioned . . . those who were without friends or servants lay unburied in the streets or in their desolate houses. . . . Corpses were placed aboard ships and these abandoned to the seas . . . Physicians could not tell which cases were light and which severe, and no remedies availed.

Despite the detailed study of the effects of the bubonic plague upon its victims, none of these learned men, nor even the historian Edward Gibbon, writing in his *History of the Decline and Fall of the Roman Empire* in the late 18th century, had discovered that it was spread through rats, fleas, and bacilli. Gibbon wrote:

... during three months, five and at length ten, thousand persons died each day at Constantinople.... No restraints were imposed on the free and frequent intercourse of the Roman provinces: from Persia to France, the nations were mingled and infected by wars and emigrations; and the pestiential odour which lurks for years in a bale of cotton, was imported, by the abuse of trade, into the most distant regions.

RUSSIA
TYPHUS EPIDEMIC
1812

Nearly 100,000 Russian and French soldiers died from typhus in 1812 in Russia. Napoleon's retreating soldiers then spread the epidemic over parts of Germany, Austria, Switzerland, and France.

Typhus, the persistent microorganism that is halfway between being a louse-born virus and a louse-born bacteria, (see PLAGUES AND EPIDEMICS, Rome, 541–590, above) has been described by Columbia University professor Hans Zinsser as ". . . the inevitable and expected companion of war."

This was the case when Napoleon invaded Russia in 1812. The Russians employed a scorched earth policy, and thus, deprived of even basic sanitation, French troops suffered severely from dysentery and diarrhea on the way to Moscow. It might have weakened them. At any rate, by the time they reached Smolensk, on August 14, 1812, they were no match for the typhus infection that beset them in that city. By mid-September, when they reached Moscow, the French army was riddled with the disease, and, according to Friedrich Prinzing, in his book, *Epidemics Resulting from Wars*, ". . . when Napoleon's army withdrew from the city it left behind several thousand typhus-fever patients, almost all of whom died—only the stronger patients were taken along on wagons." But, predictably, the epidemic did not respect boundaries of nationality. Sixty-two thousand Russian soldiers, who pursued the French army, were also felled by typhus.

On December 8, the remaining bits and pieces of the French Army crossed Lithuania at Vilna, and 30,000 soldiers died there, most of them of typhus. These soldiers spread the disease to civilians, who in turn spread it through much of Russia west of an imaginary line running north and south through Moscow.

It did not stop there. Pressing on in their retreat, the French troops fanned out over a large part of Germany, France, Switzerland, and Austria, leaving typhus behind wherever they went.

UNITED KINGDOM (AND WESTERN EUROPE)
MAD COW DISEASE/CREUTZFELDT-JAKOB DISEASE
1986

Mad cow disease, or bovine spongiform encephalopathy (BSE), and its human manifestation, Creutzfeldt-Jakob disease (CJD), have as of this writing claimed 147 lives in the United Kingdom and three in France since 1986. Undetectable in its early stages and untreatable, this fatal disease has the potential of becoming a major pandemic.

Bovine spongiform encephalopathy (BSE), popularly known as mad cow disease, first appeared in the United Kingdom in 1986. Since that time, 180,778 cattle in Britain have suffered and died from this terrifying, always fatal affliction that turns the brain into a sponge-like mass and destroys the ability to see, hear, or stand.

Its origins extend back to the 1700s, when sheep first began to suffer from scrapie, the ancestor and equivalent of BSE. The first BSE cases in cattle arose after the cattle, in order to accelerate their growth, were fed the protein-rich remains of scrapie-infected sheep. The first reported cases of infected humans, in the 1920s, came from eating beef from contaminated cattle. Thus, unlike the plagues of history, mad cow disease was man-made.

BSE in humans is known as Creutzfeldt-Jakob disease (CJD), so called because of the work in the 1920s of two German physicians, Hans Gerhard Creutzfeldt and Alfons Jakob, who independently reported the first examples of human infection from BSE.

In humans, this currently incurable and untreatable disease begins with failing memory, behavioral changes, visual disturbances, and a lack of coordination.

This is followed by involuntary and progressively more violent jerking motions. As the disease spreads, the brain is literally eaten away, until the afflicted person has lost the ability to hear, see, speak, and to understand written and spoken language. Eventually, he or she dies, usually in less than a year after the onset of symptoms.

BSE has an insidious quality: It is a self-perpetuating disease. Cattle are fed contaminated sheep offal and contract BSE; contaminated cattle are slaughtered and their offal is fed to uncontaminated cattle, who become carriers or victims of the disease. In the 1980s, it was discovered that contaminated cattle also passed the disease into their offspring.

In 1986, some 10,000 cattle were discovered to be infected in the United Kingdom. But nothing was done about it, allegedly because of budgetary considerations. In 1988, scrapie-contaminated animal feed was banned in the UK, but the disease continued to spread until a wholesale slaughter of contaminated cattle was finally ordered.

But by this time, BSE had entered the world food chain. In 1989, the first BSE case was reported outside of the United Kingdom, in France. And now, panic and scientific study began to increase.

It was discovered that BSE was transmitted to humans through a prion, which is an infectious protein that seems to "persuade" other proteins to imitate its abnormal folds. The problem is that there is no known diagnostic method to discover the disease in its very early stages. By the time it is diagnosed, it is already too late, as the 80 people in the United Kingdom and the three in France who have died from it have proved.

As of the publication of this edition of *Natural Disasters,* mad cow disease has been largely confined to the United Kingdom, though sizable numbers of infected cattle have been detected in France (257), Ireland (614), Portugal (503), and Switzerland (365). Lesser numbers have shown up in Belgium (22), Denmark (3), and Germany (31).

The spread has been enough to cause the sale of beef to plummet in Europe and the sale of horse meat to rise by 30 percent in France. This, however, is only a practical reaction to the presence of a threat to public health and safety.

In contrast, the onset of any potential epidemic produces wildly divergent reactions. Possibly the most bizarre one was the spectacle of Britain's agricultural minister, John Gummer, feeding his four-year-old daughter a hamburger on TV in 1990 to prove to the world and his people that British beef was safe. It was a particularly reckless move, since Creutzfeldt-Jakob disease has, like AIDS, a notoriously long incubation period, anywhere from six months to 20 years.

At the other, cautious end of the spectrum of reaction are ongoing fears that BSE can be transmitted by blood transfusions. To this end, the United States banned blood donations from anyone who lived in Britain for a period of six months between 1980 and 1996, and in February of 2001 was publicly contemplating the banning of blood donations from travelers to France and Portugal for a total of 10 years beginning with 1980.

The World Health Organization (WHO) has made the discovery of diagnostic evidence for the presence of CJD a priority. Dr. David L. Heymann, the executive director in charge of communicable diseases for WHO, told reporters, "We have to make recommendations based on limited information, but it's better to be on the conservative side and change the rules later."

To this end, WHO has convened nine scientific consultations on issues related to animal and human examples of mad cow disease infection since 1991, and from 1997 to 2000, the organization has conducted a series of training courses in recognizing the disease worldwide, particularly in developing countries.

WHO's recommendations include a prohibition of the entrance into any human or animal food chain of any parts or products of any animals that have shown signs of the disease, and the banning in all countries of the use of ruminant tissues in ruminant feed.

Since some human and veterinary vaccines are prepared from bovine materials, and so may carry the risk of transmitting the infection, WHO has also warned the pharmaceutical industry to use only those materials from countries that have a surveillance system for BSE in place, and that report either zero or only sporadic cases of BSE.

In January 2001, a Purina Mills plant in Texas recalled 22 tons of feed in which cow meat and bone meal was mixed and placed on a truck that delivered it to cattle ranches. One thousand, two hundred and twenty-one head of cattle were quarantined as a result of the mixup, which Purina Mills discovered internally and reported to the Food and Drug Administration.

However, on December 23, 2003, the U.S. Department of Agriculture (USDA) announced the first confirmed U.S. case of mad cow disease, and quarantined a 4,000-member herd in Mabton, Washington. The cow, born in 1997 in Alberta, Canada, was exported to Washington State in September 2001 along with a herd of 81 dairy animals.

A week later, the USDA issued a series of bans on "downer," or sick cattle, being introduced into the human food supply chain, along with several other safeguards against the introduction of CJD into meat directed for human consumption. Simultaneously, Japan banned the importation of U.S. beef until early 2004. In May 2003, a case of CJD was diagnosed in Canada, and in March 2004, Canada admitted that contaminated feed from two Canadian feed mills was to blame for the infection of the Washington cow.

Although earlier in the year, Japan had lifted its ban on U.S. beef, yet another cow in Alabama exhibited symptoms of mad cow disease. By June 10, the United States confirmed that the cow did indeed have CJD and Japan reinstated its ban on imported beef from America.

By late 2006, no new outbreaks had occurred in the United States, but a disquieting development in American regulation of feed for cattle destined for slaughter came about, despite the fact that 65 nations in the rest of the world had enforced full or partial restrictions on importing U.S. beef products because of fears that U.S. testing was not rigorous enough. The USDA ignored this concern as well as the statistics that showed that the value of U.S. beef products had declined from $3.9 billion per year in 2003, before the first mad cow had been detected in the United States, to $1.4 billion in 2005. The USDA, in March 2006, went on to invoke an obscure 1913 law that blocked companies from selling CJD testing kits to beef producers. Furthermore, the USDA in 2006 only tested 1 percent of the roughly 100,000 cows slaughtered daily in the United States, citing budget problems, and in December of that year reduced its mad cow-testing program by 90 percent.

Mad cow disease has not yet become a worldwide epidemic. But it trembles on the brink of possibility. In a 1998 paper delivered at the University of Melbourne, Australia, Dr. Kynette J. Dumble, a senior research fellow, observed that "A worst case scenario–sized CJD epidemic will smash rather than stretch every available human resource."

UNITED STATES
CALIFORNIA
BUBONIC PLAGUE
1899

The third pandemic, bubonic plague, which began in China and India (see pp. 208–211), was brought to the West via San Francisco, in 1899.

The third pandemic, which began in China in 1892, traveled to San Francisco in 1899, where it flourished and spread. There is evidence that the third pandemic is still with us (see PLAGUES AND EPIDEMICS, the world, 1855–Present, p. 220).

On June 27, 1899, the S.S. *Nippon Haru* arrived in San Francisco harbor from Hong Kong via Honolulu. Two cases of plague had occurred on the ship, which carried plague-infected rats, much like the Genoese ships that spread the Black Death to Europe in the 1300s (see pp. 208–211). There were two Japanese stowaways aboard the *Nippon Haru,* who apparently jumped overboard before the ship reached quarantine on Angel Island. When the stowaways' corpses were fished from the bay two days later, they were found to be riddled with plague bacilli.

All was quiet until nine months later, when the body of a Chinese man, Wing Chut King, was discovered in the basement of the Globe Hotel in the heart of Chinatown. An autopsy revealed that he had died of bubonic plague.

The Board of Health of San Francisco immediately cordoned off the 12 square blocks of Chinatown, and a search was launched among the 25,000 inhabitants for more corpses. However, San Francisco at the time was controlled by conservative business interests and railroad and lumber magnates, who saw the plague scare as a threat to their profits.

Working through sympathetic newspapers—which included every newspaper except William Randolph Hearst's *Examiner*—they launched a campaign to minimize the danger. The *San Francisco Bulletin* skewered Board of Health head Dr. J. M. Williamson in verse:

> Have you heard of the deadly bacillus,
> Scourge of a populous land,
> Bacillus that threatens to kill us,
> When found in a Chinaman's gland?

Under fire from all sides, health officials lifted the quarantine of Chinatown two days later. Now, the fight between the Board of Health and the business interests escalated, while the plague, unattended, grew. Extremism blunted the arguments of both sides. Governor Henry T. Gage insisted that there was no plague in his state, and dismantled the Board of Health, but not before the board had announced a plan to remove all Chinese in the city to detention camps on Angel Island and to demolish Chinatown.

Finally, U.S. Surgeon General Edward Wyman and President William McKinley took a hand, and, beginning on April 8, 1901, saw to it that Chinatown was scrubbed clean. The new governor of California, George C. Pardee, was a practicing physician, and he soon cleaned out the corrupt health practices of his predecessor.

But the damage had been done. The plague was spreading. On February 29, 1904, a 33-year-old woman died of the plague in the town of Concord, California, northeast of San Francisco. Until that time, there had been 121 reported cases in San Francisco and only five outside the city. The mortality rate was 97 percent.

For years, the books were then closed on the so-called San Francisco epidemic. But, according to some scientists, this was merely the beginning of the third pandemic, one that would claim millions.

UNITED STATES
PENNSYLVANIA
YELLOW FEVER EPIDEMIC
1793

Four thousand residents of Philadelphia died in a yellow fever epidemic in 1793.

Yellow fever arrived in Cuba from Veracruz in 1761. The following year, British forces besieged Havana, contracted the disease in large numbers, and brought it back to Philadelphia. There was a moderate outbreak in the city that year, but then, for 30 years—from 1763 to 1793—no cases of yellow fever were reported.

Then, in August 1793, it reappeared in Philadelphia and from that month until the October frost, when it abated, 4,000 people would die of the disease.

According to publisher Mathew Carey, an eyewitness,

> The consternation of the people . . . was carried beyond all bounds. Dismay and affright were visible on the countenance of almost every person. Most people who could . . . fled the city. Of those who remained, many shut themselves in their houses and were afraid to walk the street. . . . Many were almost incessantly purifying, scowering, and whitewashing their rooms. Those who ventured abroad, had handkerchiefs or sponges impregnated with vinegar or camphor at their noses. . . . The corpses of the most respectable citizens, even if they did not die of the epidemic, were carried to the grave, on the shafts of a chair, the horse driven by a negro, unattended by a friend or relation, and without any sort of ceremony. People shifted their course at the sight of a hearse. . . . The old custom of shaking hands fell into such general disuse, that many were affronted at even the offer of a hand.

Two theories raged in Philadelphia almost as furiously as the epidemic. The contagionists, led by Oliver Wilcott, Jr., the comptroller of the U.S. Treasury, argued that it was transmitted overland as easily as by ship, and that quarantining would do little to avoid outbreaks of yellow fever.

The anti-contagionists, led by James Hutchinson (himself a victim of the epidemic) and Noah Webster, also argued against quarantining, and cited instead filthy conditions on the wharves—a recent outbreak, they argued, had begun in damaged coffee that had putrefied on one of the wharves, and in a proclamation, declared that the only cure could be "extensive sanitary measures, including sewer construction, waste removal, broad streets planted with trees, numerous open squares, large house lots, and an end to overcrowding—in short, comprehensive city planning, sanitation and housing reform."

It was good advice that went unheeded. Quarantine still remained Philadelphia's chief weapon against yellow fever. And so, the epidemic inched northward to New York City and killed 2,000 there in 1798.

THE WORLD
PLAGUE
1200 B.C.E.

....

The first recorded plague struck the Philistines in 1200 B.C.E., and they, bearing the Ark of the Covenant, spread it throughout the known world.

The first recorded plague is the one which beset the Philistines in 1200 B.C.E., and which is recorded in the Bible in the Book of Samuel. The Philistines in this year defeated an army of nomadic Hebrews at Eben-ezer, captured the sacred Ark of the Covenant, and carried it in triumph to Ashdod, a city near Mediterranean Sea.

But their triumph was immediately tainted, according to 1 Samuel 5:9: "the hand of the Lord was against the city with a very great destruction: and he smote the men of the city, both small and great, and they had emerods [swellings] in their secret parts."

The description makes it clear that bubonic plague had invaded the army of Philistines, probably from a stricken ship. If it had originated in the Ark of the Covenant, as the Bible notes, it would have been mentioned in the Old Testament.

Wherever they took the Ark, the Philistines took plague, too. They moved from Ashdod inland to Gath, then to Ekron. The plague followed them. Terrified, they trundled the Ark of the Covenant into a cart pulled by two milk cows. If the cows took the Ark to the Hebrew border town of Beth-shemesh, they reasoned that the Lord of Israel was responsible for the plague, and had indeed smitten them.

The cows took the Ark into the field of the Beth-shemite Joshua, stopping alongside a huge stone. Israel rejoiced, but not for long. In 1 Samuel 6:19, the Bible chronicles the inexorable progress of the plague: "And he smote the men of Beth-shemesh, because they had looked into the ark of the Lord, even He smote of the people fifty thousand and three score and ten men: and the people lamented because the Lord had smitten many of the people with a great slaughter."

THE WORLD
CHOLERA EPIDEMIC
1846–1860

....

The 15-year cholera epidemic that roamed the world from 1846 to 1860 killed millions.

For 15 years, cholera swept the world. When it began in India, in roughly 1840–41, it was generally conceded that great plagues and epidemics were either forms of supernatural punishment or the result of "miasmic vapors." However, by 1849, John Snow, a pioneer in inhalation anesthesia, proved in a highly conclusive paper that won him a 30,000-franc prize from the Institute of France, that cholera was waterborne and taken into the system by mouth.

The epidemic spread throughout India in the early 1840s, completely covering the country by 1846. From there, it spread to the Philippines and China in one direction and to Persia in the other. By 1848–49, it had consumed Europe. From there, it leaped to Canada and the United States via immigrant ships from Europe. New York, New Orleans, and ports along the Mississippi, Arkansas, and Tennessee Rivers were infested. By early 1849, all of the United States east of the Rockies was infected, and tens of thousands died monthly.

By the 1850s, Snow's theories had become widespread, and efforts were made to decontaminate water supplies. But by that time, the cholera had spread from Texas south to Mexico, to the West Indies, Cuba and Jamaica, the Bahamas, Puerto Rico, Venezuela, and Brazil. There would be no corner of the world that would escape this scourge, which John Snow, according to Norbert Hirshhorn and William Greenough's account in *Scientific American* in 1971, carefully mapped London and ". . . determined that most of the people who had died had drunk water from the Broad Street pump. The water, drawn from the Thames River, had the taste and odor of sewage. Snow advised the guardians of the parish to remove the pump handle, and the epidemic quickly waned. . . ."

Snow's contaminated water theory was soon dramatically confirmed. In 1854, London was struck by another severe cholera outbreak. In the intervening years, one of the two private water companies serving the city had changed its water source, drawing from a clean area of the Thames instead of the sewage-laden area. Among the users supplied by this company, relatively few persons succumbed to cholera. There were 461 deaths in a 14-week period, or 2.6 per 1,000. In the population served by the other company, which was still drawing from the sewage-laden waters, there were more than 4,000 deaths from cholera, or 15.3 per 1,000.

THE WORLD
PLAGUE
1855–Present

A modern theory proposes that the third pandemic of bubonic plague, which began in 1855, is still with us today. According to these theorists, it has been confined to Africa since 1959, but this is only a temporary confinement.

According to a convincing case made by Charles T. Gregg in his book *Plague!*, the horror of bubonic plague, which accounted for the Black Death that once decimated one-third of the known world (see pp. 208–211), is still with us.

According to Gregg, the so-called third pandemic wound down in 1959 when the total of world deaths dropped from nearly 5,000 in 1953 to a little more than 200 in 1959.

But more recent incidents of plague have been found to occur in areas characterized by poor sanitation practices, famine, or starvation. The carriers of the disease, rats and fleas, seem to thrive in these conditions. Outbursts of plague have occurred in Madagascar, where 132 cases were reported from 1953 to 1958 with a 72 percent mortality rate, and between 1960 and 1970, when another 143 cases were reported with a fatality rate of 52 percent. In the Zaire outbreak in 1960, the fatality rate was 80 percent, and in Nepal in February 1966, 12 out of 13 cases were fatal.

THE WORLD
INFLUENZA EPIDEMIC
1918–1919

Over 22 million people died in the mammoth influenza epidemic of 1918–19. Its true source has never been accurately traced, but it has been established that returning servicemen from World War I helped to spread it.

The monumentally devastating influenza epidemic of 1918–19 killed more than 22 million people in practically every corner of the globe. This was more than twice the total killed in World War I, which was at least partially responsible for the rapid progression of the pandemic.

One theory has it that the virus began in Fort Riley, Kansas, and was spread through Europe by American soldiers. But throughout history, epidemics have generally followed an east to west pattern, and this theory has generally been discounted.

Other historians have claimed that it was introduced into Europe by Chinese labor battalions that landed on the coast of France. Still others blamed it on Russian soldiers arriving from Vladivostok, others said it developed from an earlier bronchitis noted in Spain ber of 1918, one out of every five soldiers stationed in the United States was felled by influenza. Twenty-four thousand died. (Thirty-four thousand were killed in World War I from other causes.)

Makeshift emergency hospitals were set up in major cities and at major military encampments, but they were short-staffed because of the war. Fortunately, the outbreaks of the disease were short-lived. There would be one or two weeks of rapid dissemination of the disease, followed by two or three weeks of deaths, and then the epidemic would rapidly subside. The third week of October was the peak of the epidemic in the nation's major cities, particularly New York, New Orleans, and San Francisco. In all three cities, the mortality rate was roughly 28 percent.

The rest of the world did not fare quite so well. A staggering 12 million died in India alone. In Argentina, the mortality rate was a low 120 per 100,000 people, in England and Wales it climbed to 680 per 100,000, in South Africa it was 2,280 per 100,000.

What was astonishing about this pandemic was its ability to reach into remote sections of the world. In cloistered islands of the South Pacific, where respiratory diseases were all but unknown, whole populations were wiped out. The *Sydney Daily Telegraph* described the situation in Samoa: "As at one time 80 or 90 percent of the people were lying helpless, many died from starvation who might probably have recovered, for even when rice milk and other items were sent out and delivered, the survivors were too weak to prepare and apportion the food."

New Zealand recorded 6,717 influenza deaths, and some settlements in Alaska were completely eliminated.

The conductor on a Seattle, Washington, streetcar wears a surgical face mask distributed by the Red Cross during the worldwide influenza epidemic of 1918. (American Red Cross)

(it was generally called the "Spanish Influenza"), others blamed it on a confluence—and presumably an inter-infection—of American with European and African troops in the north of France.

Thus, the true source of the great influenza epidemic will probably never be accurately known (see the color insert on p. C-7). But whatever the sequence of events and the source, the pandemic made its first dramatic appearance in America on August 28, 1918, in Boston via a sailor on a transport tied up in the port. According to medical historian Henry A. Davidson, "It infected New England like a forest fire. In Massachusetts alone it killed 15,000 civilians in four months, plus an unknown number of others whose deaths were erroneously classed as 'pneumonia,' 'encephalitis,' 'meningitis' or masked under other rubrics."

Within days, the epidemic had traveled the length of the East Coast, but that was only the beginning. Some of the sailors on the same boat were transferred to Michigan and Illinois, and in September and Octo-

THE WORLD
AVIAN (BIRD) FLU
1959–Present

Avian flu, known genetically as virus subtype H5N1 and under its nickname bird flu, made its first tentative appearance in Scotland in 1959, disappeared in 1997, and has, since its reemergence in 2003, infected 256 people and killed 152. As of this writing, experts in the field continue to consider that its latest outbreak could become a possible pandemic, capable of killing anywhere from 2 million to 1,000 million people.

The ancestor of avian (bird) flu was first recorded in Italy in 1878. The cause of massive infections in poultry, it was then known as "Fowl Plague." Thousands of birds were slaughtered, and it disappeared, but, in characteristic fashion, reappeared during the world-wide "Spanish flu" epidemic in 1918. In 1924–25 and again in 1929, it surfaced once more, in the United States. No humans were infected in either outbreak, and once again, the disease disappeared until 1959.

By then, scientists had determined that an influenza virus was fundamental to its identity, had isolated the strain, and described it genetically as H5N1. The disease had struck large numbers of chickens on farms in Scotland where the poultry had perished from the disease.

Its next appearance was in England in 1991, where it decimated the populations of a large number of turkey farms. After this outbreak, the strain of flu seemed to disappear. But it was not gone for good. It developed, adding new genes that would turn it more virulent and dangerous.

In 1997, in Hong Kong, the first known case of avian flu infecting humans occurred: in May, a 3-year-old boy died in great respiratory distress in a Hong Kong hospital. Investigations after his death revealed that he had attended a daycare center where the teachers kept chicks as playthings for the children. Some of the birds had died from avian flu.

By December, more cases in humans surfaced, and an enormous slaughter of chickens, ducks, geese, quail, and pigeons had begun. Every market and farm in Hong Kong was cleared of both ill and healthy poultry: 1.5 million birds, in all, were killed. In March 1998, after 18 people had been reported stricken by avian flu and six had died, the epidemic was pronounced over.

A pattern developed: Most people who contracted the disease received it from contact either with infected poultry or surfaces contaminated with secretions or excretions from sick poultry. As the disease has developed, it has become clear that the spread of the viruses from one ill person to another has been reported very rarely, and transmission has not been observed to continue beyond one person.

The symptoms in humans have ranged from typical human flulike symptoms—fever, cough, sore throat, and muscle aches. But other manifestations can occur—eye infections, pneumonia, and severe respiratory distress, can lead to life-threatening complications.

From 1997 to 2003, no cases were reported, and then, in 2003, the disease reappeared with a vengeance. In January, a family visiting Fukien (Fujian) Province in China contracted the disease from poultry. Two members of the family died and the epidemic began once again. It had been, from the beginning, primarily a disease that began and mostly ended with birds. Tens of millions of birds, mostly in Asia, contracted the disease, and out of that number, only 256 people would fall victim to it. Until 2006, there would be no cases of it being transferred from human to human.

International agencies now conducted intensive studies, and found that the virus was most contagious among domesticated or farm chickens, ducks and turkeys. These birds shed the virus in their saliva, nasal secretions and feces, and so, susceptible birds become infected when they have contact with contaminated secretions or excretions—sometimes directly, and sometimes from surfaces such as dirt, cages, water, or feed that have been contaminated.

Infected wild birds—at least so far—carry the viruses worldwide in their intestines, but usually do not become sick from them.

Asia was the logical cradle of avian flu, since, in rural areas, families and livestock live in close proximity, often under the same roof. Later investigation found that there were outbreaks of the disease in Asia during 2003, but they were not reported. It was not until December 2003 that 140 tigers in a zoo in Thailand died after eating infected chicken carcasses—an incident that led to the discovery that the disease had already been transmitted to household cats as well.

Humans in 2003 seemed to escape the disease: Only 4 cases—one in China and three in Vietnam—were reported. However, this escape was short lived. In 2004, Thailand and Vietnam reported a mass infection in their poultry industries, which produced a total of 46 cases in human beings, resulting in 32 deaths.

In January 2004, a 13-year-old boy was hospitalized in Ho Chi Minh City for flu symptoms: a high fever, trouble breathing, and diarrhea. The boy lived near a live-poultry market and handled birds at cockfights. Within nine days, he was dead of avian flu.

In September, an 11-year-old girl was hospitalized with what was recognized as avian flu in Kamphaeng Phet, Thailand. The girl played and slept where chickens were kept. Her mother, who lived in Bangkok, where she had no exposure to birds, came to the hospital and stayed at her daughter's bedside for 16 hours, kissing and wiping the child's mouth. The child died and within the next two weeks the mother also died.

The epidemic was clearly increasing. Within weeks of the outbreak in Thailand and Vietnam, 10 other countries and regions in Asia, including Indonesia, South Korea, Japan, and China, reported evidence of a new and more virulent strain of the disease. Waterfowl were found to be directly spreading the pathogenic infection to crows, pigeons and other birds. More than 100 million birds were killed during 2004, which resulted in a brief cessation in the spread of the disease. However, in October 2004, scientists found that domestic ducks could act as silent reservoirs, excreting the virus without becoming sick themselves.

The incidents of infection in humans increased from 46 in 2004 to 97 in 2005. It was found that there was no evidence of the virus in birds that had been cooked. But in Vietnam, the virus was found to have transmitted to at least two persons through the consumption of uncooked duck blood.

Of the four South Asian countries contaminated by avian flu—Cambodia, Indonesia, Thailand, and Vietnam—Cambodia is the poorest, and poverty played a role in the forward march of the disease in 2005. In all of South Asia, chickens still wander freely in and out of homes and even apartment buildings, mixing constantly with people to an extent not found in most other countries.

This was the case in the village of Prey Rognieng, just west of Cambodia's capital, Phnom Penh. In the summer 2004, half-starved chickens began to die in the streets, and shortly afterward, some of the village's children came down with flulike symptoms. Their parents took them to a clinic in Phnom Penh, where they were found to have not avian, but ordinary flu.

Some other village residents, however, believed that the outbreak came from witchcraft, practiced by the only village resident who had not been born there. Fifty-three-year old Som Sorn had arrived in the village eight years before.

On a day when Som Sorn had gone into the jungle to cut wood and his wife had begun cooking rice over a fire on the dirt floor of her hut, a local man with a machete appeared at the hut's entrance. "[He] grabbed her hair, pulled her head back and cut her throat," Ya Phreorng, the village leader later told Western reporters. "Her neck was almost completely severed."

After finishing his task, the man collected $30 from the woman's grateful neighbors. When he was later arrested, he was reportedly astonished that anyone would identify him to the police and equally surprised to be arrested and sentenced to 15 to 20 years in prison. The neighbors, who paid the killer, after making further payments to the police, were not prosecuted.

The conditions of this tragedy reflected, according to U.N. officials, the difficulty of preventing a global human epidemic of bird flu, since the disease is most prevalent among poultry and wild birds in impoverished rural areas of Southeast Asia in which the inhabitants have low levels of literacy, high levels of superstition, and very little health care.

Now, the disease began to proceed out of Asia, westward into Europe. In October 2005, avian flu appeared in Turkey and Romania, the first time the disease had been reported in Europe. No one died from the disease in Turkey that year, but in 2006, 12 people would be infected, and four of them would die.

In November 2005, in Toronto, Canada, nearly 35 wild ducks tested positive for bird flu but results showed that their flu was of a different, less lethal variety. It was the second scare for Canada. In 2004, approximately 17 million birds had been slaughtered in British Columbia when rumors of the presence of avian flu had surfaced.

In January 2006, a vaccine was developed that proved 100 percent effective in mice and chickens. But it was a long way from ready for use in humans. While experimentation on the vaccine continued, avian flu intensified, as migratory birds were now carrying the infection, and spreading it westward. Winter 2006 was unusually cold in Europe, and the great annual northern migration from Africa began early. Poultry had been infected with the virus in Nigeria and Niger, and there was testing for the virus in dead poultry in Kenya, Ethiopia, Gabon, Gambia, and Sierra Leone. In March, the beginning of the spring migration of waterfowl reached France where more than 400 turkeys were killed in a matter of hours, infected by a duck that had migrated west from the Black Sea. More ducks arrived, excreting the virus into ponds from which other birds drank. Dozens of wild waterfowl were subsequently found dead in the Dombes wetlands near Joyeux, in the southeastern part of France.

In Sweden, two ducks were found dead on the Baltic Sea coast. Both contained the flu virus. A cat on the Baltic island of Rugen in Germany tested positive for the virus, and previous discoveries of cats infecting other cats opened the possibility of transmission of the disease between and among mammals. In Azerbaijan, several children who were involved in removing feathers from wild dead swans became infected, but recovered. In Turkey, two teenage siblings died shortly after playing catch with the heads of dead chickens. In Indonesia, other evidence of human-to-human spread of the disease surfaced. The World Health Organization (WHO) reported that Avian Flu infected eight people in one rural family. The first family member became ill through contact with infected poultry, and she infected six other members of the family. One of these six was a child who in turn infected her father. And there, the circle of infection closed, strengthening the view that human-to-human infection only proceeds to one person.

Studies have shown that the virus can latch on only to cells deep in the respiratory system in humans, too far down to be coughed up or sneezed out to infect other people. This, however, according to Dr. David Nabarro, could change in the future, considering the constant mutations of the disease. "If bird flu ever gained the ability to spread easily among humans [one] patient would . . . infect thousands before diagnosis," he told reporters.

Epidemiologists forecast another problem in 2006: Since humans have not (to the point of this writing) been affected in huge numbers yet, if the disease graduated to the status of a pandemic, they would have no resistance to the infection.

The developed vaccine has been used in Vietnam to vaccinate chickens, but it has yet to advance to human use. Antiviral drugs, particularly Tamiflu and Relenza, when taken within 24 hours of the onset of the disease, have proved to make the illness less severe and life threatening, but they do not prevent its spread. Vaccine is the answer, but it is not yet available, and the supply of flu drugs at present cannot stop a pandemic if it starts (see the color insert on p. C-7).

So for the moment, at least, containment has been the method of choice in efforts to avert a pandemic. One method is the practice of "culling"—slaughtering infected birds, hopefully with as little spilling of blood as possible, and then disposing of the bodies so that they cannot spread the virus. This has proved to be an enormous job, since if one bird is infected on a farm the entire flock and birds in a large circle around it must be exterminated. Farmers, particularly in poor countries, have been loath to cooperate, and demand compensation for their lost flocks. In Turkey, farmers get up to $3.50 a chicken; Vietnam pays $1; Indonesia pays 80 cents; and China pays 60 cents. Still, farmers bribe cullers to save their flocks, and breeders of fighting cocks, worth up to $5,000, sneak their chickens away, unwilling to sacrifice the cash or the bloodlines.

And still, the number of cases of avian flu increases worldwide. From 2005 to 2006, the number of cases of human infection jumped from 97 to 109, and the deaths from 42 to 74. From 2003 to 2006, 256 cases of human infection have been reported, with a consequent toll of 152 deaths, mostly in people whose lungs simply gave out, but in the cases of many children, through the development of encephalitis, since the virus had attacked their brains.

To some epidemiologists, this is only the beginning. Migrating birds, bird smuggling, and travelers jetting from one country in Asia to other countries throughout the world could—and perhaps already have—spread the virus worldwide. Many scientists say it is just a matter of time before avian flu is detected in birds in North America, since migratory birds from Europe and Africa, which are natural hosts for the virus, mingle with North American birds in Greenland and Canada.

If avian flu becomes a pandemic, millions—some scientists say fully half of the world's population, in the thousands of millions—will become infected. As of this writing, only one-fifth of the world's countries have a pandemic-response plan, according to the WHO, and those plans vary widely in comprehensiveness. Fewer than 10 nations have domestic vaccine companies trying to produce an avian-flu vaccine, mainly because producing vaccines is notoriously unprofitable. Nevertheless, the United States is considering a plan to spend $10 billion on stockpiling vaccines. "We are seeing the unfolding of a pandemic in slow motion," Dr. Klaus Stohr of WHO said in a speech to business leaders in 2005. "We can reduce the damage, but we cannot avoid it."

THE WORLD
AIDS EPIDEMIC
1980–Present

From the first reported case in the United States, in 1980, AIDS has spread worldwide.

AIDS (acquired immune deficiency syndrome) is the possible pandemic of the 21st century. "Anyone who has the least ability to look into the future," noted Dr. Ward Cates of the U.S. Centers for Disease Control in 1989, "can already see the potential for this disease being much worse than anything mankind has seen before."

Epidemiologists are fairly certain that the AIDS virus originated in Central Africa, where it was originally harbored by green monkeys. The generally accepted theory states that the virus was mutated in a way that made it able to attack humans. According to Dr. Myron Essex, of the Harvard School of Public Health, a mechanism may have evolved in these monkeys to control the virus that infects them, so that they remained healthy, while those that they bit became infected with the virus.

Generally speaking, the level of health in Africa deteriorated in the 20th century while the health of the more affluent nations of the world improved. The transfer of the virus from human to human was unwittingly accomplished in the late 1970s and early 1980s through health workers using a single needle to inject a number of patients with penicillin and other drugs. In 1986, *Newsweek* reported that 10 percent of the blood stocks in Zambia were contaminated with AIDS; it was easy to contract it during transfusions.

But the largest early transmission of the disease in Africa is thought to have been sexual. Since multiple heterosexual partners (some Africans infected with AIDS were reported to have had an average of 32 sexual partners) and visits to prostitutes constituted a way of life in Central Africa, the disease spread quickly and alarmingly.

From Africa, some theorists state that the disease was spread to the Caribbean by natives, and was then picked up by American homosexuals, many of whom had made Haiti a vacation spot of choice in the 1970s. Haiti hotly contests this, although medical investigators from the Centers for Disease Control and the University of Miami Medical School, as well as doctors from Cornell University, traced the first cases of AIDS-related diseases to Port-au-Prince in 1979, two years before the first well documented cases of what was later called AIDS were found in San Francisco and New York.

In 1981, a young male homosexual in San Francisco was found to be suffering from a severe fungal infection to which he had little immune reaction. Shortly afterward, he developed *pneumocystis carinii* pneumonia (PCP). At roughly the same time, a dermatologist in New York encountered two cases of Kaposi's sarcoma (KS) in one week, both in young male homosexuals.

It would be 1982 before the virus was given the name HIV, and there are no records of who named it, but in 1980–81, Dr. Michael S. Gottlieb of the University of California at Los Angeles noted that several men he had examined had contracted *pneumocystis carinii* pneumonia, or PCP—a rare type of pneumonia that is caused by a small protozoan organism of a mere two to four nanometers in size (a nanometer is one-billionth of a meter). Discovered in 1955, it first appeared in malnourished infants housed in displaced person camps after World War II, and then only in cancer patients, or in organ transplant recipients, whose immune systems had become weakened.

At roughly the same time, Dr. Alvin Friedman-Kien of New York University Medical Center, diagnosed a young, gay man with Kaposi's sarcoma, a particularly uncommon, slow-growing cancer customarily seen among elderly men of Mediterranean extraction. The cancer, rare in the United States but prevalent in equatorial Africa, is a tumor of the blood vessels. In this disease, the lining of the body's small vessels becomes stippled with irregular tumor cells which cling to the inner walls. When the lymph nodes and internal organs are affected, the tumor can clog narrow blood vessels; thus, limbs that are affected may become swollen and internal organs congested and enlarged.

In older patients, this process takes anywhere from eight to 30 years to develop. In young AIDS patients (the vast majority are in their thirties) it becomes fatal in three.

At first, AIDS was thought to be a strictly homosexual disease, transmitted sexually. Then, it was discovered that the transmission was through the blood, and intra-venous drugs users and hemophiliacs, who must receive constant blood transfusions, were added to the list of people at risk. By early 1983, heterosexual recipients of blood transfusions were also thought to be potentially at risk. Then, in Africa, heterosexual transmission of the disease was discovered to be a common occurrence.

It would be 1984 before the AIDS virus would be isolated by two researchers, Dr. Robert C. Gallo of the U.S. National Cancer Institute, and Dr. Luc Montagnier at the Pasteur Institute in Paris. (In 1990, Dr. Gallo admitted that he had used some of Dr. Montagnier's early findings to reach his breakthrough and acknowledged that the French research team actually isolated the virus.)

The findings revealed that the virus reproduces itself, killing, in an accelerating fashion, T-4 lymphocytes, a subgroup of white blood cells that play a major role in defending the body against infections like PCP and some cancers.

Unfortunately, the discovery of AIDS coincided with a resurgence of conservative political and religious philosophy in the United States. The fact that the disease first seemed confined to homosexuals caused the government to do little about it. "The poor homosexuals," Nixon speechwriter and Reagan staff member Patrick Buchanan said in 1983. "They have declared war on nature, and now nature is exacting an awful retribution." The Reverend Jerry Falwell and his moral majority preached like sermons, and other conservative spokespersons advocated everything from 20th-century leper colonies to summary executions for victims of the nascent pandemic. Historical perspective will undoubtedly compare the behavior of modern-day American politicians with the panic that surrounded the Black Death in medieval Europe (see pp. 208–211).

By the time Surgeon General Dr. C. Everett Koop had shaken the U.S. administration into a grudging acknowledgment that an epidemic cannot be controlled by moral pronouncements, the epidemic had burst full force upon the world. In the United States the victim count grew from one in 1980, to 12,000 in 1985, and 38,312 in 1987, and of that figure, 22,057 died. By 1989, the estimate of infected individuals in New York State alone—the state with the largest number of AIDS cases—would rise to 400,000, and of those 400,000, scientists further estimated that 88 percent to 90 percent of them were still untested, and therefore probably freely spreading the disease. The National Institute of Health in 1989 estimated that up to 1.5 million people throughout the United States were already infected with AIDS, and that by 1992, the number of deaths would total 263,000.

Europe's statistics were scarcely less alarming, with constantly escalating figures in every country, ranging, in 1987, from 1,980 in France to 1,298 in Germany.

As the 1990s unfolded, AIDS continued to proliferate. By the mid-1990s, the World Health Association would estimate that 14,000,000 people in the world were infected with the virus, most of them as a result of heterosexual sex. And, as preventive and educational efforts faltered, the future began to look exceedingly bleak.

In 1993, the Centers for Disease Control redefined the disease, adding three illnesses—invasive cancer of the cervix, pulmonary tuberculosis, and recurrent bacterial pneumonia—to the list of 23 diseases already used to classify a person as having AIDS. The redefinition pinpointed women and drug users who had previously not been contained under the epidemic's umbrella.

By 1993, AIDS could no longer be defined, either, as an underground affliction. Dancer Rudolf Nureyev and tennis star Arthur Ashe died of complications related to AIDS and suddenly it belonged to the mainstream of humankind.

And now, AIDS began to spread to areas of the world that heretofore had been thought to be free of it. Latin America, in denial for years because of social and religious forces, finally admitted that the virus had reached its countries and was growing. The reluctance of men to discuss homosexual and heterosexual affairs with their wives allowed these wives to become infected, and the disease, in turn, spread to their unborn children.

In Brazil, a growing legion of infected women pushed the total of infected people to roughly 1,000,000 in a population of 150,000,000, making it second, in 1993, to the United States in AIDS cases. Mexico's total rose to 500,000. Colombia admitted to 200,000 and Argentina, 2,754.

Haiti had its AIDS plague, a heterosexually transmitted disease in this poor country in which inhabitants simply could not afford drugs. Argentina, on the other hand, with its wealth and ability to buy drugs, had a higher percentage of AIDS cases traceable to contaminated needles.

Africa continued to be the largest repository of AIDS in the world. Uganda's government estimated, in 1993, that 9 percent of the country's 16.7 million people carried the AIDS virus.

In the early 1990s, AIDS swept southward from its equatorial epicenter to South Africa. Three hundred thousand black heterosexuals were diagnosed with the HIV virus by the middle of 1993, and estimates were that it was spreading to at least 300 new carriers every day. As in other African countries, ignorance of AIDS was abnormally high.

The figures for those infected with AIDS in North Africa and the Middle East climbed to 75,000 by 1995, and in sub-Saharan Africa, the figure was a numbing 8,000,000. Until 1996, Zaire had been an exception. A vigorous prevention campaign waged by its government had kept it from the statistical map. But as political and social unrest hit the country in 1996, AIDS began to spread, and it appeared that Zaire would soon join its neighbors Uganda, Rwanda, and other East African countries with staggering numbers of AIDS sufferers.

In 1996, too, AIDS exploded into Asia. Thailand, with its permissive sexual laws, was the first to feel the full effect of its invasion. Though the first cases reported in Asia went back to the late 1980s—a decade after the United States outbreak—it would take until 1996 before it reached epidemic proportions. In 1995, 50,000 people died of AIDS in Thailand alone. Cambodia, Myanmar, Malaysia, and Brunei all reported huge increases in HIV infection.

But the most severely affected Asian nation was India. In 1997, a focus on 5 million truck drivers whose habit it was to buy sex every day, sometimes several times a day from local women and teenage girls for as little as 10 rupees (28 cents), revealed widespread HIV infection.

As a result, fanned by the flames of denial and taboo, AIDS began to rampage through India, enough so that the United Nations predicted that by the end of the decade, India would replace Africa as the center of the AIDS plague, with 10,000,000 infected with the HIV virus—a quarter of all the worldwide projected infections.

In a reversal, in 1996, Uganda became the first African nation to exhibit a decline in AIDS cases. Preventive methods, once considered too expensive and involved to work in African nations, seemed to take hold in Uganda. Prenatal clinics reported a sharp decline in HIV infection from 30 percent to between 15 percent and 20 percent.

Still, the scale of infection in Uganda was enormous. Between 1.2 million and 1.5 million people were infected, and each year, an estimated 150,000 to 200,000 Ugandans died from the disease.

To balance this, there was further encouraging news. By 1996, strides began to be made in therapies to treat AIDS. In addition to AZT, a class of drugs known as protease inhibitors attached to the old drugs, were hailed as potent forces against the disease.

Experiments began in mid-1996 of combining AZT, 3TC, and a third drug that attacks the protein enzyme known as protease. The combination, the so-called cocktail, made it difficult for the HIV virus to sidestep the drugs, as it had with AZT and 3TC alone.

It was an encouraging situation. Conservative U.S. governments in the 1980s had first ignored the disease, then condemned it, then blocked, through the FDA, testing of new drugs. The Clinton administration reversed this practice, and canceled other inhibiting regulations, thus opening the door for research into not only treatment but a potential cure. And although there were significant improvements in U.S. support of AIDS research and prevention during these years, the advent of a new, conservative administration in Washington, D.C., in 2000 did little, at first, to help international efforts to prevent the spread of the pandemic.

In January 2001, on his first day in office, President George W. Bush reinstated the Reagan era "global gag rule" on international family planning assistance.

In May 2002, U.S. administration representatives at the United Nations Children's Summit opposed the use of condoms for HIV/AIDS prevention. In July, the administration withheld from the U.N. Population Fund $34 million in funding for birth control, maternal and childcare, and HIV/AIDS prevention. In August, it withheld more than $200 million in funding to support/HIV-infected women.

Even later, in 2003, when the purse strings were—in a State of the Union speech given by the president—abruptly and unexpectedly loosened, and a $15 billion, five-year campaign to control AIDS in Africa was announced, the campaign and the promise came with a caveat: One third of the funds spent on prevention programs—approximately $130 million—could only promote abstinence before marriage and could not support condom use.

The restriction was immediately seized upon by Ugandan President Yoweri Museveni, who deemed condoms "inappropriate for Ugandans," and banned them, despite his country's record as having one of the highest rates of HIV infection in the world.

The entire $15 billion initiative also fell prey to restrictions and bureaucracy. Instead of utilizing The Global Fund, founded in 2001 to pool funds from many nations, organizations, and religious institutions to fight AIDS, the U.S. administration offered a $1 billion contribution to the fund, but withheld $88 million of this because of a bookkeeping difference in the fiscal years of various countries, and a calculation that the U.S. economy is only 30 percent of the world economy. The effect was to leave a group of programs unfunded. A partial amount of the remainder was delivered to 15 nations individually.

A further problem arose because of the U.S. administration's refusal to buy and send pharmaceuticals that lacked approval from the U.S. Food and Drug Administration (FDA), thus closing the door to nonbranded, generic AIDS therapies. In addition, the United States severely curtailed its participation in the 2004 U.N. World AIDS Conference, thus forcing the cancellation of meetings, workshops and subsidiary conferences. The negative impact was regretted by Peter Piot, of the United Nations AIDS program, because, as he noted, "The largest group in the world in terms of AIDS expertise comes from the U.S."

Reaction worldwide and at home to the gap between promises and delivery and the restrictions on needed pharmaceuticals forced another about face, and in early 2004, the U.S. Administration directed the FDA to license generics for use in U.S. Global AIDS program, although these generics could not be sold in the United States.

Still, further participation in the sixteenth World Aids Conference in Toronto in 2006 continued to be minimal from the United States.

The gains, however, were offset by the expense of the new drugs. Patients elated by the prospect of staving off the former death sentence of AIDS ran into a wall of resistance from HMOs. The cost of the drugs amounted to totals of from $10,000 to $15,000 a year, which some patients would have to spend indefinitely. The insurance companies held fast to their $3,000 per year ceiling, and law suits erupted around the United States utilizing provisions of the Americans With Disabilities Act of 1990.

At least Americans had that possibility. The gigantic cost of the new drugs to combat AIDS, while producing a 12 percent decline in deaths from the disease in the United States in 1997, caused hardly a ripple in the rest of the world. In Africa, affording treatment was as impossible as flying.

And now Russia, absorbing the habits of the West, was the latest victim of the pandemic. Kindled by a surge in the use of an easily contaminated liquid form of heroin, and fed by poverty and unemployment, HIV spread like a forest fire through Russia. It was estimated that 75 percent of the HIV infections in Russia came from contaminated needles. In the year 2000, 2,000 new cases of AIDS were reported.

And in Africa, the pandemic continued. In 1998, almost 3,000,000 South Africans—12 percent of all adults—were infected by HIV. Zimbabwe, Botswana, Kenya, and Nigeria had similar percentages.

The 12th World AIDS Conference, held in Geneva in July of 1998, in contrast to the euphoria of the Vancouver meeting of 1996, in which the discovery of new drugs was announced, was somber. There was no breakthrough with a vaccine that would end the AIDS pandemic. Tested on monkeys, the vaccine gave them the disease, rather than curing them of it. And some new drugs had proved to be failures. While some patients had left their death beds and returned to work, others had not responded, and still others had suffered extreme side effects. The total number of people with AIDS worldwide had climbed to 34,000,000.

But as the 1990s ran out, more research did produce new drugs. Antiretrovirals proved to be constantly successful in preventing deaths from AIDS. But access to this treatment was confined to the United States and other wealthy countries. In Africa, where 25 million people were infected by the year 2000, fewer than 25,000 received the therapy that could avert their deaths. By 2006, this situation had hardly improved, despite a dramatic expansion of antiretroviral therapy. More then 1 million people were receiving antiretroviral treatment by June 2006, a tenfold increase since December 2003. But the sheer scale of need in Africa meant that only 20 percent of individuals clinically qualified to receive antiretroviral therapy were receiving these lifesaving medicines and the 80 percent who did not receive them were estimated to die within the next year.

The first negotiations between world health organizations and pharmaceutical companies over global access to AIDS drugs began in Geneva in 1991. They went nowhere. And so, patients in poor countries, through the 1990s, died in six months or less. The cost of the new drugs put them beyond all but the imagination of most people in the world.

"The brutal fact," health economist William McGreevey told a World Bank audience on May 22, 1998, was that "those who could pay for Africa's AIDS therapy—the pharmaceutical industry, by way of price cuts and rich country taxpayers, by way of foreign aid—are very unlikely to be persuaded to do so." His words did not fall on entirely deaf ears, but in sub-Saharan Africa, it would seem so. In 2000, 2.4 million people died of AIDS in Africa alone. By 2006, 24.7 million people—almost two-thirds of all persons infected with HIV—were living in sub-Saharan Africa. In that year, an estimated 2.8 million adults and children became infected with HIV, more than in all other regions of the world combined. In fact, the 2.1 million AIDS deaths in sub-Saharan Africa represented 72 percent of global AIDS mortality.

Throughout this region, women would bear a disproportionate part of the AIDS burden. Not only were they more likely than men to be infected with HIV but also in most countries they were also more likely to be the ones caring for people infected with the disease.

The one country in southern Africa that experienced a declining adult HIV presence in 2006 was Zimbabwe, where both the prevalence and incidence fell, apparently in relation to a combination of factors, especially reductions in casual sex relations with non-regular partners, along with increases in condom use and later sexual debuts. Nevertheless, approximately one in five adults in Zimbabwe continued to live with HIV and young women from 15 to 25 years old were four times more likely to be HIV-infected than young men.

In contrast, North America in 2005 reported 350,000 cases of HIV/AIDS were attributed to women out of 1.2 million cases. The reason, the United Nations concluded, was that in developing countries, women are often unable to refuse sex or to dictate protective practices such as the use of condoms.

In addition, the number of new HIV cases varied only slightly from the late 1990s to 2006, and the widespread availability of antiretroviral treatment increased the longevity of those infected. Still, women and minorities in both the United States and Canada remained at a significantly higher risk for contracting HIV, which reflected larger systematic biases in health care and prevention efforts. In the United States, 50 percent of newly reported infections from 2000 to 2006 were among African Americans, although the group represented only 12 percent of the population. Thus, their HIV prevalence was as much as 12 times higher than that of whites.

African-American women accounted for an increasing proportion of new infections. Moreover, HIV infection in 2006 was the leading cause of death for African-American women aged 25–34. Many of these women did not engage in high risk behavior, but were contracting HIV through sex with their long-term male partners, a significant proportion of whom also had sex with men or injected drugs.

To add to the divide, a new U.N. statistic in 2006 revealed that Hispanics in the United States were experiencing significantly higher rates of HIV infection than whites. Latinos comprised 18 percent of new HIV diagnoses in 2005, while accounting for only 14 percent of the total U.S. population.

The AIDS pandemic, if it continues at its present pace, will, by 2010, have eclipsed both the 14th-century Black Death and the 1918 flu pandemic, thus making it the deadliest epidemic in human history. As of December 2006, there were 39.5 million people living with HIV, including 2.3 million children under the age of 15. The number of deaths continued to climb. In 2004, the total was 2.7 million and in 2006, it had risen to 2.9 million.

On a more optimistic note, by 2006, the global AIDS pandemic, marking its 25th anniversary, had begun to show signs of maturation. It had become, finally, as editor Chinua Akukwe, an editor at www.worldpress.org, wrote, ". . . a mainstream political issue, engaging policymakers in the United Nations and at the highest national governments, bilateral and multilateral institutions, and regional organizations. HIV/AIDS also became universally accepted as a national and international security issue."

Despite this good news, the fight to stem the growth of the pandemic is far from over. Approximately 4.3 million people worldwide contracted HIV in 2006. As of this writing, less than 50 percent of young people, according to the Joint United Nations Programme on HIV/AIDS, have "comprehensive knowledge" of HIV preventive strategies, in spite of the goal stated in the 2001 U.N. Declaration of Commitment on HIV/AIDS to reach 90 percent of young people by the end of 2005. Only 9 percent of pregnant women worldwide have received antiretroviral therapy that can prevent HIV infection in newborns and infants. To this date, there is no comprehensive remedial effort to meet the needs of 15 million AIDS orphans worldwide.

Most of the progress against the disease has been composed of emergency actions to stave off imminent disaster. This is all well and good, but the only cure that will stop the growth of this monster pandemic is, if history is the teacher, a vaccine. But in 2007 United States, there is still a lack of federal support for research. Prevention and education measures are still fought by conservative groups and the FDA has blocked the testing of some new drugs, including those designed to alleviate the symptoms of AIDS. As a result, some desperate patients and their families travel from the United States to Mexico and Europe in search of miracle cures.

Researchers are cautiously optimistic about finding a vaccine. Still, the arrival of it can only accelerate if, instead of unkept or modified promises, a sustained universal increase in both unfettered financial support and preventive education comes about. If so, this, the grimmest, cruelest, and most cataclysmic of all pandemics yet may be alleviated within the lifetime of some of the present world population.

THE WORLD
SARS
November 2002–May 2005

The short-lived epidemic of SARS (Severe Acute Respiratory Syndrome)—which began in November 2002 in China, where its presence was kept silent for months—infected 8,096 persons and killed 774 of them before it was pronounced "eradicated" in May 2005.

The secretive policies of the government of the People's Republic of China underwent a necessary change with the advent of SARS, or Severe Acute Respiratory Syndrome. The disease originated in China's Kwangtung (Guangdong) Province in November 2002 where 305 residents of the province contracted the disease, but

Chinese authorities, while making an effort to control the outbreak, restricted media coverage of it—reportedly to preserve public confidence in their ability to protect that public. But information, even in repressive societies, has a way of leaking out, and in April 2003, Dr. Jiang Yanyong—at great personal risk—exposed the repression and the Chinese government confessed to the World Health Organization (WHO) that it had withheld information about the epidemic.

But by then, the world was already aware of SARS. In February 2003, an American businessman traveling from China came down with pneumonialike symptoms while on a flight to Singapore. The plane stopped in Hanoi, Vietnam, and the man was transferred to a hospital, where he died. Several of the doctors and nurses in that hospital soon came down with the same disease, prompting the WHO to issue a global alert on March 12.

Also in February 2003, SARS had spread to Hong Kong. Of the first 45 people who contracted it, most had been in the company of employees of the Prince of Wales Hospital. And those people, in turn, had had contact with a smaller circle who had either treated or visited a 26-year-old male patient who had been diagnosed with a nonspecific fever (one of the symptoms of SARS, along with headaches, body aches, a dry cough, and shortness of breath, is a fever of 100.4 degrees or higher).

This patient in turn had visited a friend on the ninth floor of the Metropole Hotel in Hong Kong's Kowloon district in February. Six other people who had stayed on the same floor of the hotel between February 12 and March 2 had also contracted SARS. One of them might have had a more important role in the spread of the disease than the patient: He was a 64-year-old doctor from Kuang-chou (Guangzhou), the capital of Kwangtung. Members of Hong Kong's health ministry concluded that he might have been the bearer of SARS from Kwangtung and had infected the others at the Metropole Hotel. This theory was strengthened when it was found that another man from Kwangtung took SARS with him to Spain, and infected people there. At the same time, several Metropole victims also set out into the world to Singapore, Vietnam, and Canada, infecting other people who in turn spread the disease even farther afield.

By the end of February, the WHO and the United States Centers for Disease Control and Prevention (CDC) issued a global alert. Both launched, in concert with a number of disease labs, a coordinated effort to understand the illness as quickly as possible.

By mid-March, the University of Hong Kong announced that a strain of corona virus, possibly never seen before in humans, was the infectious agent responsible for the spread of SARS, and by April the WHO stated that a corona virus was the official cause of the disease. It was agreed that SARS was spread by the inhalation of droplets expelled by an infected person when coughing or sneezing, or possibly by contact with secretions on objects. Antibiotics were ineffective in treating the disease. The only strategies that seemed to work were the transmission of antipyretics, supplemental oxygen, and ventilatory support.

More than 1,200 suspected cases were put under quarantine in Hong Kong, 977 in Singapore, and 1,147 in Taiwan, and in Canada, thousands were quarantined. In Singapore, schools were closed for 10 days, and longer in Hong Kong. On March 24, Singapore invoked the Infectious Diseases Act, which mandated a 10-day home quarantine for anyone who might have come in contact with SARS patients. In addition, SARS patients discharged from Singapore hospitals were given a 21-day home quarantine, enforced by telephone surveillance.

On March 27, 2003, the WHO recommended the screening of airline passengers, particularly those traveling to and from Asia, for symptoms of SARS, and on April 23, the organization advised against all but essential travel to Toronto, noting that a small but potent number of persons from Toronto appeared to have exported SARS to other parts of the world. Toronto public health officials questioned this, and by April 30, the advisory was withdrawn. But damage to the Toronto tourism industry had already been done by the WHO's announcement. The Rolling Stones and other rock groups organized a massive "Molson Canadian Rocks for Toronto" concert, entitled *SARSstock,* to revitalize the city's tourism trade.

It helped somewhat, but the U.S. Library of Congress officially excused itself from attending the American Library Association convention in Toronto in the summer of 2003. Most conferences and conventions scheduled for the city were cancelled, and the production of at least one movie was moved out of Toronto. The hotel occupancy rate was reported by the Canadian Broadcasting Corporation on April 22 to be only half the normal rate.

In other parts of the world, similar conditions occurred. The 2003 FIFA Women's World Cup, scheduled to be held in China, was moved to the United States, and the International Ice Hockey Federation cancelled the 2003 IHF Women's World Championship tournament which was to take place in Beijing. Hong Kong merchants withdrew from an international jewelry and timepiece exhibition in Zurich. North American airline bookings to Hong Kong plunged more than 85 percent. Personal bankruptcies in Hong Kong rose 74 percent, and retail sales plunged 50 percent. In the Chinatowns of New York and San Francisco, customers of Chinese cuisine decreased by as much as 90 percent. And the Catholic Church of Singapore suspended confessions in booths and instead granted "general forgiveness" to its followers. In Ontario, Canada, worshipers were asked to refrain from kissing icons, dipping their hands in holy water, and sharing Communion wine.

Finally, in May 2005, the disease was declared "eradicated" by the WHO. It therefore became only the second disease in history, with smallpox, to receive this label. In its brief life, SARS had been contracted by 8,096 people in 28 countries. Some of these countries—Germany, Mongolia, Thailand, France, Malaysia, Sweden, Italy, the United Kingdom, India, South Korea, Indonesia, South Africa, Macau, Kuwait, New Zealand, Ireland, Romania, Russia, Spain, and Switzerland—only had single digit cases (the United States had 27). Others, like China, which had 5,327, Hong Kong which had 1,755, Canada which had 432, Taiwan, which had 346, and Singapore, which had 238, were more severely hit. Of these cases, 774 died, a fatality percentage of 9.6 percent, which was higher than that of the 1918–19 Spanish influenza pandemic (see p. 220).

CYCLONES

· ·

THE WORST RECORDED CYCLONES

* Detailed in text

Australia
    Darwin (1974)

Bangladesh
    * (1970) (worst disaster of century)
    * (1985)
    * Southeastern Coast (1991)

Burma
    (1926)

China
    Hong Kong (1906)

India
    * Andhra Pradesh (1996)

* Backergunge (1876)
* Bay of Bengal (1737)
Bay of Bengal (1942)
Bombay (1882)
Calcutta (1833)
* Calcutta (1864)
Ceylon (1964)
* Coringa (1789)
* Coringa (1839)
Cuttack District (1971)
* Orissa (1967)
* Orissa (1999)
Pondicherry (1916)

Madagascar
    Antalaha (2004)

Pakistan
    (1960)
    Chittagong (1965)
    Ganges Delta (1965)
    Hyderabad (1964)
    Jessore (1964)

Persian Gulf
    (1925)

United States
    Carolina Coasts (1881)

CHRONOLOGY

* Detailed in text

1737
    October 7
      * Bay of Bengal, India

1789
    December
      * Coringa, India

1833
    May
      Calcutta, India

1839
    November
      * Coringa, India

1864
    October 5
      * Calcutta, India

1876
    October 31
      * Backergunge, India

1881
    August 27
      Carolina Coasts

1882
    June 5
      Bombay, India

1906
    September 18
      Hong Kong

1916
    December 1
      Pondicherry, India

1925
    October 23
      Persian Gulf

1926
    May 28
      Burma

1942
    October 16
      Bay of Bengal, India

1960
    October 31
      Pakistan

1964
    April 12
      Jessore, Pakistan
    June 13
      Hyderabad, Pakistan
    December 23
      Ceylon, India

1965
    Chittagong, Pakistan
    May 11
      Ganges Delta, Pakistan

1967
    October 12
      * Orissa District, India

1970
    November 12
      * Bangladesh

1971
    October 29
      Cuttack District, India

1974
    December 25
      Darwin, Australia

1985
    May 25
      * Bangladesh

1991
    April 30
      * Southeastern Coast,
      Bangladesh

1996
    November 8
      * Andhra Pradesh, India

1999
    October 29
      * Orissa, India

2004
    March 8
      Antalaha, Madagascar

CYCLONES

Cyclone is the generic name given to storms that rotate around a core of low pressure in a counter-clockwise direction in the Northern Hemisphere and in a clockwise direction in the Southern Hemisphere. This circular movement is caused by a combination of two forces: (1) the contrast between the low core, or axis of atmospheric pressure and the relatively higher pressure surrounding it, and (2) the Coriolis effect, which, simply stated, is the tendency for any moving body on or above the Earth's surface to drift sideways from its course because of the rotation of the Earth. In the Northern Hemisphere, this deflection is to the right of the motion; in the Southern Hemisphere, the deflection is to the left. The combination of these two forces sets up what is termed a "cyclonic pattern."

This establishes the pattern of the storm; further characteristics of low pressure systems contribute to the destructive forces within the heart of the cyclone. The force of air moving over the Earth's surface is affected by objects in its path or variations in the Earth's surface. For instance, on or near the surface of the Earth, there is a frictional drag on air currents, which causes them to spiral inward toward areas of low pressure, and this in turn builds up cyclonic forms. This is compensated for by currents that rise upward from the center of this bowl of low pressure, forming a chimney of rising air. These currents, in turn, eventually cool at high altitudes, which then increases the humidity of the air. Thus, in any region of low pressure, clouds and high humidity come into being, and are characteristic not only of cyclones but of storms in general.

As cyclones intensify, they often spread outward, sometimes reaching a radius of 500 miles or more—although this is a relatively unusual size.

Finally, cyclones generally fall into two major categories: middle latitude and tropical.

Middle latitude cyclones, which can form over either land or water, are sometimes associated with waves or ripples along polar fronts and generally move from west to east along with the prevailing winds.

Tropical cyclones, which occur over warm, tropical oceans, in their formative stages usually move toward the west with the flow of the trade winds, and when mature, curve towards the poles.

A tropical cyclone that has matured to severe intensity is called a hurricane when it occurs in the Atlantic Ocean or its adjacent seas, a typhoon when it occurs in the Pacific Ocean (or adjacent seas), and a cyclone when it occurs in the Indian Ocean region.

BANGLADESH
November 12, 1970

The cyclone of November 12, 1970, in Bangladesh is widely considered to be the worst natural disaster of the 20th century. Between 300,000 and 500,000 were killed by a combination of wind and water.

On November 12, 1970, five months before it became Bangladesh, East Pakistan experienced the worst disaster of the 20th century. The combination of a killer cyclone and a tidal wave said to be 50-feet (15.2-m) high caused a death toll of between 300,000 and 500,000. Winds of up to 150 MPH (241.4 km/h) lashed the East Pakistan Coast, the Ganges Delta and the offshore islands of Bhola, Hatia, Kukri Mukri, Manpura, and Rangabali.

Bangladesh is roughly the size of Wisconsin with an enormous population of over 95 million people. It borders on India and Burma and consists mainly of a low plain cut by the Ganges and Brahmaputra Rivers and their delta. A Hindu portion of India until 1947, the area became part of East Pakistan and remained so until its declaration of independence from Pakistan in March 1971. Alluvial and marshy along the coast, with hills breaking this monotony only in the extreme southeast, it has one of the rainiest climates in the world, and is thus a breeding ground for tropical monsoons.

Ironically enough, this, the worst disaster of the century, was first dismissed as a false alarm. Barely a month earlier, on October 23, 1970, a small cyclone had frightened the inhabitants of the Ganges Delta into evacuating the environs, and only minimal damage resulted. But this false alarm brought with it a false sense of confidence, and when, on November 11, 1970, an American weather satellite warned of a giant cyclone heading toward the same region, Radio Pakistan ignored it. Uninformed, the huge population slept blissfully as this meteorological monster pounded inexorably toward it.

The storm hit in the middle of the night of November 12. Cyclonic winds pushed a tidal wave of at least 20 feet (6.1 m)—some said, 50 feet (15.2 m)—in height towards islands whose highest ground level rested a maximum of 20 feet (6.1 m) above the sea's surface. Thus, when the wave curled over and crashed upon the thatched roof houses and paddies of these islands, it absolutely consumed them, and only the second stories of the manor houses of a few well-to-do farmers were saved.

Most houses were smashed into piles of soaked straw, and the sleeping inhabitants were swept out to sea by the roaring current. Moments later, the storm made landfall with winds of 150 MPH (241 km/h). Houses, hospitals, power lines were instantly collapsed, cutting off all communication with the outside world. It would be two days before the rest of Pakistan would know of the calamity, and by then the tragedy would have climbed to monumental proportions.

More than 20,000 inhabitants of one island alone disappeared into the sea without a trace; corpses covered the land like grim cobblestones. They were scooped up from the islands and thrown into the sea, where they floated toward the land. There, inhabitants lined the beaches, shoving the beached corpses back out to sea with bamboo poles.

Disease spread rapidly. Cholera ravaged the island of Rangabali. Rice paddies turned the color of blood. Vultures circled constantly, and the smell of death and decaying corpses hung like a sickly sweet mist over the entire area. Water was unobtainable; food supplies were spoiled or tainted by disease.

Within two days, medical supplies, personnel and food began to be airlifted into the region from the rest of the world. America and Great Britain ferried in the largest amount of supplies and engineers to reconstruct the transportation and health-support systems. But air drops of food supplies caused further misery in the form of food riots.

It would be months before the dead would be collected from the streets of the demolished city of Patuakhali and its surrounding fields and paddies, and months more before the International Red Cross would be able to stem the rampant spread of cholera and typhoid in the region.

BANGLADESH
May 25, 1985

Ten thousand people perished in the cyclone that hit Bangladesh on May 25, 1985, and destroyed 80 percent of the dwellings on the seven islands at the mouth of the Meghna River.

Bangladesh is no stranger to cyclones and tidal waves, nor to monumental loss of life because of both. In 1942, when it was still part of India, 40,000 people lost their lives in a cyclone; in 1963, when it had become Pakistan, 22,000 perished in a windstorm; in 1965, 30,000 and 10,000 died respectively in two more windstorms; in 1970, a staggering 300,000 died in the worst cyclone of the century (see p. 235).

On May 25, 1985, a cyclone struck, preceded by a 15- to 20-foot (4.6–6.1 m) wall of water that ravaged seven islands at the mouth of the Meghna River, part of the delta system in the Bengal basin. The disaster killed 10,000 people, 500,000 head of cattle, and flattened 80 percent of the area's dwellings.

The 400-square-mile (643.7 sq. km) area affected embraces approximately one-eighth of Bangladesh's 95 million people, contains nearly 1,000 islands (some of which disappeared entirely in this storm)

Survivors of the Bangladesh cyclone of May 25, 1985, huddle together, waiting for shelter and food. (Rudolph von Bernuth, CARE)

Families were separated and decimated by the 1985 Bangladesh cyclone. (Rudolph von Bernuth, CARE)

the island of Hatia under three feet of water. Miraculously, only seven people in the city of 40,000 were killed.

The next day, relief aid began to arrive from Europe and the United States. But most of these supplies did not arrive before much of the populace had been forced to survive on rancid food and salt water.

BANGLADESH
SOUTHEASTERN COAST
April 30, 1991

Nearly 139,000 residents were killed on the southeast coast of Bangladesh and the islands along it when a giant cyclone with winds up to 145 MPH struck on the night of April 30, 1991. A storm surge 20 feet high inundated homes and fields, and drowned 5,000 fishermen, caught at sea. Four million people were made homeless, and the total damage estimate totaled $3 billion.

"No country has got a worse natural and environmental deal than Bangladesh," noted an editorial in the *Dhaka Courier* in 1991. "One must seriously consider whether there is a curse on this land or not."

The writer was reacting with suitable bleakness to the appalling destruction left in the wake of the cyclone of April 30, 1991, whose 145 MPH winds and 20-foot (6.1-m) high waves killed 139,000 hapless residents of the islands and the southeast coast of this overpopulated country, which exists a bare five to six feet above sea level.

Situated on the delta of the Ganges, Brahmaputra, and Meghna Rivers, Bangladesh is, as well as overpopulated, one of the world's poorest countries. Its 110 million people, whose average yearly income is $170, are crammed into a space roughly the size of Wisconsin. Thus, a number of them are forced to inhabit the coast and the silt islands, which are most vulnerable to the frequent cyclones that roar in from the Bay of Bengal.

Not since 1970, however (see p. 235) has a cyclone of the dimensions of the 1991 storm ravaged this blighted country in which, even without cyclones, over 260,000 children under the age of five die each year of nothing more than diarrhea. Respiratory infections kill another 157,000, and measles claim the lives of 60,000 children annually.

There was no shortage of warnings about the coming of the 1991 storm; an efficient program of cyclone warnings had been instituted after the 1970 calamity. On the island of Manpura, cyclone warnings were

and is located southeast of Dhaka, the capital. The most severely battered islands were Sandwip, Hatia, Mahash Khali, Bhola, Ulir Char, Char Clerk, and Dhal Char.

This part of Bangladesh is beset by monsoon rains twice a year, from the east in the fall, and from the west in the spring. Half of these monsoon rains coincide with the annual May rice harvest, and, in May 1985, the population of 10 million was swelled by 300,000 migrant workers brought in to assist in the harvest.

Although the cyclone was detected three days earlier in the Indian Ocean by satellite, there is no record of a warning being sent to Bangladesh. Thus, workers and dwellers on the islands were swept into the sea by the sudden, monumental waves that roared across the islands. Only a few survivors made it to the concrete cyclone shelters.

Winds of over 100 MPH ripped the roofs from thatched-roof houses and submerged the main town on sounded over megaphones and by the beating of drums. But 11 false alarms in a row had produced casualness bordering on disbelief in the populace. As a result, tens of thousands were caught directly in the path of the gigantic cyclone that kicked up high waves on April 29, as it headed straight for the Bangladesh coast.

At 1 A.M., on the 30th of April, at the precise moment of high tide under a full moon, the storm made landfall just south of Chittagong, the country's second largest city. The winds to the left of the storm were clocked at 50 to 70 MPH, enough to inundate the islands of Sandwip, Bhola, and Hatia. To the right of the storm, winds reached 145 MPH, devastating the islands of Kitubdia, Moiscal, and the seaside resort of Cox's Bazar.

Mufizur Rahman recalled seeing waves "as high as mountains" emerge from the sea and roar toward him before he blacked out. When he awoke, his wife, his son, and his three daughters had all been swept away by the 20-foot (6.1-m) high storm surge; his village of Vijandya had been leveled beyond recognition.

The storm was relentless; it pounded the area for eight hours, drenching it with rain, flooding it with the storm surge, swelling its rivers so that they continued to flood the countryside after the storm abated.

Over 10 million people, or one-tenth of Bangladesh's population, lived in the area that was in the direct path of the cyclone. Eighty percent of the straw huts that housed most of these people were blown away or crushed by the storm surge.

Three million residents did heed the warnings, and were moved to concrete buildings erected on higher ground following the 1970 storm. And the wealthy of Bangladesh, insulated from the yearly cycle of cyclones by their wealth and the stone dwellings in which they lived, survived universally.

But elsewhere, chaos and terror reigned. Storm rains set off a flash flood in the Meghna River, engulfing the rail station and a majority of the buildings in the town of Chandpur. The entire airport at Chittagong was under three feet of water, and nearly 5,000 people were trapped on the rooftops of the airport buildings. In the city itself a food storage depot collapsed under cyclone winds, burying the workers within it.

Casualty figures began to pour in, shortly after the storm abated, but they were untrustworthy. The bloated bodies that were washing ashore daily were countable; those of the Bangladeshi who were washed out to sea were not. Five hundred fishing vessels containing over 5,000 fishermen had simply disappeared. Six larger vessels were missing and presumed sunk. A passenger ship from India with 800 passengers and crew aboard was stranded, powerless, in the Bay of Bengal.

The Bangladeshi air force, which consisted of 17 helicopters and three or four fixed-wing aircraft, began to drop food and supplies helter-skelter through the afflicted area, on rooftops to waving survivors. And journalists began to arrive.

"I saw deaths, devastation, agony and misery of a magnitude I have never seen before," said an AP photographer, used to covering tragedies.

As days passed, it became clear that nearly 115 million people had been uprooted by the disaster, which was continuing and growing. Ninety-five percent of the houses in and around Cox's Bazar were totally destroyed. Along the southeast coast of Bangladesh, brackish water had ruined drinking supplies. Roads and bridges were tangled and useless wrecks. And most important, a rice crop that was on the verge of being harvested had been flooded with salt water. Not only was this crop spoiled but the land would be unworkable for two to three years.

Shrimp farms, salt pans, and fishing fleets and a fledgling oil refinery were all wiped out. Hospitals and schools, normally shelters for the homeless, had disappeared.

A week after the storm struck, the islands were still under water. Three C-130 cargo planes dropped dry food and clothing in plastic containers, but a large percentage of these containers burst upon landing, rendering their contents unobtainable or inedible.

Stories circulated of survivors dying from snakebite as they tried to grab floating banana trees, while tiny children began to gather along roads with empty water pots and plastic bags, begging for food or handouts of any kind.

Four days after the storm struck, a combination of continuing squalls, waterborne diseases, and starvation began to loom as further killers. The nearly penniless Bangladeshi government, under Prime Minister Khaleda Zia, appealed to the rest of the world for immediate help. But the United States, Japan, and Western Europe were concentrating their aid moneys on Kurdish and African calamities. The major task of relief fell to UNICEF and CARE, which began to distribute plastic sheeting for shelters, oral rehydration mixes, water-purification and basic medicine kits, intravenous drip for cases of severe dehydration, and vaccine for measles.

But even this was hampered by foul weather, which moved over the disaster site and slowed the missions under way. As high ground began to disappear again, refugees competed with poisonous snakes for this unflooded territory.

All of this was complicated by the government of Bangladesh's apparent inability to coordinate its efforts. Chaos began to layer itself over tragedy as tens of thousands of carcasses of livestock began to rot in the tropic heat, adding to the threat of epidemic disease.

Five days after the storm struck, Charles H. Larsimont, the head of the United Nations Development Program in Bangladesh, lashed out in frustration. "I keep insisting on a common list of needs," he told reporters. "Let us have one common master. One list of needs. One list of donors. One point of entry. Otherwise we are not going to be able to do much."

Democracy had only lately come to Bangladesh, and with it, came opportunists who cashed in on relief agency buying of rice, and raised its price 30 percent, which placed it out of reach for the hundreds of thousands of Bangladeshi poor. Nails cost 10 times more than they did a week before the storm.

The airdrops did not go well. Along the southeastern coast, some packages of food were airdropped into the sea alongside floating bodies. Some plastic containers of drinking water were dropped from a level of 500 feet (152.4 m) and thus burst upon impact.

And now, steady rain and gusty winds grounded even the few helicopters and planes of the Bangladesh air force. Winds in the nine northern districts as well as the Noakhali region, on the edge of the disaster area, were measured at 60 to 70 miles an hour.

Roads became rapidly flooded, but not before looting began. Three trucks were attacked and divested of their rice by groups of hungry men in Faujdarhat, 135 miles southeast of Dhaka.

On May 6, a relief plane carrying officials and journalists attempted to land in the area but was unable to find enough dry land. Nizam Ahmed, a local journalist aboard, reported that "It looked like the area had been carpet-bombed with damaged houses and craters created by the tidal surge. People were either half-naked or wearing torn and dirty clothes and appeared to be in shock."

On May 7, a tornado and thunderstorms struck the area, piling new misery on the survivors. The tornado struck in Tungi, an industrial suburb 13 miles north of Dhaka, and so out of the original disaster area. But the thunderstorms pounded Chittagong, one of the hardest hit areas.

Finally, Pakistan and Britain added helicopters to the relief flights, and the Royal Fleet Auxiliary ship *Fort Grange*, which had helped supply the allied fleet during the Persian Gulf War, was dispatched to the area to distribute rice, molasses, bread, and drinking water.

As the weather momentarily cleared, relief workers were able to reach some, but not all villages. Bridges that had been destroyed were replaced temporarily by bamboo poles, lashed together. Workers found cyclone survivors living in makeshift shelters of tin and plastic scavenged

from debris. The huts they called home had collapsed into nonentity, often with their loved ones inside.

Most of the fishing boats in the area were destroyed in the storm, and the fishermen had no money to buy new ones. Farmers were in an equal situation. Not only were their fields contaminated by salt water, but the dikes that once protected these fields from the ocean had been broken through. All the way from Chittagong south to Cox's Bazar, dikes were destroyed, and now, twice a day, more salt was deposited onto the fields.

And all of this occurred a mere few weeks before the monsoon season, with its relentless rain and higher than normal tides.

Some fields were spared contamination, but the farmers had no money to buy seed, fertilizer, and equipment destroyed by the storm. And so they waited for the government to save them, by giving them seed and fertilizer, as the fishermen waited for the government to give them nets.

Some refugees had made their way to Chittagong, where they stood outside government offices, hoping for food. On several nights, small altercations broke out, during the distribution of rice. Local young people climbed onto a building and hurled bricks and stones at the police during one of these incidents, but they were easily controlled. Hunger apparently took away not only their strength but their will.

Finally, on May 11, 12 days after disaster struck, the United States dispatched nearly 8,000 marines and dozens of helicopters, along with the amphibious assault ship *Tarawa*, which had been headed home to California from the Persian Gulf. Preventive medicine teams and Navy seabees landed, to help with water purification and the rebuilding of homes and bridges.

Meanwhile, other nations, led by Saudi Arabia, began to make donations to the Bangladeshi government, which now put the damage estimate at over $3 billion. Four million people had been rendered destitute by the storm.

Rebuilding began. One hundred tons of food a day were ferried to remote parts of Bangladesh, as well as 30,000 gallons (113,562 l) per day of drinkable water. A threatened cholera epidemic did not materialize.

The death toll, however, was staggering. The official figures totaled nearly 139,000, not as cataclysmic as the 300,000 to 500,000 of the 1970 storm, but bad enough.

And still, this terribly poor country, built upon shifting sands, the recipient of yearly floods that begin in the Indian Himalayas, and battered by yearly storms that eat away at the embankments that protect its often below-sea-level agriculture, continues to grow at the rate of 2.4 million people per year. And as the population grows, more and more peasants are forced to live on the silt islands that were never meant for human habitation.

With the inevitable increase in global warming, the future looks bleak and troublesome for this precariously situated, poverty-stricken, and overpopulated country. With the rising of sea levels, it is almost certain that Bangladesh will disappear like Atlantis beneath the sea before the century ends.

INDIA
ANDHRA PRADESH
November 6, 1996

....

A cyclone with winds of 100 MPH struck the south-eastern Indian state of Andhra Pradesh on the night of November 6, 1996. One thousand six hundred people were killed and tens of thousands were rendered homeless.

The state of Andhra Pradesh, on the southeast coast of India, is a plateau, which slopes down to the waterline. At its center, two major rivers, the Krishna and the Godavari, empty into the Bay of Bengal on wide deltas. The upward slope makes it extraordinarily vulnerable to the periodic tropical storms that gather force in the Bay of Bengal, then roar inland. In 1977, over 10,000 people were killed by a cyclone that slammed into the coast and pushed a 50-foot (15.24-m) high tidal wave eight miles inland.

The cyclone that struck the same area on November 6, 1996, was not as large, but in many respects it was equally as vicious. First of all, it was unexpected. Weather forecasts spoke of a storm that was headed for the neighboring Krishna district, where residents were alerted. But at the last minute, it veered off course, and headed to the Godavari River delta.

Secondly, the storm struck at night with winds that climbed to nearly 100 MPH. It slammed full force into the coast, and sent a storm surge crashing inland at express train speed. "The wind came like a thundering airplane and shook the houses," said Venkat Reddy, a high school principal. "Everybody locked themselves in. We could hear the trees falling. Nobody dared to move out."

Apparently, Mr. Reddy was in a brick house when the storm struck. Residents of mud homes had neither the time nor the choice to lock themselves in. Ten thousand mud homes were collapsed as if they were toys. "Except for houses made of brick and cement, nothing [was left] standing," said Chandrababu Naidu, the state's chief minister, as he later surveyed the scene from a helicopter.

The walls of the mud houses that formed the usual sort of dwelling place in the state fell in on occupants, crushing them to death, then were dissolved in the roaring waters. Roads of escape were flooded, and refugees drowned in them. Telephone poles were uprooted, cutting off communication with a large portion of the state. One and a half million acres of rice crops, and more than 770,000 acres of bananas, coconuts, and sugar cane were destroyed by the salt water, which swirled into the waters still remaining from a flood that had swept through the area a mere three weeks before, killing 350 people.

Seventy-two hours after the cyclone hit, the roads that were free of flooding had clothes, bedding, pillows, and grain lined along them, drying. Vultures circled over bloated carcasses of cattle, floating in a village canal swollen by floodwater. This harbinger of water contamination was given the impact of fact when seven people were admitted to a hospital with symptoms similar to cholera. Fortunately, none of them actually had cholera, and no epidemic began.

Conforming to the rules of their religion, survivors built funeral pyres to cremate their dead. Since wood was too scarce and expensive, bodies were placed on top of car and truck tires and ignited with the gasoline and kerosene dispensed by the government to help head off the spread of disease.

The fourth day after the storm was supposed to be a day of Hindu celebration—Diwali, the Festival of Lights. Though some few devotees salvaged and lit candles, most were busy with relief workers, disseminating or receiving food, drinking water, and medicine.

Families of fishermen were beside themselves with grief. Some 1,000 fishermen from Balusutippa, a village of about 10,000 near the Godavari River delta, had ignored warnings and set off into the bay hunting for shark. They had apparently disappeared and were presumed drowned.

However, as time passed, 162 fishing boats returned, carrying 400 of the 1,000 fishermen believed missing. The final death toll reached 1,600, 600 of them fishermen.

INDIA
BACKERGUNGE
October 31, 1876

....

Two hundred thousand residents of the Indian city of Backergunge died from wind, water, and cholera caused by the wreckage and pollution of a cyclone that struck the city on October 31, 1876.

One hundred thousand people were killed when a gigantic cyclone roared into the Indian city of Backergunge on October 31, 1876. Sitting at the mouth of the Megna River, Backergunge is also serviced by the Bay of Bengal, which is a well known sluiceway for cyclones.

The warning signs of this tragedy formed a classic pattern. For several days prior to the arrival of the storm, above average tides had flooded the shores of the Megna River. Thus, when a huge tidal wave, pushed by the storm, broke over these shores and the outlying islands, they were instantly inundated under water that reached 40 feet (12.19 m) in depth at certain points.

The death toll from drowning in Backergunge was swollen within weeks by another 100,000, who succumbed to disease—probably cholera—caused by the pollution of the water supply by the storm.

More than 50,000 were killed when this cyclone tore through Calcutta, India, on October 5, 1864. (Illustrated London News)

INDIA
BAY OF BENGAL
October 7, 1737

A cyclone which struck the Bay of Bengal on October 7, 1737, killed 300,000 residents of that thickly populated part of India.

The tidal wave caused by the cyclone of October 7, 1737, wrecked 20,000 ships that were anchored in the harbors of the Bay of Bengal, and, crashing over the shorelines, particularly at the mouth of the Hooghly River, went on to claim 300,000 lives.

Densely populated in the 18th century because of the huge amount of shipping that originated within the Bay of Bengal's fairly sheltered environs, this area was easy prey to a storm that claimed the greatest death toll up to that time in that region.

INDIA
CALCUTTA
October 5, 1864

Striking at high tide, a cyclone devastated Calcutta on October 5, 1864, killing over 50,000 people. Nearby cities were untouched.

The great killer cyclone of 1864, which instantly drowned more than 50,000 inhabitants of the bustling city of Calcutta on the morning of October 5, was a selective storm. Sinking more than 200 ships, sending a 40-foot (12.19-m) tidal wave over the docks and into the streets of the city, it seemed to localize itself solely in and around Calcutta. Nearby coastal sites such as Contai remained blissfully dry and happily unaffected.

The high water level of the northern part of the Bay of Bengal makes cities like Calcutta particularly vulnerable to storms at times of high tide. (The 1833 cyclone claimed 50,000 lives, and 300 of the surrounding villages were destroyed.) October 4, 1864, was a time of extremely high tide, thus aiding the 40-foot (12.2-m) storm wave, when it slammed into the city.

In moments, the city was almost entirely submerged, which accounted for the great number of drownings. Disease, brought on by the polluted water supply, killed another 30,000 during the following weeks.

INDIA
CORINGA
December 1789

Three tidal waves and a cyclone battered Coringa, India, in December 1789, killing 20,000.

Located at the mouth of the Ganges River, the city of Coringa was completely destroyed and 20,000 of its inhabitants were drowned by three tidal waves spawned by a cyclone that hit the city at the end of 1789.

Although detailed information is sparse, it was apparently the first of the three waves that did the major damage, sinking most of the ships moored in the harbor, flooding every alley and small street, crushing businesses, government buildings, and the thatched roof dwellings that housed much of the city's population.

The second wave, less intense than the ones before and after it, hit the area a glancing blow, saving most of its fury for the countryside surrounding the city.

The third wave followed the path of the first, adding measurably to the depth of the water in the city and the surrounding countryside. It went on to hit Yanaon, a town nearby.

Enormous quantities of mud and silt choked the entrance to the mouth of the river, hampering the search for bodies and cloaking the ruins of a once bustling city now reduced to one building and the shattered remains of its smashed docks.

INDIA
CORINGA
November 1839

Three hundred thousand people lost their lives in a cyclone and tidal wave that assaulted Coringa, India, in November 1839. The city was never entirely rebuilt.

It took the rebuilders of Coringa many years to reconstruct their ancient city after the triple waves of the cyclone of 1789 (see preceding entry). And then, 40 years later, a gigantic 40-foot (12.19-m) tidal wave, spawned by an enormous cyclone, once again wiped out the city, destroying 20,000 vessels in its harbor and killing 300,000 people.

It was one of the worst cataclysms ever to hit this cyclone-ravaged area, and the ancient city of Coringa was never wholly rebuilt after it.

INDIA
ORISSA
October 12, 1967

Hardly anyone survived the cyclone that struck the state of Orissa, of India on October 12, 1967.

Orissa, a primitive state consisting of small villages and located some 200 miles (321.8 km) south of Calcutta, keeps no records, and so the precise number of fatalities from the killer cyclone of October 12, 1967, remains unknown. Suffice it to say that only a small percentage of humans survived.

All but a handful of the buildings—most of them thatched palm roofed huts held together by mud—were flattened as if a giant fist had driven them into the earth. Mingled with the wreckage of the huts were banana, palm, and banyan trees, all of them uprooted and crushed.

On the one-lane roads, steel I-beams that had been used to hold electric lines were bent into U-shapes. The rice crops that ordinarily rimmed the roads were destroyed without a trace.

One unique circumstance makes this particular tragedy outstanding and strange. Ordinarily, a mass destruction of livestock is taken care of by nature, for vultures strip the carcasses clean. In the Orissa cyclone, every bird, as well as every animal was killed. There were no vultures to clean the thousands of decaying animal carcasses, which permeated the disaster site with their fearsome stench for months.

INDIA
ORISSA
October 29, 1999

Over 9,500 residents of the state of Orissa, India, perished in a giant cyclone that slammed into India's coast with wind gusts up to 190 MPH on October 29, 1999. The resulting floods were among the worst in India's history.

A monstrous cyclone, later classified by meteorologists as a "supercyclone," churned across the Bay of Bengal on October 29, 1999, heading straight for an 85-mile stretch of coastline on the eastern Indian state of Orissa, with its population of 35 million people. Waves of up to 15 feet (4.57 m) in height and winds that eventually gusted to 190 MPH (309 km/h) crashed ashore. It was the second cyclone in two weeks to hit at precisely the same spot, but experience in this case was apparently not a good teacher. Thousands, including the fishermen in several dozen fishing boats, either ignored the warnings of the national meteorological service, or did not understand the complex jargon used by its meteorologists.

The first cyclone killed 100 people and injured 1,000. This one immediately cast its earlier cousin into stark shadow, as it flattened homes, ripped up telephone lines, uprooted railroad tracks, smashed buildings, and drowned humans and livestock alike.

For two days, the storm pounded the area mercilessly, isolating it from the rest of India and the world. Torrents of water aided the wind in destroying all communication lines into and out of the state. Two days after the storm abated, over a million homeless people huddled in improvised campsites on plateaus of higher ground, while raging waters frothed below them, eroding more and more land.

The state's capital, Bhubaneshwar, was particularly flattened. Ninety percent of its trees were uprooted, billboards were torn to pieces, and its electric and telephone lines, detached and frayed from their supports, snapped in the breeze. Its slum districts were almost entirely swept away by flood waters. Two hundred thousand people—one of every six residents—lost their homes.

The districts within Orissa that were most cruelly ripped asunder by the storm were Kendrapara, Jagatsinghpur, Puri, Suttack, and Jaipur. Here, days after other portions of the state received army personnel, who cleared roads and brought relief supplies, survivors depended upon air drops of dried food and fresh water.

By November 1, navy ships carrying medicine, water and generators positioned themselves near the port of Paradip, near Kendrapara, but there was not enough cleared land to receive the supplies.

As time wore on, and rain continued, families squatting under thin sheets of plastic became restless and desperate. Groups of starving people began to attack trucks carrying emergency supplies. Buses were stopped and ransacked for food. A helicopter carrying Defense Minister George Fernandes, sent to survey the area, was attacked at one point by a protesting mob of hungry survivors.

Refugees trapped on housetops and hilltops surveyed a landscape dotted with bodies hanging from trees and floating through villages. Indian troops, traveling by boat, arrived in the county of Ambiki, four and a half miles inland and 85 miles from Bhubaneshwar, to a scene that one of them described as "a hellish sight." Rotting bodies were everywhere. Six villages that were once the home to 3,000 people had disappeared, leaving little trace that they had ever existed.

As time passed, starvation and cholera seeped into the isolated villages, some of which never received relief supplies, some of which contained people too poor to pay for the food that entrepreneurs were selling at a 300 percent markup over pre-storm prices.

A month after the cyclone struck, a train loaded with medical and sustenance supplies was dispatched from New Delhi. For tens of thousands, it was far too late. Almost all the rice paddies and thousands of acres of farmland had been destroyed. Four hundred thousand head of livestock drowned.

Urban destruction was equally horrendous. Over 11,000 schools alone were destroyed or heavily damaged. Hospitals were swept away. "This is the worst flooding in 100 years," said Asim Jumor Vaishnov, the chief administrator of Baleshwar. "I would say it's the worst in India's history."

It was certainly nearly that. Over 9,500 people were killed, 2.5 million people were rendered homeless, and the damage estimates reached $3.5 billion.

HURRICANES

....

THE WORST RECORDED HURRICANES

* Detailed in text

Barbados
    (1694)
  * (and Martinique, St. Lucia, and
    St. Eustatius) (1780)
    (1782)
  * (1831)

British Honduras
  * Belize (1931)

Cape Santa Cruz
    (1481)

Caribbean
  * Hurricane Charley (2004)
  * (and Mexico and Texas) Hurri-
    cane Gilbert (1988)
    Hurricane Gilbert (1997)
  * Hurricane Inez (1966)
    Hurricane Ivan (2004)
  * Hurricane Janet (1955)
  * Hurricane Joan (1988)
    Hurricane Luis (1995)

Cuba
  * (1926)
    (1933)
    Havana (1768)
  * (and Florida) Havana/Florida
    (1944)
  * Hurricane Flora (1963)
  * Hurricane Georges (1998)
  * Santa Cruz del Sur (1932)

Curaçao
    (1877)

Dominican Republic
  * Hurricane Georges (1998)

El Salvador
    (1934)

England
  * (1703)

Grenada
  * (1956)

Guadeloupe
  * (1666)
  * (1928)

Guatemala/El Salvador/Mexico
  * Hurricane Stan (2005)

Haiti
  * (1909)

* Hurricane Cleo (1964)
* Hurricane Flora (1963)
* Hurricane Georges (1998)
* Hurricane Hazel (1954)
    Tropical Storm Gordeon (1994)
* Hurricane Jeramie (1935)

Hawaii
  * Hurricane Iniki (1992)

Holland
  * Leyden (1574)

Honduras
    (1941)
  * (1955)
  * Hurricane Hattie (1961)

Honduras (and Nicaragua)
  * Hurricane Mitch (1998)

Jamaica
  * (1784)
  * Hurricane Charlie (1951)

Leeward Islands
    Twin hurricanes (1747)

Lesser Antilles
    Basseterre (1650)

Martinique
    (1680)
  * (1695)
  * (1881)
    (1891)

Mexico
  * (1909)
  * Hurricane Dean (2007)
    Hurricane Henriette (2007)
  * Hurricane Pauline (1997)
  * Hurricane Wilma (2005)
    Manzanillo (1959)
  * Tampico (1933)
  * Tampico (Hurricane Hilda) (1955)
  * Yucatan (Hurricane Gilbert)
    (1988) (see HURRICANES, Carib-
    bean, Hurricane Gilbert)

Montego Bay
  * (1780)

Newfoundland
  * (1775)

Nicaragua, Honduras
  * Hurricane Felix (2007)

Nova Scotia
  * (1873)

Puerto Rico
  * (1533)
  * (1825)
  * (1899)
  * (1932)
  * Hurricane Donna (see HURRI-
    CANES, U.S., Florida)
  * Hurricane Georges (1998)
    Hurricane Jeanne (2004)

St. Croix
  * (1772)

Santo Domingo
  * (1780) (see HURRICANES, Montego
    Bay)
  * (1834)
  * (1930)

United States
  Alabama
  * Mobile (1819)
  * Mobile (Hurricane Frederic)
    (1979)
  Bahamas, Florida, and Louisiana
  * Hurricane Andrew (1992)
  East
    (1849)
    (1944)
  * Hurricane Bob (1991)
  * Hurricanes Connie and Diane
    (1955)
  * Hurricane David (1979)
  * Hurricane Donna (1960)
  * Hurricane Eloise (1975)
  * Hurricane Floyd (1999)
  * Hurricane Hazel (1954)
  Florida
    (1758)
  * Keys (1906)
  * Keys (1935)
  Key West (Hurricane Donna)
    (1960)
  Florida (Louisiana, Mississippi)
  * Hurricane Katrina (2005)
  Florida (Northwest, Georgia, and
    Alabama)
    Hurricane Alberto (1994)

Florida (Northwest and Alabama)
Hurricane Opal (1995)
Miami
* (1926)
* Hurricane Cleo (1964)
* Okeechobee (1928)
* St. Jo (1841)
* Straits (1919)
Georgia
* (1893)
Louisiana
* (1812)
* (1893)
* (and Mississippi) (1909)
(1918)
* Last Island (1856)
Mississippi
* Hurricane Elena (1985)

New England
* (1773)
* (1815)
* (1841)
* (1938) (see HURRICANES, New
York [Long Island] and New
England)
* (1944)
* Hurricane Carol (1954)
New York (Long Island) and New
England
* (1938)
North Carolina
* (1713)
Hurricane Emily (1993)
* Hurricane Fran (1996)
Hurricane Isabel (2003)

Puerto Rico (North and South
Carolina, Virgin Islands)
* Hurricane Hugo (1989)
Southern U.S.
* Hurricane Audrey (1957)
* Hurricane Camille (1969)
* Hurricane Juan (1985)
Texas
* Galveston (1900)
* Galveston (1915)
* Indianola (1886)
West Indies
(and Florida) (1928)
* Hispaniola (1495) (first reported
hurricane, by Columbus)
Hispaniola (1509)
* Trois-Ilets (1766)

CHRONOLOGY

· · · · · · · · ·

* Detailed in text
1495
June
* Hispaniola, West Indies (first
reported hurricane, by
Columbus)
1509
July 10
Hispaniola, West Indies
1533
July 26–August 31
* Puerto Rico
1574
October 1
* Leyden, Holland
1650
Basseterre, Lesser Antilles
1666
August 4
* Guadeloupe
1680
August 3
Martinique
1694
September 27
Barbados
1695
October 20
* Martinique
1703
November 27
* England
1713
September 16–17
* North Carolina

1747
October 24
Leeward Islands (twin
hurricanes)
1758
Florida
1766
August 12
* Trois-Ilets, West Indies
1768
October 25
Havana, Cuba
1772
August 31
* St. Croix
1773
August 14
* New England
1775
September 9
* Newfoundland
1780
October 3
* Montego Bay
October 10
* Barbados
1782
Barbados
1784
July 30
* Jamaica
1812
August 19
* Louisiana

1815
September
* New England
1819
July 27–28
* Mobile, Alabama
1825
July 26
* Puerto Rico
1831
August 10
* Barbados
1834
September 23
* Santo Domingo
1841
September
* St. Jo, Florida
October 3
* New England
1849
October 6
Eastern United States
1856
August 13
* Last Island, Louisiana
1873
August 24
* Nova Scotia
1877
September 21
Curacao
1881
August 18
* Martinique

1886
> August 19
> > * Indianola, Texas

1891
> August 18
> > Martinique

1893
> August 27
> > * Georgia
> October 1
> > * Louisiana

1899
> August 8
> > * Puerto Rico

1900
> September 8
> > * Galveston, Texas

1906
> October 18
> > * Florida Keys

1909
> August 23
> > * Haiti
> August 27
> > * Mexico
> September 10–20
> > * Louisiana and Mississippi

1915
> August 16
> > * Galveston, Texas

1919
> September 9–10
> > * Florida Straits

1926
> September 17
> > * Miami, Florida
> October 20
> > * Cuba

1928
> September 10
> > West Indies (and Florida)
> September 12
> > * Guadeloupe
> September 16
> > * Okeechobee, Florida

1930
> September 3
> > * Santo Domingo

1931
> September 10
> > * Belize, British Honduras

1932
> September 26
> > * Puerto Rico
> November 9
> > * Santa Cruz del Sur, Cuba

1933
> September 1
> > Cuba
> September 24
> > * Tampico, Mexico

1934
> June 8
> > El Salvador

1935
> September 2
> > * Florida Keys
> October 22
> > * Haiti (Hurricane Jeramie)

1938
> September 21
> > New York (Long Island) and New England

1944
> September 8–16
> > * New England
> September 14
> > Eastern United States
> October 13
> > * Havana, Cuba (and Florida)

1951
> August 17
> > * Jamaica (Hurricane Charlie)

1954
> August 26–31
> > * New England (Hurricane Carol)
> October 12
> > * Haiti, Eastern and Southern U.S. (Hurricane Hazel)

1955
> August 4–18
> > * Eastern United States (Hurricanes Connie and Diane)
> September 19
> > * Tampico, Mexico (Hurricane Hilda)
> September 22
> > * Caribbean (Hurricane Janet)
> September 27
> > * Honduras (Hurricane Janet)
> October 12
> > * Haiti (Hurricane Hazel)

1956
> September 22
> > * Grenada

1957
> June 27–30
> > * Southern United States (Hurricane Audrey)

1959
> October 27
> > Manzanillo, Mexico

1960
> September 4–12
> > * Eastern United States (Hurricane Donna)

1961
> October 31
> > * Honduras (Hurricane Hattie)

1963
> September 30–October 9
> > * Cuba (Hurricane Flora)

1964
> August 22–27
> > * Florida and Haiti (Hurricane Cleo)

1966
> September 24–29
> > * Caribbean (Hurricane Inez)

1969
> August 17
> > * Southern United States (Hurricane Camille)

1975
> September 22–27
> > * Eastern United States (Hurricane Eloise)

1979
> August 31–September 8
> > * Eastern United States (Hurricane David)
> September 11
> > * Mobile, Alabama (Hurricane Frederic)

1985
> August 30–September 2
> > Mississippi (Hurricane Elena)
> October 27–November 5
> > * Southern United States (Hurricane Juan)

1988
> September 12–19
> > * Caribbean (Hurricane Gilbert)
> October 17–23
> > * Caribbean (and Mexico and Texas) (Hurricane Joan)

1989
> September 17
> > * North and South Carolina, Puerto Rico (Hurricane Hugo)

1991
> August 19
> > United States East Coast (Hurricane Bob)

1992
> August 24
> > * Florida, Louisiana, the Bahamas (Hurricane Andrew)

September 11
* Kauai, Hawaii (Hurricane Iniki)

1993

August 31
North Carolina (Hurricane Emily)

1994

July 8
Florida, Georgia, Alabama (Hurricane Alberto)
November 14
Haiti (Tropical Storm Gordon)

1995

August 3
Florida (Hurricane Erin)
August 11
Caribbean (Hurricane Luis)
September 16
Virgin Islands (Hurricane Marilyn)

1996

July 12
North Carolina (Hurricane Bertha)
September 6
* North and South Carolina (Hurricane Fran)

September 10
Puerto Rico (Hurricane Hortense)

1997

September 20
Caribbean (Hurricane Gilbert)
October 9
* Mexico (Hurricane Pauline)

1998

September 20
* Dominican Republic, Leeward Islands, Puerto Rico, Cuba, Haiti, Key West, Mississippi, Louisiana (Hurricane Georges)
October 26
* Honduras and Nicaragua (Hurricane Mitch)

1999

September 17
* United States East Coast (Hurricane Floyd)

2003

September 18
North Carolina (Hurricane Isabel)

2004

August 13
Caribbean (Hurricane Charley)

September 8–16
Caribbean (Hurricane Ivan)
September 18
Puerto Rico, Haiti, Florida (Hurricane Jeanne)

2005

August 25–30
* Florida, Louisiana, Mississippi (Hurricane Katrina)
October 1–5
* Guatemala, El Salvador, Mexico (Hurricane Stan)
October 15–25
* Mexico, Cuba, United States (Hurricane Wilma)

2007

August 13–23
* Mexico, Belize, Caribbean (Hurricane Dean)
August 31–September 5
* Nicaragua, Honduras (Hurricane Felix)
September 4–6
Baja, Mexico (Hurricane Henrietta)

HURRICANES

A hurricane is a tropical cyclone that takes place over the North Atlantic Ocean and is characterized by windspeeds greater than 75 MPH.

There are four stages in the development of a full blown hurricane: (1) tropical cyclone, (2) tropical depression, (3) tropical storm, (4) full hurricane. These storms generally form over the tropical North Atlantic, often off the west coast of Africa, and mature as they move westward.

Occasionally, hurricanes develop off the west coast of Mexico and move northeast, thus threatening the coastal areas of Texas.

The intensity of hurricanes is measured on the Saffir-Simpson scale, which ranks them according to wind speed from a low-intensity category 1 to a very severe Category 5. A Category 1 storm has wind speeds of 74–95 MPH (119–152.9 km/h), a Category 2 storm 96–110 MPH (154.5–177 km/h), a Category 3 111–130 MPH (178.6–209.2 km/h), a Category 4 131–155 MPH (210.9–249.4 km/h), and a Category 5 above 156 MPH (251.1 km/h).

Generally, the hurricane season is considered to run from June 1 through November 30, with the peak activity between mid-August and October, when the sea surface temperatures (SST) are high enough to form tropical storms.

Hurricanes generally have a lifespan of from one to 30 days. They thrive over tropically heated water and are transformed into extratropical cyclones after prolonged passage over the cooler waters of the North Atlantic. Once they make landfall, they decay rapidly.

This dispassionate description may convey a benign picture of hurricanes. But just the opposite is true. In fact, in an average hurricane, the release of latent heat from the condensation of water vapor provides as much energy as the detonation of 400 20-megaton hydrogen bombs. Fortunately for those in a hurricane's path, only 2 percent to 4 percent of this heat energy is translated into the kinetic force of the winds' motion—still enough of a percentage to cause immense damage from hurricane winds, and secondary damage from the flooding that results from the coastal storm surge and tropical rains that generally accompany the storm.

The precise conditions necessary to spawn a hurricane are not entirely known, although Project Stormfury, a U.S. government program designed to develop ways of defusing hurricanes at their source is currently studying this in depth. What is known is this: A mature hurricane is nearly symmetrically circular in form, sometimes extending to a diameter of 500 miles (804.7 km). Within a chimney of superheated, tropical air, is a so-called eye—a space of calm and often blue sky that is usually 20 miles (32.2 km) in circumference. Surrounding this is the "eye-wall," the place of greatest danger and turbulence. This is where the inward-spiraling, moisture-laden air is shot aloft, causing condensation and the release of dangerous, latent heat—the source of the energy of the storm. After reaching tens of thousands of feet of altitude, this energy is expelled toward the storm's periphery. But at the location of the wall, the upward velocity of the air, mixed with condensation, forms a combination of maximum high winds and furious precipitation.

Clouds spread outward from this wall in spiral bands parallel to the wind direction, thus forming the hurricane's characteristic shape, and giving way from persistent rain at the storm's core to tropical showers along its outer rim.

Hurricanes generally move at speeds of about 10 MPH on their westward track, often picking up speed as they curve toward the north pole, usually at 20 to 30 degrees north latitude, but often in a more complex and unpredictable pattern. Predicted or unpredicted, hurricanes are capable of causing enormous destruction and staggering loss of life.

And although that loss of life has declined as forecasting methods have improved, the cost of replacing the physical destruction hurricanes cause has correspondingly risen, and not merely because the cost of everything has risen at the turn of the 21st century.

The South Carolina coast of the United States, for instance, is a striking example of the rush to seaside development at the end of the 20th century. Spurred by an economic boom in the industrialized world, more and more people are buying more and more property that is in great danger of damage or extinction by natural forces, and particularly by hurricanes.

As of 2000, approximately 63 million people lived within 50 miles (80.5 km) of the Atlantic Ocean or the Gulf of Mexico, despite the fact that, for years, environmentalists had argued that much of the building on fragile barrier islands should have been prohibited. The environmentalists argued that these coastal areas are not only too vulnerable to storms but building upon them contributes to their erosion.

The monstrous loss of life and property caused by Hurricane Katrina in 2005, (see pp. 290–293) is a prime example of building and living in the face of potential disaster. The entire Gulf Coast of the United States is particularly vulnerable to repeated visits of strong hurricanes, and yet housing developments, resorts, oil rigs, and floating casinos continue to proliferate on the coast, and often in its most environmentally threatened locations.

Ironically, this chancy building boom coincides with an increasing awareness of the presence and implications of global warming. With climate changes factored in, the potential from increased damage from more and more ferocious hurricanes has gone from possibility to near certainty.

It has long been known that storms tend to be stronger during times in which the surface temperature of the ocean is higher, and in 1987, Dr. Kerry Emanuel of the Massachusetts Institute of Technology linked the strength of tropical storms and global warming.

Statistics seem to support his theory: Sea surface temperatures (SSTs) tend to fluctuate naturally over time in a process called Atlantic Multidecadal Oscillation (AOM), which causes the SST to oscillate between 0.1 and 0.2 degrees centigrade. Thus, over the period from 1995 to 2005, a common explanation for an escalating series of strong storms during the hurricane season has been an "upturn" of this cycle. However, the SST of the Atlantic Ocean has, from the early 1990s to 2004, increased 0.5 degrees centigrade, which is two to five times higher than the temperature increase historically associated with AOM fluctuations.

Matching this with other increases in world temperatures, Thomas R. Knutson and Robert E. Tuleya of the government-funded Geophysical Dynamics Laboratory concluded that the recurrence of the strongest category of storms, with maximum wind speeds and minimum central pressures suggested a systematic increase in the strength and intensity of tropical storms.

Their study was published in *Journal of Climate* in September 2004. According to the study, an 80-year buildup of atmospheric carbon dioxide (thought to be a major culprit in global warming) at a rate of 1 percent per year would result in a one-half category increase in hurricane intensity.

The very next year, Katrina, with its gigantic devastation of the Gulf Coast, seemed to give credence to the link between global warming and the rising intensity—if not frequency—of hurricanes, and, in March 2006, another study conducted by researchers at the Georgia Institute of Technology in Atlanta reached the same conclusion after collecting data from satellite observations of hurricanes in six ocean basins. In explanation, NASA climatologist James Hansen noted that month that ". . . an increase of ocean surface temperature provides more 'fuel' to drive strong storms." Hansen further noted that global warming has also driven up ocean temperatures at intermediate depths, thus reducing the ability of this cooler water stirred up by the storm to check the hurricane's strength.

It all seems to add up to a stormy future.

BARBADOS (AND MARTINIQUE, ST. LUCIA, AND ST. EUSTATIUS)
October 10, 1780

In an eight-day trip through Barbados, Martinique, St. Lucia, St. Vincent, and St. Eustatius, the hurricane that entered the area on October 10, 1780, killed 20,000.

The great hurricane of 1780 roared through Barbados on the 10th of October, 1780, totally leveling the entire island, killing 6,000 on the island itself and a total of nearly 20,000 on its eight-day rampage through Barbados, Martinique, St. Lucia, St. Vincent, St. Eustatius, and Puerto Rico.

The intense destruction of Barbados resulted from the full fury of the storm passing directly across it. An entire day of increasing rains and rising winds preceded the final destruction of every tree and every building on the island, including its Government House, which, with its three-foot-thick walls, was heretofore thought to be hurricane-proof. (Major General Cunningham, governor of Barbados, and his family escaped unharmed by hiding in a storm cellar.) The terror was increased by the fact that the eye of the storm passed over the island at night, and thus whole families were buried in the rubble of their dwellings as they slept.

Once the storm crossed Barbados, it smashed an entire British fleet anchored off St. Lucia, then barreled on to Martinique, which it also flattened, destroying a French fleet of an estimated 40 ships and drowning 4,000 French soldiers and sailors. On St. Lucia, 9,000 persons lost their lives to the storm.

After the devastation of Barbados, Martinique, and St. Lucia, the storm reeled over open water, sinking a score of other ships and eventually blowing itself out someplace past Puerto Rico.

BARBADOS
August 10, 1831

Fifteen thousand people died in the hurricane of August 10, 1831, in Barbados.

A huge hurricane, pushing a series of tidal waves before it, was first sighted as it roared into the island of Barbados on August 10, 1831. Compounding the human toll of 1,500 was the incredible loss of property: $7.5 million worth, which in current value totals hundreds of millions.

As in the 1780 event (see preceding entry), trees were uprooted, and scarcely a dwelling was left standing. In addition, millions of dollars worth of crops were laid to waste, inundated by water and stripped clean by wind.

A day later, the same storm slammed into Haiti, raced on to Cuba, totally submerged several small islands, careened through the Gulf of Mexico and eventually made landfall at New Orleans, after which it blew itself out over the mainland of Louisiana.

BRITISH HONDURAS
BELIZE
September 10, 1931

A small but powerful hurricane slammed into Belize, British Honduras, on September 10, 1931, killing more than 1,500 people.

With winds exceeding 132 MPH (212.41 km/h), a compact but deadly hurricane ran through the Caribbean from Barbados on September 6, 1931, to British Honduras on September 10. Its full force slammed into the small city of Belize, where it picked up a score of ships and flung them through the streets of this normally busy port.

A coastal surge pushed inland from the storm filled the streets of Belize, drowning over 1,500 inhabitants and causing over $7 million in damage.

CARIBBEAN (AND MEXICO AND TEXAS)
HURRICANE GILBERT
September 12–19, 1988

The gigantic Hurricane Gilbert killed over 350 people in the Caribbean, Mexico, and Texas between September 12 and 19, 1988. The National Hurricane Center established it as the most violent storm in its annals, and the worst hurricane ever to occur in the Western Hemisphere up to that time.

The worst hurricane ever to hit the Western Hemisphere cut a 2,500-mile- (4,023.31 km) wide path of stupendous destruction as it crossed the Caribbean westward from September 12 to 19, 1988. It wreaked nearly $10 billion worth of havoc and killed more than 350 people, as it roared across the Caribbean with winds of up to 200 MPH.

In the Caribbean, Haiti, the U.S. Virgin Islands, the British Virgin Islands, St. Lucia, Puerto Rico, the Dominican Republic, the Cayman Islands, and, particularly, Jamaica suffered everything from mild to extreme damage. Mexico's Yucatán Peninsula, which is normally spared hurricanes, suffered huge damage. Brownsville, Texas, where the storm eventually ran ashore, suffered minor damage, but the storm continued to move upward from Texas, spinning off 41 tornadoes that killed three people in Texas and Oklahoma, flooding land near rivers in both states, and spreading moisture as far north as Chicago. The last casualty of the storm was a 50-year-old pilot from Santa Fe, New Mexico, who was killed on September 17, when his plane broke apart while flying near Muskogee, northeast of Oklahoma City, in heavy rain produced by the storm.

Gilbert began its short but destructive life off the coast of Africa in late August and traveled the trade winds, arriving in the eastern Caribbean at the beginning of September 1988.

On Sunday, September 11, it grazed the southern tip of Haiti, drowning 10 people and engulfing hundreds of head of livestock. Roads were completely cut off to this peninsula and the banana crop was destroyed. Simultaneously, its winds struck the Dominican Republic and the U.S. and British Virgin Islands. Damage was slight to the Virgin Islands; utility poles were toppled, there was some livestock and crop loss on the British Virgin Islands, and there were numerous power outages. In the Dominican Republic, however, the flooding and crop damage were far more widespread. Five died on the island, 100 families were rendered homeless, and the main electricity relay station was blown down, blacking out much of the capital city of Santo Domingo.

Squarely in the path of Gilbert lay the island of Jamaica. It had not had a visit from a hurricane in 37 years, but this one would hit it head on, at 9 A.M. on the morning of September 12. By this time, Gilbert's winds had climbed to 145 MPH. It slammed into Kingston, swept across the banana plantations and livestock farms, and lifted the roofs from 80 percent of the homes on the island. Trees were leveled as if a cosmic buzzsaw had rushed through the countryside. Four out of every five homes were rendered uninhabitable; once the storm had passed, more than a million people would be affected, and 500,000 made homeless.

In Kingston, all 82 patients from the National Chest Hospital were moved hastily after the hurricane chewed through 50 feet (15.2 m) of roof, spewing the remnants of solid wooden beams like toothpicks across the hospital garden.

At another public hospital, the storm crushed its kitchen and flooded the maternity ward. Two days later, doctors and nurses there were still standing in rubber boots in water an inch deep delivering babies in a hallway.

Seventy percent of the dwellings in Kingston, home to a fourth of Jamaica's 3.4 million population, were either damaged or destroyed.

Twenty-five people died on Jamaica; the banana and poultry crops of the entire island were completely wiped out. Communication with the island would be impossible for days. All in all, Jamaica suffered $8 billion in damages, making it the greatest natural disaster in that island's history.

But the storm had not finished its destruction. Back over the tropical waters of the Caribbean, it intensified still further, and both American and Soviet hurricane watchers (the Soviets were based in Cuba) flying into it were astonished by the pressure in its eye: 26.13 inches—the lowest ever recorded for a hurricane in the Western Hemisphere. (The only lower pressure in a hurricane, 25.69 inches (63.5 cm), was recorded on October 12, 1979, in the eye of Hurricane Tip as it rocketed across the Pacific Ocean between Luzon in the Philippines and Iwo Jima in the western Pacific.)

Something else was odd about Hurricane Gilbert. Ordinarily, the eyes of hurricanes measure from 20 to 25 miles (32.5 to 40 km) across. Gilbert's eye measured a mere eight miles in diameter. Thus, it resembled a tornado more than a hurricane, and this, according to Frederick J. Gadomski, a climate analyst at Pennsylvania State University, probably accounted for the extremely violent winds—now clocked at 200 MPH—within its eyewall.

By Tuesday night, September 13, the storm was upgraded to a force 5 storm—the highest force rating possible. Accorded the title of the century's fiercest storm, it headed due west, passed 20 miles (32.5 km) south of the Cayman Islands, where it leveled trees and disrupted electricity and telephone service and headed straight for Mexico's Yucatán Peninsula. With wind gusts of 218 MPH and a storm surge that sent 23-foot- (37-m-) high waves curling over beachfronts, Gilbert slammed into the Yucatán at daybreak on Wednesday, the 14th.

The night before, 20,000 residents of the coastal areas of Yucatán state, including 6,000 tourists—90 percent of them American—were evacuated from the beachfront resort areas of Quintana Roo State and housed in schools, hospitals and government buildings inland.

Well that they had been, for, as it had in Jamaica, the storm began systematically to peel the roofs from homes and other buildings, uproot trees, smash docks, and topple communications towers. The storm cut a 90-mile- (144-km-) long swath across the Yucatán, wrecking the banana crops, flattening thatched roof houses, and rendering small villages on the edge of the fabled Yucatán jungle, where Mayan ruins vie with the undergrowth, ghost towns. The devastation in the city of Cancún was extreme; businesses were wrecked, debris was everywhere. Seventeen people were dead; 300,000 people were made homeless. Ninety percent of the corn and fruit crops were destroyed. In Cancún and Cozumel, posh resort hotels were severely battered. The Club Med lost half of its installations.

Meanwhile, the storm roared on. Honduras, Nicaragua, and Guatemala all reported deaths, mostly by drowning, and extensive damage to crops and animal herds. Sheets of rain pelted Brownsville and Galveston, Texas. Over 175,000 people were evacuated from northern Mexico. City officials on Galveston Island asked the city's 70,000 residents to leave. In Houston, space agency officials delayed the first space shuttle launching since the *Challenger* disaster in January 1986 because the hurricane threat in Houston was interfering with the work of engineers on the project. Prisoners from four Houston area prisons were moved to facilities farther north for the duration of the storm. A hurricane warning was posted from Tampico in northern Mexico to Port O'Connor near the middle of the Texas coast, and a hurricane watch extended as far north as Louisiana.

Dozens of tornadoes, spawned by the storm, whirled through the Lower Rio Grande Valley of Texas and brought wind gusts of up to 82 MPH on the Texas coast at South Padre Island, ripping tin roofs off stores and homes, overturning mobile homes and automobiles. Tornadoes ripped the roof off an apartment building in Brownsville.

Off the Texas coast, the Coast Guard received rescue calls from three shrimp boats that tried to outrun the storm but wound up caught in rough waters. Helicopters quickly found one boat, the *Ki II* of Venice, Louisiana, and picked up her crew of five. A sister boat, the *Ki I*, ran aground, and its three-man crew survived, as did the third shrimper and its men.

An estimated 20,000 people stayed at shelters in the Brownsville area, and roughly that number were housed in shelters in Matamoros, across the border, including 3,000 who stayed in railroad cars.

The storm, whose winds had weakened over the Yucatán Peninsula, and whose eye had widened from eight miles to 50, made its final landfall at 5:30 P.M. Friday, September 16 (Mexico's Independence Day), just south of the border city of Matamoros, Mexico. Its sustained winds were still 120 MPH, enough to topple trees, devastate houses, and cause the most tragic occurrence of its 2,500 mile (4,023.4 km) trek.

Four buses, loaded with a reported 200 passengers stalled near the overflowing Santa Catarina River near Monterey. Trying to avoid the flooding, the buses detoured unwisely and soon found themselves mired on a two-lane highway, engulfed by the raging rain-swollen waters of the Santa Catarina.

Panicked passengers soon clambered to the roofs of the buses. For six hours they remained there, buffeted by winds and rain, while the waters slowly rose around them and rescuers tried to reach them—to no avail. Four police officers drowned when the tractor they were using as a rescue vehicle overturned in the water and they were swept away. Two civilian rescue workers in a boat fell into the floodwaters.

Finally, all four buses toppled over, spilling their passengers into the raging waters, and sweeping them downstream, past their rescuers. Only 13 of the 200 passengers were saved; some of the drowned were fished from the water as far away as Cadereyta, 35 miles (56.3 km) from the site of the tragedy. Some of the bodies were never found.

All in all, more than 350 people lost their lives to Hurricane Gilbert, 750,000 were made homeless, and at least $10 billion in damage was caused by this, the most ferocious storm of the century.

If anything positive can be gleaned from the hurricane, it is this: Modern, up-to-date forecasting techniques and the willingness of populaces to follow the warnings undoubtedly kept the statistics from matching the size and tragic intensity of the storm. Had it occurred 75 years earlier, the casualty figures would have been a hundred times worse.

CARIBBEAN
HURRICANE INEZ
September 24–29, 1966

More than 2,500 people were killed, 3,000 were injured, and 250,000 were made homeless by Hurricane Inez's five-day stay in the Caribbean, from September 24 to 29, 1966.

Over 2,500 residents of Haiti, Cuba, and the Dominican Republic died in the onslaught of Inez, the Caribbean-spawned hurricane that raged for five days (September 24–29, 1966), laying waste to over $100 million worth of crops, flattening villages, and injuring an estimated 3,000 people. Over 250,000 were forced to evacuate their homes—150,000 in Cuba alone.

A particularly furious storm, with winds clocked at 160 MPH, Inez hit Haiti the hardest, turning some of its countryside into a valley of death, with corpses piled high enough to dominate the devastated landscape. The town of Jacmel, a hitherto peaceful enclave on the southwest coast of Haiti, reported a death toll of 1,000. Accurate figures were hard to come by from the dictator, François Duvalier, but huge crop damage was reported.

The Dominican Republic suffered complete destruction of all crops on the island. Homes were completely flattened (in the town of Oviedo the only building that remained standing was the town hall). More than 1,000 people were reported injured, and 200 were known dead.

The coastal areas and riverbanks of Cuba were overrun by the coastal surge of Inez, sending 150,000 persons fleeing from their inundated homes and farms. Hundreds more, caught in the path of waves and wind, drowned or were killed by falling trees and houses.

CARIBBEAN
HURRICANE JANET
September 22, 1955

Five hundred people died and 60,000 were rendered homeless by Hurricane Janet as it carved its destructive path through the Caribbean on September 22, 1955.

The Caribbean season of 1955, with 11 hurricanes, was the most extraordinary and disastrous to date, and Hurricane Janet was one of the worst of them.

Packing winds of 114 MPH (83 km/h), Janet, an immense storm that originated off the coast of Venezuela, claimed over 200 lives in Honduras and 300 in Mexico, made over 60,000 homeless and caused millions of dollars in damage to the coastal areas of Mexico, Honduras, Veracruz, Tampico, and the U.S. Navy Hurricane Observation Station at Swan Island, located off Cape Gracias a Dios.

Janet claimed its first lives in the first recorded wreck of a hurricane hunter plane. The Navy Neptune flew out of Jacksonville on September 26, 1955, with nine crewmen and two Canadian news photographers. Early in the morning of September 27, Lieutenant Commander G. B. Windham reported penetrating the eyewall at an altitude of 700 feet (213.3 m) and announced that winds in the eyewall were clocked at 150 MPH. That was the last message received from the plane, which was never found.

The second victims were also personnel, stationed, along with U.S. Weather Bureau personnel, on Swan Island. The center of the storm passed over the tiny island, destroying all of its buildings with the exception of its mess hall, whose refrigerator exploded. When the winds reached 150 MPH, the roof of this one remaining building was ripped off, soaking seven men with 500 gallons (1,892.8 l) of diesel fuel from a roof tank.

All of the foliage on the island was destroyed, including 10,000 coconut palms which were snapped off at a uniform height of 15 feet (4.57 m). The beaches were carpeted with dead fish, almost all with their eye popped out. Fortunately, there was no loss of human life.

From here, the storm proceeded to Honduras, where it killed an estimated 200 inhabitants of Corozal, a small Honduran city. Exiting Honduras, it menaced the coast of Mexico, ricocheting off the beaches, demolishing huts and buildings in the ancient city of Chetumal in the state of Quintana Roo, as well as other sites in the Yucatán.

But its most horrendous fury was saved for last. Refueled by the tropical waters of the Caribbean, it crashed into Tampico with unabated, monstrous force, ripping countless homes asunder and rendering over 60,000 people homeless.

A grim aftermath occurred when the Panuco and Tamesi Rivers, swollen by the coastal surge and tropical downpours, overflowed their banks, disgorging countless poisonous rattlesnakes into the nearby villages, and increasing the total number of dead from the storm.

CARIBBEAN
HURRICANE JOAN
October 17–23, 1988

The first Atlantic hurricane to cross over land and continue as a Pacific tropical storm, Hurricane Joan killed 500, made over 60,000 homeless, and destroyed scores of villages and towns in the Caribbean from October 17 to 23, 1988.

The second deadly Atlantic hurricane in six weeks (see HURRICANES, Caribbean, Hurricane Gilbert, p. 253) in the Caribbean, Joan was also the first recorded Atlantic hurricane to cross over and continue as a tropical storm in the Pacific. After devastating parts of Nicaragua on October 22, it headed over the Pacific, assuming a new name and a new identity. Weakened by its crossland passage, it became Pacific tropical storm Miriam and threatened San Salvador, Guatemala, and Mexico.

Nearly 500 people lost their lives to Joan in Nicaragua, Costa Rica, Panama, Colombia, and Venezuela as the hurricane wandered for six days across the Caribbean and finally, with winds of 155 MPH, slammed into the coast of Nicaragua at Bluefields. This city of 60,000 was 90 percent destroyed, "practically disappear[ing] from the map," according to the Managua daily *El Nuevo Diario*.

Off the coast of Bluefields, a deepwater port that Bulgarian engineers had been building was 60 percent destroyed. The winds uprooted the breakwater, pulling 10-ton (9.1 metric ton) concrete sections from their moorings.

Houses were splintered into kindling; dikes and dams burst; mudslides buried entire villages. Shrimp packing plants in El Bluff, near Bluefields, were flattened. Every house was smashed on Little Corn and great Corn Islands, 45 miles (72.4 km) east of Bluefields, and Great Corn's only clinic was destroyed.

Many refugees from the Contras, who lived in tin and wood shacks along the riverbeds near Managua, were forced to flee from their homes, which were washed away by roaring, overflowing riverwaters.

In Nicaragua alone, 360 people died, 178 were seriously injured, 110 were reported missing, and 29,000 were made homeless. Seventy-six bridges were wiped out by floodwaters. "This could be the worst natural disaster in Central American history, but people don't realize it because we didn't have 50,000 victims," commented Nicaraguan social security minister Reynaldo Antonio Tefel.

In neighboring Costa Rica, eight people were killed when a dike burst in the town of Neily, 10 miles (16 km.) from the Panama border. In Costa Rica's mountains, the Terraba River rose almost five feet (1.5 m) in an hour of pounding rain, then burst its banks and flooded the village of Cortés, forcing its 2,000 inhabitants to flee. Throughout Costa Rica, 1,500 homes were destroyed, six hospitals and health centers were severely damaged, and 55,000 people were forced to evacuate their homes.

But the greatest casualties in both human life and ecological loss were centered in Nicaragua, which, because of the presence of its Communist Sandinista government, found nearby disaster aid hard to come by. Flying in the face of its normally humane attitude toward the victims of natural disasters, the Reagan administration failed to respond to Nicaragua's pleas for relief aid and sent nothing. Cuba and the Soviet Union sent 25,000 tons (22,679.6 tonnes) of food, medicine, and other necessities and 300 construction workers to rebuild 1,000 homes in Bluefields.

In terms of longterm destruction, the gutting of 70 percent of the fishing fleet, the entire crop in the south of Nicaragua, and more than 600,000 acres of its rain forest was incalculable. "To rebuild a town of 25,000 is possible with the right amount of aid," commented Interior Minister Tomas Borge, "but the ecological damage to our rivers and forests is apocalyptic."

A similar conclusion was expressed by Sandinista commentator Sofia Montenegro, who dubbed her country "The nation of Sisyphus. Damned country!" Montenegro went on, in an article in the official newspaper *Barricada:*

> Once more in ruins! How many times will we have to rebuild what has been destroyed?
>
> This new catastrophe, devastating but above all unfair for a country already worn down by war and enemy siege, gives one a sense that the effort to raise Nicaragua up is futile and hopeless. Against the background of aggression, the hurricane seems not a caprice of nature, but rather a joke, a hideous conspiracy by who knows what Indian gods against our absurd mortal efforts. It produces a strange and atavistic rebellion against our apparent destiny to plant again and again, without ever harvesting anything.

CUBA
October 20, 1926

Every window in Havana, Cuba, exploded from the force of the hurricane that raced through Cuba on October 20, 1926, killing 650 and rendering 10,000 homeless.

The monumental hurricane that battered Cuba for six hours, from early morning until the late afternoon

of October 20, 1926, packed winds that climbed to 130 MPH, strong enough to explode every window in downtown Havana, kill 650 people, render 10,000 homeless, sink dozens of boats in Havana Harbor, and cause $100 million in damage.

Roaring into the harbor of Havana in the early morning of October 20, the storm pushed a 25-foot (7.6 m) high coastal swell that swamped huge steamers like the *Puerto Tarafa,* which went instantly to the bottom with all hands. The *Maximo Gomez* was turned into a juggernaut, ramming a score of other ships and sending them under the roiling waters of the bay before an enormous swell sent it careening ashore.

Matanzas and Pinar del Río provinces were devastated; the small town of Batabano, located at the extreme southern end of Havana Province, suffered a death toll of 300, which was 15 percent of its total population of 2,000. Inhabitants of Santa María del Rosario, Fajardo, Guanabocoa, Jaruco, and Minas were trapped and crushed within their collapsing dwellings. The next day, these villages had virtually disappeared from the horizon.

A distraught officer at the Cuban army headquarters and Camp Columbia walked out to the middle of the parade field and committed suicide at the height of the storm. Other members of the army reacted with more equanimity by assisting what was left of the Havana police force in digging through wreckage for survivors.

When the coastal surge broke over Havana's sea wall, it drove trolleys, automobiles, and people ahead of it, snapping off trees as if they were twigs, crushing one four-year-old boy against a hospital wall, and drowning or maiming rescue workers under collapsed walls or automobiles.

The famous American monument to the dead of the U.S. battleship *Maine,* which was blown up in Havana harbor in 1898, was toppled, its marble columns sheared off at centerpoint, its stone eagles liberated and sent flying, never to be reclaimed.

By the time the storm, which had now centered itself inland where it destroyed crops and farmhouses, had veered out to sea and headed toward Key West, there was hardly any means left to broadcast the news of the wholesale and horrible destruction to the outside world.

CUBA (AND FLORIDA)
HAVANA/FLORIDA
October 13, 1944

Possessing an eye that was 70 miles (112.6 km) in diameter and 167 MPH winds, a deadly hurricane struck Cuba, killing hundreds.

The great freak Havana-Florida hurricane, born in the Caribbean Sea, packed incredible force. For 12 long hours, it clocked 167 MPH (268 km/h) winds, which threw huge schooners used for sponging two and a half miles inland near Batabano, where hundreds died. From here, the storm headed directly for Florida, through the Gulf of Mexico, passing, with an eye that was 70 miles (112.6 km) in diameter, through the space between Ocala and Jacksonville.

From this point, the storm skirted the coast, making no landfall as it rocketed up between Nantucket and Cape Cod and on to the colder waters of Nova Scotia, where it would ultimately die at sea.

CUBA/HAITI
HURRICANE FLORA
September 30–October 9, 1963

Hurricane Flora, which remained in Cuba for 10 days, from September 30 through October 9, 1963, killed over 1,000 and made 175,000 homeless. Proceeding to Haiti, it killed 5,000 and rendered 100,000 homeless on that island.

The paths of hurricanes can be by the book, moving westward along the trade winds and then curving northward toward the poles. Or, they can behave erratically, and this was the reason that Hurricane Flora caused such unbridled havoc on Haiti and Cuba during 10 horrendous days and nights—September 30 through October 9, 1963.

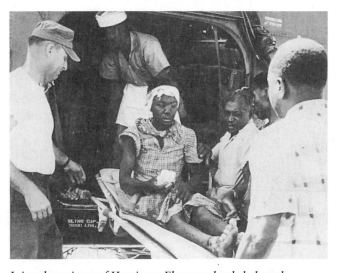

Injured survivors of Hurricane Flora are loaded aboard helicopters in Haiti in September 1963. (CARE)

Fidel Castro accused the United States of deliberately withholding information from Cuba about the imminent danger posed by the threatening storm, which wiped out one-fourth of that island country's sugar crop, 90 percent of its coffee crop, killed over 1,000 persons, and rendered more than 175,000 homeless. (Later, apparently reconsidering his hasty accusation, the Cuban leader filed for United States disaster aid.)

What increased the cataclysm in Cuba was the longevity of the storm. It spent five days crossing and recrossing Cuba, particularly over its easternmost quadrant. Here, it claimed 1,000 lives and accounted for the crushing of the homes destroyed, and erased the once abundant and rich crops of sugar and coffee that formed the basis of Cuba's economy in 1963.

François Duvalier's subjects in Haiti also fared badly, as the hurricane whipped through the island, felling forests, unceremoniously tilling fields, and swelling streams into roaring rivers. Duvalier, noted for undercounting victims, nevertheless admitted that scores of villages had simply disappeared, 5,000 persons had died, and 100,000 had been rendered homeless.

CUBA
SANTA CRUZ DEL SUR
November 9, 1932

A hurricane with 210 MPH winds killed half the population of Santa Cruz del Sur, Cuba—2,500 people—on November 9, 1932.

A ship's captain off the coast of Cuba on November 9, 1932, recorded in his log that the winds that roiled the seas around him into a furious froth and bounced his vessel like a youngster on a pogo stick were "infinite."

That may have been an exaggeration, but to the 4,000 inhabitants of Santa Cruz del Sur in Cuba that same day, the winds recorded at 210 MPH (337 km/h) might just as well have been beyond measurement. The mighty forward motion, the incredibly high winds of this monster hurricane churned up a gigantic tidal wave which curled forward and smashed into the Cuban city full force.

Within minutes, Santa Cruz del Sur was flattened beyond recognition, and 2,500 people—more than half of the city's population—had lost their lives, mostly from drowning. There were simply no shelters for the inhabitants to escape the violent fury of this storm, even if they had had adequate warning.

DOMINICAN REPUBLIC (AND LEEWARD ISLANDS, PUERTO RICO, CUBA, HAITI, KEY WEST, MISSISSIPPI, LOUISIANA)
HURRICANE GEORGES
September 20, 1998

Hurricane Georges, a slow moving, category 4 hurricane, cut a huge swath of death and destruction through the Leeward Islands, Puerto Rico, Cuba, Haiti, the Dominican Republic, Key West, and the coasts of Louisiana and Mississippi in late September and early October of 1998. The Dominican Republic and Haiti suffered most from the storm, which claimed a total of 603 deaths, nearly 400,000 homeless people, and damage estimated at $5.9 billion.

Hurricane Georges, one of the most destructive and lethal hurricanes of modern times, cut a huge path of destruction across the Caribbean in late September and early October of 1998. Georges was born off the west coast of Africa late on the afternoon of September 13, 1998. It became a tropical depression on the 15th, a tropical storm on the 16th, and a hurricane on the 17th. It roared into Antigua, St. Kitts, and Nevis on the evening of the 20th, destroying 70 percent of the living quarters on each island, and made landfall in southeast Puerto Rico on the evening of September 21, went on to particularly devastate the Dominican Republic and Haiti, glance off Key West, and finally blow itself out on the Gulf coast of Louisiana, Mississippi, and Alabama.

The storm, with sustained winds of 115 MPH (185 km/h) and gusts up to 190 MPH (305 km/h), was a killer hurricane, and Puerto Rico, which had been decimated by Hurricane Hugo in 1989 (see p. 307), prepared well. Thus, when the storm struck on September 21, though its maximum winds of 150 MPH ripped roofs from houses, collapsed others, caused huge mudslides, and pelted the island with 28.36 inches (72 cm) of rain, it caused relatively few casualties. Twenty thousand people huddled in shelters while the storm, its winds weakened by land to 115 MPH, crossed the entire length of the island.

Virtually all of Puerto Rico was stripped of electricity and nearly 80 percent of its populace found itself without water or telephone service.

In the capital, San Juan, thousands of trees were toppled by the storm, taking with them streetlights and power lines. Streets were paved with broken glass, metal signs, shards of awnings, and pieces of houses.

In the mountainous interior, the zinc roofs of residents' houses collapsed under the steady onslaught of wind and rain. The lush vegetation of the mountain town of Cayey was reduced to greenish-brown confetti by the passage of the storm; throughout the region, homes were turned into twisted, anonymous hunks of material. The carcasses of poultry dotted the backyards of scores of homes. A warehouse for the town's medical center was destroyed, along with the medicine kept inside it.

"When the eye passed over, it was silent," one resident, Frank Vega, told reporters afterward. "And then you could hear it coming again. It sounded like a tractor-trailer running down the highway."

Meanwhile, Georges continued on into the Caribbean, where it regained strength. Late in the morning of September 22, it made landfall in the Dominican Republic, approximately 75 miles (120.7 km) east of Santo Domingo. Its winds had again climbed to over 125 MPH (201.1 km/h).

The storm ran along the southern coast of the country, then veered inland. And here was where it inflicted its worst damage. The hills and mountains of the Dominican Republic, once covered with old growth trees, had been stripped, in far too many locations, by poor residents, who used the wood for fuel. Now, with torrential winds and rains, and swollen rivers surging over their banks, these hills were turned into mountains of mud that avalanched downward into villages, burying and flattening them, and killing hundreds of residents. In the town of San Cristóbal, 45 people were killed when the walls of a school being used as a shelter collapsed under an assault of water and mud. Seventy-six townspeople drowned in the overflow from a river in San Juan de la Maguana, 120 miles (193.1 km) west of Santo Domingo. Outside Santo Domingo, guards and several hundred inmates remained on the roof of the pentagonal La Victoria prison, most of which was underwater.

President Leonel Fernández described the dreadful destruction of the storm as "a national tragedy" that had left more than 100,000 people homeless (the eventual total would be 185,000) and severely damaged sugar, banana, rice, and coffee crops that were nearly ready for harvest. "We face a titanic task," he added, looking toward the coming cleanup.

The tragedy of the Dominican Republic was matched by that of Haiti, to which Georges next laid waste. The populace of Haiti had been given false alarms once too often. When a real warning was issued, few paid attention to it, and therefore, many paid a terrible price.

Homes, livestock, and crops were swept away by the swirling winds and relentless rain. As in the Dominican Republic, trees that had held hills together had been cut down, and countless floods and mudslides began to envelop and destroy numerous villages, towns, and settlements. Scores clambered aboard floating pieces of anything to escape the raging floodwaters; others climbed trees and remained there, sometimes for days.

Over 150,000 people were rendered homeless; thousands of them crowded into makeshift shelters bereft of heat, light, or water. As in Puerto Rico and the Dominican Republic, huge acreages of crops were flooded, beaten down, and destroyed. A fertile crescent of huge rice paddies and banana plantations in the country's heartland was hopelessly demolished.

And still, Hurricane Georges had only half-finished its destruction. The mountains of Haiti and the Dominican Republic weakened it somewhat, but as it entered the Windward Passage on the morning of the 23rd, Georges again began to gather strength. When it made landfall that afternoon in eastern Cuba, approximately 25 miles (40.2 km) east of Guantánamo Bay, the storm's sustained winds were nearly 85 MPH. It would take it nearly an entire day to cross the island of Cuba west-northwestward on the northern coast.

Over 370,000 people were evacuated from southern regions of Cuba at the beginning of the storm's entrance to the country, as mudslides again buried towns and crops, and rivers overflowed, drowning livestock and people. The entire eastern part of Cuba up to Cayo Coco, a small island popular with tourists north of Ciego de Ávila in central Cuba, was turned into a morass of mud.

Once more, the storm moved out over open water and intensified. At mid-morning on the 25th, its eye passed directly over Key West, Florida, which by that time had been evacuated except for a handful of youths who had never experienced a hurricane and a few habitually contrary souls. Evacuation orders had gone out for 1.4 million people from Key West to Tampa and Miami.

Georges struck Key West just before midday, coincidental with high tide. Its 105 MPH winds whipped up a snowstorm of spray and drove seven-foot (2.11 m) storm surges ahead of it. Since many of the islands in the lower Keys were barely above sea level, this caused widespread flooding. Canals rose four feet above their banks, floating boats out of the canals and into nearby yards. In Cudjoe Key, a local radio station reported a refrigerator floating down a submerged road. In Tavernier and Marathon, roofs blew off houses and from Tampa to the Keys, power was universally lost.

U.S. 1, the only road in and out of the Keys, was under five feet of water in places, and wreckage from collapsed houses choked side roads. A curfew was established in Key West, but there were few to keep indoors.

And still, Hurricane Georges plowed on, heading for its seventh and last landfall. Drawing energy from the warm waters of the Gulf of Mexico, the storm spent the 26th and 27th moving slowly but deliberately toward the Gulf coast. On the morning of the 28th, it came ashore near Biloxi, Mississippi, and, after meandering around southern Mississippi, diminished to a tropical storm that afternoon.

The winds had climbed once again to 105 MPH when Georges crashed into the Mississippi coast. Again, trees fell, power lines were severed, roofs were ripped from houses, windows were smashed. Waves from the Gulf of Mexico arced across beachfront roads along Route 90 in Mississippi as thousands crammed themselves into 250 shelters.

In Venice, Louisiana, the tidal surge topped a levee, sending eight to nine feet of water into the small fishing village. In Lake Pontchartrain, the storm surge banged against levees, now and then sending a random wave over them.

The disabled freighter *Golden Star,* under tow from Beaumont, Texas, to Tampa, Florida, wallowed in 20- to 30-foot (9.14 m) seas in the Gulf, and dragged its anchor.

New Orleans, normally a bustling tourist and convention hub, was eerily deserted, its attractions boarded up with plywood. And for good reason. A great deal of New Orleans is located below sea level, and its levees were being battered vigorously by Georges's storm surge.

But that would be the end of the storm's fury. For the next six to 12 hours, it would become quasi-stationary in a cyclonic loop over southern Mississippi. By early morning of October 1, it would become a mere frontal zone over the northeast Florida/southeast Georgia coast.

The effects of its monumentally destructive path would linger. In Haiti, the poorest country in the Western Hemisphere, despair reigned. Providing relief to the stricken farmers and restoring crops and livestock production loomed as an almost insurmountable task. For nearly a week after the storm worked its havoc, makeshift shelters in schools, churches, and civic centers continued to be overwhelmed by new refugees.

In the countryside, families who had used livestock as a savings account to pay for events such as first communions, weddings, and funerals, were suddenly bankrupt. "We're sleeping in trees like our animals," Clovis Daniel, a cattle rancher who had lost his entire herd told a reporter. Nearly a hundred others shared his living quarters.

In the Dominican Republic, residents whose homes had been demolished found that there was a shortage of building materials. Nails were almost nonexistent. A week after the storm, some thousands had still not received a single relief supply.

Cattle and horses wandered aimlessly through the devastated countryside; rivers were choked with mud, trees and crops were flattened, and houses were without roofs. Officials estimated that several hundred thousand people had been cut off from the rest of the country by the hurricane and forced to survive on whatever food and water they had before the storm struck.

With only four small helicopters, the Dominican government had slow going in ferrying relief supplies to the country's residents. Finally, after a week, military trucks began to distribute food, water, small electric generators, and tin roofing materials in both the capital and rural towns.

In these countries, and in the Leeward Islands, Georges cut a huge swath of death as well as destruction. Two hundred and three people died in the Dominican Republic alone; 400 more died in Haiti, Puerto Rico, and the Leewards.

In Key West and on the Gulf coast, the destruction was purely material. In Pascagoula, Mississippi, water reached rooftops, roads were washed out, and oyster shells were swept into yards. In Alabama, the Dog River near Mobile overflowed its banks to a depth of 11 feet, floating, sinking, or damaging over 100 boats. In northwest Florida, 28 tornadoes, spawned by the storm, touched down.

The total death toll from this extraordinarily destructive storm was 602. In Puerto Rico, a total of 72,605 homes were damaged, of which 28,005 were completely destroyed. In the Dominican Republic, 100,000 of the 185,000 people made homeless by the storm remained in shelters through the entire month of October. In Haiti, 167,332 people were rendered homeless.

In Cuba, 60,475 homes were damaged and 3,481 were totally destroyed. In the United States, the damage was considerably less. One thousand five hundred and thirty-six homes were damaged in the Florida Keys and 173 were totally destroyed.

Damage to crops was enormous. Ninety-five percent of the plantain and banana crop and 75 percent of the coffee crop in Puerto Rico was wiped out by the storm. All of the banana plantations in eastern Cuba were destroyed.

The total financial loss from the storm was $5.9 billion, and the American Red Cross spent $104 million on relief services throughout the area, making it the most expensive disaster aid effort in the organization's 117-year history.

ENGLAND
November 27, 1703

The hurricane of November 27, 1703, drowned 30,000 sailors off the coast of England, countless others on land, and crushed 5,000 houses in Gloucestershire, Sussex, Kent, Oxfordshire, Somersetshire, Northampton, Suffolk, and Worcestershire.

Daniel Defoe, the creator of *Robinson Crusoe*, must have drawn some inspiration from the raging sea caused by the enormously destructive hurricane of November 27, 1703. Slamming into the southern coast of England, it sank 300 ships, killed an estimated 30,000 sailors at sea, caused the deaths of an incredible—though unrecorded—number of cattle and people, and tumbled or crushed 5,000 houses in Gloucestershire, Sussex, Kent, Oxfordshire, Somersetshire, Northampton, Suffolk, and Worcestershire.

It was, in fact, Defoe, then a pamphleteer, who is responsible for whatever records survive of this 18th-century catastrophe.

For 14 days prior to making landfall, the storm lashed at the English coast with pelting rain and high wind squalls, softening it up for the final blow, which, when it came, virtually scoured the harbors of Grimsby, Falmouth, Bristol, Gravesend, Plymouth, and Cowes clean of the ships and seamen of the French, Russian, and English fleets which happened to be unlucky enough to be riding at anchor in these ports when the storm struck. Even ships inland, on the River Thames, were demolished as they were driven either into each other or onshore.

The wind turned tiles into flying missiles, flinging them through the air at enormous speed and driving them, according to Defoe, ". . . from five to eight inches into the solid earth." Other solid pieces of dwellings—bricks, timber, sheets of lead and iron—were lifted from buildings and sent soaring murderously above the earth.

Even after the storm careened inland, its devastation continued. The next day, a prodigious tide that was said to be six to eight feet (1.82–2.43 m) higher than normal roared over seawalls and river banks, killing still more people and cattle.

GRENADA
September 22, 1956

Two hundred people died and practically every building on the island of Grenada was destroyed in the 127-MPH hurricane of September 22, 1956.

In one five-hour nightmare of destruction, a hurricane packing winds of up to 127 MPH (204 km/h) barreled into Grenada on September 22, 1956, killing 200 persons and destroying almost every building on the 120-square-mile (193.12 sq. km) island.

The Crown Colony of England (it would become an associated state of Britain 12 years later, in 1967, and gain full independence in 1974), primarily a resort, had been amply warned of the impending storm. A day earlier, the hurricane had roared in from the Atlantic and ripped across Barbados, 98 miles (157.7 km) northeast of Grenada, killing 30 persons and flattening over 40,000 homes.

But even radioed warnings, the firing of cannons in the harbor of Saint George's, and frantically pedaling policemen on official bicycles failed to prevent the catastrophic loss of life.

Saint George's was particularly hard hit, losing not only its buildings but its warehouses and piers, the heart of its commercial and tourist area. It would take almost a year, with copious American and British aid, before Grenada would return to any semblance of stability and serenity.

GUADELOUPE
August 4, 1666

More than 2,000 French sailors were drowned when a violent hurricane struck Guadeloupe on August 4, 1666.

After surviving a minor sidehand sweep from a hurricane in 1665, Guadeloupe received the full force of what has been recorded as one of the most violent storms of the 1600s in the Western Hemisphere, on August 4, 1666. Over 2,000 French sailors alone were drowned when a troop fleet was sent to the bottom of the Caribbean by the monster storm surge that roared ahead of the hurricane.

First coursing through St. Christopher and Martinique, the storm was not abated a bit by this brief landfall and went on to slam into Guadeloupe with full force, sinking every sailing vessel either at anchor or at sea near the island. A six-foot- (1.81-m-) thick defense wall surmounted by huge cannons was crumpled like paper by the pounding of the same waves that demolished the ships it was supposed to shelter.

The sugar crop that the French had been nurturing with imported slaves was wiped out in a matter of hours. It would take years before it would be reinstated and the colony transformed into one of the world's richest growers and exporters of sugar.

GUADELOUPE
September 12, 1928

Six hundred people died without warning when a surprise hurricane hit Guadeloupe on September 12, 1928.

The same storm that meandered into Puerto Rico on September 13, killing 300 persons and causing $50 million in damage, laid waste to Guadeloupe the day before, causing the deaths of 600 people.

The mammoth storm was a slow mover, and this, in the computer-controlled forecasting atmosphere of today, would have assured a much smaller fatality figure. But the fact that most of the dead were caught in their fields or on village streets indicates that warning failures were the primary cause of the magnitude of this disaster.

GUATEMALA, EL SALVADOR, MEXICO
HURRICANE STAN
October 1–5, 2005

From October 1–5, 2005, a cataclysmic convergence of the category 1 Hurricane Stan and an extensive tropical storm system that pelted Central America with a week of downpours resulted in mudslides and drownings that killed 1,540–2,000 residents of Guatemala, El Salvador, Honduras, and Mexico.

On October 1, 2005, two storms in an unusually active season formed over Central America. One was a torrential rain system—poised over Mexico, Guatemala, El Salvador, Nicaragua, Honduras, and Costa Rica—that was loosening the earth, filling rivers to overflowing, and causing mudslides. The other was a tropical depression off the coast of the Yucatán Peninsula. By early in the morning of October 2, this depression had become Tropical Storm Stan, which would make landfall that day, weakening as it traversed the Yucatán Peninsula, but regaining tropical storm strength when it reemerged into the Bay of Campeche.

By October 4, the storm had reached hurricane status, with winds of 80 MPH (128 km/h) making it a Category 1 hurricane. That by itself would have caused minimal damage and little notice. But when Hurricane Stan became imbedded in the already torrential rain system in the area, transferring some of its energy to it and absorbing some of the rain system's energy into itself, the combination proved to be lethal.

Continuing on through the bay, Stan made landfall later on October 4, on the east-central coast of Mexico south of Veracruz. Its strength spread outward, feeding the deluge that now covered a great deal of Central America.

In Guatemala, in Panabaj, an impoverished Mayan village in the highlands near Lake Atitlán in Sololá Department, the rains fell gently before Stan became involved. On Wednesday, October 5, however, creeks and rivers began spilling their banks. As the day progressed, the waters continued to rise; a nearby mountain simply came apart, and, in an enormous wall of mud and rocks, it bore down on the village.

"I saw it coming," Pablo Gaspar, a 16-year-old firewood collector, told rescuers later. "I leaned up against the wall. Then, the wall collapsed."

"It was a roar, like an earthquake," Ramiriz Tacaxoy, a 41-year-old farmer told the rescuers. "Some of my neighbors were buried up to their necks. Some people on the edges were able to grab hold of coffee trees, and we pulled them out. But many remained underneath." Six of his family members, including both of his parents, were buried in the mudslide, which also obliterated nearby Piedra Grande, a hamlet of 1,400 people in the municipality of San Pedro Sacatepéquez. Before the rains ended on October 5, 36,559 people had fled to emergency shelters.

El Salvador's tragedy was compounded by another natural disaster. On October 1, the Santa Ana volcano, located near the capital of San Salvador, erupted, and this, plus the driving rains, caused a portion of the mountain to break away and cascade into the surrounding valley. Three hundred communities were buried, and 54,000 residents managed to flee from the disaster.

In Mexico, where Stan had come ashore, some 100,000 inhabitants of the Sierra de los Tuxtlas region on the Gulf Coast were evacuated from their homes, which in many cases had their roofs ripped away by Stan's winds. Coastal areas of Veracruz, including its port, Boca del Río, San Andrés Tuxtla, Santiago Tuxtia, Minatitlán, and Coatzacoalcos were flooded and ripped apart by hurricane winds.

Inland, areas of Chiapas, near the Guatemalan border, were also hit hard, particularly the coastal border town of Tapachula, where the river overflowed its banks, flooded the town, and destroyed all the bridges leading into and out of it.

Throughout the area, 33 rivers broke their banks and nearly 20 bridges and other infrastructures were ripped apart.

In Honduras, 7,042 people were evacuated and 2,475 homes were shattered or washed away when the Nacaome River overran its banks and washed away the populace and its dwellings.

By October 10, military and international aid groups had arrived, and were helping in supplying water and food to survivors and aiding in the efforts to dig out the bodies buried under mud that was, in some places, 40 feet (12.2 m) high. It was a hopeless task, and Guatemalan vice president Eduardo Stein finally gave towns legal permission to declare the buried areas cemeteries. Some of these towns, such as Tacana, had already done this, and had spread lime over the mud to prevent disease.

The true count of fatalities will probably never be precisely known because of the hundreds counted missing beneath the caking mud. Estimates were that there were 80 deaths directly caused by Hurricane Stan and between 1,540 and 2,000 fatalities caused by the combination of Stan and the rain system. The total damage was set at $1 to $2 billion.

HAITI
August 23, 1909

(See HURRICANES, Mexico, 1909, p. 268.)

HAITI
HURRICANE CLEO
August 22, 1964

(See HURRICANES, United States, Florida, Hurricane Cleo, p. 295.)

HAITI
HURRICANE FLORA
October 9, 1963

(See HURRICANES, Cuba, Hurricane Flora, p. 258.)

HAITI
HURRICANE HAZEL
October 12, 1954

Villages were buried and cities devastated, 1,000 were killed, and over 40,000 were made homeless when Hurricane Hazel visited Haiti on October 12, 1954.

(See also HURRICANES, United States, East, Hurricane Hazel, p. 287.)

Sometime before dawn on October 12, 1954, Hurricane Hazel smashed into Haiti's southern coast with 125 MPH winds. Before it left the island, it would kill nearly 1,000 people and render over 40,000 homeless. Soil and crops of coffee, sugarcane, beans, rice, and corn were washed out to sea.

Near the capital, Port-au-Prince, torrential rains at the heart of the storm washed tons of mud over the village of Berly, burying alive all but two of the town's 260 inhabitants. In Jacmel, a city of over 8,000 people, the wind sheared off the faces of buildings, exposing the interiors to the soaking storm. Monumental tides caused by the storm surge of the hurricane pummeled the coast, washing over the seashore villages of Les Cayes on the tip of the southern peninsula.

Heading northeast, the storm seemed to blow itself out against Haiti's mountains, after leaving its destructive aftermath of tuberculosis, typhoid, and typhus. U.S. Navy observation planes and Weather Bureau experts pronounced it tamed. But Hazel, heading for the Carolinas, was about to be reborn.

HAITI
HURRICANE JERAMIE
October 22, 1935

Over 2,000 died when one of the first named hurricanes, Hurricane Jeramie, the aptly named "hairpin hurricane," hit Haiti on October 22, 1935.

More than 2,000 farmers, most of them caught at work, were killed when Hurricane Jeramie, one of the first named hurricanes, slammed into Haiti on October 22, 1935.

Jeramie was nicknamed "The Hairpin Hurricane" because of the abnormal path it followed. Ordinarily, hurricanes travel predictably along the trade winds and then turn toward the polar north. This hurricane originated in the western Caribbean, traveled eastward, and slammed into Jamaica, where it wiped out over $2 million in sugar plantations, and then veered, not north, but south.

Thus, neither weather forecasters nor the population of Haiti were at all prepared for the fatal onslaught of the storm, which hit most devastatingly in the agricultural regions of the southwest, killing farmers, crops and livestock.

HOLLAND
LEYDEN
October 1, 1574

Twenty thousand occupying Spanish soldiers were drowned when a hurricane destroyed dikes near Leyden, Holland, on October 1, 1574.

The siege of Leyden had gone on for many days before a gigantic wind and rain storm of hurricane force hit this lowland city, located on the old Rhine River in Holland. Spanish troops, warring against William the Silent, were in the process of starving out the populace of this ancient, walled city.

They were having little luck. The plucky Netherlanders were, according to historian A. H. Godbey, "... devouring roots of grass, old leather, offal, anything that could in the least aid to sustain life. . . ." But the Spanish soldiers, who ringed this city that dated back to Roman times, held fast.

On the night of October 1, 1574, a hurricane arrived, lashing against the coast some distance beyond the encamped Spanish soldiers. The dikes gave way before the storm surge, flooding the lowlands and drowning an estimated 20,000 Spanish troops. The siege of Leyden was over. (See also FLOODS, Holland, Leyden, p. 164.)

There is an equally persuasive version of this same incident that states that there was no storm, and that William the Silent was the one who smashed the dikes, thus allowing the fleet of the Beggars of the Sea (crews of patriotic privateers chartered in 1569 by William to harass Spanish shipping) to sail across dry land and subdue the Spanish. Belief in either depends upon a choice of historian.

HONDURAS
September 27, 1955

(See HURRICANES, Caribbean, Hurricane Janet, p. 255.)

HONDURAS
HURRICANE HATTIE
October 31, 1961

Carrying winds clocked at 200 MPH, Hurricane Hattie struck Honduras on October 31, 1961, killing 400.

Hurricane Hattie, the eighth hurricane of an ordinary season in 1961, proved to be a distinctly out of the ordinary storm when it slammed into Belize, the capital of Honduras, with winds that were gauged from 160 to 200 MPH (321 km/h). Two hundred seventy-five people died in this city, most from drowning when a series of 10-foot tidal waves crumbled seawalls and piers, and rose to the third stories of buildings near the waterfront. Days later, debris and mud still clung to the third-story floors and walls.

After wreaking its havoc on Belize, the storm plummeted on toward smaller villages, leaving only three buildings standing in the town of Stann Creek. The waves continued to roll in, inundating such islands as Caulker and Turneffe Cays, and burying many inhabitants under mud and collapsed homes. The final tally of dead reached 400, with countless others missing and presumed swept out to sea.

HONDURAS AND NICARAGUA
HURRICANE MITCH
October 26, 1998

Hurricane Mitch, considered the worst natural disaster to strike Central America in the 20th century, pelted Honduras and Nicaragua with six days of relentless rain, causing monster floods and mudslides that killed 11,000 people, and left nearly 20,000 missing and presumed drowned or buried. The homeless total for the two countries reached 1.5 million and the cost of the storm, which totally destroyed Honduras's infrastructure, was $5 billion.

"Apocalyptic" was the word that Nicaraguan president Arnoldo Aleman used to describe the damage to his country caused by Hurricane Mitch, the deadliest Atlantic hurricane since the Great Hurricane of 1780 killed over 25,000 people in Montego Bay, Martinique, and St. Eustatius (see p. 272).

Ironically, although Hurricane Mitch was a category 5 storm with sustained winds of 190 MPH and gusts well over 200 MPH, its greatest devastation was caused after it had diminished in intensity, slowed down, and pumped relentless rains onto Honduras and Nicaragua.

Born as a tropical depression on October 22, Mitch matured into a monster storm quickly. By 9 P.M. on October 26, its winds had climbed to 180 MPH, and they remained there for 33 hours—the longest continuous period for a category 5 storm since the 36 consecutive hours by Hurricane David in 1979 (see p. 283).

Squalls that ran out from the 350-mile- (563.3-km-) wide goliath first threatened Jamaica, forcing churches to cancel services and flooding streets in the capital city of Kingston. It skirted the Cayman Islands, then headed westward toward Honduras on October 27. Thousands of people were evacuated on that day from coastal regions of Belize, Mexico, and Cuba.

Later that night, Mitch positioned itself 35 miles (56.3 km) north of Honduras, and began a slow movement west-southwest, parallel to the Honduran coast. As it did so, its winds dropped from 180 to 140 MPH, a category 4 speed.

Near midnight of the 27th, Mitch veered toward the mainland of Honduras, but still remained tantalizingly 25 miles (40.2 km) offshore. However, its winds, now 120 MPH, continued to pump rain and wind on the coastal regions of the country, sweeping away bridges, collapsing trees and buildings, and flooding coastal cities and villages.

As the rain continued, more than 50 rivers in Honduras began to overflow their banks. The government evacuated 45,000 people from low-lying coastal areas. Communication was severed between the mainland and the Bay Islands, off the coast.

On the Yucatán Peninsula, Mexico moved residents and tourists from beach resorts and cut back on oil pumping in the Gulf of Mexico.

Mitch remained stationary for three days, as its winds diminished to 75 MPH. All the while, relentless rain, averaging an astonishing two feet a day, continued to pound Honduras, first on the coast, then farther and farther inland.

The *Fantome*, a four-masted schooner, one of the boats of the Windjammer Barefoot Cruises, was reported missing with a crew of 31 aboard. Repeated searches by the Coast Guard failed to find the ship, and after the storm cleared, its wreckage was spotted. Families of the crew sued the company, charging that the crew had been sent by the company on a "suicide cruise."

Now the center of the storm finally moved slightly southeast and made landfall near Limón, and the rainfall began to take its toll across the entirety of Honduras and into Nicaragua. In some places, two feet of water fell in six hours, turning small streams into rivers and rivers into raging, out of control torrents. Soaked into a state somewhere between solid and liquid, the ground began to shift, and mudslides began to swallow trees and landscapes, then houses and eventually villages.

In the city of Comayagua, in central Honduras, the Salistroso River burst through its banks and swept 150 houses ahead of it. Mudslides consumed houses and people in the southern city of Choluteca. In the capital,

Tegucigalpa, a prisoner was shot to death and two were wounded when they tried to escape during the evacuation of the Central Penitentiary.

Neighboring Nicaragua suffered identical horrors. Mudslides buried entire communities—dwellings and their occupants—without warning. On the northwestern border with Honduras, mudslides avalanched upon communities, burying them and their inhabitants without mercy in minutes.

"It is like a desert littered with buried bodies," one rescue worker commented, as he worked to extricate the few who remained alive after countless mudslides in the country.

Back in Honduras, army helicopters began to rescue some flood victims clinging to rooftops in the capital. In coastal communities, rescuers plucked people from roofs by floating up to them in boats.

The rain continued ceaselessly, and as it did, more bridges fell, more mudslides occurred, more communities were buried. The Choluteca River, beyond control, swept away neighborhoods, vehicles, trees, power lines, and livestock. As people took refuge on roofs, looters took advantage of the chaos and raided stores and homes.

"The capital has been leveled," Mayor Cesar Castellianos told reporters. "Blocks and blocks of middle class and poor neighborhoods, shops—they have all been completely demolished." Later that same day, Mayor Castellianos and three aides were killed when their helicopter crashed during an inspection flight.

"This is a catastrophe beyond measure," General Rodolfo Pacheco, the chief of the Honduran air force observed. "It's incredible. The entire nation is in danger."

Still the storm remained nearly stationary, blocked by a strong front over the Gulf of Mexico. And as it did, drownings and burials under mud increased. "I believe that since the earthquake of 1972 (see p. 78) we have not suffered so much loss of human life as the misfortune experienced in recent days," President Arnoldo Aleman of Nicaragua said in a televised address. "There are corpses everywhere—victims of landslides or the waters," Honduras's President Carlos Flores Facusse echoed.

After six days of torrential rains, rescue workers began to infiltrate the afflicted areas. The scenes that greeted them were, in the words of one worker, "a vision out of Dante," and in the words of another "a deluge of Biblical proportions."

There were bodies everywhere—some half-buried in mud, others pinned against tree trunks or tree limbs protruding upward out of the muck. "It would be appropriate to declare the area of the tragedy a national cemetery and seal it off," declared another rescuer.

The rotting bodies of humans and livestock produced a growing danger of epidemics of cholera, dengue fever, and other diseases. And to add to the misery, there was a growing shortage of food and water in rural areas cut off from the rest of the country. Warehouses that had contained foodstuffs were destroyed, roads that would have allowed the transport of goods and supplies were impassable, and crops of beans, bananas, vegetables, and coffee had been washed away.

One of the worst tragedies occurred in the villages on the slopes of the Castas volcano in northwestern Nicaragua. The pileup of foot after foot of falling rain swelled the volcano's crater lake until, with a roar that survivors said sounded like a fleet of helicopters arriving, it burst through the mountain slopes. Ton after ton of mud, like a lava flow, roared down the mountainside, crushing everything and everyone in its path.

Some farmers and their families, in the fields that day, reported that they saw a wall of dirt mixed with trees, rocks, and roofs more than 20 feet (6.1 m) high coming toward them, with a herd of frantic oxen running ahead of it.

Rolando Rodríguez, a farming village of 164 houses, was completely demolished. Only a field of mud and one house remained after the mudslide, from which horribly disfigured bodies and pieces of houses and trees protruded. Six hundred and ten bodies were recovered and 1,900 people were never found and were presumably buried under the wreckage. All the rescuers could do was to douse the area with gasoline and set it afire to prevent the spread of disease.

In both Honduras and Nicaragua, nearly every bridge was reported to have suffered damage. Most were unusable and would have to be totally replaced. The entire infrastructure of Honduras was destroyed. Bridges, roads, communications, ports, hospitals, and schools were annihilated or so decimated that they were beyond restoring. Entire villages and farms were erased from the map, as if they had never existed. Three-quarters of the country's crops were ruined or simply swept away.

Some surviving families waited for days on rooftops or perched in trees, waiting to be rescued. There were reports from the northern provinces of Honduras of desperate parents tying small children to the limbs of trees to protect them from drowning. All of the major cities in Honduras were like islands, cut off from each other and the rest of the country.

"This is the worst," concluded Delmer Urbizio, the minister of government and justice in Honduras. "This has no precedent in the history of the country, or even in the history of Central America."

Mitch, meanwhile, meandered northward. It struck the western side of Mexico's Yucatán Peninsula, which weakened it to a tropical depression, then regained tropical storm status as it crossed the Gulf of Mexico. On the night of November 4, it struck the Florida Keys with 55 MPH winds, knocking over trees and power lines, ripping roofs from houses, and spawning several tornadoes, but causing only seven injuries and no fatalities. Within 24 hours, it had crossed the state and headed out to sea. Near the Dry Tortugas, a fisherman was drowned when his boat capsized in the seas caused by Mitch before it eventually blew itself out past the Bahamas.

Meanwhile, in Central America, the suffering continued. Large bodies of stagnant water were becoming breeding grounds for malaria, dengue fever, and other mosquito-borne diseases. Cases of cholera began to be reported.

The industrialized world responded by sending rescue teams and supplies and forgiving debts to Honduras and Nicaragua, two of the world's poorest nations.

The statistics were staggering, and made Mitch the most deadly natural disaster to strike Central America in the 20th century. A total of 11,000 were killed, most of them in Honduras and Nicaragua. Honduras's toll was 6,500; Nicaragua's, 3,800. In addition, nearly 20,000 were reported missing and presumably drowned or buried by the storm. Up to 1.5 million people were displaced and homeless. The crop losses were estimated at $900 million. The cost of the storm was set at $5 billion—a staggering amount for these poor nations, which faced a total of from $4.5 billion to $5 billion in lost future production. Restoring Honduras's infrastructure alone would be a 15- to 20-year task.

And all this from a storm whose eye never came ashore.

JAMAICA
July 30, 1784

The only known record of the Great Cape Verde Hurricane of July 30, 1784, was kept by American poet Philip Freneau, who favored imagery over statistics.

The Great Cape Verde Hurricane of 1784 was given an odd sort of immortality when it was recorded in verse by American poet Philip Freneau, a witness and near victim of this catastrophic, meteorological monster.

On July 30, Freneau, sailing as supercargo on the brig *Dromelly,* heading for Jamaica, found himself suddenly surrounded by darkness, roaring winds, and mountain-high waves. The ship nearly went to the bot-

tom, which would have eclipsed the early career of this journalist turned seaman.

Shaken, he later set down the following—the only known—account of the storm, which he titled "Verses, made at Sea, In a Heavy Gale." Short on statistics, it is abundant in imagery and terror:

> Happy the man who, safe on shore,
> Now trims, at home, his evening fire;
> Unmoved, he hears the tempest roar,
> That on the tufted groves expire.
> Alas! on us they doubly fall,
> Our feeble barque must bear them all.
>
> While death and darkness both surround,
> And tempests rage with lawless power,
> Of friendship's voice I hear no sound,
> No comfort in this dreadful hour—
> What friendship can in tempests be,
> What comforts on this raging sea?
>
> The barque, accustomed to obey,
> No more the trembling pilots guide.
> Along she gropes her trackless way,
> While mountains burst on either side—
> Thus, skill and science both must fall;
> And ruin is the lot of all.

JAMAICA
HURRICANE CHARLIE
August 17, 1951

Fifty-four people were killed, 2,000 injured, and 50,000 made homeless on the island of Jamaica by Hurricane Charlie on August 17, 1951.

Hurricanes were first given women's names in 1953. Two years before that, in 1951, they were noted alphabetically by the Navy as Able, Baker, Charlie, etc. Thus, Charlie was the third of the 12 hurricanes of that season. It began life gently enough in the Atlantic, and then in the Lesser Antilles. But, by the time it finally blew itself out, 54 people died, 2,000 were injured, and 50,000 were rendered homeless.

On August 16, Jamaica forecasters warned that the storm was on a course running east and north, which would allow it to miss Jamaica. But by 8:30 P.M. it took an unexpected shift, crashing into the old town of Morant Bay, located on the coast of St. Thomas Parish. Every coco palm that had hitherto lined that town's streets was flattened in a matter of minutes. Every house was smashed, and the Yallahs Valley was stripped of topsoil and crops, which were washed out to sea.

Winding up now, with winds of 125 MPH, the storm spun on toward Port Royal, the former pirate hangout, where it demolished the waterfront, sweeping wharves, warehouses, and rum shops out to sea.

Heading inland, the storm lay waste to 12 parishes, wiping out 90 percent of the banana export trade and a large percentage of the coconut industry. Thousands of coco palms fell, cocoa crops were inundated and washed away by torrential downpours, and 154 people either drowned, were buried under fallen houses and debris, or were crushed by crazily blown timbers.

The storm wore itself out at sea near Tampico, but the devastation it left behind would number in the millions of dollars.

MARTINIQUE
October 20, 1695

An incomplete but colorful account of the great hurricane of October 20, 1695, on Martinique was left by Father Jean-Baptiste Labat.

The monstrous hurricane of October 20, 1695, was recorded in sketchy form by Father Jean-Baptiste Labat, a scourge and a legend to the native population of Martinique.

Father Labat, a torturer of natives who, when he caught them practicing their black magic, whipped them and then rubbed the raw wounds with lime juice and red pepper, still survives in legend as a figure galloping through hurricanes, living on as an evil spirit responsible for the ills of the people he both hated and oppressed.

Certain that hurricanes were caused by divine intervention he wrote, ". . . this compression and rarefaction of the air (brought on by either the movement of Earth around the sun or the sun around Earth) is the cause of the winds. They are not here by accident."

Caught in the rising winds of October 20, Father Labat galloped on horseback to his church and snatched the Holy Sacrament from the altar. Tucking it under his arm, he remounted his horse, and, through winds so strong he had to clutch at the neck of his horse, galloped across meadows that resembled a sea of waves. Thus, the legend and the longevity of Father Labat.

MARTINIQUE
August 18, 1881

Seven hundred people were killed in a deceptively small hurricane that hit Martinique on August 18, 1881.

Hurricanes that seem small by comparison to others can often outdo the damage wrought by their bigger brothers. Through a combination of bad reporting and complacency, a compact, self-contained hurricane hit Martinique a crippling blow in the early evening of August 18, 1881. The huge storm surge sent waves crashing into this volcanic, windward island, flooding the streets of Fort-de-France and inundating the vast sugar fields that, since 1654, had governed the economy of this island and covered 80 percent of its land area.

The storm continued to batter the island for four hours, and when it finally smashed up against Pelee, the island's volcano, it split and veered out to sea. It left behind 700 dead and over $10 million in damage to livestock, buildings, and flooded sugar crops.

MEXICO
August 27, 1909

Over 1,500 people died in the assault upon Tampico, Mexico, by 1909's worst hurricane, on August 27 of that year.

In 1909, multiple hurricanes hit the Caribbean. The worst hurricane of the year began its life to the east of the Windward Islands on August 20. For two days it roiled the waters of the Caribbean, first hitting Puerto Rico a glancing blow, then completely leveling the city of St. Nicholas, in Haiti.

On August 24, with winds clocked at 60 MPH, Cuba received a moderate brushing from the storm.

But that was a mere respite. Drawing more energy from the superheated waters of the Yucatán Channel, the hurricane's winds cranked themselves up to speeds exceeding 100 MPH in the next two days, sideswiped the coast of Texas, and barreled full force into the northeastern provinces of Mexico. A combination of high winds and torrential rains swelled this region's rivers, and by the time the hurricane had passed on August 28, over 1,500 people had been killed.

MEXICO
BELIZE AND CARIBBEAN
HURRICANE DEAN
August 13–23, 2007

Hurricane Dean, the fourth-named storm and the first hurricane of the 2007 Atlantic hurricane season, was a monster the size of Texas that reached the top of *the Saffir-Simpson scale of hurricanes at a Category 5. The first hurricane to make landfall in the Atlantic Basin at Category 5 intensity since Andrew (see p. 280) 15 years earlier, Dean raked through the Windward Islands (particularly St. Lucia, Martinique, and Dominica), the Leeward Islands, Puerto Rico, the Dominican Republic, Haiti, Jamaica, the Yucatán Peninsula, Belize, Nicaragua, and central Mexico. At least 42 died and there was $3.8 billion in damage from the storm.*

The prophecies of scientists sometimes come true slowly and subtly. But the warnings issued by an international group of experts at a meeting of the Intergovernmental Panel on Climate Change (IPCC) in February 2007 came true with a dramatic impact. The 2007 Atlantic hurricane system began with not one, but two Category 5 hurricanes—Dean and Felix.

A Category 5 hurricane rests at the top of the Saffir-Simpson Hurricane scale and is defined as capable of causing "catastrophic damage." Anything more would, it is supposed, be apocalyptic. The appearance of Hurricanes Dean and Felix were, in a sense, the predictions of the panel come true. "Numerous long-term changes in climate have been observed," the report stated. "These include changes in . . . the intensity of tropical cyclones." In the North Atlantic, the report continued, "[fiercer hurricanes] are correlated with increases of tropical sea surface temperatures."

Dean was the first major hurricane of the 2007 season—the ninth most intense Atlantic hurricane since records began to be kept in 1850 and the third most intense Atlantic hurricane ever at landfall. Only the "Labor Day" hurricane of 1935 (see p. 289) and Hurricane Gilbert in 1988 (see p. 253) outpaced it in intensity, but not by much.

As is the case with most Atlantic hurricanes, Dean was born off the coast of Africa on August 11 as a tropical wave. Conditions were favorable for nurturing this wave, and by August 14, it was upgraded to Tropical Storm Dean. From that point forward, the intensification was steady and rapid. By August 15, it had become Hurricane Dean and at 5:00 A.M. on August 16, it was classified as a Category 2 storm. Picked up by strong moving westerlies, it headed directly for the Caribbean, and specifically toward the Antilles, Lesser and Greater.

On August 16–17, Dean graduated from a Category 2 to a Category 4 storm. As it neared settled land, warnings and watches went up for all the windward and leeward islands along its path. Jamaica, Haiti, and part of the Dominican Republic battened down. Oil rigs were evacuated, airports were shut down, a dozen cruise ships altered their itineraries, and NASA cut

short its *STS-118* mission as a precaution in case Dean veered toward the Lyndon B. Johnson Space Center in Houston, Texas. The mission's final space walk was shortened by 2 hours and the vehicle touched down a day earlier than originally planned.

Dean entered the Caribbean through the Saint Lucia Channel between St. Lucia and Martinique on August 17, killing three people, flattening buildings and devastating the agricultural economies of St. Lucia, Martinique, and Dominica.

The Greater Antilles were spared direct contact with Dean, but its outer bands caused flooding and landslides on the island of Jamaica. By this time, it had strengthened to a Category 4 hurricane and its size, which would eventually become as large as the state of Texas, left a trail of destruction and death. Five died in the Lesser Antilles, a man in St. Lucia drowned in a river while trying to save a cow, a woman and her son died in a landslide on Dominica, and two more perished on Martinique.

Haiti and the Dominican Republic were spared much of the force of Dean as it passed 170 miles (270 km) to their south. Nevertheless, ancillary winds and flooding killed 15 people on the island of Hispaniola, and hundreds of homes were destroyed. As far away as Nicaragua, a 4-year-old girl drowned on a boat that sank during high winds and waves on the Kukra River.

And now, Dean had achieved its final identity, as a Category 5 monster, carrying maximum sustained winds of 160 MPH (257 km/h) and wind gusts of 200 MPH (320 km/h). It smashed into the Quintana Roo coast of the Yucatán Peninsula near Majahual at 3:30 A.M. EDT on August 21. Warnings had been heeded, Chetumal, the capital of the state of Quintana Roo, had been evacuated of 13,000 of its residents, and the tourist-crammed coastal cities of Cancún and Cozumel were largely emptied.

Dean made landfall in the sparsely populated jungle that was home to largely separated Mayan communities, who would be cut off from civilization for weeks after its passing. Troops evacuated more than 250 small communities, 8,000 people took refuge in 500 shelters, but others turned away soldiers with machetes and refused to leave.

In the nearly deserted city of Chetumal, diehards spent a nightmare night with windows shattering, billboards catapulting down streets and heavy water tanks flying off rooftops. Sirens wailed for hours, and by midday the next day, the Federal Electricity Commission reported 90,000 customers without power.

Majahuai, a fishing village turned into a cruise ship destination over which Dean's eye passed, suffered extraordinary damage—trees ripped up, power lines downed, roads demolished, crumpled steel girders and the loss of half of the immense concrete dock that had berthed scores of cruise ships. Still, a handful of people chose to ignore military orders to evacuate. "It wasn't minutes of terror. It was hours," Catherine Morales, a one year resident of the settlement told reporters afterward. "The walls of our house felt like they were going to explode." Winds of 156 MPH (250 km/h) with gusts of 200 MPH (320 km/h), faster than the take-off speed of many passenger jets, blew out the windows of the house and shredded major parts of the roof. Morales was wading in waist high water as she gave the interview.

Further south, in Belize City, officials closed the hospitals and urged people to head inland, saying the town's shelters were not strong enough to withstand the hurricane.

The cataclysmic power of Dean began to weaken slightly as it roared over the Yucatán Peninsula, and headed toward the Bay of Campache, the home to more than 100 oil platforms and three major exporting ports. Before the storm left the Yucatán peninsula, Mexico's state-oil company Petróleos Mexicanos (PEMEX) had evacuated its workers and shut down all production on its offshore rigs. Earlier, PEMEX had shut down production in the Gulf of Mexico and by Monday, the day the storm was ravaging the Yucatán Peninsula, 18,000 workers were evacuated from the remaining platforms. This reduced the worldwide production of oil and natural gas by 2.65 million barrels and 2.6 billion cubic feet per day.

Now it was the turn of the mainland of Mexico to receive the onslaught of Dean. Its trip over land on the Yucatán Peninsula had weakened it to a Category 1 storm, but the warm waters of the Gulf of Mexico replenished its energy, and at 11:30 A.M. CDT, it crashed as a Category 2 hurricane into the coast of Mexico near Tecoluta, Veracruz, south of Tuxpan.

Fatalities continued to mount. In Jalisco, a mudslide consumed 10 houses, killing one of the occupants. One person died in Puebla after a wall fell on him in his house. Another individual, doing roof repairs on his house in Veracruz touched a power line and was electrocuted. In Michoacán, a man was struck by lightning under a tree and two people died in Hidalgo when the roof of their house collapsed.

There was no land impact on the United States, but heavy surf and rip currents were reported on Florida beaches. One person drowned and at least 35 people were rescued from Siesta Key due to these rip currents.

On April 23, Dean disintegrated over land in Mexico and disappeared, leaving a total of 42 dead and approximately $3.8 billion in damage.

MEXICO
HURRICANE PAULINE
October 9, 1997

Hurricane Pauline struck the south coast of Mexico on October 9, 1997, with 115 MPH winds and up to 16.5 inches (41.9 cm) of rain. Floods and mudslides killed 230 people. Over 2,000 were reported missing, all in the countryside and the poor sections of Acapulco.

Hurricane Pauline was born off the western coast of Africa on September 16, 1997, but instead of following the usual Atlantic track into the Bahamas and up the eastern coast of the United States, it crossed northern South America as a tropical depression and began to gather force in the eastern Pacific Ocean near Panama. On October 6, it intensified to a category 3 hurricane.

On the evening of October 8, Pauline, pushing 30-foot (9.1-m) waves ahead of it, headed toward the mountains of Oaxaca, Mexico. But then, defying forecasters, it tore through the resorts of Puerto Escondido and Huatulco on the Oaxaca coast, went out to sea, where it gathered more power, and then, with winds of 115 MPH (185 km/h), slammed into the posh resort of Acapulco at 6:30 A.M. on October 9.

In three hours, it dropped 16 inches (40.64 cm) of rain, which brought about enormous mudslides that demolished the flimsy houses that clung to hillsides above the resort. And that was the unique feature of the damage of the storm: It left the high-rise resorts populated by tourists unscathed, but wreaked death and destruction on the poor sections of the city of 600,000.

"We don't have a single broken window," boasted the president of the Acapulco Association of Hotels and Tourist Businesses after the storm cleared. "However," he admitted, "the hurricane punished the poor people of the city and the neighborhoods where our hotel employees live very, very cruelly."

It was an understatement. Red Cross ambulances were unable to get through streets blocked with debris and were forced to shout first aid instructions to those who could hear them from a distance.

The mayhem was caused by floodwaters fed by Pauline's rains, that cascaded down the hills upon which Acapulco's working families lived. Tons of mud descended upon the thinly, sometimes hastily, built houses on the sides of these hills.

Ironically, then, the low ground in Acapulco was the safest ground; by the time the descending floodwaters reached the beach resorts, their force had been dissipated into mere annoyances.

But it was another world in the hillside neighborhoods that had grown explosively in the decade from the mid-1980s to the mid-1990s. Poor Mexicans crowded into these neighborhoods in the hope of making a good living from the growing tourist industry.

Now, these neighborhoods were muddy tombs. Of the 230 people killed by the storm, three-quarters of them drowned or were buried by mud in Acapulco. For days, rescue workers unearthed bodies from the waist-high mud that coated the area.

In the storm's path, six main bridges were damaged, cutting off key highways. Cars were tossed about like toys, some stacked on top of each other. The two main aqueducts that carried water to the city's water purification plant were damaged, and within days, water rationing was instituted in both the city and the resorts. Within a week, eight cases of cholera were reported.

It was, in many ways, the price paid by a location that had, in two decades, grown from a sleepy tourist port to a hill-straddling metropolis. In its burgeoning growth, its government had failed to provide a drainage system for its steep streets, which, even in a light rainstorm, turned into surging streams.

Rescuers continued to dig for bodies, but some 2,000 missing people would never be unearthed. Meanwhile, as some residents bathed in the same brown, sewage-laden river that had carried away dozens of people, food and water remained scarce, and, in some neighborhoods, unavailable. It would be months before Acapulco and the coastlines of Oaxaca and Guerrero returned to a semblance of normality.

MEXICO
TAMPICO
September 24, 1933

There is no official record of the human loss from the hurricane that hit Tampico, Mexico on September 24, 1933, causing $5 million in property damage.

In September 1933, at the depth of the Depression and the height of the hurricane season, two major storms ravaged the city of Tampico within the space of 10 days.

No estimate of fatalities was kept from this killer storm, largely because thousands of unemployed persons from outside of Tampico had gravitated to the city, swelling its population and muddling the statistics.

What is known is that the hurricane approached from the southwestern Gulf of Mexico, pushing a strong storm surge of sea water ahead of it. By the time the waters had receded and the winds had ceased, over $5 million in property and agricultural damage had been caused.

MEXICO
TAMPICO
HURRICANE HILDA
September 19, 1955

Three hundred people died when Hurricane Hilda hit Tampico, Mexico, on September 19, 1955.

Three hurricanes battered Tampico in the season of 1955: Gladys, Hilda, and Janet (see HURRICANES, Caribbean, Hurricane Janet, p. 269). However, it was Hilda that did the most damage and which, with its awesome power, was described by Mexico's then-president Adolfo Ruiz Cortines as ". . . the worst disaster in the city's history."

The bustling city of 110,000, located on the Panuco River a few miles inland from the Gulf of Mexico, was wearily cleaning up the debris from Gladys when Hilda roared into its docks, smashing its oyster and shrimp fleet and the piers that contained loads of cattle, hides, and other agricultural products saved from Gladys's fury.

The strong storm surge swelled the Panuco River to its greatest height in 30 years, funneling it into the streets of the city. Before it left, Hilda had flooded half the city and drowned 200 people.

A month later, Janet would deliver the third strike, drowning another 100 people and bringing the total of property destruction for the three storms to $42 million.

MEXICO (AND FLORIDA)
HURRICANE WILMA
October 15–25, 2005

Wilma, the 21st hurricane of the 2005 hurricane season, the third most costly storm in U.S. history, and the most powerful hurricane ever to form in the Atlantic Basin, careened across the Caribbean Sea and the Gulf of Mexico from October 15 to 25, 2005, killing 63 people in Haiti, Cuba, Jamaica, Mexico, and Florida, and causing $28.8 billion in damages.

On October 15, 2005, a tropical depression formed in the Caribbean Sea. By the 17th, it had acquired the name of Wilma, making it the 21st storm of the 2005 hurricane season and the most powerful hurricane ever to form in the Atlantic Basin, with winds reaching 175 MPH (281.6 km/h) and a barometric pressure that plunged to 28.04 inches (0.7 m).

The giant storm took a northwesterly path from its origin, which headed it directly toward the tip of the Yucatán Peninsula, where multiple resorts are located. It would take the storm four days to reach Mexico, and on its unsteady trip, it would kill 13 people in Haiti and Jamaica. By the time it reached Mexico, it had weakened slightly from a category 5 to a Category 4 hurricane. The winds within its 500-mile (804.7-km) width were still swirling at 145 MPH (233.4 km/h) as it sped over Cozumel Island. It would stall for the next two days directly over Cancún and Playa del Carmen upon which it would dump an unprecedented 5 feet (1.5 m) of rain.

As it approached, Mexican authorities evacuated 72,000 people from Quintana Roo and Yucatán States. But tens of thousands of tourists and residents remained in hotel ballrooms, schools, and gymnasiums. Eleven pregnant women went into early labor and were ferried to hospitals through water that had reached five feet (1.5 m) in height and, driven by the high winds, had smashed through barricaded windows in the hotels. Rough seas consumed the white sand beaches, and nearly 50 hotels were evacuated. Cars were crushed under falling trees, waterfront huts were washed away, buildings blew apart, and major hotels, occupied or evacuated, suffered major damage.

"I never in my life wanted to live through something like this," Guadalupe Santiago, a 27-year-old cook at the Xbalamque Hotel in Cancún, told an Associated Press reporter.

By 2:00 P.M. on October 22, the eye of Wilma was inland over northeastern Yucatán about 10 miles (16.1 km) southwest of Cancún and about 400 miles (643.7 km) southeast of Key West, Florida.

And meanwhile, in Florida, tourists were ordered out of Key West, as Wilma headed for a collision course with a cold front blowing across the Gulf of Mexico, forcing it to make a right hand turn toward Florida.

Shortly after dawn on Monday morning, October 24, after absorbing the energy of tropical storm Alpha, Wilma hit the southwest Gulf Coast of Florida near Marco Island as a Category 3 hurricane with winds of up to 125 MPH (201.2 km/h). Seawater, pushed by the hurricane, inundated much of the Florida Keys and knocked out power to an estimated 3.4 million people.

In Everglades City, a village of about 700 fishermen and tourist boat operators, a storm surge flooded streets and yards with up to four feet (1.2 m) of water,

but because most homes were built on stilts or mounds of earth, the damages were minimal.

On its way to Florida, the hurricane roiled the waters of the Gulf of Mexico and brought about heavy flooding in Cuba. Several low-lying neighborhoods were evacuated in Havana, where waves poured over the Maleceon, the historic waterfront promenade, and inundated streets.

The damage in Florida was less severe. Wilma's rapid spanning of the state cut the rainfall amounts down to eight inches (20 cm) in Miami and 6.5 inches (16.6 cm) in Naples which nevertheless bore the brunt of the state's damage. Hundreds were made homeless, many of them in mobile homes, and the downtown area suffered over $2 million in tree and water damage. Mobile homes were ripped apart in several areas. Windows were blown out of high rises in Miami, and lampposts, pieces of wood and bits of power transformers floated in Biscayne Bay.

It was a considerably calmer scene than that in Mexico, however, where hundreds tried, with varying degrees of success, to flee Cancún after the storm had passed. The normally happy resort had become a city beset by looters, rumors of vandals, and a scarcity of food and water. Bands of young men wielding machetes broke into homes, sprayed walls with graffiti and stole food and appliances.

Thousands of tourists tried to leave for Mérida in rented vans, only to be turned around at Leona Vicario, about 30 miles (48.3 km) outside of Cancún, where a section of road was under more than 5 feet (1.5 m) of water. Some tourists ended up returning to the same hotels they left. Others in large buses hired by tour companies were able to leave. Along the way, poor people set up roadblocks with stones and logs, trying to extract a toll from passing cars.

A total of 63 people died in the wandering wake of Wilma, which spread rainfall over 11 countries—more than any other hurricane in recent history. The damage was estimated at over $28.8 billion, which made it the third costliest storm, after Camille (see p. 311) and Katrina (see p. 290), in U.S. history.

MONTEGO BAY
October 3, 1780

The first hurricane to be assiduously tracked, the great Atlantic hurricane of October 3, 1780, killed over 25,000 people at Montego Bay (on Jamaica), Martinique, and St. Eustatius, and was felt as far north as New England.

The greatest Atlantic hurricane of the 18th century is also the first hurricane in history to have been tracked. The day-to-day logbooks of a British fleet of 27 ships that, had it not been decimated by the storm, might have carried out a plan to capture Puerto Rico, collected more minutiae on this storm than hundreds that came after in supposedly more meticulous and scientific times.

The storm first struck the Jamaican town of Savanna-la-Mar at 1:00 P.M. on October 3. The flimsily constructed English houses, crowded below their tide marks, were easy targets for the storm surge, and by evening, the undertow had sucked the supporting sand away and many of these houses were swung, with people in them, out to sea, never to be found again.

By morning, there were no homes left below the high water mark, and the only visible survivors were on ships that were hurled up on the marshes far inland.

Meanwhile, the storm coursed northward toward Montego Bay, where four ships of the Royal Navy were anchored. When the full fury of the hurricane hit, crashing across the island, it either smashed the ships to bits in the harbor or foundered them at sea. Only a handful of men survived, as their ships were thrown against the rocks of Cabo de la Cruz on the coast of Cuba.

The storm, spawning ancillary cyclones, continued to whirl through the Indies, sending its effect as far north as Newport and Rhode Island, pushing tides up into bays and toppling windmills, loose shutters, and washlines, thus earning it the title in New England of the "line storm."

Meanwhile, the main storm whirled over Barbados, reducing it to mud, debris, dead cattle, and corpses. The few survivors were given the unprecedented sum of 80,000 pounds in relief aid by Lord Rodney, the British Lord Protector.

The center continued, passing over St. Lucia, which was flattened. The only cheerful news out of that battered place was that the storm drowned a plague of ants that had been eating the sugar cane. Three British ships—the *Vengeance,* the *Deal Castle,* and the *Thunderer*—were either smashed on the rocks or sent spinning out to sea, never to be heard from again. As far away as Grenada, 15 Dutch ships were slammed against the rock-strewn shore.

The storm crashed into Martinique, demolishing ships, towns, churches, houses, and a hospital, and killing 9,000 people. On the Dutch island of St. Eustatius, every building was smashed and washed out to sea and 5,000 people were drowned.

On the morning of October 15, the storm moved up the Mona Passage, its rain squalls extending to Puerto Rico and Santo Domingo. It swallowed the island of Mona whole and disabled two British ships, the *Ulysses* and the *Pomona,* which had survived the first battering a few days earlier.

From here, this monster storm curved eastward, soaking the East Coast of the United States and tormenting ships at sea to the extent of shearing off their topsails. Finally, spending itself out over open water, it reached the latitude of Great Britain, after at one point covering, in its gigantic width, the distance from Newfoundland to St. George's Channel.

NEWFOUNDLAND
September 9, 1775

Four thousand people died in the hurricane that struck Newfoundland on September 9, 1775. Accounts of the storm's passage and effect conflict.

On September 2, 1775, while British redcoats and American militiamen fought in North Carolina, an Atlantic hurricane swept ashore, killing 150 people and wiping out the corn crop in Pasquotank County. Some reports say the storm continued on through Virginia and blew itself out over Pennsylvania. Others say that it curved out to sea again, hitting Newfoundland on September 9, and wiping out, in one awful afternoon, its entire fishing fleet.

Whether the September 9 hurricane was a new storm or the North Carolina one, it was devastating for Newfoundland. Unwarned, hundreds of fishing boats were caught offshore when the storm struck, and every one of them was lost. Four thousand sailors and fishermen lost their lives that day.

On shore, roofs tore away, chimneys disintegrated, and entire houses were lifted from their foundations and transported by the winds until they eventually either exploded or collapsed. In all, the loss of ships, land, buildings, and produce was estimated to total a staggering 140,000 pounds, or $350,000 in today's dollars.

NICARAGUA, HONDURAS
HURRICANE FELIX
August 31–September 5, 2007

Hurricane Felix smashed into the Mosquito Coast region, an area between Nicaragua and Honduras, on September 4, 2007, making it the second Category 5 hurricane to make landfall within the first two months of the 2007 Atlantic hurricane season, an occurrence that had not happened since the keeping of records of hurricanes were begun in 1850. At least 133 died, mostly from drowning.

For the first time since records began to be kept in 1850, two Category 5 hurricanes—Dean and Felix—made landfall in one Atlantic season. Both of these storms climbed to the very top of the Saffir-Simpson scale appeared at the beginning of the 2007 season and arrived within a month of each other.

Hurricane Dean (see p. 268) began the season by entering the Caribbean on August 17, 2007. Felix made landfall just south of the border between Nicaragua and Honduras on September 4. Though Dean maintained a fairly steady identity as a Category 5 storm before it made landfall, Felix had a more episodic history. It began life as a barely perceptible tropical wave off the coast of Africa on August 24 and it would not be until August 31 that it would gain identity as Tropical Depression Six.

The next day, shortly after passing over the island of Granada, it was upgraded to Tropical Storm Felix. And from this point, it grew rapidly " [as] . . . one of the more rapid deepening rates we have observed," according to the U.S. National Hurricane Center. Within a day, it had become a Category 4 hurricane, and on September 3, it had earned its Category 5 credentials by carrying winds of 165 MPH (263 km/h). Later in the day, however, it would weaken to a Category 4 storm.

At this point, Felix was predicted to track west-northwestward to the Yucatán Peninsula and Belize, much as Dean had proceeded. But instead, its course sent it due westward, toward the coast of Nicaragua. And that same day, for the second time, it once again climbed to Category 5 status, with winds of 160 MPH (260 km/h). And it was at this strength that it slammed into the extreme northeastern coast of Nicaragua, just below the border with Honduras.

As Felix tracked toward its landfall, hurricane watches and warnings went up and down through the Caribbean. Heavy rain fell across the Windward Islands and mudslides and overflowing rivers caused damage on Trinidad and Tobago. But Felix continued at sea on its westward course, and when it became apparent that it would hit Honduras and Guatemala, evacuations began in both countries. By the time it hit, 20,000 people had left the site of its landfall, and 140,000 pounds (63,000 kg) of meals and emergency supplies had been flown in to a nearby location.

It was just before dawn on September 4 when Felix roared ashore with 160 MPH winds (257 km/h) and a storm surge of 18 feet (5.6 m) in an isolated, swampy jungle part of Nicaragua, where its inhabitants—some 150,000 Miskitos, descendents of Indians, European settlers, and African slaves—live in hamlets of wooden shacks and coconut groves which are remote even in good weather, and reachable only by air or flat-bottomed boats.

Shortly before Felix arrived, Marco Burgos, the national commissioner of the Honduran emergency response center sent a radio message to the area. "I'm asking people, no, I'm ordering them to leave their wood homes and head to shelters," he said in the announcement. "Not one wooden house is going to survive such a hit."

No word was received from the Miskito, but in Puerto Cabezas, Nicaragua, Claudio Vanegas, a Red Cross official, told the Associated Press "the winds send roofs flying through the air, so we aren't going outside because it is too dangerous."

The next day dawned to a scene of heavy rain curtaining limitless devastation. Residents hacked at fallen trees with machetes, trying to uncover the remnants of their crushed homes. Reporters descended and gathered stories from survivors: Teresa Flores had ridden out the storm at a neighbor's house after her wooden home collapsed, injuring her husband and 3-year-old son. No sooner had the storm passed than looters had entered the area taking everything from Mrs. Flores's home. "They took the clothes, even the barrels where we keep water," she said. "Now we have nothing to drink. We lost all our food, even the toilet. Nothing works. We are pretty much in the street."

In the sea, more than 150 Miskito Indians, clinging to buoys, canoes, and slabs of wood, were finally rescued by Honduran authorities. The next day, the bodies of 25 dead fishermen, from a group of 109 Miskito Indians who sought refuge from Felix in canoes were fished from the sea near their settlements. At least 52 other Miskitos found something to hang onto, fighting for hours to stay alive as huge waves and lightning crashed around them.

The exhausted survivors told rescuers that the storm caught them by surprise, flooding the tiny islands used by lobster fishermen off the Nicaraguan coast and forcing them to spend 16 hours clinging to anything that would float.

Fernando Pereira told Associated Press reporters he clung to a piece of wood for 12 hours, despite a dislocated shoulder, and washed ashore at the village of Sandy Bay. "I felt horrible," he said. "I was drinking salt water, and I thought I was going to die." Jelivaro Climax told reporters he had to swim through enormous waves to reach shore. "Lightning flashed through a pitch black sky," he said. "I don't know how I survived. I swam with everything I had, and I was sure the sea would take me." Many of the survivors suffered dehydration and received medical care in the seaside town of Villeda Morales, on the Nicaraguan border.

Across the border, in Honduras, the city of Puerto Cabeza was mutilated. During the storm, tin roofs flew through the air like straight razors, slashing trees and collapsing structures. Felix snapped steel cables that guided a small ferry carrying people and cars from Puerto Cabezas to the village of Wawahum. Seventy six shelters in and around the city were filled to overflowing. The city's hospital was completely flooded, and a makeshift medical center was set up in a nearby university building.

Crops were flooded and destroyed; 70,000 pounds (31.8 kg) of food supplies dropped into the area were washed away. And meanwhile, Felix, now a tropical depression, continued to dump countless gallons of rain water on Honduras, El Salvador, Nicaragua, and Guatemala. Thousands of refugees stayed anxiously away from shaky hillside slums and swollen rivers. In San Pedro Sula, in northern Honduras, one shantytown filled with water after a river erupted over its banks. Police rode bulldozers to evacuate slum dwellers from water that was waist deep and rising, but many of the residents refused to leave, fearing their remaining possessions would be stolen.

Nicaraguan President Daniel Ortega declared the coastal region a disaster area and the Honduran military flew in sheets, mattresses, food, and first aid. Still, officials in Honduras were anything but hopeful. "Most of our municipalities have no firefighters, no ambulances," Marco Burgos, national commissioner of the Honduran emergency response center, told reporters. "The people have to save themselves."

At least 133 people died in the storm. At the time of this writing, the cost of the damage had not been estimated.

NOVA SCOTIA
August 24, 1873

At least 600 died and 1,032 ships foundered in the hurricane that devastated Nova Scotia on August 24, 1873.

While it may be true that the death toll of pre-20th century hurricanes is abnormally tilted toward perished sailors, it is also true that the military kept far better records in those days than civil authorities. Thus, the fact that the roll of an estimated 600 lost lives in the great Nova Scotia hurricane of 1873 is composed mostly of ships' crews tells as much about record keeping as it does about the location and the impact of the storm.

Roaring in from the sea lanes, where it left in its wake 190 ships or boats either foundering or ripped from stem to stern by its enormously high winds, this hurricane shot up the Gulf of St. Lawrence on August 24, 1873. Its winds and storm surge engulfed no less than 1,032 ships, anchored in the supposedly protected ports of Nova Scotia, Newfoundland, and Cape Breton. The ships were either driven ashore, split asunder, or sent to the bottom. Property damage from the storm, which eventually blew itself out in the Canadian provinces, was estimated at a staggering $5 million.

PUERTO RICO
July 26–August 31, 1533

Nearby 2,000 people—almost all of them slaves—died as a result of three successive hurricanes that assaulted Puerto Rico from July 26 to August 31, 1533.

Puerto Rico had been settled for only 25 years (Ponce de Leon had named it Puerto Rico in 1508, although Columbus had actually discovered it in 1493) when the triple hammer blows of three major hurricanes battered this rich paradise on July 26, August 23, and August 31, 1533.

The Spanish had all but swept the island clean of gold in the early 1530s and had executed most of the native Arawak Indians. Boatload after boatload of slaves from Africa were brought in to plant and work the sugar plantations that would replace the gold deposits.

Thus, when the first of the three major storms hit, it claimed its greatest loss of life among these slaves, many of whom had not been provided with adequate shelter.

When the second storm struck, even the poor shacks being erected to replace those demolished by the first storm were not yet in place, and still more slaves were drowned.

By the morning of September 1, 1533, nearly 2,000 people, almost all of them slaves, had been killed by the three storms, and the sugar cane fields lay flooded and useless. Overlooking the casualties, the Spanish conquerors sent envoys back to Spain to request more slaves to rebuild and work the gutted plantations and repopulate the island.

PUERTO RICO
July 26, 1825

Three hundred seventy-four people died and nearly 1,200 were made homeless on Puerto Rico as a result of the hurricane of July 26, 1825.

Residents of Puerto Rico have become accustomed, if not inured to storms. But Santa Ana, the monster hurricane of July 26, 1825, was a particularly memorable one.

The storm first laid waste to Guadeloupe, then bore down on Puerto Rico with winds strong enough to collapse 7,000 buildings over the entire length of the island. Three hundred seventy-four people, most of them trapped and crushed in buildings or buried under falling walls or flying debris, lost their lives to the storm, and nearly 1,200 were rendered homeless.

PUERTO RICO
August 8, 1899

The entire island of Puerto Rico was laid waste by the hurricane of August 8, 1899, called the worst Caribbean storm of the 19th century. More than 3,000 died.

More than 3,000 people were killed by the most ferocious Caribbean storm of the 19th century, which literally laid waste to the entire length of the island of Puerto Rico from early in the morning until late at night on August 8, 1899.

The only city that escaped the wrath of this particular monster was San Juan. Winds were a mere 90 MPH (145 km/h) by the time San Ciriaco, as the storm was called, reached there. When it roared into the port of Humacao, its first landfall, its winds were clocked at 125 MPH (200 km/h)—enough to send multiple tidal waves crashing over wharves and through streets. Farther inland, Aguadilla received winds of over 100 MPH (161 km/h), which peeled the tiles from roofs and turned metal stripping into flying knives which decapitated some of those fleeing from the storm. Twenty million dollars in property damage seems slight compared to the monumental loss of life in this killer storm.

PUERTO RICO
September 26, 1932

Two hundred twenty-five died, 3,000 were injured, and 75,000 were made homeless by the killer hurricane that struck Puerto Rico on September 26, 1932.

Winds measuring 120 to 150 MPH toppled the weather tower in San Juan during the night of September 26, 1932. San Ciprian, as this storm was called, made landfall at Ceiba, then tore swiftly through the island of Puerto Rico, killing 225 people, injuring 3,000,

rendering 75,000 homeless, and causing $30 million in property damage. Its tenure was short but devastating; within a day it had blown itself out over Mexico.

PUERTO RICO
HURRICANE DONNA

(See HURRICANES, United States, East, Hurricane Donna, p. 284.)

ST. CROIX
August 31, 1772

Hundreds were killed and the course of history was changed by the hurricane that struck St. Croix on August 31, 1772.

If a few—a very few—hurricanes can be categorized as ill winds blowing some mite of good, the hurricane of August 31, 1772, would certainly qualify as one of those select storms.

Alexander Hamilton, confidential aide to George Washington and later America's first secretary of the treasury, was born illegitimately in 1757 on Nevis, one of the small British islands of the West Indies, to a 20-year-old mother divorced from an older husband. In 1772, at the age of 15, he was employed in Nicholas Crueger's counting house on the nearby Danish island of St. Croix. He was hardworking, a financial whiz, a sometime manager of Mr. Crueger's entire operation, and therefore headed for wealth and comfort as a business manager on St. Croix.

But the hurricane of August 31—a giant that wrecked all of the ships anchored in Dominica, killed hundreds, and destroyed 2.5 million pounds of prime muscovado sugar in unroofed warehouses—slammed into St. Croix, Hamilton's life, and American history, all at once. The following day, still shaken by the fury of the storm, Hamilton wrote, in part, to his father the following summation of it:

> I take up my pen just to give you an imperfect account of one of the most dreadful hurricanes that memory, or any records whatever, can trace, which happened here on the 31st ultimo at night.
>
> It began about dusk, at north, and raged very violently till ten o'clock, then ensued a sudden and unexpected interval, which lasted about an hour. Meanwhile the wind was shifting round to the southwest point, from whence it returned with redoubled fury and continued so till near three

o'clock in the morning. Good God! What horror and destruction—it is impossible for me to describe—or you to form any idea of it. It seemed as if a total dissolution of nature was taking place. The roaring of the sea and wind—fiery meteors flying about in the air—the prodigious glare of almost perpetual lightning—the crash of the falling houses—and the ear-piercing shrieks of the distressed were sufficient to strike astonishment into Angels. A great part of the buildings throughout the island are levelled to the ground—almost all the rest very much shattered—several persons killed and numbers utterly ruined—while families running about the streets unknowing where to find a place of shelter—the sick exposed to the keeness [sic] of water and air—without a bed to lie on—or a dry covering to their bodies—our harbor is entirely bare. In a word, misery in all its most hideous shapes spread over the whole face of the country—a strong smell of gun powder added somewhat to the terrors of the night, and it was observed that the rain was surprisingly salt. Indeed, the water is so brackish and full of sulphur that there is hardly any drinking it . . .

Hamilton's father, no doubt feeling a tingle of paternal pride, submitted the letter to the editor of the *Royal Danish-American Gazette*, which printed it in its October 3, 1772, issue. A group of wealthy planters read it and decided that the boy had promise. Together, they raised enough money to send young Alexander Hamilton away from the limited future of the islands to the promise of King's College (now Columbia University) and a college education.

SANTO DOMINGO
October 3, 1780

(See HURRICANES, Montego Bay, p. 272.)

SANTO DOMINGO
September 23, 1834

Thousands were killed by the so-called Padre Ruiz Hurricane that struck Santo Domingo on September 23, 1834.

The hurricane of September 23, 1834—known as the "Padre Ruiz Hurricane" because it occurred simultaneously with the burial of a highly respected and popular priest—flattened a large portion of Santo Domingo during its ravaging stay there.

Thousands were killed in the entire country; Dominica, the small, lush, volcanic island to the west of Santo Domingo, had been mangled by the storm three days earlier, and 200 of its inhabitants lay dead as a result.

In addition to the enormous loss of life brought about by the storm, huge tracts of forest and crops ready for harvest along the Ozama River were crushed and flooded. Fishing and transportation boats on the river sank, and their crews, trapped by the force and suddenness of the storm, went to the bottom with their boats.

SANTO DOMINGO
September 3, 1930

Four thousand people died, 5,000 were injured, and every wooden building was destroyed in Santo Domingo during the 200-MPH hurricane of September 3, 1930.

The numbing effects of the hurricane of September 3, 1930, upon Santo Domingo, the capital of the Dominican Republic, were swift but telling. In a mere four hours, 4,000 people died, 5,000 were injured, and 9,600 buildings—every one of the wooden structures in the city—were demolished. When the skies cleared that night, a mere 400 stone buildings remained, stark sentinels in the midst of $40 million worth of damage.

The eye of the storm passed directly over the city, giving its inhabitants a 30-minute respite before the storm surge that would be whipped up by the sudden wind flow from the opposite direction.

The hurricane struck Santo Domingo at almost exactly noon, when the city was most crowded. Tourists and natives alike sought shelter in the hotels, which were thought to be the most solid structures. And they did stand—minus their shutters, windows, doors, and skylights.

The dictator Rafael Trujillo narrowly missed being one of the storm's victims when his executive offices began to come apart around him. Accompanied by his plenipotentiary, he scampered swiftly through the chaotic streets, finally taking refuge in Fort Ozama.

When the sun broke through the next morning, prisoners were turned loose from the jails to assist in the grim cleanup, which included burning bodies in huge funeral pyres. Martial law was declared to protect the remaining food supplies. The water supply was totally destroyed by the storm, and the casualty figures were swollen by hundreds who died of typhoid fever, the result of drinking contaminated water.

An entire area of Santo Domingo is flattened in the hurricane of September 13, 1930. (American Red Cross)

UNITED STATES
ALABAMA
MOBILE
July 27–28, 1819

More than 200 people were killed by winds and life-threatening debris—including snapping turtles and alligators—washed up by a hurricane that hit Mobile, Alabama, on July 27, 1819.

Possibly one of the oddest hurricanes of the 19th century struck Mobile Bay on the night of July 27, 1819. The storm was odd not because of its size, which was described by one observer as "the most severe and strongest that ever blew on this coast since I came to it," and not because of the number of ships it sent to the bottom of the Gulf, nor its death toll of over 200. It was unique in the fact that many of its casualties were caused not by the storm itself but by some of the debris

washing up into the streets of Mobile at the height of the storm: hordes of snapping turtles and alligators.

The hurricane roared up Mobile Bay on the evening of July 27, 1819, taking four hours to wind up and hit the city of Mobile its most devastating blow, when the eye passed over the city around midnight. The sudden shift of wind caught residents by surprise. As is often the case, they were lulled into false security by the momentary calm of the storm's eye. The sudden rush of wind from the opposite direction carried high tides into the city, pummeling buildings and collapsing them, and carrying a lethal load of snapping turtles and alligators through the city streets. Fully half of the 200 casualties were blamed on alligator and turtle bites. Huge ships were floated onto dry land. Dauphin and Water Streets became the docking point for an immense brig; other smaller vessels moored at other locations.

A few captains had the good sense to weigh anchor and escape up the Pearl River, where it was possible to ride out the storm. Others, like the *Favorite,* were thrown against reefs at Henderson Point, which splintered the ship's hull but did not effect casualties in her crew. The man-of-war *Firebrand* was not so lucky. Her captain and seven crewmen were in New Orleans when the storm struck. The first mate in charge, left to protect the ship and crew, apparently either made the wrong decision or no decision at all, and when the captain and his group returned on the morning of July 28, they found the ship with its capsized hull jutting up from a sandy shoal known as "The Square Handkerchief." The crew, confined to its quarters, apparently drowned.

Along the banks of Mobile Bay, whole herds of cattle were swept from the shore. A mere three buildings were left standing at Pass Christian, and water six feet deep churned through the streets of Bay St. Louis (both Mississippi cities), undermining and collapsing scores of buildings, some with inhabitants inside.

Although the storm reached its fingers out as far as Biloxi, Mississippi, and New Orleans, damage was slight there, and, though it must have made itself felt in Pensacola, that Florida city was still under the control of Spain, and no reports were released from it.

UNITED STATES
ALABAMA
MOBILE
HURRICANE FREDERIC
September 11, 1979

Hurricane Frederic, the worst storm ever to strike the Mobile, Alabama, area, caused mass evacuations—nearly 450,000 people were moved out before it struck—and this kept the casualties low. One hundred ten people died.

Hurricane Frederic plowed ashore at beleaguered Dauphine Island (see HURRICANES, United States, Mississippi, p. 299) 25 miles south of Mobile, Alabama, on the night of September 11, 1979, with winds of 100 MPH. It was the worst storm ever to hit the Mobile area, one that, it was predicted, would carve up and destroy Dauphine Island, a small, 16-mile- (25.7-km) long spit of populated sand that rose only 10 to 15 feet (3.04–4.57 m) above sea level. Since the tides were expected to climb at least that high, experts figured that the end was in sight for this Alabama resort haven.

But Dauphine Island would survive, and so would Hurricane Frederic, but not before turning Mobile and Dauphine Island into twin war zones.

The storm, with an eye 50 miles (80.46 km) in diameter, had been closely tracked by the U.S. Weather Bureau. As a result, 400,000 people were evacuated from coastal areas along a 400-mile (643.7-km) strip from Pensacola, Florida, to Biloxi, Mississippi. The air force evacuated 9,000 service personnel and dependents stationed at the Tyndall Air Force Base near Panama City and 1,000 Florida National Guardsmen assisted in that state's evacuation plan. As a result, the loss of life—110—was low, considering the fury of the storm, which, when it hit Mobile shortly after midnight, had narrowed its eye to 20 miles in diameter, thus increasing its power and wind velocity to 130 MPH.

The wind pummeled the city like a giant fist. The south wall and part of the roof of the City Hall collapsed. Other buildings were flattened as if a bomb had gone off inside. Broken power lines snaked across streets, setting fires before the power went off. Three stores in a major shopping area went up in flames during a downpour, and the sky was lit by blue flashes as power transformers blew up.

In the surrounding towns, some evacuation centers had their roofs ripped away, forcing refugees to flee. Inshore, along the Alabama-Mississippi state line, the low-lying towns of Bayou La Batre, Theodore, and Grand Bay received the full force of the storm. At Theodore, a house containing 15 people was demolished, leaving them stranded in an open field at the height of the storm.

At Bayou La Batre, Herbert McKinney and his family spent the night in a gravel pit 30 miles (48.3 km) inland because they could not afford a motel. When they finally returned, by sawing their way through roads barricaded with fallen pines and monster oak limbs, they found only battered remnants of their mobile home. Around them, groves of pecan trees were flattened in rows as militarily correct as those of a cornfield.

At sea, the 230-foot (70.1 m) freighter *Mary's Ketch,* out of Grand Cayman Island, went dead in the water during the height of the storm. It survived waves ". . . as high as the mast . . ." according to a radio message the Coast Guard received from its captain.

Seventy-five-foot shrimp boats were driven ashore as if they were rowboats. A stacking effect of rising water occurred when the storm surge rushed into Mobile Bay, which was already swollen by the storm's heavy rains.

And what of Dauphine Island? The causeway to it from the mainland was wiped away like a toy construction. Twenty-seven stubborn residents rode out the storm on the island and survived by huddling together in an army radar bunker. The island was a sea of crushed homes. Deputy Sheriff Bud Wilson, a resident of the island, gathered other homeowners around him on the roadside leading to the causeway and said to his neighbors, "Why don't we all gather around and cry?"

Fortunately, few people lost their lives during Frederic, although the damage in Alabama was $493.3 million and in Mississippi $175.5 million. Between David (see p. 283) and Frederic, the two brotherly storms that hit roughly the same area within a period of two weeks in September 1979, 1,567 died, 600,000 were made homeless, and $3 billion in damages were caused.

UNITED STATES
BAHAMAS, FLORIDA, AND LOUISIANA
HURRICANE ANDREW
August 24, 1992

Hurricane Andrew, the most expensive catastrophe in U.S. history, slammed into southern Dade County, Florida, on the morning of August 23, 1992, and later in the day hit the bayou country of southern Louisiana. Sixty-nine people in Florida, Louisiana, and the Bahamas died in the storm; 250,000 were rendered homeless. Damage from the storm was estimated at $25 billion—enough to cause several large insurance companies to abandon the state of Florida and enough to force others into bankruptcy.

"If you don't go, you're risking suicide," director of Dade County's Office of Emergency Management Kate Hale told residents of Miami on Sunday, August 22, 1992. But it was sunny, a benign 90-degree day. Just right for the surfers who were riding the admittedly larger than usual waves, eminently fine for the eternal sunbathers on the beaches.

And besides, Andrew was the first storm of the season, an early arriver. And early arrivers usually blew themselves out. Besides, the last hurricane to hit the Miami area was Hurricane Betsy, 27 years ago.

This was the attitude of hundreds of young residents of the area. Tourist season had not yet arrived, and they weren't about to give up their time on the uncrowded beaches and streets.

But older inhabitants, some of them survivors of Betsy and therefore experienced in the ways of storms, heeded the warning. Occupants of high-risk condominiums on the beach boarded buses that took them to already overcrowded shelters. Northbound highways were jammed with traffic; hardware stores and supermarkets were overflowing with tens of thousands of shoppers loading up on food, bottled water, batteries, and plywood for storm sashes.

By 9:30, Miami International Airport had shut down its operations, and late leavers were forced to bed down in the terminal's waiting room.

By midnight, Dade County shelters were full to overflowing. In Broward County, north of Miami, 21,000 people filled its shelters, some to twice their capacity.

Andrew had begun its lethal trip toward the mainland as a tropical depression in the mid-Atlantic. By the time it had traveled north of Puerto Rico on Friday, it had turned into a tropical storm, and at 5 A.M. Saturday, the 22nd, it graduated from a tropical storm to a hurricane, with sustained winds of 74 MPH (119 km/h).

By Sunday afternoon, Andrew's winds had climbed to 120 MPH (143 km/h), and the storm slammed into the Bahamas with frightening force. Fierce waves washed over the island, as most of its 255,000 residents sought shelter. Early on Sunday, 1,200 tourists climbed aboard the liner *Scandinavian Dawn*, sent empty from Fort Lauderdale to evacuate them. Four deaths were reported from the Bahamas as a result of the storm.

Meanwhile, Miami battened down. Lumberyards made record profits, as the price of plywood leaped from $22 to $40 a sheet in an hour. Some few elder residents were unable to leave their houses or condominiums. Eighty-seven-year-old James L. Greenhill, confined to a walker, remained in his third floor apartment. "I'm staying," he told a reporter. "I don't know where the hell to go. The shelters are awful." His wife, who had long since given up on moving him, added, "If the house goes, we go. I hope it's not painful."

Young Miamians stocked up on cases of beer. Shanna Ochery, a 29-year-old unemployed cocktail waitress, did just this. "Hey," she said to a reporter, "you gotta die sometime. It can't be that bad."

It was, but not for Miami. Andrew sideswiped the resort and instead, plowed into the southern end of Dade County, 15 miles (24.1 km) south of Miami, in the Homestead-Kendall-Cutler Ridge area. It was a part of Florida occupied equally by the wealthy and the poor, migrant workers, young families, and retirees, and Andrew roared into its center at approximately 9 A.M. on Monday, August 24.

It literally flattened the landscape, yanking towering palm trees from the ground, pockmarking lawns, hacking down cypresses, pines, and palms. Roofs were peeled away like the covers on sardine cans; even shelters were rendered roofless.

Homestead Air Force Base was evacuated; most of its F-16 fighters and C-130 transports were sent to bases in South Carolina and Georgia. They would remain there. After the storm passed, Toni Riordan, spokesperson for the State Community Affairs Department, made a succinct evaluation: "Homestead Air Force Base no longer exists," she said.

All electricity failed in Miami. Store windows disintegrated; a storm surge filled the streets. A 7 P.M. to

7 A.M. curfew was instantly instituted to curb looting. But this was benign compared to the gigantic destruction in the Homestead area. Entire colonies of homes were reduced to splinters. Gas and water mains erupted. Roads were either blocked by cat's cradles of downed trees or ripped up. "I never saw anything like that," Florida governor Lawton Chiles observed after a tour of the area. "It's like an air bomb went off."

Meanwhile, Andrew, after slamming into the east coast of Florida with winds said to be close to 150 MPH, exited the west coast and, over warm water in the Gulf, gathered speed and energy again. Its winds, dissipated somewhat over land, resumed their 145 MPH intensity, as the storm headed directly for New Orleans.

But just as it had missed Miami, Andrew veered again, bypassing New Orleans and smashing into the sparsely populated towns along U.S. 90 from Morgan City to New Iberia, flooding the surrounding marshlands and ripping up the sugar cane fields of southern Louisiana. From here, it rapidly broke up, drenching Louisiana with rain and blowing itself out over Mississippi.

The wind gauge in Franklin, Mississippi, registered 140 MPH before it broke and the wind blew the skylights in on people huddled in Franklin High School, its designated shelter. Rain poured into every room but the gymnasium, where the refugees finally huddled in terror.

"It sounded like two trains crashing," said Marsha Repouchon, one of the 400 people in the shelter. "People were singing, crying, chanting, praying. Children were screaming every time there was a sudden noise. It was awful."

Trailer homes were decimated. In New Iberia, Louisiana, the storm ripped a chunk off the roof of a shelter. In LaPlace, a tornado caused one death and 30 injuries. Grand Isle and other coastal communities were deeply flooded.

Meanwhile, in Florida, the toll of the homeless rose to staggering heights. By August 26, it was generally agreed that 250,000 residents of the area south of Miami were now homeless, their residences totally uninhabitable, and in some cases, no longer identifiable as homes.

Florida Power and Light began the laborious task of restoring power to approximately 1.4 million homes and businesses. In Key Biscayne, 80-year-old retiree George Grim described his first view of destruction as he opened his storm shutters: "We saw complete destruction all around us," he said, "All the foliage was gone. You could see windows blown out, parts of roofs missing . . . We were like an island out of time, completely cut off. It was an extremely eerie feeling."

In Homestead, University of Miami freshman wide receiver Jermaine Chambers recalled riding out the storm in a friend's house: "We stayed on the second floor in the bathroom," he said. "I kept thinking that this was it. My life was over. I said, 'when is it going to stop?' But the wind just kept coming."

At one point, the house's roof was whipped off by the wind. "I didn't realize it until I opened the door at the end," he concluded. "I looked out at the sky and said, 'Geez, no house left.'"

He was not alone. At least one of every 10 of the 2 million residents of Miami and points south were left without a home.

President George H. W. Bush promised troops to help feed these homeless residents, but they were slow in coming. There was growing discontent at the apparent stasis in relief for the thousands uprooted by the storm. There were stories of refugees calling emergency numbers and getting either endless busy signals or officials who brushed them off.

In Louisiana, on August 27, 200,000 people were still without electricity. In scores of places, all commerce was shut down, food supplies began to run out, and entire towns were without running water or sewage disposal. Most of the bayou country of Louisiana, where Andrew struck, is either at or below sea level. A seven-foot tidal surge swept over this part of the world, laying waste to nearly everything in its path. An accompanying tornado took care of the rest. And yet, there was only one death in that state that was attributable to the storm, compared to 13 in Florida. The greatest loss was to its sugar cane crop, which was destroyed, and its shrimp beds, which were severely disturbed. A large number of shrimp boats were either sunk or ruined.

Leroy Hose, a worker in the sugar cane fields was interviewed by a *New York Times* reporter in a field in Charenton, Louisiana. As on other farms in the state, the crop lay in a twisted, broken, and muddy mess.

"If Uncle Sam doesn't take care of the boss, the boss doesn't get a profit," Mr. Hose reasoned, in perfect synchronicity with the current trickle-down economics practiced in Washington. "If there's no profit, there's no work, and if I don't work, the kids don't eat."

In both places, the statistics began to reveal that Andrew was the most expensive disaster ever to strike the United States. Sixty-three thousand homes were completely destroyed and 250,000 people were rendered homeless in Dade County alone. The cost was estimated at $30 billion, with insurers picking up $7.3 billion of it. Over 100 oil platforms in the Gulf of Mexico were damaged.

Finally, on August 28, army and marine troops began to arrive with food supplies, fresh water, generators, tents for shelter for the homeless, and equipment to set up military kitchens to feed them.

By August 30, 275,000 people in Florida were still without power and 150,000 had still not found shelter. Over 2,000 were living in 13 Red Cross shelters. The military effort was sputtering. Facilities had still not been put into place, despite the arrival of nearly 20,000 troops.

And now, the insurance adjusters arrived, and the shock renewed itself. "At points," said Joe House, a regional vice president of the USAA Property and Casualty Group, "it looks like a nuclear detonation, where the buildings just exploded."

Agricultural businesses began to tally their losses. Although the main Florida citrus crop was located north of the storm's path and so escaped unscathed, its $20.3 million-a-year lime crop suffered a direct hit. It would take five to eight years before new trees, planted immediately, would bear fruit. Avocado trees, reduced to broken sticks, would take longer to regenerate. In addition, the state's $150 million-a-year houseplant industry, located in southern Dade County, was also virtually wiped out.

By August 31, the possibility of a health crisis began to loom. The Homestead branch of South Miami Hospital, closed for a week, reopened on the 31st, but the six other hospitals in south Dade County remained closed, leaving makeshift clinics to care for afflicted adults and children.

Garbage and debris remained, and infestations of mosquitoes and rats became a particular problem for the thousands living in the remains of their homes. Bill Carruthers, the deputy director of an emergency medical team, observed that "at this point, we have to be on the alert for a whole range of diseases that flourish in unsanitary conditions, from diphtheria and dysentery to malaria and hepatitis."

The threat seemed to dissipate as more hospitals opened, and within a week, the warnings were discontinued. It would be months—for some, years—before the semblance of a normal life would return for the residents of Florida and Louisiana whose lives and livelihoods had been shattered by Andrew.

The total death toll in Florida rose to 38; one person died in Louisiana and 30 in the Bahamas. The homeless total remained at 250,000.

UNITED STATES
EAST
HURRICANES CONNIE AND DIANE
August 4–18, 1955

Occurring back to back, hurricanes Connie and Diane killed 310 people and caused $1.5 billion in property damage from August 4 to 18, 1955, in large areas along the East Coast of the United States.

Considered to have created the worst storm impact upon the eastern seaboard of the United States in this century, the back-to-back fury of hurricanes Connie and Diane killed 310 people and caused $1.5 billion in property damage to coastal areas in the Carolinas, Virginia, Delaware, Maryland, Pennsylvania, New Jersey, New York, Connecticut, and Massachusetts.

Connie, the first of the fatal sisters, was first sighted by a Navy Neptune weather plane in the southeastern Atlantic. The plane invaded the storm's eye and measured winds of 125 MPH, which caused the Weather Bureau and the Red Cross to call for an evacuation of the Carolina beaches. On the 11th of August, 10-foot (3.04-m) high rollers thundered into Myrtle Beach, South Carolina, crushing it, and then roared on to Wilmington and Pamlico Sound.

Thousands warned by the Weather Bureau evacuated the area, but in New Bern, North Carolina, 41 residents met their deaths through drowning, auto accidents, or electrocution from live wires that fell into the swirling flood waters.

By the 12th of August, Connie had sideswiped New York with gale winds and had blown itself out over Pennsylvania. But Diane was right behind her, following the same path until she got to the vicinity of New Jersey. There, instead of veering inland, Diane curved northeastward, smashing into coastal areas from New Jersey to Boston, where the storm finally took a right turn into the North Atlantic.

Monumentally high tides, brought on by the storm surge, overflowed the banks of Connecticut's Naugatuck, Still, and Mad Rivers. Winsted, Connecticut, was inundated with 12 feet (3.6 m) of brackish water. In Putnam, a magnesium plant was broken apart and hundreds of barrels of burning magnesium floated through that town's streets, exploding sporadically and sending cascades of white-hot metal 250 feet (76.2 m) in the air.

Diane blundered on, tearing up Revolutionary War–era coffins in the cemeteries of Woonsocket, Rhode Island, and grimly floating them down the swollen rivers. In Massachusetts, the roaring, swollen waters of the Concord River ripped apart the historic bridge where the "shot heard round the world" was fired.

Six eastern states were declared disaster areas by the federal government immediately after Diane departed, and local, state, federal, and private relief supplies and money began to flow into the area.

UNITED STATES
EAST
HURRICANE DAVID
August 31–September 8, 1979

Fourteen hundred fifty-seven people died and hundreds of thousands were made homeless by Hurricane David on its path from the Caribbean island of Dominica, through Puerto Rico, the Dominican Republic, and the East Coast of the United States from Florida to New England, from August 31 to September 8, 1979.

Hurricane David first hit the tiny eastern Caribbean island of Dominica on August 31, 1979, killing 22. It then lunged toward Puerto Rico, where it claimed 16 lives and made 1,500 homeless. From here, it blundered directly into the Dominican Republic, where it killed over 1,000 people and rendered 150,000 homeless, 90,000 of them in Santo Domingo alone.

Its 150 MPH winds and huge storm surge ripped chunks of concrete loose from sidewalks and crushed cars waiting on a dock. Hotels were smashed. Power poles were broken off at their bases and palm trees were uprooted. A long concrete walkway along George Washington Avenue in Santo Domingo was shredded and its remains hurled across the street. Waves broke directly into the lobbies of ocean-front hotels.

In the tiny town of Padre las Casas in the Ocoa Mountains 75 miles (120.7 km) southwest of Santo Domingo, 400 Dominicans who had sought refuge in a church and school drowned when flood waters from the Yaque River swept through both buildings. The few who escaped death in that village were forced to climb to the top of the church steeple. Fifteen people were killed when winds collapsed a concrete block church in San Cristobal, 12 miles (19.3 km) west of Santo Domingo.

Ninety percent of the nation's crops were destroyed—although the nation's precious sugar crop, which had already been harvested and bagged, escaped unharmed—and damage estimates exceeded $1 billion, including $350 million damage to agriculture, $130 million to business and industry, $30 million to housing and $30 million to the electrical system. A week after the storm's passing, 150,000 people were still homeless and uncounted thousands were in need of food, medicine, clothing, and water.

While aid was being organized to help the Dominican Republic, which was already under the threat of the visitation of a second storm, Hurricane Frederic, David headed toward Florida. It had weakened somewhat when it made landfall near Palm Beach; its winds were clocked at between 75 and 90 MPH. Still, the storm spanned a 220-mile (354.06 km) stretch of coastline, stretching as far north as the Carolinas.

The hardest hit area appeared to be South Melbourne Beach in Brevard County, Florida, where a tornado associated with the storm ripped the roof and rear wall from the Opus 21 condominium, sucking furniture and appliances from the building.

Fortunately, most of the area's 220,000 residents either left themselves or were evacuated, although some refused to leave. (A police dispatcher in Daytona Beach admitted in frustration, "they're just laughing at us," and 27 diehard drinkers remained in the Sand Spur Bar, on the beach at Cocoa Beach, Florida. Fortunately, their hurricane party had a happy, if hungover, ending.)

Ultimately, only four people lost their lives in Florida in the storm, including two who suffered heart attacks while shuttering their homes, one killed when his car over-turned on rain-swept Interstate 95, and another electrocuted when his sailboat mast hit a power line as he towed his boat out of the Keys. The property damage was in the millions.

After smashing through Florida, David once more headed out to sea, gathered energy, and then, on the

The remains of a luxury Florida condominium after Hurricane David's assault on the eastern United States, August 31, 1979
(CARE)

afternoon of September 4, slammed into the Georgia–South Carolina coast. Its first landfall whipped the barrier islands near the border of Georgia and South Carolina with 70- to 90-MPH winds, then hit Savannah, Georgia, a slashing blow.

The downtown streets of this colonial city, population 120,000, were quickly rendered impassable, choked with the remains of stately trees and other debris. One large oak was split in two by a lamp post and traffic signal hurled by the storm.

In the lush resort of Hilton Head, South Carolina, 30 miles (48.3 km) north of Savannah, vacationers gleefully ignored evacuation orders and became isolated when the James F. Byrne Causeway, linking the resort to the mainlands, disappeared beneath David.

In Charleston, South Carolina, immense waves broke over the High Battery wall at 7:15 P.M., flooding East Bay Street and sending rolling water onto the lawns and over the porches of century-old homes. Undeterred, a crowd of about 50 people cavorted in the spray, yelling and drinking beer, apparently oblivious to the danger.

Outside of Savannah, farms and yards were covered with sea water. On the 17 small islands lying off Savannah, which housed 18,000 people, wind damage was devastating. The 70-year-old lighthouse on Tybee Island was abandoned early by the coast guard, but five men and a woman holed up in it with beer and sandwiches, determined to tough out the storm. They fled when the pounding of waves threatened to crush in the lighthouse's seaward door.

And David had still not ended his destruction. Down-graded to a tropical storm, it was nevertheless capable of claiming eight lives and leaving more than 2.5 million residents of the northeastern United States without power.

A combination of abnormally high tides brought on by a rare total eclipse of the moon, residual winds of 55 MPH, and accompanying tornadoes caused huge breakers to slam into the coastal northeast, uprooting trees, inundating highways, toppling power lines, and forcing yet another mass evacuation. Sixteen more people died, among them, policeman Alvin Williams who dove into a raging torrent in Woodbridge, New Jersey, in an attempt to save two 10-year-old boys from drowning. Instead, all three were sucked into a storm sewer, never to be found again.

The storm wove northward, knocking out power for more than 300,000 people in Connecticut, bowling over trees and more power lines in New York's Mohawk Valley. Ultimately, it would blow itself out over Newfoundland.

The death toll from the storm in the United States would be set at 19. But the lives that were disrupted by David would number in the millions.

UNITED STATES
EAST
HURRICANE DONNA
September 4–12, 1960

One hundred forty-three people died as a result of Hurricane Donna, which originated in the mid-Atlantic, traveled through the Leeward Islands to Puerto Rico, on to Florida, and up the East Coast of the United States to the Gulf of St. Lawrence, from September 4 to 12, 1960.

Hurricane Donna was a long-distance traveler. From the time it originated in the mid-Atlantic until it blew itself out in the Gulf of St. Lawrence, ultimately claiming 143 lives, it had covered more than 5,000 miles (8,047 km).

Its path took it across the Leeward Islands and through Puerto Rico, which suffered a death toll of 106 from the 150 MPH winds.

From here, Donna directed its roaring fury on the Florida Keys, slamming into Key Largo with winds that had climbed to 180 MPH. Roofs flew through the air, boats were ripped from their moorings and sent surfing out to sea. Trees were felled by the hundreds, and scores of residents barricaded themselves within the cement walls of the Florida Keys Electric Cooperative Association, where they were bombarded by sparks from the generator before it finally went dark, too.

Leaving the Florida Keys, the storm skirted the East Coast. But by now, enough sophisticated weather data had been gathered from U.S. Weather Bureau airplanes to warn coast dwellers to leave their homes. Thus, only 22 persons were killed by the storm in the United States.

UNITED STATES
EAST
HURRICANE ELOISE
September 22–27, 1975

Hurricane Eloise killed 45 and made hundreds of thousands homeless as it ripped up the East Coast of the United States between September 22 and 27, 1975, spawning destructive tornadoes, which caused additional damage.

Eloise was a double-barreled storm that caused as much damage after it had supposedly died than it did at the peak of its youth. First sighted in the Caribbean as a tropical storm that claimed 42 lives on two Caribbean islands, it was upgraded on September 22 to a full hurricane when its winds climbed to 115 MPH. It proceeded to bear down on a 150-mile (241.4-km) long, crescent-shaped stretch of Gulf of Mexico coastal lands in Louisiana, Mississippi, Alabama, and Florida.

Landfall finally occurred on the morning of September 23, 1975, on a 40-mile (64.37-km) stretch of beach between Fort Walton Beach and Panama City, Florida. The hurricane's winds were now 130 MPH (209 km/h), making it the worst storm to hit the area in almost half a century. What was worse, Eloise, like a great many hurricanes, had spawned along its rim a group of killer tornadoes which raced over other parts of Florida, Alabama, and Georgia, causing floods and extensive wind damage.

Three early morning twisters struck Fort Walton Beach, Hartford, Alabama, and an area near Jacksonville, Florida, collapsing homes and businesses, knocking out 90 percent of Fort Walton's electricity, and causing $20 million in property damage. In nearby Panama City, which suffered $50 million in damages, motels were flattened and the roof of a shoe store was picked up and blown away, as winds scattered shoes like confetti.

That there was only one death attributed to Eloise along the Gulf was not a result of the wise judgment of its residents. Many waited until the very last minute before evacuating their beachfront homes. When 15-foot (4.6-m) high waves driven by the 130 MPH winds began to cut through the dunes and the wind to peel roof after roof from houses, the inhabitants finally crowded the causeways to the mainland. J. D. Hembree, a construction supervisor for a new condominium project at Destin, said later, "I decided to ride this thing out, but when that first break crested those dunes, I thought, 'J. D., ole boy, you've made a mistake.'"

Although only one person in this area died, hundreds were injured, 17,000 were rendered homeless, and a total of more than $150 million in damages was left behind by Eloise, which headed northward, where it was soon to fire its second barrel.

A quirk of the atmosphere sent the storm due north, through the Washington, D.C., area, Pennsylvania, and New York state and city. A blocking system of high pressure kept it in place for five days, during which time it dumped 7.89 inches (20.04 cm) of tropical precipitation on that area. Streams, rivers, and sluiceways were turned into raging, roaring tides, and in Pennsylvania, 20,000 people were driven from their homes as the Susquehanna and Juniata Rivers overflowed their banks.

In the Washington, D.C., metropolitan area, thousands were affected, as the worst rainstorm in 41 years—actually the backlash of Eloise—flooded every major commuter route in and out of the city. Approximately 600 persons were evacuated from Laurel, Maryland, midway between Washington and Baltimore, when flood gates on a reservoir were opened to keep it from collapsing.

In the area of Alexandria, Virginia, boats evacuated many of the 400 who had to move out. Refugees were taken from the tops of cars, from trees, and from partially submerged homes and businesses in Maryland.

Industries, homes, and schools around Elmira, New York, were evacuated. In New York City, all of the major access routes were flooded, as were basements and first stories of homes in Queens, Brooklyn, and Staten Island. A state of emergency was declared by Mayor Abraham Beame.

The fall crops in rural New York were ruined and the property damage was estimated to be in the "tens of millions." Ironically, there were only two deaths reported from Eloise's second coming: One 83-year-old motorist was swept to his death when he ventured from his car on the Hutchinson River Parkway in White Plains, New York, and an 11-year-old boy drowned in an overflowing brook near Mt. Pleasant, New York, thus bringing the fatality count to a total of three.

UNITED STATES
EAST
HURRICANE FLOYD
September 14, 1999

Hurricane Floyd, a category 4 storm, raked the Bahamas on September 13, 1999, then went on to hit North Carolina, parallel the New Jersey coast, and dump large amounts of rainfall on the northern and western suburbs of New York City. Fifty-seven people died in the storm, which caused upward of $6 billion in damage, most of it by floods caused by its rainfall.

For the early part of its existence, Hurricane Floyd remained a strong tropical storm. But on September 12, 1999, as it entered the eastern Caribbean, it began to strengthen, and by the 13th it was at the top end of category 4 intensity, and aimed directly at the central Bahamas.

Early in the evening of the 13th, Floyd changed direction and its eye passed just 20 to 30 miles (32.2–48.3 km) northeast and north of San Salvador and Cat Island. On the morning of the 14th, it spanned Eleuthera, and, weakening somewhat that afternoon, struck Abaco with full force.

Roofs were ripped off a large number of homes, utility poles were toppled, and thousands of trees were uprooted on all of the islands. A coast guard plane flying over Sandy Point on Little Abaxo spotted large swaths of land covered with trees that had been stripped of their leaves. Almost every home was damaged. Some had collapsed inwardly upon themselves like card houses; others were skeletons devoid of their roofs.

From Cedar Harbour to Foxtown on Abaco, cars and boats were strewn in crazy profusion, and roads were carpeted in dirt and rocks. A radio tower collapsed near a grove of palm trees bowed double by the storm's winds.

Now Floyd turned northward, paralleling the central Florida coast, and at 9 A.M. on September 15 passed approximately 95 miles (152.9 km) east of Cape Canaveral. Its winds were still recorded at 115 MPH, and although it was nearly a hundred miles offshore, it sheared off nearly 100 feet of the century-old Daytona Beach Pier and felled trees and utility poles along its periphery.

Heeded hurricane warnings in Florida and Georgia posed another, unexpected problem: Evacuation orders, released before the storm veered off the coast, clogged roads and filled shelters in Georgia with refugees from Florida and South Carolina. Hundreds of thousands—290,000 from Florida and 300,000 from Georgia and South Carolina—jammed emergency shelters, hotels, and motels, while their homes, for the most part, suffered only shingle loss and an absence of electric power.

The roads were hopelessly clogged along the east-west escape route. The disk jockey on WCGA, an AM radio station near Brunswick, Georgia, offered dirt road shorcuts. It was necessary, according to one driver: "Maximum speed was five miles per hour. People were walking along the highway to get exercise."

Along the way, fast food stops soon ran out of burgers, fries, and pizzas.

Grain trucks and Greyhound buses were pressed into service to evacuate thousands of nursing home residents, their medications, health records, and beds.

In a near-21st-century twist, some Floridians set up Web cams and beamed pictures of Floyd in action to their personal Web sites. Ralph Bonnell, a resident of Stuart, Florida, pointed his web camera out of his bedroom window, turned it on, then evacuated his house.

At Cape Canaveral, space shuttles, poised on their pads and valued at $2 billion each, escaped damage. And the space center itself suffered only some minor flooding and dislodged sheet metal.

Finally, downgraded to a category 2 storm, Floyd came ashore near Cape Fear, North Carolina, at 6:30 A.M. on September 16. It continued on a northeast path, passing over extreme eastern North Carolina on the morning of the 16th and over the greater Norfolk, Virginia, area by 3 o'clock that afternoon.

In a final move, Floyd, now a tropical storm, moved rapidly along the coasts of the Delmarva Peninsula and New Jersey on the afternoon and early evening of September 16. Heavy rainfall preceded Floyd over the mid-Atlantic states, and rainfall amounts were recorded in North Carolina and Virginia as high as 15 to 20 inches (38.1–50.8 cm). In Maryland, Delaware, and New Jersey, the rainfall amounts reached 12 to 14 inches (30.48–35.56 cm).

As the storm worked its way northward, power lines were downed, businesses were shuttered, and schools and airports were closed. In Washington, D.C., the federal government remained open for business, but the House of Representatives closed one day before its traditionally long weekend. A state of emergency was declared in Maryland, and schools were closed. Tens of thousands of residents lost power. In New Jersey, flooding, wind damage, and power outages increased through mid-afternoon. At least 10,000 customers experienced power outages. In New York City, businesses closed early, creating an early commuter rush hour. The stock markets remained open, but the bond market and the New York commodity exchange closed early. Amtrak shut down its Northeast Corridor service between Washington and Boston after mudslides, flooding, and electrical problems trapped several trains for several hours.

Floyd took a unique path when it reached the New York metropolitan area. Instead of crossing Long Island, as it was predicted to do, it veered west and cut a path of destruction across the counties bordering the southern reaches of the Hudson River. Thousands were forced to evacuate, as rivers overflowed, swamping cars in a sea of felled branches and debris. Over 300,000 homes lost power.

It's like one of these scenes in Nebraska when the tornadoes hit," one observer noted. "We've had a bridge collapse up in Old Tappan; we had mudslides in Oakland. All the major roadways have been closed down due to heavy flooding . . ."

For days after Floyd wandered over Prince Edward Island and Newfoundland and turned into an extratropical low over the North Atlantic, rivers and creeks in its path through New Jersey and the western and northern suburbs of New York City overflowed. Sixty-two MPH winds and a rainfall of over a foot in some places had swollen them over their banks, and it would be weeks before the flooding diminished.

Bound Brook, in central New Jersey, was particularly hard hit. The swollen currents of the Raritan River crested at a record 42 feet (12.8 m), 20 feet (6.09 m) above flood levels. Cars, homes, and the downtown business area were submerged under 15 feet (4.5 m) of water, which created a horseshoe-shaped lake of brisk, muddy water, a mile long and a quarter of a mile wide.

Over a thousand people were evacuated from the small, blue-collar community. A flotilla of boats, from canoes and small fishing craft to oversized police launches, crisscrossed the floodwater above the village's Main Street, rescuing hundreds of residents from second-floor windows and rooftops where they were trapped. Hundreds of National Guard troops, police officers, and firefighters manned six helicopters and the boats in the rescue effort.

In New York's Dutchess, Putnam, Westchester, and Rockland counties, mudslides continued to be fatal hazards. Two days after the storm, a nine-year-old girl was swept away by a rain-swollen creek in the village of Wingdale in Dutchess County.

It was the same in North Carolina. Three days after the storm passed, almost two-thirds of the state remained paralyzed. Rains continued, and entire towns were inundated; 300 roads, including long stretches of Interstates 40 and 95, were shut down, and the death toll mounted.

There were 67 deaths attributable to Floyd, 56 in the United States and 11 on Grand Bahama Island. Most of the deaths were due to drowning. Total damage estimates ranged from $3 billion to over $6 billion.

UNITED STATES
EAST
HURRICANE HAZEL
October 12, 1954

....

Hurricane Hazel killed over 1,300 in its passage from Haiti up the East Coast of the United States on October 14, 1954.

Hurricane Hazel gave plenty of warning before it slammed into the coast of the Carolinas. Two days earlier, it had blasted Haiti with 125 MPH winds, ripping off the fronts of buildings in Jacmel, washing tons of mud over the village of Berly, and burying all but two of the village's 260 inhabitants. Sweeping over the southern end of the peninsula, it killed another 1,000 people and rendered 40,000 homeless, toppling banana trees and washing away fields of beans, rice, coffee, sugar, and corn.

That was devastation enough for one storm, and, to the relief of the local residents, Hazel, after crossing the island, headed out to sea. However, a huge high pressure area—an anticyclone—settled down during the night of October 12, blocking Hazel's retreat. She re-formed, and, armed with an extremely small eye (Naval weather planes had great difficulty remaining within it and were constantly threatened by the 150-MPH winds of the eyewall), headed straight for the outer banks near Cape Fear.

All night on October 1, warnings were broadcast to the remaining inhabitants of this area to evacuate, and most did. On the morning of the 15th, the storm hit with full force, packing a 20-foot (6.1-m) storm surge (added to a 10-foot [3.04-m] tide). Miles and miles of houses literally exploded as the waves and wind struck. The ground was pushed into the inland waterway. Of Long Beach's 377 buildings, 352 were absolutely destroyed. At Holden Beach, 200 were erased. Ocean Isle was swept bare; Myrtle Beach, Windy Hill, Crescent Beach, Cherry Grove, Robinson Beach, Colonial Beach, and Wrightsville were horribly damaged. One-half of the taxable wealth of Carolina Beach was swept away. Shrimp boats were washed onto the verandas of some of the remaining dwellings. Nineteen people lost their lives, and over $136 million of damage was done to property, chiefly by sea water.

But Hazel wasn't through yet; on she careened, up through Virginia, the District of Columbia, Maryland, Delaware, Pennsylvania, New Jersey, and New York State, mowing down power lines, trees, crops, poultry houses, and tobacco barns. The losses due to power failure and the flooding of dams, rivers, and lakes were astronomical.

By the time the storm had reached upstate New York, Hazel had diminished to an extratropical cyclone with heavy rains and wind gusts of 90 MPH. It was on the cusp between hurricane and cyclone, and should have blown itself out in the Alleghenies. But again, a blocking system, this time a cold front over Chicago, redirected it toward Toronto, a city unaccustomed to hurricanes.

Hazel cranked itself up again. Its rains increased and its winds rose in gusts to 120 MPH. Above Toronto, north of Lake Ontario, four rivers carried more than six billion gallons of water to the city and the land surrounding it. Over this area, Hazel collided with the cold front from Chicago. The resulting water vapor condensed into an explosion of rain.

In Toronto, hurricane warnings were lifted after 7 P.M. on the night of October 15. What weather bureau experts failed to anticipate was the effect of the torrential rains of Hazel. Upriver of Toronto and its environs, 113 tons of water per inch of rainfall fell for six hours. The Humber River suddenly overflowed its banks in a black wall of water. The Little Don River soon followed, washing out roads, uprooting houses, and sending them spinning downstream. Forty concrete bridges were wiped out. The sewage plant in the town of Kitchener was gutted. Every house was surrounded by water and many washed away in Woodbridge.

The next morning, an army of rescue workers was marshaled to rescue families trapped on the roofs of floating houses throughout a silt-covered countryside that had once been green and populated.

High tides and storm surges, whipped to a froth by Hurricane Hazel, demolished waterfront homes in Swansboro, North Carolina, on October 15, 1954. (American Red Cross)

Hazel had by now blown northeastward, out of North America, beyond Greenland. Her effects would eventually be felt off the coast of Norway. In Toronto alone, she would leave over $100 million in property damage.

Altogether, the three hurricanes of 1954—Carol, Edna, and Hazel—would cause $1 billion in property damage and claim over 150 lives in the United States, and thousands more in the Caribbean.

UNITED STATES
FLORIDA
KEYS
October 18, 1906

One hundred twenty-nine people died during the hurricane that struck the Florida Keys on October 18, 1906.

The Overseas Highway and the Florida East Coast Railroad were under construction when the first important hurricane of the 20th century barreled into the little group of islands called the Florida Keys.

A day before, the hurricane had hit Cuba with minimal winds of 50 MPH—an indication to local weather forecasters that Florida had nothing about which to be concerned.

But on the night of the 17th, the storm picked up power from the moisture of the Caribbean, and by the morning of the 18th, when it slammed into Sand Key, its winds had increased to minimum speeds of 75 MPH. The storm surge washed across the key, demolishing the houseboats that sheltered the workmen, smashing at the rock embankments, and drowning 124 workmen and five others.

The partially constructed road and railway held firm. But 29 years later, they would be destroyed by another hurricane.

UNITED STATES
FLORIDA
KEYS
September 2, 1935

....

Gusts of wind clocked at 250 MPH drove a monster hurricane into the Florida Keys on September 2, 1935. Four hundred people, many of them veterans of the 1932 Washington Bonus March, were killed.

The great Labor Day hurricane of September 2, 1935, was a true monster, packing winds of from 150 to 200 MPH and gusts of an unbelievable 250 MPH. It left behind 400 dead and injured hundreds more. The villages of Matecumbe, Tavernier, Rock Harbor, and Islamorada were totally destroyed.

But it was the death of 121 of the 716 inhabitants of Camp Number 5, located near Snake Creek on lower Matecumbe Key, that drew world attention to this storm and its victims. Camp Number 5 housed World War I veterans, survivors of the famous 1932 Bonus March on Washington, during which members of the Third U.S. Cavalry, under General Douglas MacArthur, rode full herd onto their fellow veterans, whose only crime was protesting Congress's failure to pass a bill for the payment of their bonus certificates.

Two marchers in this so-called Bonus Army lost their lives that August day. Three years later, what remained of the Bonus Army was spread out over the country in work camps, such as Camp Number 5, tucked away from public view. Their task: to build a government road, for which they received 30 dollars a month and all the food they could eat.

This relief train was overtaken by the September 1935 hurricane and its storm surge in Key West, Florida. Better planning and less bureaucracy might have saved hundreds of lives. (Mattrock photo, American Red Cross)

Most of them, by this time, had become outcasts. Some were drunks, and some suffered from shell shock. Some were hardened misfits. All were an embarrassment and a shame to the country that had conscripted them to fight and then had abandoned them.

Most of the natives of the Keys distrusted and disliked this ragtag army. Only Ernest Hemingway, it seemed, who knew some of them, pleaded their cause in *The New Masses*: ". . . they were all that you get after a war," he wrote, "but who sent them there to die?"

And die they would, most of them from a hurricane that was first recorded on August 31 northeast of Turks Island. It was a small hurricane, as hurricanes go— merely 10 miles (16.09 km) across. But when it reached Andros Island on September 1, it was already a giant in force, if not in size. Northeast storm warnings were posted from Fort Pierce to Fort Myers, and caution was advised for the Florida Keys.

The next day, September 2, was Labor Day. Three hundred fifty veterans from the Keys camp were in Miami, attending a ball game. The wind was high in Miami, but not nearly so high as it was in the Keys by 12:15 that afternoon, when Captain Edney Parker telephoned the Florida East Coast Railroad. An evacuation train that was supposed to be ready and waiting at Homestead in case of just such a storm wasn't there, nor had it been prepared. It was finally made up in Miami and departed at 4:25, arriving in Homestead after 5 P.M. From there, it slowly backed down the track to evacuate the veterans, who by now had abandoned Camp Number 5 and were huddling along the railway embankment.

Storm waves began to crash over the banks. Windborn sand drew blood as it knifed into their faces. As night began to fall and the train still didn't arrive, some of the men went back to the camp.

Sometime after 8 P.M., J. A. Duncan, the keeper of the Alligator Reef Light, went out onto the observation platform to check the light. A black mass of water suddenly loomed over the lighthouse. He jumped for the ladder and held fast as tons of sea water crashed over him, shattering the glass and the lenses. Duncan reported that the wave seemed to be 90 feet (27.41 m) high. In reality, it was 20 feet (6.09 m) high and powered by 250-MPH gusts of wind.

On over the Keys the wave rushed, slicing roofs off houses, floating homes and beds and their inhabitants out to sea. At 8:30, it hit the 10 cars of the rescue train, which had finally arrived, uprooted the track, smashed the trestles, and knocked over every car but the engine. It roared on into Camp Number 5, leveling every building, sending timbers and roofs and bodies swirling out to sea.

The eye, with its brief calm center, passed over Lower Matecumbe Key, plummeting the barometer to 26.35 inches, the lowest reading yet recorded for a Western Hemisphere hurricane.

By morning, in the 10-mile (16.09-km) wide swath cut by the storm, hardly anyone was left alive. Everything was destroyed—buildings, docks, roads, viaducts, trees, the railroad and its bridges. Many of the dead had been washed out to sea and would never be seen again. Bodies were found hanging among overturned and stripped mangroves, buried in sand and debris, or in sunken wrecks of boats. Hardly any survivors were without injury.

All communication with the Keys was cut off. Some of the injured, hanging in trees, died of thirst, without help. Cisterns were choked with debris and fouling salt water. Finally boats reached the mainland at Homestead, and help arrived in the form of Coast Guard amphibians and cutters bearing supplies, administered by the National Guard and boys from the Miami CCC (Civilian Conservation Corps) camp.

The National Guard command ordered all of the bodies to be cremated. The Red Cross set up headquarters, telling people who had been left with nothing that to get help they had to submit "plans for rehabilitation." A storm of bitterness broke, the WPA (Works Projects Administration) ordered an investigation to settle blame for the tragedy, particularly the reason that the train did not arrive until it was too late for it to have any role in rescuing the veterans. Eventually, tempers cooled, orders were rescinded, and the investigation abandoned. The veterans group was fragmented and those who had survived the storm were quietly sent elsewhere.

UNITED STATES
FLORIDA (LOUISIANA, MISSISSIPPI) HURRICANE KATRINA
August 25, 2005

Hurricane Katrina, the monster storm of the staggeringly active 2005 hurricane season, laid waste to the Gulf Coast of the United States and particularly to New Orleans, the only city in the United States ever to be so devastated by a natural disaster. In total, 1,277 people lost their lives, the homeless numbered 374,000, and the cost of the destruction was estimated at more than $200 billion.

In June 2002, the New Orleans *Times-Picayune* ran "Washing Away," a five-part series that predicted the effects a Category 4 or 5 hurricane might have on New Orleans. Levee breaches, uncontrolled flooding, and thousands of people stranded or dead were laid out in precise detail. Who could have known, in August 2005, that, in an eerie scenario, each of the newspaper's predictions would come horribly true?

It all began on August 23, with a disturbance gathering itself into a tropical depression—number 12 of the season—in the Caribbean. A day later, when the depression was 135 miles (217.3 km) east of the coast of Florida, it would gain the name of Tropical Storm Katrina. By August 25, it would detonate into a Category 1 hurricane which traveled across Florida, causing little damage and downgrading into a tropical storm.

But then, the unexpected happened. Drawing energy from the warm waters of the Gulf of Mexico, Katrina rapidly expanded, in three days of travel across the Gulf, into a monster Category 5 hurricane with winds up to 175 MPH (281.6 km/h) playing across a cloud bank 1,000 miles (1,609.3 km) wide, and pushing ahead of it a storm surge 28 feet (8.5 m) high.

And then, just as the 2002 story predicted, the storm crashed into New Orleans, a city that is 70 percent below sea level, surrounded by water on three sides and protected by levees lower than 25 feet (7.6 m). An unimaginable catastrophe followed: NASA weather expert Dr. Jeffrey Halverston told reporters as the storm approached New Orleans, "This is one of the worst case scenarios. It's kind of a doomsday scenario. Very rarely have we ever seen nature conjure a storm this powerful in the past 100 years. It ranks right up there with storms such as Camille [see p. 311] and Andrew [see p. 280] and Galveston [see p. 313]. There is no way to soften the blow of this storm whatsoever. It's packing the worst possible energy in terms of ocean atmosphere, and all unimaginable energy."

On August 28, as Katrina neared, Dr. Max Mayfield, the director of the National Hurricane Center, did something he had not done in years: He called the governors of Louisiana and Mississippi and Mayor Ray Nagin of New Orleans to warn them of the severity of the storm. States of emergency were declared throughout coastal Louisiana, Mississippi, and Alabama. Mayor Nagin ordered a mandatory evacuation of New Orleans and opened ten shelters, including the Superdome. Federal Emergency Management Agency (FEMA) representatives held briefings for the press stating that all was well: they had practiced and planned for just such a disaster.

At 6:10 A.M. on August 29, Katrina made landfall near Grand Isle, Louisiana, as a Category 4 storm with winds of 140 MPH (225.3 km/h) and a storm surge of 24 feet (7.3 m). The eye was over Biloxi, Mississippi, and then moved up through Hattiesburg. It seemed as if New Orleans had missed the worst of it, though 120 MPH (193.1 km/h) winds had hit the city, ripping

buildings apart, thrusting refrigerators, air conditioners, and stoves into the streets and rendering signs and pieces of roofing and siding airborne (see color insert on p. C-8). In the 12-hour assault, centuries-old trees had fallen, dragging power lines down with them. Survivors, feeling that New Orleans had dodged the bullet, began to appear in the streets, and refugees who loaded their cars to leave the city began to turn their cars around.

What they did not know—and why they were not told would become a subject of intense investigation—is that from 2:00 P.M., the levees had been breached. As darkness fell, water in the city began to rise at the rate of half a foot an hour. Those who did not evacuate from their homes began to move to upper stories and into attics, as the water level climbed. Within the Superdome, chaos began to rise: There was no power, toilets would not flush, and in the streets, it was total darkness, lit only by Coast Guard flares dropped from helicopters to identify survivors. Rain from the storm continued to fall.

Lake Pontchartrain, swollen to six feet (1.8 m) above sea level, began to pour over the 17th Street levee; two other breaches opened along the Industrial Canal, and intense flooding poured into the lower Ninth Ward and began to flow south toward the French Quarter.

The news of the levee break spread quickly. And then the very worst happened: Hundreds of thousands of gallons (378 l) of lake water per second cascaded into New Orleans without letup, breaching the floodgates that were designed to keep it out of the city. In an indication of the confusion gripping the upper echelons of those in charge, Senator David Vitter, Republican of Louisiana, held a press conference, in which he stated, calmly, "I don't want to alarm everybody that, you know, New Orleans is filling up like a bowl. That's just not happening."

But that is just what was happening. And to compound the confusion, FEMA and the state stumbled over each other, sometimes sending busloads of survivors to the Superdome, sometimes suddenly canceling the buses. By the afternoon of August 30th, nearly 30,000 people filled the Dome, where there was no drinking water and no noticeable assistance.

Katrina had moved on into Tennessee, Georgia, Kentucky, and the Carolinas, spawning tornadoes along the way. In New Orleans, pumps that were designed to pump out overflow had broken down, and the water, named a "Toxic Gumbo" by one local wag, had begun to accumulate chemical waste, gasoline, and the contents of the above ground coffins in the city's cemeteries.

The next day, an official and military plan to drop, from helicopters, 1,500-pound (680.4-kg) concrete blocks and 3,000-pound (1,360.8-kg) sandbags filled with gravel into the breach in the 17th Street Canal levee was scrapped in favor of conducting a rescue in a church (see color insert on p. C-8). The Superdome turned violent, with roaming gangs of teenagers assaulting and robbing weaker survivors. People died and were left under blankets, on cots, and in wheelchairs. In the flooded streets, looting began, and the National Guard joined the New Orleans police in trying to control it. Appeals were made to Washington for help, but these appeals were routed to various agencies, where they disappeared.

The infrastructure into and within the city all but vanished. Part of Interstate I-10 was destroyed—the twin span bridge over the Mississippi had disappeared. An evacuation of the refugee population of the Superdome, no longer safe or able to sustain its temporary residents, began. Chinook helicopters were employed at first, but after reports that shots had been fired at one of them, they were discontinued, and buses were substituted. Meanwhile, in Biloxi, Mississippi, 20,000 people were in shelters, 185 were reported dead, the necessities for living were dwindling, and FEMA had not appeared.

In New Orleans, FEMA was waiting in staging areas. As of September 1, it had not entered the city. Four ships had left Norfolk, Virginia, with 6,000 National Guard troops; there was an impressive array of military MedEvac helicopters at the ready. But there was no evidence that they had arrived in New Orleans, except for one National Guard helicopter, which on the afternoon of September 1st dropped enough food for 25 people for the thousands stranded at the Convention Center. Confronted by this, Homeland Security director Michael Chertoff told NPR radio he was unaware that there were people in the Convention Center. Four days after the hurricane struck, reports arrived that FEMA was blocking relief efforts by other organizations and private parties.

No one seemed to be in charge except for the Army Corps of Engineers who had managed to close 50 percent of the breach at the 17th Street Canal. Those in charge of federal assistance seemed to be overwhelmed and confused. And in the city, anarchy reigned: A mall burned to the ground and hundreds of patients lay unattended—some of them dying—on the floor at the International Airport. Food had all but disappeared from the city. Gangs rampaged through the streets, setting fires, looting, robbing, and shooting unarmed survivors. Snipers fired from rooftops. And the police, stretched beyond the breaking point, were unable to control the chaos.

Finally, on Friday, September 2nd, a convoy of 50 trucks containing combat troops of National Guard soldiers fresh from Iraq rolled into the city, bearing food and supplies, and what was equally important, the assurance of a beginning of law and order. But it was only a beginning. The rancid floodwaters still remained. An exhausted and demoralized police force, 30 percent of whom had abandoned their jobs, was doing what it could, but it was clearly not enough.

Promised helicopters to evacuate the patients at Charity Hospital did not arrive, as medical supplies began to run out, and nurses and doctors struggled to keep their patients alive. Some 50,000 survivors still remained on rooftops, in shelters, in attics, or in the shells of ravaged houses. "The people of our city are holding on by a thread," announced Mayor Nagin. "Time has run out. Can we survive another night and who can we depend on? God only knows."

There was still mostly silence from the federal government, despite Congress's passing of an emergency measure providing $10.5 billion for aid, which the president had signed. Accusations of racial overtones in the lack of help from above began to surface based on the fact that New Orleans is the ninth poorest city in the United States with a population of more than 70 percent African Americans. At that time most of the 80 percent of the city that remained underwater was centered in poor areas which underlined the fact that Louisiana had the fourth highest poverty rate in the country.

Finally, on the seventh day after Katrina hit, a more organized evacuation process took form. Helicopters took off every fifteen minutes, ferrying survivors to Houston, Dallas, San Antonio, and other safe havens (see color insert on p. C-8). President Bush went on national television to reassure the nation that the situation was improving in New Orleans.

The next day, refrigerated trucks began to arrive, to begin retrieving the hundreds of dead bodies floating in the floodwaters or hidden in attics. For days, they had been bloating, preyed upon by rats and starving animals and continuing to foul the water upon which they floated (see color insert on p. C-8).

Army engineers continued to work to repair the levees, and install pumps to drain the water from the streets of New Orleans. In eastern New Orleans, pumps were started to drain water at 700 feet (213.4 m) per second from St. Bernard Parish. In Plaquemines Parish, the Corps began to notch levees with gaps up to 100 feet (30.5 m) wide so that water could flow back into the Gulf. Once the water was drained, the Corps promised to plug the gaps, but it predicted that it would be from 26 to 80 days before that could take place.

FEMA now chartered three Carnival Cruise Line liners to help house up to 7,000 evacuees for the next six months. The USS *Bataan*, anchored in the Gulf, had for days offered its extensive hospital facility, its doctors, its six operating rooms, 600 hospital beds, food and water supplies, and the ability to produce 100,000 gallons (378 l) of clean, fresh water each day. But federal authorities had never made use of the facilities.

FEMA, under increasingly angry criticism, tried to lay blame on everyone from the local government to the National Weather Service. The New Orleans *Times-Picayune* countered with an open letter to the president: "Dear Mr. President: We heard you loud and clear Friday when you visited our devastated city and the Gulf Coast and said, 'What is not working, we're going to make it right.' Please forgive us if we wait to see proof of your promise before believing you. But we have good reason for our skepticism . . . We're angry, Mr. President, and we'll be angry long after our beloved city and surrounding parishes have been pumped dry. Our people deserved rescuing. Many who could have been were not. That's to the government's shame."

But on Labor Day, September 9, the tide literally began to turn. Not by much, but at least pumps had begun to obviously drain the water from New Orleans; the water grew, if that were possible, more filthy and toxic, a black mixture of gas and mud, oil and excretions, garbage, and human remains. Mosquitoes bred in it, bacteria multiplied in it. The mayor continued in his plan of forced evacuation of the 10,000 remaining citizens who were persisting in staying in their homes, or—and this was one sign of the levity for which the Big Easy was known and noted—hanging out at Johnny White's Bar on Bourbon Street. Advertising itself as "the bar that never closes," it had lived up to its advertising, and had never closed. Nor did the locals stop drinking, and it became a gathering place for those who had decided to hold out against evacuation.

Over 750,000 residents of New Orleans, however, had removed themselves, some possibly forever, to 20 other states, and as far away as Los Angeles.

Meanwhile, in Mississippi, the destruction was appalling, and the help from federal agencies nearly nonexistent. In Gulfport, the police headquarters no longer existed. Most of Biloxi remained without either water or power, and its casinos were overturned and swept into the Gulf of Mexico. In the beach town of Waveland, a former thriving community of 7,000, nearly every building had been leveled. The worst damaged of all was at Pass Christian, which was hit dead center by Katrina, along with a 23-foot (7-m) tidal surge which destroyed nearly all 2,500 homes in the town and rendered 8,200 of its 8,500 residents homeless. "Bodies are still being pulled out of trees, out of houses," one resident told a reporter. "You drag somebody out of the rubble you grew up with, it's pretty tough."

Still, an atmosphere of rudderlessness hung over the city, and, in fact, the entire Gulf Coast. Though 5,000 sailors and 500 marines had landed on the beaches of Biloxi, Mississippi, 20,000 FEMA trailers, supposedly destined for survivors, sat idle in Atlanta, Georgia, held prisoner by miles of red tape. Aaron Broussard, president of Jefferson Parish, vented his anger and frustration on the CBS's *Morning Show*, "Bureaucracy has murdered people in the greater New Orleans area," he said. "And bureaucracy needs to stand trial before Congress today. So I'm asking Congress, please investigate this now. Take whatever idiot they have at the top of whatever agency and give me a better idiot. Give me a caring idiot. Give me a sensitive idiot. Just don't give me the same idiot."

Some heads did roll: Michael Brown, praised by President Bush on the president's first visit to the area ("You're doing a heck of a job, Brownie"), was forced to resign as head of FEMA. But, as New Orleans gradually recovered, some memories shortened, though not all. Perhaps the memories that are longest belong to the first responders, the volunteers and residents of New Orleans who, against insurmountable odds, natural and man-made, helped the survivors of Katrina, and were therefore the tragedy's true heroes.

The official toll was staggering:

- the death toll numbered 1,277
- the number of lives saved by the Coast Guard totaled 33,000
- the total number of homeless was 374,000
- more than 1 million people were evacuated from New Orleans
- the official bill for damage to the entire Gulf Coast was more than $200 billion.

UNITED STATES
FLORIDA
MIAMI
September 17, 1926

The hurricane of September 17, 1926, struck Florida at the height of the great land boom. Five hundred people died, 25,000 were made homeless, and over $760 million of property damage resulted.

The great Florida land boom had just passed its fevered peak when the Miami hurricane of September 17, 1926, struck. When the final toll was counted, the property damages in Miami alone totaled a staggering $760 million. One hundred fifteen people lost their lives in Miami; 300 were killed in Moore Haven and dozens in Fort Lauderdale; 25,000 people were rendered homeless.

The storm originated in the Cape Verde Islands on September 15, brushed Puerto Rico, and roared into Miami at approximately 7:30 P.M. on September 17. There was no warning. The Weather Bureau reported a tropical storm nearby, but issued no warnings for the steadily increasing northeast winds.

By midnight, waves were crashing over the breakfronts and the wind had climbed to 120 MPH, blowing away house roofs and the instrument shed at the top of the Weather Bureau. People trying to crawl to safety were crushed by trees or bricks, and water was foaming everywhere.

The devastation was horrendous. Concrete office floors weighing five tons were broken up and carried for 50 feet (15.2 m) by the flood waters. Trees, automobiles and entire families were buried under the drifting sand.

Salon Maritime, a palatial real estate office in Coral Gables, lost its roof and filled first with sand, then with 500,000 gallons (1,892,705.9 l) of fresh water when the Roman Pool next door split in two. By dawn, a policeman and a real estate agent still clung to its chandelier, as sea water mixed with fresh water and rose beneath them. Finally, they floated through the door on a piano.

At 6:47 A.M., the eye of the storm passed overhead. Unwarned, people poured out of shelters, and when the gale suddenly leaped anew out of the south at 128 MPH (which made this storm one of the fiercest in Florida history) many were instantly killed by blown timbers or falling, weakened walls. The tide crashed in from the opposite direction, sucking the two men on a piano back inside the wreckage of the real estate office. They would climb from here onto the oversize mantel and remain there until they were rescued, 10 hours later.

On the storm raged, obliterating the boom town of Moore Haven on Lake Okeechobee, hitting Fort Lauderdale, falling on Pensacola. Shacks, flimsy buildings, and billboards were blown entirely away.

The cleanup would take weeks. More than 2,000 workmen were hired for the job, cutting up fallen trees,

A devastated beachfront in Miami, Florida, following the hurricane of September 17, 1926. Striking at the height of the Florida land boom, the hurricane killed 500, made 25,000 homeless, and set the boom back considerably. (Library of Congress)

Shacks are turned into lean-tos after Hurricane Cleo rampaged through Haiti on August 22, 1964. (CARE)

burning wreckage, and collecting rotting fish. The whole-sale carnage brought about by the collapse of flimsy, expensive buildings built during the boom led Miami to revise its building code, which would later become a model for all hurricane endangered cities in Florida.

UNITED STATES
FLORIDA
MIAMI
HURRICANE CLEO
August 22–27, 1964

One hundred thirty-three people died when Hurricane Cleo traveled from Guadeloupe to Miami from August 22 to August 27, 1964.

Spawned in the Caribbean, where it located itself for four days, August 22–26, 1964, Hurricane Cleo killed 120 people in Guadeloupe and Haiti with 100 MPH winds and enormous storm surges.

On August 27, its fury was turned on Miami. By this time, its maximum winds had climbed to 135 MPH, and

the waves it sent roaring over breakwaters of the famous resort's oceanfront sank, smashed, or overturned hundreds of boats, from dinghies to six-figure yachts.

The so-called Gold Coast, that strip of beachfront property that housed the poshest hotels and vacation homes, was particularly hard hit. More than 1,200 homes and offices suffered wind and/or water damage. For weeks, streets were buried under debris and mud. The final damage accounting totaled $200 million in property loss. Fortunately, thanks to modern forecasting methods, only 13 people lost their lives in Miami. If the storm had struck 20 years earlier, the loss might have been catastrophic.

UNITED STATES
FLORIDA
OKEECHOBEE
September 16, 1928

Six hundred people died in Guadeloupe, 300 in Puerto Rico, when a major hurricane ripped across these islands in early September 1928. But its most tragic visit was to the Florida Everglades, where 2,500 people drowned from the storm surge of Lake Okeechobee.

In 1928, there was no flood control along Lake Okeechobee's shores, protecting the lake towns of the Everglades and their surrounding farms from being destroyed by an overflow from this, the second largest fresh-water lake wholly within the United States. It took the hurricane of September 16, 1928, in which a staggering 2,500 people—some said more—drowned, to convince both the state and federal governments that drastic flood control measures were necessary if the Everglades were ever to be a safe, habitable, and agriculturally profitable area.

The hurricane was a monster, a vicious killer right from its start, somewhere off the African coast. It crashed across Guadeloupe on September 12, leaving over 600 dead. On the next day, which was the feast of San Felipe, its 235-mile (378.2 km) wide presence made itself fatally felt in Puerto Rico, where it caused $50 million in damage, killed another 300 people, and rendered 200,000 homeless.

On the night of September 16, moving with a kind of terrible grandeur, it hovered over and smashed its way through West Palm Beach, where the barometer plunged to 27.43, a new low up to that time. Word spread that the Everglades, particularly the area around the 730 square miles (1,174.8 sq km) of Lake Okeechobee, was directly in the path of the storm. Moore Haven would not have to be warned; it had been flattened two years earlier by another killer storm. But border places like Bare Beach, Lake Harbor, Belle Glade, Pelican Bay, Pahokee, and Canal Point were active resort areas on the afternoon of Sunday, September 17.

By 6 o'clock, the winds were strong enough to overturn automobiles and strip roofs from shacks and houses. In Belle Glade, 500 people crowded into the Glades Hotel and 150 crammed themselves into the Belle Glade Hotel. On Lake Okeechobee itself, a group of women and children huddled on a barge.

When the full fury of the storm hit, the lake was completely emptied of water, which in turn was hurled towards an inadequate mud dike and the towns below it. Needless to say, the dike melted under the force of the water, and the land south of the lake was completely inundated. No trees, telephone poles, canals, or houses were visible above the surface of the water from there to the sea, 30 miles (48.28 km) away.

In Belle Glade, the Belle Glade Hotel was torn from its foundations, but it held together and continued to shield its refugees from the projectile-riddled winds. The Glades Hotel took in water up to its second floor, but the inhabitants there just moved above the high-

water mark and survived. The remainder of the town was demolished. Fifty houses were swept away.

In Pelican Bay, everyone drowned, either in their homes or trying to escape on the roads. On Ritta Island, many who had climbed to the roofs of their houses to escape the flood waters were crushed by falling trees. Others, climbing trees, were killed by the fatal bites of water moccasins swarming up the trunks or along the tops of ridges that jutted up above the roaring waters.

The women and children on the barge miraculously survived.

The next day, the water receded only slightly. Militia men and boy scouts, pressed into rescue service, fished for bodies, both human and animal, from the brackish waters that covered everything. Later, they tied the dead together and towed them behind launches to railroad platforms. Seven hundred were buried in a long trench in West Palm Beach. Finally, after four days of hot sun, the only solution was to burn the dead in huge funeral pyres.

It would take two years of lobbying to convince the Hoover administration in Washington to approve a bond issue of up to $5 million to support a plan for state flood control over the 12,000-square-mile (19,312.1 sq. km) Okeechobee-Everglades area.

UNITED STATES
FLORIDA
ST. JO
September 1841

Nearly 4,000 people—the entire population of St. Jo, Florida—were killed when the town was totally destroyed by a hurricane in September 1841.

A few miles from Apalachicola, on the banks of the Apalachicola River, the high-living boomtown of St. Jo reveled night and day. A collection of saloons, shops, and small businesses, it was a thriving town of 4,000 in 1841. The only unhappy citizens were its local preachers, who rode around the countryside denouncing St. Jo as a hell-hole of wickedness, destined to be destroyed by fire or flood.

Half of that destiny came dramatically true in September 1841, when a colossus of a hurricane slammed into the boomtown on its way eastward across the state from the Gulf of Mexico. Within hours, every single building in St. Jo was demolished. Almost every person in town was either drowned or crushed under the falling buildings or flying debris. Today, the only evidence of the storm is a cluster of deteriorating gravestones to the east of the town.

UNITED STATES
FLORIDA
STRAITS
September 9–10, 1919

Seven hundred seventy-two people died in a hurricane that whipped up the Florida Straits from September 9 through 10, 1919.

What was reputed to be the largest, most destructive hurricane of the 20th century began as a minor disturbance near Santo Domingo on September 2, 1919. Before it finally subsided near Corpus Christi, Texas, 12 days later, 772 people would lose their lives to the storm.

After Santo Domingo, the hurricane made landfall at Key West, Florida, where its 80 MPH winds stripped the weather service of some of its recording instruments. The full fury of the storm slammed into this city, collapsing buildings, flooding streets, uprooting trees, overflowing mud dams, and ultimately causing $2 million in property damage.

Settling into the Florida Straits, the storm laid waste to the sea lanes, capsizing 10 vessels. There were survivors from most of these ships, but the *Valbanera*, a Spanish cruise ship, was unceremoniously and tragically swamped before safety precautions could be taken, and it went rapidly to the bottom of the Gulf of Mexico near the Dry Tortugas, drowning 400 vacationing passengers and 88 crew members.

The killer storm maintained its size and power when it blundered ashore at Corpus Christi. Its outer winds, holding at an even 60 MPH, lashed at Galveston and Miami. But the fury of its center was saved for Corpus Christi, where it sent storm surges up to 16 feet high through the city streets, toppling tall buildings and pulverizing houses, farms, and businesses. When a watery, calm dawn finally broke over the city on September 15, $20 million in damage had been caused, and 284 people lay dead from the horrendous destructive force of the storm.

UNITED STATES
GEORGIA
August 27, 1893

Over 1,000 were killed by the hurricane that struck Georgia and North Carolina on August 27, 1893.

By the time they reach the southeastern coast of the United States, most hurricanes curve northeastward toward the pole. The hurricane of August 27, 1893, however, behaved differently. Packing huge winds, it produced a roaring, consuming tidal wave that totally inundated islands off the Georgia and North Carolina coast.

Property damage was estimated to approach $10 million, and more than 1,000 people were reported killed by the storm.

UNITED STATES
HAWAII
KAUAI
HURRICANE INIKI
September 11, 1992

Hurricane Iniki, with winds of 160 MPH, slammed into the 30-mile (48.3-km) wide island of Kauai on September 11, 1992. Three people died, 98 were injured, 8,000 were made homeless, and there was $1 billion in damage to the island's dwellings, buildings, and farming and tourist industries.

Kauai is one of the nearest examples to a pristine paradise left on Earth. This circular, 30-mile (48.3-km) wide island in the Hawaii congregation is bursting with spectacular vistas—hillsides with frozen waterfalls of jungle greenery punctuated by wild orchids and scarlet bougainvillea; Waima Canyon, dubbed by Mark Twain the Grand Canyon of the Pacific; crescents of beaches; and, in contrast, the Na Pali coast, where mountains and cliffs with plummeting waterfalls directly border the sea. And then there is Mount Waialeale, bearing the title of the wettest spot on Earth because of the 450 to 600 inches (1,143 to 1,524 cm) of rain that fall yearly on its cloud-shrouded summit.

But in the early afternoon of September 11, 1992, Hurricane Iniki (Hawaiian for, appropriately, "piercing as pangs of wind or love") roared in from the southwest, with winds up to 160 MPH. Roofs of homes that sheltered the island's 51,000 inhabitants were ripped loose and became flying missiles, buildings were flattened, 90 percent of the island's power lines were downed, and trees fell like matchsticks.

The destruction was horrifying and unselective. The only contact with the outside world became the signals from an amateur radio operator. All light aircraft was grounded or destroyed, and Lihue Airport's control tower was heavily damaged. Norfolk pines, koa, and eucalyptus trees were reduced to scattered rubble. The Na Pali coast changed its topography; huge chunks of cliffs were eroded away by the fierce storm surge. Seven thousand homes were either wiped out or made uninhabitable.

The island's tourist industry—its industrial backbone—was decimated. Before the storm, 45 percent of Kauai's economy depended upon its $1 billion tourist business. Now, nearly all of the 7,646 hotel rooms were wrecked. In Poipu, the Stouffer Waiohai Resort was completely washed out; the Sheraton Kauai Beach Resort was severely damaged as was the Sheraton Princeville Hotel. All resorts estimated that it would be six months at least before they could reopen.

Exotic species of birds and plants were feared to be exterminated in one afternoon: The Hawaiian duck, the Hawaiian coot, the Lauai Alialoa, and Kauai thrush, nearly extinct already, were pushed to the very edge of existence.

The western edge of the island, the site of a projected test of the Strategic Defense Initiative, or "Star Wars," was severely damaged, to the delight of a percentage of the islanders who opposed its planned launching of Polaris missiles southwest across the Pacific.

And Steven Spielberg's final day of shooting *Jurassic Park* on location in Kauai, was disrupted.

Natural wonders like the Coco Palms grove and the Tunnel of Trees, three-quarters of a mile of eucalyptus trees flanking the road to Koloa, were stripped of their leaves and broken down in sections. Nature trails simply disappeared.

President Bush, criticized for his slow response to Hurricane Andrew (see p. 280), immediately sent cargo planes with supplies to the stricken island. Landing military personnel found 8,000 homeless, hungry residents.

As the days lengthened after the storm left, it soon became obvious that the farm industry as well as the tourist industry had been seriously injured by Iniki. Francis Furtado, a crane operator at a sugar mill, outlined this to a *New York Times* reporter. "Sugar has been slowly going down and they've been trying to fill in by planting coffee and macadamia nuts," he said, "But now everything's been affected. Papayas have been wiped out. Bananas are wiped out. We lost pineapples already. With sugar down, it's pow! And if the tourists stop coming, we're done for. We'll have to move to the mainland."

Some looting was reported on a small island with a small population, but the young men who were involved were soon identified and stopped.

Still, the devastation remained: 10,000 homes had sustained major damage, and a week after the storm 8,000 residents were housed in Red Cross shelters. Not only were sugar cane plantations flattened but the macadamia nut industry was no more. The water system on Kauai was totally ruined, and the only drinkable water came from the military importations. Broken palm fronds and thousands of roof tiles carpeted the ground.

A brick smokestack at the Lihue Plantation, the island's largest sugar refinery, was snapped in half. Kauai's mayor, Jo Anne Yukimura, described the damage to the Na Pali coast: "It's like the coast aged a century in one day," she said.

Three hundred troops assisted in the relief operations, distributing generators to hospitals and police stations.

Governor John Walhee III called Hurricane Iniki "probably the worst disaster we've had in the state of Hawaii." It was considerably worse than the damage caused by Hurricane Iwa, 10 years before, when the population had been half of its present number and its tourist industry had been merely nascent. The final damage figure for Iniki totaled $1 billion.

Ninety-eight people were injured; there were three deaths, including a 91-year-old woman whose house collapsed on her, and a 76-year-old man who was killed by flying debris.

"Ginger, eggplant, cabbage, lettuce, carrots, papaya, bananas, tangerines, onions, corn, orchids, you name it, I had it," said Ciceilio Decay, a 68-year-old farmer. "It's all gone," he concluded, "But the people are still here . . . We've just got to start all over again."

UNITED STATES
LOUISIANA
August 19, 1812

Fatality figures were not kept, but hundreds were known to have died in one of the first hurricanes to be recorded in Louisiana, on August 19, 1812.

One of the first hurricanes to be recorded in Louisiana slammed into the shore area west of the Mississippi River and followed its entire length on August 19, 1812.

Its first impact flattened buildings in New Orleans, turned streets into streams, and collapsed the historic Market House with its supposedly indestructible 24-inch- (60 cm) diameter columns.

Settlers along the river were routed from their homes and fields or drowned in them. Fifty-three ships either anchored in or sailing upon the Mississippi were, as one writer of the time put it, "crushed to atoms." Others sank with their crews aboard.

Farther up the river, at Fort St. Philip, a contingent of American volunteers was billeted in anticipation of a British attack. Flood waters roared through the fort, and the wind shattered its buildings, all except the barracks, which were fashioned from thick logs. Some survived here but 45 others perished, adding to the unrecorded fatalities that occurred from New Orleans to Fort St. Philip.

UNITED STATES
LOUISIANA
October 1, 1893

Two thousand people were killed by a small, forceful hurricane that struck Louisiana on October 1, 1893.

Small hurricanes, if wound up tight, can cause as much damage and loss of life as more grandiose and far reaching ones. Such was the compact hurricane that caromed into the coast of Louisiana between Port Eads and New Orleans on October 1, 1893.

The storm, unannounced by the primitive weather forecasting methods of the time, pushed a 12-foot (3.6 m) high storm surge ahead of it, and it was this that drowned, according to one report, 2,000 people working the fields and the fishing boats of that rich area of Louisiana.

Boats that had not been made fast were swept out to sea with their crews aboard; the fishing shacks and curing sheds that lined the docks were swept away in minutes, or crushed in seconds.

The storm swept inward, then outward again toward the coast, laying waste to millions of dollars worth of property throughout the South Atlantic states before it finally headed out to sea and self-extinction.

UNITED STATES
LOUISIANA (AND MISSISSIPPI)
September 10–20, 1909

Three hundred fifty people died in the hurricane that hit Louisiana and Mississippi in September 1909.

A giant hurricane, that developed off the Leeward Islands, traversed the Caribbean and sideswiped Cuba. Ten days later it slammed into the Louisiana coast 50 miles (80.5 km) west of New Orleans, crumbling carefully constructed dikes at the mouth of the Mississippi River with a moderate storm surge.

The winds of the storm were not enormously strong in comparison to other storms that had already hit the region (see two previous entries). But combined with high tides in the Gulf, the storm swelled the Mississippi a full three feet (0.9 m) above the high-water mark.

Lake Borgne, located outside of New Orleans, overflowed its banks and united with the onrushing waves from the Gulf of Mexico. All the wire supplying electricity to the city and carrying communication both in and out of it was swept away.

The large storm continued on, devastating coastal Mississippi and parts of Florida. At Pensacola, ships safely moored in the harbor were overturned by the storm surge and scores of people, trapped in such supposedly secure vessels as the steamer *Romanoff,* perished. Scores of small towns simply disappeared along with their populations as the storm swept over them.

In Louisiana and Mississippi, 350 people were killed and over $5 million in property damage was sustained by this long-lasting and extensive hurricane.

UNITED STATES
LOUISIANA
LAST ISLAND
August 13, 1856

The entire population of Last Island, Louisiana—several hundred people—drowned when a huge hurricane submerged the island on August 13, 1856.

Last Island, off the Louisiana coast, was pelted with nearly 14 inches (35.6 cm) of rain during the week preceding the arrival of the colossal hurricane that would eventually destroy it.

These meteorologically correct conditions set up a classic scenario for disaster on August 13, 1856. At approximately 2 P.M. on that day, huge waves produced by the storm surge began to pound the island. Residents were blinded by both the rain and wind-driven sand. Those who chose to row out to the steamer *Star,* which was anchored in the bay, would survive, although by the end of the storm the ship would be stripped back until only her hull and boilers remained.

Those who decided to climb to the highest point on the island were less lucky, for within hours, the entire island would be submerged, the waters of the Gulf of Mexico closing over it like a green and murky blanket. Only the nearly destroyed ship with its survivors could attest to the island's existence.

UNITED STATES
MISSISSIPPI
HURRICANE ELENA
August 30–September 2, 1985

Although there were no fatalities, over 1 million people were either evacuated or affected by Hurricane Elena, which wandered through Louisiana, Alabama,

Mississippi, and central and northern Florida from August 30 to September 2, 1985.

Like a drunken sailor, Hurricane Elena wove and dipped around the Gulf of Mexico from August 30 to September 2, 1985, threatening New Orleans, Alabama, Mississippi, and central and northern Florida.

On August 30, it appeared that the 300-mile (482.8-km) wide hurricane would come ashore somewhere near Pensacola, Florida, and some 125,000 people were evacuated from Louisiana and Florida coastal areas, while oil companies airlifted 20,000 people from their offshore oil rigs. Governor Edwin Edwards of Louisiana declared a state of emergency for 14 parishes in Louisiana, and 15,000 people in Alabama evacuated their homes, condominiums, and mobile homes. The National Guard moved into the Florida panhandle.

But instead of hitting the coast where it logically should have, Elena stalled and churned over the warm Gulf waters about 50 miles (80.5 km) west-southwest of Cedar Key in central Florida. Rains and tidal surges eight to 12 feet high battered the coast on August 30 and 31, and high tides closed the three main bridges crossing Tampa Bay. The storm was behaving like 1950's Hurricane Easy, which had lingered off Florida's Gulf coast, dumping 38 inches (96.52 cm) of rain in 24 hours—a national record.

At midday on August 31, two tornadoes spun off the main storm. One hit the Kennedy Space Center at Cape Canaveral, causing no damage. The second injured seven people when it slammed into a mobile home community near Leesburg, destroying 54 trailers and damaging 83 others.

On August 31, Elena intensified, doubled back, and once again threatened the Gulf Coast east of New Orleans with winds of up to 125 MPH. By changing its course, the hurricane, which had been snaking and bobbing across the Gulf of Mexico since the previous Thursday, now menaced many of the same people who had been given an all clear late on Friday. Hundreds of thousands of people were again ordered to flee inland.

Still remaining offshore, the storm rolled up the coast of northwest Florida, battering seaside towns from Apalachicola to Fort Walton Beach. As the eye passed near Panama City, winds on the fringe of the storm caused power failures as far away as Pensacola. At Port St. Joe, Bill Eagle, a police dispatcher, said he had just heard from a neighbor. "Bill, your roof just blew right by my house," the neighbor reported. There was a short pause. "Bill," the neighbor continued, "your pump house just whipped by, too."

Now, its upper-air currents apparently diminishing, the storm wobbled towards New Orleans, simultaneously climbing on the Saffir-Simpson scale to a grade 3 hurricane.

But again, the predictors were wrong. On September 1, instead of making landfall in Louisiana, the storm came ashore farther north, near Biloxi, Mississippi. The storm's winds, which exceeded 110 MPH, peeled roofs off homes, toppled trees and power lines, and demolished structures in Biloxi and Gulfport.

Dauphine Island, a resort community southwest of Mobile, Alabama, was particularly hard hit. A 12-foot (3.8-m) tidal surge crashed over the west end of the island, sweeping away 50 homes, uprooting trees, and floating debris at lethal speeds inland.

In Gulfport, a tornado ripped into a storm shelter, injuring one person and forcing another 380 to flee into the night. A fire in the same city destroyed an apartment complex, leaving 130 people homeless. The roof of a center for the elderly in Biloxi was collapsed by a tornado, forcing paramedics to crawl through the rubble to rescue an estimated 200 people.

All in all, over a million residents of the wide area either hit or threatened by Elena had their lives and often their living spaces disrupted by the storm. The ultimate loss in homes and agriculture would rise to $543.3 million, making it the costliest storm on record until that time.

UNITED STATES
NEW ENGLAND
August 14, 1773

There are no casualty statistics for the hurricane that cut a wide path through New England on August 14, 1773. Widespread damage is known to have resulted from the storm.

Cows grazed on the Boston Common in 1773, but they were sent home early on the night of August 14, as a slashing rain began to whip across the salt marshes and the 2,000-foot (609.6-m) length of Long Wharf. All up and down the New England coast, the wind had been rising, and by nightfall, waves were breaking over the docks and smashing into warehouses and ropewalks.

In Charleston, waves cracked the wharves as they smashed down upon them. Bunker and Breed's hills were eroded and Cambridge was flooded. Houses were shuttered and safe in Boston itself, but the farmhouses along the Merrimack River were ripped from their foundations as the river overflowed its banks. Houses, barns, and windmills collapsed, and apple orchards were stripped and broken, cornfields turned into vistas of mud, and cattle drowned or felled by flying planks.

Salisbury Point, Amesbury, and Haverhill were ripped by tornadoes that grew along the edge of the storm, and ships in Boston and a score of other harbors along the coast were torn from their moorings and thrown up against the rocks of the shore, where they burst apart or simply capsized, often with all hands aboard.

The hardy colonists started rebuilding as soon as the sun rose on August 15. No casualty statistics survive. There are only accounts of wagons rumbling in from Roxbury and Dorchester to the maritime settlements, laden with supplies and men to repair homes, raise sunken ships, and get the thriving trade with the Indies and Africa back in business. These efforts, however, were shortlived; a year later, King George would shut down the port of Boston, with even greater effect than the hurricane of 1773.

UNITED STATES
NEW ENGLAND
September 1815

Eighty-five vessels were sunk in the "September equinoctial gale of 1815," which was obviously a hurricane. No casualty records were kept, but it is safe to assume that many drowned aboard the sunken ships.

Originating near the island of St. Bart's, the "September equinoctial gale of 1815," as it was called, slammed into Providence, Rhode Island, in September, pushing a storm surge ahead of it that raised the tides in Narragansett Bay 12 feet (3.65 m) above normal, and swept away the bridge that connected the two parts of Providence. Thirty-five vessels, anchored at the head of the bay, were completely demolished. Water rose to the tops of second-story windows in some houses; men from a wrecked brig on the river crawled over floating debris until people taking refuge on the roofs of nearby houses tossed them lines and pulled them to improvised safety.

Sixty vessels went to the bottom of Boston Harbor. Roofs were ripped from houses in Cambridgeport, Salem, Gloucester, Reading, Newburyport, Danvers, Saugus, even as far north as Wells, in Maine.

Point Judith light was destroyed. In Connecticut, New London, Groton, Norwich, and Stonington suffered wind and water damage; sea water was picked up by the roaring winds and flung more than 40 miles (12.19 km) inland, scalding the sides of buildings and trees with salt.

No casualty figures were kept, but the lesson of the storm prodded the residents of much of New England, and especially Providence, to burn and replace the wrecks of dilapidated wooden warehouses with brick ones. In fact, as a result of the storm, the entire architecture of the city underwent a generous, noticeable facelift.

UNITED STATES
NEW ENGLAND
October 3, 1841

Hundreds of fishermen drowned in the hurricane that struck New England on October 3, 1841.

There was no warning of the storm that whistled and screeched all night on the third of October 1841, felling trees, smashing windows, collapsing walls and buildings. In the harbors, boats simply disappeared. The next morning, it was as if winter had come early. Trees were stripped or felled in meadows white with salt. Every village and farm was piled high with debris and dead livestock.

But the worst and most tragic damage was done to the Cape Cod fishing fleet, which had headed out on October 1 for the Georges Bank, where the cod were running. Between 40 and 50 vessels foundered and split apart in the roiling sea whipped by the storm. They washed up on the beaches from Chatham to the Highlands the next morning. Gloucester, Dennis, and Hyannis Port all lost some of their finest schooners and their proudest men. At Cape Ann, 14 of the fishing fleet were washed ashore at Pigeon Cove, where fish houses, fish flakes, 60 barrels of mackerel, 200 barrels of salt, 300 empty barrels, and a catch that would have brought $50,000 to its fishermen and their families were completely wiped out by the storm.

These were the bald and horrible facts. Probably from the safety of an inner-city shelter, Oliver Wendell Holmes immortalized the storm in verse that hardly plumbed its terrible depths:

Lord how the ponds and rivers boiled
   And how the shingles rattled
And oaks were scattered on the ground
   As if the Titans battled
And all above was in a howl,
   And all below a clatter,—
The earth was like a frying pan
   Or some such hissing matter.
I lost—oh, bitterly I wept
I lost my Sunday britches.

Tragedy, as well as beauty, sometimes seems to rest in the eye of the beholder.

UNITED STATES
NEW ENGLAND
September 21, 1938

. . . .

(See HURRICANES, United States, New York [Long Island] and New England.)

UNITED STATES
NEW ENGLAND
September 8–16, 1944

. . . .

Three hundred eighty-nine people were killed in the New England hurricane of September 8–16, 1944, which followed a unique inland path from North Carolina northward.

Largely following the path of the September 21, 1938, hurricane (see HURRICANES, New York [Long Island] and New England), a monster storm, born off the West Indies, would ultimately claim 389 lives and cause $50 million in damage along a 500-mile front.

After bringing slight damage to Puerto Rico, the Bahamas, and the Florida Keys, the storm swerved inland at North Carolina. At this point, Colonel Lloyd B. Wood, deputy chief of the Army Air Corps' Weather Division, took off from Washington, D.C., piloting an A-20 Havoc light bomber. Colonel Wood flew straight through the incredible 140-MPH winds of the storm to its eye. His reports made the killer hurricane one of the best charted storms in history up to that time, thus keeping the death toll down.

Despite the fact that it had made landfall, the hurricane maintained powerful winds and barreled northward across the Carolinas, Maryland, Delaware, and then into New Jersey. At Sea Isle, New Jersey, only 300 of the resort's 700 dwellings escaped without severe damage. In Asbury Park, 200 feet (60.9 m) of that city's municipal fishing pier was torn up and washed away. In Atlantic City, the famous boardwalk was ripped apart.

By September 14, the storm, now packing winds of 95 MPH, slammed into New York City, skipping the villages of eastern Long Island that had been so severely devastated by the 1938 storm. Skyscrapers swayed and pedestrians were whipped by stinging rain and flying debris, but the metropolis remained fairly undamaged.

Not so New England, which felt the storm's final fury. Tens of thousands of ancient, towering trees were toppled, bringing down electrical lines—which in some cases caused fires—and wiping out telephone service for more than 300,000 residents. Bridges crumbled,

waterfront houses were washed away in a moment, salt spray whitened the land along the coast, but few people lost their lives. Most of the fatalities in New England took place at sea, where scores of vessels, both military and civilian, foundered with their crews aboard.

After covering a total area of close to 1,000 square miles (1,609 sq. km), the storm finally blew itself out off the coast of Newfoundland.

UNITED STATES
NEW ENGLAND
HURRICANE CAROL
August 26–31, 1954

. . . .

Five thousand tourists and fishermen were stranded on the tip of Long Island by Hurricane Carol, in late August 1954. Forty-five people died as the storm continued on to ravage New England.

Born in the so-called horse latitudes well east of Jacksonville, Florida, Hurricane Carol made an inauspicious entrance into the world on August 26, 1954. If a convergence of forces—a Bermuda high and a western high pressure area that arrived from the Rockies—hadn't created a sluiceway up the Atlantic cost, Carol might have remained an inconsequential storm.

But such was not the case. Unlike the 1944 hurricane (see previous entry), the Weather Bureau failed to issue hurricane warnings to the crowds enjoying Labor Day weekend on the Long Island beaches. Fortunately, Carol struck Long Island at low, rather than high tide. No huge tidal wave hit, but the storm surge was strong enough to strand 5,000 tourists and fishermen at Montauk, at Long Island's tip, where the surge scoured out roads and railroad tracks, isolating that community.

By the time the storm hit the coast of Connecticut, the Weather Bureau had still not issued hurricane warnings. The storm crest hit Narragansett Bay at high tide, smashing into Misquamicutt Beach and sweeping away 200 homes.

Pushed by 120-MPH winds, the surf exploded against the beaches of Block Island, Rhode Island, and Martha's Vineyard, Massachusetts, filling freshwater ponds with salt water. On the island of Martha's Vineyard, Vineyard Haven, West Chop, Edgartown, and particularly Menemsha were flattened by the winds and buried under piles of wreckage.

In Newport, Rhode Island, the venerable casino collapsed. Travelers in cars were swept off coastal highways by sudden waves. In South Kingston, eight carloads of people who had fled from their summer cottages were stranded, and many drowned. A 35-foot wave overwhelmed Westerly and Watch Hill, Rhode Island. The center of Providence was flooded up to the 1938 high water mark.

In Massachusetts, New Bedford harbor was inundated by waves, broken boats, docks, yachts, draggers, sailboats, trawlers, tugs, and houses. Fishermen were washed off boats and drowned. A fire in Wareham was fought by firemen struggling through chin-high water.

Finally, hurricane warnings filtered through from the Weather Bureau, just as the wooden spire of Boston's Old North Church cracked, shuddered, and fell. It had fallen once before, exactly 150 years ago in the New England hurricane of 1804.

Worcester County suffered severe crop damage, and in the city of Worcester, one Harry R. Davis was blown to his death from the 10th story of a downtown office building when he opened a door on a fire escape.

Maine suffered tree, boat, and dock damage, as, unannounced to its inhabitants, Carol swept northward into Canada, where it would blow itself out. Behind her, she would leave 45 dead and millions of dollars in property damage.

UNITED STATES
NEW YORK (LONG ISLAND) AND NEW ENGLAND
September 21, 1938

Four hundred ninety-four people died, over 100 were missing, and 1,754 were injured in the September 21, 1938, hurricane that hit Long Island, New York, and ultimately blew itself out over New England.

A normally tree-lined country road in Westhampton Beach, Long Island, was turned into a lumberyard after the hurricane of September 21, 1938. The wreckage consisted of the remains of once palatial beach houses smashed to splinters by a 20-foot (6.1-m) storm surge. (Quogue Historical Society/Pat Shuttleworth)

The Long Island hurricane of September 21, 1938, was born some 600 miles (965.6 km) northeast of Puerto Rico (called Porto Rico in 1938) four days before, on September 17, traveling rapidly along the prevailing trade winds, west-northwest. Two days later, it was headed on a collision course with Miami, and hurricane warnings were posted for the Florida coast.

The storm then veered northward, and by the night of September 20, was 400 miles (643.7 km) due east of Jacksonville and picking up speed.

The conditions for this hurricane to make landfall over Long Island and New England were just right. A long tongue of warm, moist air hung in a line from the Carolinas to Long Island, where it had been raining heavily for two days. But no such storm had hit Long Island since October 6, 1849—nearly a hundred years before.

That morning, fishing boats went out, as usual. The weather forecast was for showers. But the situation deteriorated rapidly during the day, and when the National Weather Bureau finally issued gale warnings, radio stations were giving priority to a speech by Adolf Hitler being shortwaved from Europe.

When the storm inched its way into Long Island, around noon, few people believed it was more than an autumn gale. But by mid-afternoon, when it slammed full force into Westhampton Beach, where the eye would pass, tide levels had risen 18 feet (5.48 m) above normal, and waves measured another 12 feet (3.65 m). Telephone poles were picked up like match sticks and became flying projectiles. A tidal wave, estimated to be 20 feet (6.09 m) in height, broke over the dunes and swept away hundreds of beachfront homes. Water rushed over the mainland, as survivors floated miles inland on pieces of houses, parts of boats, and even oil and water tanks. Steeples were stripped from the tops of churches; the 100-year-old cupola from the top of the Presbyterian Church in Sag Harbor was lopped off as if a sword had sliced through the tower.

By evening, the storm had blown itself out over the mountains of Vermont, where, a week later, records from the town hall in Southampton were found strewn across the New England countryside. State troopers were brought in to restore order in the devastated vacation land of the south shore of Long Island. Bridges that

The storm surge of the Long Island/New England hurricane inundated roads a mile and a half inland on the south shore of Long Island. (Quogue Historical Society/Pat Shuttleworth)

Main Street, Westhampton Beach, Long Island, during the 1938 Long Island/New England hurricane (Quogue Historical Society/ Pat Shuttleworth)

connected the mainland to the barrier beach no longer existed. Looters had already appeared and by midnight, patrols of local citizens, armed with clubs, walked the battered beaches and darkened towns. Country clubs were turned into temporary morgues; school buildings became combination hospitals and shelters.

It would be months before the silt-covered villages and towns would recover and years before the multimillion-dollar reconstruction would approach completion.

UNITED STATES
NORTH AND SOUTH CAROLINA
September 16–17, 1713

Seventy people died in the onslaught of a furious hurricane that hit North Carolina on September 16, 1713.

In the sparsely colonized area of the coast of the Carolinas at the beginning of the 18th century, a fatality figure of 70 was impressive and calamitous. That it was brought about by a hurricane whose winds reached 100 MPH makes this early storm notable for its viciousness and toll.

Charleston, Cape Fear, and Port Royal were the hardest hit areas, although the storm did slice over three miles inland, across the salt marshes surrounding these settlements.

One of the first casualties of the storm was the 80-foot- (24-m-) high lighthouse on Sullivan's Island. Able to withstand the screaming winds during the earlier parts of the storm, it was no match for its final, accumulated fury, and after swaying precipitously, it finally snapped in two like a dried branch.

Seaside cottages were undermined and swept out to sea with their inhabitants, who thought—as most did in those times—that home was the safest haven. Most drownings, in fact, took place in this way. Farmers in their fields, pedestrians on the streets of Charleston, and crews aboard the schooners and sloops that were driven inland or capsized in the harbors also drowned.

UNITED STATES
NORTH AND SOUTH CAROLINA
HURRICANE FRAN
September 6, 1996

Twenty-two people died, hundreds were injured, and thousands were made homeless by Hurricane Fran, which hit the North and South Carolina coasts on September 6, 1996, with 120 MPH winds. Possessed of a 25-mile-wide eye and measuring 100 miles (160.9 km) in width, Fran cut a wide swath of destruction through the Carolinas and into Virginia, where intense floods were caused by the storm's wind-driven rainfall.

The coasts of North and South Carolina had come in for their share of battering from Hurricane Bertha in the summer of 1996. The damage had been moderate, but psychologically, the people were not ready for the arrival of a major hit in the same season. And then there it was: Hurricane Fran, arriving on the night of September 6, 1996.

A category 4 storm, with winds up to 120 MPH, Fran slammed into Cape Fear, North Carolina, just before 7 P.M. on the 6th. An hour later, its eye passed over the coast. A little before 10 P.M., the storm reached Wilmington, North Carolina.

The most unique characteristic of Fran was its girth. With a 25-mile (40-km) wide eye, its overall width measured over 100 miles (160 km), thus spreading its destruction over a huge area of North and South Carolina. Along the coast, several rivers had their courses reversed, flowing upstream as a 13-foot (3.9-m) storm surge changed nature's normality. Destruction on the coast was monumental.

The North Carolina barrier island that includes Topsail Beach, Surf City, and North Topsail Beach was hit hardest. Hundreds of houses and hotels were crushed or lifted off their stilts and deposited elsewhere. Roofs were stripped from waterside condos, leaving them open to the elements.

A 10-block stretch of Kure Beach, North Carolina, was flooded by the storm surge, as was the north end of Carolina Beach, where eight feet of water remained in the streets and two to three feet of water stayed in many houses and businesses long after Fran had roared through.

In Carolina Beach, automobiles were turned into driverless boats, floating past the Breakers, a condominium complex. A group of residents of the Breakers who had opted to ride out the storm called 911 in terror, reporting that the building was collapsing. The noise was later discovered to have been made by floating cars slamming into the building.

Authorities evacuated villages and towns in the storm's path as it moved toward the border between South and North Carolina. Highways were clogged; shelters in local schools filled quickly with local residents and tourists. "We're going to need a vacation from our vacation," a tourist remarked as he entered a shelter.

As Fran moved inland, the winds dropped to a maximum of 80 MPH—still enough to cause considerable damage. On it trekked, through Fayetteville, North Carolina, at 1 A.M. on the 7th, spinning off tornadoes and soaking the land in its path with five to 10 inches (12.7–25.4 cm) of heavy, wind-driven rain.

Church steeples and roofs were blown off in Wilmington, North Carolina, and tens of thousands lost their power as soon as the storm neared.

As Fran finally crossed the Virginia border, it diminished to a tropical storm. But its destructive powers, while diminished, remained. Creeks and rivers in the Blue Ridge Mountains and the Shenandoah Valley flooded their banks. Ten thousand people were evacuated in Rockingham County, near the north fork of the Shenandoah River. Over 300,000 people were without power on the day following the storm in Virginia. In North Carolina, 1.5 million were without light, heat, or water.

Rescue and maintenance workers turned their attention and efforts to assessing and cleaning up the destruction. Raleigh, North Carolina's state capital, suffered some flood damage, and the tobacco and vegetable fields of eastern North Carolina received immense damage.

In Kenansville, on State Road 24 from Fayetteville east, pine trees, pecan trees, and 200-year-old oaks littered roads, while highways were shrunk into roads less than the width of a small car.

But it was the crops that the populace mourned most. Immature corn and soybean crops suffered markedly. "We had the prettiest crop," Violette Phillips, who owned a farm in Kenansville told a reporter. "My son Paul said early on this was just too pretty."

Along the coast, thousands were rendered homeless. Municipal water systems in the Carolinas and Virginia became contaminated, and U.S. Army units brought in drinkable water and generators. From North Carolina through Virginia, rivers continued to crest in floods, isolating hundreds.

But it was on the coast of the Carolinas that the full force of the hurricane's fury was most felt, and where lives were most disarranged. If the strength of the storm stirred up the sand on the sea floor off the coast of Long Island, New York, enough to bury valuable evidence in the wreckage of TWA flight 800, its strength around its eye was manifold. A huge number of homes and tourist attractions on the Carolina coast would not be rebuilt for over a year.

Inland, near Raleigh, homeowners began the massive job of cleaning and patching up immediately. One, Joe Daniels from the suburb of Cary, told a reporter, "We were lucky. We don't have anything compared to a lot of people. And—" he paused, "Thank God for insurance."

UNITED STATES
PUERTO RICO,
NORTH AND SOUTH CAROLINA
HURRICANE HUGO
September 17–23, 1989

Seventy-one people died and 50,000 were made home-less by Hurricane Hugo, the 10th-strongest storm to hit the United States in the 20th century. Its path of destruction extended over 1,500 miles (2,414 km) in five days, as the storm laid waste to seven islands in the Caribbean, then smashed into the mainland at Charleston, South Carolina.

Hurricane Hugo was a devastating traveler. From the first warnings that were issued on September 17, 1989, for a hurricane headed toward Guadeloupe, to September 21, when it hit Charleston, South Carolina, full force, Hugo laid waste with a fearsome 125 MPH power to Antigua and Barbuda, Guadeloupe, Dominica, Montserrat, the U.S. and British Virgin Islands, Puerto Rico, and North and South Carolina. And yet, because of early warnings, the death toll was astonishingly low for the 10th most intense hurricane to hit the United States in the 20th century—a dangerous, category 4 storm that was the most intense since Camille in 1969 (see p. 311).

Hugo began its 350 mile (563 km) trek through the Caribbean on September 17, when it hurtled, with 140 MPH winds, into the tiny British-owned island of Montserrat. A formerly verdant, tree-covered volcanic island, Montserrat was stripped of its greenery and most of its structures. Ninety-nine percent of its 12,000 residents were instantly made homeless. The island's only hospital was destroyed, as was its police station and two of its schools. The control tower at Pointe-à-Pitre was demolished and power and communication with the outside world were simultaneously extinguished. No buildings escaped damage, and hundreds were injured, but only nine people died.

On nearby Guadeloupe, five were killed and 3,000 were left homeless when Hugo slashed through the island, toppling telephone poles and severing all power lines on the island, and flattening the resort town of St. Francois. Thirty percent of the island's roads were made impassable.

A 20-foot (6.09-m) high storm surge washed over Nevis, inundating its coastal settlements. Wind and water demolished the homes of 99 percent of the island's population, and four died.

Thirty percent of the 100,000 people on Antigua were rendered homeless and the main business area in the capital of St. John's was flooded.

By September 18, Hugo's winds had diminished to 115 MPH, and it had ripped across St. Croix and St. Thomas, smashing residences and businesses, injuring scores of residents, but killing none. On St. Thomas, 80 percent of all structures were destroyed or damaged.

Dominica, the British island, lost almost all of its vital banana crop.

Now, Hugo turned its wobbly course west-north-west, and headed straight for Puerto Rico. Early in the morning of the 18th, containing winds of 120 MPH, it collided with the eastern coast of the island, blowing out scores of windows in high-rise apartment and office buildings, decapitating palm trees, collapsing the fragile homes of the poor. Over land, Hugo's winds steadily diminished, but never dropped below 75 MPH, with gusts up to 110 MPH.

It was San Juan's turn at mid-morning of Monday, the 18th. Ocean water, whipped by the wind, broke over sea walls. Telephone poles tumbled, and the power to the main city flickered and went out.

The night before, tourists in the major hotels, on vacation and therefore largely immune to reality, greeted the coming of the storm festively. But that was when the lights and the air conditioning were functioning. By dawn on the 18th, the tourists were considerably subdued. Guests in the Caribe Hilton were roused from their rooms and herded into the main ballroom, whose windows had been boarded up.

But halfway through the morning, a huge gust of wind caved in the plywood covering the ballroom's plate glass windows, shattering the windows and spraying glass on the terrified tourists. "It was like a crystal wind," 29-year-old John Kim told a reporter, as hotel employees wrapped the frightened guests in blankets to shield them from the glass that was still detaching itself in shards from its metal frames.

Led into the restaurant, which was protected by metal sheeting, the hotel guests hunkered down on the floor while the wind roared and whistled around the trembling hotel.

When the storm passed, they returned to a scene of sobering damage and devastation. Part of the lobby ceiling was scattered in pieces on the floor, a shattered chandelier lay in the middle of it, and water cascaded through holes in the lobby's walls and windows.

Outside, 10,000 residents of the island had been deprived of their homes by the storm; 25,000 remained in the shelters set up by the Red Cross. Coast Guard helicopter pilots reported that 80 percent of the roofs were blown off the houses between San Juan and Sajardo, 25 miles to the east. The same figures came from the offshore islands of Culebra and Vieques.

Shantytowns at the base of a steep hill in Old San Juan were wiped out, and the mansions and high rises in fashionable Condado Beach suffered the same amount of damage. Looting began in its expensive stores. Some fire hydrants were labeled as sources of drinking water, but the water emerging from them was brown. When power was restored, officials warned that all water should be boiled before drinking.

In the fields of the island, devastation was dramatic. Eighty percent of the coffee crop, located around Utuado, in the center of the island, was ruined. Virtually the entire ornamental plant industry, which supplied the mainland with poinsettias and orchids, was wiped out.

Roads were flooded or made impassable by fallen trees. In Humacao, the third, fourth, fifth, and sixth stories of the local hospital were reduced to rubble. Alert staff members had moved all of the patients on these floors to the first two before the damage began.

The San Juan airports were inoperable, their control devices destroyed by the storm, which was once again gathering strength as it now headed for the southeastern coast of the United States.

Meanwhile, on St. Croix, terror and anarchy reigned. The St. Croix prison was damaged enough to let all 220 of its inmates loose. Armed gangs moved through the streets, shooting in the air, terrorizing residents, smashing what windows still existed on store fronts, and looting at will. Policemen and National Guardsmen, who should have been maintaining order, joined in the looting. "We saw a National Guard truck filled to capacity with all kinds of stuff in it," one rattled resident told an Associated Press reporter later.

On Thursday, September 21, President George H. W. Bush ordered over 1,000 military police to St. Croix and St. Thomas to restore order. They were met by a crowd of hysterical tourists, crying, "Please help us! They're looting. We've seen police looting. We've seen National Guard looting. There's no law and order here!"

"You should have seen them," oil refinery worker Hector Vidaz told a *New York Times* reporter. "Fifty or 60 of them. They smashed the windows of the sportswear shop, then started passing out T-shirts and boxes of sneakers. Here was 50 people sitting in the street, trying on sneakers and passing them around."

"I saw National Guardsmen reaching into the windows and handing jewelry and stuff to people," another man related.

Other activity was more dangerous, and threatening to white tourists particularly. St. Croix's population was three-quarters black, and crowds of black residents surrounded terrified tourists. "They followed us down the road shouting, 'Whitey go home!'" one tourist told a reporter.

Gary Williams, a San Juan newspaper reporter, told the AP that he flew in a helicopter over the Sunny Isle shopping center in Christiansted (St. Croix), and reported that he saw at least 1,000 people in the parking lot, walking in and out of shattered shops with their arms loaded with loot. "We saw three men and a woman walking out with garbage bags loaded with stuff," he recalled.

The looting continued even after the soldiers arrived. Grandmothers with toddlers, pregnant women, and men of all ages carried away paint, hardware, slide projectors, and other items overlooked in the earlier rounds of pilfering.

Some looters were hungry people, picking over groceries that were about to rot anyway. They openly carried their contraband through streets tangled with snapped utility poles and wires. The air was permeated by the stench of rotting debris, and the sun shone mercilessly through trees bereft of leaves.

Chaos of a different sort was present on other islands that had been raked by Hugo earlier as relief workers reached them. Montserrat was a horrific mess. Because the jetty for receiving ships had disappeared, crates of generators, boxes of water purification chemicals, and medicines had to be brought in by landing craft. Sailors and marines came ashore from the frigate *Alacrity*, anchored off shore, set up a bivouac in a soccer field, and began to clear the island's roads of debris.

Though the airport was demolished, its runways were miraculously saved. The Royal Air Force thus was able to land its C-130 cargo planes, packed with supplies.

Still, coordination became difficult. One woman from Tallahassee, Florida, arrived on the island to see her mother with a golf bag full of purification tablets. "Every conceivable thing you can think of has been touched by damage," one worker said, in despair, but another, riding in a truck, waved cheerfully at reporters. "We all carpenters now, mon," he shouted.

Meanwhile, Hugo was gathering deadly and increasing force. Having caused 26 deaths and deprived 50,000 people of their homes in its five-day, 1,500-mile (2,414-km) trip through the islands, it now plunged northeastward, toward the mainland. Coastal towns from Georgia through the Carolinas were evacuated. Over half a million people boarded up their beach houses and fled, as a storm surge 12 to 17 (3.6 to 5.2 m) feet above normal crashed into their towns.

Shortly before midnight on Thursday, September 21, Hugo came ashore, at the worst possible moment for Charleston, South Carolina. The extreme low pressure in its eye created a storm surge 12 to 17 feet (3.6 to 5.2 m) high, just before high tide. Powered by 135 MPH winds, the full force of the storm's wind, rain, and storm surge hit the city dead center.

"My ego is the only thing bigger than Hugo," bragged a cocky radio announcer, just before his station lapsed into sustained silence. All electricity was abruptly cut off in the city as utility poles were torn from the earth and turned into lethal, hurtling weapons. One smashed through the plate glass window of a prominent shoe store; others destroyed benches, front porches, and cars.

A rancid odor of burnt powder hung over the city from fires caused by parting power lines and lightning strikes. A large, jagged crack was opened in the roof of the Charleston City Hall, a 199-year-old brick building that had weathered other hurricanes without damage. Now, deprived of part of its roof, it became a receptacle for torrents of rainwater, which poured into its paneled council chamber.

Over half of Charleston's 355,000 residents were moved inland that afternoon. By sunset, most emergency shelters were filled to capacity. Most of the remainder of the population, as adventurous or foolish as the radio announcer, regretted their decision to stay. The city was blanketed by darkness. In places, streets were ankle deep in glass shards. The air was salty and thick and smelled of fish. The only sounds were the sirens of the emergency vehicles, the rush of the wind, and the periodic pop of exploding windows.

Roofs were peeled back by the wind; water damage was immense; boats, once in the harbor (one of them 50 feet (15.2 m) long), came to rest on city streets. Mud was everywhere, five feet high in some buildings. A large percentage of the ancient trees that helped to give Charleston its peaceful charm were split as if by some giant wielding a monster axe. In all, 30 major downtown buildings were destroyed, and hundreds of buildings were rendered uninhabitable. Several people were rescued by National Guard troops from a collapsed condominium complex.

One family who stayed in their home in the historic district reported later that the fireplace exploded in the wind. A refrigerator, shoved up against the back door, was the only obstacle that kept the door from flying from its hinges.

Horrific damage was done to historic Fort Sumter, the fortification in Charleston Harbor where the first shots in the Civil War were fired. A 17-foot (5.2-m) wall of water pounded the compound, crushing and scattering its artifacts. It would take over a million dollars in repairs to restore it.

The damage to the Charleston Air Force Base was termed "catastrophic" by its commander. "We have on our hands," Charleston's mayor said, "a degree of physical destruction that is unprecedented in everyone's living memory."

By evening, Hugo was headed inland, and at 6 A.M., on September 22, it plowed into Charlotte, North Carolina, 200 miles (321.8 km) inland. Tamed a bit by land, its sustained winds were now 70 MPH, but constant gusts of 90 MPH toppled the city's magnolias and willow oaks, and drove three and a half inches of rain into shattered windows and through the ripped up roofs of houses.

On it proceeded, through the Appalachians of West Virginia, Ohio, and Pennsylvania, where it finally blew itself out.

And now, the cleanup and assessment began. In St. Croix, 60 prisoners were returned to the St. Croix Prison, 20 more were paroled, and 140 remained at large. Ten thousand homes, three-quarters of the dwelling places on the island, had been totally destroyed. There was no power on the island; 90 percent of its utility poles lay in splintered pieces on the ground.

Michael DeLorenzo, a part-time resident, returned to find his home totally destroyed, and wrote feelingly about the post-storm situation on the island: "No jobs. No money. Checks and credit cards are of no value. The banks are closed. No electricity and no estimate of how many months until it is restored. No telephone service. The only communication with the rest of the country is by short-wave radio.

"No running water. No toilet facilities. No schools. No garbage disposal. No hospitals. Scarcely a leaf or flower on the island. Bees are biting everyone because of the loss of flora and fauna."

In a part of the world in which there was 30 percent unemployment, the devastation was magnified.

On Vieques, off Puerto Rico, the situation was even worse. Unemployment there had been 40 percent before the storm; afterwards, it soared to 70 percent. Three thousand homes, nearly half the homes on the island, were rubble.

In Comerio, in central Puerto Rico, the $150 million in agricultural losses was personified by the poultry ranches, where tens of thousands of hens lay dead.

Once spectacularly green hills had become piles of dirt. The island's government estimated that 25,000 to 53,000 of Puerto Rico's 750,000 families were homeless. Shattered boats, the vehicles that allowed the livelihood of the island's fishermen, were everywhere. Some harbors looked like forests after a fire, with the masts of sunken boats congregated near the wreckage of the docks. In Culebra alone, 300 boats were lost. The total loss to businesses in Puerto Rico from Hugo would total over $1.4 billion. Four hundred million dollars of the $1.1 billion offered by Congress to Puerto Rico was immediately put to use making the water on the island safe to drink again.

And now, heavy rains returned to the Carolinas, making the cleanup that much more difficult. Water, food, money, and gasoline were in short supply. Many in Charleston wrote $3 checks for lunches at the few delicatessens that were open, while others waited on line at gas stations to pay $2.50 a gallon to fuel their cars.

On Sullivan Island and the Isle of Palms, the rain that followed the storm unleashed flooding and poisonous snakes. "I've never seen snakes as big as I've seen in the past few days, every time we pick up a board," said police officer Tom Buchanan to a reporter. In the islands and on the mainland, rain poured through gaps in plastic sheeting that had been drawn across the space once occupied by the roofs of homes.

When the weather cleared, a grim reality set in. There was no electricity for 750,000 people and 75,000 were in shelters. Charlotte, North Carolina, was in a state of emergency, patrolled by 350 National Guard troops. Schools were closed until October 2.

Everywhere in the storm's path, from Charleston to Myrtle Beach to towns a hundred miles from shore, homes from expensive restored Colonials to trailers and shacks were damaged beyond habitation.

On September 27, aid trucks began to reach these battered towns and cities. Middle schools were turned into diners; in Mount Pleasant, facing Charleston across the harbor, volunteers served survivors dinners of hamburgers, beans, and fruit. For weeks afterward, some of these refugees slept in the cabs of trucks or on mattresses in the help centers.

Some homes had seemingly just disappeared. Stephen Blanchard, the owner of a seaside restaurant, came back to Ocean Boulevard in Charleston, where the house he rented once stood. All that was left was a blue and white water heater and a few cinder blocks. The house itself was 120 feet (36.6 m) north, in a neighbor's backyard.

Supply trucks arrived, but dry places for them to unload their supplies were in short supply. "They give you enough for a day," one of the recipients told a reporter, while holding a wet box containing cereal, canned beans, and bottled water. "So you come here every day." She was accompanied by her three children, dressed in plastic garbage bags against the steadily falling rain.

The devastation extended to crops. Over $100 million in damage was done to overturned peach and pecan trees and washed out fields of cotton, tobacco, and vegetables. Miles of wooden triangles were formed by pine trees snapped halfway up the trunk, with their tops dug into the ground beside the trunks. Seventy-five percent of South Carolina's timber was felled, at a loss of $1 billion.

The weather worsened. Floods and a tornado added to the misery of the survivors and rescue workers.

As September gave way to October, shellfish harvesting, the livelihood of thousands of fishermen, was banned and warnings were issued against swimming in ocean water and rivers that were choked with untreated sewage. The plants that had treated this sewage were only partly operational; only 30 percent of pollutants were being removed from river water.

Ultimately, the death toll for Hugo was put at 71 people, (32 on the mainland). Some drowned, but many were electrocuted, some during the cleanup. The damage in South Carolina alone was set at $3.7 billion. Hugo destroyed 3,785 homes and 5,185 mobile homes. An additional 27,211 homes suffered major damage. Over 292,000 people were thrown out of work, and 230,937 households were forced to go on food stamps.

"We were battened down, really pretty well prepared," the governor of Montserrat told reporters after the storm had ravaged his island. "But you can't plan for a hurricane of maximum force. There's a rule of thumb that for every additional 20 miles (32.2 km) an hour, you double the damage. I've been through four hurricanes myself, and I've never seen anything like this."

UNITED STATES
SOUTH
HURRICANE AUDREY
June 27–30, 1957

Hurricane Audrey, the out-of-season hurricane that struck the southern United States on June 27, 1957, killed 534, injured thousands, and destroyed 40,000 homes.

Ordinarily, the hurricane season begins in August. Thus, on June 27, 1957, when a tropical depression in the Gulf of Mexico, 350 miles (563.3 km) southeast of Brownsville, Texas, unexpectedly developed into Audrey, the first hurricane of the season, residents of the Texas and Louisiana bayou country refused to believe early Weather Bureau warnings.

Such was not the case with residents of Galveston and Port Arthur, Texas. Hearing radio reports that the 105-MPH winds of Audrey had already claimed their first casualties—nine men killed when a 78-ton (70.76-metric ton) fishing boat was hurled into an oil rig—the inhabitants of these two cities evacuated their gulfside homes, and casualties there were relatively minimal.

After brushing these two Texas cities, Audrey flung itself toward the Louisiana coastal villages of Grand Chenier, Creole, and Cameron. The storm slammed into these thickly populated villages at 97 MPH, shoving a storm wave ahead of it that inundated streets, uprooted trees and houses, and sent automobiles whirling through streets.

Cameron, Louisiana, was the worst hit. Homes were demolished and strewn across the countryside. The entire village virtually disappeared beneath the roiling waters of the Gulf. A tidal wave consumed the entire countryside as if it had been swallowed whole. One doctor remained in the local medical center, tending to the injured, while his own family was swept away. His children were never found; his wife was discovered clinging to a piece of driftwood 20 miles (32.2 km) from Cameron.

When the storm finally blew itself out over Ohio and Pennsylvania, it had left behind 534 dead—mostly from Cameron—and 40,000 destroyed homes.

UNITED STATES
SOUTH
HURRICANE CAMILLE
August 17, 1969

The largest recorded storm ever to hit a heavily populated area of the United States, Hurricane Camille, with winds of close to 200 MPH, killed at least 360, injured thousands, and left tens of thousands homeless.

Hurricane Camille, described by Dr. Robert H. Simpson, then head of the National Hurricane Center at Miami, Florida, as ". . . the greatest recorded storm ever to hit a heavily populated area of the Western Hemisphere," began its life off the coast of Africa and was first sighted by hurricane watch planes on August 14, 1969, in the Caribbean, 480 miles (772.5 km) south of Miami. From then until its demise over Newfoundland four days later, the storm wreaked incredible havoc—in fact, causing more fatalities in an aftershock than it had in its initial assault upon Louisiana and Mississippi.

That more people weren't killed by this killer storm can largely be credited to the advanced methods of hurricane reporting. From the time it was first sighted, Camille was closely tracked. It brushed by the western tip of Cuba with no effect except accelerated winds. By Friday, August 15, it had entered the Gulf of Mexico, with gale-force winds spreading 150 miles (241.4 km) ahead of it.

On August 16, as pressure dropped within the storm and its winds rose, weather forecasters predicted landfall in southern Florida. But that night, the storm changed direction and headed towards New Orleans. An Air Force plane, entering the eye, clocked winds of nearly 200 MPH and a barometric reading of 26.61—the second lowest barometric pressure ever recorded. The lowest—26.35—was recorded in the eye of the great Labor Day hurricane of 1935 over Key West. (See HURRICANES, United States, Florida, Keys, 1935, p. 289.)

By noon the next day, Camille had aimed her fury directly at the bayou country of Mississippi and Louisiana. Two hundred thousand people boarded up their houses and headed for higher ground. The area just west of Gulfport, Mississippi, received the full force of Camille—winds of 190 MPH and a storm surge that sent awesome 25-foot (7.6-m) waves roaring through the tiny resort city of Pass Christian. That city's Trinity Church literally exploded in a gust of 200-MPH wind, burying 11 people who had taken shelter there.

Beach houses were swept out to sea. In one luxury condominium, the steel and concrete Richelieu Apartments, 12 people gathered for a hurricane party, ignoring the pleas of local police to evacuate the area. All 12 perished when the building collapsed on them. Slightly up the beach from the Richelieu, a motel was flattened when several oceangoing ships riding the crest of the 25-foot (7.6-m) waves plowed into it.

By 2 A.M., the wind had subsided, and local police, accompanied by a fleet of amphibious Navy DUCKS manned by Seabees plied the waist-deep waters, looking for survivors. Every house in town, they discovered, had been either destroyed or badly damaged. Bodies floated on the water; natural gas lines were smashed; electricity was gone; the water was contaminated. When the sun rose, hundreds of deadly cottonmouth snakes swarmed out of the swamps and into town, trying to escape the flooding.

In all, 137 people died and 500 were injured in the area in and around Pass Christian, which was soon put under martial law while nearly a million pounds of food, fresh water, and medical supplies poured in from as far away as Arizona.

But meanwhile, Camille was far from dead. For reasons not quite clear to meteorologists, the storm had either conserved some of her original cargo of rain or had in some manner acquired fresh quantities. Instead of blowing out over the landmass of Tennessee and Kentucky, she seemed to gather new power, and by the time she had inexplicably turned east and crossed the Blue Ridge Mountains into Nelson County, Virginia, she was still a menace. Thirty-one inches of rain fell during the night of August 19—more than 400 million gallons (15,141,647,201 l) of water within a period of

A wrecked house left amid the detritus from Hurricane Camille in Biloxi, Mississippi, in September 1969 (American Red Cross/Ted Garland)

four hours. Scientists afterward noted that rain of this magnitude occurs on an average of only once in a thousand years.

Now, mudslides were added to the devastation caused by Camille. At Holcomb Rock the James River rose from eight feet on August 18 to 47.5 feet (14.5 m) on August 19. Families, sleeping through what they thought was just another rainstorm, were buried under tons of mud. Nearly 200 people lost their lives in Nelson County, and $100 million worth of property was simply washed away. The Tye and Rockfish Rivers overflowed; trees were uprooted; bridges washed away and highways and railroad right of ways were undercut and disappeared.

After Virginia, Camille went out to sea and eventually blew herself out over the open waters of the Atlantic near Newfoundland.

UNITED STATES
SOUTH
HURRICANE JUAN
October 27–November 5, 1985

More than 60 people were killed, hundreds were injured, and thousands were made homeless by Hurricane Juan, when it struck the southern coast of the United States on October 27, 1985.

Hurricane Juan was a wanderer, causing as much damage after it was downgraded to a tropical storm as it did when it qualified as a full-fledged hurricane.

As a hurricane, with winds of up to 85 MPH, Juan first roared into the Gulf of Mexico on October 27, 1985, swamping boats and oil rigs and stranding some

1,400 people in Grand Isle, Louisiana, when they failed to evacuate over the causeway. By nightfall, the only power in town was at City Hall and the high school gymnasium, which, lit by a portable generator, was turned into a storm shelter for the 70 percent of the villagers who chose to stay on the island.

In the Gulf, waves that often reached heights of 18 feet, driven by 75- to 80-MPH winds, pummeled oil rigs. Ten people were stranded on a semi-submersible offshore rig whose supporting legs partly collapsed after catching fire. *Miss Agnes,* a boat used to ferry workers to the offshore oil rigs, capsized, drowning two crewmen. Other crewboats ran aground.

Juan made landfall shortly before dawn on October 28 along the south-central Louisiana coast, then bounced out into the Gulf again, where it knocked down two more oil rigs. At 3 P.M. on October 29, it again made landfall near Lafayette, Louisiana, about 100 miles (160.9 km) west-northwest of New Orleans. Waves over 20 feet (6.1 m) high and tides 10 feet above normal rushed inland, flooding areas in New Orleans and on the west bank of the Mississippi River and forcing over 2,000 people from their homes in Louisiana and from 5,000 to 6,000 people in Mississippi.

Tornadoes spun off the edges of the hurricane, causing isolated damage in Mississippi, Alabama, and Florida, and dumping 7.25 inches (18.4 cm) of rain on southeastern Texas.

By October 30, damage from Juan was estimated at $1 billion. Seven deaths had been recorded, 50,000 homes were flooded in Louisiana, and $110 million in damage had been done to the Louisiana sugarcane crop. Several multimillion-dollar oil rigs were either capsized or badly damaged and three bodies were recovered from a 52-foot (15.8-m) long drilling barge that capsized in a marshy ship channel 20 miles (32.2 km) east of New Orleans. Six people were rescued from a foundering shrimp boat 150 miles (241.4 km) off the coast of Galveston, Texas.

Now packing winds of 65 MPH and downgraded to a tropical storm, Juan continued to hit the coast, dumping 10 inches (25.4 cm) of rain on New Orleans and eight inches at Pensacola, Florida; Beaumont-Port Arthur, Texas; and Alexandria and Boothville, Louisiana. By the night of the 30th, approximately 2,500 people who had been evacuated from their homes were still waiting in 23 Red Cross shelters for floodwaters to drop in 10 southern Louisiana parishes.

Moving back into the Gulf of Mexico, Juan briefly cranked itself up to near-hurricane status, then weakened as it came ashore near Gulf Shores, Alabama, alongside the Florida state line, and headed towards Georgia. In some frustration, M. K. Renfroe, the

Escambia (Florida) County civil defense coordinator, commented: "It was predicted there would be more than the usual number of storms this year, but they didn't say they would be weird."

In Louisiana, floodwaters kept 50,000 homeless, and by now, coffins were floating out of mausoleums.

Juan continued northward, spreading devastation everywhere. Twenty-six people lost their lives in Virginia and Maryland, and thousands more fled their homes when swollen rivers, fed by the rains from the last, tattered remnants of Juan, pummeled the mid-Atlantic states. Six one-ton cannisters of poisonous chlorine gas were washed into the James River from a plant near Lynchburg, Virginia, and they roared like torpedoes past concrete bridges that, had they been hit by the cannisters, would have been demolished in fiery, explosive splendor.

Governor Arch Moore of West Virginia mobilized the National Guard; up to 60 barges ripped loose from moorings on the Monongahela River and evening rush hour traffic was tied up in Pittsburgh as the police closed bridges while some of the barges swept past.

Towns in Pennsylvania were inundated by floodwaters that had not reached such levels in the past 100 years. Twenty people were listed missing, 10 more reported dead, more than 2,000 more were rendered homeless in Virginia, West Virginia, and Pennsylvania. The James River flooded approximately 40 square blocks of warehouses and small businesses in low-lying areas of Richmond, Virginia, and historic sites around Washington, D.C., were sandbagged against the expected rise of the Potomac River.

By the time Juan had blown itself out, over 60 people would be dead, thousands homeless, and the total damage would reach nearly $2 billion.

UNITED STATES
TEXAS
GALVESTON
September 8, 1900

Six thousand people died, 6,000 were injured, and tens of thousands were rendered homeless by the huge hurricane that hit Galveston, Texas, on September 8, 1900. It was one of the last untracked storms in the history of the United States.

Six thousand people—one-seventh of the population of Galveston—were killed and an equal number were injured by the monster hurricane that struck the city on September 8, 1900. This event is an example not

Six thousand people—one-seventh of the population of Galveston, Texas—drowned or were buried under falling debris in the great hurricane of September 8, 1900. Six thousand more were injured, and tens of thousands were left homeless. (Library of Congress)

merely of the deadly force of tropical hurricanes but of the dramatic contrast in weather reporting methods between 1900 and today.

In 1900, the U.S. Weather Bureau was unable to detect storms at sea; in fact, it was not until five years later, in 1905, that ships began to send weather reports directly to the Washington Weather Bureau. Of course, whether or not the Washington Weather Bureau would evaluate this data correctly is another story. In 1900, Isaac and Joseph Cline, two brothers who manned the Galveston branch of the Weather Bureau, had reason to doubt it would be.

In hindsight, it seems that the storm was born somewhere off Venezuela in late August. It blew down telegraph lines on Antigua on August 30, raked Haiti on September 2, and dumped 12.5 inches (31.7 cm) of rain on Santiago, Cuba, on September 3.

But once the storm left Cuba, Washington lost track of it. Meteorologists assumed that it would curve north and hit Florida. But the storm defied the pattern and instead headed due west, toward Galveston. The Weather Bureau still predicted that it would make landfall east of Galveston.

By the afternoon of Friday, September 7, the booming waves hitting the beach could be heard all over town. Dr. Isaac Cline and his brother Joseph determined that the wind was coming not from the northwest, as it would if the storm were to hit east of them, but from the northeast.

By 5 A.M. on Saturday, September 8, the tide had risen high enough to flood four blocks of the city—and at a time when the wind was blowing against the tide. People began to wade through ankle-deep water to work (in 1900 almost everyone worked on Saturday), and by 8 A.M., the rains began.

By 10 A.M., the waves had increased in force and size. Groups of curiosity seekers, huddled under umbrellas, went down to the beach to watch the surf as it smashed into beach cottages and other obstructions, sending plumes of spray 20 feet (6.09 m) into the air. By this time Dr. Cline had mounted a horse-drawn cart to ride up and down the beach, warning sightseers and residents to move to higher ground. The tide was already five feet above the high water mark and rising at a rate of nearly a foot an hour.

At 11:15, Joseph Cline received word from Washington to change the wind warnings from northwest to northeast. The storm would stay out at sea, the U.S. Weather Service confidently predicted. Against regulations, Dr. Cline wired hurricane warnings up and down the coast. He wrote later, ". . . neither emergency nor hurricane warnings for the disaster were received from the forecaster in the Washington office."

By the afternoon, half the city of Galveston was under water, and all the telegraph lines were down but one. Joseph Cline sent one last message to the National Weather Bureau, stating that the storm was upon them. Two bridges to the mainland had fallen, and a causeway was inundated. The city was completely cut off.

By early evening, the wind had risen enough to shear off roofs and hurl timbers through the air at enough velocity to kill hundreds of people trying to wade to safety through the flooded streets. The anemometer on the roof of the Weather Bureau blew off after registering 120 MPH.

The wooden barracks at Fort Crockett, in the southwestern part of the city, washed away, carrying with them seven soldiers who had remained behind to guard it. Huge rafts of debris—battered and crushed remains of houses and people—planed through the streets, carried by the current out toward the Gulf. A thousand people crowded into the Tremont Hotel lobby, climbing the stairs just ahead of the rising water.

Dr. Cline found 50 people besides his family huddled in his house that evening. A moment later, water roared into the first story of the house, reaching Dr. Cline's armpits. Moments after all 50 had climbed to the second story, a railroad trestle collapsed on the house, splitting it asunder and spilling its inhabitants out into the raging floodwaters. Joseph Cline grabbed two of Dr. Cline's children and escaped through a window; Dr. Cline grabbed his baby and clambered aboard some floating wreckage. His wife and 31 of their friends were trapped in the wreckage that sank beneath the water.

Over 700 people sought shelter in the city hall. It collapsed, killing 50 and injuring hundreds more. The Rosenberg school caved in on a score of people huddled inside it. St. Mary's Infirmary collapsed on hundreds, leaving only eight survivors.

The cemeteries were scooped clean of coffins. One metal one weighing 200 pounds floated as far as Virginia Point. An ocean liner was torn from its moorings. It rocketed down the channel, ramming all three bridges to the mainland and ripping them apart, thus cutting off three major avenues of escape.

At dawn, the sun rose on a scene of incredible carnage. Limbs of trees, pieces of buildings, piles and piles of rubble stretched as far as the eye could see. Nothing but the larger buildings of Galveston were left standing, and one of them, fortunately, was the Levy Building, which contained the record of the storm.

In addition to the flooding, two billion tons of rain had fallen on this hapless city during a 24-hour period. Six thousand people died in the city; another 2,000 more bodies were scattered for miles along the coast on either side of it. Many bodies had been swept out into the Gulf. Others were piled into funeral pyres and burned, for health reasons.

Chaos reigned. Some survivors went insane. Several shot themselves when they discovered the bodies of loved ones. But the worst was yet to come. Hundreds of looters scurried into the city, many arriving by boat from the mainland. The militia was called in, and over the next two days hundreds of looters were executed on the spot.

The tally from the storm was ruinous. Once the fourth wealthiest city in the nation, Galveston had lost half of its taxable property in the storm—a loss of over $20 million.

Still, two lessons were learned and put to immediate use: First, a seawall a foot higher than the highest waves of the 1900 storm was built between the city and the Gulf. It still stands, with one major modification (see following entry).

Second, the Weather Bureau of the United States appointed Dr. Isaac Cline to direct the nation's first scientific study of hurricanes, so that no hurricane would ever again surprise a city quite the way this one did.

UNITED STATES
TEXAS
GALVESTON
August 16, 1915

Two hundred seventy-five people were killed when the Galveston Seawall, built after the 1900 hurricane, collapsed under the force of the storm surge set in motion by the hurricane of August 16, 1915.

Human beings still try, in spite of experience, to erect structures designed to thwart nature's furies. This futile effort resulted in the construction of the giant seawall at Galveston, Texas, after the city was devastated by the great storm of 1900 (see previous entry). The 11-mile (17.7 km) long structure was built of granite, timbers, and concrete, 16 feet (4.9 m) thick at its base and between 17 and 19 feet (5.2 and 5.8 m) high and designed in a concave curve to give it added strength. It was thought to be impermeable to anything natural forces could fling against it.

But like all seawalls, it proved to be no match for nature.

The hurricane of 1915 was first sighted off Cape Verde on August 10, 1915. Barely more than a gale, with winds of 60 MPH, it continued on a track that took it past Guadeloupe and Dominica. Fifteen years earlier, the U.S. Weather Bureau would have lost track of it at this point. But, thanks to the efforts of Galvestonian Dr. Isaac Cline, the bureau since 1905 was pinpointing hurricanes with increasing accuracy.

This one cranked its forces up, turned into the Gulf of Mexico, and now, brandishing winds of 120 MPH, headed straight for Galveston. Both local and national Weather Bureaus warned of its coming. But the residents of Galveston, for the second time in 15 years, went about their normal business, confident that their new seawall would hold.

When the storm hit, during an abnormally high tide, the waves it pushed ahead of it rose to 21 feet (6.40 m), easily broaching the wall and in some places penetrating it. Water roared into the business district, funneling furiously through its streets and covering them to the second stories of most buildings. Two hundred seventy-five people were drowned or crushed by the force of the water and falling debris, and a record $50 million in damages was caused before the storm roared inland, where it quickly dissipated.

The game survivors of the second Galveston disaster in a decade set about elevating the seawall to a uniform 19.2 feet (5.85 m). No other hurricane for more than 75 years has conquered this man-made structure. Not yet, at least.

UNITED STATES
TEXAS
INDIANOLA
August 19, 1886

One hundred seventy-six people lost their lives when a hurricane hit Indianola, Texas, on August 19, 1886, absolutely destroying it.

On August 18, 1886, Indianola, Texas, was a thriving town. On August 20, 1886, it was totally demolished— and later deserted. The cause was a hurricane that slammed into it on August 19, flattening every house and killing 176 people, including its weather observer,

I. A. Reed, who, while fleeing from the storm, was struck from behind and decapitated by flying timber.

The giant hurricane moved in from the Gulf, first making landfall between Lavaca and Matagorda Bays. Its wind velocity was only 72 MPH, but its concentrated force, colliding with Indianola in the form of both winds and a storm surge of building-crushing force, flattened, crumbled, or gutted every structure in town. What it didn't destroy it washed away, scattering the remains of these structures on the plains beyond.

The following day, the small number of survivors picked through the wreckage of homes, personal possessions, and dead animals, which were strewn everywhere. Nothing was salvageable. The town would have to be rebuilt from the bottom up, and rather than do that, its inhabitants decided to leave and settle elsewhere. To this day, only the rubble left from the passage of the storm remains.

WEST INDIES AND FLORIDA
September 16, 1928

(See HURRICANES, United States, Florida, Okeechobee, pp. 295–296.)

WEST INDIES
HISPANIOLA
June 1495

Christopher Columbus recorded the hurricane that slammed into Hispaniola in June 1495 and sank the Pinta and the Santa Maria. He did not record human casualties.

The first hurricane to be documented was recorded by none other than Christopher Columbus. In Hispaniola to rout out and enslave the Indians and establish a colony, Columbus was preparing to leave for home when the hurricane surged out of the Atlantic, crossed the lesser Antilles and slammed into Hispaniola, the second largest island of the West Indies, located between Cuba and Puerto Rico. Discovered and claimed for Spain by Columbus in 1492, the island was a rich trove of subtropical forests and fields, ideal for growing coffee, cocoa, and sugarcane.

The native Indians were used to the frequent hurricanes, and Columbus and his crew had encountered one at sea in September of the famous first voyage. Still, he was as helpless as the horrified settlers when this giant storm roared into the island, uprooting trees from the rain-softened ground, toppling them in windrows. Huts were flattened and buried under the whirling mud; some settlers were buried in their crushed dwellings.

In the harbor, ships strained at their anchors, broke loose, turned over, and sank. The *Pinta* and *Santa Maria* both went to watery graves, with some hands aboard. Only the *Nina* remained afloat.

It would take 10 months for Columbus to salvage enough from the wreckage of the two sunken caravels to build another, which he called the *Santa Cruz*. During that time he also built more forts on the island, exacted tributes of gold from the Indians under threat of slavery or death, and, besting the record of the hurricane, killed or hunted down with dogs thousands of native Indians.

WEST INDIES
TROIS-ÎLETS
August 12, 1766

The ruination caused by the hurricane of August 12, 1766, on Trois-Îlets, a small village on Martinique, set in motion a succession of events that would eventually unite Napoleon Bonaparte and his future empress, Josephine.

In the French West Indies, 1765 and 1766 were bad years for hurricanes. And the small village of Trois-Îlets, which clings to the rocky southern coast of Martinique and is composed of a few narrow houses, a cobbled square, and a church, was hit particularly hard in the middle of the second season.

History would probably have passed by Trois-Îlets had it not been for the fact that in 1763, in this tiny town, the wife of Lieutenant Joseph-Gaspard Tascher de La Pagerie gave birth to her second girl, Marie-Josephine-Rose, who would later, as the wife of Napoleon Bonaparte, be known as Empress Josephine of France.

On the night of August 12, 1766, a storm slammed into Trois-Îlets, collapsing many of the town's houses, uprooting trees, flooding the sugarcane fields, and collapsing the slave quarters on La Pagerie, the sugar plantation owned by Lieutenant Tascher.

As the fury of the storm mounted past midnight, slaves escaped from their smashed huts and huddled in the mansion with the Tascher family.

But the mansion proved no match for the storm. Tiles blew off the roof, then the entire roof was lifted off and the screaming wind and torrential rain rushed in, forcing both family and slaves to flee to the shelter of an old stone sugar house whose walls were two feet thick.

The following day, the plantation lay in ruins. Tascher was forced to sell his slaves, and for the next 10 years, Josephine would grow up in the stone sugar house. She was eventually bundled off to France, where she married a nobleman who was decapitated during the Revolution. She returned to Martinique and the old sugar mill, and finally, as a merry widow, returned to France with a ticket to immortality.

ICESTORMS AND SNOWSTORMS

THE WORST RECORDED ICESTORMS AND SNOWSTORMS

* Detailed in text

ICESTORMS

Asia/Northern India
Bangladesh (2003)

France
* Chartres (1359)

India
Moradabad (1853)
* Moradabad (1888)

United States
Central (1951)
South (1948)
Southeast (1994)

SNOWSTORMS

Bulgaria
(1936)
China
* North (1999)
Europe
* (1956)
Himalayans
* (1950)
Iran
(1972)
Japan
(1931)
Sweden
(1719)
Turkistan (Djetisusisk)
(1928)

United States
Atlantic Coast (1922)
California (1952)
* Truckee Pass (Donner Party)
   (1846–47)
* Colorado (1997)
* East (1947)
East (2005)
* (1958)
Massachusetts (Boston) (2003)
New England (1898)
* Northeast (1888)
* Northeast (1958—March)
Northeast (1993)
* Northeast (1996)
* Northwest (1996)
Northwest (1891)
Oklahoma to New England (2002)
Southwest (1886)

CHRONOLOGY

ICESTORMS

1359
   * Chartres, France
1853
   Moradabad, India
1888
   April 30
   * Moradabad, India
1948
   January 24
   Southern United States
1951
   November 2
   Central United States
1994
   February
   Southeastern United States
2003
   December
   Bangladesh (Asia/Northern
   India)

SNOWSTORMS

1719
   Sweden
1846–1847
   October 28, 1846–April 21, 1847
   * Truckee Pass, California (Don-
   ner Party)
1886
   January 6–13
   Southwestern United States
1888
   March 11–14
   * Northeast United States
1891
   February 8
   Northwest United States
1898
   January 31
   New England, United States
1922
   January 27–28
   Atlantic Coast, United States

1928
   January 12
   Turkistan (Djetisusisk)
1931
   January 12
   Japan
1936
   February 11
   Bulgaria
1947
   December 26
   * Eastern United States
1950
   November–December
   * Himalayas
1952
   January 13
   California
1956
   January–February
   * Europe
1958
   February 7–10
   * Eastern United States

February 15–16
* Northeastern United States
March 19–22
Northeastern United States

1972
February 10
Iran

1996
January
* Northeast United States

December 26–29
* Northwest United States

1997
October
* Colorado, United States

1999
October
* North China

2002
December 23–25

United States (Oklahoma to
New England)

2003
December 14–18
United States, Boston,
Massachusetts

2005
January 22–23
Eastern United States

ICESTORMS

Icestorms may precede or follow snowstorms and are usually composed of a combination of sleet and hail.

Sleet is precipitation of small, partially melted grains of ice. These grains of ice are formed in two ways, either by drops of rain falling through a layer of air with a temperature below freezing, or from snowflakes that have fallen through a layer of air with a temperature above freezing. Unlike hail, which can fall at any time of the year, sleet only falls in winter.

Although sleet is an annoyance, it rarely causes the wholesale disruption that hail does, and thus, the casualties and damage figures in the following section will be entirely attributable to the effects of hailstones.

Hail is precipitation in the form of pellets of ice or a combination of ice and snow. Usually, hailstorms occur during the passage of a cold front, or during a thunderstorm.

Small hailstones are simple structures formed when the surfaces of snow clumps melt and refreeze, or become coated with water droplets which subsequently freeze. Thus, they have a soft center and a hard outer coating.

Large hailstones, which have diameters of from one-half inch to—in rare cases—five inches (12.7 cm), are more complex, and the theories on their formation vary. Generally, they are composed of alternating layers of hard and soft ice. One theory has it that they are formed in clouds, when supercooled raindrops freeze on dust particles or snowflakes. These tiny hailstones are then blown upward and downward, repeatedly, by winds within the cloud. Each time they descend and pass through a region whose temperature is above freezing, they collect moisture, and each time they are blown upward into a region in which the temperature is below freezing, they either freeze or accumulate a layer of snow. The stones continue to grow until they reach a weight that the winds will no longer support, after which they fall to the ground.

Another theory suggests that hailstones fall through varying pockets of air, gaining layers by passing through air containing varying amounts of water.

Whatever the method of formation, falling hailstones have caused astonishing damage and loss of life.

FRANCE
CHARTRES
1359

One thousand men and 6,000 horses in the army of Edward III were killed by a destructive hailstorm in 1359 near Chartres, France.

In 1359, if old chronicles are to be believed, one of the most prodigious and devastating hailstorms of all time struck the army of Edward III, just outside of Chartres.

It occurred during the 22nd year of the Hundred Years' War. Edward, claiming the French crown, had already captured Calais and Crecy and approached Chartres heading an elite corps of horsemen and soldiers.

Suddenly, a fierce rain and lightning storm struck the army. Within moments, raindrops had turned to hailstones the size of goose eggs, which, in a matter of moments, pounded some 1,000 men and 6,000 horses to death.

Reversing his trek towards Chartres, Edward escaped unharmed and went on to sign the treaty of Calais with Philip of Burgundy.

INDIA
MORADABAD
April 30, 1888

Two hundred fifty people died in India's worst hailstorm on April 30, 1888.

In the same year that the famous blizzard of '88 pounded the northeastern United States with high winds and snow (see p. 331), the worst hailstorm in the history of India killed 250 people, caused widespread havoc among livestock, and atomized windows, glass doors, and unreinforced roofs.

Older residents of India were no strangers to damaging icestorms. In 1853, a hailstorm pounded 84 people and 3,000 head of cattle to death, and in 1855, another storm killed dozens of farmers in the fields of Naini Tal.

But for sheer fury and damage, the 1888 storm was considered to be the worst, since it was accompanied by a giant windstorm that lifted roofs, uprooted porches, shook walls and drove hailstones said to be the size of eggs and oranges into buildings, animals, and people.

When it was over, ridges of ice one- to two-feet deep remained on the ground, and buildings looked as if they had been pummeled with giant lances. The bodies of farmers and livestock beaten to death by the hail laid partially buried in the white carpet of ice that would soon melt in the spring heat.

SNOWSTORMS

Snow in its most benign form is what children and romantics wait for from the first turn of the calendar towards November and the first drop of the thermometer below freezing. Snow has a way of softening the sharp edges of urban environments and providing the tools of play for imaginative youngsters.

But in its less benign form, when it roars into our lives in the form of a blizzard, it can be a killer.

Snow itself is precipitation formed by the sublimation of water vapor into solid crystals at temperatures below freezing. This sublimation generally takes place around a dust particle, in the same way that rain drops are formed, except that snowflakes emerge as symmetrical, hexagonal crystals, no two of which are alike. The difference in size and shape is the result of the matting together of several crystals, caused by the descent of these snowflakes through air warmer than that of the cloud in which they originated.

On average, 10 inches (25.4 cm) of snow is equivalent to 1 inch of rain, and the factors determining rainfall are roughly those affecting snowfall.

Blizzards, then, are winter storms that are characterized by low temperatures, high winds, and driving snow, as a hurricane is characterized by tropical temperatures, high wind, and driving rain. According to the U.S. Weather Bureau, which evolved the parameters of blizzards in a 1958 definition, a snowstorm becomes a blizzard when the winds reach 35 MPH, and the temperature dips below 20°F. In the United States, blizzards are most common in the northern Great Plains states, but they also occur as far south as Texas and as far east as Maine.

CHINA
October 1999

Particularly heavy snow crippled the northern provinces of China from October 1999 through January 2000. Three feet of snow from continuous storms collapsed structures, destroyed infrastructures, and killed nearly 2 million livestock and 11 people.

Persistent snowstorms choked the northern provinces of China, in particular Ch'ing-hai (Qinghai), Inner Mongolia, and Sinkiang (Xingjiang), from October 1999 through January 2000. Transportation between population centers was severely curtailed, and food supplies dwindled in some areas. The threat of flooding when the inevitable thaw occurred was acute, though the worst predictions failed to materialize.

In Ch'ing-hai Province, slightly over two feet of snow fell from October 1 through 21. In 58 of the most hard-hit villages, over 50,000 people and close to 1,000,000 livestock were affected. Some livestock were lost; others were buried under the snow; others perished from the constant cold.

In Inner Mongolia, the snowfall began at the end of October and concluded on January 6, 2000. Snow fell continuously during this time in the towns of Hulunpeierh (Hulunbeier), Hsilinkuole (Xilinguole), Wulanch'apu (Wulanchabu), and Hsinkan (Xingan) and in the cities of Ch'ihfeng (Chifeng) and Tongliao. By December, the snow level averaged three feet. In the 260 townships of this province, 300,000 herdsmen were deprived of their livelihoods and 15,000 livestock were killed by the storms.

A blanket of snow three feet deep also covered the northern and eastern portions of Sinkiang Province, from two blizzards that struck in both October and January. Ch'angchi (Changji) and the Haomi area of eastern Sinkiang were the two areas most seriously crippled by the snow. Eleven people were killed, 100,000 livestock were killed, and more than 1,000,000 people were affected. Farm buildings collapsed under the snow; vegetable crops were buried and lost; fruit trees withered under the assault of the snow.

The infrastructure of this province was particularly ravaged. Electrical power, roads, and communication facilities underwent enormous, heavy damage.

EUROPE
January–February 1956

Nearly 1,000 died in a record series of snowstorms that blanketed Europe in January and February 1956.

One of the fiercest winters on record gripped the entire European continent in January and February 1956. Close to 1,000 people died as a result of prolonged cold and a repeated series of blizzards that roared out of Siberia and blanketed Europe from Norway to the Mediterranean with repeated, paralyzing snowstorms. The French Rivera received its first snow in years, and the first snow in memory blanketed parts of North Africa.

Adding to the danger of frostbite and exposure, roving bands of wolves and wild boar, forced from their lairs to forage for food, wandered into villages. Near Kalambaka in central Greece, a pack of wolves attacked children returning home from school. In Italy, Germany, and Belgium, similar attacks occurred.

The capital cities of Paris, London, Madrid, and Athens were brought to a standstill by the numbing cold and ferocious blizzard conditions.

While the northern countries such as Norway, Denmark, Sweden, West Germany, and Czechoslovakia merely suffered the effects of a particularly severe winter, the Mediterranean area experienced extreme economic damage because of the annihilation of its crops.

In Italy, fruit and vegetable crops were severely damaged. Spain suffered millions in agricultural damage. Citrus crops, hit with the cold before they could be picked, were totally destroyed. Fifty thousand people employed in the citrus groves of the southeast Mediterranean seaboard were thrown out of work.

Olive groves in both Spain and Greece were ravaged. In Greece alone, frost and flood damage to crops, trees, livestock, and installations such as bridges, dams, and culverts was estimated at millions of dollars.

In Turkey, the effects of the cold produced huge floods, particularly in Thrace and along the Ceyhan River in southern Turkey.

In mid-February 1956, President Dwight Eisenhower sent millions of dollars in supplies and seed for replanting crops to stricken countries who requested aid.

HIMALAYAS
November–December 1950

Over 500 perished in the horrific trek of several thousand refugees from Communist China through the high passes of the Himalayas in November and December 1950.

The background of the story begins in 1948, when Chinese Communists conquered most of Northern China. By 1949, the Chinese Nationalists were driven off the Chinese mainland, and the People's Republic of China was established, with Mao Tse-tung as chairman and Chou En-lai as premier and foreign minister.

By 1950, the Chinese Communists were in firm control, joining North Korea in pushing U.N. forces under General Douglas MacArthur back from the Manchurian border to the 38th parallel.

These are the neat lines of history. Contained within them are the human stories of refugees and loyalists to the Nationalist regime who were forced to flee from their destroyed homes.

In early November 1950, several thousand Muslim Chinese set out with food, provisions and what they could salvage of their possessions to immigrate over the Himalayan passes into India. Like the Donner Party (see following entry), they were not seasoned travelers, and numbered among them wealthy dignitaries. In this case, the dignitaries—Isa Yusuf, secretary-general of the former Nationalist government of Sinkiang, and Colonel J. A. Sabri, formerly of the Nationalist army garrison in Sinkiang's capital, Urumchi—may have unwittingly contributed to the tragedy, for when 50 caravans reached the China-India border in mid-November, Communist soldiers detained them for weeks, forcing them to surrender arms, extra clothing, and valuables.

Yusuf and Mohammed Amin, deputy governor of the province, were imprisoned, and Yusuf was hanged by his wrists from a rafter in the makeshift prison for an entire night. The remainder of the refugee party was forced to remain nearby, and in so doing, consumed half of their food supply.

The party finally set forth toward the Himalayan passes in early December—far too late in the year for safe passage through trails that often climbed to 18,000 feet. Yusuf and Amin finally bribed their guards and escaped, joining the remainder of the refugees as they set out to climb into swirling snowstorms which coated ice-plated trails.

All of the travelers began the passage on horseback. But as they reached higher elevations the snow accumulated to a depth of six feet (1.82 m) and the horses died.

The thin air made the blood gush from the nostrils of humans and horses alike, and headaches were constant and debilitating.

As the horses began to perish, some of the refugees were forced to wait behind. The first of the band began to die from the exposure and cold.

At night, the caravans attempted to make camp on the glassy ice, but many had no tents and lay exposed to the elements. Mothers with nursing children found their bones frozen into hunched positions as they tried to shelter the infants in their arms. Extremities—both hands and feet—began to freeze.

As the bands of refugees plodded through the treacherous ice and snow in a desperate race against dwindling supplies, they were forced to abandon relatives and friends. Those who stayed behind to help died by the side of the trail.

One man left his wife in the snow to push on with their three small children. Half an hour later, he was able to obtain a horse. He mounted it and went back for his wife, and found her frozen to death in the snow. By the time he had returned to his children, they were suffering from frostbite.

Finally, near Christmas, the survivors of the trek reached Leh, the 12,000-foot (13,657.6 m) high capital of Ladakh province in Kashmir. Two hundred remained there, some 700 miles (1,126.5 km) from their starting point. The remainder went on to Srinagar, where winter deepened, darkening the plight of those who were still suffering from frozen extremities. Secretary-General Yusuf brought his family of five through safely only to see his seven-year-old daughter die in Srinagar on Christmas Day of tetanus resulting from frozen feet. Days later, his eight-year-old son's frostbitten foot had to be amputated.

Altogether, nearly one-quarter of the several thousand who began the trip through the Himalayas perished, and those who survived would be forever scarred by the experience.

UNITED STATES
CALIFORNIA
TRUCKEE PASS
October 28, 1846–April 21, 1847

Approximately 100 members of the Donner Party, a group of pioneers ill-suited for pioneering, perished in the snows of Truckee Pass, California, in the winter of 1846–47.

The Donner Party, about which volumes have been written, was a group of slightly more than a hundred businessmen and their families who had decided to quit the relatively humdrum existence of Springfield, Illinois, for the untested promise of California.

Led by Jacob Donner and James F. Reed, they began their westward journey on April 14, 1846, but through a series of mishaps and misfortunes, did not reach the most treacherous terrain of their trip until October. Some accounts state that they were warned not to attempt to cross the Sierras until spring; others that they were promised safe passage by guides who later abandoned them. At any rate, around 90 of the party reached Truckee Lake (now known as Donner Lake) high in the Sierras, on October 28, 1846. And there, the snow began to fall, in great, blustering abundance.

The group hastily began to erect shelters, which, from their very concept, were inadequate. There was no time to build cabins; tents and lean-tos were all they could manage. An account by one of the survivors states, "The snow came on so suddenly that we had barely time to pitch our tent, and put up a brush shed . . . one side of which was open. This brush shed was covered with pine boughs, and then covered with rubber coats, quilts, etc. . . ."

The storm continued, day and night. All of the passes through the Sierras were blocked by the steadily accumulating snow. Hunting parties that went out for food supplies came back empty-handed. From October to mid-December, several unsuccessful attempts were made to drive men and cattle through the waste-high snow to the summit. All failed.

Food supplies continued to dwindle. First cattle, then horses, and finally dogs were killed, dried, and eaten. Finally, on December 16, a party of 17 men and women, led by poet-woodsman C. T. Stanton, set off in desperation to scale the summit and bring help to the slowly starving inhabitants at the Donner camp. They were apparently the first to revert to cannibalism when their meager supply of six days' worth of dried beef gave out, and their fellow explorers died around them. Stanton, their leader, was one of the first to perish. Three days after the beginning of the trek, he went snowblind, and two days after that, he built a tiny fire, wrote a poem to his dead mother, and died.

On the survivors battled, through sudden blizzards and waist-high drifts. Two Indian guides were killed and eaten by the dwindling band of semi-delirious travelers.

Thirty-two days later, on January 17, 1847, seven survivors reached an Indian camp on the western slopes of the Sierras. Gaunt, ravaged, sometimes unintelligible, they pleaded with whomever would listen to them to send rescue parties to bring back the remaining men, women, and children who might still survive at the camp on Truckee Lake.

One of the men who immediately set out with a rescue party was James Reed, one of the original organizers of the party, who had been exiled from it at the beginning of October when he had killed another man in a fight. Ironically, Reed had reached California safely, while his wife and children remained at the top of the pass.

Back at Truckee Lake, Jacob Donner, the party's other organizer, was dead. The only food available was green rawhide, which was boiled down to a barely edible, glue-like substance. But even that gave out, and, when the children began to die, these survivors also turned to cannibalism, stripping the flesh from those who were dying around them each day.

The misery mounted; the snows continued. Finally, the first of three rescue parties sponsored by the philanthropic California pioneer Captain J. A. Sutter broke through the summit snow and reached the horror of the Donner campsite on February 19, 1847. What they saw was indescribable. Skeletons, their flesh half ripped away, and the remnants of shelters littered the snow. In the midst of this, looking scarcely different from the corpses, was a gaunt band of 48 survivors, some of them already insane.

James Reed arrived in the second party. His wife and two daughters had survived, and would write about their experience later. Trials were held, murder charges were brought against those who had reverted to cannibalism, and then dropped. There was simply no precedent for the horrors experienced by the Donner Party.

UNITED STATES
COLORADO (AND NEBRASKA AND KANSAS)
October 1997

Fourteen people and 50,000 head of cattle perished in a giant blizzard that blanketed eastern Colorado and parts of Kansas in late October of 1997. Snow depths reached four feet and drifts were 15 feet (4.57 m) high on most farms.

The eastern plains of Colorado were swept by a ferocious blizzard that contained snow and shrieking 50-MPH wind gusts in October 1997. Four feet of snow piled up in most places, and drifts 15 feet (4.57 m) high were commonplace.

Fourteen people lost their lives, some from automobile accidents, others from being lost and perishing in the snow, like wanderers in the Arctic or the high Himalayas.

This is cattle country, and the storm could not have struck at a worse time. It hit at precisely the moment that calves were being weaned from their mothers and cows from the south were entering feedlots. There was no way to continue this since the blizzard arrived on October 25. It was all that farmers could do to secure their own lives and buildings.

And so, some animals suffocated under the snow. Others wandered over snow-covered fences and drifted for miles through the falling snow until, desperate for water, they walked into irrigation canals, fell through the ice, and drowned.

"I've never seen anything like this," said one rancher after the storm cleared. "And we've been ranching for years and years. Out of 350 cows, calves and yearlings, we lost 110."

The overall total of dead cattle reached 50,000 in Colorado and Kansas, the heaviest toll since a spring blizzard in 1957.

In neighboring Kansas, for the National Byproducts Company, a rendering outfit that ground cow carcasses into bone meal and sold hides for low-quality leather, the toll of the storm meant extra business. "It's so big I can't skin them all, and I have men skinning 600 a day," confessed James Davis, the company's general manager.

Cattle were not the only casualties. Two Colorado State Patrol planes and two helicopters from the Fort Carson army post, and army vehicles and snowmobiles swept the highways for survivors. On Interstate 25, between 700 and 1,000 cars were abandoned during the height of the storm.

Over two feet of snow paralyzed Denver at the height of the storm. Because of the loss of power, lost heat was a problem in the sub-freezing temperatures. Red Cross shelters were opened in Denver and in Omaha, Nebraska, and they served over 1,000 of the estimated 100,000 people without power in the two states.

UNITED STATES
EAST
December 26, 1947

Fifty-five died in a colossal snowstorm that paralyzed New York City and its suburbs, Connecticut, and New Jersey on December 26, 1947. The storm established a new record for snowfall in the area.

An extraordinarily vicious snowstorm swept in from the Atlantic at 5:25 A.M. on December 26, 1947, descending upon the eastern United States, and eclipsing the previous snowfall record of the fabled Blizzard of '88 (see p. 331) in 12 hours. It took 30 hours for the 1888 storm to accumulate 20.9 inches (53 cm); during the 1947 storm, from 5:30 A.M. until midnight of the 26th, 25.8 inches of snow paralyzed New York City. At times, the snowfall exceeded three inches an hour.

The only benign aspect of this storm was the fact that it was not, as in 1888, accompanied by hurricane force winds. Thus, it did not qualify as a full-fledged blizzard. Still, 99 million tons of snow fell on New York City alone and brought the huge city and other parts of the eastern seaboard as far north as northern Connecticut and as far south as Washington, D.C., to an absolute, silent standstill.

Fifty-five people died, some of exposure, some of heart attacks from shoveling snow, some in fires ignited by makeshift heating devices invented when power lines weighted with ice were severed. One hundred eighty fires were reported in the first 24 hours after the snowfall ended in New York City, further straining the 25,000 police and firemen already battling the snow. In Philadelphia, five brick buildings in the downtown warehouse and wholesale district burned out of control, causing an estimated $1 million in damage. Four hundred firemen, hampered by icy streets and residual snow, battled the blaze for nearly five hours before bringing it under control. Two of them received leg fractures when a brick wall collapsed on a fire engine; eight others were less seriously hurt. Only the fact that the buildings were commercial properties, unoccupied during the weekend, accounted for the relatively low number of casualties.

Public transportation on the eastern seaboard ground to a muted stop for two days. Mayor William O'Dwyer, in California for the Christmas holidays, could not return to New York until Sunday, three days after the storm struck. He immediately went on the radio, assuring the public that his deputy mayor, Vincent Impelliteri, had kept the wheels of the city moving, and that food and fuel supplies would soon be getting through to the snow-locked city.

Seventy-three freighters, loading or unloading on the then-busy docks of New York City and Hoboken, New Jersey, were forced to suspend activities. Because of low visibility, decks banked high with snow, and the absence of docking crews, five passenger liners either did not leave on time or docked late. The *Queen Mary* postponed a scheduled sailing for 12 hours, awaiting most of her 860 passengers, who could not get to the dock. Snow prevented the Coast Guard from maintaining its watch at Breezy Point for three days.

Some commuter trains were stalled for up to 14 hours on runs that normally took under an hour to accomplish. Abandoned cars tied up main thoroughfares for a week.

Some price gouging was reported, with some New York City cabdrivers charging the equivalent of weekly salaries for rides. In one part of the Washington Heights section of Manhattan a grocer was cited for charging 35 cents for a quart of milk and 24 cents for a loaf of bread.

Still, tales of heroism abounded. Volunteers from the wartime Civilian Defense Corps, the army, navy and Red Cross put themselves at New York City's disposal. State troopers of the Hawthorne Barracks in Westchester County, north of New York City, wearing snow shoes, hauled toboggans loaded with milk, bread, and other foods to marooned residents.

An unidentified young sailor rescued a 50-year-old bridge operator who had been trapped in a snow-covered automobile for 21 hours, a half mile away from his job at the Mill Basin Bridge in Brooklyn.

A policeman delivered a baby in Maspeth, Queens, when the family doctor could not make it through the drifts in time.

Police cars were used for ambulances when ambulances could not get through snow-clogged streets. The Morro Limousine Service in Brooklyn could not free its 18 limousines for ambulance service, and began to receive an average of 25 calls an hour to transport sick persons to hospitals. Radio station WNEW broadcast an appeal for horse-drawn sleds to transport the sick, and the need was met.

In one bizarre case, policemen who pursued a larceny suspect in Harlem had to carry him to a hospital after they had shot him in the legs. No ambulance was available.

Metropolitan Opera star James Melton, who lived on a farm six miles (9.65 km) from Westport, Connecticut, shoveled snow from his backyard on the Saturday afternoon after the storm so that a chartered helicopter could land and take him to LaGuardia Field. From here, a limousine scooped him up and got him to his engagement on the Metropolitan's Opera Quiz of the Air. Still wearing his raccoon coat, Melton dashed to the broadcasting room. Boris Goldovsky, who had been pinch-hitting, shifted over and the panting singer expressed his opinion about American opera and volunteered an answer about off-stage music in *Don Giovanni*.

Also admirable was the performance of the Post Office, which, demonstrated its slogan, "Neither sleet, nor snow, nor dark of night shall stay these couriers. . . ." Mail deliveries took place on time, even at the height of the storm.

UNITED STATES
EAST
February 7–10, 1958

A monumental blizzard swept through the entire United States, from the Rocky Mountains to the eastern seaboard, from February 7 through 10, 1958. Sixty died and cities were immobilized from upstate New York to Jackson, Mississippi.

One of the most severe blizzards on record swept through the United States from the Rocky Mountains to the eastern seaboard for three days in February 1958, killing 60 people, stranding tens of thousands, and causing economic damage in the hundreds of thousands of dollars.

Upstate New York was hit the hardest, when a series of new storms swirled in on February 9, blowing 40 inches (101.6 cm) of snow into drifts up to 15 feet (4.51 m) high, snarling transportation, marooning hundreds, immobilizing industries, and closing schools and businesses. Twenty-six inches (66 cm) of snow fell overnight between February 9 and 10, stranding 50,000 workers in the Syracuse-Utica area. The General Electric Company shut down all four of its plants in Utica. By early afternoon on Monday, all of the airports in the region had suspended activity.

The entire New York Thruway was closed off, and stalled vehicles dotted the roads in most of New York State. In Baldwinsville, on the outskirts of Syracuse, the faint sound of an automobile horn led rescuers to a car buried beneath a huge drift. It had been there for two days. Inside the car was a woman with both of her legs frozen. Her husband lay dead beside her.

At Augusta, in central New York, state police abandoned a stalled patrol car and trudged 15 miles (24.1 km) on snowshoes to aid a marooned farm wife and her eight children. The family was taken to shelter on toboggans.

Milk farmers in Ontario and Tompkins counties were forced to dump their milk because they could not deliver it. One dairy that normally shipped 84,000 quarts a day was only able to deliver 9,000 at the height of the storm.

The effects of the blizzard and cold extended offshore and into the South. The Cunard liner *Saxonia* battled fierce headwinds and ferocious seas for 36 hours before finally limping into port in New York City. Up to two inches of snow fell as far south as Vicksburg and Jackson, Mississippi. In that state, bus service was suspended, schools were closed, and motorists were warned to stay off highways.

UNITED STATES
NORTHEAST
March 11–14, 1888

Four hundred people died in the Blizzard of '88, which raged in the northeastern United States from March 11 through 14, 1888.

The famous "Blizzard of '88" was a four-day, raging snow-and-wind storm that claimed 400 lives and $7 million in property damage in an area that extended from Maryland to Maine. Its fury was particularly felt in New York City, where wind gusts climbed to nearly 100 MPH, the snowfall exceeded 20 inches (50.8 cm), and drifts climbed to heights of nearly 20 feet (6.09 m).

Occurring at a time when weather forecasting was decidedly unsophisticated, the storm was at first greeted with the innocent joy most snowstorms receive. But by midnight, most East Coast cities had lost not only power, fire alarm systems, and transportation facilities but also hope that the snow would ever end.

By the second day of this roaring blizzard, sustained by hurricane force winds, the frozen bodies of homeward bound office workers began to dot the drifts. Those foolhardy enough to venture out into the raging storm sank up to their armpits in snow. By the end of the second day, Boston, New York, and most large cities had been brought to a virtual standstill. Hotels were packed to capacity, their lobbies carpeted with sleeping refugees from the storm, which continued to howl into its third day.

Telephone and telegraph lines were downed and communication, even within major cities, was impossible. The major stock exchanges were closed for three days. Without power, without light, families built fires to survive, often igniting their homes in the process. Fire companies could not be contacted, or, if reached, could not get through the towering drifts.

Equally destructive were the winds. Fishing boats off the coast and even in harbors were swamped and sunk. Walls and roofs of houses were torn away, and the families within the homes froze to death, or, after finally being rescued and given shelter, died from the effects of prolonged exposure to the cold and the elements.

By the second day of the storm, all manner of improvised shelters were full to overflowing.

On March 13, a bridge of ice appeared, spanning the East River between Manhattan and Queens. Some foolhardy adventurers tried to cross it and were pitched into the icy waters when a change of tide shattered the bridge.

A bustling New York City was brought to a standstill by the famed blizzard of 1888. The corner of Pierpoint and Fulton Streets, normally awash in activity, is a quiet landscape on the day following the storm. (Library of Congress)

The winds calmed and the snow diminished to flurries on March 14, and a combination of bonfires and an emerging, warm March sun melted the mountains of snow quickly, but the cleaning up, restoration of power, and the counting of the dead would go on for months.

UNITED STATES
NORTHEAST
February 15–16, 1958

Only five days after a monster blizzard struck the eastern United States, an even more severe snowstorm blanketed the Northeast on February 15 and 16, 1958.

Over 500 died and the entire Northeast, from Washington, D.C., to Maine was immobilized.

The worst snowstorm in U.S. history from the viewpoint of human casualties struck the Northeast on February 15, 1958. More than 500 persons were killed, and millions of dollars in crop and property damage resulted from this two-day blizzard that brought the nation's capital to a complete standstill, and left only the janitor to answer the telephone at the U.S. Weather Bureau.

Various cities and their populations experienced various depths of snow and cold. It was 48 degrees below zero in Wisconsin and 26 degrees below zero in Bismarck, North Dakota. Lebanon, New Hampshire, recorded a record snow depth of 60 inches, and

Michigan City, Indiana, was buried beneath 54 inches (137.2 cm) of closely packed snow. For the first time in history, the locks at Lake Michigan were frozen shut by ice blocks eight feet thick.

New York City was virtually closed down, as its new snow-fighting equipment, purchased after the 25.8-inch (65.5 cm) snowfall of 1947 (see p. 328), failed to open public thoroughfares.

The Beltway outside of Baltimore was completely shut down when two tractor trailers jackknifed across its entire width. Some of the passengers and drivers in the thousands of cars halted by the accident died of frostbite.

At nearby Bowie racetrack, over 4,000 fans were stranded by the onset of the fast-moving storm. As quickly as state troopers in half-tracks and jeeps opened the access road to the track, gale winds closed it behind them. Eventually, they too were stranded with the 4,000 sports enthusiasts at the track. It would be another day before all would be rescued by emergency Pennsylvania Railroad trains.

On February 16, Washington, D.C., resembled a ghost town. Over 80 percent of government employees did not report for work. No transportation—public or otherwise—operated in the paralyzed seat of national government until late on the second day of the storm.

UNITED STATES
NORTHEAST
January 1996

A blizzard that shattered snowfall records throughout the northeastern United States immobilized the area on January 7 and 8, 1996. Roaring up the Atlantic Seaboard with winds gusting as high as 60 MPH, the storm dropped snow as deep as 48 inches in West Virginia and 30.7 inches (78 cm) in Philadelphia, Pennsylvania. One hundred eighty-seven people died in the storm.

A record-shattering blizzard roared into the northeastern United States on January 7, 1996. A classic Nor'easter, which drew moisture from the sea and hurled it over Arctic air, shot up the Atlantic Seaboard and buried coastal cities from Norfolk, Virginia, to Boston, Massachusetts, with snow totals that, in nearly every location, far outdistanced the 1947 blizzard (see p. 328), and the notorious Blizzard of 1888 (see p. 331).

Philadelphia was hit hardest—paralyzed by 30.7 inches (78 cm) of driving snow, thus shattering the records kept for 125 years by the National Weather

Bureau. "There have been other storms with higher winds or more snow in other areas," Penn State University meteorologist Fred Gadomski told reporters, "But if you focus on snow and hazardous conditions in the highly populated areas this is it."

New York, Washington, Baltimore, Philadelphia, New Haven, and Boston were brought to a virtual standstill on the night of January 7. As each city was affected by the northward movement of the storm, state governors declared states of emergency and dispatched teams to rescue stranded motorists on highways or residents in isolated homes. In New Jersey, state police closed state highways to all but emergency vehicles.

In all parts of the Northeast, roads turned perilous. Tractor-trailer rigs and buses were ordered off highways in some of the states as whiteouts cut visibility to zero and packed-down snow and ice produced spinouts and collisions. Overtaxed snowplow operators tried their best, but all of their energies were devoted to major roads. Secondary roads turned impossible. New York City mobilized every one of its 1,300 snowplows and 350 salt spreaders, but could not keep up with the fiercely falling snow.

Traveling by mass means also became futile, except for the rails. Amtrak trains in the Northeast Corridor were jammed and late, but running. "South of Philadelphia, we were the only show in town," an Amtrak spokesman boasted.

The airports were madhouses. Kennedy, LaGuardia, and Newark airports closed, and so did National and Dulles in Washington and the Baltimore-Washington and Philadelphia airports. Practically all flights were cancelled throughout the area.

The mercury plunged just before the storm to record lows. In Saranac Lake, New York, 118 miles (190 km) north of Albany, the mercury plummeted to 37 degrees below zero. As the storm came ashore, winds howled at up to 45 MPH (72.4 km/h) with gusts up to 60 and visibility was measurable only in feet.

Police and rescue workers rounded up the homeless and the heatless, which numbered in the thousands. In New York City, 7,200 beds in 39 homeless shelters were filled, and people sprawled on floors and chairs in the shelters. In Grand Central Terminal and Penn Station, the homeless were allowed to remain overnight. At LaGuardia, Kennedy, and Newark airports, 1,800 people were stranded overnight.

High overnight tides posed further threats. Coastal flood warnings were posted in New York, New Jersey, and Connecticut, and nursing homes and dwellings in vulnerable areas were evacuated by police and national guardsmen. A change in the wind, however, suppressed coastal flooding. In Connecticut, the Connecticut River overflowed, flooding and icing shorelines and properties on the river.

The following day, the snow totals were tallied: An all-time record snowfall in New Jersey was established when 35 inches (88.9 cm) was measured at Whitehouse Station in northeastern Hunterdon County. The depth of snow in Central Park in New York City was 20.2 inches (51.3 cm), which made it the third highest snowfall ever there. Twenty-seven inches (68.6 cm) of snow fell on Staten Island; LaGuardia Airport recorded 24 inches (60.9 cm) of snow; Dulles Airport in Northern Virginia had 24.6 inches (62.5 cm).

And, though Philadelphia held the record for cities, Pocahontas County in West Virginia had the greatest snow depth—between 40 and 48 inches (101.6 and 121.9 cm).

For the next two days, the entire area remained pretty but immobilized. The federal government shut down, except for emergency services, including mail deliveries. Schools were shut; places of entertainment were shuttered; department stores, shopping malls, specialty shops, retail merchants, banks, and the New York Mercantile and Commodity Exchanges were closed. The stock market was open, but trading was nearly nonexistent. Commuter buses and trains ran rarely, if at all.

New York harbor was brought to a standstill, except for the Staten Island ferries, which continued to ply their way through the black and bitter water.

One of the busiest corridors in the United States gradually came back to life, and major roads were intermittently opened. It would take weeks before the last body of the last victim of the storm was unearthed from one of the cars buried along a major highway. The toll from the storm was sobering: 187 people died.

UNITED STATES
NORTHWEST
December 26–29, 1996

Twin snowstorms buried the northwestern United States under three feet of snow during the week of December 26, 1996. Four people died and an enormous amount of property damage was caused by snow, icing, and flooding.

The day after Christmas 1996 was no day to celebrate for the residents of Washington and Oregon. A combination of snow and freezing rain descended upon the area, cutting off the power to more than 300,000 homes and businesses, shutting down bus and plane service, blocking highway passes in the Cascades, and killing at least four people.

Ice-crusted trees collapsed on homes, streets, and power lines, and both Seattle, Washington, and Portland, Oregon, were turned into ghost towns, bereft of traffic and electricity.

Snow levels were moderate, ranging from six inches in downtown Seattle to over a foot in suburban Mountlake Terrace. Two feet of snow closed the two major routes across the Cascade Range in Washington: Interstate 90 over Snoqualmie Pass and U.S. 2 over Stevens' Pass.

But it was the layer of ice that descended on the snow that caused the most damage. In Oregon, more than two inches of freezing rain coated the eight inches of snow already on the ground. In Port Orchard, across Puget Sound west of Seattle, the combined weight of the snow and ice collapsed the roof of a marina, sinking eight boats and putting 70 others in jeopardy.

If that were all that occurred, the December 26 storm would have been that and nothing more. But nature had a second round waiting. On December 29, another storm arrived, and dumped two more feet of snow on top of the snow already on the ground, then coated all of this with yet another layer of ice, which melted as the sleet turned to rain and raised fears of flooding.

This time, roofs caved in and avalanches closed roads. All three routes across the Cascade Range in Washington—the two aforementioned ones and Route 12 over White Pass—were closed down by avalanches, which also shut down a 45-mile (72.42 km) stretch of Interstate 84 on the Oregon side of the Columbia River Gorge.

The combined weight of the snow and ice collapsed roofs all over the area. In Seattle, a Drug Emporium and a Kmart and the roof of the Entiat High School gymnasium fell in, injuring no one. The 610-foot (185.9 m) Space Needle, Seattle's prime tourist attraction, was shut down when missiles of snow and ice began to loose themselves from its upper reaches and plummet dangerously to earth.

Puget Sound Power and Light, which was western Washington's largest utility, again lost power, this time for 132,000 customers; Seattle City Light left 1,000 customers without power, and Portland General Electric in Oregon was unable to deliver electricity to 27,000 customers.

The effects of the twin storms were felt as far away as Nevada, where, in Reno, wild winds toppled truck rigs and snapped power lines. Snow fell at the 7,300-foot (2,225-m) line above Lake Tahoe, which meant that when it melted, it would overflow the lake and cause flooding around it.

TORNADOES

THE WORST RECORDED TORNADOES

* Detailed in text

Africa
  Congo (2003)
Bangladesh
  (1972)
  (1973)
  * Tangail (1996)
England
  * Widecombe-in-the-Moor (1638)
India
  (1936)
Madagascar
  (1951)
Sierra Leone
  Freetown (1913)
United States
  * Alabama, Georgia, Tennessee
    (1994)
  * Arkansas (1945) (see TORNADOES,
    Oklahoma, Arkansas, and Mis-
    souri, 1945)
  Delaware (1995)
  Florida, Georgia, and Maryland
    (2004) (see HURRICANES, Hur-
    ricane Ivan)
  Georgia
  * Gainesville (1903)
  Illinois
  * Chicago (1920)
  * Illinois, Indiana, and Missouri
    (1925)
  * Indiana (1925) (see TORNADOES,
    Illinois, Indiana, and Missouri,
    1925)

Kansas
  * Irving (1879)
  * Udall (1955)
  Kansas (1992)
  Kentucky (1996)
  Maryland (2002)
  Massachusetts
  * Worcester (1953)
  Michigan
  * Flint (1953)
  Midwest
  * (1965)
  * (1992)
  * Midwest and South (1917)
  Midwest and South (2003)
  Minnesota
  Odessa (1881)
  Mississippi
  Beauregard (1883)
  * Natchez (1840)
  Natchez (1842)
  Natchez (1971)
  Missouri
  * (1925) (see TORNADOES, Illinois,
    Indiana, and Missouri, 1925)
  * (1945) (see TORNADOES, Okla-
    homa, Arkansas, and Missouri,
    1945)
  * Marshfield (1880)
  * Poplar Bluff (1927)
  * St. Louis (1896)
  * St. Louis (1927)
  Nebraska
  * Omaha (1913)
  * North and South Carolina (1984)

North Dakota to Oklahoma
    (2004)
  * Northeast (1985)
  Northern Illinois (1990)
  Ohio
  Xenia (1974)
  Oklahoma
  * Oklahoma City (1942)
  * Oklahoma City (1970)
  * Oklahoma, Arkansas, and Mis-
    souri (1945)
  South
  * (1884)
  * (1917) (see TORNADOES, Midwest
    and South, 1917)
  (1920)
  * (1924)
  (1932)
  * (1936)
  (1951)
  (1974)
  South Carolina
  * Charleston (1811)
  * Charleston (1938)
  * Southwest (1947)
  Tennessee (1995)
  Texas
  * (1947) (see TORNADOES, South-
    west [Texas, Oklahoma, and
    Kansas])
  (1967)
  Saragosa (1987)
  * Waco (1953)
  Virginia (1993)
  Wisconsin (1994)

CHRONOLOGY

* Detailed in text

1638
  October 21
    * Widecombe-in-the-Moor,
      England

1811
  September 10
    * Charleston, South Carolina

1840
  May 7
    * Natchez, Mississippi
1842
  June 16
    Natchez, Mississippi
1879
  May 30
    * Irving, Kansas

1880
  April 18
    * Marshfield, Missouri
1881
  July 15
    Odessa, Minnesota
1883
  April 22
    Beauregard, Mississippi

1884
> February 19
> > * Southern United States

1896
> May 27
> > * St. Louis, Missouri

1903
> June 1
> > * Gainesville, Georgia

1913
> March 23
> > * Omaha, Nebraska
> December 11
> > Freetown, Sierra Leone

1917
> May 26–27
> > * Midwest and South
> > United States

1920
> March 28
> > * Chicago, Illinois
> April 20
> > Southern United States

1924
> April 29–30
> > * Southern United States

1925
> March 18
> > * Illinois, Indiana, and Missouri

1927
> May 9
> > * Poplar Bluff, Missouri
> September 29
> > * St. Louis, Missouri

1932
> March 21–22
> > Southern United States

1936
> April 2–6
> > * Southern United States
> October 30
> > India

1938
> September 29
> > * Charleston, South Carolina

1942
> June 12
> > * Oklahoma City, Oklahoma

1945
> April 12
> > * Oklahoma, Arkansas, and
> > Missouri

1947
> April 9
> > * Southwestern United States

1951
> January 4
> > Madagascar
> June 8–10
> > Southern United States

1953
> May 11
> > * Waco, Texas
> June 8
> > * Flint, Michigan
> June 9
> > * Worcester, Massachusetts

1955
> May 25
> > * Udall, Kansas

1965
> April 11
> > * Midwestern United States

1967
> September 20
> > Texas

1970
> April 30
> > * Oklahoma City, Oklahoma

1971
> February 21
> > Natchez, Mississippi

1972
> April 2
> > Bangladesh

1973
> April 13
> > Bangladesh

1974
> April 3
> > Xenia, Ohio

1984
> March 29
> > * North and South Carolina

1985
> May 31
> > * Northeast United States

1987
> May 22
> > Saragosa, Texas

1990
> June 16
> > * Midwest United States

1991
> April 28
> > Kansas

1992
> May 12
> > Oklahoma
> June 16
> > * Midwest

1993
> August 6
> > Virginia

1994
> March 26
> > * Alabama, Georgia, North and
> > South Carolina
> August 27
> > Wisconsin

1995
> May 13
> > Tennessee
> May 20
> > Delaware

1996
> May 13
> > * Tangail, Bangladesh
> July 20
> > Wisconsin

1998
> April 9
> > Mississippi, Alabama,
> > Georgia

1999
> January 17
> > Tennessee and Arkansas
> May 3
> > * Oklahoma and Kansas

2002
> April 28
> > Maryland

2003
> May 1–11
> > Midwest and South
> June 23
> > Africa (Congo)

2004
> May 29
> > North Dakota to Oklahoma
> September 15
> > Florida, Georgia, and
> > Maryland

TORNADOES

In contrast to a cyclone, a tornado covers a relatively small area of ground. But it is also much more violent and destructive. In fact, for sheer, concentrated energy, there is very little else in nature that compares to this kind of storm.

There is no doubting a tornado by those who see it coming, or hear its explicit, roaring freight-train sound. Surrounded by rain, thunder, and lightning, a dark, funnel-shaped cloud extends from the cumulonimbus cloud from which it was formed to the ground. Within this funnel, air rotates violently. The tornado rises and falls, twists about, and wherever it touches earth, it causes instant and enormous destruction.

The diameter of a tornado ranges in size from a few feet to a mile. Within it, there are two types of violent wind: the winds rotating in its walls can achieve speeds of 200 to 300 MPH; the winds in its updraft at the center may reach 200 MPH.

The atmospheric conditions necessary for the formation of a tornado include high humidity, great thermal instability, and the convergence of warm, moist air at low levels with cooler, drier air aloft—a condition that explains the frequent presence of tornadoes within or accompanying hurricanes.

Tornadoes can travel distances from a few feet to a hundred miles, usually in a northeasterly direction at a speed of between 20 and 40 MPH. In the United States, the greatest concentration of tornadoes is usually over the central and southern plains and the Gulf states.

BANGLADESH
TANGAIL
May 13, 1996

A tornado that sped through the northern district of Tangail in Bangladesh on May 13, 1996, caused unimaginable havoc, killing 600 people, injuring 50,000, and making 100,000 homeless.

Bangladesh, whose inhabitants are yearly victims of horrendously destructive monsoons and floods, was attacked by a tornado on the night of May 13, 1996. The twister touched down in the northern district of Tangail, 45 miles (72.4 km) north of Dhaka, the country's capital, while its inhabitants slept.

Although tornadoes are common in the tropical delta of Bangladesh during April, May, and June, leading up to the annual monsoon season in July, the northern part of the country is generally spared their destruction. Not so this night. In less than half an hour, the tornado utterly destroyed 80 villages. The not very well built houses were universally flattened, but pieces of them—tin roofs and even tree trunks—were turned into lethal missiles by the 125 MPH winds of the storm.

The district of Tangail received the worst of the storm. In the village of Bashail, 120 died, many of them in a boarding school building that collapsed on top of the children sleeping inside.

"Tin roofs were cut into pieces and were flying everywhere," Reazuddin Ahmed, a weaver from the village, told a *New York Times* reporter. Mr. Ahmed had gathered up his family when he heard the storm approaching, and they huddled near a cement wall alongside a road near their home. There, they could see the tornado toss the village's wooden houses into the air and reduce several larger buildings, including a movie house, to rubble.

Fifty-five of their neighbors died, and were lined up in a makeshift morgue at the soccer field that had previously been used by the dead students from the boarding school.

Local hospitals were overwhelmed by the sudden influx of the injured. One 200-bed hospital accepted more than 1,000 wounded, many of whom lay on its floors. Another hospital in Tangail turned away four pickup trucks, some filled with bodies, others with injured, because the hospital had run out of space.

One nurse related how her hospital ran quickly out of bandages, medicine, and surgical equipment. At least 50 people died because they could not be treated, she added.

Finally, on May 15, the government sent teams of soldiers to transport tens of thousands of injured people to hospitals in other towns. The death toll was 600, though there were hundreds that were reported missing and never found. There were 50,000 injured, and over 100,000 were made homeless by a storm that touched down for only 30 minutes.

ENGLAND
WIDECOMBE-IN-THE-MOOR
October 21, 1638

Fifty members of a congregation were killed and 12 were injured when a tornado entered a church in Widecombe-in-the-Moor, England, on October 21, 1638, during a Sunday service.

The tiny hamlet of Widecombe-in-the-Moor earned itself the sobriquet of "The Evil One's Residency" when a tornado struck it on Sunday morning, October 21, 1638. Local authorities blamed the devil for timing the arrival of the storm to coincide with the commencement of the morning worship service, when the entire population of the town was gathered within the hamlet's only church.

Lightning often accompanies tornadoes, and in this case, a fireball was unleashed as lightning struck the church just as the storm's winds arrived. According to records left by parishioners, ". . . most part of [the congregation] fell downe into their seates, and some upon their knees, some on their faces, and some one upon another, with a great cry of burning and scalding, they all giving themselves up for dead . . ." and one Mistress Ditford reported that she had ". . . her gowne, two wastcoates, and linnen next to her body burned cleane off."

Far worse, however, seemed to be the effect of the twister itself, which apparently entered the church, flinging the congregation against the stone walls of the structure, killing 50 people and injuring 12 more.

UNITED STATES
ALABAMA, GEORGIA,
NORTH AND SOUTH CAROLINA
March 26, 1994

A series of 13 tornadoes roared down "Tornado Alley" from Alabama up the Appalachians to North Caro-

lina on Sunday morning, March 26, 1994. Forty-one people were killed, 250 were injured, and nearly 1,000 homes were destroyed or damaged.

Rain and hail began to pelt the Goshen Methodist Church near Piedmont, Alabama, during Palm Sunday service on March 25, 1994. The brick and cinder block church was in the middle of farmlands and pine forests, and the 140 people who crowded the church that Sunday morning came from the surrounding fields.

In one corner of the church, the children of the congregation were gathered to present their Palm Sunday play, "Watch the Lamb." They began a hymn as the lights began to flicker.

Suddenly, a window blew out. Someone yelled "Hit the floor!" and the roof ripped off. A moment later, the tornado collapsed the wall against which the children were gathered. Bricks flew into the sanctuary. Beams crashed down from the ceiling.

When it was over, 20 people, most of them children, lay dead among the ruins of the church.

It was part of the yearly visit of storms in "Tornado Alley," so named because tornadoes regularly ripped through its countryside. But never before on Palm Sunday, and never when there had been so many people congregated in one place.

The pack of tornadoes tore through Alabama, Georgia, South and North Carolina that day. Two tornadoes touched down 11 times in Pickens County, Georgia, killing nine people, including one family whose trailer was picked up and flung, with them in it, more than 30 yards across a country road. Fifty-nine other homes were exploded by the storm that day in Pickens County and 200 were seriously damaged. Thirteen other tornadoes touched down in Bartow, Lumpkin, White and Habersham counties, killing 15 people.

Back in Alabama, the miracle of selectivity also occurred. The nursery of the Goshen Methodist Church, where children played who were too young to participate upstairs, served as a storm cellar and remained undamaged.

A few miles away, in Rock Run, 75 worshipers in the Union Grove Methodist Church heard the tornado coming, and took refuge in the church's basement. The church was demolished above them, but they emerged unscathed.

Power and telephone service was interrupted along the path of the tornadoes; cars were flung around on country roads; a man died in Tennessee when he drowned in a creek swollen by runoff from the storms. All told, 41 people died and 250 were injured, all within hours of each other on one Palm Sunday morning.

UNITED STATES
GEORGIA
GAINESVILLE
June 1, 1903

Two hundred three people were killed and thousands were injured when a tornado struck Gainesville, Georgia, at the beginning of the shift in the local cotton mill on June 1, 1903.

The worst storm disaster in the southern United States to that date occurred on the morning of June 1, 1903, when a twister slammed into Gainesville, Georgia. The characteristic dark and funnel-shaped cloud appeared suddenly, with little warning, and most of the fatalities occurred in the downtown area, where people were instantly buried under collapsing and exploding buildings.

The storm arrived at precisely the hour that 750 employees of the city's largest cotton mill punched in for the day. The twister hit the mill broadside, crushing parts of it, and flinging other sections scores of feet into the air. In the mill and in the city 203 persons died, and nearly a thousand were seriously injured. (See also TORNADOES, United States, South, 1936, p. 353.)

UNITED STATES
ILLINOIS
CHICAGO
March 28, 1920

The worst tornado to strike Chicago killed 28, injured 325 and destroyed 113 buildings on March 28, 1920.

Twenty-eight people lost their lives when the great tornado of March 28, 1920, slashed through Chicago. Three hundred twenty-five people were injured, 113 buildings were demolished, and over $3 million in property damage made this the most destructive of the six major tornadoes to strike this city. (Others hit on May 25, 1896; April 6, 1912; May 18, 1926; April 1, 1929, and May 1, 1933.)

The storm remained chiefly in the northwestern portion of Chicago, centering on Melrose Park, in which 10 people died. It was an unusual storm that included huge hailstones, along with the thunder, lightning, and rain that accompanied its path. Power lines were snapped like toothpicks. Animals were flung crazily about. Streetcars and automobiles were overturned. A freight car, packed with 1,500 pounds (680.4 km) of unidentified cargo, was lifted 40 feet (12.2 m) in the air and deposited on the roof of a train station.

The storm lasted only an hour, from 12:15 to 1:15 P.M. But to those caught in it, it seemed to last for decades.

UNITED STATES
ILLINOIS, INDIANA,
AND MISSOURI
March 18, 1925

Six hundred eighty-nine people died, 2,000 were injured, and over 15,000 were made homeless by the tornado that rolled through Illinois, Indiana, and Missouri on March 18, 1925.

Ordinarily, tornadoes give warning of their coming by their characteristic funnel-shaped clouds. Ordinarily, tornadoes move at approximately 45 MPH for roughly 13 miles (20.9 km) in a path that is usually less than a quarter of a mile wide (1.21 km). But the famous tri-state tornado, a storm that has been called ". . . perhaps the worst tornado that has ever occurred anywhere in the world . . . ," defied all of these rules. Its path was three-quarters of a mile wide (1.21 km). It traveled for 219 (352.4 km) miles. And worst of all, its covering cloud was so close to the ground that hardly anyone saw the hidden funnel within it. Thus, the 689 who were killed, the 2,000 who were injured, and the more than 15,000 who were rendered homeless in the three hours and 18 minutes of the storm's fury probably never knew what hit them.

Formed by the collective massing of a cold low pressure system from Canada and a warm and wet low pressure system from the Gulf of Mexico, the tri-state tornado was one of eight that formed on the morning of March 18, 1925.

It first touched the ground at about 1 P.M. in Reynolds County, Missouri, and then began its 60 MPH race toward the northeast. Eleven people died, buildings were collapsed, and trees uprooted as it roared toward the Mississippi River.

Across the river, Gorham, Illinois, was totally obliterated by the storm. Ninety people were killed and 200 were injured.

The same fate awaited the larger city of Murphysboro, which was also left in ruins seconds after the storm struck. Over 150 of the 200 blocks that comprised the city were totally destroyed; 234 people were crushed by falling buildings; 800 were injured; 11 steam engines lay on their sides in the railroad yard. The fires, some of which accounted for the fact that more than 8,000 people would be homeless by nightfall, raged without control. The town's water supply had also been destroyed by the storm.

DeSoto, Illinois, felt a particular fury from the tornado, which was described by resident F. M. Hewitt as a ". . . seething, boiling mass of clouds whose color constantly changed. From the upper portion came a roaring noise as of many trains. Below this cloud was a tapering dark cloud mass reaching earthward."

Wooden houses were blown apart as if dynamited. A mother and her two-week-old baby miraculously survived when falling roof timbers formed a protective tent over their bed. The 100 who died in DeSoto died violently, many of them through dismemberment or from splinters of wood that were driven into them by the wind with the force of bullets.

The storm seemed to feed on destruction, gathering power and speed. Near Zeigler, Illinois, it lifted the Illinois Central railroad bridge from its concrete base. Sixty-four houses were destroyed within two minutes, and 127 people were killed. A two-by-four plank was driven through the side of a metal railroad car. Nearby, a large plank was driven so far into a tree that a man could later stand on the end of the plank without budging it.

As the storm continued on, it picked up a huge grain binder and carried it for a quarter of a mile; further on, it gathered up a touring car and blew it 225 feet (68.6 m) through the air before dropping it to the ground. A barber chair was discovered later in a field, miles from the nearest barber shop. A school building was destroyed, but its 16 pupils were picked up and deposited, unhurt, 450 feet (137.1 m) away.

Griffin, Indiana, was totally destroyed. Thirty-four people were killed, and every house was so badly damaged that there was no shelter left in the town for the survivors.

Finally, at 4:18 P.M., the tornado blew itself out near Princeton, Indiana, but not before cutting a grim path of destruction through that town. It demolished buildings, flung furniture and belongings over the countryside, and in one case, popped open the doors of a car containing four miners, pulled the passengers and driver from the car, dropped them unhurt by the side of the road, and tore the car into little pieces, scattering its remains along several miles of road.

UNITED STATES
KANSAS
IRVING
May 30, 1879

Fifty people were killed and the entire town of Irving, Kansas, was leveled by a multiple-barreled tornado on May 30, 1879.

One normal-size tornado hit the tiny town of Irving, Kansas, on the late afternoon of May 30, 1879. It caused minimal damage and no loss of life, and the inhabitants were just picking themselves up and brushing themselves off when the sky turned an unearthly black, shot through with red and purple flashes that seemed to give off smoky, sulfurous fumes. It was a mass of whirling, superheated air that contained in its two-mile-wide swath not one, but a dozen tornadoes.

Within minutes, the bundle of storms had slammed into the little town, leveling or blowing apart every building, leaving behind them the sort of scene that only a cosmic steamroller could have created. Records on the frontier were casual in 1879; around 50 people were killed, according to reports, but no estimate was made regarding either the injured or property damage.

UNITED STATES
KANSAS
UDALL
May 25, 1955

One hundred one people died and at least 500 were injured when a tornado struck Udall, Kansas, on May 25, 1955.

Throughout the United States, 1955 was a particularly bad year for tornadoes. Normally, about 150 of them occur in any 12-month period. In 1955, nearly 900 were recorded. Without touching the ground, one of these killed the 15-man crew of a B-36 bomber flying over Sterling, Texas.

One of the most destructive of them to come to earth entirely destroyed the farming town of Udall, Kansas, 22 miles (35.40 km) south of Wichita, killing or injuring half its population and scarring the rest with the horrible memory of the black night of May 25, 1955.

At 9:30 P.M., the twister smashed through Blackwell, Oklahoma, near the Kansas border. Five hundred homes were destroyed there; 19 people were killed and 493 were injured. A refrigerator was flung with so much force into a tree that refrigerator and tree fused into one inseparable mass. The pavement on the streets of Blackwell was peeled up like an apple skin. Automobiles were flung to the top of two-story buildings.

At 10:30, hail began to fall on Udall, lit by flashes of brilliant lightning, intermixed with rain driven almost parallel to the ground by a roaring, teeming wind. The twister struck with such force and rapidity, most local residents had no time to get to their storm cellars.

Eighty-two people died; more were buried under the rubble, or struck by pieces of windows, walls, and roofs that were turned into flying missiles when the sudden drop in outside pressure exploded houses. Every home was damaged. The Community Center was destroyed, although those who were attending a meeting there that night escaped injury by crowding against the west wall. As the pressure outside the building dropped, the roof exploded upward and the wind carried away the walls on the north, east and south sides of the building. Only the west wall held, protecting the 12 women and children who huddled against it.

The town's huge cement and metal grain elevator was demolished. Every car in town was damaged so badly that none could be started. A visiting policeman filled his car with injured and drove them to the hospital in Belle Plain, nine miles away.

Word spread quickly about the destruction of Udall, and within hours, dozens, then thousands of rescuers descended upon the scene. The Kansas National Guard and the air force sent detachments, beds, bedding, a crane, a wrecker, and food. A dairy in Wichita sent in a gasoline-powered generator and searchlights. The Red Cross and Salvation Army brought in food. A car dealer from Mulvane sent all of his station wagons to serve as ambulances. The rebuilding began quickly.

UNITED STATES
MASSACHUSETTS
WORCESTER
June 9, 1953

The worst tornado in New England's history killed 62, injured more than 1,200, and destroyed 4,000 buildings in Worcester, Massachusetts, on June 9, 1953.

The tornado that struck Worcester, Massachusetts, at 5 P.M. on June 9, 1953, remains the worst tornado ever to strike New England. Moving at 36 MPH, the twister traveled through 46 miles (74 km) of Massachusetts, roaring through Holden and Shrewsbury and claiming 28 lives in these two towns before fusing with another tornado and slamming into Worcester.

Sixty-two people perished in this cataclysmic collision of forces which twisted several steel towers that had been built to withstand winds of 375 MPH. More than 4,000 buildings were destroyed; over 1,200 people were injured.

One man, in a newspaper report, related that his wife went out to rescue the dog as the storm struck.

The worst tornado in New England's history was captured as it slammed through Holden, Massachusetts, on June 9, 1953. Sixty-one people died, 1,200 were injured, and more than 4,000 buildings exploded. (Library of Congress)

"She opened the door and that is the last we saw of her," he said. "She just went up into the air and out of sight."

The property damage estimate of $52 million was the highest ever recorded in the United States for tornado destruction.

UNITED STATES
MICHIGAN
FLINT
June 8, 1953

One hundred sixteen people were killed and 386 homes were totally destroyed when a twister hit Flint, Michigan, on June 8, 1953.

In terms of grim tornado statistics, 1953 challenged even 1955. From January to June of that year, 250 twisters rampaged through Michigan and Ohio, killing 28 and injuring over 350. But it was the 251st tornado that was by far the worst.

It began around 8:30 P.M. on June 8, 1953, in Flushing, Michigan, a suburb of Flint. In the manner

Forty-five drive-in moviegoers were killed and scores were injured when a tornado tore through Flint, Michigan, in June 1953. (American Red Cross)

of most twisters, its path was contained but viciously ravaging. Within minutes, it had flattened buildings, exploded others, and ripped the roofs from the suburban settlements along the Coldwater Road and the Flint River. Three hundred eighty-six homes were totally obliterated, 525 others were severely damaged. But the worst of it was the human toll: 116 victims were trapped indoors and crushed to death by this killer tornado.

UNITED STATES
MIDWEST
April 11, 1965

On April 11, 1965, 37 tornadoes tore through the Midwest, killing 271 and injuring over 5,000.

Palm Sunday, April 11, 1965, will be long remembered by survivors of what has been called the "worst tornado epidemic in history." On that day, an amazing total of 37 tornadoes formed from the warm, humid thunderstorm conditions that prevailed over a wide area of the Midwest, including Kansas, Missouri, Illinois, and Iowa. Before the day ended, 271 people were dead, more than 5,000 had been injured, and the property damage totaled $300 million.

UNITED STATES
MIDWEST
June 16–19, 1992

Seven people died and millions of dollars in damage occurred during a three-day rampage of 160 tornadoes from June 16 through 19, 1992, in 11 Plains and Midwest states.

In the biggest three-day outbreak of tornadoes since 1984 (see p. 349), 160 tornadoes swirled across 11 Plains and Midwest states from June 16 through 19, 1992.

Ohio was the first to be hit, with 55 MPH winds uprooting trees and damaging mobile homes.

Spikes of trees and shards of houses in Russiaville, Indiana, are all that remain after the passage of one of 37 tornadoes that tore through six states on Palm Sunday, 1965. (American Red Cross/Jack Shere)

McFarland, Wisconsin, was damaged extensively. On the 16th, a midday tornado blew apart 33 homes and 200 others were damaged, but miraculously, no one was seriously hurt. "People weren't home, thank God," a local official told reporters.

The sign on Tim Sweeney's mailbox in McFarland read "Garage Sale." But there was neither a garage, nor the roof nor several walls of his house, left after a twister passed through.

Four counties in Minnesota were pummeled by rain, hail, and tornadoes, and cottonwood trees—one of them 322 feet (98.15 m) in circumference, with a 90-foot (27.41 m) spread at its crown—were destroyed.

A total of seven people died in the three-day period, which was a testament to the effectiveness of local warning systems, which also caught a following tornado near Pleasant Hill, Ohio, the following day. It was all part of a weather system that brought thunderstorms to Indianapolis and that caused the collapse of a canal supplying much of the city's water.

UNITED STATES
MIDWEST AND SOUTH
May 26–27, 1917

Three hundred ninety-seven people died in a series of tornadoes that traversed Illinois, Louisiana, Mississippi, Missouri, Kansas, Tennessee, Indiana, Kentucky, Alabama, and Arkansas on May 26 and 27, 1917.

An unusual, perhaps freakish series of tornadoes invaded seven states in two days, from May 26 to 27, 1917, killing 249 and causing more than $5.5 million in property damage. Illinois, Louisiana, Mississippi, Missouri, Kansas, Tennessee, Indiana, Kentucky, Alabama, and Arkansas fell victim to a collection of killer storms that—in the usual behavior pattern of tornadoes—demolished some buildings while sparing others nearby, killed some people and spared others only feet away.

In Konts, Indiana, one of the largest, most opulent estates in town was transported more than a city block, without so much as unseating the curtains hung in its many windows. A Catholic church in Andale, Kansas, standing directly in the path of another twister, remained intact when the storm took an abrupt turn.

Other locations were less lucky. A tornado that, in its 293-mile (471.5 km) journey from Missouri to Indiana established the record as the tornado that traveled the longest continuous distance ever, left a wide swath of smashed farms, toppled grain elevators, and mangled automobiles. The northern section of Mattoon, Illinois, was virtually eliminated from the map, and 110 persons lost their lives there. Charleston, Illinois, suffered a similar fate when a different tornado slammed into it, demolishing 221 homes and killing 38 residents of that medium-sized town.

UNITED STATES
MISSISSIPPI
NATCHEZ
May 7, 1840

Three hundred seventeen people died and 109 were injured when a tornado struck Natchez, Mississippi, on May 7, 1840.

Nine inches of rain punctuated by huge hailstones preceded a tornado that followed the Mississippi River northward and slammed into Natchez, Mississippi, at 2 P.M. on May 7, 1840. In the five minutes the tornado took to traverse the city, it chewed up the riverfront, demolished two residential areas, killed 317 people, injured 109, and caused a staggering—for 1840—$1.26 million in property damage.

Pieces of houses were later found 30 miles (48.3 km) away from the city. Sheets of tin from roofs were scaled through the air like scimitars with a similar, murderous effect. Houses, their windows shut, exploded when the pressure around them suddenly plummeted.

The most tragic occurrences, however, took place on the river, near the docks. Two large steamers crowded with passengers and 60 flatboats were simultaneously capsized by the storm, drowning a large percentage of their passengers and crews.

UNITED STATES
MISSOURI
MARSHFIELD
April 18, 1880

The collision and consequent fusion of two tornadoes killed 101 and injured 600 in Marshfield, Missouri, on April 18, 1880.

Two tornadoes collided in Marshfield, Missouri, on April 18, 1880, and that cataclysmic meeting of forces combined to erase the town from the face of the Earth.

Not one structure was left unaffected. All were splintered, pounded into the earth or exploded sky high. One hundred one people died; 600 were injured seriously. Over $2 million in damages was recorded—an incredible amount in 1880, in a thinly populated corner of the United States.

UNITED STATES
MISSOURI
POPLAR BLUFF
May 9, 1927

One hundred forty-two people were killed by a tornado that struck Poplar Bluff, Missouri, and parts of Oklahoma, Texas, and Kansas on May 9, 1927.

Ninety-two residents of Poplar Bluff, Missouri, were killed by one of a series of tornadoes that left another 50 dead in Oklahoma, Texas, Kansas, and other parts of Missouri on May 9, 1927.

The storm, which caused over $1 million in damage to the small city, slammed into the middle of its business district at 3:15 P.M., while children in the East Side School were finishing their afternoon classes. Several children were killed; many more were trapped and injured in the collapsed school.

Most of the fatalities in the downtown part of Poplar Bluff took place in its three largest hotels. Its newest structure, the Melbourne Hotel, was hammered into the ground, trapping and killing those inside it. The four-story Drucker Hotel suffered a similar fate, but some of its inhabitants were saved by brick walls. They had to wait a number of hours before being rescued. The Harris Hotel—a wooden structure—did not collapse entirely, but was swept by a huge fire that roared out of control, threatening the rest of the business district. Even before the fire department could reach the Harris, an enormous rain squall generated by the storm doused the flames.

Relief arrived soon after the storm in the form of troops to prevent looting and a trainload of doctors, nurses, and medical supplies.

UNITED STATES
MISSOURI
ST. LOUIS
May 27, 1896

A tornado containing 560 MPH winds killed 306 residents of St. Louis, Missouri, and injured 2,500 on May 27, 1896.

If the estimate of weather expert Frank H. Bigelow is to be believed, the winds within the vortex of the immense killer tornado that ravaged St. Louis, Missouri, on May 27, 1896, were traveling at an amazing speed of 560 MPH.

The evidence of horrendous ruination, superphenomenal in its scope, gives credence to Bigelow's estimate: 306 killed, 2,500 injured, and more than $12 million in property damage, and all of it accomplished in 20 minutes. Straws were driven with the force of a rifle bullet six inches (15.24 cm) into a tree trunk. A six-by-eight wooden plank was pounded four feet straight into the ground. A farmer's spade burrowed six inches (15.24 cm) into another solid tree trunk.

This, the most violent storm ever to visit St. Louis, was preceded by spectacular thunder and lightning, first observed, according to an official of the United States Weather Bureau, ". . . at 5:00 P.M., an hour before the tornado occurred. This continued with short intermissions until 5:45 P.M., when it became almost continuous. . . . At 6:00 P.M., when the tornado occurred, the whole west and northwest sky was in a continuous blaze of light. Intensely vivid flashes of forked lightning were frequent, being outlined in green, blue, purple and bright yellow colors against the dull yellow background of the never ceasing sheet lightning."

A period stereo view captures the havoc wrought by the 560-MPH (901 km/h) winds of the tornado that struck St. Louis, Missouri, on May 27, 1896. The view is of the corner of Rutger and Seventh Streets, where seven of the 306 residents who were killed by the storm lost their lives. (Library of Congress)

Somewhere within this web of color and light was the tornado, which, uncharacteristically, was not formed in a funnel, but rather as a mass of black clouds. At 6:10 P.M., it roared into the city limits, rocketed across six miles (9.65 km) of densely populated parts of the city, exploding homes, driving trees like medieval pikes through thick walls, crushing, with its awful velocity, entire city blocks.

And then it was gone, across the Missouri River, leaving behind it raging fires, the cries of the wounded and the dying, and thousands of bewildered, sometimes crazed survivors.

UNITED STATES
MISSOURI
ST. LOUIS
September 29, 1927

Eighty-five people died and over 1,300 were injured when a tornado struck St. Louis, Missouri, on September 29, 1927.

One of the most horrendous happenings in the five-minute passage of the tornado that killed 85, injured over 1,300, caused $100 million in property damage, and virtually wiped out a large part of St. Louis's southwest section, was the presence of mysterious sheets of flame that scores of inhabitants swore they saw, whipping like grim, fluttering curtains down streets and abruptly disappearing against walls of buildings.

In contrast, one of the most inspiring happenings was the conduct of the two radio stations, KMOX and KFVE—KFWF had suffered a demolished antenna as a result of the storm and so was rendered silent—in directing rescue efforts and appealing for outside aid. Stationing himself on the roof of the Chase Hotel, KFVE broadcaster Patrick Convey, following a game plan that had been evolved two years earlier by the local disaster committee, was able to direct units of the Red Cross, Fire Departments, National Guard, American Legion, and Boy Scouts to specific disaster sites. At ground level, KMOX's George Junkin kept in further contact with rescuers and, working around the clock, appealed for relief supplies and money. Because of this, it took only hours for medical attention, food, clothing, and money to arrive in the city, thus preventing the spread of fires and the seemingly inevitable aftermath of looting.

The twister announced its approach shortly before 1 P.M. on September 29, 1927. The noontime air suddenly thickened, thunder growled in the distance, and lightning flashes illuminated the fast approaching black and orange clouds.

The full impact of the twister struck at 1:05 P.M., exploding homes, shearing off chimneys, hurling building blocks and automobiles through the air, collapsing streetcars, felling trees and telephone poles, ripping live wires from their protective moorings.

Survivors on the ridges of the city reported that the twister churned up the earth as if some sort of cosmic plow had touched down. In East St. Louis, Granite City, Madison, and Venice, Illinois, and throughout St. Louis itself, over 1,800 dwellings were either exploded or slammed into the ground.

UNITED STATES
NEBRASKA
OMAHA
March 23, 1913

A giant tornado hit Omaha, Nebraska, on March 23, 1913, killing 115, destroying 600 homes, and seriously damaging 1,100 others.

The worst tornado ever to smash into Omaha, Nebraska, was documented on its way toward its grim mission of destruction by *Topeka Daily Capital* reporter Milton Tabor, who was picnicking at the time: "The tornado," he wrote, "whirled furiously over our heads. We looked up into what appeared to be an enormous hollow cylinder, bright inside with lightning flashes, but black as blackest night all around. The noise was like ten million bees, plus a roar that beggars description."

Tabor must have witnessed this sometime between 5:30 and 6:00 P.M. on the late afternoon of March 23, 1913. At approximately 6:00 P.M., the tornado touched down at Forty-fourth and Frances streets, then tumbled on northeastward to Center Street, near millionaires' row in West Farnum. Before it would leave, 12 minutes later, it would claim 115 lives, destroy 600 homes, seriously damage 1,100 others, and rob the economy of this midwestern city of $3.5 million in property damage.

The path of the tornado was limited to a small area—one-quarter of a mile in width and five miles in length—but within that path, horror reigned.

Barn roofs were lifted and sent flapping through the air. On Mayberry Street, a family was crushed at the dinner table when their home folded in upon them. At Forty-eighth and Leavenworth streets, a streetcar was lifted off its tracks and capsized, with all of its passengers inside. A baby was killed in its father's arms by flying glass.

The huge stone palaces built by millionaires in the Bemis Park District were no match for the combined forces of wind, hail, sleet, and rain generated by the storm. Roofs were peeled away, windows were exploded

A ruined landscape reflects the passage of the worst tornado to ever hit Omaha, Nebraska. The 1913 twister killed 115, destroyed 600 homes, and seriously damaged 1,100 others. (Library of Congress)

outward, and monumental stones were flung like baseballs from yard to yard.

Not only the wealthy were affected. At Twenty-fourth and Grant Streets, the Idlewild Poolhall collapsed upon 20 pool-playing patrons, crushing them to death.

And even the dead were not left undisturbed. In West Lawn cemetery, cadavers were ripped from their graves, spun skyward within the tornado's eyewall, and flung yards from the graveyard.

The continuing rain extinguished fires set by the tornado, but that was only the beginning of a massive cleanup effort that would take months to complete.

UNITED STATES
NORTH AND SOUTH CAROLINA
March 29, 1984

A freak series of 24 tornadoes ripped through North and South Carolina on March 29, 1984. Sixty-one people were killed, 1,000 were injured, and nearly 3,000 were made homeless. New York and New England were also affected.

An unusual set of atmospheric events on March 29, 1984, produced an immense storm system that would send 24 tornadoes through North Carolina, scatter thunderstorms accompanied by hurricane-force winds and blinding snow into New York, topple a 103-year-old lighthouse in Nantucket, and raise wind levels to over 115 MPH on Mount Washington, New Hampshire.

The gigantic storm system was born when a vast body of warm, moist air from the Atlantic was driven over cold, dry air descending from Canada. At the same time, an unusually active jet stream blew across the top of the storm system, sweeping the topmost clouds downward into an anvil shape and setting up huge updrafts of air. These factors converged to form, according to Joe Friday, deputy director of the National Weather Service, ". . . one of the most severe general cyclonic storms in recent history."

The worst damage came in North and South Carolina, where 61 people were killed, 1,000 were injured, nearly 3,000 were rendered homeless, and damage estimates climbed to $200 million. The tornadoes, racing along a 260-mile (418.4-km) path across both of these states, missed most thickly populated urban areas, but wreaked havoc in rural spots, tossing mobile homes about like pieces of cardboard and shearing the roofs off homes and businesses alike.

Red Springs, North Carolina, was one of the few towns struck dead center by a tornado, which touched down in the town's main street, blowing apart businesses, flattening a school, and knocking down hundreds of 200-year-old oak and pecan trees that had lined the elegant, residential area of the city. A three-year-old boy was killed when the roof of a church collapsed upon him and other refugees from the approaching tornado, and more than 300 houses and 100 mobile homes were severely damaged.

In the northern part of South Carolina, the storm hit the towns of Anderson, Winnsboro (where one person was killed), Newberry, and Bennettsville (where seven people perished when a twister slammed into an apartment building and shopping center).

The group of twisters then entered North Carolina's sandhills and tore through the coastal plain, including the towns of Maxton, Shannon, Red Springs, Parkton, and Mount Olive. People and homes were blown ". . . all over the fields . . ." according to Mark Tartis of the Emergency Division of Scotland County, North Carolina.

Before the giant storm system would blow itself out over New England, hundreds of thousands of homes would be robbed of power and some of their roofs. But the greatest siege of terror occurred in North and South Carolina, which one North Carolina survivor described to a reporter as ". . . the hand of God reaching down to crush us."

UNITED STATES
NORTHEAST
May 31, 1985

Ninety-three people were killed and over 2,000 were injured when more than 24 tornadoes traversed the northeastern United States on May 31, 1985.

Ninety-three people lost their lives and over 2,000 were injured when more than two dozen tornadoes slashed through central Ontario to parts of Ohio, Pennsylvania, and New York on the evening of May 31, 1985. A combination of warm, humid air from the Gulf of Mexico, cooler air from the north, and a cold front appearing from the west sent giant hailstones hurtling through the air in Ohio and Pennsylvania and spawned the tornadoes—which churned along at 200 MPH with a deafening noise that witnesses compared to freight trains or jets.

In Ontario, where the tornadoes first touched down, the heaviest damage occurred north of Toronto, were 12 died and 150 were injured. Eight people died in Barrie, the site of the Barrie Racetrack, where cars and trucks were strewn crazily around the track and terrified horses, freed from their partially demolished stalls, ran wild over the countryside.

In Ohio, over $4 million in damage was caused as the twisters virtually destroyed the village of Newton Falls, crushing or exploding 300 homes, turning much

A house in North Sewickley, Pennsylvania, was no longer for rent after one of the series of tornadoes that raged through western Pennsylvania, southwestern New York, and eastern Ohio on May 31, 1985. (American Red Cross/Joseph Matthews)

of the downtown area into rubble, and setting off numerous fires. Nine people were killed in Youngstown, when a tornado cut a 200-foot (60.9-m) swath across three and a half miles (5.6 km) of the town.

New York State suffered only property damage, but neighboring Pennsylvania was hit extremely hard. The state suffered hundreds of millions of dollars of damage—more than in any other area visited by the storm. In Atlantic, Pennsylvania, five people were killed, dozens of homes, a grain mill, and a 312-foot (95 m) microwave tower were pulverized. In Wheatland, a steel plant that employed 500 people simply disappeared. Mayor Helen M. Duby and her family escaped to their cellar, and when they emerged, the town of 1,122 was gone. "It was like a nightmare," she said.

In Butler County, eight women at a Tupperware party tried to make it to a storm cellar outside the house. Seven succeeded. The eighth was found in a creek two miles away, dead.

Albion, in northeastern Pennsylvania, was struck shortly after 5 P.M. Martha Sherman, the Red Cross disaster coordinator, noted that a warning appeared on her television screen only seconds before the tornado struck her block. The storm roared through, turning tall maples trees into missiles, sending them through the air and dropping them half a mile from where they had been rooted. Automobiles were flung about like plastic toys. When the storm passed, there was little left of the town, and what remained was unrecognizable.

In the wake of the tornado, roads were sealed off to the stricken towns to prevent looting, and the National Guard hauled 500-gallon (1,892.7-l) water tanks to many areas, since many private wells were now inoperable.

UNITED STATES
OKLAHOMA
OKLAHOMA CITY
June 12, 1942

Thirty-five people died and 30 were injured on June 12, 1942, in still another Oklahoma City tornado.

Sometime during the night of June 12, 1942, a tornado touched down in a residential section of Oklahoma City, Oklahoma, killing 35 sleeping citizens in their collapsing and exploding dwellings. Thirty more were taken to hospitals the following morning. More than 100 homes were blown apart by the nocturnal twister, and more than $300,000 in damage was caused to both the residential and neighboring business district of this city that—more than any other city in the United States—has been the victim of constant assaults by tornadoes.

UNITED STATES
OKLAHOMA
OKLAHOMA CITY
April 30, 1970

Because of the population's response to early warnings, there was no human loss when a tornado hit Oklahoma City on April 30, 1970.

The area of frequent tornado activity that runs from Texas northward through Oklahoma, Kansas, and Iowa is known to its residents as "Tornado Alley." Perhaps because the people in this area pride themselves in "sensing" tornadoes before television and radio reports warn of them, perhaps because they know, through experience, to heed forecasters' warnings, the twin twister that hit Oklahoma City at around 2 A.M. on April 30, 1970, caused monumental damage but killed no one.

A torrential thundering deluge had hit at 11:30 P.M. the night before, filling storm drains and flooding parts of the city. Up until 2:20 A.M., when the first house collapsed, cleanup crews were busy clearing flooded drains and streets. But during that time, radio and television reports constantly warned of approaching tornadoes. Shortly after 2 A.M., the radar at the Emergency Operating Center noted a funnel 12 miles (km) from the city moving toward it at nearly 50 MPH.

At 2:20, the two funnels smashed into the northwestern section of the city, while warning sirens wailed. The tornadoes hit the earth, bounced, and hit again. And again. Telephone lines snapped, showering glistening displays of sparks skyward and earthward. Houses and other buildings exploded. A semi-trailer was flung onto its side at Interstate 40 and Council Road.

In 20 minutes, the storms were gone. Emergency crews fanned out over the affected area, a thickly populated Oklahoma City suburb. Twenty miles of it had been scoured clean by the storm, and millions of dollars of property damage was recorded. But, again, because of training and reporting, there were no serious injuries and no deaths.

UNITED STATES
OKLAHOMA, ARKANSAS, AND
MISSOURI
April 12, 1945

One hundred twelve people were killed when a series of tornadoes traveled through Oklahoma, Arkansas, and Missouri on April 12, 1945.

A series of tornadoes raged through Oklahoma, Arkansas, and Missouri on April 12, 1945, and, although damage was extensive in Arkansas and Missouri, Oklahoma suffered the majority of the 112 deaths and hundreds of thousands of dollars in damage.

In Oklahoma, Oklahoma City was sideswiped by the storms and four other nearby towns received great damage: Rowland, Boggy, Red Oak, and Greenwood Junction. But the worst damage and the greatest loss of life occurred in Antler, Oklahoma, the town first hit by the storms. A small hamlet of 3,000, it lost all of its power and its residential area was torn asunder, with scores of houses flung wildly apart by explosions and the flattening force of the twister. Sixty-three people were killed—including the owners of the town's only funeral parlor—and hundreds were injured.

UNITED STATES
OKLAHOMA AND KANSAS
May 3, 1999

The most expensive tornado outbreak in U.S. history struck Oklahoma and Kansas in the early evening of May 3, 1999. Seventy-six tornadoes roared through central Oklahoma and the Wichita, Kansas, areas, killing 42, injuring 748, and causing nearly $1.5 billion worth of damage.

A violent storm system containing 76 tornadoes roared through central Oklahoma and the Wichita, Kansas, areas on the evening of May 3, 1999. Tornado warnings were well posted in 44 of Oklahoma's 77 counties, and so Oklahoma was ready—or as ready as any place in the world can be for the extreme forces of nature.

The first storm originated around Fort Sill, formed from three to five tornadoes, and headed directly for Oklahoma City, but it was only a few seconds away from some of the other tornadoes, which spread fanlike throughout all of central Oklahoma. Some were born and died almost instantly. Some fused with each other. At one time, storm chasers were able to see 14 tornadoes at once.

The most intense tornado began near Chickasha, about 45 miles (72.4 km) southwest of Oklahoma City. Its winds were clocked at one moment at 300 MPH. A mile wide, this monster first struck the bedroom community of Bridge Creek, 30 miles (48.3 km) southwest of Oklahoma City, where it damaged 3,000 homes, injured 30 people and killed 11.

A few minutes later, it plowed into Moore, flattening 836 buildings and damaging 285 others. Seventy-five per cent of the local elementary school was destroyed;

the local high school was heavily disfigured, and Moore Industrial Park received $2,000,000 worth of damage.

And now the tornado entered Oklahoma City proper. At 6:50 P.M. the Oklahoma Red Hawks game at Bricktown ballpark was stopped, and some 2,000 fans and employees were herded into an underground area outside the locker rooms. The storm uprooted trees, flung cars around, blew out the windows in businesses, and damaged or destroyed 1,000 homes.

Although almost all of the human beings in the city escaped injury, cattle, horses, and smaller family pets were killed. The body of a gray horse was lifted up by the storm and dropped onto a high school parking lot.

In some areas, softball-size hail accompanied the winds, and they became destroyers of property. In fact, at the end of the following day, it was estimated that there was enough debris in Oklahoma City alone to fill 4,700,000 18-wheelers.

Although the storm struck after rush hour, some commuters were still on the road, heading home from work. The scene on Interstate 40 was described by a local paper as "horrific—rush hour transformed into an auto junk yard." Cars and trucks, some of them with their drivers still in them, were tossed about like toys. Several car dealerships, struck by the tornado, added their stock to the general junkyard scene.

As the storm cell raged on, it killed several people, flattened all 53 stores in the Tanger Mall in Stroud, Oklahoma, southwest of Tulsa, and damaged or destroyed every single building in Mulhall, Oklahoma. The elementary school there was demolished, the grocery store was eliminated, and three churches were damaged. The town's 75-foot (22.8 m) high, 400,000-gallon (1,514,64.7-l) water tank collapsed, causing a cascade of water to push a nearby house 20 feet (6.1 m) off its foundation.

There were 748 injuries from the storm treated at area hospitals. Forty-two people were killed. Estimates of damage were $1.485 billion, making this the most expensive tornado outbreak in U.S. history.

UNITED STATES
SOUTH
February 19, 1884

Over 800 people died and over 10,000 were made homeless when 60 tornadoes traversed the southern United States on February 19, 1884.

More than 60 tornadoes tore through the southern states on February 19, 1884, flattening dwellings, killing more than 890 people, and rendering over 10,000 homeless. The entire town of Davisboro, Georgia, was

Stripped trees, bodies, and a war-torn landscape were all that remained of multiple areas of the southern United States after a series of 60 tornadoes ripped back and forth across the area on February 19, 1884. (Frank Leslie's Illustrated Newspaper)

destroyed, as well as several other small farming villages in six southern states. The twisters, of unusual ferocity, arrived in split seconds, with almost no warning in certain cases. Reports, mostly contained in Frank Leslie's *Illustrated Newspaper,* were rife with accounts of entire families being crushed within houses when one of the 60 twisters struck.

UNITED STATES
SOUTH
April 29–30, 1924

One hundred fifteen people died and more than 1,500 were made homeless when 22 tornadoes raised havoc in the southern United States on April 29 and 30, 1924.

Twenty-six towns in South Carolina, Georgia, Alabama, Arkansas, Louisiana, and Mississippi were obliterated by 22 tornadoes that touched down in the southern United States during the two-day period of

April 29 and 30, 1924. Riverside Mill, South Carolina, was particularly hard hit, losing its entire business district, half of its residences, and a $500,000 mill. Over 1,500 became homeless in that town alone.

A total of 115 people lost their lives and a conservative estimate of the storm's damage reached $10 million.

UNITED STATES
SOUTH
April 2–6, 1936

Multiple tornadoes killed 421 and injured 2,000 in the southern United States from April 2 through 6, 1936.

On the five days of April 2, 3, 4, 5, and 6, 1936, a veritable firestorm of tornadoes touched down in six southern states—Alabama, Arkansas, Georgia, Mississippi, North Carolina, and Tennessee. Four hundred twenty-one people lost their lives, 2,000 were injured, and property damage was estimated in excess of $25 million. Not since the notorious twister of June 1, 1903, in Gainesville, Georgia, in which 203 people died and nearly a thousand were injured, did inhabitants remember such devastation.

Ironically, Gainesville was also hit the hardest in this series of storms; its entire downtown section was demolished by the tornado that roared into this town 55 miles outside Atlanta at 8:45 A.M. on the morning of April 6. Without warning the tornado slammed into the Cooper Pants Manufacturing Company, destroying it and killing 125 people instantly.

In the middle of town, the hotel, the First Baptist Church, and virtually every one of the downtown area's 630 buildings was either flattened or damaged, and 33 businesses were whisked off their foundations. Only the new post office and federal courthouse remained unscathed.

Five days before, at the same hour of 8:45, the first of the tornadoes had sideswiped Gordo, Georgia, and roared through Cordele, Georgia, and Greensboro, North Carolina. In Cordele, a 12-mile (19.31-km) long, 400-foot (121.9-m) wide swath was cut through the town's center, 23 people were killed, 600 were injured, and 289 buildings were lost.

Between the 2nd and 6th of April, Coffeyville, Booneville, and Auburn, Mississippi; Columbia, Fayetteville, and LaCrosse, Arkansas; Elkwood and Red Bay, Alabama; and Acworth, Woodstock, and Lavonia, Georgia, felt the storms' furies.

On the last day of the storms' multiple crisscrossings of the six states, Tupelo, Mississippi, was hit a devastating blow. The small town with a population of 10,000 suffered a loss of 195 lives in the space of a few tumultuous moments. Office buildings were turned into hospitals as the town's only hospital was filled beyond capacity. Fires threatened to add to the damage, but fortunately, rains appeared shortly after the storm passed and extinguished what the local fire department could not.

UNITED STATES
SOUTH CAROLINA
CHARLESTON
September 10, 1811

Over 500 people were killed by the noon-hour tornado that struck Charleston, South Carolina, on September 10, 1811.

A southerly gale blew through Charleston, South Carolina, through most of the morning of September 10, 1811. But by noon, the winds had calmed. Citizens went about their ordinary tasks, only mindful of an odd pinkness that seemed to paint the underside of the clouds that remained clustered over the city.

Deathly calms are notoriously the harbingers of storms to come, and this was precisely the case in Charleston that day. At approximately 12:30 P.M., the air began to thicken, and a telltale dark, funnel-shaped cloud bore down directly upon the city. Within moments, it had cut a swath of devastation through the crowded noontime streets, uprooting and twisting trees into odd shapes, blowing up buildings, and killing more than 500 people.

UNITED STATES
SOUTH CAROLINA
CHARLESTON
September 29, 1938

Five tornadoes struck Charleston, South Carolina, simultaneously on September 29, 1938, killing 32 and injuring 150.

Five twisters at once slammed into Charleston, South Carolina, on the morning of September 29, 1938. In less than two hours more than $2 million in property damage was sustained, 32 people were killed, and 150 more were hospitalized.

The first tornado was a relatively mild one, setting up a feeling of complacency in the public—a blueprint for disaster. The second and third, touching down a mere two miles apart, caused the most damage and loss of life. While the second twister hit the city directly, the third, more powerful one first hit land on James Island, and then ran into the city for another 1,600 feet (487.68 m), destroying everything in its path. The fourth and fifth merely splintered what had already been demolished.

UNITED STATES
SOUTHWEST (TEXAS, OKLAHOMA, AND KANSAS)
April 9, 1947

The most destructive and widest tornado ever to hit the Texas panhandle killed 169 and caused millions of dollars in property damage throughout the southwestern United States on April 9, 1947.

The tornado of April 9, 1947, was a record breaker. It traveled in the widest path ever recorded—one and a half miles (2.4 km) (the width of a tornado's path is usually measured in feet)—and was the most destructive tornado in the entire history of the Texas panhandle.

Its first recorded touchdown was in the small (population 500) town of White Deer, Texas. Normally, tornadoes only sound like freight trains. This one struck a freight train just outside of White Deer, lifting it from the tracks as if it had been made by Lionel. Fortunately, only three people were injured in this small town, but a goodly portion of its few buildings were demolished.

The still smaller hamlet of Glazier, Texas (population 125), was the next stop for the tornado. It flattened every building in town except the schoolhouse, killed 16 people, and transported two 200-pound (90.7 kg) men—C. S. Wright and Art Beebe—30 feet (9.1 m) in the air for nearly 300 yards before dropping them, unhurt, into the debris.

On the twister proceeded, demolishing Higgens, Texas, where it killed 45 people and set fires that even rain and golf ball-sized hailstones failed to extinguish.

After destroying the town of Shattuck, Texas, the tornado crossed the border into Oklahoma and wreaked its greatest havoc in the town of Woodward, which was also the largest town on its itinerary. One hundred blocks of Woodward were blown off the face of the Earth, 95 people were killed, and for the few minutes the tornado took to rage through the town, the air was filled with flying pieces of houses, roofs, roadsigns, and automobiles.

It was the last location to suffer a human toll. White Horse, Oklahoma, received a sideswipe, and Kansas, while suffering $200,000 in property damage, reported no deaths.

In all, 169 persons died and more than $15 million in damage was caused in the 221-mile (355.7 km) path of destruction left by the April 9, 1947, Southwest Tornado.

UNITED STATES
TEXAS
WACO
May 11, 1953

One hundred fourteen people were killed and 500 injured when a fierce tornado hit Waco, Texas, on May 11, 1953.

In 1953, weather scientists were just beginning to realize the value of radar in tracking tornadoes and other

Rescuers probe through the wreckage left by the tornado that struck Waco, Texas, on May 11, 1953. Failure to heed warnings resulted in the deaths of 114 people. (Smithsonian Institution)

violent storms. The Air Force Weather Service had developed a radar network that ran from Omaha, Nebraska, to Oklahoma City's Tinker Field, and a week before May 11, it had tested it to the point of satisfaction. It could, evaluators felt, become a useful tool to supplement the tornado watch alerts issued by the U.S. Weather Bureau.

Late in the afternoon of May 10, 1953, a tornado watch alert was issued by the U.S. Weather Bureau for central Texas, warning that a mass of warm, moist air was moving northward from the Gulf of Mexico and was expected to meet a large mass of cold, dry air that had moved in from the north somewhere between Waco and San Angelo.

The population of San Angelo heard and heeded the warning. The population of Waco did not. The resultant tragedy was a primer in preparedness.

Patrolmen of the Texas Department of Public Safety spotted the storm three hours before it hit San Angelo. The twister was a slow mover. Its forward speed was judged to be 15 MPH, and it stuck closely to the Concho River valley.

By 2:15 P.M., when the elephant trunk of the tornado's cone touched down in San Angelo, schoolchildren in the Lake View residential section had moved into inside hallways or under desks away from the windows, which had been opened on the side of the building away from the storm. Although the roof of the school lifted off and the windows blew out, no one was injured.

Police began rescue work even before the wind had begun to lessen. Local ambulances arrived upon the scene within 10 minutes. By 3:15 P.M., National Guard and Goodfellow Air Force Base personnel had set up a command post. Still, 11 people died and another 66 suffered serious injury, but the toll could have been far higher. There was hardly anything left of San Angelo's business section.

Now, the storm gathered speed and headed toward Waco. Here, in contrast to San Angelo, there was a minimum of preparation. Hardly any of the 90,000 residents of the city had heard the news about San Angelo. Perhaps because many believed the Indian legend that the "dancing devil" could never dance between the hills that circled the town—a legend accepted so unquestioningly that few Waco residents carried wind damage insurance on their homes and automobiles—the city was caught absolutely unaware.

The tornado touched down on the city's southwest side at approximately 4:45 P.M. It cut a grim swath into the business district, demolishing 83 buildings in a matter of minutes. At the Joy Theater, late afternoon patrons were buried when the arched roof collapsed, and the front of the building exploded into the street.

Steel mailboxes were turned into deadly flying objects; bricks and chunks of cement shot through the air. Automobiles tumbled end over end down city streets. Apparently descending directly on top of the Dennis Furniture Building, the tornado wrenched the top four floors off the ground floor, exploding them into dust and scattering them over a five-block area.

Twenty-five high school boys and older men were playing pool in the Torrance Recreation Hall behind the Joy Theater. Tons of debris from the Dennis Building arched through the air and landed on the Recreation Hall, collapsing its roof and burying all of the men and boys within. None survived.

A steady rain fell after the storm passed on. By 5 o'clock, police, fire department, National Guard, and army rescue teams were at work. Human chains were formed to unearth people from beneath mountains of debris.

In the few moments that it took the tornado to traverse Waco, it destroyed two square miles of the city's business district, killing 114 people, injuring 500, and causing $50 million in damage.

Ironically, a month later, on June 8, another tornado struck the same area, bypassing Waco, but killing 11 people in the nearby town of Flint.

STORMS

THE WORST RECORDED STORMS

* Detailed in text

Bahrain
   (windstorm) (1959)

England
   * London (fog) (1952)
   * London (fog) (1962)

English Channel
   * (windstorm) (1974)

India
   * (monsoon) (1955)
   (monsoon) (1972)
   (monsoon) (1983)

Italy
   Calabria (windstorm) (1951)

Japan
   Honshu (rain) (1953)
   Kyushu Island (rain) (1957)

Malagasy Republic
   (rainstorm) (1959)

Thailand
   * (monsoon) (1983)

Tunisia
   * (rain) (1969)

Western Europe
   * (windstorm) (1999)

CHRONOLOGY

* Detailed in text

1951
   October 21
      Calabria, Italy (windstorm)

1952
   December
      * London (fog)

1953
   June 26
      Honshu, Japan (rain)

1955
   October 7–12
      * India (monsoon)

1957
   July 25–28
      Kyushu Island, Japan (rain)

1959
   February 28
      Bahrain (windstorm)
   May 26–27
      Malagasy Republic (rain)

1962
   December 3–7
      * London (fog)

1969
   September–October
      * Tunisia (rain)

1972
   June
      India (monsoon)

1974
   January 16
      * English Channel (windstorm)

1983
   September 1–18
      India (monsoon)
   September–December
      * Thailand (monsoon)

1999
   December 26–28
      * Western Europe (windstorm)

STORMS

The following storm disasters are collected together because they occur too infrequently to justify a section of their own.

All are self-explanatory, with the exception of monsoons, which possess distinct and individual characteristics.

Simply stated, monsoons are winds that change direction with the change of seasons in India and Southeast Asia. In the winter, they bring no moisture; in the summer they carry with them torrential rains, which, when unleashed, cause large amounts of human and crop damage.

The change in wind direction results from changes in air temperature over land masses and oceans. In winter, the wind change that causes monsoons results from an area of high pressure that develops over south-ern Siberia, out of which dry winds cross India from northeast to southwest, and Southeast Asia directly from north to south. The wet, or summer monsoon formation is a little more complex: Its source is an area of low pressure which develops over Southeast Asia as the land mass slowly warms. As the low pressure area deepens, moisture-heavy air from the oceans is drawn toward it. Then, this air cools as it ascends the slopes of mountain barriers in Asia and India, and, losing its ability to retain moisture, releases it in torrential rainfall.

Since the destruction from monsoons almost always manifests itself in floods, consult the separate section on floods (p. 137) for additional information on the effects of monsoons in India and Southeast Asia. Only two monsoons are included in this section.

ENGLAND
LONDON FOG
December 1952

London's deadliest fog occurred in December 1952. Twelve thousand people died from ingesting soft-coal smoke, sulphur dioxide, chemical fumes, and gasoline fumes contained in the fog.

London's fogs are both legendary and picturesque—to all but those who have experienced them. Formed by a combination of stalled warm air masses and the smoke given off by thousands of soft-coal stoves and factory chimneys, they have, for centuries, literally strangled a large part of the populace of this city, bringing its commerce to a stop, hospitalizing many and killing more.

Even the interiors of theaters, restaurants, and other public places whose doors to the street opened and closed frequently were not exempt from the notorious fog. The dense, pungent, opaque air would roll into movie theaters, obscuring the screen and making the auditorium uninhabitable.

The deadliest fog ever settled over London in December 1952. Soft-coal smoke, sulphur dioxide, chemical fumes, and gasoline fumes collected in the atmosphere, intensifying as the warm mass lingered. Four thousand people died from ingesting the fumes, and another 8,000 eventually died from its aftereffects.

It was this catastrophic fog that brought about the four-year study that would produce England's Clean Air Act.

ENGLAND
LONDON FOG
December 3–7, 1962

One hundred thirty-six people died and 1,000 were hospitalized in the last lethal London fog, from December 3 through 7, 1962.

Great Britain's Clean Air Act of 1956 had scoured the London air of most pollutants. But certain industries were still sending great quantities of sulphur dioxide into the air in and around England's capital city.

Thus, when another air inversion brought on a heavy fog on December 3, 1962, 136 elderly persons succumbed to the fumes and another 1,000 were hospitalized.

The *New Yorker's* London correspondent gave the following colorful description: "The fog floated visibly, like ectoplasm, into theaters, cinemas, and shops, where sales assistants stood around gossiping among the sparsely populated wastes of Christmas presents. In the evenings, the West End was as empty as on a night in the blitz. . . . If you had felt the urge to do so, you could have danced in the eerie middle of Piccadilly in safety, if not in comfort."

That, apparently, was after the first casualties of the fog conditions, and after London's 5,000 red double-decker buses were pulled from the streets after two of them collided, injuring 13 passengers. A train motorman was drowned when he stepped from his cab and, not realizing he had stopped on a bridge, plummeted into 40 feet (12.2 m) of water. Scores of pedestrians were run down by motorists, driving blind.

Perhaps the only positive advantage of this fog was the use it was put to by the British Ministry of Aviation. It successfully tested a new blind flying system at London Airport.

ENGLISH CHANNEL
WINDSTORM
January 16, 1974

A freak windstorm with 100-MPH gales killed 36 people and sent numerous ships to the bottom of the English Channel on January 16, 1974.

Completely outside the hurricane season and without the snow that would qualify it as a blizzard, a freak windstorm with 100-MPH gales whipped up 55-foot (16.8 m) high waves in the English Channel on January 16, 1974.

The 2,088-ton (1,894-tonne) Cypriot freighter *Prosperity* swirled out of control in the middle of the channel in the midst of the storm, which finally drove it aground on a reef off the island of Guernsey. Eighteen of its crew members drowned.

The *Merc Enterprises*, a 781-ton (708 tonnes) Danish freighter, also went down in the storm, taking with it eight crew members.

The storm then struck the coast of Brittany and battered the 2,447-ton (2,219 tonnes) freighter *Marta*,

10 miles (16.1 km) off the coast of Ostend, Belgium, causing the craft to catch fire. Hours afterward, the *Marta*, missing one of her seamen who had leaped with his clothes aflame into the water, limped into Ostend harbor, which by now was calm.

Other fatalities occurred in different locations. Six fishermen in Brest and Brittany were lost when their boats disappeared. At Dawlins, in Devon, two boys were swept off a seawall and into the Channel. Almost instantly, a wave scooped up one of the boys and swept him safely ashore. A 16-year-old girl was lost from a Brest ferry when wind-powered waves washed across the ferry's decks.

INDIA
MONSOON
October 7–12, 1955

One of India's worst natural disasters occurred as a result of floods generated by the monsoons of October 7–12, 1955. Seventeen hundred people were killed and 28,000 villages were totally destroyed.

When monsoon rains struck Orissa province from October 7 through 12, 1955, the population was caught unprepared.

The results were horrendous. According to the estimate of Dr. François Daubenton of the International Red Cross, 1,700 persons lost their lives and 28,000 villages were wiped off the face of the Earth, obliterated by the roaring flood waters from eight rivers; 3,500 square miles (5,632.7 sq. km) of cropland were flooded and rendered useless.

In India and Pakistan, thousands of people clung to trees, their only refuge above the roaring waters fed by downpours and whipped by the monsoon winds. Over 300,000 persons were evacuated by Indian paratroopers. The International Red Cross and other international relief agencies were quick to converge upon the disaster site. The United States alone sent 10,000 tons (9,071.9 tonnes) of wheat to the devastated area. (See also FLOODS, p. 137.)

THAILAND
MONSOON
September–December 1983

Ten thousand people died, 100,000 contracted waterborne diseases and 15,000 were evacuated as a result

of the monsoons that raged throughout Thailand from September through December 1983.

The three months of monsoon rains that engulfed Thailand from September through the middle of December 1983, caused more than $400 million in damage and 100,000 people to contract water-borne diseases. Ten thousand died in floods, and 15,000 were evacuated. Many of those who stayed cut their feet on submerged glass and other obstacles, and one hospital reported 20 cases and two deaths from leptospirosis, a disease once common in the trenches of World War I.

The storms caused the worst flooding in 40 years in the capital city of Bangkok, and brought to the attention of city planners the uncomfortable fact that, like Venice, Bangkok was sinking.

For two centuries, the city, built on low ground, had been drained by a network of canals. As urban sprawl took over, the canals were blocked or simply paved over, thus causing the massive flooding following the monsoon.

Thus, after floods, some city streets resembled riverbeds, with gravel and boulders littering them; others were like the aftermath of some recent battle.

Finally, after an exchange in which Admiral Tiam Makarandanda, the governor of Bangkok, accused the interior minister and his Irrigation Department of relieving flooded farmlands east of the city by pumping water into the suburbs, the interior minister replied—with what he believed to be supreme logic—that both men, after all, had the people's best interest at heart. An impatient and truly logical king Phumiphol Adulet asserted, in a national address, that unplanned construction had blocked Bangkok's drainage.

Cowed by the king's knowledge of hydraulics and stung by his criticism, the Cabinet of Thailand approved a $40 million series of drainage schemes designed to build flood-gates, dredge canals, and install permanent pumps. It had taken the carnage of a monsoon to bring this about. (See FLOODS, Thailand, 1988, p. 177.)

TUNISIA
RAIN
September–October 1969

Thirty-eight consecutive days of rain produced flooding over 80 percent of Tunisia in September and October of 1969. Five hundred forty-two people were drowned, 100,000 were made homeless, and 250,000 became refugees in other countries.

An unprecedented 38 consecutive days of rain fell over Tunisia in late September and early October 1969. Every river in the country was swollen beyond its banks, and over 80 percent of the country was flooded, from the Gulf of Bou Grara to the Gulf of Tunis.

Many of the rivers crested without warning in the middle of the night. Once unleashed, the floodwaters rampaged over the countryside, rising 36 feet above their normal levels, ripping apart 35 major bridges, and washing away a $7 million irrigation project.

One hundred-ton concrete blocks were shoved out of the way like children's toys, and millions of tons of valuable loam were washed into the Mediterranean. One hundred thousand head of cattle and other livestock drowned.

The human toll was staggering: 542 people drowned; more than 100,000 rendered homeless; and a quarter of a million people temporary or permanent refugees from the storm and the floodwaters.

Starvation was the immediate effect of the catastrophe. Villagers, wild with hunger, ripped the door off one U.S. Air Force helicopter that landed, some weeks later, in the remains of one gutted town. All told, 84 nations sent relief funds and supplies to alleviate the human toll and the $200 million in damage brought on by a deluge that was just two days short of the biblical one, and, to the human beings caught in it, nearly as devastating.

WESTERN EUROPE
December 26–28, 1999

Two monster windstorms roared across western Europe between Christmas of 1999 and New Year's Day of 2000, killing over 140, injuring thousands, and displacing thousands more. France received the worst damage, particularly to its electricity grid, its transportation system, and some of its greatest artistic treasures.

The storm of the century in western Europe was a doubleheader that chose France as its battleground. A monster storm that spanned the Atlantic and roared across the northern part of Europe just before dawn on the day after Christmas of 1999, it packed sustained winds of 95 MPH and gusts up to 110 MPH. "Since I was born, I've never seen anything like that storm," said David Chezaud, a not very young worker who was given the job of blocking off sidewalks and buildings that were littered with tons of debris in Paris.

Though all of northern France was battered by the extraordinary winds, which ripped trees from the ground, collapsed walls, crumbled buildings, and ripped roofs from their moorings, Paris, with its artistic and historic treasures, had the most visible damage.

Tens of thousands of Parisians who battled their way back from the Christmas holiday were greeted with immense scenes of destruction. Broken glass and bricks were strewn through the streets, uprooted trees blocked access to hundreds of blocks. Hundreds of the fabled Paris chimneys were turned to piles of brick and mortar; giant construction cranes toppled, decapitating the tops of buildings, kiosks, and streetlights. Traffic lights were bent at crazy angles, making them useless. Newspaper kiosks were overturned, and their contents strewn for blocks.

Notre Dame cathedral lost enormous slabs of its roof, which were lifted off and flung into the courtyard before it. Flying masonry shattered centuries-old stained glass windows in the 13th-century Saint-Chapelle. The roofs of the Bibliotheque Nationale and the Pantheon were ripped open, exposing priceless treasures to the elements.

Balzac's home, the site of the Musée Balzac, was cruelly collapsed when a huge tree smashed through its roof. Miraculously, the priceless sculpture within the museum escaped destruction.

The two principal city parks, the Bois de Boulogne and the Bois de Vincennes, were scenes of terrible carnage. Seventy thousand of their 300,000 trees were felled, and thousands more were stripped of enough of their branches to kill them.

Outside the city, the palace of Versailles and its meticulously planned and kept gardens resembled a battlefield after a war. Portions of the castle's roof were torn open, and windows were smashed. Fortunately, groundskeepers and guards managed to preserve the antiquities from the wind and rain that roared through the openings.

The grounds fared far worse. Ten thousand trees were ripped from the ground, including 200-year-old cedars planted at the time of the French Revolution. The trees that hid Marie Antoinette's *petite ferme* fell like matchsticks, exposing the centuries-old buildings to the elements.

"This is an ecological disaster for Paris," Françoise de Panafieu, Paris's deputy mayor, told reporters. "We'll need three years to reconstruct."

Ordinary citizens lost much in the storm. If hundreds of trees fell in Paris, more did outside of the city, and frequently on homes whose roofs had been peeled away by the 110 MPH gusts. By nightfall, 1.4 million people were without power. Roads were clogged with fallen trees and mudslides. Air traffic was universally cancelled, and train travel was brought to a standstill. In Lille, Lyon, and other cities, passengers spent the night in emergency lodgings in sports halls or youth hostels, where Red Cross volunteers supplied them with breakfast.

Off the coast of Britanny, gigantic waves were whipped up in the Atlantic, and the tanker *Erika,* which had foundered two weeks previously, gushed forth more oil, which was floated at express train speed toward 250 miles (402.3 km) of the Britanny coast.

France was not the only country to be impacted by the monster windstorm. In Germany, 17 people, most of them in the state of Baden-Wurttemberg, across the Rhine from France, died from falling trees and collapsing houses. In Bavaria, 1,000 homes lost power.

In Switzerland, falling trees killed 13, and two skiers plunged to their deaths when a tree crashed into a lift cable and sent the gondola in which they were riding spiraling hundreds of feet downward, into a ravine.

In northern Spain, walls, loosened by repeated wind gusts, collapsed upon workers at construction sites. In Italy, 7,000 passengers were stranded at Milan's Malpensa airport, highways were closed, and power was lost to thousands of homes and businesses. In Belgium and in Great Britain, rivers overflowed their banks, flooding several villages.

And then, just as the cleanup in Paris and throughout the storm-shredded countryside began to make a difference, a second, equally ruinous storm slammed into the southwest coast of France, in a region that had been spared the fury of the first storm. On the night of December 27 and well into the morning of the 28th, gale-force winds of up to 125 MPH ravaged the Atlantic coast and Bordeaux, moved to the French Riviera, then crossed the continent, laying waste to the countries along its path.

In La Rochelle, luxury yachts, tied up in safe harbors, were lifted by floodwaters and flung far inland. In the Vendee region, over 2,000 people were evacuated from their homes as floodwaters ravaged coastal towns. In Rouen, portions of the roof of the famous medieval cathedral were ripped loose, letting gusts of rain into the sacristy.

In Paris, two spires on Notre Dame cathedral, spared by the first storm, were severely damaged—one was decapitated by the wind, the other knocked at a crazy angle. At Versailles, more trees, loosened by the first storm, succumbed to the second. The Virginia tulip tree that Marie Antoinette had nurtured was ripped from its roots and sent tumbling down a muddy hill. The Corsican pine that Napoleon had ordered planted nearby was splintered asunder.

In Calvados, 100 people were rescued from flooded areas in which entire houses were submerged under raging river waters.

In the French Alps, snow, whipped up by the same storm system, buried Alpine villages and created chaos in ski resorts. The Savoy region was particularly hard hit with avalanches and snow-barricaded roads. In

Austria, driving winds and snow created avalanches that killed 12 people.

Rail traffic that had just begun to return to a semblance of normalcy was once more wildly disrupted. "Road, rail and air routes are affected," France's transport minister, Jean-Claude Gayssot, announced. "We are really in great difficulty."

Now, at least 3.4 million homes in France were without electricity, and 400,000 were without telephone communication with the world. Heat and water shortages made life miserable for tens of thousands. Without electricity, power tools needed to clear debris were silent and unusable.

On December 29, 9,000 French soldiers were deployed to the most severely damaged areas of the country to deliver drinking water and food supplies. Crews from Germany, Belgium, Italy, and Spain arrived to aid the French national electricity company as it replaced one-quarter of the country's electricity grid, and restored power to the millions who had lost it.

The destruction left by the twin storms was staggering, exceeding $4 billion by the time the second storm had blown itself out. It had all come almost without warning, crossing the Atlantic in 24 hours instead of the usual 48, then exploding upon the western coast of France. "The problem with this storm is that it came like a cannon shot, very fast and very straight, so there wasn't much time to collect anything," explained Patrick Galois, a forecaster with Metro, the French national weather service.

Pennsylvania State University meteorologist Todd Miner told the Associated Press that the first storm ". . . was a one-in-50-year event, but its quick succession by an equally powerful storm in nearly the same area could be expected only once in a hundred years."

Gradually, life returned to normal. Over 140 people had been killed by the storms, 88 of them in France. Thousands were injured and thousands more lost their homes. "France has been wounded and many French are faced with cruel hardship just when they were about to celebrate the end of the year and the millennium," President Jacques Chirac announced in the French press.

But they would celebrate. The 11 giant Ferris wheels placed in the middle of the Champs Elysees were unaffected by the storms, and at midnight on New Year's Eve, hundreds of Parisians rode them into the new century.

TYPHOONS

THE WORST RECORDED TYPHOONS

* Detailed in text

China
(1926)
Canton (1862)
Fukien (Fujian) (Typhoon Iris) (1959)
Haifung (1881)
* Hong Kong (1841)
* Hong Kong (1906)
Hong Kong (1947)
Hong Kong (1964)
Swatow (Shant'ou) (1922)
Typhoon Aere (2004)
* Typhoon Fred (1994)
* Typhoon Rananim (2004)
* Typhoon Saomai (2006)
* Typhoon Wanda (1956)

Formosa
* Taito (1912)

Guam
* Typhoon Omar (1992)

India
Barisol (Ganges Delta) (1941)

Japan
* Hakodate (1954)
Ichinoseki (1948)
Kiashuni (1281)
Kyushu (1927)
Kyushu (1951)
Loochow Island (1930)
* Osaka (1934)
* Tokyo (1918)
Typhoon Ida (1958)
Typhoon Toage (2004)
Typhoon Vera (1959)

Korea
(1936)
(1949)
Typhoon Billie (1967)
* Typhoon Rusa (2002)

Philippines
(1937)
(1952)
(1964)
* (1970)
Leyte (1929)
* Luzon Islands (1944)

Manila (1928)
Manila (1929)
Manila (1936)
Mauban (1934)
Mayon (1776)
Mayon (1875)
Quezon Province (1970)
* Typhoon Ike (1984)
* Typhoon Ruby (1988)
* Typhoon Angela (1995)
Visayas (1946)
* Typhoon Florita/Tropical Storm Bilis (2006)
* Typhoon Thelma (1991)

Philippines and Korea
Micronesia, Philippines and Korean Peninsula (2002)

Thailand
(1962)

Vietnam
* (1953)
* Typhoons Chip, Dawn, and Elvis (1998)
* Typhoon Linda (1997)

CHRONOLOGY

* Detailed in text

1281
July 17
Kiashuni Island, Japan

1776
Mayon, Philippines

1841
July 21–22
* Hong Kong, China

1862
July 27
Canton, China

1881
Haifung, China

1875
Mayon, Philippines

1906
September 18
* Hong Kong, China

1912
September 16
* Taito, Formosa

1918
September 30
* Tokyo, Japan

1922
August 2–3
Swatow (Shantou), China

1926
September 30
China

1927
September 13
Kyushu Island, Japan

1928
November 17
Manila, Philippines

1929
May 27
Leyte, Philippines
September 3
Manila, Philippines

1930
November 14
Loochow Island, Japan

1934
September 21
* Osaka, Japan
November 14
Mauban, Philippines

1936
August 28
Korea
October 11
Manila, Philippines

1937
August 5
Philippines

1941
May 25
Barisol (Ganges Delta), India

1944
December 17–18
* Luzon Islands, Philippines

1946
November 13
Visayas, Philippines

1947
October 7
Hong Kong, China

1948
September 17
Ichinoseki, Japan

1949
December 8
Korea

1951
October 14
Kyushu, Japan

1952
October 22
Philippines

1953
September 25
* Vietnam

1954
September 26
* Hakodate, Japan

1956
August 2
* China (Typhoon Wanda)

1958
September 27–28
* Japan (Typhoon Ida)

1959
Fukien (Fujian), China
(Typhoon Iris)
September 26–27
* Japan (Typhoon Vera)

1962
October 28
Thailand

1964
June 30
Philippines
September 5
Hong Kong, China

1967
July 9
Korea (Typhoon Billie)

1970
September 11
Quezon Province, Philippines
October 14–15
* Philippines

1984
September 2
* Philippines (Typhoon Ike)

1988
October 24
* Philippines (Typhoon Ruby)

1991
November 5
* Philippines (Typhoon Thelma)

1992
August 28
* Guam (Typhoon Omar)

1994
August 21
* China (Typhoon Fred)

1995
November 3
* Philippines (Typhoon Angela)

1997
November 2
* Vietnam (Typhoon Linda)

1998
November 11–26
* Vietnam (Typhoons Chip,
Dawn, and Elvis)

2002
July 16
Philippines and Korea
August 22–September 1
* Korea (Typhoon Rusa)

2004
August 12
* China (Typhoon Rananim)
August 24–25
China (Typhoon Rananim)
October 20
Japan (Typhoon Toage)

2006
July 7–15
* Philippines, Taiwan, Southern
China (Typhoon Florita/
Tropical Storm Bilis)
August 4–11
* China (Taiwan) and the
Philippines (Typhoon
Saomai)

TYPHOONS

For a detailed description of typhoons, consult the section on hurricanes, p. 245. In the North Atlantic, this meteorological phenomenon is considered a hurricane, in the Indian Ocean, a tropical cyclone, and in the Western Pacific Ocean, a typhoon. Thus, all of the storms described in this section have taken place in that land area bordering upon the Western Pacific Ocean.

CHINA
HONG KONG
July 21–22, 1841

One thousand people were killed in the typhoon that struck Hong Kong harbor on July 21–22, 1841. Practically all of the casualties were Chinese; British naval personnel knew of the approaching storm, but neglected to inform the Chinese residents.

The typhoon that roared into Hong Kong Harbor on the night of July 21, 1841, wrecking or sinking every boat except the British frigates anchored there, claimed an estimated 1,000 lives, most of them Chinese fishermen and inhabitants of the junks that dotted the harbor. According to Royal Navy reports, some waterfront dwellings were also swept into the harbor under the incessant battering of winds, rain, and sea.

There was ample warning of the approaching storm. For hours, the telltale purple clouds that signal the onset of a typhoon wandered and brooded over the hills that ring the harbor, and there was continuous lightning and thunder. By 10:30 on the night of July 21, the Royal Navy's barometers registered 28.50.

However, there seems to be no record of Royal Navy efforts either to rescue or warn the Chinese inhabitants. According to the diary of a British officer: "The Chinese were all distracted, imploring their gods in vain for help; such an awful scene of destruction and ruin is rarely witnessed, and almost every one was so busy in thinking of his own safety as to be unable to render assistance to anyone else. . . ."

CHINA
HONG KONG
September 18, 1906

An absence of warning again caused a huge death toll among the Chinese of Hong Kong from a typhoon that struck on September 18, 1906. Of the 10,000 dead, only 20 were Europeans.

In the harbor of Hong Kong is a "Typhoon Gun"—a cannon that is fired to signal the impending approach of a typhoon. The fact that the gun was not discharged until 8:40 P.M. on the night of September 18, 1906, a mere 20 minutes before a typhoon roared into Hong Kong Harbor, may account for the ghastly death total of 10,000.

It was a horrendous storm, with winds of 100 MPH or more, and a storm surge that was powerful enough to sink 22 medium-sized steamers, 11 heavy ships, and more than 2,000 junks and sampans.

Within a half-hour after the gun exploded, the 2,000-ton (1,814.4-tonne) American steamer *S. P. Hitchcock* was blown ashore, and the 1,698-ton (1,540.4-tonne) German steamer *Petrarch* was catapulted by wind and waves onto the steamers *Emma Luyken* and *Montreagle*. The hulls of both of these smaller steamers were crushed by the *Petrarch*, which then careened into the Kowloon wharf, demolishing a large chunk of it.

The *San Cheung*, a heavy steamer, did tremendous damage when it snapped its mooring cable and was flung around the harbor like a menacing toy ship, ramming and sinking dozens of other ships. Its captain, trying to grab and steady the ship's wildly spinning wheel, broke his fingers in the attempt.

In the Kowloon District, hundreds of houses were turned into fragmentary, lethal weapons, as their roofs were torn off, and the bamboo walls splintered and knifed through the air, killing or injuring hundreds of people. Rickshaws, some with passengers inside, were flung crazily through the air.

The worst toll of human life and property fell on the native Chinese. Of the 10,000 dead, only 20 were Europeans whose brick homes withstood the storm, and afterwards, there was justifiable criticism for their refusal to take in Chinese victims of the storm.

Still, there were acts of heroism. A large wooden drawbridge spanned the waterway leading to the boat club lagoon, a logical shelter for the wildly pitching junks and sampans and other small boats. Thousands of coolies raised this bridge by hand at the height of the storm, and hundreds of small boats scurried for shelter. Unfortunately, the fury of the storm drove into the lagoon, too, wrecking every boat in it. The next morning, it was possible to walk across the lagoon on the wreckage of sunken boats.

At the height of the storm, dozens of Chinese men braved the collapsing piers to thrust bamboo poles into the raging waters, allowing hundreds of drowning boatmen and their families, dumped from sampans and junks, to climb to safety.

H. R. Bevan, a legendary soldier of fortune, spotting a coolie who was picked up by the wind, rammed into a lamppost, and then flung into the water, joined forces with an Indian constable. The constable removed his turban, unravelled it and held onto one end while Bevan looped the other around his waist and leaped into the water, rescuing the coolie as he was about to go underwater for the last time. The coolie survived, as did Bevan, who, in true soldier of fortune fashion, disappeared before he could be either rewarded or thanked.

CHINA
TYPHOON FRED
August 21, 1994

Typhoon Fred decimated the coastal province of Chekiang (Zhejiang), China, for 48 hours from August 21 through 23, 1994. Heavy rain and wind killed 700 people and caused $1.6 billion in damage.

The summer of 1994 had been one brimming with natural disasters for China. In July, 128 people were killed and 70 injured in floods that roared through the three northeastern provinces and Inner Mongolia. In seven other provinces of China, severe drought caused suffering for 27 million people. And then there was typhoon Fred.

The eastern China province of Chekiang borders directly on the East China Sea, making it vulnerable to typhoons that migrate from the Indian and Pacific Oceans. Typhoon Fred was one of these storms that traveled down from Taiwan and pounded the coast of Chekiang Province for 48 hours, dumping eight inches of rain in some places and causing floods of up to three feet. In Wenchow (Wenzhou), a 20-year record for high tides was broken.

Throughout the province, trees were downed, homes were smashed into unrecognizable piles of lumber, and crumbling river embankments unleashed floodwaters into rich agricultural lands.

The infrastructure of the province was thrown into chaos. Train tracks were severed, roads were washed out or covered by mudslides, and port facilities were floated out to sea as storm erosion bit at the coastline.

Over 8,000,000 people were affected by the storm, and 700 died. The damage was estimated at $1.6 billion.

CHINA
TYPHOON RANANIM
August 12, 2004

From August 12 to 14, 2004, 164 residents of mainland China were killed and 1,800 injured by Typhoon Rananim, the strongest storm to hit China in 50 years. Some 42,000 homes were destroyed in a path of destruction that reached from the southeast coast to as much as 150 miles (241.4 km) inland.

Typhoon Rananim, the strongest storm to devastate China in 50 years, slammed into the country's southeastern coast on the night of August 12, 2004, with winds of more than 100 MPH (161 km/h). In a positive sense, it relieved a drought and heat wave that had been affecting much of eastern China. But it was at a price: The town of Wenling, near the landfall of Rananim, was deluged with huge amounts of rain and wind, which immediately cut off its power supply.

"Shop signboards were flying out and hit people's arms and legs like knives," a doctor at No. 1 People's hospital in Wenling told Reuters reporters. "The wind was really very, very strong and we have rarely seen this."

Trees were uprooted and small houses collapsed as the storm moved west into inland Kiangsi (Jiangxi) Province. The tropical deluges filled lakes in the Lake District to overflowing, and caused 410,000 people, many from rural villages, to flee ahead of them. Behind the refugees, the flooding destroyed huge amounts of cropland, and killed thousands of livestock. In the major city of T'ai-chou (Taizhou), the loss of power also deprived millions of people of water and phone service.

As Rananim moved inland, it weakened to the force of a tropical storm, but it still caused extensive damage as far inland as 150 miles (241.4 km) in the provinces of Fukien (Fujian) in the south and Anhui (Anhui) in the northwest. Over 4,000 houses were destroyed, thousands more were damaged, and over 1,800 people were injured by the storm. A recorded 164 died.

CHINA (AND THE MARIANA ISLANDS, PHILIPPINES)
TAIWAN
TYPHOON SAOMAI
August 4–11, 2006

Dubbed "Super Typhoon Saomai" by forecasters, Typhoon Saomai, in August 2006, traversed basically the same path that Typhoon Florita/Tropical Storm Bilis had laid down a month earlier. The monster storm killed 441 and caused $1.5 billion in damage.

Coined "Super Typhoon Saomai" by the Joint Typhoon Warning Center (JTWC), Typhoon Saomai laid waste to sections of the Mariana Islands, the Philippines, Taiwan, and the east coast of China in August 2006. Some 441 people were killed in what forecasters described as the most powerful typhoon to strike China since Typhoon Wanda in 1956.

Passing over the Mariana Islands on August 6, Saomai whipped Guam with rain and wind, but caused little damage. But its track over the Philippines was a different story: As with Florita/Bilis a month earlier (see p. 379), its outer rain bands soaked and caused mudslides in the Philippines, causing a storm surge that washed over and destroyed over 400 homes, and killed two people.

Again, as with Florita/Bilis, Saomai nicked a corner of Taiwan, bringing heavy rain and wind which disrupted traffic and cancelled flights from Taipei, but caused little damage.

Its major landfall was in China's Chekiang (Zhejiang) Province, where major highways were cut, more than 18,000 homes were torn down, and 87 people were killed. In the adjoining province of Fukien (Fujian), 138 people lost their lives to the storm from a storm surge that flooded several coastal communities and crushed 8 people to death by collapsing an evacuation shelter in which they had taken refuge. Some 37,000 houses

were demolished and 93 acres (37 ha) of farmland were flooded and rendered useless by the storm.

Even after Saomai passed by, it caused tragedy. A man was washed away in floodwaters while inspecting damage, and another was killed when a damaged building collapsed, trapping him inside.

CHINA
TYPHOON WANDA
August 2, 1956

Typhoon Wanda killed 2,000, injured 1,200, and made 38,000 homeless when it hit the Chinese coastal provinces of Anhwei (Anhui), Chekiang (Zhejiang), and Kiangsu (Jiangsu) on August 2, 1956.

Nearly 2,000 people lost their lives, 1,200 were injured, and 38,000 homes were demolished in the gigantic typhoon that ravaged Anhwei, Chekiang, and Kiangsu Provinces on August 2, 1956.

The provinces, located on the China coast, are widely separated, but this made no difference. The typhoon struck on a wider front than any other typhoon in recent history, tearing up crops, wiping out villages and farms, and drowning livestock and people alike.

Warned of the imminent approach of the storm, peasants near Shanghai attempted to harvest the rice crops, already battered by the worst heat wave in a century. Their efforts proved futile; one-third of the rice crop of the three provinces was destroyed, as well as 40 percent of the cotton crop.

FORMOSA
TAITO
September 16, 1912

A typhoon with winds of over 200 MPH killed 107 and injured 293 in the city of Taito, Formosa, on September 16, 1912

Over 200,000 houses in Taito, on the island of Formosa, were flattened or blown apart by the winds of an immense typhoon that struck the island on September 16, 1912. According to sketchy records, winds of over 200 MPH roared into this large island separated from the mainland of China by the Strait of Formosa. These monstrous winds pulverized the fishing fleet, and the consequent storm surges inundated the rice and sugar fields, wiping out that island's two major agricultural industries.

One hundred seven people were killed, either by debris unleashed by the furious wind or from drowning. Another 293 were severely injured.

GUAM
TYPHOON OMAR
August 28, 1992

Typhoon Omar ripped across the island of Guam on August 28, 1992, rendering nearly 5,000 of the island's 32,000 buildings uninhabitable or nonexistent. No one was killed, but 4,000 were made homeless. The cost of the damage was more than $250 million.

Typhoon Omar, packing winds of 150 MPH (241 km/h), with gusts up to 184 MPH, roared across the island of Guam on Friday, August 28, spreading an extraordinary amount of havoc. When the swift moving storm had departed, 2,000 of the buildings on the island had been totally destroyed and 2,300 others had been damaged enough to defy rebuilding.

The brunt of the storm smashed through central Guam, from near Agana to Andersen Air Force Base in the north. The resort hotels that stretched along Tumon Bay were especially hard hit. Trees in awesome numbers were split or flattened; floods gouged out roads and farmland; the radar system at the international airport was downed early, stranding 5,000 tourists.

Severe structural damage was done to the air force base, and two warships, the *Niagara* and the *White Plains*, broke loose from their moorings and ran aground in Apra Harbor.

Miraculously, no one was killed, but several hundred people were injured and treated at Guam Memorial Hospital. Over 3,000 people were forced to remain in shelters for weeks after the typhoon passed, and nearly 4,000 were rendered homeless.

JAPAN
HAKODATE
September 26, 1954

Sixteen hundred people died in the typhoon that hit Hakodate, Japan, on September 26, 1954, 1,000 of them in a ferryboat capsized by the storm.

Statistics never tell the entire human story, but in the case of the killer typhoon of September 26, 1954, the casualty figures completely obscure the bizarre balance struck by this storm. Sixteen hundred people died. But over 1,000 of these 1,600 drowned in one place in a matter of minutes.

The September 26 typhoon was one of 15 which raged over Japan in 1954. Some were slight; others immense. None took the toll of human life that this storm did.

Other storms glanced off Japan's outward islands. This storm hit the mainland squarely, roaring in with extremely high winds and extraordinarily high tides.

The 4,300-ton (3,900.9 tonne) ferry *Toyu Maru*, which used to transport people between Hokkaido and Honshu, made its customary trips that day. Unwisely, perhaps, it left Hokkaido as the storm approached. Midway there, the storm struck, its eye passing directly across the ferry's path. Buffeted by the winds within the eyewall of the storm and swamped by the heavy seas, the ferry's engines gave way in the Tsugau Straits.

Through some herculean effort, the captain and crew of the ferry managed to keep her afloat. Powerless against the currents, the ferry was swept against a reef at the entrance to Hakodate harbor. The boat overturned, spilling its 1,000 passengers, including 50 Americans, into the sea, where they all drowned.

JAPAN
OSAKA
September 21, 1934

Four thousand people died when a monster typhoon struck Osaka, Japan, on September 21, 1934.

Despite the devastation caused by countless typhoons throughout Japan's history, the builders of Osaka, because they had not felt the fury of these ravaging storms for more than a generation, constructed buildings not meant to withstand storms—including schools, hospitals, and even an insane asylum. These vulnerable structures stood directly in the midst of a classic storm path, one taken by the furious typhoon that bore directly upon the city on September 21, 1934.

The outer fringes of the monster storm containing winds of 50 MPH struck Kyoto, Kobe, and Tokyo glancing blows. The typhoon hit Osaka dead center, with winds exceeding 125 MPH. Osaka's major industry was textiles. In one afternoon, 3,082 factories were destroyed, literally blown to pieces, as hundreds of workers were killed in the disintegrating buildings.

The Sorojima Leper Hospital was ripped apart, killing 200 of its 560 inmates. The insane asylum was flattened, and 50 more died under its wreckage.

Eighty-seven schools were wrenched from their foundations during the middle of the school day; 420 children were killed in their classrooms and another 1,000 were seriously injured when pieces of the buildings shattered and became decimatingly airborne.

All in all, 4,000 people died in the storm, most of them residents of Osaka.

JAPAN
TOKYO
September 30, 1918

One of the worst typhoons in Japan's history slammed into Tokyo on September 30, 1918, killing 1,619 and making 139,000 homeless.

In Japan, as in the rest of the world, weather forecasting methods and warning systems were in a relatively primitive state in 1918. Thus, when the residents of Tokyo went to bed on the night of September 30, they settled in with a fair amount of serenity, thinking they were only receiving another rainstorm.

By 3 A.M., however, the wind velocity had reached 100 MPH, and the fragile structures in the poorer sections of Tokyo had begun to blow away, tumbling against each other and crushing inhabitants both inside and outside the dwellings.

The Kasai and Sunamura neighborhoods, near the waterfront, were turned into raging sluiceways with nine-foot-high waves roaring over the waterfront and rushing through the streets. Hundreds drowned.

Even such sturdy structures as the Seiyoken Hotel and the Ebisu Brewery collapsed against the assault of wind and waves. Ancient pine trees that had withstood centuries of lesser typhoons were uprooted. Most astonishingly, the windows of some houses seemed to explode outward, suggesting an abrupt drop in air pressure that usually only occurs in tornadoes.

The devastation was immense. Six hundred twenty-nine people died within the city. Another 990 deaths occurred on its outskirts. When 200,000 houses both in and outside Tokyo were demolished by this severe typhoon, 139,000 were rendered homeless.

JAPAN
TYPHOON IDA
September 27–28, 1958

Over 600 were killed, thousands were injured, and 1,200 were made homeless by Typhoon Ida, which hit the Izu Peninsula near Tokyo, Japan, on September 27 and 28, 1958.

More than 600 people lost their lives, thousands were injured, and 1,200 were reported missing in the onslaught of Typhoon Ida, which slammed into the Izu Peninsula, 70 miles (112.7 km) south of Tokyo, on September 27, 1958. This plot of land is rich in rice paddies, supplying Japan with a goodly portion of the staple of its diet. The storm destroyed most of the Izu Peninsula's crop that year, flooding 120,000 acres (48,562 ha) of rice paddies.

The Eguro, Kano, and Arakawa Rivers, swollen by torrential rains and wind-driven storm surges, overflowed their banks, taking down 244 bridges, washing away 1,000 homes, and creating 1,875 muddy landslides. More than 10,000 residents of the area were rendered homeless by the combination of winds and floodwaters.

JAPAN
TYPHOON VERA
September 26–27, 1959

Japan's worst typhoon, Typhoon Vera, hit central Honshu Province on September 26, 1959. Five thousand people died, 15,000 were injured, and 400,000 were made homeless.

The worst typhoon in Japan's history struck central Honshu Province on September 26, 1959, killing 5,000 people, injuring more than 15,000, rendering 400,000 homeless, and causing $750 million in damages.

Powered by 135-MPH winds, the storm slammed into the coastline from Nagoya to Tokyo on the night of September 26, flinging the 7,412-ton (6,124.1 tonne) British-owned freighter *Changsha* onto the beach at Nagoya.

Yokohama and Tokyo were hit first; 92-MPH winds roared down the canyons of their urban streets, forcing floodwaters ahead of them. From here, the storm flung itself inland across central Honshu Province, then turned north and traversed the island, wrecking the provinces of Toyama, Yamagata, Akita, and Niigata.

The city receiving the worst of the storm was Nagoya, which was celebrating its 70th anniversary. An early casualty of the storm was a series of golden dolphins affixed to its 350-year-old castle in commemoration of the anniversary.

The Nagoya lumber yards, heavily stocked with logs waiting to be cut into planks, were turned into

375

ammunition dumps by the storm. Floated loose by the floodwaters, and propelled by the winds, the logs shot down streets in Nagoya, ramming in doors and walls.

As the floodwaters rose, hundreds of people were swept out of their homes before the homes themselves collapsed. Over 5,800 houses in Nagoya alone were destroyed by the winds or the floodwaters. Even modern apartment buildings fell, often burying their inhabitants under the rubble.

On the day following the storm, 25,000 people wandered or swam over their devastated homes. Thieves were so plentiful and persistent they often robbed homes that were still occupied. A police force of 1,100 men was finally formed to control the looting.

Further tragedy occurred when dysentery and typhus, spread by polluted water, claimed more lives.

KOREA
TYPHOON RUSA
August 22–September 1, 2002

Typhoon Rusa, the worst typhoon to hit South Korea since 1959, slammed into the southeastern coast of the country on August 31, 2002. Its torrential rains and 80 MPH (128.7 km/h) winds caused flooding and landslides that killed 113 people, submerged 17,000 houses, displaced 27,474 residents, and created $6 billion in damage.

Typhoon Rusa began on the eastern periphery of the Pacific monsoon trough east of Pohnpei on August 19, 2002. By August 22, it had strengthened to a storm with winds of 135 MPH (217.3 km/h). But by August 31, it had weakened, and a low pressure system over Japan eroded it still further, so that when it hit South Korea that day, its winds were peaking at 80 MPH (128.7 km/h)—still strong enough to gain it the reputation of the worst typhoon to strike South Korea since 1959.

Over a two-day period, nearly 36 inches (91.4 cm) of rain were dumped on the southern and eastern provinces that resulted in ravaging floods that submerged thousands of homes, tore apart bridges, ripped up sections of railway, inundated and destroyed hundreds of acres of farmland, and cut off fresh water to 400,000 people.

The hardest hit areas were Gangwon, on the east coast, where 128 people were killed or missing, and its seat, Gangneung, a city of 220,000 people. "Nothing is more miserable than this," Kim Bun-hee, a 61-year-old housewife, told reporters as she waited on line for water from a fire truck after the storm had passed northward. "I would rather die."

Just outside of the city, hundreds of graves were washed away in a landslide that destroyed a large part of a public cemetery. People flocked to the site after the rains subsided and shoveled the leveled ground, trying to locate the missing tombs of their loved ones. Nearby, trucks and cars remained buried, the roads beneath them buckled under the weight of repeated landslides. Cows from flooded farms stood in what was left of local roads.

All told, 17,000 houses and buildings in low-lying areas were submerged, forcing 27,474 residents to take shelter in public buildings and schools.

PHILIPPINES
October 14–15, 1970

Three successive typhoons killed 1,500 and rendered 500,000 homeless when they struck the Philippines on October 14 and 15, 1970.

Three typhoons in a row ravaged southern Luzon, Mindanao, and smaller islands in the Philippines during two terror-filled days, on October 14 and 15, 1970.

There was no discernible pattern to the destruction; the storms crisscrossed the islands, striking with an average velocity of 100 MPH and completely destroying an astonishing 90,000 homes in 24 provinces. In all, 1,500 died and over 500,000 were made homeless by these rampaging typhoons. The islands' crops were destroyed, cities were rendered powerless, and the damage was reckoned in the millions.

The Philippine government was overwhelmed by the cleanup job. Red Cross disaster relief supplies and other aid quickly arrived from China, Canada, and the Soviet Union.

PHILIPPINES
LUZON ISLANDS
December 17–18, 1944

The worst storm disaster to strike the United States Navy hit Admiral "Bull" Halsey's fleet off Luzon in the Philippines on December 17 and 18, 1944. Seven hundred ninety people drowned, three ships and 156 airplanes were destroyed, and 28 ships were crippled.

During World War II, the weather warning apparatus for the U.S. Navy in the Pacific was located in Honolulu. It was efficient and up to date for the time, but when, on December 16, 1944, a 300-mile (482.8 km) wide, monster typhoon bore down on Admiral Halsey's Third Fleet, which was cruising off the Luzon Islands of the Philippines, the Honolulu weather station was unable to chart the typhoon's course.

Thus, the worst storm disaster to strike the United States Navy approached without warning. Seven hundred ninety seamen would drown, three ships and 156 airplanes would be lost, and 28 ships would be crippled.

The Third Fleet had just come from assisting General Douglas MacArthur and his troops, who had landed on the main island of Luzon. It was resting and refueling just east of the Philippines when the storm struck, on December 17, with 150-MPH winds and waves that rose 100 feet (30.4 m) above equally menacing troughs in the sea.

The fleet had already been battered by war. Most of its ships had been in continuous service since Pearl Harbor, three years before. Thus, the 24 hours of repeated pounding by immense waves, pelting rain, and enormously powerful wind sent some to the bottom and ripped rivets, rigging, and armor plate from others. One hundred fifty-six airplanes were torn from their moorings on the decks of the fleet's aircraft carriers.

Flagship Command attempted, in the early hours of the storm, to maintain order. But the power of the storm was such that this soon became impossible, and the fleet scattered, each ship seeking its own means of survival.

The 1,500-ton (1,360.78-tonne) destroyer *Monohan*, a veteran of 12 battles, keeled over early, pounded repeatedly by gigantic waves. She was driven to the bottom with 256 officers and crew aboard; only six survived.

The *Hull*, a sister ship to the *Monohan*, was likewise thrown over on its side, whipped from quadrant to quadrant by waves and wind, pummeled by huge rollers, and eventually sent to the bottom.

The 2,000-ton *Spence* tipped over as a 100-foot (30.4 m) high wave slammed it down into a seemingly bottomless valley of churning water. One officer and seventy enlisted men were flung into the water, and, although the wind and waves ripped lifejackets from some of them, all 71 managed to survive.

The magnitude of the disaster drove the Naval Command to establish another weather control station on the island of Guam to oversee storm conditions in that part of the Pacific.

PHILIPPINES
TYPHOON ANGELA
November 3, 1995

Typhoon Angela slammed into the main Philippine island of Luzon on November 3, 1995, killing 500, displacing 40,000, and causing $30 million in damage to crops, homes, and the infrastructure of the island.

Ever since the gigantic eruption of Mount Pinatubo in 1991 (see p. 417) the central Luzon province of Pampanga, 35 miles north of Manila, had been plagued by annual avalanches of mud and volcanic debris, triggered by the annual rain and windstorms that visited the area. But none of these yearly visits by tropical depressions would compare with the devastation caused by typhoon Angela on November 3, 1995. Mudflows 10 to 12 feet (3.04 to 3.65 m) high and tons of volcanic ash roared down Pinatubo's slopes that day, burying farms, fields, and livestock. Fortunately, the people of the province, used to the threat of the volcano, had evacuated the area after a week of rainstorms that had caused minor mudslides—presages, the inhabitants knew, of possible damage to come.

Not so fortunate were the rest of the inhabitants of Luzon, the Philippines' main island. The rice fields were ready for harvest at the beginning of November 1995. But the rice would never be harvested. Up to 20 feet of water inundated the fields from the rivers and lakes that overflowed that November day of furious, 130-MPH winds and rain that fell like sharpened needles.

The storm's gentle name of Angela was a misnomer. Its effects were anything but angelic. It cut a huge swath of destruction, felling, crushing, or exploding everything in its path. Power lines whipped loose from their supports, which crashed to earth. Trees and entire roads were uprooted and left at crazy angles. In the capital of Manila, buildings collapsed in on themselves, walls were torn down, and roofs were turned into flying wedges.

Manila Bay and the South China Sea were whipped into a frantic froth by the winds, which shut down the international airport and kept all ships in port. Inland, transportation ground to a complete halt as road after road disappeared. Nearly a hundred power transmission lines were downed, and it would be a week before power was restored to the province, which included the Philippines' normally teeming capital.

As many as 40,000 people evacuated their homes and crowded into schoolhouses, churches, and government buildings. The death toll would rise to 500, with 280 missing and presumed dead. The damage to crops alone was estimated at $8 million and that to bridges, buildings, and roads at $22 million.

PHILIPPINES
TYPHOON IKE
September 2, 1984

Typhoon Ike, packing winds of 137 MPH, struck the Philippines on September 2, 1984, killing 1,363, injuring over 300, and making 1.12 million people homeless.

Typhoon Ike, the worst recorded storm of this century to hit the Philippines, slammed into the southern part of these Pacific islands on September 2, 1984, with winds of 137 MPH. Ike killed an estimated 1,363 people, injured more than 300, and rendered 1.12 million homeless.

The storm uprooted trees, darkened the provincial capital of Mindanao, 450 miles south of Manila, and flattened homes and businesses across six southern provinces. In Mainit, a lake overflowed, drowning 200 people, and on Negros Island, near Cebu, 59 people died when the storm surge completely engulfed the island. Surigao del Norte and del Sur provinces were the worst hit by the storm. Ninety percent of the crops in the Surigao region were destroyed, and the day after the storm struck, Constantino Navarro, the mayor of Surigao City, appealed for immediate food assistance, stating that rice stocks in his area would last for only 10 days.

At first, President Ferdinand Marcos refused aid from outside of the Philippines, declaring that "... everything is under control ..." and "The policy has been to rely on our own resources."

But the $4.4 million in government aid released by Marcos soon proved inadequate, and the Philippine president requested aid, which was immediately forthcoming from the United States and Japan.

Meanwhile, Ike roared on to the China Sea, whipping the southern China coast. Thirteen people were reported missing at sea off the city of Bethai after enormous winds swept the coast of the Guangxi-Zhuang Autonomous Region (Kwangsi Province) on September 6, 1984. Factories and houses collapsed in Bethai, Ch'inchow (Qinzhou), Fangcheng, and the area around the regional capital of Nanning.

PHILIPPINES
TYPHOON RUBY
October 24, 1988

Three hundred fifty-six people died—240 of them aboard a capsized ferry—and 110,000 were made homeless when Typhoon Ruby hit the Philippines on October 24, 1988.

Typhoon Ruby, the strongest typhoon to hit the main island of Luzon in the Philippines since Typhoon Patsy in 1970, slammed into the coast 50 miles (80.5 km) east of Manila on October 24, 1988. Three hundred fifty-

six people died in the storm—240 of them aboard the ferry *Dona Marilyn* that sank in 140-MPH winds with 500 passengers and crew aboard at the height of the typhoon. This monster storm made 110,000 Philippine residents homeless and caused $45 million in damage. The islands were saturated with nine inches of rain in a 24-hour period, causing massive flooding, mudslides, and washed out bridges.

Twenty-six people died in central Antique province when a bus went off a bridge and into the Sibalom River during the storm.

The dramatic *Dona Marilyn* disaster was the second such sinking within a year in the Philippines, following the horrendous *Dona Paz* collision with an oil tanker on December 20 off the central part of the islands.

The 2,846-ton (2,581.8-tonne) *Dona Marilyn* was carrying fewer than half its capacity of 1,279 passengers when it left Manila at 10 A.M. Sunday, October 23. It was scheduled to arrive about 24 hours later in Tacloban, the provincial capital of Leyte.

But at 11:30 A.M. on Monday the 24th, the vessel's crew radioed Manila to say they were changing course because of choppy water. At 1:10 P.M., they radioed Manila again. This time it was a distress signal stating that they had encountered large waves and were taking on water. It was the last message received from the sinking ferryboat, which went down in the Visayan Sea just short of its destination of Leyte Island.

Later that day, two survivors were rescued from the water, eight were rescued from a nearby island, and five from another. The next day, 135 more were discovered alive on a third island. Seventy-six other people lost their lives on the Philippine mainland from drowning or injuries received from falling debris.

PHILIPPINES
TYPHOON THELMA
November 5, 1991

The port city of Ormoc, on Leyte in the central Philippines, was devastated by Typhoon Thelma on November 5, 1991. Mudslides and floods, unleashed from mountainsides that had been illegally logged, killed between 3,400 and 5,000 people. Fifty thousand were made homeless; 700,000 were affected by the storm and its resultant floods.

For years, with the complicity of the government and military leaders, illegal clearcutting of the forests in the mountains of Leyte, in the center of the Philippines, left the hills bereft of the tree roots that keep the soil in place during heavy rains.

It was a preparation for a disaster, and on November 5, 1991, the disaster arrived, shortly after breakfast time. The rains that preceded typhoon Thelma had been falling all night. And then, in midmorning, the storm descended upon the island with a vengeance. Winds uprooted trees, collapsed houses, and scattered roofs and refuse throughout the area and on the adjacent islands of Cebu and Negros.

But the worst, the terrible tragedy took place in the city of Ormoc, on Leyte. Ormoc sits on a flat plain at the bottom of a once thickly forested mountain. Bereft of trees, the soil of the mountainside turned swiftly into mud and, fed by the pounding rain of the storm, roared down the mountain and into the city. Trees surrounding the city were uprooted as if a giant hand had reached down and ripped them from the earth. Cars were flushed down streets and wooden buildings were blown apart by the roaring water, which gathered up everything, including people, in its passage.

Ninety percent of the buildings in the city were totally destroyed. Over 3,000 people—some estimates said 5,000 of the city's 120,000 residents—were drowned, many of them swept out to sea, never to be recovered.

"People were in a panic," one 48-year-old survivor told a Reuters reporter. "They were scampering into the streets. Some children climbed up trees yelling for help. People were screaming as they ran from their houses."

A large percentage of the dead were children who were swept away from their parents by floodwaters. Christina Guillotes told reporters, "We clung to a piece of wood from the house, rolling in the current. Then we got separated." Her three sons, aged two, four, and five were never found.

Porfirio Labugnay, the captain of an inter-island vessel, said, after the storm had passed, "I saw bodies and animals, cows, pigs and household appliances floating in the sea off Ormoc City. There were cars that I could see under water at the port area and divers trying to retrieve bodies trapped inside them."

Bodies were stacked on sidewalks; the courtyard of Ormoc's Health Center was strewn with bodies, left for identification by residents holding scarves to their noses. Trucks loaded with bodies drove to improvised cemeteries, where backhoes dug mass graves.

The living picked their way through the fetid and muddy streets. Local officials were overwhelmed, as food and water rapidly ran out in the city. "Bread is what we need first," one official said. "People who didn't die in the storm are dying of hunger."

Over 700,000 people were affected by the storm; 50,000 were made homeless. The death toll hovered between 3,400 and 5,000, which included 2,000 who were never found and presumed swept out to sea. Some

bodies were never identified, since entire families were drowned in the vicious flood that the typhoon loosed from the mountain.

"These people died from a natural disaster," Father Ben Bacas, a parish priest, said in a mass for the victims of the tragedy. "But they died because of human greed, too. Now, too late, we see the effects of illegal logging."

PHILIPPINES (AND TAIWAN, SOUTHERN CHINA)
TYPHOON FLORITA/TROPICAL STORM BILIS
July 7–15, 2006

In July 2006, during an unusually active typhoon season, 637 people lost their lives in the Philippines, Taiwan, and Southern China when a giant storm hit, alternately named Typhoon Florita and Tropical Storm Bilis.

The enormous storm that struck the Philippines, Taiwan, and the southern portion of China in July 2006 had two names. While the Joint Typhoon Warning Center qualified it as Tropical Storm Bilis, the Philippine Atmospheric, Geophysical and Astronomical Services Administration bumped its status upward, and named it Typhoon Florita. Whatever its name, Florita/Bilis caused a huge amount of damage and human suffering in its wake as it thundered through three countries from July 13 through July 15.

Born in the Pacific northeast of the island of Yap on July 7, it wandered northwest, passing north and west of the Philippines and dumping huge amounts of rain and causing landslides there that killed 20 people in Baguio City and the Manila area. Continuing on, it made its first landfall in northern Taiwan on July 13 and sped across the northern coast of the island, leaving behind relatively minor casualties. Two fishermen from Mainland China were drowned and one man was electrocuted in the city of Taipei.

Florita-Bilis saved its greatest force for the southeastern mainland of China, coming ashore in Fukien (Fujian), where it killed 43 in the flood that resulted from its rains and winds, and closed schools and tourist attractions. It went on to kill 35 people in Kwangsi (Guangxi) and 106 in Kwangtung (Guangdong). In Yunnan, it caused a flash flood that drowned eight road workers and floated away their huts.

Florita-Bilis traveled on, inundating roads and causing flooding and landslides that blocked several sections of the Neijing-Guangzhou railway, a main line. One train was surrounded by floodwaters in Lechang, and its passengers were evacuated to a nearby school.

Finally, Florita/Bilis smashed into Hunan province, where it did its worst damage. Mudslides, and some of the worst flooding to occur in 100 years wiped out the village of Zixing, and in it and other sections of Hunan, crushed 31,000 homes and caused 345 deaths. In the end, this twice-named storm killed 637 people, most of them in China and caused over $2.5 billion worth of damage.

VIETNAM
September 25, 1953

Over 1,300 were killed, over 100,000 were made homeless, and the U.S. Air Force Base in Otsu, Japan, was almost destroyed by a typhoon that struck Vietnam on September 25, 1953.

A typhoon that killed 21 persons and caused $1.9 million in damage to the United States Air Force Base in Otsu, Japan, slammed into the central coast of Vietnam on September 25, 1953. Its 90-mile (144.8-km) wide swath cut through the central part of the country, wiping out the entire rice crop of the two provinces traversed by the storm.

In the countryside, 1,300 people either drowned or were crushed by collapsing huts or flying debris. In the capital city of Hue, 325 miles (523 km) north of Saigon, massive numbers of people died, and a large percentage of the city's structures were flattened.

At the storm's end, over 100,000 would be left without homes or shelter.

VOLCANIC ERUPTIONS AND NATURAL EXPLOSIONS

THE WORST RECORDED VOLCANIC ERUPTIONS AND NATURAL EXPLOSIONS

* Detailed in text

Africa
    Congo, Nyingongo (2002)

Chile
    (1960)

Colombia
    (1985)
    Nevada de Huila (1994)

Ecuador
    Cotopaxi (1741)
    Tunquraohua (1794)

Greece
    * Santorini: Atlantis (1470 B.C.E.)

Guatemala
    Aqua (1549)
    * Santa Maria (1902)

Iceland
    * Laki (1783)

Indonesia
    Kelut (1909)

Italy
    * Etna (1169)
    * Etna (1669)
    * Sicily, Etna (1226, 1170, 1149, 525 B.C.E.)
    * Vesuvius (79 C.E.)
    * Vesuvius (1631)
    * Vesuvius (1779)
    * Vesuvius (1793–94)
    * Vesuvius (1906 and 1944)

Japan
    * Unsen (1793)

Java
    * Krakatoa (1883)
    * Keluit (1919)
    * Merapi (1931)
    * Papandayan (1772)
    * Miyi-Yama (1793)
    * Tamboro (1815)

Martinique
    * Pelée (1902)

Mexico
    * Paricutín (1944)

Montserrat
    * Soufrière Hills (1995–97)

New Guinea
    * Lamington (1951)

Nicaragua
    * Coseguina (1835)
    Leon (1995)

Papua New Guinea
    * Rabaul (1994)

Philippines
    * Mayon (1766)
    * Mayon (1814)
    * Mayon (1897)
    * Mount Pinatubo (1991)
    * Taal (1911)

Sumatra
    Toba Megavolcano (73,000 B.C.E.)

United States
    Alaska Spurr (1992)
    * Mount St. Helens (1980)

USSR
    * Kamchatka, Bezymianny (1956)

West Indies
    * La Soufrière (1902)

CHRONOLOGY

* Detailed in text

73,000 B.C.E.
    * Sumatra (Toba)

1470 B.C.E.
    * Greece (Santorini)

1226 B.C.E.
    * Italy (Etna)

1170 B.C.E.
    * Italy (Etna)

1149 B.C.E.
    * Italy (Etna)

525 B.C.E.
    * Italy (Etna)

79 C.E.
    August 24
    * Italy (Vesuvius)

1169
    * Italy (Etna)

1549
    * Guatemala (Aqua)

1631
    December 16
    * Italy (Vesuvius)

1669
    March 11
    * Italy (Etna)

1741
    Ecuador (Cotopaxi)

1766
    October 23
    * Philippines (Mayon)

1772
    * Java (Papandayan)

1779
    May–August
    * Italy (Vesuvius)

1783
    January–June
    * Iceland (Laki)

1793
    * Java (Miyi-Yama)
    February 1793–July 1794
    * Italy (Vesuvius)
    April 1
    * Japan (Unsen)

1794
    Ecuador (Tunquraohua)

1814
    February 1
    * Philippines (Mayon)

1815
    April 5–12
    * Java (Tamboro)

1835
    January 22
    * Nicaragua (Coseguina)

1883
    August 26–27
        * Java (Krakatoa)
1897
    June 23–30
        * Philippines (Mayon)
1902
    May 8
        * Martinique (Pelée)
    May 7 and 8
        * West Indies (La Soufrière)
    October 24
        * Guatemala (Santa Maria)
1906
    April 4
        * Italy (Vesuvius)
1909
    Indonesia (Kelut)
1911
    January 30
        * Philippines (Taal)
1919
    * Java (Keluit)

1931
    December 13–28
        * Java (Merapi)
1944
    March 18
        * Italy (Vesuvius)
    June
        * Mexico (Paricutín)
1951
    January 21
        * New Guinea (Lamington)
1956
    March 30
        * Kamchatka, USSR
            (Bezymianny)
1960
    May 21–30
        Chile
1980
    May 18
        * United States (Mt. St. Helens)
1985
    November 13
        Colombia

1991
    June 10–15
        * Philippines (Mt. Pinatubo)
1992
    June 27
        Mt. Spurr, Alaska
1994
    June 6
        Nevada de Huila,
            Colombia
    September 19
        * Rabaul, Papua New
            Guinea
1995
    July 19
        * Soufrière Hills,
            Montserrat
    December 18
        Leon, Nicaragua
2002
    January 17–18
        Congo (Nyingongo)

VOLCANIC ERUPTIONS AND NATURAL EXPLOSIONS

If drama and spectacle were the stuff of natural disasters, then volcanic eruptions would be their benchmark, for there is probably nothing more terrifying or splendid than the eruption of a volcano. It is cataclysmic, often climactic, and frequently instantaneously horrible in its consequences. Cities, civilizations, and cultures have been either wiped out or forever changed—and in one case, that of Vesuvius's hermetic sealing off of Pompeii, preserved.

There is also something primeval and mysterious about this spurting forth of the interior contents of the Earth upon which we live. For the moments of the eruption of a volcano, we are plunged backward in time to that sort of prehistory of which legends are made, that time when the solid ground upon which we walk and base our assumptions of reality was a shifting, amorphous mass of matter, and the Earth as we know it was establishing itself in a very theatrical way.

Volcanic eruptions are still, to this day, somewhat of a mystery. Volcanologists can study only the aftereffects of a volcanic explosion. The cause and sources of such an event are buried too deeply within the Earth—perhaps down to its very core—to explore scientifically. And they are likely to remain that way for the remainder of human existence on this planet.

Still, we do know more about the 516 active volcanoes on the Earth at this moment than, say, Pliny did when he described the eruption of Vesuvius in 79 C.E.

First of all, the true, real volcano is merely a vent, or a series of vents, in the Earth's crust, through which the forces of the Earth's interior are released into the atmosphere, often violently. The mouth of the vent is referred to as the crater. The large, more or less circular depression—if it is more than a mile in diameter—formed either by collapse or by explosion of the volcano, is the caldera. Crater lakes sometimes form in these calderas.

The vent is like a chimney, connecting the atmosphere with a reservoir of molten matter known as magma. The smoke that seems to come from these vents is actually condensed steam, frequently mixed with dust particles until it is dark in color. The so-called fire of a volcano is the reflection of the red-hot material on the vapor clouds above the volcano. And the lightning observed playing around and through this vapor cloud is actually St. Elmo's fire, caused by a characteristic abundance of static electricity in the atmosphere around a volcanic explosion.

Volcanic mountains—which we often mistake for volcanoes—are formed after a series of explosions have deposited cooled matter blasted forth from beneath the Earth's crust around the vent. In fact, the highest mountains in the world were formed, over millions of years, through the accumulation of volcanic detritus.

There are three varieties of material expelled from volcanoes: liquid (lava), fragmental (pyroclastics, such as ash, mud and rocks), and gaseous—usually sulphur dioxide if it comes from beneath the Earth, but sometimes a hybrid formed from contamination of the atmosphere.

And lastly, the explosions from a volcano do not always shoot straight up from its cone, as 19th-century romantic artists have depicted them. On the contrary, the most lethal and cataclysmic of volcanic explosions occur in the sides of volcanic mountains.

Volcanologists have been able to plot the major belts of volcanoes on the Earth: the Mid-Atlantic range, going through the West Indies; the Cascade Range in the Pacific Northwest; the Hawaiian Range; the unnamed string of mountains that occurs along the north shore of the Mediterranean and extends eastward through Asia Minor into the Himalayas and beyond.

The theory of plate tectonics (see EARTHQUAKES, Introduction, p. 33) established some of the causes of the eruption of volcanoes. In one theory, volcanoes appear where two plates are pulling apart, thus releasing a surge of magma from beneath the plates. The Mid-Atlantic Ridge, for instance, exists along the seam at which the Eurasian and American plates begin to move in opposite directions. Here, volcanic islands such as Iceland, the Azores, and Tristan da Cunha have been formed.

The second cause for the formation of volcanoes, according to this theory, comes where plates collide, or slide over or under each other. When one plate slides beneath the other, it melts and churns as it descends into the bowels of the Earth. Molten material simultaneously rises to the surface and erupts there, forming new land. These collisions or grindings created the Caribbean Islands, the volcanic peaks of Central America, and the Cascade Range.

The third type of volcanic formation occurs in the middle of a plate. According to Earth scientist Dr. Robert Decker, "Somehow, a hot spot in the Earth's mantle melts a hole through the middle of a plate, allowing molten material to spill out on the surface. The Hawaiian Islands are a perfect example of this process—they are right in the middle of the huge Pacific plate."

In February 1993, while mapping the sea floor 600 miles (965.6 km) northwest of Easter Island in the South Pacific, scientists discovered what appears to be the greatest concentration of active volcanoes known on Earth.

In this, the world's last unexplored frontier, 10 to 300 miles (16.1 to 482.8 km) west of the East Pacific Rise, between 16 degrees and 19 degrees south latitude and 113.5 degrees and 117 degrees west longitude, these scientists, using sonar scanning devices, came upon an astonishing 1,133 seamounts and volcanic cones, all in an area of 55,000 square miles (88,513.9 sq. km)—about the size of New York State—and at a point where the Pacific and Nazca plates are separating.

A large number of the volcanoes rise more than a mile above the ocean floor; some are almost 7,000 feet (2,133.6 m) tall, with their peaks 2,500 to 5,000 feet (762 to 1,524 m) below the sea's surface.

At any given moment, the discoverers of this undersea mountain range reason, two or three of the volcanoes could be erupting.

More exploration must be done, but these scientists believe strongly that this enormous field of active volcanoes—twice the recorded number aboveground—might be the long sought-after explanation for the origin of the El Niño phenomenon.

Volcanologists have classified volcanoes into four categories, ranging from the least to the most destructive, and those of the Hawaiian type are considered the least destructive—in fact, Mauna Loa, the fabled mountain of Hawaii, does not appear in this section, nor does Fujiyama in Japan because neither have had catastrophic explosions, at least in recorded history.

In the Hawaiian category of volcano, there is a relatively quiet effusion of basaltic lava unaccompanied by explosions or the ejection of fragments.

The Strombolian type of volcano (named for Stromboli, in the Lipari or Aeolian Islands north of Sicily) is characterized by continuous but mild discharges, in which viscous lava is emitted in recurring explosions. The ejection of incandescent material produces luminous clouds.

The Vulcanian category of volcano, typified by Paricutín (see p. 410), is more explosive. Here, the magma accumulates in the upper level of the vent but is blocked by a plug of hardened lava that forms at the orifice between consecutive explosions. When the explosive gases have reached a critical pressure within the volcano, masses of solid and liquid rock erupt into the air, and clouds of vapor form over the crater, but they are not incandescent, as in the Strombolian type.

The most violent and destructive type of volcano is the Pelean type, named after its prototype, Pelée (see p. 405). Here there is an emission of fine ash, hot gas-charged fragments of magma, and superheated steam in an incandescent cloud that moves downhill with the speed of a major hurricane. Thus, thousands of people are often caught absolutely unaware and are simply ground under by superheated mud, steam, and molten, falling ash and rocks.

Volcanoes rarely erupt without warning. The problem in the past has been the disregard of warning signs, namely earthquakes. Even in modern times, people situated near volcanoes ignore common sense and the warnings of experts.

· ·

SUMATRA
TOBA MEGAVOLCANO
73,000 B.C.E.

In 2006, three scientists, who were unaware of each other's discoveries, and a NASA computer expert discovered evidence that one of the greatest natural disasters ever to occur on Earth was the eruption of a megavolcano on what is now Sumatra in 73,000 B.C.E. The eruption, the scientists concluded, could explain the reasons and origin of a prehistoric ice age.

In a startling coincidence of 21st-century exploration, three scientists with no connection to each other except for their findings, and a NASA computer expert working by himself, uncovered evidence that one of the greatest natural disasters ever to occur on Earth happened when magma that had accumulated for a million years under a megavolcano (or supervolcano) erupted in 73,000 B.C.E.

The story of the discovery reads like a detective novel. First, on a routine mission close to the North Pole, researchers for the Greenland Ice Sheet Project, led by climatologist Greg Zilinksi of the University of Massachusetts, drilled into a glacier that was composed of the accumulation of more than 100,000 years of snowfall. The snow had captured chemicals in the atmosphere, which were in turn compressed into the surviving ice sheet. During this routine drilling exercise, Zilinksi made a startling discovery at the 75,000-year level: His instruments indicated a huge leap in concentrated sulfuric acid, a particularly noxious chemical, indicating the presence of a cataclysmic event in which billions of tons of sulfuric acid had found their way into the Earth's atmosphere. If this had happened, it would mean that the acid would have taken on the form of a poisonous, yellow haze that, because of its volume, would have blanketed the Earth.

Meanwhile, several thousand miles away, geologist Michael Rampino, similarly studying the history of the planet's climate, was tracking color changes in tiny sea creatures called Foraminifera, in ocean cores, or layers of sediment, at the sea bottom. The creatures who absorb oxygen from seawater can be read to calculate ocean temperatures over many thousands of years. In this particular experiment, Rampino discovered an abrupt drop in 10°F (12.2°C) over a few thousand years—an eye blink in the Earth's 4.5 billion year history. "Climactically, that's very surprising," he told PBS later. "That's very, very fast. That's very catastrophic."

Still unknown was what caused these coincidental evidences of a single catastrophe. The two scientists communicated, and decided that the cause would have to be either an asteroid hitting the Earth, or a volcanic explosion that was thousands of times more powerful than any in recorded history.

In 1990, John Westgate, a quaternary tephrochronologist, who identified the location of volcanoes by the chemical composition of volcanic ash, found that he was receiving boxes of ash from as much as 4,000 miles (6,437.3 km) apart that had the same chemical composition. And all were from 73,000 B.C.E. The highest number of samples came from Southeast Asia, a region with over 70 known volcanoes.

In 1994, Westgate received a sample taken from the shore of Lake Toba in a tropical jungle in Sumatra. Lake Toba is an enormous and extremely deep lake. And further exploration by Craig Chesner, a geologist from Eastern Illinois University, turned up evidence that the lake was really the caldera of a super, active volcano, only the fourth recorded in the world today (the other three are Long Valley in eastern California, Taupe in New Zealand, and Yellowstone National Park). Toba seemed to be the biggest, holding nearly 1,800 cubic miles (7,502.4 cu. km) of magma, which is enough to fill 200 Grand Canyons.

Now the other two scientists became involved, and all three determined that the magma that accumulated in the Toba volcano must have been a million years

in the making. When it finally erupted, it flung enormous ash clouds tens of miles (more than 16 km) into the atmosphere and spewed a long succession of hot avalanches of gas, ash, and rock down its slopes. "The island actually slid down off the walls of the crater during the eruption," Chesner said.

At the time of the eruption, the Earth was teeming with animal life, and recent fossil evidence has suggested that *Homo sapiens* had begun the migration from Africa and had reached Asia well before the Toba eruption. But both animals and humans would have been annihilated by the catastrophe, as would plant life. The animals and humans would have died excruciating deaths from inhaling volcanic ash.

But more than this. The resulting, Earth-embracing cloud of sulfuric acid and volcanic ash would have blotted out the Sun, causing a volcanic winter. "It's [the] mist of drops of sulfuric acid that present the veil that cuts out the sunlight," explained Michael Rampino, "that scatters the sunlight back to space, that keeps the sunlight from getting to the surface of the planet to warm the surface of the planet."

So, the evidence from Rampino's discovery of an abrupt drop in temperature near the North Pole indicated major cooling of the climate. "Based on our present information," Rampino told PBS, "Toba had a role, a major role, in causing 1,000 year climate cooling and created a mini-ice age that lasted at least 1,000 years, perhaps longer."

Now, NASA computer expert, Drew Shindell, joined the three scientists and constructed a computer model based upon their information. Shindell and the computer confirmed their conclusions. The more snow fell, the more reflection bounced back into space, and as the temperature dropped, ice formed into glaciers, which advanced, causing the temperature of the oceans to fall, which confirmed the findings of Greg Zilinksi. "Ocean surface temperatures started to cool," Zilinski said. "And by doing that, you're looking at a major part of our entire climate system, considering that the ocean covers 75 percent of the Earth's surface." And so, with the drop in temperature, all vegetation withered and died, and the animals and humans who had survived the initial toxic cloud starved to death when the food chain was disrupted.

The question that haunted the scientists was: Will it happen again? The answer was probably yes, but not in the lifetimes of the members of Earth's current population. Toba, they discovered, seems to be on a roughly 400,000-year cycle. But the eruption of any supervolcano would be as catastrophic as that of Toba. Michel Rampino laid out the scenario: "A supervolcanic eruption would affect all aspects of modern life, of modern civilization," he told *Nova*, the PBS science program. "Food resources, climate, water availability

... the water would become polluted by the ash, the crops would fail, the livestock would die. Machinery would get clogged by the ash so that automobile filters would clog and you can't drive. The jet engines on commercial airliners would clog up, and so worldwide air traffic would come to a standstill."

Not a very cheerful future. But a realistic, if distant one, given the potential power of the four megavolcanoes still accumulating magma beneath the surface of the Earth today.

GREECE
SANTORINI
1470 B.C.E.

Santorini, in the Aegean Sea, erupted cataclysmically in the summer of 1470 B.C.E. Experts theorize that it was the cause of four major prehistoric events recorded by Plato and documented in the Bible.

The overnight disappearance of Atlantis. The parting of the Red Sea. The plague of darkness that allowed the Children of Israel to leave Egypt. The disappearance of the Minoan Civilization. If the theories of Professor George A. Galanopoulos, the director of the Seismological Laboratory of the University of Athens, are to be believed, all of these legendary happenings are inextricably linked by a single, cataclysmic cause: the monumental eruption of the volcano Santorini, located in the Aegean Sea, 131 miles (210.8 km) southeast of Athens and 68 miles (109.4 km) north of the island of Crete.

Santorini (a medieval Italian corruption of Sant' Irene, the patron saint of the volcanic island of Thera) is one of a series of volcanoes that form an island arc bordering the former Aegean land mass. According to Dr. Galanopoulos, Santorini's first subterranean explosions took place in the Pleistocene Age, after which the cone of Santorini, along with a number of accompanying cones, grew to a height of 5,300 feet (1,615.4 m) above sea level.

This growth was apparently uneventful. Then, in the summer of 1470 B.C.E., Santorini erupted with

The Irazu Volcano, located in Turrialba, Costa Rica, is typical of the Central American chain of volcanoes that have erupted periodically from earliest recorded history to today. (U.N. photo)

incredible force, enough to atomize its summit, collapse the cones of several nearby volcanic mountains and send an enormous geyser of molten rock skyward, over the islands of the Mediterranean, Crete, and part of Egypt. An area of 77,200 square miles (124,241.4 sq. km) was completely covered with the ash from this gigantic explosion, which sent a cloud of poisonous vapors and sun-blotting darkness over Egypt and the eastern Mediterranean that lasted for days and possibly weeks.

The caldera (the depression formed by the volcano's explosion) of Santorini was enormous, more than three times the area of the caldera of Krakatoa. If Plato and Dr. Galanopoulos are to be believed, it was, until the moment of the eruption, the lost colony of Atlantis.

When the top of Santorini blew off in 1470 B.C.E., it took with it all of the structures and all of the people of this fabled empire, and deposited them on the floor of the Mediterranean Sea.

The legend and reality blend convincingly. First, although Santorini initially erupted in the Pleistocene Age and was active enough to build itself to a height of 5,250 feet (1,600 m), it probably was inactive enough for a civilization to have been built on its summit. Second, this summit measured 32.05 square miles (51.6 sq. km)—not enough space for a huge civilization, but certainly large enough for one of the size of either Athens or Sparta, comparable civilizations of the time.

The story of Atlantis—the advanced, island empire that sank beneath the sea in a single day and night—is told by Plato near the opening of *Timaeus* and in more detail in *Critias*. The tale is credited to Critias, an Athenian political figure in the Socratic circle, who in turn heard it when he was 10 years old from his 90-year-old grandfather, who had heard it from his father, a friend of Solon, the founder of Athenian democracy.

Solon, it seems, was a progressive and a free thinker, who believed in outlawing contracts where personal freedom was involved. This brought about his 10-year exile in Egypt, where he learned from the priests of Sais, one of the ancient cities of the Nile delta, the story of an island empire larger than Libya and western Asia combined, located outside the Pillars of Hercules (now the Straits of Gibraltar), which, 9,000 years earlier, had disappeared in a single day and night beneath the water.

Two problems arose for some historians. One was the small size of the caldera that supposedly contained the civilization; the other was the 9,000-year figure recorded by Solon. However, in 1956, Dr. Galanopoulos, examining the refuse from a severe earthquake on Thera, concluded that, among other misinterpretations of history, a misplaced decimal would make the 9,000 figure 900, and thus locate the disappearance of Atlantis at 570, approximately the time of the explosion and decimation of the summit of Santorini. Dividing the land area of Libya and Western Asia by 10 also diminishes the area to that of the caldera of Santorini.

Fascinating. But the possibility that the eruption of Santorini was also the destruction of Atlantis is only one-fourth of this intriguing theory. The disappearance of the first true civilization in the Mediterranean area, the Minoan civilization, which developed in Crete and the adjacent islands, also happened around 1400 B.C.E. At the same time, the Mycenaean civilization, which contained a multitude of Minoan traditions, suddenly appeared in southern Greece.

It seems reasonable to assume, then, as the Irish scholar K. Y. Frost did in a 1939 newspaper article titled "The Critias and Minoan Crete"—and as did the Greek archaeologist S. Marinatos, who carried out excavations at the Minoan palaces of Crete—that the Minoan civilization was not destroyed by foreign invaders, but was, instead, buried under the pumice falling from a gigantic, cataclysmic natural catastrophe of an unparalleled violence. The cataclysm? The eruption of Santorini, 68 miles (109.4 km) to the north of Crete.

The conclusions of both Dr. Marinatos and Dr. Galanopoulos—who, while investigating a mine shaft after the earthquake of 1956, discovered the ruins of a stone house in which he found two small pieces of wood and some human teeth, which radiocarbon dating proved to have come from approximately 1400 B.C.E.—were that a giant tsunami, caused by the eruption of Santorini, destroyed most of the Minoan civilization, and its survivors went to Greece, where they founded the Mycenaean civilization. This, too, was a responsible conclusion: Tsunamis were not uncommon in the region around Crete (recorded ones from earthquakes submerged parts of Alexandria, Egypt, in 365 C.E., and the southeast coast of the island of Amorgas, about 40 miles (64.3 km) southeast of Santorini, was inundated in 1956), so the theory could be true.

And if that was not enough, the conclusions of these two gentlemen had been prepared for in the 19th century by French engineers who built the Suez canal and mined the thick volcanic ash layer on Thera for the manufacture of cement used in the construction of the canal. Under the ash layer, they found remains of a civilization that was clearly pre-Greek, but whose age was then unknown.

So much for Atlantis and Minoa. Now, Galanopoulos, fired by his first discoveries, went on to speculate upon other effects of this monster eruption.

Continuing with his tsunami tracing, he used it to explain the biblical parting of the waters of the Red Sea, which allowed the Children of Israel to escape the pursuing forces of the Pharaoh. The parting, accord-

An engraving depicting the 1866 eruption of Nea Kameni, Santorini, Greece (National Geophysical Data Center)

ing to Professor Galanopoulos, was brought about by the withdrawing of the sea a half hour or so before the tsunami struck, thus exposing a large area of the sea bottom near the shore.

All of this gains credibility when one considers that the date of approximately 1450 B.C.E. is usually assigned by biblical scholars as the time of the exodus of the Children of Israel from Egypt. Noticing this, Professor Galanopoulos also concluded that the great plague of darkness brought by the Lord to compel the Pharaoh to let these children go was the same volcanic cloud that darkened all of the region after the explosion. "And the Lord said unto Moses," it says in Exodus 10:21–22, "stretch out thine hand toward heaven, that there may be darkness over the land of Egypt, even darkness which may be felt. And Moses stretched forth his hand towards heaven; and there was a thick darkness in all the land of Egypt for three days."

Considering the fact that there was total darkness for 22 hours at a distance of 130 miles (209.2 km) and for 57 hours at a distance of 50 miles (80.4 km) from the collapse of the caldera of Krakatoa in 1883 (see p. 400), it is not only possible but also probable that darkness continued for at least three days over Egypt in 1470 B.C.E.

Thus, four legends from four sources might very well have had their basis in the fact of one calamitous natural explosion—one that, if the theories that link these legends to it are even nearly true, is extraordinary, perhaps monumental—not only in its persuasive actuality but also in its effect upon the history of the world, its legends, its civilizations, and its beliefs.

GUATEMALA
SANTA MARIA
October 24, 1902

Six thousand people were killed when Santa María, a 12,361-foot (3,767.6-m) volcano in western Guatemala, erupted on October 24, 1902. An area of 125,000 square miles (323,749 sq. km) was covered by volcanic ash and stone.

Santa María, a volcano rising 12,361 feet (3,767.6 m) above the flat coffee plantations of western Guatemala, erupted with a roar that was heard 500 miles (804.6 km) away in Costa Rica on October 24, 1902.

It was an explosion of enormous proportions. An entire side of the mountain blew away, sending farmhouses into the air and crushing those which remained with horrendous showers of gigantic boulders. In fact, most of the 6,000 people who perished were crushed to death in their homes when this lethal rainfall of rocks, some of them weighing a ton or more, descended upon them.

The rest were killed by the lava flow whose pumice stone and ashes eventually covered an area of 125,000 square miles (323,749 sq. km) to a depth of eight inches (20.3 cm).

The eruption cloud lingered for weeks and climbed to a height of 18 miles (28.9 km) before dissipating. The eruption would rank as one of the worst in the entire history of volcanic explosions.

ICELAND
LAKI
January–June 1783

Twenty thousand people—one-third of the population of Iceland—were killed by the monumental, multiple eruptions of Laki, a 15-mile- (24.1-km-) long series of 100 craters and vents near Lakagigar, from January to June, 1783.

A small, thinly populated country abutting the Arctic Circle, Iceland seems like anything but a volatile place. And yet Iceland is one of the most volcanically active places in the world. Straddling the mid-Atlantic Ridge, it has, during its 1,100 years of recorded history, experienced and eruption on the average of once every five years.

In 1783, it was the site of a monumental volcanic explosion that produced the world record for casualties. It took place over a span of six months and consumed nearly 20,000 lives, or one-third of the population of the country.

"Basically," Dr. Sigurdur Thorarinsson, the dean of Icelandic geologists, told National Geographic writer William R. Gray in 1978, "Iceland is part of the ocean floor. We simply happen to be sitting over a plume where more magma [hot rock] has been released and therefore more terrain built."

The residents of Iceland had ample warning of this cataclysmic disaster. First, there was the long-term

knowledge that Iceland is situated on the northern rim of one of the major concentrations of volcanoes in the world. In a more immediate sense, they experienced, prior to the eruption, a long series of seismic upheavals over several weeks of escalating ground movements and a roiling sea that erupted periodically in enormous geysers, whirlpools, and water spouts which intensified the tides enormously.

The seas literally steamed in the few days before Laki's eruption, and the 50,000 inhabitants of Iceland, most of them farmers and families of farmers, had become so thoroughly terrified from the weeks of upheavals that they had retreated within their homes, where they locked the doors and windows.

Laki, near Thingvellir, is a 15-mile (24.1-km) long slash in the Earth, with over 100 craters and vents pockmarking it. The center of activity was near Lakagigar, one of its largest craters, in a practically deserted part of Iceland. On June 11, 1783, the mountain called Skaptar Jokul, near Lakagigar, erupted with a cataclysmic roar, splitting open the entire 15-mile (24.1-km) long fissure. The gas pressure and rock-melting heat spewed forth a powerful and prodigious lava flow. Within hours, the normally placid and huge River Skapta was dried up and replaced by a lava flow 200 feet (60.9 m) wide and from 400 to 600 feet (121.9 to 182.8 m) deep. The moving lava, fueled by repeated eruptions that continued like grim, deafening echoes, moved relentlessly and purposefully to the sea, where it turned the water into steam, killing the fish for miles.

Ashes filled the air and blotted out the daylight. The grasslands for 50 miles (80.4 km) around the mountain were poisoned, and grazing cattle and sheep died by the thousands. According to one report, "The hills were dotted with decaying carcasses. The air was filled with horrible stench." Seventy percent of Iceland's livestock would die within the next two years.

Once put in motion, the volcanic forces did not ease. The enormous lava flow, the largest in history, poured forth from this 15-mile (24.1-km) long fissure, in various places, alternately creeping and racing across a staggering 23,400 acres (9,469 ha) of land, laying waste to it, and creating cloud upon cloud of noxious fumes from the decaying vegetation.

It was the largest lava flow ever recorded, covering an estimated 20 cubic miles (32.2 km). To early volcanologist Professor E. Bischoff, it was ". . . the greatest eruption of the world—the lava, piled, having been estimated as of greater volume than is Mont Blanc."

The lava streams crossed and recrossed the landscape, swallowing up entire villages, bypassing some and then returning to consume them on a second or third pass. Two major streams of running lava, one 40 (64.4) and the other 50 miles (80.5 km) long, roared along at an average depth of 100 feet (30.4 m). Eventually, they would reach widths of 15 (24.1) and seven miles (11.1 km), thus totalling an amount of molten lava that was sufficient to cover 1,000 square miles (2,590 sq. km) to a depth of 150 feet (4.6 m). More than 9,000 villagers perished of asphyxiation from the poisonous fumes or were buried and incinerated by the lava flow.

Those who reached the sea were in no less danger. The lava flow reached the beaches and inched into the sea. But before that, the heavy cloud of ashes descended upon it. According to Professor Bischoff, "The ashes fell in such volumes into the ocean that the fish deserted the coast. The flying clouds of dust spread to Europe. The appalling horror of the scene can hardly be imagined. Death stalked aboard in his most repulsive form."

While the eruptions continued, those who had escaped the lava flow and its immediate lethal effects died of starvation when both crops and fish were eliminated by the volcanic fallout.

Some sense of the enormity of this disaster can be imagined when it is compared to that of Vesuvius in its most angry times. At its worst, Vesuvius's lava flow covered only a square mile to a depth of 25 feet (7.62 m). Thus, its most extensive lava field was only .006 as extensive as that from this eruption of Skaptar Jokul, in the volcanic ridge of Laki.

ITALY
ETNA
1169

Catania was again destroyed by the eruption of Mt. Etna in 1169, and over 15,000 of its inhabitants died. In Messina, hundreds of refugees were drowned in an ancillary tsunami set in motion by the eruption.

Catania was again (see next entry) the object of Etna's falling ash and charging lava in 1169. According to legend, when Etna erupted and sent huge rivers of lava tumbling toward the town in 251, terrified residents went to the tomb of St. Agatha, removed the veil that covered the remains of the saint, held it up against the lava flow and thus stemmed this furious fulmination of nature just before it would have consumed them.

And so, when Etna exploded with a gigantic, ear-throcking roar in 1169, thousands of peasants crowded into the great cathedral of Catania, while the bishop and 44 Benedictine monks again employed St. Agatha's veil to protect them. But this time, the immense combination of an earthquake and the blowing out of one side of Etna's cone in a huge blast of molten ash and fire overwhelmed the city built on the slopes of a natural time bomb.

Almost instantaneously, more than 15,000 citizens of Catania, including everyone in the cathedral, were either crushed to death by falling buildings or burned by falling ash and advancing lava.

All over Sicily, and particularly in Syracuse and Ajo, fountains first turned brackish and then bloody. The earth pitched and rocked all over the island, as hundreds of citizens of Messina, running to the beach to escape both the earthquake and the eruption, were met by an enormous tsunami, which roared across the beaches, flooded half the city, and drowned scores of refugees and residents.

ITALY
SICILY
ETNA
1226, 1170, 1149, 525, 477, 396, 122 B.C.E.

The written records of the earliest eruptions of Etna, while vivid in imagery, are lacking in statistics. Nevertheless, they provide a picture of a constantly erupting volcano that repeatedly destroyed the city of Catania.

Myth has surrounded Mt. Etna as certainly as the volcanic clouds that have frequently obscured its 10,870-foot (3,313.2-m) peak. Overlooking the Mediterranean from the eastern coast of Sicily, Etna is Europe's highest, grandest volcano, and one of its only active ones today.

Scientists, and particularly volcanologists, have never ceased to be interested in Etna, since, in its dormant moments, it provides a moment of frozen time, a relic from the prehistory of the Earth.

Etna may also represent the height of man's foolishness. Its base, which spans some 87 miles (140 km) in circumference, is also the location of dozens of villages and a major city, Catania. Catania has been buried and dug out, destroyed and rebuilt, over and over. And yet, human beings still choose to live there, knowing full well that Etna could at any time explode and bury buildings, people, and landscape.

The early eruptions of Etna were explained in religious terms by the Greeks who witnessed them. The first three recorded ones, those of 1226, 1170, and 1149, were attributed to the expulsion of Hercules from Sicily and the wrath of Zeus. Another theory postulated that Etna was the workshop of Hephaestus and Cyclops—a direct parallel to the Icelandic myth that blamed the eruptions of Heclas on the workshop activity of the fire god Thor.

No less a figure than Pythagoras recorded an explosion that took place in 525 B.C.E., and Thucydides set to tablet another major eruption in 477 B.C.E.: "In the first days of this spring the stream of fire issued from Etna as on former occasions, and destroyed some land of the Catanians who live upon Mount Etna which is the largest mountain in Sicily. Fifty years, it is said, had elapsed since the last eruption, there having been three in all since the Hellenes have inhabited Sicily."

The most detailed and literary reportage of the volcanic activity of Etna during Greek times was by the poet Pindar. In 474 B.C.E., he brought to center stage Typhon, whom he claimed was manacled and imprisoned in the mountain by Zeus. "He is fast bound by a pillar of the sky," wrote Pindar, "even by snowy Etna, nursing the whole year through her dazzling snow. Whereat pure springs of unapproachable fire are vomited from the inmost depths; in the daytime lava streams forth a lurid rush of smoke; but in the darkness a red roiling flame sweepeth rocks with uproar to the wide, deep sea. . . . That dragon-thing [Typhon] is what maketh issue from beneath the terrible, fiery flood."

Virgil waxed eloquent over Etna, too:

But Etna, with her voice of fear,
In weltering chaos thunders near.
Now pitchy clouds she belches forth
Of cinders red, and vapor swarth;
And from her caverns lifts on high
Live balls of flame that lick the sky;
Now with more dire convulsion flings
Disploded rocks, her heart's rent strings.
And lava torrents hurls today
A burning gulf of fiery spray . . .

Lucretius, on the other hand, chose a more scientific approach.

"I will explain in what ways yon flame, roused to fury in a moment, blazes forth from the huge furnace of Etna," wrote Lucretius.

First the nature of the whole mountain is hollow underneath, underpropped throughout with caverns of basalt rocks. Furthermore in all caves are wind and air, for wind is produced when the air has been stirred and put in motion. When the air has been thoroughly heated, and raging about, has imparted its heat to all the rocks around, whenever it comes in contact with them, and to the Earth, and has struck out from them fire burning with swift flames, it rises up and then forces itself out on high, straight through the gorges; and so carries its heat far, and scatters far its ashes, and rolls on smoke of a thick, pitchy blackness, and flings out at the same

time stones of prodigious weight—leaving no doubt that this is the stormy force of air.

Again the sea to a great extent breaks its waves and sucks back its surf at the roots of the mountain. Caverns reach from this sea as far back as the deep gorges of the mountain below. Through these you must admit that air mixed up with water passes; and the nature of the case compels this air to enter in from that open sea, and pass within, and go out in blasts and so lift up flame, and throw out stones, and raise clouds of sand; for on the summit are craters, as they name them in their own language, what we call gorges and mouths.

An inaccurate report, it nevertheless attempts an explanation of one of the most splendid and terrifying of all natural disasters.

When Rome absorbed Sicily, the mythology of the mountain was turned over to the likes of Seneca, who turned the attention of his readers and listeners to the apocryphal tale of two brothers involved in the very real eruption of Etna in 477 B.C.E., an eruption that almost totally destroyed the city of Catania. According to Seneca, these two brothers carried their aged parents on their backs and fled from the city and the lava. But they could not outrun natural forces, and the lava overtook them. And then, like the Red Sea, the ocean of lava parted on either side of the family: "The flames blushed to touch the filial youths, and retired before their footsteps," writes Seneca. "On their right hand fierce dangers prevailed; on their left were burning fires. Athwart the flames, they passed in triumph. . . . The devouring flames fled backward and checked themselves around the twin pair. At length they issued forth unharmed, and bore with them their deities in safety."

After the eruption of 477 B.C.E. Catania was rebuilt and would withstand the eruption of 396 B.C.E., which obliterated the town of Naxos and killed all but a handful of its 500 inhabitants.

But when Etna again erupted in 122 B.C.E., huge, tumbling clouds of ash caved in the roofs of hundreds of houses, and prompted Rome to grant its Sicilian colonies a stay of taxation for 10 years—thus establishing the first recorded example of disaster aid from a government.

ITALY
VESUVIUS
August 24, 79 C.E.

Practically all of the inhabitants of Pompeii and those unable to evacuate Herculaneum perished as a result of the horrific eruption of Mt. Vesuvius on the afternoon of August 24, 79 C.E.

One of many volcanoes in the Naples area, Vesuvius first began life as a submarine volcano in the Bay of Naples, then emerged as an island and finally was joined to the land by the filling and upbuilding of its eruptive products. Its first eruptions are believed to have occurred after the retreat of the last Ice Age, which would make it roughly 10,000 years old.

However, since the ancients made no record of Vesuvius' volcanic activity, the mountain was most likely dormant for hundreds or perhaps thousands of years before the cataclysmic explosions of 79 C.E., which destroyed the cities of Pompeii and Herculaneum.

The district's earliest known inhabitants called themselves Oscans, and they were primitive cattle herders. Around 800 B.C.E., the Greeks arrived and settled in eagerly. There were forests full of game. The soil around the mountain, loaded with ancient deposits of mineral-rich ash, was amazingly fertile. Apples, pears, figs, cherries, melons, almonds, and pomegranates prospered. It was possible to grow two or three crops of wheat, barley, and millet annually, and vegetables and grapes flourished everywhere.

Small wonder then that the twin cities of Pompeii, on the slope of Vesuvius, and Herculaneum, at the foot of the mountain and on the rim of the bay, grew and flourished, and when Rome conquered Greece in 88 B.C.E., Pompeii became a jewel in the imperial crown, with its own council, Temple of Jupiter, and a forum. Gaius Quinctius Balbus, a Roman speculator, erected an amphitheater capable of accommodating 16,000 spectators, the entire population of the city, while they watched gladiatorial games. Ironically, these games would be in progress on the afternoon of August 24, 79 C.E.

In 62 C.E., nature offered a warning to the residents of the slopes of Vesuvius. A gigantic earthquake shook the area, tumbling public buildings, wiping out villages, and reducing a large part of Pompeii and Herculaneum to rubble. Despite this event, for the next 17 years, new temples, better, more grandiose baths, houses, taverns, and theaters rose in both cities.

And there they remained until the afternoon of August 24, 79 C.E., when, with a stupendous roar, Vesuvius woke from a long sleep and erupted, sending incandescent ash skyward in a huge, seething cloud. Lightning flashed through this geyser of ash that blew outward and upward at unbelievable speed, bombarding the countryside and the two cities with showers of stones and fragments of pumice. Ultimately, the western wall of the volcano blew away and fell into its widening crater.

At the bottom of the mountain at Misenum on the westernmost point of the Bay of Naples, Pliny the Elder, who, as Gaius Plinius Secundus, commanded the oar-powered war galleys in the bay, was roused by the spectacle. His nephew, Pliny the Younger, then a lad of 17, wrote two letters to the Roman historian Tacitus, one of his uncle's experiences and one of his own, and they constitute the first recorded eyewitness accounts of a volcanic explosion:

> About one in the afternoon, my mother deserted him [Pliny the Elder] to observe a cloud which had appeared of a very unusual size and shape . . . a cloud, from which mountain was uncertain at this distance, was ascending, the form of which I cannot give you a more exact description of than by likening it to that of a pine tree, for it shot up to a great height in the form of a very tall trunk, which spread itself out at the top into a sort of branches . . . it appeared sometimes bright and sometimes dark and spotted, according as it was either more or less impregnated with earth and cinders. . . .

Pliny the Elder ordered his ship to proceed toward the cloud. As they drew closer, hot cinders fell onto

An artist's depiction of Pompeii during the cataclysmic eruption of Vesuvius in 79 C.E. (Library of Congress)

the ships. Pliny also describes a "sudden retreat of the sea," which indicates that a tsunami might also have occurred that afternoon.

Pliny the Elder disembarked. Climbing the side of the mountain, he made contact with his friend Pomponianus.

> They consulted together whether it would be most prudent to trust to the houses, which now rocked from side to side with frequent and violent concussions as though shaken from their very foundations; or to fly to the open fields, where the calcined stones and cinders, though light indeed yet fell in large showers and threatened destruction. In this choice of dangers they resolved for the fields. They went out then, having pillows tied upon their heads with napkins; and this was their whole defense against the storm of stones that fell round them.

Pliny the Elder now reached the shore, with the idea of putting out to sea, but the sea was roiling with waves and steam, and he laid down to sleep. Then, "He raised himself up with the assistance of two of his servants, and instantly fell down dead; suffocated, as I conjecture, by some gross and noxious vapor. . . ."

Pliny the Younger's conjecture has been challenged by historians. His uncle, they conclude, must have died of a heart attack, not asphyxiation from volcanic gases; otherwise, his companions would have all died, too.

Meanwhile, Pliny the Younger, left with his mother at Misenum, recorded his own observations for Tacitus. Setting out from his tottering home, he noticed that:

> A panic-stricken crowd followed us, and (as to a mind distracted with terror every suggestion seems more prudent than its own) pressed on us in dense array to drive us forward as we came out . . . we stood still in the midst of a most dangerous and dreadful scene. The chariots, which we had dared to be drawn out, were so agitated backwards, and forwards, though upon the most level ground, that we could not keep them steady, even by supporting them with large stones. The sea seemed to roll back upon itself and to be driven from its banks by the convulsive motion of the earth; it is certain . . . the shore was considerably enlarged, and several sea animals left upon it.
>
> It now grew rather lighter, which we imagined to be rather the forerunner of an approaching burst of flames (as in truth it was) than the return of day; however, the fire fell at a distance from us; then again we were immersed in thick darkness, and a heavy shower of ashes rained upon us, which we were obliged every now and then to stand up to shake off, otherwise we should have been crushed and buried in the heap. . . . At last this dreadful darkness was dissipated by degrees, like a cloud or smoke; the real day returned, and even the sun

shone out, though with a lurid light, like when an eclipse is coming on. Every object that presented itself to our eyes (which were extremely weakened) seemed changed, being covered with deep ashes as if with snow.

It was thus the morning of the third day since the first eruption of Vesuvius.

Pliny did not describe the fate of Pompeii and Herculaneum. For years, the two cities would remain buried and silent—Pompeii under 25 feet (7.6 m) of pumice and ash, and Herculaneum under a mud flow 65 feet (19.8 m) deep.

The world ignored the two metropolises. A new city, Resina, rose above the site of Herculaneum, and grape vines grew where Pompeii once was. Ironically, it would take almost 1,600 years and another cataclysmic eruption of Vesuvius before the twin towns would be unearthed, and the story of that fateful day would be reconstructed.

But there would be no shortage of descriptions of the last hours of Pompeii once excavators and archaeologists began to dig through the hardened ash and pumice which killed at least 2,000 of its inhabitants.

Vesuvius's first major eruption following the 79 C.E. cataclysm occurred in 1631 (see p. 397). Shortly after that eruption, workers excavating civic water systems and irrigation reservoirs came upon a few Roman coins, which they pocketed. Shortly after this, the site was looted by scavengers until the beginning of the 18th century, when a well digger unearthed a cache of alabaster and marble.

From then until 1860, various high-level looters arrived, unearthing statues, artifacts, and pieces of buildings. Among them was Sir William Hamilton, the British envoy to Naples from 1764 to 1800, and an amateur archaeologist who sold artifacts to the British Museum for a small fortune.

It would remain for the 37-year-old archaeologist Giuseppi Fiorelli to bring responsibility to the search for the lost city and the events of that August afternoon in 79 C.E. What he—and some before him—discovered was that the particular combination of pumice and ash had in fact hermetically sealed the city against the destruction of time. He unearthed an amphitheater along with the skeletons of gladiators, discovered in positions they must have been assuming when the first fall of molten ash occurred. A sealed bakery oven contained blackened loaves of 1,800-year-old bread. Wall murals were preserved with their colors intact.

The bodies of those killed by the eruption had of course decomposed soon after they were buried by the falling ash, but their imprints in the ash were absolutely intact. Fiorelli gave the last hours of Pompeii a tragic immediacy by pumping liquid plaster of paris into these perfect molds, thus reproducing the people they had contained as they had looked at the very moment of death.

Two Roman soldiers who were evidently being punished at the time of the eruption were found intact, locked in stocks. The temple of Isis was unearthed, and in it, on a pedestal, was a statue of the goddess draped in purple and gold. In the next room was the body of a priest holding an axe in his hand, and in another room was a priest sitting at dinner, apparently confident that the temple would protect him from the deadly ash falling in the streets of the city. In another area, a man was discovered in an upright position, a sword in his hand and his foot resting on a pile of silver and gold. Around him lay the bodies of five would-be looters he had apparently killed in defense of his possessions.

Outside of the city proper, in the suburban villa of Diomed, 18 adults, a boy, and an infant were discovered in a vault. "To the skulls of the children cling their long, blonde hair," stated one historian, who went on to detail a scene in which the family sought shelter in the cellar of the villa, while the servants tried to make off with the family's valuables. Near the gate of the villa, two skeletons were uncovered, one holding a bunch of keys and a large sum of money, and the other clutching two silver vases.

"In the house of Faun," reconstructed another anonymous excavator/historian, "stood the skeleton of a woman, her hands raised over her head. Her scattered jewels lay about the floor. Endeavoring at length to leave the house, she found the doorway blocked with ashes. The flooring of the upper rooms began to fall, and she lifted her arms in a vain attempt to stay the crumbling roof. Thus she was found."

A large number of residents of both Pompeii and Herculaneum were discovered with their hands or cloths to their mouths, apparently trying to keep out the lethal gases. Numerous bodies were found near the sea, often with their possessions clutched in their hands. They were obviously struck down while trying to escape.

Excavators decided that Pompeii maintained a high standard of living before the cataclysm that snuffed out its life. There were 118 bars and taverns offering food, drink, and diversion to traders and merchants from all over the Roman Empire. Some had gaming rooms and others showed evidence of having provided prostitutes. A thermopolium (a tavern specializing in hot wine) had walls inscribed with the names of women—possibly the sort of Andalusian dancing girls who, in the words of Martial, ". . . with endless prurience swing lascivious loins." "Lucilla sells her body," trumpeted one piece of graffiti, while another, left by one Livia about someone named Alexander, chided, more ironically and pro-

phetically than she could have dreamed, "Do you think I would mind if you dropped dead tomorrow?"

Into the 20th century, excavations unearthed new, preserved remains of Pomeii's doomed populace. On August 29, 1991, workmen clearing space for a service hut discovered the mummified bodies of eight people, one a teenage girl wearing a ring, indicating that she was a slave. Their faces, wrapped in tunics to protect them from the gases that may have killed them, were turned toward the town wall and the sea, a mere 200 yards away.

Herculaneum, excavated from under its gigantic mausoleum of mud, would yield very few bodies, leading to the assumption that the muddy mass of fine ash, lava fragments, and pumice which invaded every nook and cranny of the city had not reached there in the first cataclysmic moments of the eruption, but probably crept in sometime during the three days of darkness. Thus, most of the populace had a chance to escape the lethal wrath of Vesuvius.

ITALY
ETNA
March 11, 1669

The worst eruption in the long and explosive history of Mt. Etna occurred on March 11, 1669. There were at least 20,000 dead, although other estimates have run from 60,000 to 100,000.

The worst explosion in Mt. Etna's long history (see pp. 392–393) was preceded by three days of monstrous earthquakes, in which the entire town of Nicolosi was swallowed up. On March 8, 1669, a whirlwind swept across the summit of the mountain, blotting out the sun, and the earth shook with increasing ferocity. The climactic explosion occurred at dawn on March 11, resulting in a giant fissure 12 miles (19.3 km) long and six feet (1.8 m) wide in the side of the mountain.

An astonishing red-white light emitted from the crevice as six craters erupted great gouts of flame and smoke. By the end of the day, a seventh crater had appeared, shooting white-hot stones, sand, and ashes over 90 square miles (233.1 sq. km) of the Sicilian countryside.

The seventh crater was also the source of an immense lava flow, and it was this, combined with the poisonous vapors and rain of ashes, the blasts of fire and showers of boulders caused by repeated earthquakes and the splitting of the great central crater from the mountain, that caused over 20,000 deaths (some reports set the figure at anywhere from 60,000 to 100,000).

Fifty towns were buried under the rivers of molten lava, including Pietro, Camporotondo, Mascalia, and Misterbianco. As the lava flow inched toward Catania, scooping up cornfields and vineyards which boiled eerily on its surface, prayers were offered up by priests clutching the veil of St. Agatha. One man, Diego de Pappalardo, attempted a single-handed salvation of his doomed city.

Gathering 50 cohorts, clad in wet cowskins to protect themselves from the lava's intense heat and wielding long iron bars, de Pappalardo set out to divert the lava stream away from Catania by breaking down the lava-made levees of cooling substance at the sides of the flow. The game group smashed a hole in one levee, high above their city, and a tributary of molten lava worked its way through the breach, slowing the main stream.

However, this tributary was pointed directly at the village of Paterno, 11 miles northwest of Catania. Its citizens were justifiably horrified. Some 500 set out, armed with any weapon they could find, to undo the work of Diego de Pappalardo and his workers. This they did in short order, and the levee breach soon clogged and the lava stream, restored to full and formidable volume, resumed its descent directly towards Catania.

There, it came upon a 60-foot (18.3-m) high stone wall, erected by Catania to withhold future lava flows. The wall held for three days, as tons of lava built up against it. On the third day, a weak spot gave way and with a horrendous roar, the river of lava poured through the streets of Catania.

An English ambassador, in Sicily to report on the destruction, wrote, "I could discern the river of fire to descend the mountain, of a terrible fiery or red color, and stones of a paler red to swim thereon, and to be as big as an ordinary table ... of twenty thousand persons which inhabited Catania, three thousand did only remain; all their goods are carried away, the cannon of brass are removed out of the castle, some great bells taken down, the city gates walled up next the fire, and preparations made by all to abandon the city."

Once more, Etna destroyed Catania, and once more, its residents stubbornly rebuilt it.

ITALY
VESUVIUS
December 16, 1631

Estimates of between 4,000 and 18,000 dead resulted from a monumental eruption of Vesuvius on December 16, 1631.

Vesuvius erupted from time to time after the cataclysmic explosions of 79 C.E. that buried Pompeii and Herculaneum (see pp. 393–396), but, from all accounts, these were minor outpourings, causing little damage or loss of life. Eruptions were recorded in 203, 472, 512, 685, 993, and 1036. The eruption of 472 is noteworthy, since it is reported to have spread ashes over all of Europe and caused alarm as far away as Constantinople. The 1036 explosion marked the first time that lava flowed out of Vesuvius. Even the great catastrophe of 79 C.E. consisted entirely of fire and ash.

But the cataclysm of 1631 ranked in fury and effect alongside that of 79 C.E., with the added dimension of death and destruction by lava.

There was also an added similarity to the 79 C.E. eruption in that the volcano had been quiescent for nearly 500 years—long enough for the sides of the crater to be covered with brushwood and for a plain upon which cattle grazed to exist at its bottom. It was, in other words, the same soft scene that preceded the earlier holocaust, and which in part accounted for the monstrous loss of 18,000 human lives (more conservative estimates have put it at 4,000) and 6,000 animal lives in 1631.

The mountain began to erupt on December 16, 1631, with echoing explosions and the rising of a huge pink cloud which brought instant darkness and a rain of lethal cinder and ash. Some inhabitants of the villages that had sprung up along the mountain's slopes were burned to death by falling flames and red hot cinders; the survivors fled down the mountainside to Naples.

On the morning of the 17th, two fissures opened on the southwest side of the cone and floods of molten lava erupted from them, breaking into seven streams and rushing down the slope at enormous speed, invading the towns of San Giorgio a Cremano, Portici, Pugliano, La Scala, and the western portion of Torre del Greco. Tongues of lava slipped into the sea, causing it to boil and send gouts of steam skyward.

On the evening of the 17th, mud began to rain down on Naples. Huge mudflows invaded not only the villages on the slopes of Vesuvius and the north slope of Mt. Somma but also San Giorgio a Cremano, Portici, and Resina—the town built upon the site of Herculaneum. A long peninsula of land was formed and extended nearly a mile into the sea. More lava spit from the fissures on the side of the mountain, accompanied by strong explosions, adding to the stygian darkness that covered the area.

More villages were destroyed by lava and mudflows. Six and a half feet (2 m) of ash and cinders fell in Naples, and then, to add to the destruction, tremendous deluges of rain descended during the last week of

December, particularly on the north slope of Vesuvius, where the ash fall was the heaviest.

The cone of the mountain was reduced in height by 51 feet (15.6 m), and the crater was enlarged to more than twice its diameter before the eruption. From that year forward, Vesuvius's periods of repose would never extend beyond seven years.

ITALY
VESUVIUS
May–August 1779

Close to 1,000 people died in the eruptions of Mount Vesuvius from May through August 1779.

Various cyclic eruptions marked the continuation of Vesuvius as an active volcano, particularly in the latter part of the 18th century. In May 1779, a fissure opened on the northeast side, and lava poured from it throughout the month of June, covering the countryside and destroying several villages.

On August 8, huge explosions rocked the mountainside and a liquid fountain of fire rose from the crater to a height of two miles. This blazing, undulating column of fire directed itself toward the town of Ottaiano. As it fell, it covered the side of the cone with a band of red-hot scoria (a basaltic type of volcanic ejecta) more than a mile in width. Because they could not see the crater which was hidden by the rim of Mt. Somma, the inhabitants of Ottaiano were caught unawares, and were unable to escape the bombardment of red-hot scoria and cinders.

Sir William Hamilton, who had recorded earlier eruptions of Vesuvius, was an eyewitness to this:

> The sight of the place [Ottaiano] was dismal, half buried under black scoria and dust, all windows towards the mountain broken, some of the houses burnt, the street choked with ashes—in some narrow streets to a depth of four feet, so that roads had to be cut by people to reach their own doors. During the tempestuous fall of ashes, scoria and stones, so large as to weigh a hundred pounds, the inhabitants dared not stir out—even with vain protection of pillows, tables, wine casks, etc. on their heads. Driven back wounded or terrified, they retreated to cellars and arches, half stifled with heat and dust and sulphur, and blinded by volcanic lightning—through 25 minutes this horror lasted; then suddenly ceased, and the people took the opportunity of quitting the country, after leaving the sick and bedridden in the churches. One more hour of this frightful visitation and Ottaiano would have been a buried city like Pompeii.

Nearly a thousand people died.

ITALY
VESUVIUS
February 1793–July 1794

Nearly 1,000 residents of Torre del Greco, at the base of Mount Vesuvius, were killed in a series of explosions between February 1793 and July 1794.

Earthquakes shook the earth near Vesuvius beginning in February of 1793, and finally, on June 15, strong explosions opened a fissure between Resina and Torre del Greco low on the southwest side of the old Somma cone. Along this fracture, which was a half mile in length, lava poured in a single stream. In a matter of six hours, from 10:00 P.M. until 4:00 A.M., this molten river invaded Torre del Greco, flowing down its main street in a path 1,200 to 1,500 feet (366–457 m) wide.

Again, Sir William Hamilton recorded the scene:

> Huge masses of white smoke were vomited forth by the disturbed mountain and formed themselves at a height of many thousands of feet above the crater into a huge, evermoving canopy, through which, from time to time, were hurled pitch-black jets of volcanic dust, and dense vapors, mixed with cascades of red-hot rocks and scoriae. The rain from the cloud canopy was scalding hot.
>
> As the lava rushed forth from its imprisonment, it streamed a liquid, white and brilliantly pure river, which burned for itself a smooth channel through a great arched chasm in the side of the mountain. It flowed with the clearness of honey in regular channels, cut finer than art can imitate and glowing with the splendor of the sun.

Nearly 1,000 people, mostly residents of Torre del Greco, were killed.

ITALY
VESUVIUS
April 4, 1906 and March 18, 1944

Nearly 2,000 were killed in Vesuvius's two final eruptions, on April 4, 1906, and March 18, 1944.

Close to 2,000 people lost their lives in the last two paroxysmal explosions of Vesuvius, in 1906 and 1944.

The 1906 eruption began early in the morning of April 4, when a fissure opened on the southwestern side of the cone—which had added to its height until it had reached 4,338 feet (1,322 m) above sea level. Lava began to flow from this fissure, and toward midnight, another flow, some 1,200 feet (365.8 m) lower, began.

The April 4, 1906, explosion of Mount Vesuvius was the next-to-last major eruption of that dramatic volcano. Three new fissures opened on the volcano's slopes in this thunderous eruption, which climaxed in a giant, 15-hour gas blow-off on April 8. (Illustrated London News)

Meanwhile, roaring activity began to build in the crater, and suddenly, with a violent report, the top of the cone blew off, sending enormous blocks of old lava soaring into the sky. Dubbed "lava bombs" by observers, these boulders of ash and stone weighed up to two tons (1.8 tonnes) and collapsed homes and churches with their occupants inside. In the village of San Giuseppe, 105 parishioners were crushed to death as they sought shelter from the fiery eruption in the local cathedral.

On the morning of April 6, a third fissure tore open a slash in the side of the mountain, spouting molten lava which descended toward and eventually covered the town of Torre Annunziata. For the next two days, fresh lava flows and enormous quantities of fresh scoria and rocks covered the countryside and parts of the towns of Terzigno, Ottaiano, and San Giuseppe.

The culmination of the eruption came with a great "gas blow-off" which began at 3:30 A.M. on April 8, and continued for 12 to 15 hours. The tremendous outrush of gas, which by its force enlarged the crater and

carried away the upper part of the mountain's cone, was a fearsome spectacle.

There was so much ash left by the explosions of Vesuvius this time that "hot avalanches" occurred for days afterwards, and when heavy rains fell toward the end of April, mudslides caused extensive damage in and near Ottaiano.

The eruption of March 12–21, 1944, occurring at the height of World War II, did extensive damage too, but accounted for little loss of life. The heaviest destruction was from the lava flow which began on March 18. Somma, Atrio de Cavallo, Massa, and San Sebastiano were inundated by the molten mass, which was 100 yards (91.4 m) wide and 30 feet (9.1 m) deep. As it passed through the towns, it left only a smoking lava field and the ruins of some of the houses. Other lava streams poured from the west side of the cone, cutting and blocking the railway and the Funicolare Vesuviana. At the same time, explosive activity in the crater increased in violence, throwing out huge quantities of ash and scoria.

Milton Bracker, there to report the war for the *New York Times,* was aghast:

> Those who watched Vesuvius in action this morning will never forget it. The crater, from which alternately oozed or spurted the fiery volcanic matter, was forgotten in the presence of a prong of lava . . . [which] was like the monstrous paw of an even more monstrous lion, slowly inching forward toward his prey.
>
> The lava was white-hot; it was orange-gold, with occasional black patches, undulating like waves. As the stream advanced, great boulders cracked off and tumbled down, setting fire to small fruit trees . . . the general sound was like that of an infinite number of clinkers rolling out of a furnace, but sometimes a great chunk of rock bent rather than broke. Its effect was like that of the devil's own taffy being pulled and twisted to suit his taste. . . .

JAPAN
UNSEN
April 1, 1793

Unsen, the volcano which gave the island of Unsen its name, blew itself apart on April 1, 1793, sinking the island and drowning all 53,000 of its inhabitants.

The 20-year period between 1780 and 1800 was one of extreme volcanic and earthquake activity. Dozens of islands were formed. The Satsuma Sea was rife with ripples of volcanic ridges asserting themselves, rising and forming islands.

At the same time, established volcanoes exploded with extraordinary force. Sakurajima, according to records of the time, ". . . blew out so much pumice material that it was possible to walk a distance of twenty-three miles upon the floating debris in the sea." Asama, according to these same records, ". . . ejected many blocks of stone—one of which is said to have been 42 feet in diameter—and a lava stream 42.5 miles in length."

But the most cataclysmic explosion occurred on the island of Unsen on April 1, 1793. Huge earthquakes and violent, summit-smashing eruptions from Unsen, the volcano, erupted with an ear- and earth-shattering roar on that day, sending ripples through the side of the mountain and great lava-spitting fissures all the way to the sea.

The entire island, mountain and all, disappeared into the sea, taking an estimated 53,000 people to their deaths. (See EARTHQUAKES, p. 25.)

JAVA
KRAKATOA
August 26–27, 1883

Krakatoa, on Java, exploded with the loudest sound yet known to man on August 26, 1883, in the most ruinous and lethal of all volcanic eruptions. Two hundred thousand died from fire, molten lava, descending ash and rocks, or the 120-foot (36.576-m) high tsunami caused by the deafening detonation.

What may have been the greatest cataclysm in the history of the world took place on August 27, 1883, when Krakatoa, the volcanic island located in the Sunda Strait between Sumatra and Java, blew itself to pieces in a massive eruption. Five cubic miles of rocks and ash and a seven-mile (11.26 km) high gout of steam were shot into the atmosphere, exploding with the loudest sound known since man evolved. The ensuing

A drawing of the ash cloud from the 1883 eruption of Krakatoa, Indonesia. One of the largest volcanic eruptions in history, the explosion was heard more than 2,500 miles (4,000 km) away and atmospheric effects were recognized around the world. (National Geophysical Data Center)

shock waves circled the Earth seven times, and created a 120-foot (36.5-m) high tsunami, or tidal wave, that drowned 36,000 people.

The ultimate death toll would reach an estimated 200,000, and it would probably have been greater, had Krakatoa been an inhabited island. It was, instead, the uninviting stump of an ancient volcano that might well have erupted in the same way in some prehistoric time. The string of islands that formed the Kandanga range along the southeast coast of Java may have been part of one vast volcano. Certainly, the linked volcanic upthrusts of Perboewatan, Danan, and Rakata were all part of the prehistoric caldera, or summit rim, of an ancient, immense volcano, encompassing Krakatoa.

The mountain had been somnolent from 1680 to 1883, and the 1680 explosion had been merely a flow of andesitic obsidian, or volcanic glass, from the vent of Perboewatan. Still, it was from this same vent that the initial explosions occurred on May 20, 1883. They were small ones, so unimpressive, in fact, that a group of inquisitive Europeans chartered a steamer on May 27 to visit the island and observe what one adventurous member of the party described as a "vast column of steam issuing with a terrific noise from an opening about 30 yards in width," near the crater of Perboewatan. The group also noticed that Rakata and Verlaten were covered with fine ash and that the vegetation, although unburned, had been killed.

From then until June 19, the area was quiet. Then, the small explosions resumed, and the countryside underwent a metamorphosis of deterioration.

On August 11, 1883, Captain Fercenaar, chief of the topographical survey of the neighboring island port of Bantam, stepped ashore on the island of Krakatoa. It trembled beneath his feet. And worse: It looked like some alien world. Twenty inches (50.8 cm) of dust covered the entire island. Three vapor columns reached skyward, and 11 new eruptive foci that had not been there before May were spurting steam and ash. The thoroughly terrified captain left after a few hours of record taking.

For 15 days, until August 26, activity quieted, and then the rumbles resumed and became explosions. At 1:00 P.M. on August 26 the first of those detonated from the bowels of Krakatoa, rattling windows on nearby islands, rippling the earth for a hundred miles (160.9 km) in each direction. At 2:00 P.M., an immense, black cloud ascended above Krakatoa to a height of 17 miles (27.3 km). It crackled with St. Elmo's fire. On a ship 40 miles (64.4 km) distant from the volcano, a Captain Wooldridge noted, "Krakatoa was a terrifying glory . . . it looked like an immense wall, with bursts of forked lightning darting through it, and lazing serpents

playing over it. These bursts of brilliancy were the regular uncoverings of the angry fires. . . ."

Other volcanoes in the Java chain, once part of that one prehistoric mountain, now began to erupt. Krakatoa's explosions increased in violence until 5:00 P.M., when the first tidal wave formed, crashing against nearby islands and inundating fishing villages, drowning their inhabitants and washing away their boats.

All night and into the morning hours of August 27, the rumblings and explosions continued. Concussions collapsed stone walls, shattered lamps, and tore gas meters from their sockets in nearby islands. All over Java and in Batavia, 100 miles (160.9 km) away, the noise was loud enough to keep everyone awake, as the china in their closets rattled and their houses shook as if heavy artillery were being fired nearby.

Between 4:40 A.M. and 6:40 A.M., several large tidal waves spread outward from Krakatoa, probably because of the further collapse of the northern part of the island.

By 10:00 A.M., the rehearsal was over, and the main event was ready to occur. Two observers watched, and recorded it. A Dutch scientist, sent by his government to make seismic observations, was climbing a mountain near the town of Anjer on the west coast of Java that morning. According to R. Hewitt, in his book, *From Earthquake Fire and Flood*:

> . . . looking towards the island of Krakatoa, about thirty miles away in the Sunda Straits, he suddenly saw a movement among the little boats in the bay. It seemed as though a magnet pulled them from their quiet positions of rest, and they all streamed in the same direction, impelled like the Flying Dutchman by an invisible hand. In a moment they disappeared, swallowed up in a mighty boiling chasm of fire and water. Right across the Straits it appeared that a line of flame stretched towards the islands. The crust of the Earth beneath the bay had split, and it seemed as though all the fires of hell burst through the surface waters. The sea poured into the chasm, carrying all the little boats to their doom. Hissing steam added to the inferno. . . .

"It grew darker and darker," wrote R. J. Dalby, a seaman on a Liverpool clipper. He continued:

> The already loud rumblings grew louder, and now they seemed all around us. The gusts of wind increased to such a hurricane as no man aboard had ever experienced before. The wind seemed a solid mass thrusting everything before it, and roaring like an enormous engine and shrieking through the rigging like a demon in torment. The darkness became intense, but the vivid lightning, which almost blinded us, seemed everywhere. The thunder was deafening. . . .

... When we could get a glimpse of the heavens we could see there a terrible commotion: the clouds whirled round at terrific speed, and I think most of us thought that we were in the vortex of a cyclone. But as the noise grew louder and louder, I at any rate reckoned it was something volcanic, especially when, at about noon, it rained a continuous downpour of dust. This seemed a sulphurous, gritty substance, and as we wore only two cotton garments, we were soon completely smothered; burned, choked and almost blinded.

Visibility at this time was about a yard. I seemed isolated, and felt my way about the deck, never loosing my hold of anything handy. You cannot imagine the force of that wind. Occasionally I met others on the same lay as myself, but quite unrecognizable, just moving grey objects in the darkness. Ropes and lines were lashing across the decks like whips. Once I saw two terrified eyes, the eyes of a poor old coolie, peeping from under the boat.

None of us will ever be able to describe the noise, especially one great bang about noon, which is supposed to have been the loudest sound ever heard on earth.... It was the very top of Krakatoa going up into the skies.... The whole heavens seemed a blaze of fire and the clouds formed such fantastic shapes as to look startlingly unnatural; at times they hung down like ringlets of hair, some jet black, others dirty white....

The seaman was almost right. Powered by the buildup of steam beneath the surface of the Earth, the cone of Krakatoa had been blown to bits, shooting approximately five cubic miles of material into a cloud that extended 50 miles (80.4 km) into the atmosphere. Meanwhile, rock fragments from the former cones of Danan, Perboewatan, and Rakata collapsed, sinking beneath the sea and turning it into a boiling cauldron.

The sound of this detonation was so immense, it was heard nearly 3,000 miles (4,828 km) away, and the shock waves would circle the Earth seven times. On the island of Rodrigues, in the Indian Ocean, 2,968 miles (4,776.5 km) from Krakatoa, an alert Coast Guard observer recorded the sound exactly four hours after the explosion occurred. In central Australia, 2,250 miles (3,621 km) southeast of Krakatoa, the noise was also recorded. Flocks of sheep stampeded at the sound of the eruption on the Victoria plains in West Australia, 1,700 miles (2,735.9 km) away. It was as if Pikes Peak had erupted, and the sound was heard all over the United States.

Now, a stygian darkness settled over the area, plunging places 275 miles (442.6 km) away into an early night. One hundred thirty miles (209.2 km) from the explosion, the darkness lasted for 22 hours; 50 miles (80.5 km) away it lasted for 57 hours. Ships 1,600 miles (2,574.9 km) distant reported that dust began to fall on the decks three days after the eruption.

As if they were being conducted, all of the volcanoes in Java began to erupt. Papandayan ripped apart, and seven fissures hurled steaming lava down its slope. Sixty-five miles (104.6 km) of the Kandanga range in the Malay Archipelago exploded, disappearing into the sea. Fifty square miles (129.5 km²) of Java from Point Capucine to Negery Passoerang also sank beneath the ocean.

And now, one of the worst, most lethal effects of the explosion was born. Half an hour after the cataclysmic eruption, a seismic tidal wave was formed and roared into the coasts of Java and Sumatra, wholly or partially destroying 295 towns and killing 36,000 people (some authorities said 80,000), mostly by drowning.

According to N. van Sandick, the engineer of the ship *Loudon,* the scene was horrifying. "Like a high mountain," he wrote:

the monstrous wave precipitated its journey towards the land. Immediately afterwards another three waves of colossal size appeared. And before our eyes this terrifying upheaval of the sea, in a sweeping transit, consumed in one instant the ruin of the town; the lighthouse fell in one piece, and all the houses of the town were swept away in one blow like a castle of cards. All was finished. There, where a few moments ago lived the town of Telok Betong, was nothing but the open sea.... We cannot find the words to describe the terrifying events which left us with the sight of such a cataclysm. The thunderstriking suddenness of the changing light, the unexpected devastation which was accomplished in an instant before our eyes, all this left us stupefied....

New volcanic mountains rose from the sea; islands disappeared with all of their inhabitants. At Anjer and Batavia, 2,800 were washed out to sea by the tsunami; 1,500 drowned at Bantam. The captain of the *Loudon,* surviving the tsunami, made for Anjer to warn the Dutch fort. He found the entire garrison dead, except for one sailor who was wandering around through the thickly strewn corpses. The islands of Steers, Midah, Calmeyer, Verlaten, Siuku, and Silesi sank beneath the sea, drowning all of their inhabitants. Of the 2,500 workers in the stone quarry on Merak Island, which had previously been 150 feet (48.5 m) above sea level, only two natives and a government bookkeeper escaped drowning when the island was suddenly completely inundated by the sea. The German man-of-war *Berouw,* lying off Sumatra, was swept up in the tsunami and tossed inland a mile and three-quarters (2.7 km), where it came to rest in a forest 30 feet (9.1 m) above sea level.

Those residents of islands who escaped the water were rained on by white hot stones and lava. At Warlonge, 900 were killed. At Talatoa, 300 were buried beneath lava. Tamarang was consumed by flaming rocks and lava that set every building on fire; 1,800 perished.

Three more lesser explosions rocked the area during the night of the 27th, and the early morning of the next day, and then the mountain finally rested. The destruction was reported by seaman R. J. Dalby. "In place of luxuriant vegetation nothing but a brown sterile barrenness remained," he wrote. "The shores both of Java and Sumatra seemed to have been battered to pieces and burnt up. All manner of wreckage was flying past us. Huge masses of vegetation floated by, and on it we could see enormous frogs, snakes and other strange reptiles. And the sharks!—it was sickening to see them. As for our ship, we had been painting it and tarring the rigging, and now it looked as if we had been through a shower of mustard."

The reverberations of the explosion traveled the globe. According to Robert Ballin in his book, *Earth's Beginnings,* ". . . every part of our atmosphere had been set into a tingle by the great eruption. In Great Britain the waves passed over our heads; the air in our streets, the air in our houses, trembled from the volcanic impulse. The oxygen supplying our lungs was responding also to the supreme convulsion that took place 10,000 miles away."

For months, the sky all over the world glowed, prompting Alfred, Lord Tennyson to record the event in verse, in "St. Telemachus":

> Had the fierce ashes of some fiery Peak
> Been hurled so high they ranged round the globe?
> For day by day through many a blood-red eve
> . . . .
> This wrathful sunset glared . . .

Two-thirds of the main island of Krakatoa had disappeared, and where the land before the explosion had stood anywhere from 400 to 1,400 feet (121.9 to 426.7 m) above sea level, there was nothing but a great cavity 900 feet (274.3 m) below sea level.

Bizarre lightning and visual effects persisted for months after the explosions. In some parts of the East the sun appeared blue and the moon a vivid green. And because of the movement of dust particles left by the volcanic eruption scientists were able to identify the presence of the jet stream.

JAVA
KELUIT
1919

In a freak result of the eruption of Keluit, on Java, in 1919, its crater lake first shot skyward in a spectacular

fountain, and then drowned 5,500 residents of the villages on its slopes.

Keluit, in the chain of volcanic islands in and near Java, is unique in that its active volcano contains a lake within its crater, formed after the eruption of 1901. It has never been noted for stupendous blowouts, but the explosion of 1919 produced a singular and devastating situation.

The volcano, dormant for 18 years, erupted without warning, spewing ash and fire—and water. The entire contents of the lake shot skyward in a spectacular geyser, and then roared down the slopes of Keluit at tremendous speed, drowning 5,500 residents of the valleys at the bottom of the mountain.

Javanese engineers, realizing that a recurrence would undoubtedly produce exactly the same effect, immediately dug a tunnel into the crater, which constantly drained off the water from the lake. This tunnel still exists, and to date, no such catastrophe has recurred in or near Keluit.

JAVA
MERAPI
December 13–28, 1931

One of the worst volcanic eruptions of the 20th century, that of Merapi on Java, from December 13 to 28, 1931, killed more than 1,300.

For nearly three weeks in December 1931, Merapi, one of the largest volcanoes in the Java chain, belched forth a river of lava four miles in length, 200 yards wide, and over 80 feet (24.4 m) in depth. The white hot flow scorched the earth, set trees afire, and crushed villages in its path.

In addition, both flanks of the volcano blew out, spreading ashes over half of the island to which the mountain gives its name. Over 1,300 people lost their lives in this, one of the most severe volcanic eruptions of the 20th century.

JAVA
PAPANDAYAN
1772

Three thousand people were killed in the eruption of Papandayan, on Java, in 1772. The explosion destroyed the mountain and turned its slopes, formerly dotted with 40 villages, into a lake of lava.

A cataclysm of enormous impact occurred in Java in 1772, when the 8,750-foot- (2,667-m-) high volcano Papandayan literally blew itself to pieces and then sank into a huge, sulphurous, and boiling lake of lava.

Before this horrendous event, 40 villages and several towns with prosperous outlying plantations had existed placidly on Papandayan's slopes, while the steam and smoke from its green sulfur deposits wafted benignly into the sky.

But in 1772, without a rumble of warning, the entire mountain exploded, sucking 3,000 people, herds of cattle, and the 40 villages that had previously existed on it into its lethal lake.

"The area that was sunk down was fifteen miles long and six broad," recounted one historian. "No day of judgment painted by Angelo or Dore could ever match that actual horror of the solid mountain sinking into the earth with human beings on its slopes—its huge bulk going down as a ship goes down into the deep."

JAVA
MIYI-YAMA
1793

Fifty-three thousand people were reported to have died in the explosion of Miyi-Yama, on Java, in the Indonesian archipelago.

The Indonesian archipelago that contains the densely populated island of Java includes 130 volcanoes—a fifth of the world's total.

One of the worst tragedies ever resulting from a volcanic eruption occurred on the island of Kiousiou, off Java, in 1793. Miyi-Yama, that island's governing volcano, erupted in a huge explosion of mud and water that year, and although specific details are unavailable, the historian Engelbert Kampfer recorded an absolute inundation of all of the neighboring plains around the volcano, and a staggering death toll of 53,000 people.

JAVA
TAMBORO (TAMBORA)
April 5–12, 1815

An estimated 1.7 million tons of debris were flung into the air during the eruption of Tamboro, also known as Tambora, on the island of Sumbawa in Java, from April 5 to 12, 1815, thus causing the famous "year without a summer." Nearly 49,000 natives were killed in this monumental explosion, which blew the top off the volcano.

The most spectacular volcanic explosion in the history of the world caused the famous "year without a summer" in 1815. Quoits pitchers on the village green of Plymouth, Connecticut, wore overcoats on the Fourth of July, and laundry that was hung out to dry on June 10 in the same village froze on the line. In France crops were so scanty that grain carts on the way to market had to be guarded against bands of starving citizens, and wry New Englanders referred to the year as "Eighteen hundred and froze to death."

On April 12, 1815, the 13,000-foot- (3,962-m-) high volcano called Tamboro (or Tambora) blew itself asunder, sinking the island of Sumbawa, on which it is located, and killing all but 26 of the 12,000 natives instantly. Another 37,000 on neighboring islands died as a result of starvation and disease. It blew an estimated 1.7 million tons of debris into the sky. Most of this fell to earth in the vicinity of Sumbawa within a few hours, thus accounting for the staggering death toll. But the remainder was pulverized into a talc-like dust, light enough to remain suspended in the atmosphere. This dust was carried into the stratosphere, where it began to circle the Earth, reflecting incoming sunlight back into space, robbing the Earth of some of its heat and causing the spectacular orange sunsets·that English painter J. M. W. Turner would record for posterity.

That was the effect upon the world. The immediate impact upon the human beings directly affected by the cataclysmic action of this giant among titans was more monstrous and considerably more awesome.

It began on the evening of April 5, 1815, with preliminary explosions deep within Tamboro's crater. Sir Stamford Raffles, the lieutenant governor of Java, thinking there might be either a vessel in distress or rebels attacking one of the nearby British outposts, dispatched both troops and two boats from Batavia (now Jakarta) to search the Java Sea for survivors. He found none. The captain of the East India Company's armed cruiser Benares, 900 miles (1,448.4 km) to the east at Macassar on the island of Celebes, was also unsuccessful in locating the pirates he thought had unleashed a volley of artillery fire.

But later in the month Raffles did find a local surviving raja from Sumbawa who offered an eyewitness account, which Raffles recorded. According to the account, on April 6, the day after the first explosions, a light dusting of ash began to fall over a 400-mile (643.7-km) radius. On Sumbawa, the first eruptions of Tomboro, the huge mountain covered with vines and pastures, sent enormous rocks and quantities of red hot ash into the sky, dropping both upon hundreds of homes on Sumbawa and nearby Bima, and crushing these homes from the sheer weight of the ash.

Between April 6 and April 10, the explosions increased in violence and the ash turned to red hot, glowing boulders. And then, according to the raja, "About seven P.M. on the 10th of April, three distinct columns of flame burst forth, near the top of Tamboro mountain, all of them apparently within the verge of the crater; and after ascending separately to a very great height, their tops united in the air in a troubled, confused manner. In a short time, the whole mountain appeared like a body of liquid fire extending itself in every direction.

"Stones at this time fell very thick . . . some of them were as large as two fists, but generally not larger than walnuts."

Now, a unique phenomenon occurred. A giant whirlwind struck, ". . . which blew down nearly every house in the village of Saugar [a town 25 miles (40.2 km) from the center of the eruption]. In the part of Saugar adjoining Tamboro, the whirlwind's effects were much more violent, tearing up by the roots the largest trees, and carrying them into the air, together with men, houses, cattle and whatever else came within its influence."

Simultaneously, a huge tsunami formed, and reaching between 12 and 30 feet (3.7 and 9.1 m) in height, moved outward from the island, crashing into neighboring ones and drowning hundreds.

On and on the explosions went, tearing the mountain to bits, blowing away its top and ripping away its sides. An enormous cloud rose from this holocaust, blotting out the sun for 400 miles (643.7 km) around, and causing a three-day night that was, according to one of Raffles' correspondents, "blacker than the deepest darkness imaginable."

The mountain would continue to rumble and belch fire for three months, but by April 18, Raffles felt that it was quiescent enough to send a shipload of rice to feed the starving survivors of Sumbawa and the surrounding islands. Lieutenant Owen Phillips, the ship's commander, found a scene of heartrending desolation and disaster. The once majestic mountain was now flattened down to a plateau. Most of the island was coated with a two-foot-thick mantle of ash and mud. The sea around the island was thickly jammed with thousands of uprooted trees and immense, floating islands of pumice, which had trapped enough gas in its vesicles to keep it afloat.

Many of the 127,000 survivors of Sumbawa and the other islands were already racked by a virulent cholera, which had broken out immediately after the eruption. The rest waded despondently in the knee-deep ash, ready to trade their most valued possessions for a mere few ounces of rice. Along the roads, corpses were strewn like mileposts, and the handful of survivors in the villages foraged in the interior for the few remaining heads of edible palm and stalks of plantain.

Raffles later speculated that if all the ash given off by Tamboro had been gathered, it would have made ". . . three mountains the size of Mont Blanc, the highest of the Alps, and if spread over the surface of Germany, would have covered the whole of it two feet deep."

The ash and mud fill was not confined to Sumbawa alone, nor were the casualties. More than three feet of ash covered the entire island of Tombock, 100 miles from Tamboro. As a result, everything growing on the island died, and an estimated 37,000 people perished by starvation on Tombock. Add up the casualties and the primary and secondary destruction, and Tamboro's eruption becomes the most deadly volcanic explosion in recorded history.

There is an ironic footnote to the "year without a summer" effect of the explosion. For years, no one made a connection between the volcanic catastrophe and the weather that produced snowfall over all of New England in June and killer frosts everywhere. Some scientists suggested sunspots; others blamed it on a large collection of icebergs in the North Atlantic.

But if Benjamin Franklin had still been alive, he might very well have made the connection instantly. In 1784, he had attributed the unusually cold winter to "a dry fog" that was caused by eruptions in 1783 of Mounts Asama in Japan and Laki in Iceland. But no one remembered this, and, in fact, one theorist in 1815 actually blamed the widespread use of Franklin's lightning rods for upsetting a natural flow of warming electricity from the Earth's core, thus bringing on the worldwide frost.

MARTINIQUE
PELÉE
May 8, 1902

On May 8, 1902, Pelée, on Martinique, exploded, not at its peak, but at its sides. Thirty-six thousand people were killed instantly; hundreds more perished from numerous ancillary effects, including snakebite.

A new classification of volcanic explosions, a new name and respect for a branch of scientific study called volcanism, which had hitherto been considered a mere cavern in the greater mountain of geology, and a new record of death and devastation was set by the absolutely cataclysmic convulsion that killed 36,000 persons in a matter of moments, when Mt. Pelée blew itself to bits on the morning of May 8, 1902, and wiped out the city and population of St. Pierre, one of Martinique's primary ports.

Because not the top, but the sides of Pelée blew out in the enormous explosion of that horrendous morning, all volcanic eruptions of this type would henceforth be dubbed Peléan. As for the new respect accorded volcanism, that occurred because President Theodore Roosevelt chose to dispatch, on May 14, the cruiser *Dixie* filled with not only relief supplies but scientists and one journalist—George M. Kennan—to distribute the $250,000 worth of relief the United States supplied to Martinique after this nearly indefinable natural disaster.

The story begins, as most volcanic ones do, in prehistory, with the formation of the Lesser Antilles, a group of islands which stretch like the piers of a roadless bridge across the entrance to the Caribbean Sea in the shape of a 450-mile (724.2 km) arc bowed out toward the Atlantic Ocean, and stretching from the Anegada Passage east of the Virgin Islands almost to the coast of South America.

All are volcanic islands, but only two of them, Mt. Pelée on Martinique and La Soufrière on St. Vincent (see p. 423), have had violent eruptions, and these occurred only 18 hours apart, in 1902.

Pelée is located at the northern end of the island of Martinique of the Lesser Antilles, 4,430 feet (1,350 m) above the city of St. Pierre, the largest (population 32,000) settlement on the island.

According to the author Lafcadio Hearn, who visited the city before the eruption of Pelée, it was ". . . the quaintest, queerest, and the prettiest, withal, among West Indian cities; all stone-built and stone-flagged, with very narrow streets, wooden or zinc awnings, and peaked roofs of red tile, pierced by gable dormers. . . . The architecture is that of the seventeenth century, and reminds one of the antiquated quarter of New Orleans. . . ."

That this stone architecture would, in the space of a mere minute, be pulverized into pebbles was probably beyond the most vivid imaginations of St. Pierre's builders, and no eruption of Pelée prior to 1902 gave any indication that it was a dangerous volcano. The

St. Pierre, Martinique, lies in pulverized ruins following the eruption of Mount Pelée on May 8, 1902. The monstrous blast wiped out St. Pierre's entire populace in minutes. (Library of Congress)

explosions of 1767 had killed 16,000, but most of them had lived on its slopes. Since then, its eruptions had been of lesser and lesser degree.

But the holocaust of May had its warnings, which hardly anyone in the local government or local media chose to heed. On April 2, some steaming fumaroles, or vents in the side of the mountain, began to issue sizable quantities of steam. On April 23, ash began to fall in the streets of St. Pierre, accompanied by an undeniable odor of sulphur, and some mild earthquake shocks caused dishes to fall from shelves.

From April 25 on, more vigorous, though not truly alarming explosions began to occur, and a boiling sound was heard by those venturing near the L'Etang Sec crater on Pelée. This had become filled with a 650-foot- (198-m-) wide, steaming lake, alongside of which a 30-foot (9.1-m) cone emitted boiling water.

On April 27, the ash fall began to increase, blocking some roads and giving the landscape a wintry look. In addition, many birds and some animals were reported to have been asphyxiated by the poisonous gas rising from the ash.

The local newspaper, Les Colonies, described late April in St. Pierre: "The rain of ashes never ceases. At about half-past nine the sun shone forth timidly. The passing of carriages in the streets is no longer heard. The wheels are muffled [in the ashes]. Puffs of wind sweep the ashes from the roofs and awnings, and blow them into rooms of which the windows have imprudently been left open."

Now, the populace became uneasy. Large numbers of St. Pierre's residents left the city, but they were immediately replaced by even greater numbers of refugees from the mountain's slopes. The U.S. consul's wife, Mrs. Thomas T. Prentis, wrote to her sister in Boston, "My husband assures me that there is no immediate danger, and when there is the least particle of danger, we will leave the place. There is an American schooner in the harbor, the R. J. Morse, and she will remain here for at least two weeks. If the volcano becomes very bad, we shall embark at once and go out to sea."

After the holocaust of May 8, rescuers would find the charred corpses of both the consul and his wife, sitting in chairs in front of a window that faced Pelée. The bodies of their children would never be found.

But before that, the local cathedral became crowded day and night with praying citizens of St. Pierre. Another resident wrote on May 4, to relatives in France, "I am awaiting the event tranquilly. . . . If death awaits us there will be a numerous company to leave the world. Will it be by fire or asphyxia? It will be what God wills."

Animals were not so contained. Les Colonies reported, "In the meadows, the animals are restless, bleating, neighing and bellowing despairingly." At the Usine Guérin, a large sugar mill at the mouth of the Blanche River, just north of the city, an incredible invasion of ants and centipedes overran the premises. Horses in the yard of the mill bucked and bellowed as the ants and centipedes ascended their legs and bit them, while workers doused the horses with buckets of water, in a sidelong attempt to dislodge the creatures. Inside the mill, workers whacked at the centipedes with stalks of sugarcane, while in the nearby villa of the mill owner, housemaids killed them with flat-irons and boiling water.

Meanwhile, in one quarter of St. Pierre itself, snakes, displaced by the rain of ashes, slithered into streets and backyards, killing chickens, pigs, horses, dogs, and people who failed to get out of their way. Fifty humans and 200 animals died of snakebite that afternoon.

On May 5, heavy rains falling on the mountain unleashed floods of chocolate-colored water into all of the valleys on that southwest side of Pelée. Shortly after noon on that same day, the same sugar mill that had been invaded by ants and centipedes was destroyed and buried by an enormous avalanche of mud containing huge boulders, rocks, and immense trees. Plummeting down the mountain, it consumed the mill in boiling mud in moments, burying 150 men and leaving only one chimney visible. After inundating the mill, the avalanche continued on into the sea, creating an enormous wave that overturned two boats tied at anchor and flooding the lower portion of St. Pierre.

"A flood of humanity poured up from the low point of the Mouliage [the flooded part of St. Pierre]," according to Les Colonies. "It was a flight for safety, not knowing where to turn. The entire city was afoot. The shops and private houses are closing. Everyone is preparing to seek refuge on the heights."

And then, one of the more cynical acts of history occurred. Attempting to keep the population in St. Pierre for an upcoming election on Sunday, May 10, the French governor appointed a commission to investigate the danger from the volcano. Les Colonies reported, "[Professor Landes of the Lycée concludes that] Mt. Pelée presents no more danger to the inhabitants of Saint Pierre than does Vesuvius to those of Naples." Les Colonies went on, editorially, "We confess that we cannot understand this panic. Where could one be better off than at Saint Pierre? Do those who invade Fort-de-France believe that they will be better off there than here should the earth begin to quake?"

In order to emphasize his commission's findings, the governor ordered troops to turn back refugees at the city's outer limits; and to further reassure potential voters, the governor and his wife rode from Fort-de-France to St. Pierre, where, within a day, they would both meet fiery deaths.

Thomas Prentis, the U.S. consul, dispatched what would be his last communication to President Theodore Roosevelt. Recognizing the political manipulations of the governor, the local government, and the local newspaper, he bitterly observed, "To abandon the elections would be unthinkable. The situation is a nightmare where no one seems able or willing to face the truth."

All through May 7, Pelée continued to erupt, but the residents of St. Pierre took some cheer in the news that La Soufriére on St. Vincent, 90 miles (144.8 km) to the south, had exploded. That, they thought, as they went to bed on the last night of their lives, would relieve the pressure on Pelée.

May 8 dawned serene and sunny. A column of steam rose from Pelée to an extraordinary height, but otherwise there was nothing exceptional or different about its demeanor. At about 6:30 A.M. the ship *Roraima*, its decks coated with ash, steamed into the port at St. Pierre and tied up alongside the 17 other vessels docked at the city's simple roadstead.

And then, at 7:50 A.M. Pelée blew itself to pieces in four, deafening, gunlike reports, discharging upward from the main crater a black cloud pierced with lightning flashes. But this was not the most lethal explosion. It was the lateral ones—the ones that would forever afterward be regarded as "Peléan"—that launched fire and brimstone at hurricane speed down the mountain slope, straight for St. Pierre. In two minutes, all but four people, almost the entire population of St. Pierre—over 30,000 people—would be incinerated, suffocated, or blasted to death in an instant.

Every house, every structure in St. Pierre was blown apart or partially destroyed. Trees were stripped of leaves and branches down to the bare trunks. Cement and stone walls, three feet thick, were torn apart as if they were cardboard. The six-inch cannons on the Morne d'Orange were sheared from their moorings and a statue of the Virgin Mary, weighing at least three tons, was carried 50 feet (15.2 m) from its base. On the quays and in the warehouses at the port, thousands of barrels of rum exploded, the flaming liquid running through the streets and out into the water of the roadstead.

Small wonder that there were no survivors. The superheated volcanic gas, kept close to the ground because of its immense density and speed, permeated every corner of the city, allowing almost no one to escape. Even at 11:30 A.M., three and one-half hours after the blast, the heat from the burning city was so intense that a ship from Fort-de-France could not approach the shore.

Most of the ships in the port were overturned by the blast, and their crews and passengers perished in the boiling waters. A day before, perhaps the wisest captain in St. Pierre, a Neapolitan named Marino Leboff, captaining the Italian bark *Orsolina*, brushed past port authorities who refused him permission to sail. "I know nothing about Mount Pelée," he said, ignoring their orders, "but if Vesuvius were looking the way your volcano looks this morning, I'd get out of Naples."

Sixteen of the 18 ships in the harbor at the moment of the explosion capsized. The Quebec-line steamship *Roraima* was engulfed in the onrushing black cloud. Its assistant purser, Thompson, recalled the moments before the explosion, when the crew and the captain, who was later killed, gathered to see the sight: "The spectacle was magnificent," he recalled.

> . . . we could distinguish the rolling and leaping red flames that belched from the mountain in huge volume and gushed high in the sky. Enormous clouds of black smoke hung over the volcano. The flames were then spurting straight up in the air, now and then waving to one side or the other a moment, and again leaping suddenly higher up. There was a constant muffled roar. It was like the biggest oil refinery in the world burning up on the mountain top. There was a tremendous explosion. . . . The mountain was blown to pieces. There was no warning. The side of the volcano was ripped out, and there hurled straight towards us a solid wall of flame. It sounded like a thousand cannon. The wave of fire was on us and over us like a lightning flash. It was like a hurricane of fire, which rolled in mass straight down on St. Pierre and the shipping. The town vanished before our eyes, and then the air grew stifling hot and we were in the thick of it. Wherever the mass of fire struck the sea, the water boiled and sent up great clouds of steam. . . . Before the volcano burst, the landings at St. Pierre were crowded with people. After the explosion, not one living being was seen on land. . . .

The chief officer of the *Roraima*, Ellery Scott, recorded the moment it hit. "Then came darkness blacker than night," he wrote. "The *Roraima* rolled and careened far to port, then with a sudden jerk she went to starboard, plunging her lee rail far under water. The masts, smokestack, rigging, all were swept clean off and went by the board. The iron smokestack came off short, and the two steel masts broke off two feet above the deck. The ship took fire in several places simultaneously, and men, women and children were dead in a few seconds."

Only two of its passengers, a little girl and her nurse, survived and they were interviewed later by Dr. Thomas Jagger Jr., the brilliant Harvard geologist. The nurse's account was particularly vivid:

... the steward [who was later killed] rushed past [the cabin, where I was assisting the children in dressing for breakfast] and shouted "Close the cabin door—the volcano is coming!" We closed the door and at the same moment came a terrible explosion which nearly burst the eardrums. The vessel was lifted high into the air, and then seemed to be sinking down, down. We were all thrown off our feet by the shock and huddled crouching in one corner of the cabin.

... the explosion seemed to have blown in the skylight over our heads, and before we could raise ourselves, hot moist ashes began to pour in on us; they came in boiling splattering splashes like moist mud without any pieces of rock. . . .

A sense of suffocation came next [but] when the door burst open, air rushed in and we revived somewhat. When we could see each other's face, they were all covered with black lava, the baby was dying, Rita, the older girl, was in great agony and every part of my body was paining me. A heap of hot mud had collected near us and as Rita put her hand down to raise herself up it was plunged up to the elbow in the scalding stuff. . . .

The immense volcanic cloud covered a triangle of absolute destruction and a second zone of damage spreading over 24 square miles (62.2 km²). The cloud, formed of superheated steam and other gases, made heavy by billions of particles of incandescent ash and traveling with enough force to carry along boulders and blocks of volcanic material, reached temperatures of between 1,300°F (704.4°C and 1,800°F (982.2°C), which is enough to melt glass.

The city would not cool enough for rescuers to enter it for four days. When they did, they would find unbearable sights of horrible destruction. At the rum distillery, the burned-out tanks, which were massive containers of quarter-inch boiler plate riveted together, looked to one observer "as if they had been through a bombardment by artillery, being full of holes which vary in size from mere cracks at the bottom of indentations to great rents, 24, 30 and even 36 inches across."

Two weeks later, geologist sent by America on the rescue ship *Dixie* would witness, according to Angelo Heilprin of the Philadelphia Geographical Society, ". . . twisted bars of iron, great masses of roof sheeting wrapped like cloth around posts against which they had been flung, and girders looped and festooned as if they had been made of rope."

There were incinerated corpses of families sitting at their breakfast tables, which, because of the selective nature of the freakish cloud that had killed them, were still set with undamaged plates, cutlery, and glasses.

A charred woman was found with an entirely intact handkerchief pressed to her lips. Many bodies were naked; their clothing was blown away by the blast. In a jewelry shop, the heat had been so intense that it had fused hundreds of watch crystals into one lump, but in a kitchen not far away, there were still-intact corked bottles of water, and packets of starch in which the granules were untouched.

Only four people escaped the cataclysm of May 8, 1902. Two were outside St. Pierre, and one of these, a young girl named Harviva De Ilfrile, told a harrowing tale of being halfway up the mountain on an errand for her grandmother, looking into the "boiling stuff" coming out of the so-called Corkscrew crater. She ran to the wharf and jumped into her brother's boat, and, turning, saw her brother running ahead of the boiling tide of ash. "But he was too late," she related, "and I heard him scream as the stream first touched then swallowed him."

From here, this small girl somehow launched the boat, and got it to a small grotto in which she and her chums had once played. "But before I got there, I looked back—and the whole side of the mountain which was near the town seemed to open and boil down on the screaming people. I was burned a good deal by the stones and ashes that came flying about the boat, but I got to the cave," she concluded. Days later, she was discovered floating in her charred craft two miles at sea by the French cruiser *Suchet*.

The other survivor of St. Pierre, Auguste Ciparis, a 25-year-old stevedore who was a prisoner in an underground dungeon at the time of the blast, later told his story to George Kennan, and then went on to become a major attraction in the Barnum and Bailey Circus. Protected by his subterranean quarters, he nevertheless survived immense, horrendous burns over much of his body. Kennan called the burns "almost too horrible to describe."

Looters streamed into St. Pierre from other parts of Martinique. French soldiers from the cruiser *Suchet* shot 27 men, and then turned their attention to the nightmarish sight of incinerated heaps of bodies.

It would take months before the entire population of St. Pierre could be buried, and during that time, on May 20, Pelée exploded once again, with almost as much force as the paroxysm of the 8th, leveling whatever walls remained in St. Pierre and killing 2,000 more people—most of them rescuers, engineers, and mariners bringing relief supplies to the island.

The shipment of supplies and scientists from the United States arrived only hours later, and the scientists would stay for months, while Pelée treated them to one last, more benign spectacle. Sometime around the middle of October, a domelike mass of lava, too stiff to flow, formed in the crater of L'Etang Sec, and from its surface, a spine or obelisk began to grow. Its diameter was from 350 to 500 feet (106.7 to 152.4 m), and within a few months, it had reached a height of 1,020 feet (310.9 m) above the crater floor, or 5,200 feet (1,585 m) above sea level. It was twice as tall as the Washington Monument and as large as the great pyramid of Giza.

But it was too unstable to endure, and by September 1903 it had collapsed—though not before it had been photographed and preserved for all time by the Dixie scientists. And while it stood, it seemed to scientist Angelo Heilprin ". . . nature's monument dedicated to the 30,000 dead who lay in the silent city below."

MEXICO
PARICUTÍN
June 1944

There are no official fatality figures for the multiple explosions of Paricutín, the unique Mexican volcano "born in a cornfield." Unofficial statistics vary from "few to none" to 3,500.

Paricutín is the volcano that was described in various popular magazines in 1943 as "the volcano born in a cornfield while the owner watched." This was not quite the way it happened, although the actuality was close enough.

Located 200 miles (321.8 km) due west of Mexico City, Paricutín is named for a small Tarascan Indian village of 500 people five miles (8 km) from the vent of the volcano. It did rise from the cornfield of the nine-acre parcel of land owned by Dionisio Pulido, but this was not its first appearance. For years, that spot of land had contained a small hole around which the local children played. In fact, one of the local inhabitants recalled, 50 years before, playing near it and hearing underground noises like falling rocks and feeling from it "a pleasant warmth."

On February 5, 1943, a series of steadily intensifying earthquakes began to rumble and split the ground near this hole. On February 19, no less than 300 were felt.

On the morning of February 20, Pulido left the village of Paricutín to prepare his farm for planting, and here both the popular legend and the volcano were born. All proceeded normally for most of the day, and then, at about 4:00 P.M., Pulido, his wife Paula, his son, and a neighbor Demetrio Toral were thunderstruck by what was beginning to occur in the cornfield.

A fissure had begun to open on either side of the hole. Almost immediately, there was a sound like thunder. Nearby, trees trembled, and the ground swelled to

The cinder cone of Paricutín in Mexico (National Geophysical Data Center/USGS)

a height of about three feet (4.8 m). Smoke and a fine ash-gray dust began to rise from parts of the fissure.

Then, as the fascinated farmer and his family watched, a loud hissing sound accompanied the thunder, and the odor of sulphur wafted from the opening, followed by "sparks" which ignited pine trees 25 yards from the jagged rip in the ground. This was enough activity for the eye-witnesses, who hurriedly left the scene.

The volcano continued to grow through the night. By midnight, huge incandescent bombs were hurled into the air with a roar, and lightning flashes appeared in the heavy ash column. By 8:00 A.M., when Pulido returned to his transformed cornfield, a cone about five feet high was spitting smoke and rocks with regularity and great violence. By midday, this cone was 25 feet high and growing.

On the 21st of February, the first lava began to flow from the growing cone at a speed of roughly three feet an hour, spreading a slaglike mass of black, jagged blocks about three feet (0.9 m) in thickness all over Pulido's transformed farm.

On the evening of February 22, Dr. Ezequiel Ordóñez, a veteran Mexican geologist, arrived upon the scene and recorded it officially:

> I was witnessing a sight which few other humans had ever seen, the initial stages of the growth of a new volcano. Tremendous explosions were heard, ground tremors were felt frequently, and a thick high column of vapors with a great many incandescent rocks could be seen rising almost continuously from the center of a small conical mound then estimated to be 55 meters (16.764 feet) high. On the same night I noticed a red glow on the slope of the mound and large incandescent rocks rolled down a low rocky incline with a distinctly peculiar tinkling noise. Upon approaching this low scarp as closely as possible, it was found to be the front of a large lava flow moving from the north side of the base of the mound over a flat cornfield.

Scientists now took over and clocked the progress of the growth of the volcano. By the end of the first week, the cone was 50 feet (15.2 m) high; by the end of the first year, it had reached a height of a hundred feet (30.4 m). As it erupted and cooled, as lava flows built and waned, it would grow and shrink, but its height would remain fairly constant at 100 feet (30.4 m).

By the beginning of 1944, most of the lava activity had shifted to the southwest side of the cone, and for the next three years lava emerged from closely spaced vents (or *bocas,* as the Mexicans called them) on this side. The area of the lava vents became known as the *Mesa de los Hornitos,* or Tableland of Little Ovens. The lava flow increased, creating more vents.

Finally, in June 1944, there was a vigorous eruption, and a huge lava flow descended upon the village of Paricutín and the larger town of San Juan de Parangaricutiro. Heavy ash partially buried both communities and there were a few isolated fatalities, but the inhabitants of both towns shoveled the ash from the streets and roofs of houses with the hope of resuming their lives.

That was not to be. The lava continued to flow, slowly enough to allow the Mexican army to force an evacuation of residents who often waited until it was at their very back door before reluctantly leaving. San Juan de Parangaricutiro was destroyed by the advancing lava flow in June 1944. By the end of September, Paricutín had disappeared, set afire and buried by the river of molten rock.

One source states that 3,500 people died on June 10, the day that San Juan de Parangaricutiro was evacuated. This seems highly unlikely, since the town only contained 4,000 residents. More accurate might be the "few to none" estimation by the geologists in residence. Still there are, unfortunately, no official fatality figures.

MONTSERRAT
SOUFRIÈRE HILLS
July 1995–September 1997

The lush garden island Montserrat in the eastern Caribbean was devastated by a two-year-long series of eruptions of the Soufrière Hills volcano, which destroyed the island's port and capital, Plymouth, rendered two-thirds of the island uninhabitable, and forced 8,000 of the island's 12,000 residents to abandon it.

Of all the unlucky spots on the globe, the 39-square-mile (101.1 sq. km) island of Montserrat, in the eastern Caribbean, seems to be the target for much of Nature's fury. Pounded by hurricanes and cyclones (see Hugo, p. 307) and rocked by the thunder and fire of volcanic explosions, it underwent an increasing number of natural disasters in the last decade of the 20th century. In fact, nine of the 10 worst natural disasters recorded on the British-owned island took place between 1989 and 1997, and of those nine, three were part of the ongoing series of explosions of the Soufrière Hills volcano, which dominates the southern portion of the island.

The first of these eruptions tore the top of the mountain free of the land on July 19, 1995. Twenty people were killed and 6,000 were displaced then; it erupted again on July 29, 1996, killing no one but displacing 4,000 people.

But its worst series of explosions began on June 25, 1997, when the volcano exploded in a fire storm, sending a mushroom cloud of ash and pebbles 35,000 feet (10,668 m) into a blackened sky, and spewing a river of ash, rock, and gas, heated to a temperature of 900 degrees, that raged down its slopes at a speed of more than 100 MPH.

Villages that had once dotted the lush vegetation that had partially earned the island the nickname of "The Emerald Isle of the Caribbean" (the other source was a colony of Irish settlers in the 1600s) were buried or set afire by the white-hot pyroclastic flow. Upscale resorts that had attracted free-spending tourists for nearly a century were destroyed in an instant.

But then, the explosions lessened, the heaving of the earth quieted, and there was hope that the eruption was the last of a series.

On July 31, this optimistic scenario was blasted to ribbons. With a fierceness that topped its June convulsion, the mountain exploded again, and from then through September, it would erupt an average of once every 12 hours.

The island's capital, Plymouth, where cruise ships regularly docked, was squarely in the path of the sulfurous tide that roared with increasing frequency and scope downward from the mountain's peak. A gully filled with debris had, in the past, separated the capital city from volcanic encroachment. But not today.

With a spectacular fire show, the pyroclastic flow leaped across the gully and set the city on fire. Within hours, what was not burned to the ground was covered by gas, rock, and ash.

From that time forward, regular warnings from sirens sounded through the constant night caused by the continual eruption of Soufrière Hills. Roads, covered by ash, were as slippery as if they had been coated with ice. Cars carrying terrified refugees from the explosions slipped off roads. The air turned white, as if a blizzard had struck, and people covered their faces with surgical masks or gas masks to reach one place from another on foot.

Inexorably, the spouting refuse from the mountain covered the entire southern two-thirds of the island. This "exclusion zone," cleared of inhabitants who faced arrest if they ventured back into it, included the airport. Cruise ships, with no port in Plymouth in which to dock, were now passing the island by, and by the end of July, the last hotel, caught in the "exclusion zone," was forced to close, thus shutting down the tourist business and the livelihood for a majority of the island's inhabitants.

It was not much of an existence for the rest of those who lived on Montserrat, who had made their living from the land that was now under a coating of killing ash that harbored white-hot pockets of gas. "There's no way to go on with anything in these conditions," Thomas Farrell, a farmer, told a reporter. "Do we stay or do we go? We'd like to be done with it, one way or the other, bang or no bang."

A medical school for American students was gradually absorbed into the exclusion zone and closed. More and more schools on the island were appropriated for shelters for the displaced, and so, many parents sent their children to Britain or other Caribbean islands to continue their education.

Thousands of residents left for Antigua, Nevis or Guadeloupe, but 1,000 lived in shelters in schools and churches, some for two years. Their lives were spent in makeshift cubicles, separated from others only by curtains. Their bare subsistence was made possible by the Montserrat government, relocated to the north and supplying monthly vouchers of $45 per adult and less for children.

Small wonder that the 4,000 residents remaining of the 12,000 who had previously populated Montserrat were beginning to feel marooned. By mid-August, the government was offering an unspecified amount of money and transportation to neighboring islands, including Antigua and Guadeloupe. But there were few takers. A two-year British visa was added to sweeten the offer, but it, too, went largely unclaimed.

Montserrat was not much of an island anymore. Huge swaths of it were covered with up to 10 feet (3.04 m) of ash; other neighborhoods of Plymouth looked as if they had received a major bombing from a major war. Earthquakes regularly shook the ground beneath the survivors.

Thirty-two people had been killed by the volcanic explosions. But those who remained were either stubborn or incurably optimistic. They had recovered from Hugo, they reasoned. Why not from Soufrière?

And so, as this goes to press, Montserrat's population is still on the razor edge of disaster. The local newspaper, *The Montserrat Reporter*, still publishes, but most of its pages are filled with comment from survivors of the disaster. "Scores of small businesses are no longer in existence," one letter to the editor lamented. "Hundreds of people have lost everything they have ever had. Thousands of people have been uprooted from these 'No Go' areas to the rest of the island which is habitable. Many are living in sheltered accommodation. Families are cramped together, men, women and children alike. Some are living in churches, eating meals on gravestones. Sir, I am sure you, like me, find this abhorrent and unbelievable."

Part of a long poem by a young girl capsulized the wreckage—physical and human—left by a continuing natural disaster that is destroying, bit by bit, an entire, circumscribed part of the world:

Squatting in the corner on some turf
Wading through the horrors of life
In total darkness
In oblivion
Seems like I've got a million stories
In my head
Fact or fiction
Filled with a drumming of insomnia song
Trapped in a world of no sense, rain and thunder
It's kind of like a movie
But we're all the stars
Lights, camera, everyone get in position and
ACTION!!!
Hold up, wait a minute My skin is all ashy
Could you please pass me some lotion?
. . .
I'm in the centre of the fire and

I'm cold
I'm afraid of this
Bright, hot, verging, red, fiend
Who's running the show
Aspiring to be higher
I guess
All that glitters ain't gold . . .

NEW GUINEA
LAMINGTON
January 21, 1951

On January 21, 1951, the Pelean explosion of Laming-ton, on New Guinea, killed 2,942, most of them from nuée ardente, hurricane-force winds laden with steam, hot ash, rock fragments, and superheated mud.

"Nuée ardente" was the name given to the lethal, hurricane-force winds containing enormously hot ash and rock fragments, steam and superheated mud that characterized the Peléan-type explosion of Mt. Pelée in 1902 (see pp. 405–410). The tremendous speed (from 36 MPH to 300 MPH) with which these molten masses erupt from a volcano and descend its slopes makes escape almost impossible, as the monstrous fatality figures from Pelée attest.

The January 21, 1951, eruption of Mt. Lamington in New Guinea was precisely the same sort of explosion as Pelée's, with nuée ardente crashing down its slopes and instantly killing 2,942 people. In this case, the principle cause of death was believed to be steam-laden hot dust, so those who found protection from the main force of the nuée ardente in sealed rooms survived. These survivors described symptoms such as pains in the mouth, throat, and eyes, followed by a burning sensation in the chest and abdomen and finally, a sense of rapid suffocation—exactly those described by the four survivors of Pelée. As in Pompeii (see pp. 393–396) and St. Pierre, there was rigidity in the bodies of the fallen, caused by the coagulation of the albuminous material in the muscles. Many of those who survived had been severely scalded, but the shielding of their clothing had saved them from burning to death.

The paroxysm which caused such widespread death and destruction occurred at 10:40 A.M. on January 21, 1951, and, although it was described in newspapers as having occurred without warning, Fred M. Bullard, in his exhaustive study, *Volcanoes of the Earth*, insists that there had been plenty of warning, but that, as in the case of Pelée, the warnings had gone unheeded. "Some six days prior to the climactic eruption there was activity in the crater, which was not generally observed or if so was not associated with an impending eruption," notes Dr. Bullard.

Acknowledging that Pelée was known as a volcano and Mt. Lamington was not, Dr. Bullard goes on to chronicle numerous landslides from the walls of the crater, and a two-day emission of a thin column of vapor rising from the crater which identified it unmistakably as an active volcano.

By January 18, gas was belching from the cone, loaded with ash and rising many thousands of feet. Earthquakes increased until they were "almost incessant" at Higaturu, six miles (9.7 km) north of the crater.

By January 19, according to Dr. Bullard, there were "blue flashes, tongues of flames, sheet and chain lightning." He goes on: "On January 20, the day before the climactic eruption, a continuous ash-filled eruption cloud rose to a height of 25,000 feet (7,620 m) or more. . . . The activity of Mount Lamington during the six days preceding the catastrophic eruption is fairly typical of eruptions of the Pelean type, and, for the future, it is imperative that such activity be reported, its significance appreciated, and the warning heeded."

Dr. Bullard does acknowledge that the prevailing winds carried the ash fall away from the populated and settled areas on the northern and western slopes of Mt. Lamington, thus robbing the inhabitants of the sort of admittedly unheeded warning the residents of St. Vincent received from Pelée in 1902.

At any rate, the final fulguration occurred in the mid-morning hours of January 21, 1951, and when it did, there was no doubting the identity of Mt. Lamington. A series of huge explosions tore the top and the sides from the mountain, producing a huge, ash-filled, mushroom-shaped cloud which within two minutes rose to 40,000 feet (12,192 m) and 20 minutes later had climbed to 50,000 feet (15,240 m). The detonation was enormous enough to be heard on the coast of New Britain, 200 miles (321.8 km) away.

Bursting forth from the side of the mountain, the nuée ardente catapulted, with ferocious speed and destructiveness, down the mountainside, flattening the rain forest so that not even the stumps of the trees were left, sweeping cataclysmically through buildings so that neither floors, walls, nor people were ever found again, and only the foundations remained to give evidence that houses had been there at all.

In a radius of two miles (3.2 km) from the crater, the ground was scoured and grooved, as if a gigantic rake had roared through. Charred root ends were sometimes left in the area beyond two miles (3.2 km); and the few trees that still stood exhibited charred and abraded trunks.

This hurricane of fire seemed to act like an FM wave—it bounced off hills and ridges. But then again, there were aberrations within the cataclysmic cloud that flattened St. Vincent, also.

At Higaturu, more than five miles (8.04 km) from the crater, an automobile was left, hanging crazily from the tops of two truncated trees, in an attitude more akin to that left by a tornado than a volcanic eruption. Later, scientists concluded that the stupendous velocity reached within tornadoes are possible in nuée ardente when tongues of the wind extended ahead of the main front and a vortex developed. Thus, the fact that in the Pelée eruption and the Mt. Lamington one, clothing was stripped from some victims and not from others nearby.

Further tests led the scientists to conclude that temperatures within the lethal, roaring cloud reached 392°F (200°C) for a period of 1.5 minutes—enough to incinerate the countryside and everything that lived within it. Many of the valleys surrounding the mountain were filled with thick beds of hot ash which retained heat for months. It was not uncommon to see a log which had been buried in the ash for weeks catch fire when it was suddenly exposed to the air. In places where the ash was protected from ground water and rainfall infiltration, vapor columns rising from the ash had a temperature of 194°F (90°C) two years later.

After one more cataclysmic explosion, at 8:45 P.M. the night of the 21st of January, Mt. Lamington retreated from spectacular activity and became semi-quiescent. Within 15 years, vegetation had returned to normal, but the slopes have not become repopulated to this day.

NICARAGUA
COSEGUINA
January 22, 1835

The most violent eruption in the Western Hemisphere occurred when Coseguina, in Nicaragua, exploded on January 22, 1835. Eight hundred died.

The eruption of Coseguina in 1835 has been regarded as the most violent eruption in the Western Hemisphere. The mountain is part of a belt of active volcanoes that extends along the Pacific coast from Guatemala to the Panamanian border and is sometimes called the "Fire Girdle of the Pacific." Before the 1835 eruption, Coseguina was regarded as extinct.

Consisting of a great decapitated cone with an oval-shaped crater a mile in diameter, 2,000 feet (609.6 m) deep and occupied by a clear blue lake, this 2,850-foot (869-m) giant first announced its reawakened presence by a white cloud, like an immense plume of feathers, rising from its crest at 8:00 A.M. on the morning of January 20, 1835. The plume was easily visible to the residents of La Union, El Salvador, 30 miles (48.3 km) across the Bay of Fonseca.

Within a very short time, the cloud turned gray, then yellow, and finally crimson. By 4:00 P.M. some earthshocks were felt. By 11:00 A.M., the morning sky had been turned deep gray by the volcano's ash, and lamps were lit in La Union.

As the afternoon wore on, heavy showers of flour-like pumice began to fall and the darkness deepened. Terror and confusion followed, aided and abetted by escalating sounds and sights from the volcano. By 4:00 P.M., the darkness had spread to San Miguel, 50 miles (80.5 km) to the northeast, and by nightfall, the ash fall had reached as far north as Tegucigalpa, 80 miles (128.8 km) away, and San Salvador, 110 miles (177 km) to the northeast.

By the next morning, strong earthquakes and subterranean noises frightened people at La Union, and on Tigre Island in the Gulf of Fonseca, 20 miles (32.8 km) from the volcano, pumice fragments as large as hen eggs fell. The darkness, which continued for three days, convinced many local residents that the end of the world was at hand, and, according to Richard S. Williams, Jr., who recorded the explosion, "The terror of the inhabitants at Alancho, anticipating the approach of Judgment Day, was so great that three hundred of those living out of wedlock were married at once."

And then, on the night of January 22, the mountain exploded with an immense detonation that was heard as far away as British Honduras, 400 miles (643.7 km) distant, where the superintendent at Belize mustered his troops, thinking that there was a naval action off the harbor. In Guatemala City, the enormous explosions were so deafening, the local troops, thinking it was the cannon fire of an approaching army, hastily prepared to defend themselves.

For seven hours, the mountain spewed red hot pumice from its side and top. Absolute darkness enveloped everything and everyone within a radius of 50 miles (80.46 km). Most of El Salvador was obscured, and ash fell as far north as Chiapas, Mexico, and on places near the Costa Rican border. Guatemala City remained in darkness until the 27th, although the main eruption lasted only four hours, stopping on the 23rd.

The fall of ash and ejecta was enormous. E. G. Squier stated, in his recording of the eruption in 1859, ". . . the sea for 50 leagues was covered with floating masses of pumice, resembling the floe-ice of the Northern Atlantic."

Anywhere from 12 to 35 cubic miles (50 to 145.9 km^3) of ash were reported to have fallen, although Squire's estimate tended toward the lower figure.

Close to 800 natives living at the base of the mountain were killed by the brief and world-shaking eruption. Their violent and unexplained deaths resulted in a superstition passed down through generations of their descendants to this day. It states that the only way to pacify the mountain is to sacrifice a three-month-old infant by throwing it into the crater every 25 years.

PAPUA, NEW GUINEA
RABAUL
September 19, 1994

The twin peaks of Vulcan and Tavurvur, on the island of New Britain in Papua New Guinea, erupted in the early morning of September 19, 1994. The port city of Rabaul was demolished and a gigantic swath of destruction displaced 53,000 people. Miraculously, only four died.

With unexpected and astonishing swiftness, the two volcanoes of the Rabaul Caldera, in the eastern part of the large island of New Britain in Papua New Guinea, erupted in quick succession. Tavurvur, on the north side of the island, exploded on the morning of September 19, 1994, followed an hour later by Vulcan, located on the south side. Two gigantic plumes of fumes and ash rose in a joined mushroom cloud, which ascended to a height of 6,600 feet (2,011.7 m) over the entire area. Day turned to an ashen night over the town of Rabaul, the key port of the Bismarck Archipelago, located at the head of the bay that rests between the two volcanoes.

For the residents of the area who were old enough to remember, it was a painful reminder of the 1937 eruption that had formed Vulcan, and which had left 500 dead. Fortunately—or unfortunately for those whose homes were demolished—a series of minor earthquakes had preceded the present eruptions. As the earthquakes progressed, Tavurvur began to rumble and spout steam and ash. It was obviously about to erupt, and local officials evacuated 30,000 of the 60,000 residents of Rabaul and the surrounding area.

And then, the ground split apart under the fleeing refugees. Five vents opened up, spewing white-hot steam and molten rock over a huge area.

Rabaul itself was buried under falling ash and mud that fanned into the air as the eruptions consumed more and more land. A pilot flying over the city on September 20 reported that "Rabaul no longer exists, and is . . . covered in meters of ash." Over the next three days, up to four feet of mud and ash coated Rabaul, collapsing buildings, obscuring the town, and eventually erasing it from the landscape.

Clouds of burning ash continued to spew from the multiple vents of the volcanoes. One group of 600 villagers, fleeing down the mountainside, was overtaken by a cloud of molten ash that rained down on them, burning some of them, but not slowing them down. They rushed to the sea, dove in, and paddled out to a line of barrels used to mark a channel and clung to them to escape the ash. It would be hours before a ship would rescue them.

The clouds continued to belch into the sky, darkening it. More and more buildings and palm trees began to collapse, as rain and wind joined the bright red terror of falling fire from the volcanoes. Ash clouds rained gray mud 600 miles (956.6 km) away on the New Guinea mainland.

The explosions and the rain of ash continued for a fourth day, as 52,000 of the 60,000 residents of the area fled to shelters on the outskirts of the area, some with wontoks in the countryside, but most in Kerevat and Kokopo, where every church, school building, and chicken shed was filled with refugees.

Finally, six days after the initial explosions, relative calm returned to the area, though there were still rumblings and momentary spurts of ash and pumice from both volcanoes.

The port of Rabaul had undergone a dramatic reconfiguration. Where boats had been safely moored in gentle seas, there was now dry land. The port itself remained largely intact, but useless and far from the sea.

The worst destruction took place between Rabaul and Tavurvur. A flat plain that had contained the Rabaul airport—which had been pounded into the harbor and sunk—was totally covered by a foot-thick layer of ash and pumice. The small island of Matupit, which contained an extensive housing division, was pounded until it sank into the sea and disappeared.

Villages and once lush plantations and gardens that had been planted around Vulcan were no more.

Ultimately and miraculously, only four people were killed in the multiple eruptions. Fifty-three thousand were displaced and were forced to settle elsewhere.

PHILIPPINES
MAYON
October 23, 1766

Over 2,000 were killed in the second most destructive eruption of Mayon, on Luzon in the Philippines, on October 23, 1766.

The volcano of Mayon rises majestically, 8,000 feet (2,438.4 m) high and six miles (9.6 km) from the Albay Gulf on southeastern Luzon Island. Some have called it the most perfectly formed volcano in the world. A volcano of the Etna type, which spews lava flows that warn of an eruption, Mayon was for centuries a source of great superstition to the natives of Luzon. These natives believed that an evil-spirit dwelled in its depths, and every once in a while, in search of sacrifice, showered them with ash and rocks and flaming lava.

In 1592, a Franciscan priest, Estaban Solis, gathered several thousand people from various villages and marched them to the foot of the volcano. Their task was to watch him scale the mountain and report that there were no evil spirits in its crater. This he did, and the natives, apparently feeling that he had either proved them wrong or exorcised their devils, rejoiced and joined his church. Their conversion was shortlived. Father Solis inhaled enough sulfurous gas from Mayon to bring about his death less than a year from the exorcism, and his church immediately became depopulated.

Twenty-four years later, in 1616, Mayon erupted with great loss of life. But its second most destructive eruption took place on October 23, 1766. Villages by the score were either swept away, incinerated, or buried by an enormous 100-foot- (30.4-m-) wide lava flow that cascaded down the eastern slopes for two days.

Following an initial explosion and lava flow, Mayon continued to explode for four more days, this time flinging huge quantities of steam and liquified mud. The dun-colored rivers, which measured from 80 to 200 feet (24.4 to 60.9 m) in width, crashed down the sides of the mountain in a radius of 20 miles (32.2 km), sweeping it clean of roads, people, animals, and the towns of Daraga, Camalig, and Tobago. Over 2,000 inhabitants of the island were killed by the eruption, most of them swallowed up in either the first lava flow or the second superheated mudslide. For two months afterward, the mountain continued to spew ash and belch lava over the countryside.

PHILIPPINES
MAYON
February 1, 1814

Mayon's worst eruption occurred on February 1, 1814. Over 2,200 people were killed.

The most cataclysmic eruption of the Philippines's most lethal volcano, Mayon, on Luzon, occurred on February 1, 1814.

The entirety of northern Luzon was plunged into a stygean darkness by the enormous cloud of belched fire and ashes that roared from the depths of Mayon that day. There had been warnings; from early in the morning, earthquakes had rumbled through the countryside. This explosion did not send a great deal of lava roaring down the mountainsides, as the 1766 disaster had (see previous entry). Instead, an enormous bombardment of burning stones, sand, and ashes shot up into the stratosphere and then fell on the surrounding countryside. Cagsauga and Badiao were totally pulverized by these falling, flaming rocks. Not one building was left standing in either town.

The surrounding countryside was absolutely devastated. Trees were burned beyond redemption, rivers were dammed, and in some places palm trees were buried to their tops; in other locations the ash accumulation reached 30 feet (9.14 m) in depth.

Over 2,200 persons died in this, Mayon's most lethal and spectacular eruption.

PHILIPPINES
MAYON
June 23–30, 1897

Over 400 died in Mayon's longest eruption from June 23 through 30, 1897.

Mayon erupted in 1825, killing 1,500, and again in 1835, 1886, and 1888. Its longest eruption in history, however, began on June 23, 1897, and lasted for seven days of ceaseless, horrific volcanic bombardment. Professor Samuel Kneeland, an eminent geologist, on a steamer bound for Iloilo, observed and recorded the volcano's activity five days before the eruption: "[It has] a pillar of cloud by day and [a] pillar of fire by night," he wrote. "[It throws off] pieces of lava six feet in diameter . . . thrown from the crater to a distance of five miles (8.04 m)."

Spending his nights and days in constant observation, Kneeland was awestruck by Mayon's grim beauty:

> At night the scene was truly magnificent and unique. At the date of my visit the volcano had poured out . . . a stream of lava on the Legaspi side from the very summit. The viscid mass bubbled quietly but grandly, and overran the border of the crater, descending several hundred feet in a glowing wave, like red-hot iron. Gradually fading as the upper surface cooled, it changed to a thousand sparkling rills among the crevices, and, as it passed beyond the line of complete vision behind the woods near

the base, the fires twinkled like stars, or the scintillions of a dying conflagration. More than half of the mountain height was thus illuminated.

Had he stayed on until June 23, Dr. Kneeland might have been less poetic in his description of Mayon. On that day, it began its seven-day rain of fire and terror, tossing thousands of tons of white hot stones skyward and burying the town of Tobaco once again. These stones continually thundered to earth, while huge, steaming fissures opened in the sides of the mountain.

Once again, lava roared down the mountain. Seven miles eastward, the village of Bacacay disappeared beneath 50 feet (15.24 m) of lava. In Libog, 100 persons were burned to death by superheated steam and falling rocks, and the villages of San Roque, Misericordia, and Santo Niño became deathtraps.

For more than a week, night and day were indistinguishable as an enormous black cloud of ashes spread for nearly 100 miles (160.9 km) from the eruption. More than 400 were killed.

PHILIPPINES
MOUNT PINATUBO
June 10–15, 1991

Approximately 200 died and 100,000 were made homeless by the multiple eruptions of Mount Pinatubo, 55 miles northwest of Manila on Luzon Island in the Philippines from June 10 through 15, 1991. The eruption created 150,000 refugees and forced the closing of Clark Air Force Base and the evacuation of all dependents of both it and Subic Naval Base.

Of the 37 volcanoes in the world that are considered currently active, 29 lie on the so-called Pacific "Ring of Fire" (see VOLCANIC ERUPTIONS AND NATURAL EXPLOSIONS, Nicaragua, Coseguina, p. 414). Each of these volcanoes plunges into the deep earth under the islands, and thence to the Pacific Plate, on the floor of the Pacific Ocean. When the descending plate reaches a depth of

Volcanic mud appropriates a village on the slopes of Mount Pinatubo after its multiple eruptions from June 10 to 15, 1991. (CARE)

about 60 miles (96.6 km), it produces molten rock that rises to the surface and feeds these volcanoes.

Located on Luzon Island in the Philippines, some 55 miles (88.5 km) northwest of Manila, 4,000-foot- (1,219.2-m-) high Mount Pinatubo remained a quiescent member of the Ring of Fire for 611 years. It erupted in 1380, causing great damage, but in the interim, some 300,000 Filipinos had settled on its slopes, raising rice and animals and building cities and military installations.

And then, in 1991, the silence was broken. In April, the mountain began to show signs of activity. Puffs of smoke emerged from the crater at its crest. At the beginning of June, small amounts of steam and ash were released, and the rumblings within the mountain informed seismologists that something was about to happen. They issued warnings, and over 12,000 residents were ordered to be evacuated.

On June 10, a moderate eruption occurred, and 16,000 U.S. troops, dependents, and civilian employees stationed at Clark Air Force Base, 10 miles (16.1 km) east of Mount Pinatubo, were evacuated to Subic Bay Naval Base, 30 miles (48.3 km) to the southwest. The following day, 900 servicemen volunteered to return, to prevent looting.

And then, on the morning of June 12, at 8:41 A.M., Mount Pinatubo erupted in earnest, spewing a 15-mile- (24.1-km-) high gray-green mushroom cloud of ash and smoke into a formerly clear and cobalt sky. Rivers of searing gas, ash, and 1,800°F (982.2°C) molten rock gushed down the lush slopes at 60 MPH, setting fire to the underbrush and flooding the rice fields with molten material.

Day turned into night as far away as Manila, and the cloud and its ash fall would reach as far away as Singapore, 1,500 miles (2,414 km) to the southwest. Two inches of white ash would descend upon fields, animals, and houses, burning and burying them. Armor Deloso, the governor of Zambales Province, who tried to approach the slope after the eruption, told a *New York Times* reporter, "I could not breathe because the ashes were getting into my nose. It's raining ashes. The plants were all covered with ashes."

Roads instantly became jammed with military convoys and refugees, guiding wagons piled with belongings. In the city of Angeles, adjoining Clark Air Force Base, church bells pealed to sound the alarm, and some residents stood frozen in fear on city streets. But the city was considered outside of the danger zone, and their fear quickly subsided. Scientists, reasoning that the magma within the volcano was too thick to flow easily, felt no need to evacuate Angeles, and plans went forward with the parade commemorating Philippine Independence Day.

Late at night on June 12, and again early in the morning of June 13, the volcano erupted again with even more violence, throwing ash and fire 80,000 feet into the air. Terrified residents of the area now frantically loaded cars, trucks, and handcarts with their belongings, and jammed roads became nearly impassable.

Victor Gomez, a 53-year-old American who owned a restaurant near the gates of Clark Air Base, was just sitting down to a morning cup of coffee when he heard screams outside of his restaurant. "I rushed outside and there it was," he said. "It looked like a nuclear bomb had gone up, a huge mushroom cloud over the volcano." On the base, Sergeant Richard Johnson of Wilmington, Delaware, who had been at Clark for more than two years and was only 37 days away from reassignment in the United States, told a reporter, "We've had earthquakes, coup attempts, terrorist attacks and now a volcano. This place is cursed. Next thing we'll probably get is a tidal wave."

On June 14, eight more explosions sent ash and steam 19 miles (30.7 km) high, hurling the volcanic detritus over both Clark and Subic Bay.

A giant mudflow began to slide down the mountain toward nearby rivers. All day, the mountain continued to erupt, sending plumes of ash 12 miles (19.31 km) into the atmosphere.

And then, early in the morning of June 15, nature struck with the other fist. A typhoon, with winds of up to 80 MPH, slammed into the eastern coast of Luzon Island, drenching the countryside, adding weight to the ashes, and turning them to white mud. A serviceman was killed when his vehicle slid on the rain-soaked ash. The weight of the rain-coated debris from the volcano collapsed the roof of a bus terminal in Angeles, which was now under serious threat from falling ash and mud. Only 12 of the 100 people inside were immediately extricated, and scores were crushed.

Inch-thick rocks mixed with the falling ash and became lethal bombs, and now refugees from Angeles joined the crush on the roads of fleeing, terrified residents of the surrounding countryside.

The volcano continued to erupt over the weekend of the 15th and 16th. Mudslides and floods devoured homes. An old man on the outskirts of Angeles drowned in a torrent of thick volcanic ash and rainwater. Eight inches of ash turned to mud collapsed buildings.

The Red Cross worked feverishly. Shelters were set up in churches, schools, army camps, and gymnasiums in Manila to accommodate the tens of thousands of refugees. Sitting in the shaded bleachers of a Manila stadium, a young man, Angelo Luciano, told a *New York Times* reporter that his house had collapsed on his wife and him. "Our neighbors dug into the rubble and saved us," he marveled with a shake of his head.

Meanwhile, volcanologists warned of a 1.8-mile (2.9-km) crack that had developed between two craters along the volcano's southern side. The buildup of pressure within the mountain could be enough to blow it apart, they reasoned, much the way Mt. St. Helens had exploded in 1980 (see below).

Heeding the warning, the U.S. government evacuated the remaining 900 servicemen from Clark and ordered all of the dependents and civilian employees of the two bases to be sent home. Seven navy ships loaded the civilians and left immediately for the United States.

Earlier, the Nuclear-Free Philippines Coalition had charged that if volcanic lava hit Clark Air Force Base, there would be a potential for the release of radiation from weapons stored there. The U.S. military would neither confirm nor deny that it had nuclear arms at Clark, but denied that there would be any radiation danger. The military also claimed that the 8,200 tons (7,438.9 tonnes) of munitions at Clark were not at risk from the explosion of the mountain.

Neither of these assurances was tested. Instead of exploding, the mountain caved in on itself and gradually quieted.

But the devastation left behind was monumental. Two hundred died, most of them during the height of the explosions; 100,000 were rendered homeless. The rice crop was completely destroyed, buried beneath a foot of volcanic ash and rock. Farmers wondered if their soil would ever be fertile again.

In Olongapo, a city of 300,000 adjoining Subic Bay, looting occurred, and a vast food and water shortage developed. By June 18, the water supply had almost run out, and what was there was contaminated with ash. Residents began to break pipes to drain off what little pure water remained. Street vendors made a quick killing selling cloth surgical masks for 35 cents each to residents of Manila who were still choking on the volcanic ash in the air.

The once lush slopes of Mt. Pinatubo resembled a moonscape. In Zambales Province, the hardest hit region of all, as much as three feet of ash and volcanic rock covered everything. "There is nothing left," said Leonardo Mercavado, who owned a small farm. "It is all ash and sand now."

PHILIPPINES
TAAL
January 30, 1911

In the worst eruption of Taal in modern times, this constantly active volcano in the Philippines, killed 1,335 and injured 199, on January 30, 1911.

Taal, a Pelean-type volcano which rises from a volcanic island in the middle of Lake Taal, south of Manila, has had a number of cataclysmic eruptions in its extensive history, but the most devastating of modern times occurred on January 30, 1911.

It was a classic, catastrophic example of the Pelean type of explosion which blows out not only the crater but also the sides of the mountain—often with hurricane force winds and tornadoes—and releases practically no lava, but copious quantities of white hot ash and superheated steam.

The 1911 eruption of Taal typically began with a series of earthquakes, from the night of January 27 through the 29th, accompanied by a column of heavy black smoke that rose for miles into the atmosphere, and dropped thin but steady quantities of ash throughout the countryside.

There were minor explosions, accompanied by lightning, mixed in with these earthquakes on January 29. All of this was duly noted by a party of United States engineers camped at Bayuyungan, but none of these prepared the party for the major one, a double eruption that ripped away the floor of the volcano, opened huge fissures in the lake bed, and blasted these men from their beds at 2:20 A.M., January 30.

"Loud rumbling noise was again heard," noted one of the engineers, "and before I could get out of bed an explosion of indescribable severity took place." He continued his report:

> Smoke came out of the crater in dense clouds. The rumbling noise grew louder and louder and then came a heavy report. I then saw the mud issuing from the crater as a cloud. In a few seconds, I saw this cloud drifting across the lake toward our camp. Our camp was then swept by a heavy wind which broke the tent ropes and threw the tent into the air. This atmospheric disturbance threw me a distance of about fifteen feet. A rain of ashes fell about eight inches deep, the air was oppressive, and we gasped for breath for twenty seconds. For half a minute there was light warm rain, then cold rain for fifteen minutes. A tidal wave from the lake reached the camp a quarter mile back from the shore. We took refuge on higher ground, everything was washed away at the camp, one of the party was slightly burned about the arms with hot ashes, but otherwise we did not suffer injuries.

The fortunate fate of the engineers was not shared by the residents of 13 villages and towns which were obliterated by roaring avalanches of scalding hot mud, peppered with boulders and uprooted trees. For 10 miles in a fan-shaped swath, all vegetation was burned down to the bare earth. Trees were sliced off at the roots. No wildlife survived.

A 1965 aerial view of Taal, a volcano in the Philippines (National Geophysical Data Center/USGS)

The death and destruction in the towns and villages was enormous. For a distance of six miles, a 31-inch-thick layer of mud, accompanied by poisonous volcanic gases, destroyed all human life and buildings. An area of nearly 800 square miles would eventually be inundated by the ash fallout from the explosion.

Between 3:00 and 3:45 P.M., geologist W. E. Pratt and a photographer ventured near the crater to take pictures. They, too, were luckier than the residents of such towns as Guilot, Bugaan, and San Jose, which, along with their populations, no longer existed. "No animal life could have lived on that island," Pratt recorded. "Not a blade of grass escaped. Trees eight inches in diameter were broken, leaving stumps one to two feet high with ends shredded in splinters by the sand and small stones driven by the force of the eruption. A 600-pound (272.155 kg) boulder fell on top of the highest ridge. A breadcrust bomb of augite and esite was found, one meter in diameter."

Fifteen minutes after the two descended from the crater, the mountain blew again, with almost the same force as the first explosion, which was heard 310 miles (498.896 km) away. Hurricane winds swept away what the mud, ash, and fire had not already destroyed. A black cloud of ash spread up and into the atmosphere, darkening the skies over Manila, 40 miles (64.4 km) away, and visible for a distance of 250 miles (402.3 km).

In total, 1,335 people were killed and another 199 were severely injured. Taal remained quiescent until 1965, when it again erupted, killing 200. To this day, it is an active, dangerous volcano.

UNITED STATES
MOUNT ST. HELENS
May 18, 1980

Sixty-two people—many of them scientists studying the eruption—died in the monumental explosion of Mt. St. Helens, in Washington State, on May 18, 1980.

"I think people need to put all this devastation into perspective," observed Bill Ruediger, a wildlife biologist for the Gifford Pinchot National Forest after the immense eruption of Mount St. Helens in May 1980. "Although it's totally unique as far as our lifetimes are concerned, it's the way nature is: every couple thousand years she cleans the slate again. What we have here right now is almost a genesis situation."

Known for its pristine beauty, Mt. St. Helens is the youngest and one of the smallest of the mountains of the Cascade Range, which stretches 700 miles (1,126.5 km) south from Canada to California, and is part of the Pacific "Ring of Fire." Only 9,677 feet (2,949.5 m) in height, the mountain was named for the British ambassador to Spain, Baron St. Helens, in 1792 by Captain George Vancouver, a British navigator and explorer in search of the Northwest Passage. Viewing the peak for the first time in 1805 from what is now Portland, Oregon, explorer William Clark noted in his log of the Lewis and Clark expedition, "Mount St. Helens is perhaps the greatest pinnacle in America."

By the time it had exploded in the spring of 1980, the mountain would be reduced to something resembling a partially extracted tooth. The top seemed as though a cosmic fist had pummeled it. The once symmetrical and shapely peak was gone, and in its place, 1,313 feet (400.2 m) lower, was an amphitheater with sheer, 2,000-foot (609.6-m) walls opening to the barren area beneath it.

No other major volcanic eruption has been so painstakingly recorded. The energy released in its climactic explosion was charted as being equivalent to that of 500 Hiroshima-type atomic bombs, or 10 million tons (9,071,847.4 metric tons) of TNT. An area of 232 square miles (600.9 km^2) was scorched into a moonscape. Only 62 people were killed by the enormous blast, and some of those were scientists, there to record the very disaster that claimed them. One victim, Harry Truman, an 84-year-old veteran of the mountain and proprietor of the Mt. St. Helens Lodge on Spirit Lake, met his fate because he refused to evacuate the area when ordered out by state police. Even when a force 5 earthquake rattled objects loose from the walls of his lodge, Truman reportedly took another sip of bourbon, gathered some of his 16 cats around him, and refused to leave his lodge. "I am part of that mountain," he explained. "The mountain is part of me." His lodge was buried under red hot ash; his body was never found.

Mt. St. Helens's geological history is short by the world's standards. Explorers heard Indian reports of its early explosions, but generally dismissed them as primitive legends. However, there were a series of eruptions between 1800 and 1857 that were recorded in some detail.

But from 1857 until 1980—a period of 123 years—Mt. St. Helens was, for all intents and purposes, a very quiet volcano.

On March 20, 1980, a magnitude 4 earthquake (see EARTHQUAKES, Introduction, pp. xx–xx) shook the area. It was the first of a series of escalating seismic shocks. On March 27, 47 earthquakes of magnitude 3 intensity shook the area, and at noon that day, there was a thunderous explosion near the peak. A news team from the Vancouver *Columbian* commandeered a helicopter and explored the summit, discovering the source of the explosion. According to reporter Bill Steward, ". . . there was a hole in the snow on the north side of the peak. It was smudged with black ash, and the top, especially the north side was shattered." Long, jagged fissures underlay the snow, cracking it in concert with the opened earth beneath.

The next day, a team of scientists in another helicopter affirmed that there had been a phreatic, or steam, eruption caused by subsurface water that had trickled down until it hit hot rock and was vaporized. There was clearly a superheated furnace cranking up beneath the surface of Mt. St. Helens.

From the beginning of April, a bulge began to appear on the north face of the cone, just below the summit crater and above an outcropping of lava called Goat Rocks. The bulge covered an area nearly two miles in diameter which had grown as much as 30 feet (9.1 m). It was expanding at a rate of a foot per day, with the displacement nearly horizontal to the north-northwest. Scientists attributed the bulge to pressure from magma rising in the throat of the volcano, and postulated that either an eruption or a devastating avalanche was imminent.

Roadblocks were erected by the middle of April; the danger zones near the mountain and under the bulge were evacuated except for a few diehards like Harry Truman, and some remaining loggers, scientists, and their families.

And then, at 8:32 A.M., on Sunday, May 18, 1980, another magnitude 5 earthquake shook the mountain, triggering the climactic, lethal eruption.

Two geologists, Dorothy and Keith Stoffel, had boarded a light plane early that morning to take a look at the volcano. At 8:32, they were directly over the summit, looking down into the crater from a height of approximately 1,000 feet (304.8 m). "The whole north side of the summit crater began to move as one gigantic mass," they later recorded. "The entire mass began to ripple and churn without moving laterally. Then the

The growing lava dome in the explosion crater of Mount St. Helens in 1983, three years after the main explosion. (National Geophysical Data Center/University of Colorado)

whole north side of the summit began sliding to the north along a deep-seated slide plane."

What was happening was a tripleheaded explosion: The bulge abruptly broke away, triggering massive mudflows pockmarked with lethal debris. At the same time, a lateral blast of hot, ash-filled gases blew out of the sides of the mountain, and a vertical eruption cloud and ash fall spewed out of the mountain's cone. The geologists narrowly escaped being caught in the vertical cloud, and only outflew it by turning to the south, out of its path.

Meanwhile, an enormous avalanche—the largest landslide ever recorded—of more than two-thirds of a cubic mile of rock debris and glacial ice, turned into steam and boiling water by exploding steam and entrapped air, roared down the mountainside at a speed of 155 MPH. It splashed into Spirit Lake, struck a ridge beyond, and was then deflected into the northern fork

of the Toutle River. The lake and the valley were immediately jammed with debris in depths up to 500 feet (152.4 m). The combination of water and entrapped snow and ice, plus the volcanic debris, combined to form a huge mudflow that continued down the north fork of the Toutle into the Cowlitz River, and from there into the Columbia River.

The mudflow was like hot mortar, carrying jumbles of logs, pieces of bridges and houses, and other macabre debris. One observer reported seeing a fully loaded logging truck being carried upright on the surface, submerged only to the lower tier of logs. By the time it had reached the Columbia, the mudflow had slowed to 4.8 MPH. It would deposit more than 1 billion cubic feet of sediment into the Cowlitz River, and enough mud would flow on through the Cowlitz to block the shipping channel of the Columbia River completely.

At the same time, geysers of hot gas shot from cracks around the bulge at speeds of up to 250 MPH and overtook the avalanche as it rolled down the slope, heating it to temperatures up to 572°F (300°C). This combination of a steam blast and fluidized volcanic rock fragments both flattened and incinerated major ridges and valleys over an area of 212 square miles (549 sq. km^2). Entire trees, four to seven feet in diameter, were uprooted and swept away in the zone nearest to the explosion. Beyond this, other trees were stripped of their limbs and bark and snapped like matchsticks. In some places, they were lined up militarily with their trunks pointed away from the direction of the blast. In other places, tornado-like eddies in the blast left piles of trees in mangled heaps. The timber blowdown represented 3.7 billion board feet of lumber, enough to build nearly a quarter of a million homes. It was this avalanche and the lateral blast of hot ash-filled gas that caused most of the loss of life and property. And most of it was over by 9:00 A.M., an astonishingly short period of time for this sort of eruption.

Wildlife on the north slope of the mountain was almost totally erased. According to the Washington State Department of Game, 5,000 black-tailed deer, 200 black bear, and 1,500 elk died in the disaster along with all the birds and small mammals in the affected area. An eerie silence would settle over this part of the Earth for a long time to come.

Finally, while the avalanche was developing its full force, a huge eruption cloud, filled with ash, roared from the crater. Within 10 minutes, this cloud that pursued the Stoffels in their observation plane rose to a height of 63,000 feet (19,202.4 m). Day turned to night everywhere in the area. At Spokane, Washington, 250 miles (402.3 km) away, visibility was reduced to approximately 10 feet (3 m) in mid-afternoon, when the cloud reached there. In Yakima, 90 miles away, up to five inches of ash fell, and lesser amounts rained down on Idaho, central Montana, and parts of Colorado. The ash cloud circled the globe in 11 days. For weeks, a belt of ash would color sunsets and affect the world's atmosphere.

As in most such eruptions, a lava dome formed, 600 feet high and 2,000 feet (609.6 m) in diameter, and lava began to flow from it. Mount St. Helens experienced lesser eruptions throughout 1982. There is ample scientific evidence to indicate that its life as an active, dangerous volcano is by no means over.

USSR
KAMCHATKA
BEZYMIANNY
March 30, 1956

One of the greatest Pelean-type eruptions ever, the explosion of Bezymianny in Kamchatka, USSR, on March 30, 1956, resulted in no recorded loss of life.

The enormous explosion of Bezymianny in the Kamchatka Peninsula of the USSR has largely gone unnoticed because there was no loss of life. However, in intensity, it ranks as one of the greatest Pelean-type eruptions and, as such, deserves mention.

Considered to be an extinct volcano, with no record of explosions in historic times, the volcano was ignored when a series of earthquakes occurred in September and October of 1955. Even though the epicenters of the quakes were traced to Bezymianny, it was believed that nearby Kliuchevsky volcano, an active mountain, was responsible.

However, on October 22, 1955, a moderate eruption did develop at Bezymianny, and ash fell until March 29, 1956.

And then, at 5:11 P.M., on March 30, 1956, an enormous explosion ripped open the top of the mountain's snow-covered cone, which had heretofore risen over 10,000 feet (3,048 m) above sea level. In seconds, 600 feet (182.8 m) were lopped off the volcano and the eruption cloud rose to a height of 28 miles (45.1 km).

G. S. Gorschkov, the Russian volcanologist, described the scene from nearby Kliuchi:

> The cloud was curling intensely, and quickly changed its outlines ... It seemed to be very thick and almost tangibly heavy. Together with the cloud came also and was growing a rumble of local thunder accompanying incessantly flashing lightning. About 5:40 P.M. when the cloud had already passed the zenith, ash began to fall ... and by 6:20 P.M., it got so impenetrably dark that one could not see

his own hand, even if brought up to the very face. People returning from work were wandering about the village in search of their homes. Peals of thunder were crashing with deafening loudness without any interruption. The air was saturated with electricity, telephones were ringing spontaneously, loud speakers of the radionet were burning out.... There was a strong smell of sulphurous gas.

The hot ash deposited over an area of about 193 square miles (499 sq. km) melted the snow cover and torrential mudflows formed in the valley of the Dry Hapsita River and in valleys located on the flanks of adjacent volcanoes. The torrents swept huge boulders, some weighing hundreds of tons, through the valley, wiping out everything in their paths. Trees were swept away, decapitated, or scorched. Three weeks after the explosion, Gorschkov discovered thousands of steam fumaroles rising from the surface of a 100-foot- (30.4-m-) thick layering of ash over an area of 19 square miles (49.2 sq. km), and thus, the area was named the Valley of Ten Thousand Smokes of Kamchatka, after a similar occurrence in 1912 after the eruption of Mt. Katmai.

WEST INDIES
LA SOUFRIÈRE
May 7 and 8, 1902

On May 7 and 8, 1902, 1,565 died in the spectacular explosion of La Soufrière, on St. Vincent, 90 miles (144.8 m) south of Martinique. The explosion occurred less than a day before the cataclysmic eruption of Pelée on Martinique.

Less than 24 hours before the immense, death dealing eruption of Mt. Pelée on Martinique (see pp. 405–410), La Soufrière, a 4,048-foot (1,233.8-m) volcano that rises at the northern end of the island of St. Vincent, 90 miles (144.84 m) south of Martinique, staged its own spectacular and lethal geological fireworks display.

Although the Carib natives whom Columbus subdued on St. Vincent when he landed there in 1498 did tell tales of La Soufrière's eruptions, the mountain stayed relatively calm, erupting only twice, in 1718 and 1812. Both of these were major eruptions, and the 1812 explosion left behind a crystalline blue crater lake, 1,930 feet (588.3 m) above sea level and framed by sheer, 800-foot (243.8 m) canyon walls. For half a century, this lake was a major tourist attraction in the West Indies.

But in April 1901, the mountain seemed to grow restless. Earthquakes rumbled through it and around it, severely enough to cause alarm among the people living on La Soufrière's western slope. But with the end of April 1901, came the end of the earthquakes and the people's panic.

A year later, the earthquakes resumed with greater force, and the inhabitants on the leeward side of the volcano north of Chateaubelair began to evacuate their homes. Well that they did; otherwise, the toll of 1,565 deaths from the explosions that would follow would have been even higher.

Steam began to spit from the volcano at 2:40 P.M. on May 6, and by 10:30 A.M. on May 7, the mountain erupted with a roar. The cloud of steam stretched 30,000 feet (9,144 m) above the cone, and by that afternoon, huge, superheated stones started to fall from it.

Panic-stricken residents of the slopes of La Soufrière now commenced their evacuation in earnest, but those who attempted to cross the Tabaka Dry River, southeast of the cone, would encounter a torrent of boiling mud and water over 50 feet (15.2 m) deep, rumbling down the valley toward the sea.

By 2:00 P.M., the raging explosions increased, as did the clouds of steam and the rain of red hot ash. Suddenly, a dense cloud of hot dust and steam was expelled from the mountain and sped down the mountainside with the speed of a hurricane. One observer described it as ". . . a terrific, huge, reddish-and-purplish cloud," which swept everything and everyone in its advancing path. A nuée ardente, or steam-powered cloud of gas and red hot ash, of the Pelée-type, it tumbled, at astonishing force and speed, toward the sea, overturning trees, setting fire to the dwellings it did not flatten, incinerating forests and vegetation over one-third of the entire island.

At Georgetown, 5.5 miles (8.9 km) from the crater, red hot stones six inches (15.2 cm) in diameter fell on the populace like fatal hail. The immense cloud of ash spread out for miles, helped by prevailing winds. At 2:30 A.M. on May 8, it began to fall on the *Jupiter*, a ship 830 miles (1,335.8 km) east-southeast of Barbados, indicating to scientists that the cloud was traveling at a speed of 60 MPH.

In the valleys of St. Vincent, ash accumulated in depths up to 200 feet (61 m). The drifts of gray ash would remain hot for weeks, and steam columns emitting from them would ascend upwards for a full mile.

On the rim of the ash, steam condensed into scalding water, mixed with the ash, and formed a characteristic hot mud which invaded everywhere. Had there been a major city in the path of this flow, as St. Pierre had been in Martinique, the loss of life would have been as staggering as that of Pelée's. Still, a great number of people lost their lives on the Georgetown side of the Tabaka Dry River, perishing from inhalation of the hot dust, burns from steam and dust, and scalding blows from falling rocks.

Interestingly, 132 persons escaped injury by hiding out in an empty rum cellar in Orange Hill, two and a half miles north of Georgetown. The cellar was only partly underground but, providentially enough, the one window and door it possessed gave onto the side away from the volcano.

Following the cataclysmic 2:00 P.M. eruption of May 8, similar eruptions of La Soufrière continued through May 18.

On May 23, the rescue ship *Dixie*, with relief supplies that had not been unloaded at St. Pierre, and with some of its eminent scientists still aboard, arrived to aid the ravaged populace of St. Vincent.

SELECTED BIBLIOGRAPHY

Armstrong, Betsy, and Williams, Knox. *The Avalanche Book*. Golden, Colorado: Fulcrum, Inc., 1986.

Arnold, Eric. *Volcanoes! Mountains of Fire*. New York: Random House, 1999.

Asquith, Michael. *Quaker Work in Russia 1921–23*. London: Oxford University Press, 1943.

Atwater, Montgomery. *The Avalanche Hunters*. Philadelphia: Macrae Smith Company, 1968.

Aykroyd, W. R. *The Conquest of Famine*. New York: E. P. Dutton, 1975.

Barry, John M. *The Great Influenza: The Inside Story of the Deadliest Plague in History*. New York: Viking, 2004.

Barth, Tom F. W. *Volcanic Geology, Hot Springs and Geysers of Iceland*. Washington, D.C.: Carnegie Institution of Washington Publications 587, 1951.

Bascombe, Edward. *A History of Epidemic Pestilences*. London: John Churchill, 1851.

Becker, Jasper. *Hungry Ghosts*. New York: Henry Holt and Co., 1998.

Bell, W. G. *The Great Plague of London in 1665*. London: John Lane Bodley Head Ltd., 1924.

Boccaccio, Giovanni. *The Decameron*. Trans. by G. H. McWilliam, Middlesex, England: Penguin Books Ltd., 1970.

Bourriau, Janine, ed. *Understanding Catastrophe*. New York: Cambridge University Press, 1992.

Brindze, Ruth. *Hurricanes (Planet Earth Series)*. Alexandria, Virginia: Time-Life Books, 1982.

Brockwell, Ian. *Global Warming: The Final Solution*. Ian Brockwell, 2007.

Brooks, C. E. P. *Climate Through the Ages*. New York: McGraw-Hill, 1949.

Bullard, Fred M. *Volcanoes of the Earth*. Austin: University of Texas Press, 1984.

Conrad, Karia, et al. *Volcans et tremblements de terre*. Paris: Deux Corps d'Or, 1975.

Cornell, James. *The Great International Disaster Book*. New York: Charles Scribner's Sons, 1976.

Corti, Egon C. *The Destruction and Resurrection of Pompeii and Herculaneum*. London: Routledge & Paul Kegan, 1951.

Costa, John E. and Wieczorek, Gerald F. eds. *Debris Flows—Avalanches: Process, Recognition and Mitigation*. Washington, D.C.: Books on Demand, 1987.

Crandell, Dwight R. *Potential Hazards from Future Eruptions of Mount St. Helens Volcano*. Washington D.C. United States Government Printing Office, 1978.

Creighton, Charles. *A History of Epidemics in Britain*. New York: Barnes & Noble, 1965.

Cry, George W. "Tropical Cyclones of the North Atlantic Ocean: Tracks and Frequencies of Hurricanes and Tropical Storms, 1871–1963," *Technical Paper* no. 55, U.S. Weather Bureau, Washington, D.C., 1965.

Dana, J. D. *Characteristics of Volcanoes*. New York: Dodd, Mead and Co., 1891.

Daniels, George C. ed. *Planet Earth: Volcano*. Alexandria, Virginia: Time-Life Books, 1982.

Dauncy, Guy, and Patrick Mazza. *Stormy Weather*. Gabriola Island, British Columbia, Canada: New Society Publishers, 2006.

Deaux, George. *The Black Death 1347*. New York: Weybright and Talley, 1969.

Decker, Robert, and Decker, Barbara. *Volcanoes*. New York: W. H. Freeman, 1981.

Defoe, Daniel. *A Journal of the Plague Year*. Middlesex, England: Penguin Books Ltd., 1966.

Devereux, Stephen. *Theories of Famine*. Paramus, N.J.: Prentice-Hall, 1994.

Douglas, Marjorie Stoneman. *Hurricane*. New York: Rinehart & Company, 1958.

Duffy, John. *Epidemics in Colonial America*. Baton Rouge: Louisiana State University Press, 1953.

Ebert, Charles H. *Disasters: Violence of Nature and Threats by Man*. Dubuque, Iowa: Kendall-Hunt, 1996.

Flexner, Stuart B. and Flexner, Doris. *The Pessimist's Guide to History*. New York: Avon, 1992.

Fodor, R. V. *Angry Waters*. New York: Dodd, Mead & Co., 1980.

Fouque, Ferdinand Andre. *Santorini et ses eruptions*. Paris: G. Masson, 1879.

Foxworthy, Bruce L., and Hill, Mary. *Volcanic Eruptions of 1980 at Mount St. Helens: The First 100 Days*. U.S. Geographical Survey Professional Paper, 1249.

Fradin, Dennis Brindell. *Disaster! Famines*. Chicago: Children's Press, 1986.

———. *Disaster! Floods*. Chicago: Children's Press, 1982.

Francis, Peter. *Volcanoes*. Harmondsworth, Middlesex: Penguin Books Ltd., 1978.

Fraser, Colin. *The Avalanche Enigma*. Great Britain: John Murray Publishers Ltd., 1978.

Furneaux, Rupert. *Krakatoa*. New York: Prentice-Hall, 1964.

Galanopoulos, A. G., and Bacon, Edward. *Atlantis: The Truth Behind the Legend*. New York: Bobbs-Merrill, 1969.

Gallagher, Dale, ed. *The Snowy Torrents: Avalanche Accidents in the United States 1910–1966*. Washington: Alta Avalanche Study Center, U.S.D.A. Forest Service, 1967.

Garrison, Webb. *Disasters That Made History*. New York: Abingdon Press, 1973.

General State Printing Office, the Hague: *Malnutrition and Starvation in the Western Netherlands, September 1944/July 1945, Parts I and II*, 1948.

Gibbon, Edward. *The History of the Decline and Fall of the Roman Empire* (12 vols). London: T. Miller, 1820.

Gorshkov, G. S. "Gigantic Eruption of Volcano Bezymianny," *Bulletin Volcanologique*, ser. 2, 1959.

Grant, Michael. *The Art and Life of Pompeii and Herculaneum*. New York: Newsweek Books, 1979.

———. *Cities of Vesuvius: Pompeii and Herculaneum*. Hammondsworth, Middlesex: Penguin Books Ltd., 1979.

Gregg, Charles T. *Plague!* New York: Charles Scribner's Sons, 1978.

Gribbin, John. *This Shaking Earth*. New York: G. P. Putnam's Sons, 1978.

Gore, Al. *An Inconvenient Truth: The Planetary Emergency of Global Warming and What We Can Do About It*. New York: Viking, 2007.

Guomei, Xia. *HIV/AIDS in China*. China: Foreign Languages Press, 2002.

Halsey, D. S., Jr. *Earthquakes: A Natural History*. New York: MacMillan, 1974.

Harris, D. Lee. "Characteristics of the Hurricane Storm Surge," *Technical Paper* No. 48, U.S. Weather Bureau, Washington D.C., 1963.

Harris, Stephen L. *Fire and Ice: The Cascade Volcanoes*. Seattle: Mountaineers-Pacific Search Books, 1976.

Hays, J. N. *The Burdens of Disease: Epidemics and Human Response in Western History*. New Brunswick, N.J.: Rutgers University Press, 1998.

Hearn, Lafcadio. *Two Years in the French West Indies*. New York: Harper & Brothers, 1890.

Hecker, J. F. C. *The Epidemics of the Middle Ages*. London: The Sydenham Society, 1844.

Heidel, Alexander. *The Gilgamesh Epic and Old Testament Parallels*. Chicago: University of Chicago Press, 1949.

Heilprin, Angelo. *The Eruption of Pelee: A Summary and Discussion of the Phenomena and Their Sequels*. New York: J. B. Lippincott, 1908.

———. *Mount Pelee and the Tragedy of Martinique*. New York: J. B. Lippincott, 1903.

Herbert, Don, and Bardossi, Fulcio. *Kilauea: Case History of a Volcano*. New York: Harper & Row, 1968.

Hewitt, R. *From Earthquake, Fire and Flood*. New York: Charles Scribner's Sons, 1957.

Hirst, L. F. *The Conquest of Plague*. Oxford: Clarendon Press, 1953.

Hoehling, A. A. *The Great Epidemic*. Boston: Little, Brown and Co., 1961.

Hoyt, J. B. "The Cold Summer of 1816," *Annals of the Association of American Geographers*, Minneapolis, Number 48, 1958.

Hoyt, William G., and Langbein, Walter B. *Floods*. Princeton: Princeton University Press, 1955.

Huang Wei. *Conquering the Yellow River*. Peking: Foreign Language Press, 1978.

Johansen, Bruce. *Global Warming in the 21st Century*. 3 vols. New York: Praeger, 2003.

Johnson, Thomas P. *When Nature Runs Wild*. Mankato, Minnesota: Creative Education Press, 1968.

Johnson, Willis Fletcher. *History of the Johnstown Flood*. Edgewood, New Jersey: Edgewood Publishing Co., 1889.

Joseph, Stephen C. *Dragon Within the Gates: The Once and Future AIDS Epidemic*. New York: Carroll and Graf, 1993.

Judkins, Grant, ed. *The Big Thompson Disaster*. New York: Lithographic Press, 1976.

Karplus, W. J. *The Heavens Are Falling: The Scientific Prediction of Catastrophe in Our Time*. New York: Plenum Publishers, 1992.

Kirk, Ruth. *Snow*. New York: William Morrow & Co., 1978.

Krafft, Katia and Krafft, Maurice. *Volcanoes: The Story Behind the Scenery*. New York: Hammond, 1980.

Lacroix, A. *La Montagne Peleé et ses eruptions*. Paris: Masson et Cie, 1904.

Lane, Frank W. *The Elements Rage*. New York: Chilton Books, 1965.

Lauber, Patricia. *Earthquakes*. New York: Random House, 1972.

Leet, L. Don. *Causes of Catastrophe*. New York: Whittlesey House, 1948.

Lerouc, Marcel. *Global Warming—Myth or Reality?* New York: Springer-Praxis Books, 2005.

Link, V. B. *A History of Plague in the United States of America*. Public Health Monograph No. 26, U.S. Government Printing Office, 1952.

Longshore, David. *Encyclopedia of Hurricanes, Typhoons and Cyclones*. New York: Facts On File, 1998.

Luce, J. V. *Lost Atlantis*. New York: McGraw-Hill, 1969.

Ludlum, D. D. *Early American Hurricanes, 1492–1870*. Boston: American Meteorological Society, 1963.

Macdonald, Gordon A. *Volcanoes*. New York: Prentice-Hall, 1972.

Mallory, W. H. *China, Land of Famine*. New York: American Geographic Society, 1926.

Marks, Geoffrey. *The Medieval Plague*. New York: Doubleday & Co., 1971.

Marks, Geoffrey, and Beatty, William K. *Epidemics*. New York: Charles Scribner's Sons, 1976.

Martin, Fred. *Volcano*. Des Plaines, Ill.: Heinemann Library, 1996.

Marx, Wesley. *Acts of God, Acts of Man*. New York: Coward, McCann & Geoghegan, 1977.

Mather, Kirtley F. *The Earth Beneath Us*. New York: Random House, 1964.

Mayer, Ken M. D., and Pizer, Hank. *The AIDS Fact Book*. New York: Bantam Books, 1983.

McCullough, David G. *The Johnstown Flood*. New York: Simon and Schuster, 1968.

McGrew, Roderick E. *Russia and the Cholera, 1823–1832*. Madison: University of Wisconsin Press, 1965.

McKenzie, Nancy F., ed. *The AIDS Reader: Privacy, Poverty, Community*. New York: NAL Dutton, 1993.

McNeill, William H. *Plagues and Peoples*. New York: Anchor Press/Doubleday, 1976.

Nash, Jay Robert. *Darkest Hours*. New York: Nelson-Hall, 1976.

National Geographic Society. *Great Adventures*. Washington D.C.: National Geographic Society, 1963.

———. *Powers of Nature*. Washington D.C.: National Geographic Society, 1978.

Navarra, John Gabriel. *Earthquake*. New York: Doubleday & Co., 1980.

Nencini, Franco. *Florence: The Days of the Flood*. New York: Stein and Day, 1967.

Newby, Eric. *Slowly Down the Ganges*. New York: Charles Scribner's Sons, 1967.

Newson, Malcom D. *Flooding and Flood Hazard in the United Kingdom*. Oxford: Oxford University Press, 1975.

Palmieri, Luigi. *The Eruption of Vesuvius in 1872*. London: Asher & Co., 1873.

Phillips, John. *Vesuvius*. Oxford: Clarendon Press, 1869.

Pollitzer, R. *Plague and Plague Control in the Soviet Union*. New York: Institute of Contemporary Russian Studies, Fordham University, 1966.

Ponte, G. *Stromboli*. Catania: Associozione Internazionale di Vulcanologia, 1952.

Prinzing, Friedrich. *Epidemics Resulting from Wars*. Oxford: Clarendon Press, 1916.

Ritchie, David, and Gates, Alexander. *Encyclopedia of Earthquakes and Volcanoes, New Edition*. New York: Facts On File, 2001.

Rittman, A. *Volcanoes and Their Activity*. New York: John Wiley & Sons, 1962.

Rodwell, G. F. *Etna: A History of the Mountain and of Its Eruptions*. London: C. Kegan Paul & Co., 1878.

Rosenberg, Charles E. *The Cholera Years—The United States in 1832, 1849, and 1866*. Chicago: The University of Chicago Press, 1962.

Ross, Frank, Jr. *Storms and Man*. New York: Lothrop, Lee & Shepard Co., 1973.

Rost, Yuri. *Armenian Tragedy: An Eye Witness Account of Human Conflict and Natural Disaster in Armenia and Azerbaijan*. New York: St. Martin's Press, 1990.

Schuchert, C. *Historical Geology of the Antillian-Caribbean Region*. New York: John Wiley and Sons, 1935.

Seymour, Jacqueline. *The Wonders of Nature*. New York: Crescent Books, 1978.

Shope, Richard E. "Influenza," *Public Health Reports*. Washington D.C.: LXXIII (1958).

Shrewbury, J. F. D. *A History of Bubonic Plague in the British Isles*. Cambridge: Cambridge University Press, 1970.

Sigerist, Henry E. *Civilization and Disease*. Ithaca, New York: Cornell University Press, 1943.

Sutton, Ann and Sutton, Myron. *Nature on the Rampage*. New York: J. B. Lippincott, 1962.

Tannehill, I. R. *Hurricane Hunters*. New York: Dodd-Mead, 1955.

Taylor, James B., et al. *Tornado: A Community Responds to Disaster*. Seattle: University of Washington Press, 1970.

Thomas, Gordon and Witts, Max. *The San Francisco Earthquake*. New York: Stein and Day, 1971.

———. *The Day the World Ended*. New York: Stein and Day, 1976.

Times-Picayune staff. *Katrina, the Ruin and Recovery of New Orleans*. New Orleans: Times-Picayune, 2006.

Treaster, Joseph B. *Hurricane Force: Tracking America's Killer Storms*. New York: New York Times Books, 2007.

Trevelyan, Raleigh. *The Shadow of Vesuvius: Pompeii C.E. 79*. London: Michael Joseph Ltd., 1976.

U.S. Weather Bureau. "Summary of North Atlantic and North Pacific Tropical Cyclones," *Climatological Data, National Summary,* Annual.

Verbeek, R. D. M. *Krakatau*. Brussels: Institut National de Geographie, no date.

Vicker, Ray. *This Hungry World*. New York: Charles Scribner's Sons, 1975.

Walford, Cornelius. *A Statistical Chronology of Plagues and Pestilences as Affecting Human Life*. London: Harrison and Sons, 1884.

Ward, Barbara and Dubos, Rene. *Only One Earth: The Care and Maintenance of a Small Planet*. London: Andre Deutsch, 1972.

Ward, Roy. *Floods: A Geographical Perspective*. London: The Macmillan Press Ltd., 1978.

Warmington, E. H., ed. *Pliny: Letters and Panegyricus*. Boston: Harvard University Press, 1959.

Weart, Spencer R. *The Discovery of Global Warming*. Cambridge, Mass.: Harvard University Press, 2003.

Webster, Noah. *A Brief History of Epidemic and Pestilential Diseases*. Hartford, Conn.: Hudson & Goodwin, 1799.

Wexler, Harry. "Volcanoes and World Climate," *Scientific American,* 186, 1952.

White, Theodore H., and Jacoby, Annalee. *Thunder Out of China*. New York: William Sloane Associates, 1946.

Whittow, John. *Disasters: The Anatomy of Environmental Hazards*. Athens, Georgia: University of Georgia Press, 1974.

Williams, Chuck. *Mount St. Helens: A Changing Landscape*. New York: Graphic Arts Center Publishing Co., 1980.

Williams, Howell, and McBirney, Alexander R. *Volcanology*. New York: Freeman, Cooper & Co., 1979.

Williams, Knox, and Armstrong, Betsy. *Snowy Torrents: Avalanche Accidents in the United States 1972–1979*. Jackson, Wyo.: Teton Bookshop Publishing Co., 1984.

Winslow, Charles-Edward Armory. *The Conquest of Epidemic Disease*. Princeton, N.J.: Princeton University Press, 1943.

———. *Man and Epidemics*. Princeton, N.J.: Princeton University Press, 1952.

Woodham-Smith, Cecil. *The Great Hunger (Ireland 1845–1849)*. New York: Harper and Row, 1962.

Worth, Timothy, and Dean Abrahamson. *The Challenge of Global Warming*. Washington, D.C.: Island Press, 1989.

Ziegler, P. *The Black Death*. New York: The John Day Co., 1969.

Zinnser, H. *Rats, Lice and History*. Boston: Little Brown and Co., 1941.

WEB SOURCES

BBC News. "CO$_2$ Highest for 650,000 Years." November 24, 2005. Available online. URL: http://newsvote.bbc.co.uk/mpapps/pagetools/print/news.bbc.co.uk/1/hi/sci/tech/4467420.stm. Accessed on July 27, 2007. Informative article on the current levels of the greenhouse gases, carbon dioxide, and methane in the atmosphere, higher now than at any time in the past 650,000 years.

The Boston Globe. "WHO center is on the trail of deadly bird flu." February 19, 2006. Available online. URL: http://www.boston.com/news/world/europe/articles/2006/02/19/swiss_lab_is_on_the_trail_of_deadly_bird_flu/. Accessed on July 27, 2007. Intriguing article that traces the World Health Organization's role in tracking and identifying the spread of bird flu in 2006.

Forest Conservation Portal. "Deadly Floods in Haiti Blamed on Deforestation, Poverty." September 24, 2004. Available online. URL: http://forests.org/articles/reader.asp?linkid=35207. Accessed on July 27, 2007. Article tracing the two major reasons for the devastation of flooding in Haiti.

Guardian Unlimited. "Fears over climate as Arctic ice melts at record level." September 29, 2005. Available online. URL: http://guardian.co.uk/climatechange/story/0,12374,1580613,00.html. Accessed on July 27, 2007. Cautionary article charting the increase in Arctic ice melting.

National Geographic News. "Many Islands 'Gone,' Wetlands Gutted After Katrina, Experts Say." September 19, 2005. Available online. URL: http://news.nationalgeographic.com/news/2005/09/0919_050919_katrina_delta.html. Accessed on July 27, 2007. Informative article linking the force of levee breaks during Katrina to overdevelopment and destruction of tidal wetland islands.

The New York Times. "At the U.N.: This Virus Has an Expert 'Quite Scared'" March 28, 2006. Available online. URL: http://www.nytimes.com/2006/03/28/health/28worr.html?ex=1301202000&en=91a66f695484ab54&ei=5089&partner=rssyahoo&emc=rss. Accessed on July 27, 2007. Article outlining the fears that experts at the U.N. have over the burgeoning epidemic of bird flu.

———. "Avian Flu: The Uncertain Threat—Q+A; How Serious Is The Risk?" March 28, 2006. Available online. URL: http://query.nytimes.com/gst/fullpage.html?sec=health&res=990DEEDA1430F93BA15750C0A9609C8B63&n=Top%2FNews%2FHealth%2FDiseases%2C+Conditions%2C+and+Health+Topics%2FAvian+Influenza. Accessed on July 27, 2007. In question and answer form, major risks of infection by avian flu are assessed.

———. "Bird Flu Case May Be First Double Jump." May 24, 2006. Available online. URL: http://www.nytimes.com/2006/05/24/world/asia/24birdflu.html?ex=1306123200&en=37d1ce32e8915296&ei=5088&partner=rssnyt&emc=rss. Accessed on July 9, 2007. An analysis of the transference of avian flu from human to human, instead of bird to human.

———. "Bird Flu Reports Multiply in Turkey, Faster Than Expected." January 9, 2006. Available online. URL: http://www.nytimes.com/2006/01/09/international/europe/09flu.html?ex=1294462800&en=eb3d80daaaba3cca&ei=5088&partner=rssnyt&emc=rss&pagewanted=all. Accessed on July 9, 2007. Disturbing article about the increase of avian flu in the Middle East.

———. "Bird Flu Virus May Be Spread By Smuggling." April 15, 2006. Available online. URL: http://www.nytimes.com/2006/04/15/world/europe/15bird.html?ex=1302753600&en=bc2640e1464f857a&ei=5088&partner=rssnyt&emc=rss. Accessed on July 9, 2007. Intriguing article about bird flu being spread by illegally smuggled Chinese.

———. "Coffins and Buried Remains Set Adrift by Hurricanes Create a Grisly Puzzle." October 25, 2005. Available online. URL: http://select.nytimes.com/gst/abstract.html?res=F20C1EF83B5B0C768EDDA90994DD404482&n=Top%2FReference%2FTimes+Topics%2FSubjects%2FC%2Fcemeteries. Accessed on July 9, 2007. News story about the ghastly sight of coffins and remains set afloat by Hurricane Katrina.

————. "Danger of Pandemic Is Clear if Not Present." March 24, 2005. Available online. URL: http://www.nytimes.com/2005/10/09/national/09flu.html?ei=5088&en=24f0aa63bf094d06&ex=1286510400&partner=rssnyt&emc=rss&pagewanted=print. Accessed on July 9, 2007. Article detailing the possibility of a repeat influenza pandemic in the 21st century.

————. "Disaster Response: Watch TV, Go Home." February 19, 2006. Available online. URL: http://www.nytimes.com/2006/02/19/weekinreview/19lipt.html?ex=1298005200&en=033446dc7b9d3817&ei=5088&partner=rssnyt&emc=rss. Accessed on July 9, 2007. News story regarding an aspect of failed response in New Orleans as Katrina struck.

————. "Earthquake Reconstruction Will Cost $3 Billion, Indonesia Says." June 14, 2006. Available online. URL: http://www.nytimes.com/2006/06/14/world/asia/14indo.html?ex=1307937600&en=dff bfc0fdef06160&ei=5088&partner=rssnyt&emc=rss&pagewa nted=all. Accessed on July 9, 2007. Article regarding the cost of reconstruction after the December 26, 2004, earthquake in Indonesia.

————. "Egypt Reports Bird Flu Cases, but Virus Is Not Found in Humans." February 18, 2006. Available online. URL: http://www.nytimes.com/2006/02/18/health/18egypt.html?ex=1297918800&en=9676a 132a3eaf515&ei=5088&partner=rssnyt&emc=rss. Accessed on July 9, 2007. News story regarding the absence of transference of bird flu to humans in Egypt.

————. "Evacuees' Lives Still Upended Seven Months After Hurricane." March 22, 2006. Available online. URL: http://www.nytimes.com/2006/03/22/national/nationalspecial/22katrina.html?ex=1300683600&en=6b05253408f57a2d&ei=5088&partner=rssnyt&emc=rss. Accessed on July 9, 2007. Follow-up story on the lives of evacuees from Hurricane Katrina.

————. "Experts Unlock Clues to Spread of 1918 Influenza Virus." October 6, 2005. Available online. URL: http://query.nytimes.com/gst/fullpage.html?res=9C05E3DE1E30F935A35753C1A9639C8B63&sec=health&pagewanted=print. Accessed on July 9, 2007. A study of the reasons for the 1918 influenza pandemic.

————. "Experts Say Medical Ventilators Are in Short Supply in Event of Bird Flu Pandemic." March 12, 2006. Available online. URL: http://www.nytimes.com/2006/03/12/national/12vent.html?ex=1184126400&en=3d2c 6dd8e802a069&ei=5070. Accessed on July 9, 2007. News story about medical emergencies during the break out of bird flu.

————. "Few Found Alive After Mud Buries Town in Philippines." February 19, 2006. Available online. URL: http://www.nytimes.com/2006/02/19/international/asia/19filip.html?ex=1298005200& en=bd34ea3f79f169ba&ei=5088&partner=rssnyt&emc=rss& pagewanted=all. Accessed on July 9, 2007. News story about the aftermath of the catastrophic landslide in Leyte on February 17, 2006.

————. "French Farmers Shudder as Flu Keeps Chickens From Ranging Free." March 1, 2006. Available online. URL: http://www.nytimes.com/2006/03/01/international/europe/01flu.html?ex=1298869200&en=fec6fbf7b5638ce0&ei=5088&partner=rssnyt&emc=rss. Accessed on July 9, 2007. Intriguing article about the gastronomic dilemma caused by avian flu in France.

————. "Health Officials Urge Nations to Report Bird Flu Data Sooner." March 21, 2006. Available online. URL: http://www.nytimes.com/2006/03/21/health/21infect.html?ei=5088&en=3fe76d4939f3a932&ex=1300597200&adxnnl=1&partner=rssnyt&emc=rss &adxnnlx=1173625835-+Nea5Oz BAVtMD221dGz76A. Accessed on July 9, 2007. Article regarding the consequences of tardy reporting of bird flu data.

————. "Hitting The Flu at Its Source, Before It Hits Us." November 6, 2005. Available online. URL: http://www.nytimes.com/2005/11/06/weekinreview/06mcneil.html?ex=1288933200&en=3f803155d93eebc6&ei=5088&partner=rssnyt&emc=rss. Accessed on July 9, 2007. Cautionary article about preparedness for a possible arrival of bird flu in the U.S.

————. "How The City Sank." October 9, 2007. Available online. URL: http://www.nytimes.com/2005/10/09/arts/design/09ouro.html?ex=1286510400&en=c3 a0119c405ed386&ei=5090&partner=rssuserland&emc=rss. Accessed on July 9, 2007. Gripping narrative about the disappearance underwater of large portions of the city of New Orleans after Katrina.

————. "Human Flu Transfers May Exceed Reports." June 4, 2006. Available online. URL: http://www.nytimes.com/2006/06/04/world/asia/04flu.html?ei=5090&en=d930aa42b484ada4&ex=1307073600&partner=rssuserland&emc=rss&pagewanted=print. Accessed on July 9, 2007. Speculative article about the future of the spreading of influenza.

————. "Hurricane Puts Focus on Other Disasters That Are Waiting to Happen." November 15, 2005. Available online. URL: http://select.nytimes.com/search/restricted/article?res=F60B15FA3B5A0C768DDDA80994DD404482. Accessed on July 9,

2007. Speculative article regarding the influence of the impact of Hurricane Katrina.

———. "In a Melting Trend, Less Arctic Ice to Go Around." Available online. URL: http://select. nytimes.com/search/restricted/article?res=FB0616 F639540C7A8EDDA00894DD4 04482. Accessed on July 9, 2007. Article regarding the implications of the disappearance of Arctic ice.

———. "India Slaughters Chickens in Bid to Contain Bird Flu." February 21, 2006. Available online. URL: http://www.nytimes.com/2006/02/21/inter national/asia/21flu.html?ei=5088&en=4df95 d71c 94bdcaf&ex=1298178000&adxnnl=1&partner= rssnyt&emc=rss&adxnnlx=1145877046- ejL3m3Zvnv64x84eO/6EVA. Accessed on July 9, 2007. Disturbing news story about the gigantic slaughter of chickens to prevent the spread of avian flu in India.

———. "Indonesia Hit by Big Quake Near Volcano." May 28, 2006. Available online. URL: http://www. nytimes.com/2006/05/28/world/asia/28indo.html? ex=1306468800&en=78a49432b9574c4a&ei=508 8&partner=rssnyt&emc=rss. Accessed on July 9, 2007. News story about the May 27, 2006 earthquake in Yogyakarto, Indonesia.

———. "In Mississippi, Canvas Cities Rise Amid Hurricane's Rubble." December 20, 2005. Available online. URL: http://select.nytimes.com/search/ restricted/article?res=F50E13F93C540C738EDD AB0994DD 404482. Accessed on July 9, 2007. Article about the devastation caused by Hurricane Katrina in Mississippi.

———. "In War on Bird Flu, U.N. Looks to Recruit Killer Army." January 29, 2006. Available online. URL: http://www.nytimes.com/2006/01/29/ international/29chickens.html?ex=1296190800& en=46359438563ebdee&ei=5088&partner=rssnyt &emc=rss&pagewanted=all. Accessed on July 9, 2007. News story about the efforts of multiple nations to deal, in a grass roots way, with the spread of bird flu.

———. "Maker Calls New Bird Flu Vaccine More Effective." July 27, 2006. Available online. URL: http://www.nytimes.com/2006/07/27/health/ 27vaccine.html?ei=5088&en=4ba414387e7173fa &ex=1311652800&adxnnl=1&partner=rssnyt& emc=rss&adxnnlx=1183993326-MiXoVQMM kJwHl416atc0KA. Accessed on July 9, 2007. Article about vaccine for bird flu developed in 2006 for mice but not humans.

———. "More Western Drought but With a Twist." March 21, 2006. Available online. URL: http:// www.nytimes.com/2006/03/21/national/ 21drought.html?ex=1300597200&en=0f453108

20d10e8a&ei=5088&partner=rssnyt&emc=rss. Accessed on July 9, 2007. Article about the prevalence of wildfires caused by the extensive 2006 southwestern U.S. drought.

———. "New Bird Flu Cases in Turkey Put Europe on 'High Alert.'" January 7, 2006. Available online. URL: http://www.nytimes.com/2006/01/07/inter national/europe/07turkey.html?ex=1294290000& en=ee9905f94034ca21&ei=5088&partner=rssnyt &emc=rss. Accessed on July 9, 2007. News story regarding the spread of bird flu to Turkey in January of 2006.

———. "Number Overstated for Storm Evacuees in Hotels." October 19, 2007. Available online. URL: http://www.nytimes.com/2005/10/19/national/ nationalspecial/19housing.html?ex=1287374400 &en=5c5363a6a3fac7cb&ei=5090&partner=rssus erland&em c=rss. Accessed on July 9, 2007. Article about statistical overestimates following Hurricane Katrina.

———. "Old Ways of Life Are Fading as the Arctic Thaws." October 20, 2005. Available online. URL: http://www.nytimes.com/2005/10/20/science/ earth/20arctic.ready.html?ex=1287460800&en=c 77ac003845eeb1c&ei=5090&partner=rssuserland &emc=rss. Accessed on July 9, 2007. Fascinating story of the change in Eskimo life in the Arctic as the climate warms.

———. "On the Front: A Pandemic Is Worrisome but 'Unlikely'" March 28, 2006. Available online. URL:http://www.nytimes.com/2006/03/28/health/ 28skep.html?ex=1301202000&en=c2e7fe91908 b6dea&ei=5089&partner=rssyahoo&emc=rss. Accessed on July 27, 2007. Article calming the fears of a world pandemic of bird flu.

———. "Poverty and Superstition Hinder Drive to Block Bird Flu at Source." November 3, 2005. Available online. URL: http://www.nytimes.com/ 2005/11/03/international/asia/03bird.html?ex= 1288674000&en=d4532895897ec3d1&ei=5088& partner=rssnyt&emc=rss. Accessed on July 9, 2007. Intriguing article that traces the power of superstition and the presence of poverty in the regions of China in which bird flu began.

———. "Quake Adds to Fear That Indonesian Volcano Will Erupt." June 6, 2006. Available online. URL: http://www.nytimes.com/2006/06/05/world/ asia/05merapi.html?ex=1307160000&en=00ae03 ffb282fab4&ei=5088&partner=rssnyt&emc=rss. Accessed on July 9, 2007. A summary article about the aftermath of the 2005 tsunami disasters in Asia.

———. "Storm and Crisis: New Orleans—Residents Find Both Normality & Ruin." October 6, 2005.

Available online. URL: http://select.nytimes.com/ search/restricted/article?res=F30E11FF35540C758 CDDA90994DD404482. Accessed on July 9, 2007. News story about the conditions found by returning homeowners in New Orleans after Katrina.

———. "Studies Suggest Pandemic Isn't Imminent." March 23, 2006. Available online. URL: http:// query.nytimes.com/gst/fullpage.html?sec=health& res=9802E6D91630F930A15750C0A9609C8B63 &n=Top%2FNews%2FHealth%2FDiseases%2C+ Conditions%2 C+and+Health+Topics%2Fviruses. Accessed on July 9, 2007. An article summing up the conclusions of studies that downplay the possibility of a bird flu pandemic.

———. "Tsunami's Legacy: Extraordinary Giving & Unending Strife." December 25, 2005. Available online. URL: http://www.nytimes.com/2005/12/25/ international/asia/25tsunami.html?ei=5090&en=3 f7b202d14527334&ex=1293166800&partner=rs suserland&emc=rss&pagewanted=print. Accessed on July 9, 2007. Heartrending article capturing the devastation and the human reactions to it following the 2005 tsunami in Asia.

———. "Turks Were Slow to Respond to Reports of Bird Flu, Residents Say." January 10, 2006. Available online. URL: http://www.nytimes.com/ 2006/01/10/international/europe/10flu.html?ex=12 94549200&en=fcd0e729dfda9a63&ei=5088&par tner=rssnyt&emc=rss. Accessed on July 9, 2007. An analysis of reasons for the rapid spread of bird flu in Turkey in 2006.

———. "U.S. Not Ready for Deadly Flu, Bush Plan Shows." October 8, 2005. Available online. URL: http://query.nytimes.com/gst/fullpage.html? sec=health&res=9E07E7DB1F30F93BA35753C1A 9639C8B63. Accessed on July 9, 2007. A cautionary article analyzing the shortcomings of U.S. preparedness for an onslaught of bird flu.

———. "Why the Chicken Virus Crossed the Border." November 22, 2005. Available online. URL: http://www.nytimes.com/2005/11/22/opinion/ 22hasan.html?ex=1184126400&en=12716a 9907420c7d&ei=5070. Accessed on July 9, 2007. Intriguing article explaining the passage of bird flu from country to country.

———. "With Every Epidemic, Health Officials Face Tough Choices." March 28, 2006. Available online. URL: http://www.nytimes.com/2006/03/ 28/health/28docs.html?ex=1184126400&en=14d c5f 610143bcef&ei=5070. Accessed on July 9, 2007. An outline of the major problems facing health officials worldwide in the stemming of epidemics.

———. "A Worrisome New Front." February 12, 2006. Available online. URL: http://www.nytimes. com/2006/02/12/international/africa/12assess.htm l?ex=1184126400&en=906734eb5969d4e6&ei=5 070. Accessed on July 9, 2007. Article dealing with new outbreaks of AIDS in Africa in 2006.

Voice of America. "Many Children Among Dead in Flooding in Northeast China." June 12, 2005. Available online. URL: http://reliefweb.int/rw/RWB. NSF/db900SID/VBOL-6DBDD8?OpenDocument. Accessed on July 27, 2007. Sad article detailing the deaths of children among the 1,000 drowned in the China floods of the summer of 2005.

World Socialist Web Site. "Studies link global warming with increased hurricane intensity." September 13, 2005. Available online. URL: http://wsws. org/articles/2005/sep2005/warm-s13_prn.shtml. Accessed on July 27, 2007. Arresting article that details the effect of global warming on the intensity of hurricanes.

Your Guide To Geology. "The Sumatra Earthquake of 26 December 2004." Available online. URL: geology.about.com/od/historicearthquakes/a/aasu matra-p.htm. Accessed on December 27, 2005. A detailed account of the December 2004 Sumatra earthquake.

INDEX

A

Abaco, hurricane in 285
Abdi, Omar 174
Aberfan (Wales), landslides in 22–23
Abn-Dar (Iran), earthquake in 62
Acapulco (Mexico), hurricanes in 270–271
Aceh (Indonesia), tsunamis in *C-3*, 58, 59
Achenes (Indonesia), tsunamis in 59–60
acquired immune deficiency syndrome. *See* AIDS (acquired immune deficiency syndrome)
Action Against Hunger (AAH) 114
Acworth (Georgia), tornado in 352
Adams, Paul 96
Adana (Turkey), earthquake in 92
Adapazari (Turkey), earthquakes in 95, 96
Adda (Italy), flooding in 171
Afghanistan
    avalanches in 6
    earthquakes in *C-2*, 34–37
    flooding in 174
Africa. *See also specific countries in*
    AIDS in 224–225, 226, 227, 228
    avian flu in 223
    civil wars in sub-Sahara region of 110
    earthquakes in 37, 91
    encroachment of Sahara Desert in 111
    famines and droughts in xii, 110–116, *112, 113*
    flooding in 145–147
    as origin of many hurricanes 269
African Union, in fighting famines and droughts 115
aftershocks 35–36, 37, 45, 54, 62, 86
Agadir (Morocco), earthquake in 77

Agency for Technical Cooperation and Development (ACTED) 35
Agricultural Marketing Corporation 112
agriculture. *See also* famines and droughts
    flooding and 146, 151, 154, 156, 160, 164, 169, 171, 175, 178
    hurricane damage to 253–254, 255, 258, 259–260, 261, 262, 264, 267, 268, 271, 275, 281, . 282, 283, 287, 298, 303, 306, 308, 310
    snowstorms and 326
Agriculture, U.S. Department of
    ban on downer cattle 218
    ban on selling of CJD testing kits 218
Aguadilla (Puerto Rico), hurricane in 276
Ahmadabad (India), earthquakes in 52, 53
Ahmadullah, Mohammed 151–152
Ahmed, Amira 48
Ahmed, Reazuddin 339
AIDS (acquired immune deficiency syndrome) xii, 208, 224–229
Ain Fekan (Algeria), earthquake in 38
air inversion, fog and 362
Akbal, Celil 93
Akita Province (Japan), typhoon in 375
Akopyan, Aikaz 41
Akukwe (China) 228
Alabama
    hurricanes in 259, 278–280, 285, 299–300
    tornadoes in 340, 345, 352–353
Alagret, Jose 78
Alancho (Nicaragua), volcanic eruption at 414
Alaska
    influenza epidemic in 221
    landslides in 22

Albany (New York), snowstorm in 332
Albay (Philippines) 86
Alberta (Canada), mad cow disease in 217
Albion (Pennsylvania), tornado in 350
Alcoa xv
Aleman, Arnoldo 265, 266
Alexandria (Egypt)
    earthquake in 47
    plague in 215
Alexandria (Louisiana), hurricane in 313
Alexandria (Virginia), hurricane in 285
Algeria, earthquakes in 37–38
Allegheny River, flooding of 193
alligators, impact of hurricanes on 278
Alpine Fault (New Zealand) 33
Alps
    avalanches in xi
    Austrian 8, 20
    French 9–10
    Italian 6–7, 20
    Swiss 5, 19, 20–21
    Tyrolean 7, 10
    Dolomite 172
    windstorm in French 364
Alvarado, Juan Velasco 84
Ambato (Ecuador), earthquake in 46
American Enterprise Institute xv
American Red Cross 93
Americans with Disabilities Act (1990) 227
Amesbury (Massachusetts), hurricane in 301
Amin, Mohammed 326
Amnesty International 115
Andale (Kansas), tornado in 345
Andaman Island (Indonesia) 58
Andaman Sea, earthquake in 58

Anderson (South Carolina), tornado in 348

Andes Mountains, earthquakes in 82–84, *83*

Andhra Pradesh (India), cyclones in 240

Andros Island, hurricane in 290

Anegada Passage (British Virgin Islands) 406

Angarain (Afghanistan) 36

Angola, famines and droughts in 114

Angol (Chile), earthquake in 42

Anhui (Anhui) (China), typhoon in 373

Anhwei (Anhui) Province (China)
    famines and droughts in 118
    flooding in 155, 156, 159
    typhoon in 373

Anjar (India), earthquakes in 52, 53

Anjer (Java) 401
    volcanic eruption in 402

Anoud (Chile), tsunami in 42

Antalaha (Madagascar), flooding in 146

Antananarivo (Madagascar), cyclones in 146

anticyclones 287

Antigua
    flooding in 307
    hurricanes in 259, 307, 314
    volcanic eruption in 412

Antigua (Guatemala), earthquake in 51

Antilles, hurricane in 269

Antimano (Venezuela), earthquake in 103

antiretrovirals 228

anti-Semitism, search for scapegoats in fueling 210

antiviral drugs 224

Antler (Oklahoma), tornado in 351

Anwar, Megda 48

Anzob Mountain Pass (Tajikistan), avalanche in 21

Apalachicola (Florida), hurricanes in 296, 300

Appian Way 37

Aquino, Corazon 85

Arabian tectonic plate 62

Aragatsotn (Armenia), famines and droughts in 117

Arakawa River (Japan), flooding of 375

Arawak Indians 275

Arayat (Philippines) 86

Archidamus 213

Arctic ice pack, melting of xiii

Ardakul (Iran), earthquake in 61

Ardebil (Iran), earthquake in 63

Arequipa (Peru), earthquake in 91

Arg-e-Bam 61

Argentina, influenza epidemic in 221

Aristide, Jean-Bertrand 161

Arkadelphia (Arkansas), tornado in 193

Arkansas, tornadoes in 345, 350–353

Arkansas City (Arkansas), flooding in 190

Arkansas River (U.S.), flooding of 179, 190

Arlberg Pass (Austria), avalanches in 7

Armenia
    earthquakes in 38–41, *39, 40,* 45, 46, 90
    famines and droughts in 116–117

Army Corps of Engineers 144, 182, 183, 185, 193, 292

Arno, River (Italy), flooding of 169

Arop (Papua New Guinea), earthquake in 81

Arosa (Switzerland), avalanches in 20

Asama (Japan), volcanic eruption of 74

Asbury Park (New Jersey), hurricane in 302

Ashdod, plague in 219

Ashe, Arthur 226

Asia. *See also specific countries in*
    AIDS in 226
    avian flu in 222
    famine and drought in xii, 107
    flooding in 147–153

Al Asnam (Algeria), earthquake in 37

Asquith, Michael 132

Assam (India)
    earthquakes in 51–52
    flooding in 168, 169
    landslides in 12

Asti (Italy), flooding in 171

Atchafalaya levee (Louisiana), flooding of 190

Ates, Mustafa 95

Atitlán, Lake (Guatemala) 51

Atlantic Basin, hurricane formation in 271–272

Atlantic City (New Jersey), hurricane in 302

Atlantic Multidecadal Oscillation (AOM) 252

Atlantic (Pennsylvania), tornado in 350

Atlantis, disappearance of 388, 389

Atrio de Cavallo (Italy), volcanic eruption in 399

Auburn (Mississippi), tornado in 352

Audubon, John James 101

Augusta (Italy), earthquake in 65

Augusta (New York), snowstorm in 330

Australia
    drought in xii
    signing of Kyoto Protocol xvi
    as source of aid 78, 82

Austria
    avalanches in 7–8, 20
    typhus epidemic in 216

Austrian Alps, alvanches in 8, 20

avalanches 3–33
    conditions triggering 5, 7, 23
    death toll from 7, 8, 10, 12, 13, 15, 16, 18, 19, 21, 22–23
    earthquakes as trigger for 14–15, 16–17, 18–19
    fires from 19
    floods and 12
    glaciers as trigger for 17, 20
    mass burials in 7, 9, 23
    monsoon rains in triggering 12
    rescue efforts in 6, 7, 10–11, 12, 13, 15, 16, 18, 23
    torrential rain as trigger for 8–9, 11, 12–13, 15
    velocity of 5, 12, 22
    volcanic eruptions as trigger for 22
    warnings of 9–10
    winds in 5, 20
    worst recorded *3–4t*

Avcilar (Turkey), earthquake in 95

avian flu 221–224

Avignon (France), Black Death in 210

Ávila Mountains (Venezuela) 199–200

Avisio River (Italy), flooding of 172

Aydin, Deniz 96

Aykroyd, W. R. 121, 126

Azerbaijan, avian flu in 223

Azores 385

AZT 226

B

"Baat Wol is my starving black kid" 116

Bacacay (Philippines), destruction of 417

Bacas, Ben 379

Backergunge (India), cyclone in 240–241

Baden-Wurttemberg (Germany), windstorm in 364

Badiao (Philippines), destruction of 416–417

Bagbara (Italy), earthquake in 65

Baguio (Peru), earthquake in 85

Bahamas
cholera in 220
hurricanes in 280–282, 285, 302

Baird, John xvi

Balakot (Pakistan), earthquake in 80

Balbus, Gaius Quinctius 393

Balfour, Frederick H. 118

Bali, United Nations Climate Change Conference in xvi

Ballin, Robert 403

Baltimore (Maryland), snowstorm in 332

Baluchistan (Pakistan)
earthquake in 79
flooding in 174

Balusutippa (India), cyclone in 240

Balvano (Italy) 67

Bam (Iran), earthquakes in C-4, 59, 60–61

Bangkok (Thailand)
avian flu in 222
monsoon in 363

Bangladesh xi. *See also* East Pakistan
cyclones in 235–240, *236, 237*
flooding in *C-6,* 147–153, *149,* 167
tornadoes in 339–340

Bangladore (India), earthquake in 56

Barbados, hurricanes in 252–253, 262, 273

Barbuda, hurricane in 307

Barcelona (Spain)
earthquake in 45
flooding in 175–176

Bare Beach (Florida), hurricane in 296

Baroda (India), famines and droughts in 124

Barrie (Ontario), tornado in 349

Barrymore, John 99–100

Barton, Clara 198

Baruta (Venezuela), earthquake in 103

Bashail (Bangladesh), tornado in 339

al-Bashir, Omar Hassan 114–115

Bassac River (Cambodia), flooding of 149

Batabano (Florida), hurricane in 258

Batagram (Kashmir), earthquake in 79

Batavia, volcanic eruption and 401, 402

Baton Rouge (Louisiana), flooding in 188

Batticaola 59

Bavaria, windstorm in 364

Bay Islands, hurricane in 265

Bayou La Batre (Lousiana), hurricane in 279

Bay St. Louis (Mississippi), hurricane in 279

Bayuyungan (Philippines) 419

Bayview Bridge (Illinois) 186

Beame, Abraham 285

Beaumont–Port Arthur (Texas), hurricane in 313

Bedi, Pooja (actress) 13

Bedi, Protima (dancer) 13

Beebe, Art 353

Beggars of the Sea 264

Begum, Feroza 79

Begun, Kanakfool 148

Behar (India), famines and droughts in 124

Beira (Mozambique), flooding in 145

Belgium
mad cow disease in 217
snowstorm in 326
as source of aid 179
windstorm in 364

Belize, hurricanes in 253, 268–270, 274

Belize City (Belize), hurricane in 270

Belle Glade (Florida), hurricane in 296

Belluno (Italy), landslides in 14–15

Belouizdad (Algeria), earthquake in 38

Benares (India), flooding in 166

Bengal (India)
famines and droughts in 57, 124, 125–126
flooding in *165,* 165–166

Bengal, Bay of 240
cyclones in 241
flooding and 150, 153
salinity intrusion from 150

Bennettsville (South Carolina), tornado in 348

Benoit (Mississippi), flooding in 188

Berly (Haiti), hurricane in 264, 287

Bernard, Alma 172

Bespalenko, Olga 89

Bethai (Philippines), typhoon in 378

Beth-shemesh, plague in 219

Beulah (Mississippi), flooding in 188

Bevan, H. R. 372

Bezymianny (USSR), volcanic eruptions in 423

Bhachau (India), earthquake in 53, 54

Bhat, Rajesh 53

Bhola (Bangladesh)
cyclones in 235, 237, 238
flooding in 152

Bhubaneshwar (India), cyclone in 243

Bhuj (India), earthquakes in 52, 53–54

Bible
flooding in 200–201
Great Deluge and Flood in 200–201
Great Famine in 120
parting of Red Sea in 388, 389–390
plague in 219

Biblioteca Nazionale Centrale (Italy) 170

Big Blue River (Kansas), flooding of 182

Bigelow, Frank H. 346

Big Thompson Canyon (Colorado), flooding in 180–181

Big Thompson River (Colorado), flooding of 180

Bihar (India)
flooding in 147, 168, 169
landslides in 12

Billings, L. G. 47

Biloxi (Mississippi), hurricanes in 260, 279, 291, 292, 293, 300

biological warfare 209

bird flu 221–224
vaccination of birds in C-7

Birjand (Iran), earthquake in 61

Bischoff, E. 391–392

Bismarck (North Dakota), coldness in 331

Black Death xi, 110, 207, 208–211, 212, 220, 228

The Black Death 1347 (Deaux) 210

The Black Death (Nohl) 209

The Black Prophet (Carleton) 128

Black River (U.S.), flooding of 184

Blackwell (Oklahoma), tornado in 342

Blanchard, Stephen 310

Blanchard River (Ohio), flooding of 187

Blanche River (Martinique) 407

Blandin (Venezuela), flooding in 200

blizzards 22, 325. *See also* icestorms; snowstorms
  of 1888 324, 329, 332
  of 1947 332
  of 1988 330–331, *331*

Block Island (Rhode Island), hurricane in 302

Blons (Austria), avalanches in 7

Bluefields (Nicaragua), hurricane in 256

Boca del Río (Veracruz), hurricane in 263

Boccaccio, Giovanni 209

Boer, Yvo de xvi

Bogatá (Colombia) 9, 46

Boggy (Oklahoma), tornado in 351

Bohai, China. *See* Po Hai (Bohai) (China), flooding of

Boice, Alicia 10

Bolívar, Simón 103

Bolsena (Italy), earthquakes in 94

Bolu (Turkey), earthquake in 93

Bombay (India), flooding in 167

Bonaparte, Napoleon 110, 216, 317

Bonnell, Ralph 286

Booneville (Mississippi), tornado in 352

Boothville (Louisiana), hurricane in 313

Borge, Tomas 257

Borgne, Lake (Louisiana), flooding of 194, 299

Boston (Massachusetts)
  hurricanes in 282, 301
  influenza epidemic in 221
  snowstorm in 330, 332

Botswana
  AIDS in 227
  flooding in 146

Bou Henni (Algeria), earthquake in 38

Bouin (Iran), earthquake in 62

Bound Brook (New Jersey), hurricanes in 286–287

Bourke, Gerald 131

Bourmedes (Algeria), earthquake in 38

Bouteflika, Abdelaziz 38

Boutros, Shahata 48

bovine spongiform encephalopaty (BSE) 216–218

Boyars, Arthur 122

Boyars, Marion 122

BP America xv

Bracker, Milton 400

Brahmaputra River (India) 237
  flooding of 147, 148, 151, 153, 168
  landslides into 52

Bratico (Italy), earthquake in 65

Brazil
  AIDS in 226
  avalanches in 8–9
  cholera in 220
  flooding in 153–154

Brescia (Italy), flooding in 171

Brethren of the Cross 210

Brezhnev, Leonid 39, 40–41

Bridge Creek (Oklahoma), tornado in 351

Bristol (England), hurricane in 261

Britanny (France), windstorm in 364

British Honduras 414
  hurricane in 253

British Virgin Islands, hurricanes in 253, 254, 307

Bronson, William 99

Broussard, Aaron 293

Brown, Michael 293

Brown County (Ohio), flooding in 194

Browning, Elizabeth Barrett 169

Browning, Robert 210–211

Brownsville (Texas), hurricanes in 253, 254, 255, 310

Brunei, AIDS in 226

Brunswick (Georgia), hurricane in 286

Brush (Colorado), flooding in 180

BSE. *See* bovine spongiform encephalopaty (BSE)

buboes 207

bubonic plague 207, 208, 218

Buchanan, Patrick 225

Buchanan, Tom 310

Bucharest (Romania), earthquakes in 88–89

Bugaan (Philippines), destruction of 420

Bulet, Siena 96

Bullard, Fred M. 413

Bun-hee, Kim 376

Burghsluis (Holland), flooding in 163

Burgos, Marco 274, 275

Burundi, famines and droughts in 114

Bush, George H. W. 281, 298, 308

Bush, George W. 227, 292–293

Butler County (Pennsylvannia), tornado in 350

Buyuk (Turkan) 93

Buzuluk (Russia), famine and drought in 132

C

Cabanatuan (Philippines), earthquake in 84

Cabo de la Cruz (Cuba), hurricane in 273

Cabora Bassa (Madagascar), flooding and 146

Cadereyta (Mexico), hurricane in 255

Caffa, Black Death in 209

Cagsauga (Philippines), destruction of 416–417

Cairo (Egypt), earthquakes in 47–49

Cairo (Illinois), flooding in 188, 189, 193

Cajamarca (Colombia), earthquake in 46

Calabia, Arnel 85

Calabria (Italy), earthquakes in 64–65, *65*

Calais, Black Death in 209

Calarca (Columbia), earthquake in 45

Calcutta (India)
  cyclone in 241
  famine and droughts in 126
  flooding in 166

caldera 385, 389

Cali 46

California
  bubonic plague in 218
  earthquakes in 97–101, *98*
  flooding in 179, 195–196
  mudslides in 21
  rock slides in 196
  snowstorms in 327–328

Calmeyer (Java), volcanic eruption in 402

Calvados (France), windstorm in 364

Camalig (Philippines), destruction of 416

Cambodia
  AIDS in 226
  avian flu in 223
  flooding in 148–151

Cambridgeport (Massachusetts), hurricane in 301

Cameron (Louisiana), hurricane in 311

Campache, Bay of (Mexico), hurricane in 270

Camporotondo (Italy), volcanic eruption in 397

Camp Rapid (South Dakota), flooding in 199

Canada. *See also specific provinces in*
avian flu in 223
mad cow disease in 217
SARS in 229, 230
Canal Point (Florida), hurricane in 296
Cancún (Mexico), hurricanes in 254, 269, 272
cannibalism
earthquakes and 42
famines and droughts and 119–120, 121, 125, 127–128
snowstorms and 327–328
Cannitellla (Italy), earthquake in 65
Canta (Peru) 17
Canton (China), plague in 214
Cape Ann (Massachusetts), hurricane in 301
Cape Breton (Canada), hurricane in 275
Cape Canaveral (Florida), hurricane in 286, 300
Cape Cod (Massachusetts) 258
hurricane in 301
Cape Fear (North Carolina), hurricanes in 286, 287, 305, 306
Cape Gracias a Dios 256
Cape Verde Islands, famines and droughts in 111
Caporetto (Italy) 14
Cara, Maria Assunta 172
Caracas (Venezuela)
earthquake in 103
flooding in 200
carbon dioxide levels xiii
CARE (Co-operation for American Relief Everywhere) 110, 152, 166, 238
Carey, Mathew 219
Caribbean. *See also specific countries in*
hurricanes in 253–257, 268–270
Carib Indians 211
Carleton, William 128
Carlyle, Thomas 127
Carnegie, Andrew 196
Carolina Beach (North Carolina), hurricanes in 287, 306
Caroline Islands, famines and droughts in 128
Carruthers, Bill 282
Caruso, Enrico 99, 100
Carvile, William 129
Casano (Italy), earthquake in 65
Cascade Range 421
avalanches in 22
snowstorms in 333
volcanoes in 385, 386

Caspian Sea, earthquakes in 62–63
Casso (Italy), landslide in 14
Cassville (Missouri), flooding in 185
Castas (Nicaragua), mudslides in 266
Castellianos, Cesar 266
Castro, Fidel 258
Castroreale (Italy), earthquake in 65
cataclysm 23, 44, 86
in Cuba 258
of 1921–23 132
as trigger for avalanche 23
Catania (Italy) xi
earthquake in 65
volcanic eruption in 392, 393, 397
Caterpillar Inc. xv
Cates, Ward 224
Catholic Relief Societies 110
Cat Island, hurricane in 285
Catledge, Turner 190
Caucasus, famines and droughts in 116–117
Caulker Cays (Honduras), hurricane in 265
Cayey (Puerto Rico), hurricane in 259
Cayo Coco (Cuba), hurricane in 260
Cebu (Philippines), typhoon in 379
Cedar Key (Florida), hurricane in 300
Cehyan (Turkey), earthquake in 92
Ceiba (Puerto Rico), hurricane in 276
Celebes 404
Celebes Sea, earthquake in 86
Center for Global Development xv
Centers for Disease Control
on AIDS 226
on SARS 229
Central Africa, AIDS in 224–225
Central African Republic, famines and droughts in 115
Central America. *See also specific countries in*
decreased rain level in xii
earthquakes in 90–91
hurricanes in 262–263, 265–267
Cerro de Pasco (Peru) 17
cervix, invasive cancer of 226
Cesaroni, Romildo 170
Ceyhan River (Turkey), flood of 326
Chad, famines and droughts in 110, 111, 115
Chains of Exmoor (England) 161
Chakraborty, Atanu 53
Chambers, Jermaine 281
Chamoli (India), earthquake in 56
Chamonix (France), avalanches in 10–11

Champerico (Guatemala), volcanic eruption in 51
Chandpur (Bangladesh)
cyclone in 238
flooding in 147
Ch'angchi (Changji) (China), snowstorm in 325
Chaqualane (Mozambique) 145
Charenton (Louisiana), hurricane in 281
"Charge of the Light Brigade" (Tennyson) 127
Charikar (Afghanistan) 6
Charleston (Illinois), tornado in 345
Charleston (Massachusetts), hurricanes in 300, 305
Charleston (South Carolina)
earthquakes in 101, *102*, 102–103
hurricanes in 284, 307, 308–309, 310
tornado in 353
Charles V (Holy Roman Emperor and [as Charles I] king of Spain) 163
Charles VIII (King of France) 211
Charlotte (North Carolina), hurricane in 309, 310
Chartres (France), icestorms in 323
Chateaubelair (West Indies), earthquake in 424
Chatham (Massachusetts), hurricane in 301
Chauliac, Guy de 210
Chávez, Hugo 199
Chazov, Yevgen I. 39
Chekiang (Zhejiang) Province (China)
flooding in 160
typhoon in 372, 373
Chelvi Dam (Pakistan), bursting of 174
Chencha, earthquake in 91
Chengchow (Chengzhou) (China)
famines and droughts in 119
flooding in 154–155
Ch'eng-tu (Chengdu) (China), flooding in 158
Cherry Grove (South Carolina), hurricane in 287
Chertoff, Michael 292
Chesner, Craig 387
Chetumal (Mexico), hurricanes in 256, 269
Chezaud, David 363
Chiang Kai-shek 118
Chiaramonto (Italy), earthquake in 65
Chiatura (Russia), earthquake in 90

Chicago (Illinois)
  flooding in 186–187
  tornadoes in 341
Chi-Chi (Taiwan), earthquake in 91
Chickasha (Oklahoma), tornado in 351
Chien Chen-ying 157
Chifenerhshan (Taiwan), earthquake in 91
Chigasaki (Japan), earthquake in 75
Ch'ihfeng (Chifeng) (China), snowstorm in 325
Chikwe (Mozambique), flooding in 145
Child of the Century (Hecht) 192
Chile
  earthquakes in 41–43, *43*
  tsunamis in 42
  volcanic eruptions in 42
Chiles, Lawton 281
Chillan (Chile), earthquake in 42
Chillicothe (Ohio), flooding in 192
Chimborazo (Ecuador), eruption of 47
Chimbote (Peru), earthquake in 83
China
  avian flu in 222
  Black Death in 209
  cholera in 220
  earthquakes in 43–45
  famines and droughts in xii, xiv, 107, 117–120
  flooding in xiii, 147, 154–160
  plague in 214
  SARS in 229, 230
  snowstorms in 325–326
  typhoons in 371–373, *372*, 379–380
Ch'inchow (Qinzhou), typhoon in 378
Ch'ing-hai (Qinghai) Province (China), snowstorm in 325–326
Chirac, Jacques xv–xvi, 365
Chittagong (Bangladesh)
  cyclones in 238, 239
  famine and droughts in 126
  flooding in 148
cholera
  cyclones and 236, 240
  as epidemic 207, 212, 219–220
  famines and droughts and 124, 127
  flooding and 146, 148, 150, 152, 155, 165, 167, 177
  hurricanes and 266, 267, 271
  volcanic eruptions and 405
  as water-borne disease 207

Choluteca (Honduras)
  flooding in 265, 266
  mudslides in 265
Chor Shibola (Bangladesh), flooding in 153
Choudhury, Hamayun Rasheed 152
Christian Aid 110
Christiansted (St. Croix), hurricane in 308
Ch'uisan (Quisan) (China), flooding in 155
Chungar (Peru), avalanche in 16–17
Ch'ung/Ch'ing (Chongqing) Province (China), flooding in 147
Chungking (Chongqing) (China), flooding in 158
Chungmow (China) 154–155
Chuuk, famines and droughts in 128, 129
Ciego de Ávila (Cuba), hurricane in 260
Cimarron River, flooding of 186
Cimen, Ismail 96
Cinarcik (Turkey), earthquakes in 95, 96
Cincinnati (Ohio) 102
  flooding in 193, 194
Ciparis, Auguste 409
Circleville (Ohio), flooding in 192
Citadel of Bam 61
Ciudad Serdan (Mexico) 75
Clark, William 421
Clark Air Force Base (Philippines) 419
  volcanic eruption at 417, 418, 419
Clean Air Act (Great Britain,1956) 361–362
Clement VI (pope) 210
Cline, Isaac 314–315, 316
Cline, Joseph 314–315
Clinton, Bill 93, 134, 184, 226–227
Coast Guard 255, 290
Coatzacoalcos (Veracruz), hurricane in 263
Cocoa Beach (Florida), hurricane in 283
Coffeyville (Mississippi), tornado in 352
Coihueco (Chile), earthquake in 42
Col de la Traversette (Italy) 6–7
Colombia
  earthquakes in 45–46
  hurricane in 256
  landslides in 46
  mudslides in 9

Colonial Beach (South Carolina), hurricane in 287
Colorado
  flooding in 179–182, *180*
  snowstorms in 328
Colorado River, flooding of 144
Columbia (Arkansas), tornado in 352
Columbian School 211
Columbia River (Washington) 422
Columbia River Gorge (Oregon), snowstorm in 333
Columbia University's Lamont–Doherty Earth Observatory 56
Columbus, Christopher 316–317
Columbus (Ohio), flooding in 192
Comayagua (Honduras), flooding in 265
Comerio (Puerto Rico), hurricane in 309
communication, impact of hurricanes on 254, 261, 266, 286, 290, 297, 299
Concepción (Chile), earthquake in 42
Concord (California), bubonic plague in 218
Concord River (Massachusetts), flooding of 282
Condado Beach (Puerto Rico), hurricane in 308
Conemaugh River (Pennsylvania), flooding of 196
Congo, Democratic Republic of the
  famines and droughts in 114
  flooding in 145
Conidoni (Italy), earthquake in 65
Connecticut
  flooding in 182
  hurricanes in 284, 301
  snowstorms in 328–329, 332–333
The Conquest of Famine (Aykroyd) 121, 126
Consing, Ralph 86
Constantinople, plague in 215
Contai (India), cyclone in 241
contaminated water theory 220
Convey, Patrick 347
Conza Della Compagna (Italy), earthquake in 67
Cooch Behar (India), flooding in 150
Co-operation for American Relief Everywhere (CARE) 110, 152, 166, 238
Copacabana (Brazil), avalanches in 8
Coralville Lake (Iowa), flooding of 185

Cordele (Georgia), tornado in 352
Córdoba (Colombia), earthquake in 45
Coringa (India), cyclones in 241–242
Corinth (Greece), earthquake in 50
Coriolis effect 235
Corn Laws 127
Coronal (Chile), earthquake in 42
corona virus 229–230
Corozal (Honduras), hurricane in 256
Corpus Christi (Texas), hurricanes in 297
Cortés (Costa Rica), flooding of 257
Cortines, Adolfo Ruiz 271
Coseguina (Nicaragua), volcanic eruption in 414–415
Cosenza (Italy), earthquake in 65
Costa Brava (Spain), flooding in 176
Costa Rica, hurricanes in 256–257, 262
Costello, Jose 76
Cotopaxi (Ecuador), eruption of 47
Cowes (England), hurricane in 261
Cowlitz River (Washington) 422
Cox's Bazar (Bangladesh), cyclone in 238, 239
Coyne, Andre 162
Cozumel (Mexico), hurricanes in 254, 269, 272
Crapp, Austin 82
crater lakes 385
Creole (Louisiana), hurricane in 311
Crescent Beach (South Carolina), hurricane in 287
Crete, tsunamis in 389
Creutzfeldt, Hans Gerhard 216
Creutzfeldt-Jacob disease (CJD) 216–218
crime
    earthquakes and 46, 79
    famines and droughts and 117, 119, 121
    floods and 148, 155, 178
    hurricane damage and 281, 292, 298, 305, 308, 315
    typhoons and 376
Critias 389
crocodiles, flooding and 150
Crueger, Nicholas 276
Cuba
    cholera in 220
    hurricanes in 253, 255, 257–261, 258, 268, 271, 272, 288, 299, 316
Cucuta (Colombia)
    earthquake in 45
    volcanic eruption in 45

Cudjoe Key (Florida) 260
Culebra (Puerto Rico), hurricanes in 307, 309
culling 224
Cumberland River (U.S.), flooding of 189
Cumin, Gilbert 11
Cunningham, General James 252
Curuan Zamboang (Philippines) 86
Cuyoenango (Guatemala), volcanic eruption in 51
cyclones 107, 233–243. *See also* hurricanes
    causes of 235
    cholera following 236, 240
    contrast between tornadoes and 339
    crimes after 239
    defined 235
    floods from 145, 146, 242
    middle latitude 235
    tidal waves following 235, 236, 241–242
    tropical 235
    wind rotation in 235
    worst recorded 233t
cyclones by name
    Eline 145, 146
    Gloria 146
cyclonic pattern 235
Czechoslovakia, snowstorm in 326

D

Dacca (Bangladesh)
    cholera epidemic in 152
    flooding in 151, 153
Dalby, R. J. 401–402, 403
Dalí, Salvador 175
Dammastock (Switzerland), avalanche on 20
dams 144
    collapse of 167, 172, 174, 179, 182
    hydroelectric 170
    Leonardo da Vinci's plans for 169–170
    in preventing floods 144, 146, 156–157, 182
Danang (Vietnam), flooding in 150
Danan (Java)
    upthrusts of 401
    volcanic eruption in 402
Dance of Death 211
Daniel, Clovis 260

Daniels, Joe 306
Danvers (Massachusetts), hurricane in 301
Daraga (Philippines), destruction of 416
Darbhanga (India), earthquakes in 77–78
Darchula (India) 13
Darfur 110
    famines and droughts in 111, 115
Darjeeling (India)
    avalanches in 12
    flooding in 150
Dasht River (Afghanistan), flooding of 174
Daubenton, François 362
Dauphine Island (Alabama), hurricanes in 279, 300
Davenport (Iowa), flooding in 184, 185
Davidson, Henry A. 221
Davis, Harry R. 303
Davis, James 328
Davis, Richard Harding 99–100
Davisboro (Georgia), tornado in 351–352
Davos (Switzerland), avalanches in 20
Dawlins (England), windstorm in 362
Day, William 112
Daytona Beach (Florida), hurricane in 283
Dayton (Ohio), flooding in 192
Deaux, George 210
Debagram (India), flooding in 148–149
The Decameron (Boccaccio) 209
Decay, Ciceilio 298
Decca Plain (India), famines and droughts in 124
Decker, Robert 386
deforestation 55
    as cause of landslides 13
    flooding and 151, 152
De Ilfrile, Harviva 409
Delaware, hurricanes in 286, 302
Delmarva Peninsula, hurricane in 286
Del Norte County, flooding in 195
DeLorenzo, Michael 309
Deloso, Armor 418
del Sur Province (Philippines), typhoon in 378
Demirel, Suleyman 94
Dene, William 210
dengue fever
    flooding and 146, 176
    hurricanes and 266, 267

Denmark
  Black Death in 209
  mad cow disease in 217
  snowstorm in 326
Dennis (Massachusetts), hurricane in 301
Denver (Colorado)
  flooding in 180, *181*
  snowstorm in 328
Deschutes River (Washington), flooding of 195
Des Moines (Iowa), flooding in 185–186
Des Moines River (Iowa), flooding of 184, 185
DeSoto (Illinois), tornado in 342
Dezutter, Jean Paul 94
Dhaka (Bangladesh)
  flooding in 148
  tornado in 339
Dhal Char (Bangladesh), cyclone in 237
Dharmsala (India), earthquake in 54
Diablo Canyon nuclear power plant 196
diarrhea, flooding and 148, 152, 153, 159, 176
The Dictator (Davis) 99–100
dikes
  breaking of 163
  Communist destruction of 157
  flooding and 160, 163
  hurricane damage to 264
Diomed (Italy) 396
Dionysius of Helicarnassus 215
diphtheria, hurricanes and 282
Diseases of the Madras Family of 1877-79 (Porter) 110
displacement C-2
Disraeli, Benjamin 128
diversions in preventing floods 144
divine cause, attribution of natural disasters to xi, 268
Dndapa (Madagascar), flooding in 146
Doctors Without Borders 37
Dog River (Alabama), hurricane in 261
Dolomite Alps (Italy) 172
Dominguez, Gildardo Castaneda 173
Dominica, hurricanes in 269, 283, 307, 316
Dominican Republic
  flooding in C-6, 160–161
  hurricanes in 254, 255, 259–261, 269, 277, 283

Dong Thap (Vietnam), flooding in 149
Donner, Jacob 327, 328
Donner Lake 327
Donner Party 327–328, 336
Dora Baltea (India), flooding in 171
Dorchester (Massachusetts), hurricane in 301
Dordrecht (Holland), flooding in 164
Dort (Holland), flooding in 164
Dostum, Rashid 6
downer cattle, USDA bans on 217
Dragon Flood (2000) (Vietnam) 148
Drahman, Thomas 152
Drake (Colorado), flooding in 181
Dreisbach, John A. 129
droughts xii, xiii, xiv. *See also* famines and droughts
  El Niño and 129, 130
  invisible 107
  La Niña and 135, 136
  permanent 107
Dry Hapsita River (USSR) 423
Dry Tortugas (Florida), hurricanes in 267, 297
Dubuque (Iowa), flooding in 186
Duby, Helen M. 350
Duke Energy xv
Dumble, Kynette J. 218
Duncan, J. A. 290
Dunsmuir (California) 196
DuPont xv
Durgapur (India), flooding in 166
Dushambe (Tajikistan) 21
dust bowl (U.S.), drought in 133
duststorms 134–135
Duvalier, François 255, 258
Duzce (Turkey), earthquakes in 92–94
dysentery
  flooding and 146, 148, 150, 152
  hurricanes and 282
  in refuge camps 113
  typhoons and 373
Dzhava (Russia), earthquake in 90

E

Eagle, Bill 300
earthquakes 107, 400
  aftershocks from 35–36, 37, 45, 54, 62, 86
  causes of 33
  deaths from 33
  destructive power of 33, 35

epicenters of 33, 34
fires from 78, 88, 100–101
floods from 83–84
intensity of 33, 34
lessons learned from 41
long waves in 33
magnitude of 34
malaria and 36
primary waves in 33–34
secondary waves in 33–34
seismic waves in 33
tremors of 33
in triggering avalanches 14–15, 16–17, 18–19
in triggering landslides 44, 51, 52, 64, 74
in triggering tsunamis xi, C-2, C-3, C-4, 42, 69, 70, *70*, 81–82, 86, 173
undersea 173
worst recorded *27–31t*
Earth's Beginning (Ballin) 403
East Conemaugh (Pennsylvania), flooding in 198
Easter Island 386
Eastern United States, drought in 133–134
East Lyn River (England), flooding of 161
East Nusa Tengara (Indonesia), earthquakes in 58
East Pacific Rise 386
East Pakistan. *See also* Bangladesh
  cyclones in 235–236
East St. Louis (Illinois), tornado in 347
Eboli (Italy), earthquake in 67
Ecuador, earthquakes in 46–47
Edgartown (Massachusetts), hurricane in 302
Edward III (king of England) 323
Edwards, Edwin 300
Egeland, Jan 115
Eguro River (Japan), flooding of 375
Egypt
  earthquakes in 47–48
  famines and droughts in 120–121
Eisenhower, Dwight 326
Ekron, plague in 219
Elbe River, flooding of 162
El Bluff (Nicaragua), hurricane in 256
Elburz Mountains (Iran), earthquake in 62
Eleuthera, hurricane in 285
Elhaut, Tom 152

Eliasson, Jan 110
Elkwood (Alabama), tornado in 352
Ellis, Martin 192
Elmira (New York), hurricane in 285
Elm (Switzerland), avalanche in 19
El Niño xii–xiii, xiv, 386
    droughts and 129, 130
    floods and 130, 147
El Salvador 414
    earthquakes in 94
    hurricanes in 262–263, 275
    volcanic eruptions in 263
El-Sayed, Ahmed Maher 49
El-Sayed, Hoda 48–49
Emanuel, Kerry 252
Emmanuel, Victor 66
Empire of Dust (Svobida) 134–135
encephalitis
    as epidemic 207
    Japanese, flooding and 176
England. *See also* Great Britain;
    United Kingdom
    Black Death in 209
    famines and droughts in 121–
      122
    flooding in 161–162
    fog in 361–362
    hurricane in 261
    influenza epidemic in 221
    plague in 212
    tornadoes in 340
English Channel, windstorm in 362
Environmental Defense xv
epicenter 33
epidemic pestilence 214–215
epidemics. *See also* pandemic; plagues
    and epidemics; *specific infectious*
    diseases
    cholera as 207, 212, 219–220
    current 207–208
    encephalitis as 207
    localized 207
    syphilis as 207, 211–212
    transmission of 207
    yellow fever as 207, 218–219
Epidemics Resulting from Wars
    (Prinzing) 216
Eqtiedaii, Hassan 14
Ercan, Ahmet 97
Eritrea, famines and droughts in 113
Eritrean People's Liberation Front 113
Erzincan (Turkey), earthquakes in
    94–95
Esashi (Japan), earthquakes in 70
Escherichia coli food poisoning,
    floods and 146

Escuintla (Guatemala), earthquake
    in 50
Eskisehir (Turkey), earthquake in 95
Estes Park (Colorado), flooding in
    180, 181
L'Etang Sec crater (Martinique) 407,
    409–410
Ethiopia
    avian flu in 223
    famines and droughts in 111–
      113, *112*, 114
Etna (Italy), volcanic eruption in xi,
    392–393, 396–397
Eurasian plate 62
Europe. *See also specific countries in*
    AIDS in 225
    Black Death in 208–211
    mad cow disease in 216–218
    snowstorms in 326
    syphilis epidemic in 211–212
Evans, David John 23
Everest, Mount, avalanches on 16
Everglades City (Florida), hurricane
    in 272
Exxon Mobil xv
Eyam (England), plague in 212

F

Facusse, Carlos Flores 266
Fae (Italy), landslide in 14
Faini, Mario 15
Faizabad (Afghanistan) 36
Fajardo (Cuba), hurricane in 257
Falmouth (England), hurricane in 261
Falmouth (Kentucky), flooding in 194
Falwell, Jerry 225
famines and droughts 107–136
    cholera and 124, 127
    human factor in 107–108, 121
    kinds of 107
    length of 107
    longevity of 207
    natural causes of 107
    plagues and 208
    political and cultured factors in
      107
    reasons for 107
    scurvy and 127
    wildfires as threat in 136
    worst recorded *107–108t*
Fangcheng (China), typhoon in 378
Farahzad (India), flooding in 169
A Farewell to Arms (Hemingway) 14
Farouk, Nevine 48

Farrell, Thomas 412
Faujdarhat (Bangladesh), crime after
    cyclone in 239
Fayetteville (Arkansas), tornado in
    352
Fayetteville (North Carolina),
    hurricane in 306
Federal Emergency Management
    Agency (FEMA)
    famine and droughts and 129
    hurricane relief and 291–293
Federated States of Micronesia,
    famines and droughts in 128–129
Fernandes, George 168, 243
Fernández, Leonel 259
Feysabad (Afghanistan) 34
Fez (Morocco), earthquake in 87
Fieser, James 190
Findhorn Gorge (Scotland), flooding
    in 175
Findlay (Ohio), flooding in 187
Fiorelli, Giuseppi 395–396
Fire Girdle of the Pacific 414
fires
    from avalanches 19
    brush 112
    from earthquakes 74–75, 78, 88,
      100–101
    Great Fire of London 208, 211,
      212
    hot winds as cause of 112
    in Jamestown 136
    tornadoes and 341, 353
    wild 136
flash floods xi, 144, 145, 180, 185,
    199
Flat Creek (Missouri), flooding in 185
Flemme Valley (Italy), flooding in 172
Flint (Michigan), tornadoes in *343,*
    343–344
floods xiii, 139–201
    avalanches and 12
    causes of 143–144
    cholera and 146, 148, 150, 152,
      155, 165, 167, 177
    in cyclones 145, 146, 242
    diarrhea and 148, 152, 153, 159,
      176
    dysentery and 146, 148, 150,
      152
    El Niño and 130, 147
    flash xi, 144, 145, 180, 185, 199
    following cyclones 242
    following earthquakes in 83–84
    in hurricanes 251, 260
    lack of longevity and 207

landslides and 147, 173
La Niña and 168, 199
malaria and 146, 177
monsoon rains in triggering C-6
mudslides and 199–200
typhoid fever and 148, 152, 176, 177
worst recorded *139–142t*
Florence (Italy)
Black Death in 209
flooding in 169–171
Flores, Teresa 274
Flores Island (Indonesia), earthquakes in 57–58
Florida, hurricanes in 257–258, 270, 271–272, 280–282, 283, 284, 285, 286, 293–297, *294*, 313, 316–317
Florida Everglades, hurricane in 296
Florida Keys, hurricanes in 267, 284, 288–290, *289*, 302
Floridia (Italy), earthquake in 65
Floyd, Thomas 161–162
flu. *See also* influenza
bird 221–224
vaccination of birds in C-7
as pandemic 208, 228
viral, flooding and 176
Flushing (Michigan), tornado in 343
fog
air inversion and 362
London 361–362
Fond-Verrettes (Haiti), flooding in 161
Fonseca, Bay of 414
Fonseca, Gulf of 414
Food and Agriculture Organization (FAO) 110, 111
Food and Drug Administration (FDA), on AIDS therapy 227, 229
Food for the Hungry 130
Formosa, typhoon in 373–374
Fort Crockett (Texas), hurricane in 315
Fort-de-France (Martinique)
hurricane in 268
volcanic eruption in 408
Fort Dodge (Iowa), flooding in 187
Forte, Daniel 10
Fort Lauderdale (Florida), hurricane in 294
Fort Morgan (Colorado), flooding in 180
Fort Myers (Florida), hurricane in 290
Fort Ozama (Dominican Republic) 277

Fort Pierce (Florida), hurricane in 290
Fort Riley (Kansas), influenza epidemic in 220
Fort St. Philip (Louisiana), hurricane in 298
Fort Walton Beach (Florida), hurricanes in 285, 300
Fourth World 107
Fowl Plague 221–222
Fox River (Illinois), flooding of 187
Fracasoro, Girolamo 211, 214
France
AIDS in 225
avalanches in 9–11
Black Death in 210
earthquakes in *49*, 49–50
flooding in 162
icestorms in 323
mad cow disease in 217
St. Vitus's dance in 212
as source of aid 37, 62, 76, 93, 179
typhus epidemic in 216
windstorm in 363–365
Franklin (Mississippi), hurricane in 281
Free Bridge (Arkansas), flooding of 190
Fréjus (France), flooding in 162
French Alps
alvanches in 9–10
windstorm in 364
French Disease theory 211
French Rivera
earthquakes in *49*, 49–50
snowstorm in 326
windstorm in 364
French West Indies, hurricane in 317
Freneau, Philip 267
Frick, Henry Clay 196
Friedman-Kien, Alvin 225
Friends Relief Committee 132
Friesland (Holland), flooding in 163
Friuli (Italy), earthquake in 45
From Earthquake Fire and Flood (Hewitt) 401
Frost, K. Y. 389
frostbite
snowstorms and 326, 327
tetanus and 327
Fukien (Fujian) Province (China)
flooding in 117, 147, 156, *157*, 160
mudslide in 156
typhon in 373
Funston, Frederick 100

G

Gabon, avian flu in 223
Gadomski, Fred 254, 332
Gage, Henry T. 218
Gaggo Dam (Pakistan), bursting of 174
Gainesville (Georgia), tornadoes in 341, 352
Galanopoulos, George A. 388, 389–390
Galdy, Lewis 68
Galle (Sri Lanka), flooding in 176
Gallo, Robert C. 225
Galois, Patrick 365
Galteur (Austria), avalanches in 7–8
Galveston (Texas) 291
elevation of seawall in 316
hurricanes in 254, 297, 310, 313–316, *314*
Gambia
avian flu in 223
famines and droughts in 111
Gamu-Gofa (Ethiopia), famines and droughts in 112
Ganda Chashma (Afghanistan) 36
Ganges Plain (India), famines and droughts in 123–124, 127
Ganges River (India) 241
flooding of 147, 148, 151, 153, 166, 168
Gangneung (Korea), typhoon in 376
Gangwon (Korea), typhoon in 376
Garbuan (India) 13
Garcia, Joe 136
Gaspar, Pablo 263
Gath, plague in 219
Gayssot, Jean-Claude 365
Gaza (Mozambique), flood in 145
Gegharkunik (Armenia), famines and droughts in 117
General Electric xv
Geneva (Switzerland), World AIDS conference in 227
genocide 110
as hidden agenda 113
Geoghegan, John 150
Geophysical Dynamics Laboratory 252
Georges Bank, hurricane in 301
Georgetown (West Indies) 424
Georgia
earthquake in 101
famines and droughts in 116–117

hurricanes in 284, 285, 286, 291, 297, 313
tornadoes in 340, 341, 351–353, *352*
Gerasimov, Gennadi I. 41
Germany. *See also* West Germany
AIDS in 225
Black Death in 209
cholera epidemic in 212
flooding in 162
mad cow disease in 217
snowstorm in 326
as source of aid 93
typhus epidemic in 216
Gezhou Dam (China), flooding and 158
Ghana
brush fires in 111
famines and droughts in 111, 112
hot wind in 112
Giannini, Amadeo 101
Gibbon, Edward 215–216
Gifford Pinchot National Forest 421
Gilan Province (Iran), earthquake in 62
glaciers
calving of 20
melting of xvi
as trigger for avalanches 17, 20
Glazier (Texas), tornado in 353
Glendora (California), mudslides in 21
Glen Haven (Wisconsin), flooding in 186
The Global Fund 227
global warming xiii–xv, 148, 240
link between hurricanes and 252
Gloucester (England), hurricane in 261
Gloucester (Massachusetts), hurricane in 301
Gloucestershire (England), hurricane in 261
Goat Rocks 421
Godavari River (India) 240
Godbey, A. H. 164, 264
Golcuk (Turkey), earthquakes in 95, 96–97
Goldau Valley (Switzerland), avalanche in 19
Gold Coast (Florida), hurricane in 295
Goldovsky, Boris 329
Golovchinka, Irya 89
Gomez, Victor 418

Gomez, William Jaramillo 9
Gorbachev, Mikhail 39
Gordo (Georgia), tornado in 352
Gorham (Illinois), tornado in 341
Gorschkov, G. S. 423
Gottlieb, Michael S. 225
Goyder, High 112
Grafton (Illinois), flooding in 184, *191*
Grand Bahama Island, hurricane in 287
Grand Bay (Lousiana), hurricane in 279
Grand Chenier (Louisiana), hurricane in 311
Grand Isle (Louisiana), hurricanes in 281, 291, 313
Grand Rivière du Nord (Haiti), mudslides in 11
Granite City (Illinois), tornado in 347
Gravesend (England), hurricane in 261
Gray, William R. 391
Great Britain. *See also* England; United Kingdom
cholera epidemic in 212–213
plague in 212
Royal Society of xv
as source of aid 62, 78
windstorm in 364
Great Cayman Island, hurricane in 279
Great Charleston Earthquake (1886) 102
Great Chinese Famine (1876-78) 117–118
Great Corn Island, hurricane in 256
Great Deluge and Flood 200–201
Great Depression 135
great dust bowl (1934) 133, 134
Greater Antilles, hurricane in 269
Great Fire of London 208, 211, 212
Great Hurricane (1780) 265
Great Irish Potato Famine (1845-50) 127–128, 132
Great Labor Day hurricane (1935) 289–290
Great Lisbon Earthquake 86–88, *87*
Great Plague of London 212
Great San Francisco quake 73, 98–101, *100*
Great Wall (China) 154
Greece
earthquakes in 50
Plague of Thucydides in 213–214
snowstorm in 326

as source of aid 93
volcanic eruption in 388–390, *390*
Greenhill, James 280
greenhouse effect xiii
greenhouse gases xiii, xvi
reductions in xv
Greenland ice sheet, melting of xiv
Greenland Ice Sheet Project 386–387
Greenough, William 220
Greensboro (North Carolina), tornado in 352
Greenville (Mississippi), flooding in 188
Greenville (Tennessee), flooding in 189
Greenwood (Oklahoma), tornado in 351
Gregg, Charles T. 208, 220
Gregory the Great (pope) 215
Grenada, hurricanes in 261–262
Griffin (Illinois), tornado in 342
Grigoryan, Ruzanna 39
Grim, George 281
Grimsby (England), hurricane in 261
Groton (Connecticut), hurricane in 301
Guadeloupe
hurricanes in 262, 276, 295–296, 307, 316
volcanic eruption in 412
Guam
famines and droughts in 128–129
typhoons in 373, 374
Guanabocoa (Cuba), hurricane in 257
Guangxi-Zhuang Autonomous Region (Kwangsi Province), typhoon in 378
Guantánamo Bay (Cuba), hurricane in 260
Guatemala 414
earthquakes in 45, 50–51, *51*
hurricanes in 254, 262–263, 274
volcanic eruption in 50–51, 390–391
Guatemala City (Guatemala) 414
earthquakes in 51
flooding in 50
Guerrero (Mexico), hurricane in 271
Guillotes, Christina 379
Guilot (Philippines), destruction of 420
Guingyuan (China), flooding in 158
Gujarat (India)
earthquakes in 52–54
flooding in 166–167

Gulfport (Mississippi), hurricanes in 293, 300, 311

Gulf Shores (Alabama), hurricane in 313

Gummer, John 217

Gunji (India) 13

Guntoor Famine (1833) (India) 124

Guonsaugon (Philippines), landslides in C-1

Guptipara (India), flooding in 148–149

Gutiérrez Zamora (Mexico), flooding in 174

Guzara Darra (Afghanistan) 36

Guzmán (Mexico), earthquake in 76

Gwadar (Pakistan), flooding in 174

H

Hai He (Haihe) (China), flooding of 156

hail, in tornadoes 340, 342

hailstones, size of 323

hailstorms 323–324

hairpin hurricane 162–163, 264

Haiti 258, 258

    AIDS in 225, 226

    flooding in C-6, 160–161, 162–163

    hurricanes in 253–254, 255, 259–261, 263–264, 269, 271, 287, 295, 314

    mudslides in 11

Hakenaka, Toshiko 71

Hakodate (Japan), typhoons in 374

Hale, Kate 280

Halsey, "Bull" 376–377

Halverston, Jeffrey 291

Hambantota (Sri Lanka), flooding in 176

Hamelin (Germany), Black Death in 210–211

Hamilton, Alexander 276–277

Hamilton, William 395, 398

Hammock, John 152

Hankow (Hankou) (China), flooding in 157

Hannibal 6–7, 37

Hannibal (Missouri), flooding in 185, 191

Hansen, James 252

Harar (Ethiopia), famines and droughts in 112

Harerge (Ethiopia), famines and droughts in 113

Harnai (Pakistan), earthquake in 79

Hartford (Alabama), hurricane in 285

Hassan, Abdulkassim Salat 114

Hatia (Bangladesh), cyclones in 235, 237, 238

Hattiesburg (Mississippi), hurricane in 291

Havana (Cuba), hurricanes in 257–258, 272

Hável, Vaclav 131

Hawaii 386

    hurricanes in 297–298

Hawaiian Range 385

Hays (Kansas), flooding in 182

Hearn, Lafcadio 406

Hearst, William Randolph 218

Hecht, Ben 192

Heclas (Italy), eruptions of 393

Heilongjiang (China), flooding in 160

Heilprin, Angelo 409, 410

Heliopolis (Egypt), earthquake in 48

Helli, Samir 38

Hemingway, Ernest 14, 290

Henan Province (China), flooding in 160. See also Honan (Henan) Province (China)

hepatitis, hurricanes and 282

Hephaestus 392

Heppner (Oregon), flooding in 194

Herbert, Ted 190

Herculaneum, destruction of 395, 396, 397393

Herman (Missouri), flooding in 185

Hewitt, F. M. 342

Hewitt, R. 401

Heymann, David L. 217

Hidalgo (Mexico)

    flooding in 173–174

    hurricane in 270

Higaturu (New Guinea), volcanic eruption in 413, 414

Higgens (Texas), tornado in 353

Higgins, Harold 199

Hillister, Perry 192

Hilton Head (South Carolina), hurricane in 284

Himalaya Mountains 385

    creation of 56, 79

    snow melt in 168

    snowstorms in 326–327

Hindu Kush Mountains (Afghanistan)

    avalanches in 6

    creation of 79

    drought and famine in 34

    earthquakes in C-2, 34–35

Hindustan (India), famines and droughts in 124

Hirajima, Shizuko 72–73

Hir (Iran), earthquake in 63

Hirohito (emperor of Japan) 74

Hirshhorn, Norbert 220

Hispaniola (Dominican Republic and Haiti)

    flooding in C-6, 161, 269

    hurricane in 316

History of Rome (Niebuhr) 215

History of the Decline and Fall of the Roman Empire (Gibbon) 215–216

HIV. See AIDS (acquired immune deficiency syndrome)

Hoboken (New Jersey), snowstorm in 329

Ho Chi Minh City (Vietnam)

    avian flu in 222

    flooding in 150

Hoka (Minnesota), flooding in 187

Hokkaido (Japan), earthquake in 69–71

Holcomb Rock (Virginia), hurricane in 312

Holden Beach (South Carolina), hurricane in 287

Holden (Massachusetts), tornado in 343, 343

Holland. See also Netherlands

    famines and droughts in 110, 122–123

    flooding in 163, 164

    hurricane in 264

    Nazi occupation of 122–123

Holmes, Oliver Wendell 301

Homeland Security, U.S. Department of 292

Homestead Air Force Base (Florida), hurricane in 280–281, 290

homosexuality, AIDS and xii, 225, 226

Honan (Henan) Province (China)

    famines and droughts in 118–119

    flooding in 154, 155, 160

Honduras

    earthquakes in 78, 94, 266

    hurricanes in 254, 256, 262, 264–267, 273–275

Hong Kong

    avian flu in 222

    bubonic plague in 218

    famines and droughts in 117

    SARS in 229–230, 230

    typhoon in 371–372

Hongo (Japan), tsunamis in 69

Honshu Province (Japan), typhoon in 375

Hood, Alexander 66

Hooghly River (India), flooding of 148

Hoover, Herbert 190

Hoover Dam 144

Hopei (Hebei) Province (China)
famines and droughts in 118
flooding in 154, 155, 160

Hose, Leroy 281

hot winds 112

House, Joe 282

Houston (Texas), hurricanes in 254–255, 269

Howard, John xvi

Hsiangchow (Xiangzhou) (China), flooding in 156

Hsilinkuole (Xilinguole) (China) snowstorm in 325

Hsinkan (Xingan) (China), snowstorm in 325

Huang Ho (China), flooding in 154

Huarasca, Mt. (Peru), avalanche in, 17, 22

Huaraz (Peru), earthquake in 83

Huatulco (Mexico), hurricane in 270

Huaylas (Peru), earthquakes in 83–84

Huber, Thomas 8

Hulunpeierh (Hulunbeir) (China), snowstorm in 325

Humacao (Puerto Rico), hurricane in 276

Humber River (Toronto), flooding of 287

Humboldt (Iowa), flooding in 187

Hunan (China)
agriculture in 118
flooding in 155

Hunan Kwangsi (Guangxi) (China), flooding in 147

Hun Sen 149

Hupei (Hubei) Province (China), flooding in 155, 158, 159, 160

Hurarz (Peru), earthquake in 84

hurricanes 107, 247–317. See also
cyclones; typhoons
attribution of, to a divine cause 268
Category 1 xiv, 251, 262, 270, 291
Category 2 xiv, 251, 269, 270
Category 3 xiv, 251, 272, 300
Category 4 xiv, 251, 269, 272, 274, 285, 291
Category 5 xiv, 251, 254, 265, 269, 272, 273, 274, 291

characteristic shape of 251

cholera following 266, 267, 271

classifiction of xiv

conditions for spawning 251

crime and 292, 308, 315

defined 251

dengue fever and 266, 267

diphtheria and 282

dysentery and 282

eyes of 251, 253, 254, 255, 257, 277, 278, 279

eye-wall of 251, 256

flooding in 251, 260

forecasting 254, 255, 262, 267, 295, 311

hairpin 64, 162–163

hepatitis and 282

intensity of 251, 255

lack of longevity and 207

lifespan of 251

link between global warming and 252

malaria and 267, 282

mature 251

mudslides in 262, 263, 265, 266, 270–271, 312

naming of xiv, 267

paths of xi, 258, 264

problem of pets in C-8

season of 251

sea surface temperatures and 252

speed of 251

spin off of tornadoes from 253, 255, 281, 313

stages in development of 251

storm surges in 254, 260, 272, 273, 279, 283, 299, 302, 307, 308–309, 315, 316

tidal waves in 253, 265, 297, 304

tracking of 272–273, 279

as trigger for famines 128

typhoid fever and 264, 277

warnings for 254–255

windspeeds of 251

worst recorded 247–250t

hurricanes by name
Andrew (1992) 280–282, 291, 298

Audrey (1957) 310–311

Bertha (1996) 306

Betsy 280

Camille (1969) 272, 291, 307, 311–312, 312

Camille (Mississippi) 312

Carol (1954) 288, 302–303

Charlie (1951) 267–268

Cleo (1964) 263, 295, 295

Connie (1955) 182, 282

David (1979) 265, 280, 283–284

Dean (2007) 268–270

Diane (1955) 182, 282

Donna (1960) 276, 284

Easy (1950) 300

Edna 288

Elena (1985) 299–300

Eloise (1975) 284–285

Felix (2007) 269, 273–275, 274

Flora (1963) 258, 258, 263

Floyd (1999) 285–287

Fran (1996) 305–306

Frederic (1979) 279–280, 283

Georges (1998) 259–261

Gilbert (1988) 153, 253–255, 256, 269

Gladys (1955) 271

Great Cape Verde (1784) 267

Hattie (1961) 264–265

Hazel (1954) 263–264, 287–288, 288

Hilda (1955) 271

Hugo (1989) 259, 307–310, 411, 412

Inez (1966) 255

Iniki (1992) 297–298

Iwa (1982) 298

Janet (1955) 255–256, 271

Jeramie (1935) 264

Joan (1988) 256

Juan (1985) 312

Katrina (2005) xiv, C-7, C-8, 80, 81, 144, 252, 272, 290–293

Mitch (1998) 265–267

Padre Ruiz (1834) 277

Pauline (1997) 270–271

Santa Ana (1825) 276

Stan (2005) 262–263

Wilma (2005) 271–272

hurricane season xi, 310

Hutchinson, James 219

Hwai River (China), flooding of 155–156

Hyannis Port (Massachusetts), hurricane in 301

hydrologic cycle 143, 144

I

Ibrahim, Mohammed 48

Iceland 385
avalanches in 11
volcanic eruption in 391–392

icestorms 323–324. *See also*
    snowstorms
        composition of 323
        theories on 323
        worst recorded *321–322t*
Idaho
    flooding in 195–196
    mud slides in 195
    rock slides in 196
Ijssel River (Holland) 163
Illinois, tornadoes in 136, 341–342,
    344, 345
Illustrated Newspaper (Leslie) 352
Imam Zadeh David (mountain shrine)
    169
Impelliteri, Vincent 329
Inagaki, Michihiko 70–71
Incomati River (Mozambique),
    flooding of 145
India
    avalanches in 12
    Black Death in 209
    cholera epidemic in 219–220
    cyclones in 240–243
    earthquakes in *C-5*, 51–57,
        77–78
    famines and droughts in xiii, xiv,
        123–127
    flooding in 147–151, *165,*
        165–169
    icestorms in 323–324
    influenza epidemic in 221
    landslides in 12–13
    monsoons in xii, 362
    plague in 214
    tsunamis in 58
Indiana, tornadoes in 341–342, 345
Indianola (Texas), hurricane in 316
Indonesia
    avian flu in 223
    drought in xii
    earthquakes in 57–70
    tsunamis in *C-3*
Indonesian archipelago 404
infectious diseases. *See specific*
Infectious Diseases Act (Singapore)
    230
influenza. *See also* flu
    as epidemic 207, 220–221, *221*
    as pandemic *C-7,* 230
Inhofe, James xv
Inner Mongolia (China), snowstorm in
    325–326
Inohara, Chieko 72
Intergovernmental Panel on Climate
    Change (IPCC) xiii, xv–xvi, 269

International Committee of the Red
    Cross 177
International Federation of Red Cross
    Socities 151
International Fund for Agricultural
    Development 152
International Red Cross 38, 53, 62,
    114, 118, 150, 362
International Russian Relief
    Commission 132
International Tsunami Information
    Center 86
Inverness (Scotland), flooding in 175
invisible drought 107
Iowa, tornado in 344
Iowa City (Iowa), flooding in 186
Iowa River, flooding of 184, 185
Ipanema (Brazil), avalanches in 8
Iquiquen (Chile), earthquake in 91
Iquitos (Peru), earthquake in 83
Iran 36
    avalanches in 13–14
    earthquakes in *C-4,* 45, 60–64
    flooding in 169
Irani, Hafeez 167
Iranian Red Crescent 61
Irazu Volcano (Costa Rica) *388*
Ireland
    Black Death in 209
    famines and droughts in 127–
        128
    mad cow disease in 217
Irving (Kansas), tornado in 342
Ishikawa (Japan)
    flooding in 173
    landslides in 15
Islamabad (Pakistan) 16, 36
Islamic Fundamentalists 111
Islamorada (Japan), hurricane in 289
Isle of Palms (South Carolina),
    hurricane in 310
Ismit (Turkey), earthquake in. *See* Izmit
    (Turkey), earthquake in
Israel, as source of aid 93, 171
Issyk-Kul, Lake (Central Asia), Black
    Death in 209
Italian Alps, alvanches in 6–7, 20
Italian Rivera, earthquakes in *49,*
    49–50
Italy
    Black Death in 209–210
    earthquakes in *49,* 49–50,
        64–67, *65*
    epidemic pestilence in 214–215
    flooding in 169–172
    landslides in 14–15

Plague of Justinian in 215–216
    snowstorm in 326
    as source of aid 62, 93
    typhus epidemic in 214
    volcanic eruption in 392–400
Ithome (Greece), earthquake in 50
Izmit (Turkey), earthquake in 95–96
Izmit, Gulf of (Turkey) 96

J

Jackson (Mississippi), snowstorm in
    330
Jacksonville (Florida), hurricanes in
    258, 285, 302, 304
Jacmel (Haiti), hurricanes in 255, 287
Jacoby, Annalee 119
Jagatsinghpur (India), cyclone in 243
Jagger, Thomas, Jr. 408–409
Jaipur (India), cyclone in 243
Jakarta (Indonesia) 60
Jakob, Alfons 216
Jalpaiguri (India), flooding in 150
Jamaica
    cholera in 220
    earthquake in 68
    famines and droughts in 128
    hurricanes in 153, 163, 253, 254,
        265, 267–268, 269, 271, 272
James Island (South Carolina),
    tornado on 353
James River (Virginia), flooding of
    312, 313
Jamestown (Virginia), famines and
    droughts in 136
Jammu (India), avalanche in 12
Jamuna River (Bangladesh), flooding
    of 153, 168
Janibeg Kipchak Khan 209
Janjaweed (Sudanese militia) 110
    recruiting of 115
Japan
    ban on imported beef 217
    earthquakes in 15, 68–74
    flooding in 173
    landslides in 15
    as source of aid 62, 153
    typhoons in 374–375, 374–376
    volcanic eruption in 400–405
Japanese encephalitis, flooding and
    176
Jaruco (Cuba), hurricane in 257
Java, volcanic eruptions in xiii,
    400–403
Javalge, Priyanka 57

Javed, Sayed Ali 36
Jebul Siraj (Afghanistan) 6
Jefferson City (Kansas), flooding in 184
Jha, Beshan 78
Jha, Ishrar 78
Jiayu County (China) 159
Jimani (Dominican Republic), flooding in 161
Jiujiang (China), flooding in 159
John Day River (Washington), flooding of 195
Johnson, Clarence 182
Johnson, Richard 418
Johnstown (Pennsylvania), flooding in 196–198, *197*
Joint Typhoon Warning Center (JTWC) 373, 379
Joint United Nations Programme on HIV/AIDS 229
Joseph, Dom 88
Jospin, Lionel 10
Julius II (pope) 211
Junction (Oklahoma), tornado in 351
Juniata River (Pennsylvania), flooding of 285
Junkin, George 347
Justinian I (emperor of Rome) 208
plague of 208, 215

K

Kabul (Afghanistan) 6, 34, 35, 36
avalanche in 6
Kailash, Mount (India) 13
Kailash Mansarovar (Tibet) 13
Kalambaka (Greece), snowstorm in 326
Kalimasia, earthquake in 97
Kali River (India) 13
Kalutara (Sri Lanka), flooding in 176
Kamaishi (Japan), tsunamis in 69
Kamalyari, Baba 62
Kamchatka (Russia)
tsunamis in 69
volcanic eruptions in 423
Kampfer, Engelbert 404
Kamphaeng Phet (Thialand), avian flu in 222
Kampuchea, genocidal policies of the Khmer Rouge in 110
Kandeth, Karen 166
Kangra (India), earthquake in 54
Kangwon (South Korea), flooding in 175

Kano River (Japan), flooding of 375
Kansas
flooding in 179–180, 182–185, *183*
snowstorms in 328
tornadoes in 342–343, 344, 345, 346, 351, 353–354
Kansas City (Kansas), flooding in 183–184
Kansas City (Missouri), flooding in 181
Kansas River, flooding of 182, 183–184
Kansu (Gansu) Province (China)
earthquake in 44
flooding in 160
Kao Chen-ting 91
Kaohuang (Gaohuang) (China), flooding in 160
Kaposi's sarcoma 225
Karachi (Pakistan), flooding in 174
Karakorum Mountains, creation of 79
Kashmir
avalanches in 12, 16
earthquakes in *C-5*, 54–55, 79–81
flooding in 174
snowstorm in 327
Katmai, Mount (USSR), eruption of 423
Katmandu (Nepal)
avalanche in 16
flooding in 147
Kauai (Hawaii), hurricane in 297–298
Kaziranga National Park (India) 168
flooding in 148–149
Keithburg (Illinois), flooding in 191
Keller, Bill 41
Kelly, Mrs. Gregory 193–194
Kel (Pakistan), avalanche in 16
Keluit (Java), eruption in 403
Kempos, earthquake in 97
Kenansville (North Carolina), hurricane in 306
Kendrapara (India), cyclone in 243
Kennan, George M. 406, 409
Kent (England), hurricane in 261
Kentucky
drought in 133
hurricane in 291
tornado in 345
Kenya
AIDS in 227
avian flu in 223
famines and droughts in 114

Key Biscayne (Florida), hurricane in 281
Key Largo (Florida), hurricane in 284
Keystone (South Dakota), flooding in 199
Key West (Florida), hurricanes in 259–261, 272, 297, 311
Khan, Ihsanullah 79, 81
Khartoum (Sudan), flooding in 176
Khludnev, Colonel 41
Kholi, D. R. 166
Kiangsi (Jiangxi) Province (China)
agriculture in 118
famines and droughts in 118
flooding in 147, 155
typhoon in 372–373
Kiangsu (Jiangsu) Province (China)
flooding of 155
typhoon in 373
Killari (India), earthquake in 57
Kim, John 307
Kim Jong-il 130
Kingfisher (Oklahoma), flooding in 186
Kingston (Jamaica)
earthquake in 68
hurricanes in 254, 265
Kirakosyan, Emile 40
Kirghizia-Tajikistan border 18
Kiribati (formerly Gilbert Islands), famines and droughts in 128–129
Kirk, Mark 130
Kitchener (Ontario), hurricane in 287
Kliuchi (USSR) 423
Kneeland, Samuel 416–417
Knox, Philander C. 196
Knutson, Thomas R. 252
Kobe (Japan)
earthquakes in 71–73
typhoon in 374
Koga, Masakazu 72
Koja River, flooding of 174
Konts (Indiana), tornado in 345
Koop, C. Everett 225
Korea. *See also* North Korea; South Korea
typhoons in 376
Korovakan (Armenia), earthquakes in 39, 41
Kortgene (Holland), flooding in 164
Kosem, Ersu Berkcan 97
Kosovo, reconstruction of 97
Kosrae, famines and droughts in 128, 129
Krakatoa (Java), volcanic eruption in xiii, 173, 390, 400–403, *406*

Krishna River 240
Kristoff, Nicholas 115
Kronstadt (Russia), flooding in 175
Kruger National Park (South Africa), flooding in 147
Kuibyshev, USSR, famine and drought in 133
Kukra River (Nicaragua), hurricane in 269
Kukri Mukri (Pakistan), cyclone in 235
Kumming (China) 44
Kunlun Mountains (China) 154
Kuoshin (Taiwan), earthquake in 91
Kure Beach (North Carolina), hurricane in 306
Kwangsi (Guangxi) Province (China)
    flooding in 156
    typhoon in 379
Kwangtung (Guangdong) Province (China)
    flooding in 117, 155, 158
    SARS in 229
    typhoon in 379
Kweichow (Guizhou) Province (China), flooding in 147
Kyonggi (South Korea), flooding in 175
Kyoto (Japan), typhoon in 374
Kyoto Protocol xv, xvi
Kyrgyztan, Black Death in 209

L

Labat, Jean-Baptiste 268
Labor Day hurricane (1935) 269
Labugnay, Porfirio 379
Lacaita 64
LaCrosse (Arkansas), tornado in 352
La Crosse (Wisconsin), flooding in 187
La Fayet (Switzerland), avalanche in 20
Lafayette (Indiana), flooding in 193
Lafayette (Louisiana), hurricane in 313
La Grange (Missouri), flooding in 185
La Guaira (Venezuela), flooding in 200
La Guaya (Venezuela), earthquake in 103
Laguna Beach (California) 59
Lai Chai province (Vietnam), flooding in 150
Lakagigar (Iceland), volcanic eruption in 391

Lake Harbor (Florida), hurricane in 296
Lakehead (California), flooding in 196
Laki (Iceland), volcanic eruption in 391–392
Lamare, Malika 38
Lamar (Texas) 186
Lamington (New Guinea), volcanic eruption in 413–414
Lanao Province (Philippines), earthquake in 86
Lancang (China), earthquake in 44
land mines 146
landslides. See also avalanches; mudslides
    deforestation as cause of 13
    earthquakes as trigger of 35, 44, 51, 52, 64, 74
    flooding and 147, 168, 173
    torrential rains as trigger for 15
    worst recorded 3–4
La Niña xii, xiii
    droughts and 135, 136
    flooding and 168, 199
Laos, flooding in 148–151, 149, 151
LaPlace (Louisiana), hurricane in 281
Larantuka (Indonesia), earthquake in 58
La Rochelle (France), windstorm in 364
Larsimont, Charles H. 239
La Scala (Italy), volcanic eruption in 397
Last Island (Louisiana), hurricane in 299
Latif, Abdul 121
Latin America. See also South America; specific countries in
    AIDS in 226
La Union (El Salvador) 414
Laurel (Maryland), hurricane in 285
lava 385
lava bombs 399
Lavaca (Texas), hurricane in 316
lava flow 392
Laveder, Mario 15
La Vega (Venezuela), earthquake in 103
Lavonia (Georgia), tornado in 352
Lazzaro (Italy), earthquake in 65
League of Red Cross Societies 110
Lebanon Junction (Kentucky), flooding in 193
Lebanon (New Hampshire), snowstorm in 331–332

Leboff, Marino 408
Lebu (Chile), tsunami in 42
Leeward Islands, hurricanes in 259–261, 261, 284, 299
Leghorn (Italy), earthquake in 49
Leh (India), snowstorm in 327
Lehman Brothers xv
Lele, Dwarkanath 57
Leninakan (Armenia), earthquakes in 38–39
lenticulae 214
Leonardo da Vinci 169–170
Leona Vicario (Mexico) 272
leptospirosis 363
    flooding and 150
Les Bossons (France), avalanche in 11
Les Cayes (Haiti), hurricane in 264
Leslie, Frank 352
Les Orres (France), avalanches in 9–10
Lesotho, famines and droughts in 115
Lesser Antilles
    formation of 406
    hurricanes in 267, 269, 316
Le Tour (France), avalanches in 10–11
Leukerbad (Switzerland), avalanche in 19–20
Levane hydroelectric dam 170
levees
    break of, in New Orleans 290, 291–292
    flooding and 157, 190
Leyden (Holland)
    flooding in 164–165
    hurricane in 264
Leyte (Philippines)
    landslides in C-1
    typhoon in 378–379
Liao He (Liaohe) (China), flooding of 156
Licking River (Kentucky), flooding of 194
lightning, tornadoes and 340
Lille (France), windstorm in 364
Lima (Peru) 17
Lima, Francisco Negrão de 8–9
Limpopo River (Mozambique), flooding of 145
Limpopo Valley (Mozambique), flooding in 145
line storm 273
Lisbon (Portugal), earthquakes in 86–88, 87
Lithuania, typhus epidemic in 216
Little Corn Island, hurricane in 256

Little Don River (Canada), flooding of 287

livestock, impact of hurricanes on 253, 254, 259, 260–261, 279, 300

London 1808–1870: the Infernal Wen (Sheppard) 213

London (England)
cholera epidemic in 212–213, 220
fog in 361–362
great fire of 208, 211, 212
great plague of 212
snowstorm in 326

Longarone (Italy)
bursting of dam at 172
landslide in 14

Long Beach (North Carolina), hurricane in 287

Long Island (New York), hurricanes in 302, *303,* 303–305, *304, 305,* 306

Longku (China), flooding in 160

Long Valley (California), volcanic eruption in 387

Long (Vietnam), flooding in 149

"Lord how the ponds and rivers boiled" (Holmes) 301

Lori (Armenia), famines and droughts in 117

Los Angeles (California), earthquakes in 94, 97–98, *98*

Louisiana
hurricanes in 253, 259–261, 280–282, 285, 290–293, 298–299, 310, 313
tornadoes in 345, 352–353

Louisville (Kentucky), flooding in 193, 194

low countries 163

Lower Matecumbe Key (Florida), hurricane in 290

Lower Saxony (Germany), flooding in 162

Lower Van Norman Reservoir (California) 98

Luciano, Angelo 418

Lucretius 393

Lutfi, Gafur 63

Lutz, Johann 21

Luzon (Philippines)
earthquakes in 84–86
typhoons in 376–377, 378, 418
volcanic eruption in 415–419

Lyme disease 207–208

Lynchburg (Virginia), hurricane in 313

Lynmouth (England), flooding in 161–162

Lyon (France)
earthquake in 49
windstorm in 364

M

Maas River (Holland) 163

Mabton (Washington), mad cow disease in 217

MacArthur, Douglas 289, 326, 377

Macassar (Celebes) 404

Machu Dam (India), collapse of 167

Machu River (India), flooding of 167

Madagascar
flooding in 145, 146–147
plague in 222

mad cow disease 216–218

Madison (Illinois), tornado in 347

Madison (Wisconsin), flooding in 187

Madras (India)
earthquakes in 56
famines and droughts in 124

Madrid (Spain), snowstorm in 326

Mad River (Connecticut)
flooding of 192, 282

magma 385

Maharashtra (India) 56–57, *57*
flooding in 167

Mahash Khali (Bangladesh), cyclone in 237

Mainit (Philippines), typhoon in 378

Majahual (Mexico), hurricanes in 269–270

Makarandanda, Tiam 363

Malabang (Philippines), earthquake in 86

malaria
earthquakes and 36
flooding and 146, 177
hurricanes and 267, 282

Malawi, famines and droughts in 115

Malay Archipelago 402

Malaysia, AIDS in 226

Malda (India), flooding in 150

Maldives, tsunamis in 58

Maleceon (Cuba), hurricane in 272

Mali, famines and droughts in 111

Malik, Iqbal 13

Mallorca, Luisa 85

Malpa (India) 13

Managua (Nicaragua), hurricane in 256

Mandeville, Canyon (California), mudslides in 21

Mandi (India), earthquake in 54

Manhattan (Kansas), flooding in 182

Manila (Philippines)
earthquakes in 85–86
typhon in 378

Manjil (Iran) 62

Manpura (Pakistan), cyclone in 235, 237–238

Mansarovar Lake (India) 13

Mao Tse-tung (Mao Zedong) 118, 157, 158, 326

Mapou (Haiti), flooding in 161

Maputo (Mozambique), flooding in 145

Marathon (Florida), hurricanes 260

Marco Island (Florida), hurricane in 272

Marcos, Ferdinand 378

Margeulil River (Tunisia), flooding of 178

Margosatubig City (Philippines), earthquake in 86

Mariam, Mengistu Haile 113

Mariana Islands, typhoons in 373

Marianopoli (Italy), earthquake in 65

Marijan, Mbah 60

Marinatos, S. 389

Marmara Sea (Turkey), earthquake in 95

Marmolada (Switzerland), avalanches in 7

Marshall Islands, Republic of, famines and droughts in 128

Marshfield (Missouri), tornado in 345–346

Martha's Vineyard (Massachusetts), hurricane in 302

Martin, Margaret Sue 77

Martinique 424
hurricanes in 252–253, 262, 265, 268, 269, 272–273, 317
volcanic eruption in 268, 405–410

Martirano (Italy), earthquake in 65

Maryland
drought in 133
hurricanes in 286, 302, 313

Marysville (California), flooding in 195

Mascalia (Italy), volcanic eruption in 397

Mascara (Algeria), earthquake in 38

Mashad (Iran) 14

Maspeth (Queens, New York), snowstorm in 329

Massachusetts
flooding in 282
hurricanes in 258, 282, 300,
301, 305
tornadoes in 343, *343*
Massa (Italy), volcanic eruption in
399
Matagorda Bay (Texas), hurricane in
316
Matamoros (Mexico), hurricane in
255
Matanzas (Cuba), hurricane in 257
Matara (Sri Lanka), flooding in 176
Matecumbe (Florida), hurricane in
289
Mathura (India), flooding in 166
Mattei, Enrico 171
Matter, Stephen 20
Matupit 415
Maumere (Indonesia), earthquake in
58
Mauritania, famines and droughts in
111
Maxton (North Carolina), tornado
in 349
Mayfield, Max 291
Mayon (Philippines), volcanic eruption
in 415–417
Mayquetia (Venezuela), earthquake
in 103
Mazandarani, Shahrbanoo 61
Mazar-i Sharif (Afghanistan) 6
Mazatenango (Guatemala), volcanic
eruption in 51
Mbabane (Swaziland), flooding in 145
McAllister, Dick 187
McAllister, Peggy 187
McCord, James O. 194
McDougal, Robert 179
McDowell, Bart 17
McFarland (Wisconsin), tornado in
345
McGreevey, William 228
McKinley, William 218
McKinney, Herbert 279
McMaster, John 181
Mead, Lake 144
measles
in Bangladesh 237
in refuge camps 113
Medellín (Colombia) 46
mudslides in 9
Meghna River (Bangladesh) 236, 237,
241
flooding of 147
Mejia, Enrique 46

Mekong Delta (Vietnam), flooding in
148, 149
Mekong River (Cambodia), flooding
of 149
Melcher, Bert xv
Melcombe (England), Black Death in
209
Melda (India), flooding in 168
Mellon, Andrew 196
Melrose Park (Illinois), tornado in
341
Melton, James 329
Memphis, Fernando 85
Memphis (Tennessee), flooding in 188
Menglian (China), earthquake in 44
meningitis, in refuge camps 113
Mequinez (Morocco), earthquake in 87
Merak Island (Java), volcanic eruption
in 402
Merapi (Java), volcanic eruption in
60, 403
Mercalli, Giuseppe 34
Mercalli scale 34
Mercavado, Leonardo 419
Mérida (Mexico), hurricane in 272
Mérida (Venezuela), earthquake in
103
Merosa, Junjun 85
Merrimack River (Massachusetts),
flooding of 300
Mesa de los Hornitos 411
Meshed (Iran) 61–62
Meshkinshahr (Iran), earthquake in
63
Messina (Italy), earthquakes in 65–67
Mexico
earthquakes in *75*, 75–76
flooding in 173–174
hurricanes in 253–255, 256,
262–263, 265, 266–267,
268–272
landslides in 15–16
tsunamis in 15–16
volcanic eruptions in 410–411
Mexico, Gulf of 316
hurricanes in 311, 312–313
Mexico City (Mexico)
earthquakes in *75*, 75–76
flooding in 173
Miami CCC (Civilian Conservation
Corps) camp 290
Miami (Florida)
hurricanes in 272, 280, 290,
293–295, *294*, 297, 304, 311
revision of building code
following hurricane 295

Miami River (U.S.), flooding of 192
miasma 210
miasmic vapors 220
Michelson, Fred R. 14
Michigan, tornadoes in *343*, 343–
344
Michigan City (Indiana), snowstorm
in 332
Michoacán (Mexico), hurricane in
270
Micronesia, famines and droughts in
128–129
Midah (Java), volcanic eruption in
402
Mid-Atlantic Ridge 385, 391–392
Middle America Trench 75
Middle East. *See also specific
countries in*
AIDS in 226
famine and drought in xii
middle latitude cyclones 235
Midwest (United States). *See also
specific states in*
drought in 134–135, *135*
tornadoes in *344*, 344–345
Mikhail, Joseph S. 48
Milan (Italy), flooding in 171
Miller, William 181
Minas (Cuba), hurricane in 257
Minatitlán (Mexico)
hurricane in 263
landslides in 15–16
tsunamis in 15–16
Mindanao (Philippines)
earthquake in 86
typhoons in 376, 378
Mineo (Italy), earthquake in 65
Miner, Todd 365
Minnesota
flooding in 184, 186
tornado in 345
Minnesota River, flooding of 184
Minoan civilization 389
disappearance of 388
Min River (China), flooding of 157
Miranda (revolutionary leader) 103
Misenum 395
Misericordia (Philippines), destruction
of 417
Miskito Indans 274
Misquamicutt Beach (Rhode Island),
hurricane in 302
Mississippi
hurricanes in *C-7*, 259–261, 285,
290–293, 299–300
tornadoes in 345, 352–353

Mississippi River 144
  flooding of 184–185, 186, *187*,
    187–191, *189, 191*
Missouri
  earthquakes in 101–102
  flooding in 185
  tornadoes in 341–342, 344,
    345–347, *346*, 350–351
Missouri River, flooding of 184, 186
Misterbianco (Italy), volcanic eruption
  in 397
Miyamoto, Kaneatsu 72
Miyi-Yama (Java), volcanic eruption
  in 404
Mobile (Alabama), hurricanes in 261,
  278–280, 300
Mobile Bay, flooding of 279
Mohammadi, Ali 62
Mohawk Valley (New York),
  hurricane in 284
Mokelumne River California),
  flooding of 196
Molson Canadian Rocks for Toronto
  concert 230
Mona Passage, hurricane in 273
Monongahela River (West Virginia),
  flooding of 193, 313
monsoons xi, xii
  change in wind direction 361
  flooding and 147, 151, 152, 153,
    165, 166–167
  impact of, on rebuilt villages 81
  off-schedule, as cause of famine
    124
  in triggering avalanches 12
  in triggering floods C-6
monsoon season, length of 150
Montagnier, Luc 225
Montana, flooding in 179–180
Mont Blanc (Switzerland) 20, 391
  avalanches in 10–11
  calving of Tete Rousse glacier
    on 20
Montego Bay (Jamaica), hurricanes in
  265, 272–273
Monte Leone 65
Montenegro, Sofia 257
Montgomery, Bernard Law 122
Montreal (Canada), global warming
  meeting in xv
Montroc-le-Planet (France),
  avalanches in 10–11
Montserrat 310
  hurricane in 307
  volcanic eruption in 411–413
Moore, Arch 313

Moore Haven (Florida), hurricanes in
  294, 296
Moore (Oklahoma), tornado in 351
Moquette, Socorro 161
Moradabad (India), icestorm in
  323–324
Morales, Catherine 270
Morant Bay (Jamaica), hurricane in
  267
Moray Firth (Scotland), flooding of
  175
Morbus Gallicus theory 211, 212
More Haven (Florida), hurricane in
  294
Morgan City (Louisiana), hurricane
  in 281
Morganza Bend (Mississippi), flooding
  in 188
Morikawa, Akiko 71
Morne d'Orange (Martinique) 408
Morocco, earthquake in 77, 87
Moro Gulf (Philippines) 86
Morrell, Daniel J. 197
Morton, R. S. 212
Morvi (India), flooding in 167–168
Mosquera, Mary 9
Motagua Fault (Guatemala) 51
Mound Landing (Tennessee), flooding
  in 189, 190
Mount Olive (North Carolina),
  tornado in 349
Mount Pleasant (New York),
  hurricane in 285, 310
Mozambique
  famines and droughts in 115
  flooding in 145
Mpumalanga (South Africa), flooding
  in 146
Mubarak, Hosni Said 48
mudflow, volcanic eruptions and 418
mudslides 5. *See also* avalanches;
  landslides
  in Brazil 8–9
  earthquakes as trigger of 35
  flooding and 199–200
  in Haiti 11
  hurricanes and 263m 262, 265,
    266, 270–271, 312
  in India 12
Mugabe, Robert 145
Mulhall (Oklahoma), tornado in 351
Mulholland, William 179
Mumbai (India), flooding in 167
Muradiye (Turkey), earthquake in 45
Murat, Kemal 96
Murphy, Robert 190

Murphysboro (Illinois), tornado in
  341–342
Murshidabad (India), flooding in 150
Museveni, Yoweri 227
Muskingum River (U.S.), flooding of
  192
Muskogee (Oklahoma), hurricane in
  253
Mussis, Gabiel de 209
Muzaffarabad (Pakistan), earthquake
  in 80
Myanmar, AIDS in 226
Mycenaean civilization 389
Myrtle Beach (South Carolina),
  hurricanes in 282, 287, 310

N

Nabarro, David 223–224
Nacaome River (Honduras), flooding
  of 263
Naggar (India), earthquake in 54
Nagin, Ray 291, 292, 293
Nagoya (Japan), typhoon in 376
Nahrin (Afghanistan), earthquakes in
  34–35
Naidu, Chandrababu 240
Naini Tal (India), icestorm in 324
Nakhon Sri Thammarat (Thailand),
  flooding in 178
Nakisony, Fabian 81, 82
Nanning (China), typhoon in 378
Nantou County (Taiwan), earthquakes
  in 91–92
Nantucket (Massachusetts) 258
  tornado in 348
Na Pali coast (Hawaii), hurricane in
  297
Naples (Florida), hurricane in 272
Naples (Italy), earthquakes in 67
Naresh Dayal (India), flooding in 169
Naro (Italy), earthquake in 65
Narragansett Bay (Rhode Island),
  hurricane in 302
NASA, hurricanes and 254–255, 269,
  300
Natchez (Mississippi), tornado in 345
National Earthquake Information
  Center (Colorado) 62
National Guard
  in flood control 182, 185, 186,
    199
  in hurricane rescues 279, 287,
    290, 292, 300, 308, 309, 313
National Hurricane Center 253

National Oceanic and Atmospheric Administration (NOAA) 136
National Resources Defense Council xv
natural disasters, attribution of, to a divine cause xi, 268
natural explosions 383–424
    worst recorded 383–384t
Naugatuck River (Connecticut), flooding in 282
Nauru, famines and droughts in 128–129
Navarro, Constantino 378
Naxos (Italy), volcanic eruption in 393
Nazca (Peru) 83
Nazca plate 386
Nebraska
    snowstorms in 328
    tornadoes in 347–348, 348
Nebukawa (Japan) 74
Neftegorsh (Russia), earthquakes in 89–90
Negery Passoerang (Java), volcanic eruption in 402
Negros (Philippines), typhoon in 379
Nehru, Jawaharlal 166
Neily (Costa Rica), hurricane in 256–257
Neira (Mozambique), flooding in 145
Nencini, Frano 170–171
Nen River (China), flooding of 159
Nepal
    avalanches in 16
    earthquakes in 77–78
    flooding in C-6, 147–151
    plague in 222
Netala (India), earthquake in 56
Netherlands. See also Holland
    as source of aid 37, 179
Neva, River (Russia), flooding of 174–175
Nevada
    flooding in 195–196
    rock slides in 196
Nevis (West Indies) 276
    hurricane in 259
    volcanic eruption in 412
New Bern (North Carolina), hurricane in 282
Newberry (South Carolina), tornado in 348
New Britain (New Papua), volcanic eruption in 415
Newburyport (Massachusetts), hurricane in 301
New Delhi (India) 13, 243
    flooding in 166

New England. See also specific states in
    drought in 133
    hurricanes in 300–305, 303
    tornadoes in 343, 343, 349
Newfoundland (Canada), hurricane in 273, 275, 284, 286, 311
New Guinea, volcanic eruption in 413–414
New Haven (Connecticut), snowstorm in 332
New Iberia (Louisiana) 281
    hurricane in 281
New Jersey
    drought in 133
    hurricanes in 282, 285, 286, 302
    snowstorms in 328–329, 332, 333
New London (Connecticut), hurricane in 301
New Madrid (Missouri), earthquakes in 101–102
The New Masses (Hemingway) 290
New Mexico
    flooding in 179–180
    grass fires in 136
New Orleans (Louisiana)
    flooding in 144, 188, 190
    hurricanes in xiv, C-8, 253, 260, 279, 290–293, 298, 299, 300, 311, 313
    influenza epidemic in 221
Newport (Rhode Island), hurricane in 302–303
Newton Falls (Ohio), tornado in 349–350
New York
    drought in 133
    hurricanes in 282, 284, 285, 287, 303, 303–305, 304, 305
    snowstorms in 330
    tornado in 348, 349, 350
New York City (New York)
    AIDS in 225
    hurricanes in 285, 286, 302
    influenza epidemic in 221
    snowstorms in 328–329, 330–331, 331, 332, 333
New Zealand
    influenza epidemic in 221
    as source of aid 82
Nicaragua
    earthquakes in 78–79, 94, 266
    hurricanes in 254, 256, 257, 262, 265–267, 269, 273–275
    mudslide in 266
    volcanic eruption in 414–415

Nichkasia (Bangladesh), flooding in 152
Nicholson, Barbara 181
Nicholson, Christine 181
Nicobar Island (Indonesia) 58
Nicolosi (Italy), volcanic eruption in 396
Niebuhr, Barthold G. 215
Niger, famines and droughts in 111
Nigeria
    AIDS in 227
    famines and droughts in 129
Niigata (Japan)
    flooding in 173
    landslides in 15
    typhoon in 375
Nile River
    failure of, to provide natural irrigation 121
    flooding of 176–177
Nimad, Farouk 48
Nimis, Raymond 82
Nishinomya (Japan) 71
Nixon, Pat 84
Nixon, Richard M. 21, 199, 225
Nohl, Johannes 209
nor'easter 332
Norfolk (Virginia)
    hurricane in 286
    snowstorm in 332
Norman Conquest (1066) 121–122
North Africa
    earthquakes in 37
    snowstorm in 326
Northampton (England), hurricane in 261
North Carolina
    drought in 133
    hurricanes in 285, 286, 287, 291, 297, 302, 305–310
    tornadoes in 340, 348–349
North China Plain (China), flooding in 160
North Chungchong (South Korea), flooding in 175
Northern Marianas, Commonwealth of the, famines and droughts in 128–129
Northern Province (Mozambique), flooding in 146
North Korea. See also Korea; South Korea
    famines and droughts in xii, 129–131
    flooding in xiii
North River (China), flooding of 158

North Sea
    breaking of dikes 163
    flooding and 162, 164, 165
North Sewickley (Pennsylvania),
    tornado in *349*
North Topsail Beach (North
    Carolina), hurricane in 306
Northwest Tsinghai (Qinghai)
    Province (China) 154
Norway
    Black Death in 209
    snowstorm in 326
Norwich (Connecticut), hurricane in
    301
Noto (Italy), earthquake in 65
Nova Scotia 258
    hurricane in 275
Nowrouz-Zadeh, Gholamreza 61
Nuclear-Free Philippines Coalition 419
nuée ardente 413, 424
Nueva Ecija Province (Peru),
    earthquake in 84
Nureyev, Rudolf 226
Nyang Chu River (Tibet), flooding of
    178

O

Oakland (Nebraska)
    hurricane in 286
    mudslides in 286
Oaxaca (Mexico), hurricanes in 270,
    271
Oberammergau (Germany), Black
    Death in 210
Ocala (Florida), hurricane in 258
Ocean Isle 287
Ochery, Shanna 280
Ocos (Guatemala), earthquake in 50
O'Dwyer, William 329
Ogawa, Masayoshi 72
Oglio (Italy), flooding in 171
O'Hara, Lark 194
Ohio
    drought in 133, 134
    flooding in 186, 187, 193–194
    hurricane in 309
    tornadoes in 344–345, 349
Ohio River, flooding of 186
Ohio River Valley, flooding in 193–194
Okeechobee, Lake (Florida),
    hurricanes in 294, 295–296
Okha (Russia), earthquake in 89
Oklahoma, tornadoes in 136, 346,
    350–351, 353–354

Oklahoma City (Oklahoma)
    hurricane in 253
    tornadoes in 350, 351, *355*
Okushiri (Japan), earthquake in
    69–71
Olivera, Ricardo 17
Olongapo (Phillipines), destruction
    in 419
Omaha (Nebraska)
    snowstorms in 328
    tornadoes in 347–348, *348*, 355
Ongor, Leman 93
Ontario (Canada)
    snowstorm in 330
    tornado in 349
Ontario, Lake (Canada), hurricane in
    287
Onuk, Pinar 95
Orange Hill (West Indies) 424
Ordóñez, Ezequiel 411
Ordos Desert (China) 154
Oregon
    flooding in 194–196, *195–196*
    rock slides in 196
    snowstorm in 333
Orissa (India)
    cyclones in 242–243
    famines and droughts in 124
Orizaba (Mexico) 75
Ormoc (Philippines), typhoon in
    378–379
Ortega, Daniel 275
Osaka (Japan), typhoon in 374–375
Osborne, Sidney Godolphin 127
Oscans 394
Osormo (Chile), earthquake in 42
Ostend (Belgium), windstorm in 362
Otsu (Japan), typhoon in 380
Ottaiano (Italy), volcanic eruption in
    398, 399
Ouddorp (Holland), flooding in 164
overpopulation, effects of 110
Oviedo (Dominican Republic),
    hurricane in 255
Oxfam America 112, 152
Oxfam (Indonesia), tsunamis in 59
Oxfordshire (England), hurricane in
    261
Ozama River (Dominican Republic),
    flooding of 277

P

Pacheco, Rodolfo 266
Pacific Northwest, flooding in 195
Pacific Plate 386, 417–418

Pacific Rim, earthquakes in 91
Pacific Ring of Fire 58, 60, 417–418,
    421
Pacific tropical storm Miriam 256
Padma River (Bangladesh), flooding of
    147, 153
Padre las Casas (Dominican Republic),
    hurricane in 283
Pagliuca, Salvatore 67
Pahokee (Florida), hurricane in 296
Pakistan 36
    avalanches in 16
    drought in xiii, xiv
    earthquakes in *C-5*, 63, 79–81
    flooding in *C-6*, 147–151, 174
    humanitarian aid from 54
    monsoons in xii, 362
Paktia Province (Afghanistan),
    flooding in 174
Paladino, John 187
Palau, Republic of, famines and
    droughts in 128
Palewela (Sri Lanka), flooding in 176
Palmi (Italy), earthquake in 65
Pamir Mountains (Soviet Central Asia)
    avalanches in 18–19
    creation of 79
Pamlico Sound (North Carolina),
    hurricane in 282
Pampasinga River (Philippines),
    flooding of 85–86
Panabaj (Guatemala), hurricane in
    263
Panafieu, Françoise de 364
Panama, hurricane in 256
Panama City (Florida) 279
    hurricanes in 285, 300
Panchen Lama 178
pandemic. *See also* epidemics
    AIDS as a 208, 224–229
    avian flu as potential 221–224
    Black Death as a 208–211, 215
    bubonic plague as 208, 214, 218,
        220
    defined 208
    first 208, 215
    flu as 208, 228
    fourth 208
    influenza as *C-7*, 230
    mad cow disease as potential
        216–218
    second 208, 214
    third 208, 213, 214, 218, 220
Panuco River (Mexico), flooding of
    256, 271
Papandayan (Java) 402
    volcanic eruption in 403–404

Pappalardo, Diego de 397
Papua New Guinea
    earthquakes in 81–82
    tsunamis in 81–82
    volcanic eruption in 415
Parantij (Gujarat), famines and
    droughts in 125
Pardee, George C. 218
Paricutín (Mexico), volcanic eruption
    in 410–411
Paris (France)
    snowstorm in 326
    windstorm in 363–305
Park, Han S. 130
Parke, John G. 197
Parker, Edney 290
Parkton (North Carolina), tornado
    in 349
Parma (Italy), Black Death in 209
Pascagoula (Mississippi), hurricane
    in 261
Pasni Tehsil (Pakistan), flooding in
    174
Pass Christian (Mississippi),
    hurricanes in 279, 293, 311
Pasteurella pestis 207, 209
Paterno (Italy)
    earthquake in 65
    volcanic eruption in 397
Patterson, John H. 192
Patti (Italy), earthquake in 65
Pavy, Jean-Marie 11
Paxton, Alexander G. 189
Pearl River (Alabama) 278–279
Pelée, Mt. (Martinique), volcanic
    eruption in 268, 405–410, 413
Pelican Bay (Florida), hurricane in
    296
Pelileo (Ecuador), earthquake in 46
Peloponnesian War 213
PEMEX 270
Pendleton (Arkansas), flooding in 190
Pennington, Wayne 80
Pennsylvania
    drought in 133, 134
    flooding in 196–198, 197
    hurricanes in 282, 285, 309, 313
    tornadoes in 349, 350
    yellow fever epidemic in 218–219
Pensacola (Florida), hurricanes in
    279, 294, 299, 300, 313
Pepys, Samuel 212
Peralta, Zeferino 174
Perboewatan (Java)
    upthrusts of 401
    volcanic eruption in 402

Percy, William 189
Pereira, Fernando 274
Periera (Colombia), earthquake in 45
permanent drought 107
Pertley, David 92
Peru
    avalanches in 16–17, 22
    earthquakes in 17, 47, 82–84,
        83, 84
Pescaopagano (Italy), earthquake in
    67
Petrarch (Francesco Petrarcha) 209–
    210
pets, problem of, in hurricanes C-8
Pew Center on Global Climate Change
    xv
PG&E Corporation xv
Philadelphia (Pennsylvania)
    snowstorms in 329, 332, 333
    yellow fever epidemic in 218–219
Philip of Burgundy 323
Philippine Atmospheric, Geophysical
    and Astronomical Services
    Administration 379
Philippines
    cholera in 220
    earthquakes in 45, 84–86
    landslides in C-1
    typhoons in 373, 376–380
    volcanic eruptions in xiii, 415–
        417, 420, 420
Philippine trench 86
Philistines, plague among 219
Phillips, Owen 405
Phillips, Violette 306
Phnom Penh (Cambodia), avian flu in
    223
Phnom Penh River (Cambodia),
    flooding of 149
Phoenix, Arizona, drought in 136
Phreorng, Ya 223
Phumiphol Adulet 363
Piajo 46
Piavet River valley (Italy), landslide
    in 14
Picasso, Pablo 175
Pickens County (Georgia), tornado in
    340
Pico de Orizaba (Mexico), extinct
    volcano of 75
Piedmont (Alabama), tornado in 340
Pied Piper of Hamelin (fairy tale)
    210–211
Piedra Grande (Guatemala), hurricane
    in 263
Pietro (Italy), volcanic eruption in 397

Pigeon Cove (Massachusetts),
    hurricane in 301
Pijao (Columbia), earthquake in 45
Pillars of Hercules 389
Pilot, Peter 227
Pinar del Río (Cuba), hurricane in 257
Pinatubo, Mount (Philippines),
    volcanic eruption in xiii, 377, 417,
    417–419
Pindar 393
Pineapple Express 195
Pine Bluff, flooding of 190
Pipe Stem Reservation (North Dakota)
    184
Pirago (Italy), landslide in 14
Piscopio (Italy), earthquake in 65
Piskatee Lake (Illinois), flooding of
    187
Pittsburgh (Pennsylvania), flooding
    of 193
Plafker, George 22
Plague! (Gregg) 208, 209
Plague of Florence 209
Plague of Justinian 208, 215
Plague of Thucydides 213–214
plagues and epidemics xvi, 205–230.
    See also specific infectious diseases
    bubonic 207, 208, 218
    direct effect on human
        population 207
    in London 212
    longevity of 207
    pneumonic 207
    spread of 207
    worst recorded 205–206t
Plateinsis, Michael 209
plate tectonics, theory of 385
Plato 389
Playa del Carmen (Mexico), hurricane
    in 272
Pleasant Hll (Ohio), tornado in 345
Pliny the Elder 385, 394–395
Pliny the Younger 394, 395
Ploieşti (Romania), earthquakes in
    88–89
Plymouth (England), hurricane in 261
Plymouth (Montserrat), destruction of
    411–413
pneumocystis carinii pneumonia 225
pneumonia
    pneumocystis carinii 225
    recurrent bacterial 226
    in refuge camps 113
pneumonic plague 207
Pocahontas County (West Virginia),
    snowstorm in 333

Po Hai (Bohai) (China), flooding of 154

Pohnpei (Korea)
    famines and droughts in 128, 129
    typhoon in 376

Point Capucine (Java), volcanic eruption in 402

Pointe-à-Pitre (Montserrat) 307

Point Judith (Connecticut), hurricane in 301

Poipu (Hawaii), hurricane in 298

Poland, Black Death in 209

polar bears, extinction of xiv–xv

poliomyelitis, as epidemic 207

Pol Pot 149

Pombal, Marquês de (Sebastião de Carvalho e Mello) 88

Pompeii (Italy) 385
    destruction of xi, 393–394, *394, 395, 396, 397*

Pomponianus 395

Pontchartrain, Lake (Louisiana)
    flooding of 144, 191, 194
    hurricanes in 260, 291

Poona (India), famines and droughts in 124

Poplar Bluff (Missouri), tornado in 346

Po River (Italy), flooding of 171

Port Arthur (Texas), hurricane in 310

Port-au-Prince (Haiti), hurricane in 264

Port Eads (Louisiana), hurricane in 299

Porter, A. 110

Portici (Italy), volcanic eruption in 397

Portland (Oregon), snowstorm in 333

Port Orchard (Oregon), snowstorm in 333

Port Royal (Jamaica)
    earthquake in 68
    hurricane in 268, 305

Port St. Joe (Florida), hurricane in 300

Portsmouth (Ohio), flooding in 192

Port Sudan (Sudan), flooding of 176

Portugal
    earthquakes in 86–88, 87
    mad cow disease in 217

Post, C. W. 134

Potema (Italy), earthquake in 67

Po Valley (Italy), flooding in 171

Powell, Valerie 36

Powers of Nature (Dreisbach) 129

Pratt, W. E. 420

Prentis, Mrs. Thomas T. 407

Prentis, Thomas 408

Press, Frank 41

Prey Rognieng (Cambodia), avian flu in 223

Price, Robert 23

Prince Edward Island (Canada), hurricane in 286

Princeton (Indiana), tornado in 342

Prinzing, Friedrich 216

prion 217

Procopius of Caesarea 215

Project Storm-fury 251

protease 226

Providence (Rhode Island), hurricanes in 301, 303

Puebla (Mexico) 75
    flooding in 173–174

puerperal fever, as epidemic 207

Puerto Cabezas, hurricane in 274, 275

Puerto Escondido (Mexico), hurricane in 270

Puerto Montt (Chile), earthquake in 42

Puerto Rico
    cholera in 220
    hurricanes in 252, 253, 259–261, 262, 268, 269, 273, 275–276, 280, 283, 284, 294, 295–296, 302, 304, 307–310, 316

Pugliano (Italy), volcanic eruption in 397

Pulido, Alvaro 45

Pulido, Dionisio 410

Pulido, Paula 410

pulmonary tuberculosis 226

puncticulae 214

Purdy, Hugh 181

Puri (India), cyclone in 243

Putnam (Connecticut), flooding in 182, 282

pyroclastics 385

Pythagoras 393

Q

al-Quadi 35

Quain (Iran), earthquake in 61

Quaker Work in Russia 1921-23 (Asquith) 132

quarantine
    in containing plague 211
    in containing yellow fever 219

Quelin (Chile), tsunami in 42

Quetta (India)
    earthquakes in 57, 79

Quincy (Illinois), flooding in 186, 191

Quindio (Colombia), earthquakes in 45–46

Quinebaug River (Connecticut), flooding of 182

Quintana Roo State (Mexico), hurricanes in 254, 256, 269, 272

Quito (Ecuador), earthquakes in 46–47

R

Rabaul (New Guinea), volcanic eruption in 415

Rabbani, Ghulam 35

Rabbini, Burhanuddin 6, 36

Raccoon River (Iowa), flooding of 185

Raffaelli, Signora Ida 170

Raffles, Stamford 404–405

Rahman, Mufizur 238

rain
    lack of, as natural phenomenon 111
    monsoon xii, C-6, 12
    torrential 8–9, 11, 12–13, 15, 199

Rajasthan (India), flooding in 167

Rakata (Java)
    upthrusts of 401
    volcanic eruption in 402

Raleigh (North Carolina), hurricane in 306

Rampino, Michael 387–388

Ranco, Lake (Chile) 42

Rangabali (Pakistan)
    cholera in 236
    cyclone in 235

Rapid City (South Dakota), flooding in 199

Rapid Creek (South Dakota), flooding of 199

Rasht (Iran), earthquakes in 62–63

Rasool, Said 80

Ratnapura (Sri Lanka), flooding in 176

Rattus rattus 207

Rawai, A. S. 55–56

Rawley, Wayne 195

Reading (Massachusetts), hurricane in 301

Reagan, Ronald 112, 113, 225

recurrent bacterial pneumonia 226

Red Bay (Alabama), tornado in 352
Red Crescent 38, 150
Red Cross 36
Reddy, Venkat 240
Redfoot Lake (Tennessee), earthquake in 102
Red Oak (Oklahoma), tornado in 351
Red Sea 393
    flooding of 176
    parting of 388, 389–390
Red Springs (North Carolina), tornado in 348, 349
Reed, I. A. 316
Reed, James F. 327, 328
refugee camps 110, 111
    disease in 113
Reggio de Calabria (Italy), earthquake in 65
Reghia (Algeria), earthquake in 38
Relenza 224
Renfroe, M. K. 313
Reno (Nevada), snowstorm in 333
Repouchon, Marshal 281
rescue dogs
    in avalanche rescues 8, 10–11
    in earthquake rescues 76, 95
Resina (Italy), volcanic eruption in 397, 398
Reynolds County, Missouri, tornado in 341
Rhine River (Holland) 163
Rhode Island
    drought in 133
    flooding in 282
Richmond (Virginia) 102
    hurricane in 313
Richter, Charles F. 34
Richter scale 34, 44
Righini-Bonelli, Maria Luisa 170
"Ring around the Rosies" (nursery rhyme) 210
Rioco, Antonio 65
Rio de Janeiro (Brazil)
    avalanches in 8–9
    global warming meeting in xv
Riordan, Toni 280
Riposta (Italy), earthquake in 65
Rita Island (Florida), hurricane in 296
Rivas, Arturo Lara 76
Riverside Mill (South Carolina), tornado in 352
Robinson Beach (South Carolina), hurricane in 287
Rockfish River (Virginia), flooding of 312

Rock Harbor (Florida), hurricane in 289
Rock Island (Illinois), flooding in 185
Rock Run (Alabama), tornado in 340
Rodney, Lord 273
Rodrigues, volcanic eruption and 402
Roe, Mark 133
Rolando Rodríguez (Nicaragua), mudslides in 266
Roller, George 192
Rolling Fork River, flooding of 193
Roman Empire, plague in 215
Romania
    avian flu in 223
    earthquakes in 88–89
Rome (Italy)
    earthquake in 67
    epidemic pestilence in 214–215
    Plague of Justinian in 215–216
Romme, Lucien 81–82
Roosevelt, Theodore 406, 408
Rosen, George 212
Rossberg Peak (Swiss Alps), avalanche at 19
Roudehen (Iran), avalanches in 13–14
Rouen (France), windstorm in 364
Rousseau, Jean-Jacques 88
Roussel, Frederic 35
Roussel, Michael 10
Rowland (Oklahoma), tornado in 351
Roxbury (Massachusetts), hurricane in 301
Rudbar (Iran) 63
    earthquake in 62
Rudd, Kevin xvi
Ruediger, Bill 421
Ruef, Abe 100
Ruff, Benjamin F. 196
Rugen (Baltic island), avian flu in 223
Ruiba (Algeria), earthquake in 38
Ruiz de Isla, Rodrigo 211
Ruiz-Tolima (Columbia) 45
Russell, Lord John 128
Russia. See also USSR
    avalanches in 18–19
    Black Death in 209
    earthquakes in 89–90
    famines and droughts in 132
    flooding in 174–175
    as source of aid 62, 179
    typhus epidemic in 216
Russiaville (Indiana), tornado in 344
Rustaq (Afghanistan), earthquakes in 35, 36–37
Rwanda, AIDS in 226

S

Saavedra, Joanna 200
Sabadell (Spain), flooding in 175–176
Sabri, J. A. 326
Saffir-Simpson scale 251, 269, 274, 300
Sagami Bay (Japan), landslide into 74
Sahara, encroachment of 111
St. Bartholomew's hospital, sanitary conditions in 213
St. Gervais (Switzerland), avalanche in 20
St. Moritz (Switzerland), avalanches in 20
St. Vitus's dance in France 212
St. Agatha (Italy) 392
St. Bart's, hurricane in 301
St. Christopher (Guadeloupe), hurricane in 262
St. Croix, hurricanes in 276–277, 307, 308, 309
St. Elmo's Fire 385, 401
St. Eustatius, hurricanes in 252–253, 265, 272–273
St. Francis Dam (California), collapse of 179
St. George's Channel, hurricane in 273
Saint George's (Grenada), hurricane in 262
St. Helens, Baron 421
St. Helens, Mount (Washington)
    avalanche on 22
    volcanic eruptions of 22, 419, 420–423, 422
St. Jo (Florida), hurricane in 296
St. John's, flooding in 307
St. Kitts, hurricane in 259
St. Lawrence, Gulf of, hurricane in 284
St. Louis (Missouri) 184
    flooding in 186
    tornado in 346, 346–347
St. Lucia, hurricanes in 252–253, 253, 269
St. Nicholas (Haiti), hurricane in 268
St. Petersburg (Russia), flooding in 174–175
St-Pierre (Martinique), destruction of 406, 407, 408, 413, 424
"St. Telemachus" (Tennyson) 403
St. Thomas 308
    hurricane in 307
St Thomas Parish (Jamaica), hurricane in 267

St. Vincent (Martinique)
hurricane in 252
volcanic eruptions in 406, 408, 414

Sajardo (Puerto Rico), hurricane in 307

Sakhalin Island (Russia), earthquakes in 89–90

Sakurajima (Japan), volcanic eruption in 400

Salang area (Afghanistan), avalanche in 6

Salasaca Indians 46

Salem (Massachusetts), hurricane in 301

Salisbury Point (Massachusetts), hurricane in 301

Salistroso River (Honduras), flooding of 265

Sambava (Madagascar), flooding in 146

Samia Ragab Khalil 48

Samoa, influenza epidemic in 221

Samsun (Turkey), earthquake in 94

San Andreas Fault (California) 33, 94, 99

San Andrés Tuxtla (Veracruz), hurricane in 263

San Angelo (Texas), tornado in 355

San Cayetano (Colombia), earthquake in 45

San Ciprian (Puerto Rico), hurricane in 276

San Ciriaco (Puerto Rico), hurricane in 276

San Constantino (Italy), earthquake in 65

San Cristóbal (Dominican Republic), hurricane in 259

Sand Key (Florida), hurricane in 288

Sandu, Dumitru 88

Sandwip (Bangladesh), cyclone in 237, 238

Sandy Bay (Honduras), hurricane in 274

San Felipe (Venezuela), earthquake in 103

Sanfeo (Italy), earthquake in 65

San Fernando Valley (California), earthquakes in 97–98, 98

San Francisco (California)
AIDS in 225
bubonic plague in 218
earthquakes in 42, 98–101, 100
influenza epidemic in 221

San Francisquito Canyon (California) 179

San Gabriel Mountains, mudslides in 21

San Giorgio a Cremano (Italy), volcanic eruption in 397

San Giovanni (Italy), earthquake in 65

San Giuseppe (Italy), volcanic eruption in 399

Sangupta (Nitish) 166

San Jose (Philippines), destruction of 420

San Juan (Puerto Rico), hurricane in 259, 307, 308

San Juan de la Maguana River (Dominican Republic) 259

San Juan de Parangaricutiro (Mexico), volcanic eruption in 411

San Juan Ostancalco (Guatemala), volcanic eruption in 51

San Marcus (Guatemala), volcanic eruption in 51

San Miguel (Nicaragua) 414

San Pedro (Guatemala), volcanic eruption in 51

San Pedro Sacatepéquez (Gautemala), hurricane in 263

San Pedro Sula (Honduras), hurricane in 275

San Roque (Philippines), destruction of 417

San Salvadore 414

San Salvador (Guatemala), hurricane in 263, 285

San Sebastiano (Italy), volcanic eruption in 399

Santa Ana (Guatemala), eruption of 50–51, 263

Santa Catarina River (Texas) 255
flooding of 255

Santa Clara Valley (California), flooding in 179

Santa Cruz del Sur (Cuba), hurricane in 258

Santa Fe (New Mexico), drought in 136

Santa María del Rosario (Cuba), hurricane in 257

Santa María (Guatemala), volcanic eruption in 390–391

Santa Monica Mountains (California), mudslides in 21

Santa Teresa (Brazil), avalanches in 8

Santiago Creek (California) 21

Santiago (Cuba), hurricane in 314

Santiago (Guadalupe) 272

Santiago Tuxtia (Veracruz), hurricane in 263

Santo Domingo (Dominican Republic), hurricanes in 259, 273, 277, 278, 283, 297

Santo Niño (Philippines), destruction of 417

Santorini (Greece), volcanic eruption in 388–390, 390

Saraien (Iran), earthquake in 63

Saranac Lake (New York), snowstorm in 332

Sarat (Iran), earthquake in 62

Saroya, Paul 82

SARS (Severe Acute Respiratory Syndrome) 229–230

SARSstock 230

Satsuma Sea (Japan) 400

Saugar (Java), volcanic eruption in 405

Saugus (Massachusetts), hurricane in 301

Saurashtra (India), flooding in 167

Savannah (Georgia), hurricane in 284

Savanna-la-Mar (Jamaica), hurricane in 273

Save River (Mozambique), flooding of 145

Save the Children Fund 110, 112

Scanio, Alfonso Pecoraro xvi

scapegoats, search for, in fueling anti-Semitism 210

Scheldt River (Holland) 163

Schleswig-Holstein (Germany), flooding in 162

Schmitz, Eugene 100

Scioto River (U.S.), flooding of 192

Scio (Turkey), earthquake in 97

scorched earth policy 110

Scotland
avian flu in 221
Black Death in 209
flooding in 175

Scott, Ellery 408

scurvy, famine and drought and 127

Scylla (Italy), earthquake in 65

sea ice, melting of xiv

Sea Isle (New Jersey), hurricane in 302

seamounts 385

Sea of Japan, flooding in 173

seasonal drought 107

sea surface temperatures (SSTs), hurricanes and 251, 252

Seattle (Washington)
influenza epidemic in 221
snowstorm in 333

seawall, elevation of, in Galveston 316

Secundus, Gaius Plinius 394
Segni, Bernardo 170
seismic waves 33
seismograph 33
seismologists 33
Seminaria (Italy), earthquake in 65
Semiryeschensk, Black Death in 209
Seneca 393
Seoul (South Korea), flooding in 175
Septembr equinoctial gale of 1815 301
Sextus Quintilius 214
Seyss-Inquart (Nazi Reichskommissar) 122–123
Shadi Kor Dam (Pakistan), bursting of 174
Shakhnasaryan, Gevork 39–40
Shandong (Shantung) (China), flooding in 160
Shanghai (China), flooding in 155
Shangtung (Shandong) Province (China), flooding in 160
Shannon (North Carolina), tornado in 349
Shansi (Shanxi) Province (China), flooding in 147, 160
Shantung (Shandong) Province (China), flooding in 154, 157
Shari-Buzurg (Afghanistan) 36
Shashi (China), flooding in 159
Shattuck (Texas), tornado in 353
Shbigu, Sergei 89
Shea, Todd 187
Shehada, Ali 81
Sheik, Mohammad Harunuddin 153
el-Sheik, Omar 177
Shelley, Percy Bysshe 161
Shemaka (Armenia), earthquake in 39
Shensi (Shaanxi) Province (China) 44
famines and droughts in 119
flooding in 156
Sheppard, Francis 213
Sherman, Martha 350
Shigatse (Tibet), flooding in 178
Shihkang Reservoir (Taiwan), earthquake in 92
Shimane (Japan)
flooding in 173
landslides in 15
Shindell, Drew 387
Shinkankow (Xinkankou) (China), flooding in 159
Shirak (Armenia), famines and droughts in 117
Shoshun River Basin (China), flooding in 158–160

Shoshun River (China), flooding of 159
Shreveport (Louisiana), flooding in 188
Shrewbury, J. F. 211
Shrewsbury (Massachusetts), tornado in 343
Sicily (Italy), volcanic eruption in 392–393
Sierra de los Tuxtlas region (Mexico), hurricane in 263
Sierra Leone
avian flu in 223
famines and droughts in 114
Sierra Madre (Mexico) 75
Sierra Nevada 179
rockslides in 196
Siesta Key (Florida), hurricane in 270
Sihel, famines and droughts in 120
Sikaravadi (India), earthquake in 54
Silesi (Java), volcanic eruption in 402
Silk Road 61
Simpson, Robert H. 311
Singapore, SARS in 229, 230
Singh, Reghubir 56
Sin Ho (Vietnam), flooding in 150
Sinkiang (Xingjiang) Province (China), snowstorm in 325–326
Sins, Claude 63
Sissano (Papua New Guinea) 81
Siuku (Java), volcanic eruption in 402
Sivas (Turkey), earthquake in 94
Skaptar Jokul (Iceland) 391
Skate, Bill 82
Skopje (Yugoslavia), earthquakes in 103–104
Smith, George 201
Smolensk (Russia), Black Death in 209
snapping turtles, impact of hurricanes on 278
snow 325
Snow, John 220
snow cover, melting of xiv
snowstorms 325–333. See also icestorms
as blizzards 325
worst recorded 321–322t
soil erosion, flooding and 152
Solis, Estaban 416
Sololá Department (Guatamala), hurricane in 263
Solon 389
Somalia, famines and droughts in 114
Somersetshire (England), hurricane in 261

Somma (Italy), volcanic eruption in 397, 398, 399
Sorn, Som 223
Soufrière, La (Martinique), volcanic eruptions in 406, 408, 423–424
Soufrière Hills (Montserrat), volcanic eruption in 411–413
Southern Botswana, flooding in 145
South Africa
earthquakes in 94
flooding in 145–146, 147
influenza epidemic in 221
South America. See also Latin America; specific countries in
earthquakes in 90–91
landslides in 22
South Carolina
earthquakes in 101, 102, 102–103
hurricanes in 284, 286, 291, 305–310
tornadoes in 340, 348–349, 352–353
South Dakota, flooding in 199
Southern Mississippi River Basin, flooding in 181
Southern Oscillation xii
South Fork Dam (Pennsylvania) 196
South Kingston (Rhode Island), hurricanes in 302–303
South Korea. See also Korea; North Korea
flooding in 175
South Melbourne Beach (Florida), hurricane in 283
South Ossetia (Russia), earthquake in 90
South Padre Island (Texas) 255
South Platte River (Colorado), flooding of 180
South (United States). See also specific states in
tornadoes in 351–352, 352
South Wales, famines and droughts in 122
Southwest (United States). See also specific states in
drought in 135–136
tornadoes in 353–355
Soviet Georgia (Russia), earthquake in 90
Spain
flooding in 175–176
influenza in 221, 222, 230
SARS in 229

snowstorm in 326
as source of aid 179
windstorm in 364
Sparta (Greece), earthquake in 50
Spielberg, Steven 298
Spirit Lake (Washington) 421, 422
Spitak (Armenia), earthquakes in 41
Spokane (Washington) 422
Spurius Furius 214
Squier, E. G. 414
Sri Lanka
flooding in 176
tsunamis in 58, 59
Srinagar (India)
avalanche in 12
earthquake in 55
snowstorm in 327
Stann Creek (Honduras), hurricane in 265
Stanton, C. T. 327
Stava (Italy), flooding in 172
Stava River (Italy), flooding of 172
Steers (Java), volcanic eruption in 402
Stein, Eduardo 263
Stellendam (Holland), flooding in 163
Stepanavan (Armenia) 41
Sterling (Colorado), flooding in 180
Stevenson, J. Sinclair 125
Steward, Bill 421
Still River (Connecticut), flooding in 282
Stoffel, Dorothy 421–422
Stoffel, Keith 421–422
Stonington (Connecticut), hurricane in 301
Stony Creek (Pennsylvania), flooding in 196
storms 359–365. See also icestorms;
snowstorms
dust 134–135
hail 323–324
wind 362, 363–365
worst recorded 359t
storm surges
damage to dikes 264
hurricanes and 254, 260, 272, 273, 279, 280, 283, 299, 302, 307, 308–309, 315, 316
Straits of Gibraltar 389
Strasbourg (France), St. Vitus's dance in 212
Stromboli (Italy) 65
Stroud (Oklahoma), tornado in 351
Subic Bay Naval Base (Philippines) 418, 419

Suchow (Suzhou) (China), flooding in 155
Sudan
Dafur conflict in 110, 111
famines and droughts in 110, 111, 113, 113–114, 114
flooding in 176–177, 177
Sudavik (Iceland), avalanches in 11
Suffolk (England), hurricane in 261
Suket (India), earthquake in 54
Sukisikanyan, Anton 40
Sullivan, Dennis T. 100
Sullivan's Island (South Carolina), hurricanes in 305, 310
sulphur dioxide 385
Sultenpur (India), earthquake in 54
Sumatra
earthquakes in 58–60
tsunami in C-2, C-3
volcanic eruptions in 386–387
Sumbawa (Java), volcanic eruption in 404
Sunamganj (Bangladesh), flooding in 151
Sun Moon Lake Reservoir (Taiwan), earthquake in 91
supercyclone 242
Super Typhoon Saomai 373
Surat (India)
earthquake in 53
flooding in 166, 169
Surf City (North Carolina), hurricane in 306
Surigao City (Philippines), typhoon in 378
Surigao del Norte (Philippines), typhoon in 378
Susquehanna River (Pennsylvania), flooding of 285
Sussex (England), hurricane in 261
Sustenhorn (Switzerland), avalanche on 20
Suttack (India), cyclone in 243
Sutter, J. A. 328
Suwon (South Korea), flooding in 175
Svobida, Lawrence 134–135
Swan Island (Honduras), hurricane in 256
Swaziland
famines and droughts in 115
flooding in 145
Sweden
avian flu in 223
snowstorm in 326
Sweeney, Tim 345

Swen Family Gap (China), earthquake in 44
Swiss Alps 5
avalanche at 19, 20–21
Switzerland
avalanches in 19–21
Black Death in 209
mad cow disease in 217
as source of aid 37, 76
typhus epidemic in 216
Sykes, Lynn R. 56
syphilis as epidemic 207, 211–212
Syracuse (New York), snowstorm in 330
Szechuan (Sichuan) Province (China), flooding in 147, 156, 158, 160

T

Taal (Philippines) 86
volcanic eruptions in 419–420, 420
Taal, Lake (Philippines) 419
Tabaka Dry River 424
Tabasco (Mexico), flooding in 173–174
Tabas (Iran), earthquake in 64
Tableland of Little Ovens 411
Tabor, Milton 347
Tacana (Guatemala), hurricane in 263
Tacaxoy, Ramiriz 263
Taching (Daqing), flooding in 159
Tachou (Dazhou) (China), flooding in 156
Tacitus 394, 395
Tacna (Peru), earthquake in 91
Tadmore (Ohio), flooding in 192
Tagus River (Portugal), earthquake in 87, 88
Tahoe, Lake (Nevada), snowstorm in 333
T'ai-chou (Taizhou) (China), typhoon in 373
Taichung (Taiwan), earthquakes in 91–92
Taipei (Taiwan), earthquake in 91
Taito (Formosa), typhoon in 373–374
Taiwan
earthquakes in 91–92
SARS in 230
typhoons in 373, 379–380
Tajikistan 36
avalanche in 21
famines and droughts in 116–117

Takai, Manabu 73

Takhar (Afghanistan), earthquake in 36

Takri Tsoma, Lake (Tibet), flooding of 178

Talatoa (Java), volcanic eruption in 402–403

Taliban 6, 34

Tali (Taiwan), earthquake in 91

Tamarang (Java), volcanic eruption in 403

Tambora (Java), volcanic eruption of xiii, 404–405

Tamesi River (Mexico), flooding of 256

Tamiflu 224

Tamil Nadu (India) 59

Tamil Tigers 60

Tampa (Florida), hurricanes in 260, 300

Tampico (Mexico), hurricanes in 256, 268, 271

Tanaguarena (Venezuela), flooding in 200

Tanaro (Italy), flooding in 171

Tangail (Bangladesh), tornado in 339–340

Tangshan (China), earthquakes in 44–45, 58

Tan Lap (Vietnam), flooding in 150

Tanzania, famines and droughts in 114

Tanzawa, Mount (Japan), landslide in 74

Tapachula (Guatemala), hurricane in 263

Tapayula (Mexico), flooding in 174

Tapi River (India), flooding of 166, 169

Tarrasa (Spain), flooding in 175–176, 176

Tartis, Mark 349

Tascher de La Pagerie, Joseph-Gaspard 317

Tascher de La Pagerie, Marie-Josephine-Rose (Josephine, empress of the French) 317

Taupe (New Zealand), volcanic eruption in 387

Tavernier (Florida), hurricanes in 260, 289

Tavurvur (Papua New Guinea), volcanic eruption in 415

Tavush (Armenia), famines and droughts in 117

Tawashigad (India), earthquake in 57

Taylor, Katherine 169

Tbilisi (Russia) 117
    earthquake in 90

Tecoluta (Mexico), hurricane in 270

Tegucigalpa (Honduras), hurricane in 266, 414

Tehana (California), flooding in 195

Tehran (Iran) 14
    earthquake in 64

Tehuacán (Mexico) 75

Tekirdad (Turkey), earthquake in 95

Tempey, Terrence C-7

Tennessee
    earthquake in 101–102
    hurricane in 291
    tornado in 345

Tennyson, Alfred, Lord 127, 403

Terraba River (Costa Rica), flooding of 257

Terranova (Italy), earthquake in 65

Terre Haute (Indiana), flooding in 193

Terzigno (Italy), volcanic eruption in 399

Tesero (Italy), flooding in 172

tetanus
    flooding and 185
    frostbite and 327

Tête Rousse glacier (Switzerland), calving of, on Mont Blanc 20

Tethys Sea 39

Texas
    hurricanes in 268, 269, 310, 313–316, *314*, 316
    loss of irrigated land in xiv
    tornadoes in 346, 353–355, *354*

Thailand
    AIDS in 226
    avian flu in 222, 223
    flooding in 150, 151, 177–178
    monsoon in 362–363
    tsunamis in 58, 59

Thapra (Nepal), flooding in 147

Thenia (Algeria), earthquakes in 38

Theodore (Lousiana), hurricane in 279

Thingvellir (Iceland), volcanic eruption in 391

Thorarinsson, Sigurdur 391

Thrace (Turkey), snowstorm in 326

Three Gorges Dam Project 159

3TC 226

thrust faults 79

Thucydides 213
    plague of 213–214

Thunder Out of China (White and Jacoby) 119–120

Tibet, flooding in 178

Tibetan-Tsinghai (Qinghai) Plateau (China), flooding in 160

Ticino (Italy), flooding in 171

tidal waves
    in cyclones 235, 236, 241–242
    in hurricanes 253, 265, 281, 297, 304, 311

Tientsin (China), famines and droughts in 117–118

Tigre, famines and droughts in 113

Tigre Island (Nicaragua) 414

Tiptonville Dome (Missouri) 101

Tischenbko, Yelena 89

Tista River (India) 12

T-4 lymphocytes 225

Toba (Sumatra), eruption of 387

Toba, Lake (Sumatra), volcanic eruption in 387

Tobago (Philippines), hurricane in 274, 416

Toba Megavolcano (Sumatra) 386–387

Tokyo (Japan)
    earthquakes in 74–75
    typhoons in 374, 375

Tombre, Fabian 82

Tompkins County (New York), snowstorm in 330

Tongliao (China), snowstorm in 325

Toni (Japan), tsunamis in 69

Tonle Sap River (Cambodia), flooding of 149

Tonto National Forest, grass fires in 136

Topeka (Kansas), flooding in 182

Topsail Beach (North Carolina), hurricane in 306

Toral, Demetrio 410

Tornado Alley 340, 350

tornadoes 107, 337–355
    atmospheric conditions necessary for formation of 339
    behavior pattern of 345
    contrast between cyclones and 339
    diameter of 339
    direction of 339
    distances traveled by 339
    fires and 341, 353
    funnel-shaped clouds in 339, 341
    lack of longevity and 207
    lightning and 340
    paths of xi

spin off, from hurricanes 253,
255, 261, 281, 284, 285, 286,
300–301, 313
tri-state 341
value of radar in tracking 354–
355
tornado seasons xi
Toronto (Canada)
avian flu in 223
hurricane in 287, 288
World AIDS conference in 227
Torre Annunziata (Italy), volcanic
eruption in 399
Torre del Greco (Italy), volcanic
eruption in 397, 398
torrential rains
flooding and 199
as trigger for avalanches 8–9, 11,
12–13, 15
Torres, Lem C-8
Tottori (Japan)
flooding in 173
landslides in 15
Toutle River (Washington) 422
Toyama (city) (Japan)
flooding in 173
landslides in 15
Toyama Province (Japan), typhoon in
375
Trasimeno Lake (North Africa) 37
Treatise on the Serpentine Malady
(Ruiz de Isla) 211
Trinidad, hurricane in 274
Tripapni (Italy), earthquake in 65
Tristan da Cunha 385
tri-state tornado 341
Trois-Îlets (Martinique), hurricane in
317
tropical cyclones 235
tropical depressions 274, 275, 291
tropical storms. *See also* hurricanes
Bilis 373
Dean 269
Erin 186
Felix 274
Katrina 291
Troy (Ohio), flooding in 192
Truckee Pass (California), snowstorm
in 327–328
Truckee River (Nevada), flooding of
195
Trucker Lake (California) 327
Trujillo, Rafael 277
Truman, Harry 184, 421, 422
Tsinghai (Qinghai)-Tibet Plateau,
flooding of 159

tsunamis xi
in Chile 42
in Crete 389
earthquakes in triggering xi,
C-2, C-3, C-4, 15–16, 42, 69,
70, 81–82, 86, 173
flooding and 173
in Indonesia 58–60
in Jamaica 68
in Japan 68, 69, *70*
in Morocco 77
in Papua New Guinea 81–82
in Philippines 86
volcanic eruptions and 389
warning system for 59, 71
Tuba (Peru), earthquake in 85
Tuba City (California), flooding in
195
Tubarao (Brazil), flooding in 153–
154
Tubarao River (Brazil), flooding of
153–154
tuberculosis 114
hurricanes and 264
pulmonary 226
Tucana (Guatemala), volcanic eruption
in 51
Tuleya, Robert E. 252
Tungi (Bangladesh), cyclone in 239
Tunisia
flooding in 178–179
rain in 363
Tupelo (Mississippi), tornado in 353
TUPRAS refinery 95–96
Turin (Italy), flooding in 171
Turkey
avalanches in 95
avian flu in 223
coal mine disaster in 95
construction requirements in
96–97
earthquakes in 92–97
snowstorm in 326
Turkish Red Crescent 93
Turkistan, Black Death in 209
Turkmenistan 36
Turks Island, hurricane in 290
Turneffe Cays (Honduras), hurricane
in 265
Turner, J. M. W. 404
Turrialba (Costa Rica) *388*
Tuscarora Deep (Japan) 69
Tuxpan (Mexico), hurricane in 270
Twain, Mark xvi, 297
Tye River (Virginia), flooding of 312
Tyndall Air Force Base 279

typhoid fever
flooding and 148, 152, 176, 177
hurricanes and 264, 277
Typhoon Gun 371
typhoons xi, 369–380. *See also*
hurricanes
dysentery in 373
warning of 371
worst recorded *369–370t*
typhoons by name
Angela (1995) 377
Florita (2006) 373, 379–380
Fred (1994) 372
Ida (1958) 375
Ike (1984) 377–378
Omar (1992) 374
Patsy (1970) 378
Rananim (2004) 372–373
Ruby (1988) 378
Rusa (2002) 376
Saomai (2006) 373
Thelma (1991) 378–379
Vera (1959) 375–376
Wanda (1956) 373
typhus
as epidemic
in Italy 214
in Russia 216
hurricanes and 264
in refuge camps 113
typhoon and 373
Tyrolean Alps (Austria), avalanches
in 7, 10

U

Udall (Kansas), tornadoes in 342–343
Uganda
AIDS in 226
famines and droughts in 114
Ukai dam (India), flooding in 169
Ulir Char (Bangladesh), cyclone in
237
Umit, Onur 96
Unger, E. J. 197
UNICEF. *See* United Nations
Children's Fund
Union of Concerned Scientists xv
United Kingdom. *See also* England;
Great Britain
mad cow disease in 216–218
United Nations
Children's Fund 38, 146, 174, 238
Children's Summit 227
Climate Change Conference xvi

Declaration of Commitment on
　　HIV/AIDS xii, 229
Development Program 239
Disaster Relief Organization 177
Economic and Social
　　Commission for Asia and the
　　Pacific 151
Environmental Programme xiii
Food and Agricultural
　　Organization (FAO) 114
Framework Convention on
　　Climate Change (UNFCCC)
　　xvi
Hunger and Human Rights
　　report to 131
Office for the Coordination of
　　Humanitarian Assistance to
　　Afghanistan 35
Relief and Rehabilitation
　　Administration (UNRRA) 110
as source of humanitarian aid
　　35, 36, 37, 59, 61, 80, 114
weather scientists at 151
World AIDS Conference 227
World Food Programme (WFP)
　　131
United States
　　AIDS in 208, 225
　　avalanches in 22
　　bubonic plague in 218
　　cholera in 220
　　earthquakes in 97–103, 98, 100,
　　　102
　　famines and droughts in 133–
　　　136, 135
　　flooding in 179–199, 180, 183
　　hurricanes in 257–258, 278–316,
　　　283, 288, 289, 294, 295, 303,
　　　304, 305, 312, 314
　　mudslides in 21
　　SARS in 230
　　snowstorms in 327–333, 331
　　as source of aid 37, 62, 76, 78,
　　　93, 113, 153, 179, 362
　　tornadoes in 340–355, 343, 344,
　　　346, 348, 349, 352, 354
　　volcanic eruptions in 420–423,
　　　422
　　yellow fever epidemic in 218–219
United States Climate Action
　　Partnership (USCAP) xv
U.S. Global AIDS program 227
United States Signal Service 188
U.S. Virgin Islands, hurricane in 253,
　　254, 307
unpredictable droughts 107

Unsen (Japan)
　　earthquake in 74
　　volcanic eruption in 400
Upper Volta (Burkina Faso), famines
　　and droughts in 111
Urbizio, Delmer 266
Urumchi, snowstorm in 326
Usine Guérin (Martinique) 407
USSR. See also Russia
　　famines and droughts in 133
　　volcanic eruptions in 423
Utica (New York), snowstorm in 330
Uttar Kashi (India), earthquake in 55
Uttar Pradesh (India)
　　flooding in 153, 168, 169
　　landslides in 12–13
Uzbekistan 36

V

vaccines xi, 207. See also specific
　　infectious diseases
　　of birds in bird flu C-7, 224
Vaiont Valley (Italy) 14
Vaishnov, Asim Jumor 243
Vajpayee, Atal Bihari 52
Valdiva (Chile), earthquake in 42
Valley of Ten Thousand Smokes of
　　Kamchatka 423
Valparaíso (Chile), earthquake in 41,
　　43
Vals (Switzerland), avalanches in
　　20–21
Vancouver, George 421
Vanegas, Claudio 274
Vanimo (Papua New Guinea) 82
van Sandick, N. 402
Vardar River (Yugoslavia) 104
Varlamova, Serafina 90
Vega, Frank 259
Vega, Patricia 45–46
Velji, Panchabhai 53
Venezuela
　　cholera in 220
　　earthquake in 103
　　flooding in 199–200
　　hurricanes in 256, 314
Venice (Illinois), tornado in 347
Venice (Louisiana), hurricane in 260
Venrick, Dorothy 181
vent 385
Veracruz (Mexico) 75
　　hurricanes in 256, 263, 270
Verlaten (Java), volcanic eruption in
　　402

Vermont, hurricane in 304
Verona (Italy), flooding in 171
Versailles (France), windstorm in 364
"Verses, made at Sea, In a Heavy
　　Gale" (Freneau) 267
Vesuvius (Italy) 385
　　volcanic eruption in 393–400,
　　　394, 399
Vicksburg (Mississippi)
　　flooding in 188
　　snowstorm in 330
Vidaz, Hector 308
Vieques (Puerto Rico), hurricanes in
　　307, 309
Vietnam
　　avian flu in 222, 223
　　flooding in 149, 150, 151
　　SARS in 229
　　typhoons in 380
Vigo, Giovanni di 211
Vijandya (Bangladesh), cyclone in 238
Villadaragh (Iran), earthquake in 63
Villeda Morales (Nicaragua) 275
Vilna (Lithuania), typhus epidemic in
　　216
Vilyati Leninobad (Tajikstan) 21
Vineyard Haven (Massachusetts),
　　hurricane in 302
viral flu, flooding and 176
Virginia
　　earthquake in 101
　　famines and droughts in 133,
　　　136
　　hurricanes in 286, 306, 311–
　　　312, 313
Virgin Islands, U.S. 253, 254, 307,
　　406
virus, corona 229–230
virus subtype H5N1 221, 222
Vitter, David 291
Vittoria (Italy), earthquake in 65
Vohemar (Madagascar), flooding in
　　146
volcanic cones 385
volcanic eruptions 383–424
　　worst recorded 383–384t
volcanic mountains 385
volcanism 405–406
volcanoes xi, xiii
　　classification of 386
　　dormant xi
　　formation of 385–386
　　lack of longevity and 207
　　as trigger for avalanche 22
Volcanoes of the Earth (Bullard) 413
volcanologists 385, 386, 392

volcanology xi, xvi
Vollertsen, Norbert 131
Voltaire 88
Vorarlberg (Austria) 8
Vorkuta (Russia) xiv
Vulcan (Papua New Guinea), volcanic
 eruption in 415

W

Waal River (Holland) 163, 164
Wabash River (Indiana), flooding of
 193
Waco (Texas), tornadoes in *354,*
 354–355
Waima Canyon (Hawaii), hurricane
 in 297
Wales
  influenza epidemic in 221
  landslides in 22–23
Walhee, John, III 298
Walker, Peter 151
Wallace, Nathan 10
Wallace, Robert 39
Walla Walla (Seattle), flooding in 195
Wallow (Ethiopia), famines and
 droughts in 112
Wankaner (India), earthquake in 52
war. *See also* World War I; World War
 II
  biological 209
  plague and 208
Warlonge, volcanic eruption at 402
War on Want 110
Washington (state)
  avalanches in 22
  flooding in 195–196
  rock slides in 196
Washington, D.C.
  earthquake in 101
  hurricanes in 285, 286, 313
  snowstorm in 329, 332, 333
Washington, Mount (New
 Hampshire), tornado in 348
Washington Bonus March 289–290
Watch Hill (Rhode Island), hurricane
 in 303
water, importance of uncontaminated
 214
water-borne diseases
  cholera as 207
  floods and 146, 148
  monsoons and 362–363
Waualeale, Mount (Hawaii), hurricane
 in 297

Waveland (Mississippi), hurricane in
 293
Wawahum (Honduras) 275
Webster, Noah 219
Weingartner, Wendelin 8
Wellington (Washington), avalanches
 in 22
Wells (Maine), hurricane in 301
Wenling (China), typhoon in 372
West Alton (Missouri), flooding in
 185
West Bengal (India)
  flooding in 168, 169
  landslides in 12
West Chop (Massachusetts), hurricane
 in 302
Westerly (Rhode Island), hurricane
 in 303
Western Europe, windstorms in
 363–365
Westgate, John 387
West Germany. *See also* Germany
  snowstorm in 326
  as source of aid 37, 76, 179
Westhampton Beach (Long Island),
 hurricanes in 304, *305*
West Indies
  cholera in 220
  hurricane in 316–317
  volcanic eruptions in 423–424
West Lyn River (England), flooding
 of 161
West Palm Beach (Florida) 296
West Point (Kentucky), flooding in 194
West Virginia
  drought in 133
  hurricanes in 309, 313
  snowstorm in 333
Wheatland (Pennsylvania), tornado
 in 350
Wheeler, David xv
White, Theodore H. 119
White Deer (Texas), tornado in 353
White Horse (Oklahoma), tornado in
 354
White Pass (Washington), snowstorm
 in 333
White Plains (New York), hurricane
 in 285
Wichita (Kansas), tornado in 351
Wigen, Syd 86
Wilcott, Oliver, Jr. 219
Wilder, Andrew 36
wildfires, as threat in drought 136
Willamette River (Washington),
 flooding of 195

Williams, Alvin 284
Williams, Gary 308
Williams, Richard S., Jr. 414
Williamson, J. M. 218
William the Silent 164–165, 264
Willow Creek (Oregon), flooding of
 194
Wilmington (North Carolina),
 hurricanes in 282, 306
Wilson, Bud 279
Wilson, Peter 41
Windham, G. B. 256
windstorms
  in English Channel 362
  in western Europe 363–365
Windward Islands, hurricanes in 268,
 269, 274
Windward Passage, hurricane in 260
Windy Hill (South Carolina),
 hurricane in 287
Wing Chut King (plague victim) 218
Winnsboro (South Carolina), tornado
 in 348
Winsted (Connecticut), flooding in
 282
Wisconsin, coldness in 331
Wisconsin River, flooding of 184
Witschi-Cestari, Alfredo 35
Wolfe, Sam 99
Wood, Lloyd B. 302
Woodbridge (Ontario), hurricane in
 287
Woodson, George 181
Woodstock (Georgia), tornado in
 352
Woodward (Oklahoma), tornado in
 353–354
Wooldridge, Captain 401
Woonsocket (Rhode Island), flooding
 in 282
Worcester (Massachusetts), tornadoes
 in 343, *343*
Worcestershire (England), hurricane
 in 261
Works Projects Administration (WPA)
 290
World AIDS COnference 227
World Food Program (WFP) 35, 110
World Health Organization (WHO)
  on AIDS 226
  on avian flu 223
  on Creutzfeldt-Jacob disease 217
  on SARS 229
World Resources Institute xv
World Vision 111

World War I
>deaths from avalanches in 7
>depletion of granaries in Russia
>>in 132
>influenza epidemic following
>>220–221
>leptospirosis in 363
>Russian famine in 132
World War II
>Bengal famine and 125–126
>famine as weapon of war in
>>110–111
>Nazi occupation of Holland in
>>122–123
Wright, C. S. 353
Wright, Frank Lloyd 74
Wrightsville (North Carolina),
>hurricane in 287
Wuhan (China) 159
Wulanch'apu (Wulanchabu) (China),
>snowstorm in 325
Wyman, Edward 218
Wyoming, flooding in 179–180

X

Xai-Xai (Mozambique), flooding in
>145, 146
Xenosylla cheopis 207

Y

Yakima (Washington) 422–423
Yallahs Valley (Jamaica), hurricane
>in 267
Yalova (Turkey), earthquake in 95, 97
Yamada (Japan), tsunamis in 69
Yamagata Province (Japan), typhoon
>in 375
Yamaguchi-gumi (crime syndicate)
>72, 73
Yamamori, Ted 130
Yamuha River (India), flooding of 166
Yanamachico (Peru), avalanche in 17

Yangtze River (China), flooding of
>155, 156, 158–160
Yangtze/Shoshun River Basin (China),
>flooding in 158–160
Yanyong, Jiang 229
Yap, famines and droughts in 128,
>129
Yaque River (Dominican Republic),
>flooding of 283
Yazoo River (U.S.), flooding of 189
yellow fever epidemic 207
>in United States 218–219
Yellow River (China) xiv
>flooding of 154, 156, 157, 160
Yellowstone National Park, volcanic
>eruption and 387
Yenan Province (China). *See* Yunnan
>(Yenan) Province (China)
Yerevan (Armenia), earthquake in 40,
>90
Yezo Island (Japan), tsunamis in 69
Yingde (China), flooding in 158
Yogyakarta (Indonesia), earthquake
>in 60
Yokohama (Japan) 71
>earthquake in 74–75
>typhoon in 375
Yongwol (South Korea), flooding in
>175
Yosemite National Park (California),
>flooding in 195
Yucatán Channel (Mexico), hurricane
>in 268
Yucatán Peninsula (Mexico),
>hurricanes in 253, 254, 255, 265,
>266–267, 269, 270, 272, 274
Yueyi, Liu 160
Yugoslavia, earthquake in 103–104
Yukimura, Jo Anne 298
Yungay (Peru)
>avalanche in 17
>earthquakes in 82–84, *83, 84*
Yunnan (Yenan) Province (China)
>flooding in 160
>plague in 214
>typhoon in 379

Yusuf, Irwandi 59
Yusuf, Isa 326, 327

Z

Zacapoaxtla (Mexico), flooding in
>173, 174
Zacoalpan (Mexico), landslide in 16
Zaire
>AIDS in 226
>plague in 222
Zambales Province (Philippines)
>418
>destruction of 419
Zamberletti, Giuseppe 172
Zambesi River 143
Zambia
>AIDS in 224
>famines and droughts in 115
>flooding in 143, 146
Zamboanga City (Philippines) 86
>earthquake in 86
Zammaro (Italy), earthquake in 65
Zanjan (Iran), earthquake in 62
Zeigler (Illinois), tornado in 342
Zemin, Jiange 160
Zenmount (Algeria), earthquake in
>38
Zentner, Kasper 19
Zermatt (Switzerland), avalanches in
>20
Zeroud River (Tunisia), flooding of
>178
Zia, Khaleda 238
Zilinksi, Greg 386–387
Zimbabwe
>AIDS in 227, 228
>famines and droughts in 111
>flooding in 145, 146, 147
Zimnicea (Romania), earthquake in
>88
Zinsser, Hans 216
Zixing (China), typhoon in 380